BIOLOGY THE EASY WAY

Second Edition

Gabrielle I. Edwards

Assistant Principal Supervision Emerita
Science Department
Franklin D. Roosevelt High School
Brooklyn, New York

Barron's Educational Series, Inc.

All inquiries should be addressed to:
Barron's Educational Series, Inc.
250 Wireless Boulevard
Hauppauge, New York 11788

Library of Congress Catalog Card No. 89-18096

International Standard Book No. 0-8120-4286-7

Library of Congress Cataloging-in-Publication Data

Edwards, Gabrielle I.
    Biology the easy way / Gabrielle I. Edwards. — 2nd ed.
        p.   cm.
    ISBN 0-8120-4286-7
    1. Biology.   I. Title.
QH307.2.E38   1990
574—dc20

89-18096
CIP

PRINTED IN THE UNITED STATES OF AMERICA

345   100   987

CONTENTS

PREFACE

BIOLOGY THE EASY WAY presents an up-to-date account of the principles and concepts of modern biology. Over the past 30 years there has been an explosion of knowledge in the biological sciences, which has revolutionized our way of thinking and doing in terms of biology. As a consequence of the many new discoveries in biology, some of the traditional beliefs have been discarded, others have been modified and much new material has been added to replace the invalid theories of old. A direct result of the increase in biological knowledge has been the increase in size and weight of the modern biology textbook. Although all of the newer knowledge in biology cannot be contained under one cover, the standard biology text has become a volume of substantial weight and cost.

BIOLOGY THE EASY WAY was prepared with the reader in mind. Although it contains the important information of modern biology, the writing is concise. The content material is presented clearly in language that is easy to read and to understand. The clarity of language serves to make difficult concepts more easily understood. Technical language is avoided wherever possible. However, biology has its special vocabulary which is essential for description. The words new to the reader are presented in italics, clearly defined and used appropriately in the text material. The overall quality of the writing in BIOLOGY THE EASY WAY is lively, modern and interesting.

BIOLOGY THE EASY WAY is divided into 16 chapters. Each of these chapters represents an area of specialization in the field of biology. For each such area modern principles of biology are presented in an appealing way for the reader. In each chapter, the content discussion is accompanied by carefully placed line drawings of the organisms, organs, structures or processes under discussion. Wherever possible, summary material has been presented in tabular form, providing a quick means of study and review of a given topic.

Each chapter in BIOLOGY THE EASY WAY provides special study aids that are designed to enhance the learning and understanding of the biological principles or concepts under study. Because scientific discovery and invention cannot be divorced from the history of human life on earth, a Chronology of Famous Names has been included in each chapter. Men and women scientists representing all countries, races, and religions are included in the chronology. The contributors mentioned range in time from the ancient Greek philosopher-scientists to modern day investigators. For each name the approximate date of the discovery or invention, the contributor's nationality, and a one line summary of the accomplishment is given. Another study aid included in this book is a listing headed "Words for Study." The important words of biology introduced in the chapter are listed for review. Readers who take the time and effort to learn all of these words will measurably increase their knowledge of modern biology. At the end of each chapter are at least 30 questions designed to assess the reader's grasp of some important facts and concepts of biology; an answer key is provided to help with self-evaluation. Also, under the heading "Think and Discuss" a group of four or five questions invites reflection and discussion on the important points covered in the chapter.

By reading the Table of Contents you can see the range of topics that is included in this text. By design, BIOLOGY THE EASY WAY is versatile, having appeal and use for students in secondary school through first year college and also to adults who are not in school but who desire to learn some modern biology. A chapter may be read in its entirety or in parts. The information is presented so that one can skim or read deeply. The index provides topic-page information.

The principle aim of BIOLOGY THE EASY WAY has been to present the facts of biology as they are known today in such a way that the curiosity and interest of the reader is aroused.

Gabrielle I. Edwards

HOW TO USE THIS BOOK

BIOLOGY THE EASY WAY presents the principles and concepts of biology in language easy to understand, using illustrations that simplify the text material further. To get the most in study help from BIOLOGY THE EASY WAY, however, you must use it efficiently.

Outlined below are procedures for self-study that will enable you to learn the content of biology efficiently and easily.

FOCUS

1. Begin with Chapter 1. Look through the entire chapter by scanning each page as you turn to it. Look at the illustrations and read their captions. Scan the chapter word list. Look over the review questions at the end of the chapter. This procedure will help you develop a mindset for the material that you are about to learn.

READ

2. Begin to read the chapter. Read only about four pages at a time, stopping at some logical place. For example, in Chapter 1 read the first three and two-thirds pages, stopping at "The Work of the Modern Biologist." Review what you have just read by skimming through the material again and by looking for familiar terms in the word list. Now complete the reading of the chapter. Review by skimming. Study the illustrations in detail.

THINK

3. Think about the chapter material. The word list at the end of the chapter will help you. For each word, follow this study tip: (1) Write down the word. (2) Now try to write its meaning. Do not attempt to memorize definitions; they are easily forgotten. Learn new vocabulary in ways that are understandable to you. (3) Think about the material you have read so that it becomes meaningful to you.

REVIEW

4. Now turn your attention to the Questions for Review. Begin with Part A. Read each question carefully, and write down the answer. Now check your answers with the answer key. For wrong answers, locate the appropriate material in the chapter text and review. Repeat this procedure for Parts B and C.

Using BIOLOGY THE EASY WAY efficiently will yield learning dividends.

CHAPTER 1

BIOLOGY: THE SCIENCE OF LIFE

BIOLOGY DEFINED

Biology is the science that studies life and living things, including the laws that govern the phenomena of life.

Every aspect of life from the smallest submicroscopic living particle to the largest and most imposing of plant and animal species is included in the study of biology. Biological study encompasses all that is known about any plant, animal, microbe or other living thing of the past or present.

Biology is a *natural* science because it is the study of organic (living) nature. It is the science of fishes and fireflies, grasses and grasshoppers, humans and mushrooms, flowers and sea stars, worms and molds. It is the study of life on top of the highest mountain and at the bottom of the deepest sea. Biology is the accumulated knowledge about all living things and the principles and laws that govern life. Those who specialize in biology are known as *biologists* or *naturalists*, and it is through their observations of nature and natural phenomena that the great ideas of biology have been born.

BIRTH OF BIOLOGY

"Truth is the daughter of time" is an adage worth repeating. History has shown that only within a certain time frame of thought could the major ideas of biology have evolved. The birth of biology as a *bona fide* science was slow and painful, taking place over many centuries. The object of scientific study is to find the truth. So it was and still is that human beings have sought the truth about the nature of life.

The study of life is as old as humankind, dating back to ancient peoples who observed and wondered about the characteristics of the animals and plants around their limited sphere. The ancients used their observations to help them in such activities as hunting, food gathering and crop growing.

A great deal of credit is due to the ancient Greeks for having begun a systematic study of living things, including human beings. A brief chronology of ancient achievements follows:

Hippocrates (460–370 B.C.) founded the first medical school on the Greek island of Cos.

Aristotle (384–322 B.C.) founded the systematized study of natural history. He was a keen observer and a prolific writer-illustrator on subjects dealing with plants and animals.

Theophrastus (380–287 B.C.) founded the organized study of plants. He is called the "ancient father of botany."

Galen (A.D. 130–200) founded the science of anatomy. The accuracy of his anatomical drawings of animal bodies and the gross structure of the human body remained unchallenged for centuries.

The rise of ancient science reached a peak and then declined sharply. For several centuries interest and activity in scientific investigation waned, creating a void in scientific discovery. This depressed period in which there was little or no inquiry about nature and life is known historically as The Dark Ages; it lasted from A.D. 200 to 1200. During this time, books were scarce and authority was considered to be the source of all knowledge. Observation was regarded as impious prying.

The 14th century ushered in a revival of scientific thought and inquiry. Among the reasons for the change in attitude were the invention of the printing press, the voyages of the explorers, the expansion of ideas brought on by the Crusades and the rise of universities. All of these things contributed to a return to the study of nature and to the methods of science.

A thumbnail sketch of some important contributions to the understanding of the laws of natural science by early investigators follows:

Andreas Vesalius (1514–1564) refuted the authority of Galen and studied the human body by dissection.

Marcello Malpighi (1628–1694), an expert in plant and insect anatomy, described the metamorphosis of the silkworm.

Robert Hooke (1635–1703) discovered and named the "cells" in cork.

William Harvey (1578–1667) demonstrated the circulation of the blood in the human body.

Anton Von Leeuwenhoek (1632–1723) was the first person to see living cells.

Carolus Linnaeus (1707–1778) devised the system of binomial nomenclature by the double naming of species.

Georges Cuvier (1769–1832) founded the study of comparative anatomy.

Jean Baptiste Lamarck (1744–1829) coined the word *biology* by putting together two Greek words: *bios* meaning "life" and *logos* meaning "study."

Branches of Biology

Biology is an extensive science encompassing many life science disciplines that cover enormous areas of study and information. Estimations based on biostatistical methods indicate that the total number of living plant and animal species is probably about ten million. Scientists agree that not all living species have been discovered and perhaps only 15 percent of the total have been described. *Paleobiologists*, biologists who specialize in the study of ancient life, believe that the number of extinct species may range between 15 and 16 million.

Investigators of the 18th and 19th centuries were faced with the monumental task of trying to put some kind of order into the study of natural science. Thus, work during these centuries involved sorting, identifying, naming, describing and classifying the myriads of species that were being discovered. Biology was then divided into a few discrete areas which allowed for specialization of study.

Botany is the study of plants and plant life cycles. *Zoology* is the study of animals and their life histories. *Morphology* is the study of the structure and form of living things. There are two sub-areas of this discipline—namely, plant morphology and animal morphology. *Physiology* is that branch of biology that deals with the way living things function. It explains the way a whole organism and its parts work. The study of function may be related to plants or to animals. *Taxonomy* is the science of classification. It provides a logical and consistent method for naming and grouping the many species of living things.

TABLE 1.1. Branches of Botany

Botany is the systematized study of plants.	
Disciplines of Botany	*Study of*
bryology	mosses and liverworts
paleobotany	fossil plants
pteridology	ferns
palynology	modern and fossil pollen
plant pathology	plant diseases

TABLE 1.2. Branches of Zoology

Zoology is the systematized study of animals.	
Disciplines of Zoology	*Study of*
entomology	insects
ethology	animal behavior
herpetology	amphibians and reptiles
invertebrate zoology	animals without backbones
ichthyology	fishes
mammalogy	mammals
ornithology	birds
vertebrate zoology	animals with backbones

To accommodate the explosion of knowledge in the 20th century the number of biological science specialities increased. The modern study of life science takes into consideration levels of biological organization. Using the knowledges of chemistry, physics and mathematics, the modern biologist inquires into finite structure and function.

Molecular biology explains the roles of elements, compounds and ions in the fundamental biochemical processes of cells and their parts. *Energetics* is that branch of science used to explain the vital energy transformations that take place in living cells.

Above the molecular level, the major biological sciences present organized information about parts of organisms or about whole individuals or populations. *Morphology*, the study of structure, is now divided into three sub-fields: anatomy, histology and cytology. *Anatomy* is the study of gross structure that is visible to the naked eye. *Histology* is the study of microscopic structure on the tissue level of organization. *Cytology* is the study of cells on the microscopic level.

Taxonomy is a study of whole organisms. It is the science of classification by which living species are grouped according to observed similarities and presumed evolutionary relationships in structure and molecular organization. *Systematics* includes the identification and ranking of all plants, including their classification and nomenclature; it is the science that provides understanding about plant diversities.

Embryology is the science of embryo development in plants and animals. Embryology commences with the union of sperm and egg and continues to the completion of body structures.

Genetics is a specialty of biology that explains heredity and variation in living things. How molecules bearing genetic information are passed from parent to offspring and how they function in the cellular environment are important aspects of genetic studies.

Evolution is the science that determines the possible origins and relationships of living things and analyzes the changes that occur in species. Modern classification is based upon evolutionary relationships.

Paleontology is the study of ancient life accomplished through examination of fossil records.

Ecology is the study of relationships of plants and animals to one another and to their physical environment.

The Work of the Modern Biologist

It is through the painstaking efforts of biologists that so much information about living species of the present and past has been gathered. The methods used to gather facts are as varied as the disciplines of biological science and the myriad of scientists who work to unravel the secrets of life. Some basic principles direct the efforts and interests of those men and women who devote their lives to the demands of research in biology. Two of these principles are summarized below:

Homeostasis is the term used to describe the stable internal environment of the cell and the organism as a whole. Homeostasis is a condition necessary for life. The internal chemical balance of the cell must be maintained at a steady state to promote innumerable biochemical activities that foster the production and use of energy. The concept of homeostasis was initially developed by the 19th century physiologist Claude Bernard.

Unity is shared by all living species in that they have certain biological, chemical and physiological characteristics in common. There is unity of the basic living substance—*protoplasm*. Rudolph Virchow, a 19th century pathologist, established the concept of continuity of the germ plasm in which he stated, "All living cells arise from pre-existing living cells." All cells synthesize and use enzymes. Their genetic information is carried by DNA molecules. DNA gives cells the ability to replicate by a pattern which is followed in all cells.

THE SCIENTIFIC METHOD AND PROBLEM SOLVING

The methods which enable the biophysicist to measure particles too small to be resolved, or distinguished, by an electron microscope, the biologist to discover the cause of diseases such as toxic shock syndrome, the geologist to determine the age of the earth, and the biochemist to identify minute quantities of enzyme in a respiratory pathway are not the same as those used by the average person to solve the problems of daily living.

Scientific problem solving depends upon accuracy of observation and precision of method. Inherent in scientific thinking is orderliness of approach which invites the forming of conclusions from hypotheses, theories, principles, generalizations, concepts and laws. Scientific problem solving follows a pattern of behaviors which are collectively known as the *scientific method*.

The first step in the scientific method is *observation*. Observation is an accurate sighting of a particular occurrence. The accuracy of an observation becomes substantiated when a number of independent observers agree on seeing the same set of circumstances. For example: early in the 16th century, Vesalius presented the first accurate description of the arteries and veins in the human body. His drawings were based on first-hand observations obtained through dissection of human bodies. Over a period of 400 years succeeding generations of anatomists have attested to the accuracy of Vesalius' work. Moving away from our example, it is to be noted that although several trained observers may agree upon the same set of observations, the conclusions they draw from these observations may differ. Therefore, problem solving depends on other activities which enable the investigator to form conclusions based on accurate observations.

An observation becomes useful only in terms of the conclusion that can be drawn from it. A tentative conclusion derived from a set of observations is known as a *hypothesis*. A hypothesis is a very general statement made to tie together observational data prior to experimentation.

An *experiment* requires a set of procedures designed to provide data about a well-defined problem. The purpose of an experiment is to answer a question truthfully and accurately. A reliable experiment requires experimental conditions and control procedures. A *control* is a check on a scientific experiment and differs from the experimental procedure in one condition only. An experiment is designed to test a specific hypothesis.

A nontestable hypothesis is not useful in problem solving because it does not lend itself to experimental conditions. An example of a nontestable hypothesis follows: blow flies think that meat is an excellent site for egg laying. Because there is no way to test the thought processes of a blow fly, the hypothesis is invalid.

A classical work illustrating the effectiveness of a well-designed experiment is *Redi's Blow-Fly Experiment*. That living things arise from lifeless matter was a popular 17th century belief. In 1650, the Italian physician Francesco Redi designed and performed an excellent experiment which disproved the spontaneous generation of blow flies from decaying meat:

> In each of three jars, Redi placed the same kind and quantity of meat. The first jar was left uncovered. The second jar was covered with a porous cheese cloth. The third jar was covered with a parchment thick enough to prevent the odor of the meat from diffusing to the air outside of the jar (Fig. 1.1). Flies laid their eggs on the meat in the open jar and on the cheese cloth which covered the second jar. Eggs were not laid on the parchment covering the third jar. Redi watched as the eggs hatched into larvae (maggots) and observed the change of the larvae into adult flies. There were neither maggots nor flies on the meat in the third jar.

> On the basis of this experiment, Redi concluded that maggots or flies do not come from decayed meat, but from the fertilized eggs laid by the female fly.

Fig. 1.1 Redi's Blow-Fly Experiment. Can you explain the results?

When an hypothesis is supported by data obtained by experimentation, the hypothesis becomes a *theory*. A theory that withstands the test of time and further experimentation becomes a law or principle. A *concept* is a blanket idea covering several related principles.

THE TOOLS OF THE BIOLOGIST

Problem solving by the modern biologist requires the use of some very special tools and laboratory techniques. The scientific method is a major idea of biology upon which all research is built. All research begins with observation, a behavior that requires training, skill and concentration. To increase the accuracy of observation, the research biologist uses devices and techniques that will extend his or her vision and other senses. Because human sensing ability is limited and variable and not really reliable for seeing and recording fine and sometimes fleeting detail, modern scientists make use of devices and methods that enable them to perceive, measure and record data precisely.

Measurement is a basic requirement of all research. An investigator must answer questions concerning quantity, length, thickness, periods of time, mass, weight and the like within a hair's breadth of accuracy. Modern instruments of measurement permit astounding degrees of numerical precision. *Biometrics* is the science that combines mathematics and statistics needed to deal with the facts and figures of biology. Biologists handle enormous numbers which must be organized and simplified so that they become useful in the analysis of data. Computers aid the investigator's problem-solving tasks by accepting massive quantities of numerical data and making calculations at lightning speeds.

The laboratory or field scientist derives numerical data from using all kinds of measuring instruments designed for remarkable precision. Weighing in micro quantities requires the use of an *analytical balance* that is capable of weighing accurately to infinitessimal fractions of micrograms. The *manometer* measures the uptake of gases involved in cellular respiration and photosynthesis. *Spectrophotometers* are able to measure differences in densities of fluids by color comparisons not discernible to the human eye. Radioactivity is located and measured by the *Geiger counter*, a machine structured to feed pulses of electricity into an electronic counter. *Scintillation counters* measure light flashes imperceptible to the eye while ultraviolet rays are used to measure objects thinner than onion membrane.

Discovery in biology involves the ability to take apart the substances of cells so that the secrets of their biochemical processes can be exposed. The technique of *chromatography* permits the biologist to separate unbelievably small quantities of a substance into its component parts. For example: $\frac{1}{100,000}$ of a gram of protein can be separated into its constituent amino acids in one hour. Before the discovery of chromatography, a scientist needed 100 grams of protein and one year's time in order to make this separation. As the name implies, chromatography is "color writing," a technique that uses paper or a column of chalk as a stationary phase. The test sample, dissolved in a suitable solvent, is allowed to travel up the paper or through the chalk column. The constituents of the sample settle out at various levels as bands or dots of color (Fig. 1.2). *Electrophoresis* is a technique that uses electrical charges to separate the amino acids in proteins dotted on specially treated glass plates (Fig. 1.3).

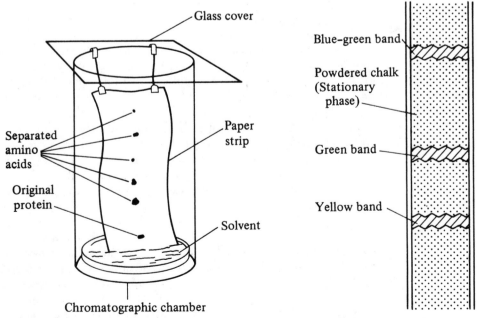

Fig. 1.2 Paper chromatography of a protein sample

Fig. 1.3 Column chromatography of green plant pigments

A biologist studying cell processes wishes to remove the organelles (cell parts) from within the cell unit. Tissue and lysing (loosening) fluid are put into an electric blender or homogenizer. Here the tissue cells are separated from each other and the cell membranes are broken open. The suspension of ruptured cells and lysing fluid is then placed in a test-tube-like device and spun in an *ultracentrifuge* at 70,000 rotations per minute. The force exerted at extreme speeds of rotation is equivalent to 300,000 to 400,000 times the force of normal gravity. Such force causes the membranes enclosing the tissue cells to split open, releasing their contents into the sucrose lysing fluid. On the basis of weight differences, the organelles of each cell, cell particles, and molecules will be distributed in the lysing fluid in layers. The heavier cell structures fall to the bottom of the tube, and the smaller and lighter molecules remain in zones at the top. The fluid that tops a zone of heavier particles is called the *supernatant*. Organelles of intermediate weight are distributed in zones between the top and bottom layers. The technique of ultracentrifugation enables the research biologist to remove layered zones and to study them separately, either by biochemical means or by observation through electron microscopy.

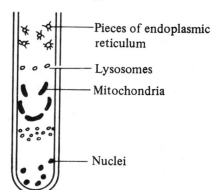

Pieces of endoplasmic reticulum

Lysosomes

Mitochondria

Nuclei

Fig. 1.4 Results of centrifugation—the heavier particles fall to the bottom of tube

A number of techniques and instruments have helped scientists see things better. *X-ray diffraction* is a procedure by which X rays sent through a crystal reveal the pattern of molecules and atoms contained in the crystal. This technique was used to determine the structure of the hereditary material DNA (deoxyribonucleic acid).

All sorts of microscopes have extended the cytologist's (cell biologist's) field of vision. *Magnification* is the measure of how much a microscope can enlarge specimens under view. The quality of a good microscope is based on its *resolving power*, that is, its ability to distinguish two closely applied points separately. The resolving power of the normal unaided human eye is 0.1 mm.

The best *light microscope* is capable of magnifying objects 2,000 times (Fig. 1.5). the *phase contrast microscope* makes transparent specimens visible, while the *darkfield microscope* or the *ultramicroscope* gives vivid clarity to fragile and transparent organisms such as the spirochetes that cause syphilis. The *ultraviolet microscope* is used for photographing living bacteria and naturally fluorescent substances. Most amazing of all is the *electron microscope*, a device capable of magnifying objects more than 200,000 times. Using electrons instead of light and magnets in place of lenses, the electron microscope has revolutionized the study of the cell

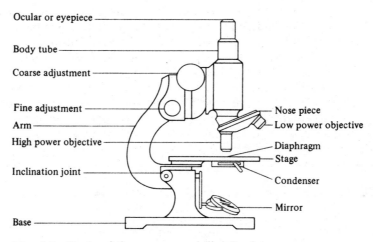

Ocular or eyepiece

Body tube

Coarse adjustment

Fine adjustment

Arm

High power objective

Inclination joint

Base

Nose piece

Low power objective

Diaphragm

Stage

Condenser

Mirror

Fig. 1.5 Parts of the compound (light) microscope

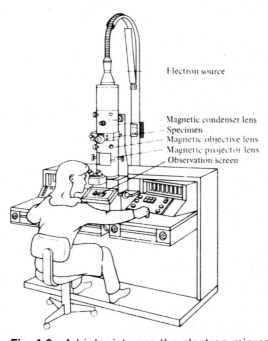

Electron source

Magnetic condenser lens
Specimen
Magnetic objective lens
Magnetic projector lens
Observation screen

Fig. 1.6 A biologist uses the electron microscope

(Fig. 1.6). The *scanning electron microscope* has improved upon the resolution of fine detail made possible by electron microscopy. A fine probe directs and focuses electron beams over the material being studied, affording quick scanning and giving finer detail than is possible with the standard electron microscope.

The *high-voltage electron microscope* is a tremendous three-story structure that outputs about 3 million electron volts. This enormous amount of energy permits specimens to be viewed without the thin slicing required by other electron microscopes. In addition, the high-voltage electron microscope produces a three-dimensional micrograph (photograph) not possible with other electron microscopes.

Mention has been made herein of only a small sampling of the kinds of tools and techniques available to the modern biologist. There are also incubators for temperature control, refrigerators for keeping things cool, autoanalyzers for separating and classifying the elements in the blood, automatic mixers and stirrers and shakers, pH meters, autoclaves for sterilizing, timers and machines that measure time to the fraction of a second. The modern biologist uses anything and everything that helps to extend his or her senses and improve observation.

Chronology of Famous Names in Biology

1895	**Wilhelm Konrad Roentgen** (German)—discovered X rays.
1903	**Mikhail Tsvett** (Russian)—discovered column chromatography.
1920	**Theodor Svedburg** (Swedish)—invented the centrifuge.
1923	**Georg von Hevesy** (English)—developed technique of using radioactive tracers.
1925	**Joseph E. Barnard** (English)—developed technique for photographing living bacteria using ultraviolet light.
1933	**Arne Tiselius** (Swedish)—developed electrophoresis.
1933	**Max Knell** and **Ernst Ruska** (German)—invented the single "lens" electron microscope.
1935	**Wendell Stanley** (American)—isolated the tobacco mosaic virus by centrifugation.
1938	**James Hillier** and **Albert Prebus** (English)—invented the compound electron microscope.
1940	**A.P.J. Martin** and **R.L.M. Synge** (English)—discovered the technique of paper chromatography.
1952	**Alfred D. Hershey** and **Martha Chase** (American)—developed the technique of using radioisotopes to study viruses.
1953	**Maurice H.F. Wilkins** (English)—developed the technique of using X-ray diffraction to decipher the structure of DNA.

Words for Study

analytical balance	centrifuge	ecology
anatomy	chromatography	electron microscope
biology	concept	electrophoresis
biometrics	controlled experiment	embryology
botany	cytology	energetics

evolution	morphology	spectrophotometer
Geiger counter	naturalist	spontaneous generation
genetics	nature study	supernatant
histology	paleobiology	systematics
homeostasis	paleontology	taxonomy
hypothesis	physiology	theory
light microscope	protoplasm	ultracentrifuge
manometer	resolving power	zoology
molecular biology	scintillation counter	

Questions for Review

PART A. **Completion.** Write in the word that correctly completes each statement.

1. The science that specializes in the study of all life is ..1..

2. The ancient who founded the systemized study of natural history was ..2..

3. The first person to study the dissected human body was ..3..

4. The word biology was coined by ..4..

5. The branch of biology that specializes in the study of ancient life is ..5..

6. The science of classification is known is ..6..

7. Vital energy changes in the cell are studied in the science of ..7..

8. The study of plants and animals in their natural environment is known as the science of ..8..

9. Ichthyology is the study of ..9..

10. Steady state of the internal environment is described by the term ..10..

11. A tentative conclusion based on observation is the ..11..

12. Redi disproved the theory of ..12.. (2 words) of flies.

13. In the modern laboratory, the device used to measure the densities of colored liquids is the ..13..

14. A spinning machine used in the modern laboratory to separate cell structures is the ..14..

15. A device used to locate and measure radioactivity is the ..15..

16. Sucrose (sugar) solution is used as ..16.. fluid.

17. An ocular is an essential part of the ..17.. microscope.

18. The word "micrograph" is best associated with the ..18.. microscope.

19. X rays were discovered by ..19..

20. The ability of a microscope to enlarge an image is called ..20..

PART B.   Multiple Choice. Circle the letter of the item that correctly completes each statement.

1. An example of a natural science is
   (a) geology   (c) physics
   (b) biology   (d) chemistry

2. The science of anatomy was founded by
   (a) Hooke   (c) Galen
   (b) Plato   (d) Tisellius

3. The "cells" in cork were named by
   (a) Hooke   (c) Galen
   (b) Plato   (d) Tisellius

4. The study of comparative anatomy was founded by
   (a) Malpighi   (c) Harvey
   (b) Lamarck   (d) Cuvier

5. The system of binomial nomenclature was developed by
   (a) Lamarck   (c) Leeuwenhoek
   (b) Linnaeus   (d) Lintel

6. The branch of biology that deals with function is
   (a) zoology   (c) physiology
   (b) taxonomy   (d) morphology

7. Included in the study of botany would be
   (a) dandelions   (c) guppies
   (b) sparrows   (d) skates

8. Inheritance and variation in living things is explained by the science of
   (a) ecology   (c) paleontology
   (b) genetics   (d) taxonomy

9. The study of insects is known as
   (a) ichthyology   (c) bryology
   (b) herpetology   (d) entomology

10. X-ray diffraction techniques were used to determine the structure of
    (a) ATP   (c) DNA
    (b) NAD   (d) PGA

11. The science of development is called
    (a) cytology   (c) systematics
    (b) embryology   (d) pteridology

12. A check on an experiment is a (an)
    (a) observation   (c) control
    (b) theory   (d) conclusion

13. A blanket idea covering several related principles is called a
    (a) theory   (c) concept
    (b) hypothesis   (d) conclusion

14. A microgram or its fraction can be weighed most accurately on a balance known as a (an)
    (a) analytical   (c) triple beam
    (b) spring   (d) double platform

15. Respiratory gases can be measured by using a device called a (an)
    (a) calorimeter   (c) barometer
    (b) manometer   (d) anemometer

16. The quality of a microscope is judged by its
    (a) volumetric capacity   (c) magnification power
    (b) light strength   (d) resolving power

17. A three-dimensional photograph of a specimen is made possible by
    the type of microscope known as
    (a) phase contrast   (c) scanning electron
    (b) high-voltage electron   (d) ultraviolet

18. To isolate individual parts of a cell, a research worker would use the
    technique of
    (a) ultracentrifugation   (c) electrophoresis
    (b) microscopy   (d) chromatography

19. The word "organelles" refers to
    (a) small body organs   (c) supernatant fluid
    (b) cell parts   (d) microscope parts

20. The value 70,000 rotations per minute is best associated with the
    (a) scintillation counter
    (b) scanning electron microscope
    (c) ultracentrifuge
    (d) pH meter

PART C. Modified True-False. If a statement is true, write "true" for
your answer. If a statement is incorrect, change the underlined word to one that will make the statement true.

1. A <u>botanist</u> is a scientist who specializes in the study of plants and
   animals.

2. The organized study of plants was founded by the ancient named
   <u>Hippocrates</u>.

3. The first person to see living cells was <u>Hooke</u>.

4. How blood circulates in the human body was first accurately demonstrated by <u>Harvey</u>.

5. Morphology is the branch of biology that studies <u>function</u>.

6. The study of tissues is known as <u>phylogeny</u>.

7. <u>Cytology</u> is the identification and ranking of plants.

8. The study of reptiles is known as <u>herpetology</u>.

9. <u>Ethology</u> is the science of birds.

10. The living substance of cells is <u>neoplasm</u>.

11. Problem solving in science is known as the scientific <u>module</u>.

12. A proven theory becomes a <u>conclusion</u>.

13. A combination of mathematics and statistics applied to facts of bi-
    ology is known as <u>energetics</u>.

14. Amino acids in proteins can be separated by using electrical charges in a process known as <u>X-ray diffraction</u>.
15. The microscope that can magnify more than 200,000 times is the <u>electric</u> microscope.
16. The ability to distinguish two closely applied points separately is known as <u>reduction power</u>.
17. Living, transparent microscopic organisms are best viewed by using either an <u>electron</u> or <u>phase contrast</u> microscope.
18. The human eye cannot resolve <u>electrons</u>.
19. The heaviest of the cell parts are the <u>mitochondria</u>.
20. A cytologist studies the <u>mind</u>.

Think and Discuss

1. Why is it reasonable to believe that the study of life is as old as humankind?
2. Why is the science of biology divided into so many fields of study?
3. What was the control in Redi's blow-fly experiment?
4. Why are numbers and mathematics so important to the modern biologist?

Answers to Questions for Review

PART A

1. biology
2. Aristotle
3. Vesalius
4. Lamarck
5. paleobiology
6. taxonomy
7. energetics
8. ecology
9. fish
10. homeostasis
11. hypothesis
12. spontaneous generation
13. spectrophotometer
14. centrifuge
15. Geiger counter
16. lysing
17. light
18. electron
19. Roentgen
20. magnification

PART B

1. b
2. c
3. a
4. d
5. b
6. c
7. a
8. b
9. d
10. c
11. b
12. c
13. c
14. a
15. b
16. d
17. b
18. a
19. b
20. c

PART C

1. naturalist or biologist
2. Theophrastus
3. Von Leeuwenhoek
4. true
5. form or structure
6. histology
7. Systematics
8. true
9. Ornithology
10. protoplasm
11. method
12. law or principle
13. biometrics
14. electrophoresis
15. electron
16. resolving power
17. darkfield
18. true
19. nuclei
20. cell

CHAPTER 2

CHARACTERISTICS OF LIFE

DEFINITION OF LIFE

As you already know, biology is the science that is concerned with life and living things. Given a collection of familiar objects, you would probably experience little difficulty in separating them into a group of living and a group of nonliving things. To determine if something is living, you would look for signs of life such as movement, response to touch, patterns of growth and ability to take in food. Let us suppose that included in the group of things you wish to classify are a mildew-covered towel, a green stain on tree bark, a large yellow turnip and slime mold on the forest floor. What signs of life would you look for in mildew, tree stain, the turnip and slime mold? Are these things living? What criteria are used to distinguish living organisms from nonliving objects?

Scientists are unable to agree on a standard definition of life because it is impossible to define life and its qualities in a sentence or two. For every definition that might be formulated, too many exceptions are found. For example, we can say that living things grow. But so can crystals be made to grow. They do so by adding on molecules of transparent quartz. We might also say that living things move. But what of the ball that rolls down a hill or a kite that flies in the air? Are these not moving? It is examples such as growing crystals and flying kites that make life a difficult concept to define. Life is best described in terms of the functions performed by living things.

Life Functions

Living things are highly organized systems. They are self-regulating and self-reproducing and capable of adapting to changes in the environment. To satisfy all of the conditions necessary for life, all living systems must be able to perform certain biochemical and biophysical activities which collectively are known as *life functions.*

Nutrition is the sum total of those activities through which a living organism obtains *nutrients* (food molecules) from the environment and prepares them for use as fuel and for growth. Included in nutrition are the processes of *ingestion, digestion* and *assimilation. Ingestion* is the taking in or procuring of food. *Digestion* refers to the chemical changes that take place in the body by which nutrient molecules are converted to forms usable by the cells. *Assimilation* involves the changing of certain nutrients into the protoplasm of cells.

Transport involves the absorption of materials by living things, including the movement and distribution of materials within the body of the organism. There are several transport methods, including diffusion, active transport and circulation. *Diffusion* is the flow of molecules from an area where these molecules are in great concentration to an area where there are fewer of them. *Active transport* is the movement of molecules powered by energy. *Circulation* is the movement of fluid and its dissolved materials throughout the body of an organism or within the cytoplasm of a single cell.

Respiration consists of breathing and cellular respiration. *Breathing* refers to the pumping of air into and out of the lungs of air-breathing animals or the movement of water over the gills of fish. During breathing, oxygen diffuses into the air sacs in the lungs, and carbon dioxide and water vapor move out of the lungs through the nose and mouth. *Cellular respiration* is a combination of biochemical processes that release energy from glucose and store it in ATP (adenosine triphosphate) molecules.

Excretion removes waste products of cellular respiration from the body. The lungs, the skin and the kidneys are excretory organs in humans that remove carbon dioxide, water and urea from the blood and other body tissues. *Guttation* is the excretion of drops of water from plants during periods of high humidity.

Synthesis involves those biochemical processes in cells by which small molecules are built into larger ones. As a result of synthesis amino acids, the building blocks of proteins, are changed into enzymes, hormones and protoplasm.

Regulation encompasses all processes that control and coordinate the many activities of a living thing. Chemical activities inside of cells are controlled by enzymes, coenzymes, vitamins, minerals and hormones. The *nervous* and *endocrine systems* of higher animals integrate and coordinate body activities. Growth and development of plants is regulated by *auxins* and other growth-control substances.

Growth describes the increase of cell size and increase of cell numbers. The latter process occurs when cells divide in response to a sequence of events known as *mitosis.*

Reproduction is the process by which new individuals are produced by parent organisms. Basic to the understanding of reproduction is the concept that organisms produce the same kind of individuals as themselves. There are two major kinds of reproduction: asexual and sexual. *Asexual reproduction* involves only one parent. The parent may divide and become two new cells, thus obliterating the parent generation. Or the new individual may arise from a part of the parent cell; in such a case, the parent remains. In either case, *replication* (duplication) of the chromosomes is involved. *Sexual reproduction* requires the participation of two parents, each producing special reproductive cells known as *sex cells,* or *gametes.* The continuation and survival of the species is dependent upon reproduction. Once a species has lost its *reproductive potential,* the species no longer survives and it becomes extinct.

BASIC CONCEPTS IN BIOLOGY

Metabolism is an inclusive term concerning all of the biochemical activities carried on by cells, tissues, organs and systems necessary for the sustaining of life. Metabolic activities in which large molecules are built from smaller ones or in which nutrients are changed into protoplasm are called *anabolic activities*, or *anabolism*. Destructive metabolism in which large molecules are degraded for energy or changed into their smaller building blocks is called *catabolic activity*, or *catabolism*.

Homeostasis, as mentioned earlier, is a fundamental concept of biology. Homeostasis literally means "staying the same." Cells within the body of an organism must keep the chemical composition of the fluid contained in them and the chemistry of the water which surrounds them stable. To stay alive cells must regulate their internal and external fluids in temperature, acid-base composition, and the amount and content of mineral salts, ions and other substances that affect the steady state of their environment. This means that the life functions of an organism (or cell) are carried out in an integrated manner that helps the maintenance of a nearly unchanging internal environment.

Adaptation refers to a trait that aids the survival of an individual or a species in a given environment. An adaptation may be a structural characteristic such as the hump of a camel, a behavioral characteristic such as the mating call of a bull frog, or a physiological characteristic controlling some inner workings of tissue cells (Fig. 2.1). Adaptations permit the survival of species in environments that sometimes seem forbidding. For example, some bacteria are able to live in hot springs that have temperatures up to 80° C (175° F). They have adaptations that permit the carrying out of metabolic functions at extremely high temperatures.

Fig. 2.1 The long neck of the giraffe is an adaptation

How Living Things Are Named

THE CONCEPT OF SPECIES

The basic unit of classification is the *species*. A species is a group of similar organisms that can mate and produce fertile offspring. The red wolf, the African elephant, the red oak, the house fly, the hair cap moss—each belong to a separate and distinct species. For example, the red wolf belongs to the red wolf species in which the male mates with the female and produces fertile red wolf offspring. Upon maturity these red wolf offspring will reproduce just as their parents did. Species is a reproductive unit, not one defined by geography. Once brought together, a Mexican male Chihuahua can mate with a female Chihuahua born in France because they are species compatible.

In rare instances, members of closely related species—horses and donkeys, for example—can mate and produce offspring. But the products of interspecies matings are not fertile and therefore cannot reproduce. When a male donkey mates with a female horse (mare), a *mule* is produced. The mule is an infertile *hybrid* (Fig. 2.2). The mating of a male horse (stallion) and a female donkey results in a *hinny*, also an infertile hybrid.

Fig. 2.2 The mule, an infertile hybrid

BINOMIAL NOMENCLATURE

In modern biology living things are named according to the groups in which they are classified. The Swedish botanist and physician Carl von Linné (1707–1788) is credited with having devised the first orderly system of classification. Under the Latinized form of his name—*Carolus Linnaeus*, he is now known as "the father of modern taxonomy." Linnaeus based his system of classification on the anatomy and structure (morphology) of organisms. He recognized that living things fall into separate groups distinguishable by specific structural characteristics. He called each separate group of organisms a *species*. To each species he assigned a two-word Latin name or a *binomial*.

The first word in the species binomial designates the *genus*. Linnaeus assigned the name genus (plural *genera*; adjective *generic*) as the next higher unit of classification after the species. In his system the genus is a classification group consisting of one or more related species.

The species name is the *scientific name*, a double name in Latin known as *binomial nomenclature*. The system of double naming is so orderly and practical that it is used in every country in the world. Thus the scientific name for organisms is the same worldwide no matter the language of the country.

The scientific or species name—*Quercus rubra*, for example—provides us with some information about the organism. First of all, the genus name *Quercus* is the Latin name of oak. Species that are classified as oaks are grouped into the genus *Quercus*. The name *rubra* (red) indicates that it is a separate and distinct kind of oak apart from *Quercus alba* (white oak) or *Quercus suber* (cork oak). See Table 2.1.

TABLE 2.1.   The Genus *Quercus*

Scientific Name	Common Name
Quercus alba	white oak
Quercus coccinea	scarlet oak
Quercus montana	chestnut oak
Quercus rubra	red oak
Quercus suber	cork oak
Quercus virginiana	live oak

Certain rules are followed in the use of the genus and species names. The genus name is capitalized; the second word in the species name is not. The scientific name (genus and species name) is always italicized in print and underlined when written in long hand. Whenever the species name is used more than once in the same paragraph, it can be abbreviated. The first letter of the genus name is used followed by the complete name of the species: *Q. rubra*. When used alone, the genus name is always capitalized and spelled out: *Quercus*.

TAXONOMIC GROUPS

The design of the classification system is a simple and practical one that easily lends itself to the addition of new names of organisms as they are discovered. The scheme makes use of a number of hierarchical groups each related to the other in terms of biological significance. Each group of organisms within the scheme is known as a *taxon* (plural, *taxa*). The classification groupings are as follows: *kingdom*, the largest and most inclusive group, followed by the *phylum, class, order, family, genus,* and *species.*

Linnaeus based his system of classification on the morphology (structure) of organisms in which those individuals that appeared most closely related were grouped into species and similar species into genera. The current practice is to group living things according to evolutionary relationships. *Evolution* is that branch of biology that studies changes in plants and animals that occur over long periods of time. Since the anatomical structures of plants and animals seem to reflect their evolutionary histories, the Linnaean system is compatible with the present day approach to classification.

TABLE 2.2.   Classification of the Human

Category	Taxon	Characteristics
Kingdom	Animalia	Multicellular organisms requiring preformed organic material for food, having a motile stage at some time in its life history.
Phylum	Chordata	Animals having a notochord (embryonic skeletal rod), dorsal hollow nerve cord, gills in the pharynx at some time during the life cycle.
Subphylum	Vertebrata	Backbone (vertebral column) enclosing the spinal cord, skull bones enclosing the brain.
Class	Mammalia	Body having hair or fur at some time in the life; female nourishes young on milk; warm-blooded; one bone in lower jaw.
Order	Primates	Tree-living mammals and their descendants, usually with flattened fingers and nails, keen vision, poor sense of smell.
Family	Hominidae	Bipedal locomotion, flat face, eyes forward, binocular and color vision, hands and feet specialized for different functions.
Genus	* Homo	Long childhood, large brain, speech ability.
Species	** Homo sapiens	Body hair reduced, high forehead, prominent chin.

* Homo = Latin: man
** sapiens: wise

Five-Kingdom System of Classification ____

The largest and most inclusive classification category is the kingdom. For many decades, living things were considered to be either plants or animals and thus were grouped into one of the two established kingdoms. Improved microscopic and other research techniques have revealed many species that cannot be classified either as plant or animal. For example: the one-celled organism *Euglena*, which has both plant and animal characteristics, was commonly referred to as a "biological puzzle" because taxonomists could not classify it to meet the satisfaction of all biologists. Other organisms such as bacteria, in which plant and animal characteristics could be demonstrated among the various species, also presented classifications problems.

In 1969, the ecologist Robert Whittaker proposed an updated system of classification in which every living thing is grouped into one of *five* kingdoms, based on the extent of its complexity and the methods by which nutritional needs are met.

TABLE 2.3.   The Five-Kingdom System

Kingdom	*Characteristics*
Monera	All monera are single-celled organisms. Unlike other cells, the monerans lack an organized nucleus, mitochondria, chloroplasts and other membrane-bound organelles. They have a circular chromosome. Examples are bacteria and blue-green algae.
Protista	Protists are one-celled organisms that have a membrane-bound nucleus. Within the nucleus are chromosomes that exhibit certain changes during the reproduction of the cell. Other cellular organelles are surrounded by membranes. Some protists take in food; other make it by means of *photosynthesis*. Some protists can move from one place to another (motile); others are nonmotile. Examples of protists are *Amoeba, Euglena, Paramecia, Punnularia* (diatom).
Fungi	Fungi are nonmotile, plant-like species that cannot make their own food. The fungi (fungus, singular) absorb their food from a living or nonliving organic source. Fungi differ from plants in the composition of the cell wall, in methods of reproduction and in the structure of body. Examples of fungi are mushrooms, water mold, bread mold.
Plantae	The plant kingdom includes the mosses, ferns, grasses, shrubs, flowering plants and trees. Most plants make their own food by photosynthesis and contain chloroplasts. All plant cells have a membrane-enclosed nucleus and cell walls that contain cellulose.
Animalia	All members of the animal kingdom are multicellular. The cells have a discrete nucleus that contains chromosomes. Most animals can move and depend on organic materials for food. Excluding the very simple species, most animals reproduce by means of egg and sperm cells.

Six-Kingdom System of Classification

In the five-kingdom classification system, all bacteria are grouped in the kingdom *Monera*. New evidence from molecular biology and electron microscopy, however, has indicated to biologists that all bacteria do not have the same evolutionary history. Therefore the kingdom Monera has now been divided into two kingdoms: the *Archaeobacteria* and the *Eu-*

bacteria. (Descriptions of the species classified as Archaeobacteria and as Eubacteria are given in Chapter 4.) The current classification model exhibits six kingdoms.

TABLE 2.4. The Six-Kingdom System

Kingdom	Characteristics
Archaeobacteria	Archaeobacteria survive only where there is no oxygen. They live in harsh environments: hot springs, salt lakes and seas, anaerobic (without oxygen) marshes, sewage plants, and the sludge of lake bottoms and seas. Examples are methane-producing bacteria, salt-loving bacteria, and heat-loving bacteria.
Eubacteria	Most species of bacteria belong to the Eubacteria. Certain kinds of these bacteria are classified by their shapes: bacillus (rod-shaped), coccus (round), spirillum (spiral). This kingdom is wide ranging and consists of both helpful and disease-producing bacteria.
Protista	Same as in Table 2.3
Fungi	Same as in Table 2.3
Plantae	Same as in Table 2.3
Animalia	Same as in Table 2.3

Chronology of Famous Names in Biology

342 B.C.	**Aristotle** (Greek)—first to try to group organisms by selecting a single outstanding feature.
1627–1705	**John Ray** (English)—introduced the word "species" to describe an organism. Catalogued about 19,000 plants of Europe.
1707–1778	**Carolus Linnaeus** (Swedish)—devised the first classification system. Wrote *Systems Natural* and *Species Plantarum*. Introduced the idea of the *type species*—typical specimen of the species.
1969	**Robert H. Whittaker** (American)—devised the five kingdom system of classification.
1975	**Carl R. Woese** (American)—divided the kingdom Monera into two separate kingdoms: Archaeobacteria and Eubacteria.

Words for Study

active transport	Fungi	order
adaptation	gametes	phylum
anabolism	genus	Plantae
Animalia	growth	Protista
Archaeobacteria	guttation	regulation
asexual reproduction	homeostasis	replication
assimilation	*Homo sapiens*	reproduction
binomial nomenclature	hinny	respiration
breathing	hybrid	sexual reproduction
catabolism	ingestion	species
cellular respiration	kingdom	synthesis
circulation	life function	taxon
class	metabolism	taxonomy
diffusion	mitosis	Tetropoda
digestion	Monera	transport
Eubacteria	mule	Vertebrate
excretion	nutrient	
family	nutrition	

Questions for Review

PART A.   Completion. Write in the word that correctly completes each statement.

1. The taking in of food is known as ..1..
2. The type of transport requiring energy is ..2.. transport.
3. The pumping of air into and out of the lungs is known as ..3..
4. Small molecules are built into larger ones in biochemical processes known as ..4..
5. Examples of excretory organs are the kidneys, the skin and the ..5..
6. The building blocks of proteins are ..6..
7. Reproduction that involves one parent is known as ..7..
8. Chromosomes are duplicated in a process called ..8..
9. Destructive metabolism is known as ..9..
10. "To stay the same" is expressed by the biological term ..10..
11. The mating of a male donkey and a mare produce the animal known as a ..11..
12. Binomial nomenclature refers to ..12.. naming.

13. *Equus asinua* and *Equus caballus* indicate that these organisms belong to the same ..13..

14. The ..14.. name is always capitalized.

15. The science of classification is known as ..15..

16. The current classification system consists of ..16.. kingdoms.

17. Animals with backbones are classified as ..17..

18. Linnaeus based his system of classification on the ..18.. of organisms.

19. Under the five kingdom system of classification, all bacteria were grouped in the kingdom ..19..

20. A bacillus is a bacterium that is shaped like a ..20..

PART B. Multiple Choice. Circle the letter of the item that correctly completes each statement.

1. Chemical changes which change large food molecules into smaller soluble ones are known as
   (a) egestion   (c) mastication
   (b) ingestion   (d) digestion

2. The general term applied to the absorption and distribution of molecules within the body of an organism is
   (a) diffusion   (c) cyclosis
   (b) transport   (d) motion

3. Energy-releasing activities carried out by cells are known as
   (a) guttation   (c) digestion
   (b) respiration   (d) assimilation

4. The wastes of cellular respiration are
   (a) CO_2 and nitrogen   (c) carbon dioxide and water vapor
   (b) urea and oxygen   (d) urea and nitrogen

5. Auxins are most correctly associated with
   (a) bean plants   (c) snails
   (b) giraffes   (d) beetles

6. Loss of water by plants during periods of high humidity is known as
   (a) excretion   (c) cremation
   (b) transportation   (d) guttation

7. All of the biochemical activities in the body are included in the term
   (a) anabolism   (c) metabolism
   (b) atavism   (d) catabolism

8. When species lose their reproductive potential, they
   (a) become extinct   (c) develop alternate systems
   (b) increase in numbers   (d) develop more energy

9. The long neck of a giraffe is considered to be a (an)
   (a) throw-back   (c) effector
   (b) affectation   (d) adaptation

10. The first word in the species binomial refers to the
    (a) kingdom   (c) genus
    (b) phylum   (d) class

11. The basic unit of classification is the
    (a) taxon   (c) kingdom
    (b) phylum   (d) species

12. A mule can properly be described as a
    (a) high breed   (c) mongrel
    (b) hybrid   (d) hinny

13. The name Linnaeus is best associated with
    (a) taxonomy   (c) paleobiology
    (b) morphology   (d) anatomy

14. The scientific name is the same as the
    (a) common name   (c) genus name
    (b) species name   (d) popular name

15. *Euglena* is best classified as a
    (a) plant   (c) fungus
    (b) animal   (d) protist

16. Tetrapoda are animals that have four
    (a) limbs   (c) fingers
    (b) legs   (d) toes

17. All bacteria that cause disease are placed in the kingdom
    (a) Fungi   (c) Eubacteria
    (b) Monera   (d) Archaeobacteria

18. Mushrooms are classified as
    (a) nongreen plants   (c) protists
    (b) bacteria   (d) fungi

19. "Anaerobic" means without
    (a) oxygen   (c) methane
    (b) carbon dioxide   (d) heat

20. All members of the animal kingdom
    (a) have legs   (c) breathe in carbon dioxide
    (b) are multicellular   (d) metabolize methane

PART C. Modified True-False. If a statement is true, write "true" for your answer. If a statement is incorrect, change the underlined word to one that will make the statement true.

1. A definition of life is quite <u>easy</u> to formulate.

2. Living things are highly <u>dispersed</u> systems.

3. The process in which nutrients are changed into protoplasm is known as <u>digestion</u>.

4. The movement of materials from place to place in the body is known as <u>locomotion</u>.

5. Processes that involve control and coordination of the activities of living organisms are known collectively as <u>metabolism</u>.

6. Increase in cell size is known as <u>growth</u>.

7. The endocrine system is a part of the <u>plant</u> body.

8. Auxins help regulate growth in <u>plants</u>.

9. Two parents are required for the process of <u>asexual</u> reproduction.

10. Anabolism is a <u>breaking down</u> process.

11. A hinny <u>can</u> reproduce others like itself.

12. The <u>genus</u> name is a binomial.

13. The largest classification group is the <u>species</u>.

14. **The branch of biology that studies change in form of living things is <u>classification</u>.**

15. **A classification group of organisms is known as a <u>taxon</u>.**

16. All Archaeobacteria live where there is no <u>water</u>.

17. Amoeba belong to the kingdom <u>Animalia</u>.

18. The name Carl Woese is best associated with the <u>five-kingdom</u> system of classification.

19. John Ray was the first to introduce the concept of the <u>genus</u>.

20. The fine detail of bacteria is best resolved through the <u>light</u> microscope.

Think and Discuss

1. Compose a definition of life. Why is life difficult to define?
2. Why must biologists classify living things?
3. Explain the meaning of homeostasis.
4. Why are bacteria not placed in the kingdom Protista?

Answers to Questions for Review

PART A

1. ingestion
2. active
3. breathing
4. anabolism
5. lungs
6. amino acids
7. asexual
8. replication
9. catabolism
10. homeostasis
11. mule
12. double (scientific)
13. genus
14. genus
15. taxonomy
16. six
17. vertebrates
18. structure
19. Monera
20. rod

PART B

1. d	8. a	15. d
2. b	9. d	16. a
3. b	10. c	17. c
4. c	11. d	18. d
5. a	12. b	19. a
6. d	13. a	20. b
7. c	14. b	

PART C

1. difficult
2. organized
3. assimilation
4. transport or circulation
5. regulation
6. true
7. animal
8. true
9. sexual
10. building up
11. cannot
12. species
13. kingdom
14. evolution
15. true
16. oxygen
17. Protista
18. six
19. species
20. electron

CHAPTER 3

THE CELL: BASIC UNIT OF LIFE

The body of a living organism is built of units called *cells*. All living things are similar in that they are composed of one or more cells. The body of a *unicellular* organism is composed of one cell. Most plants and animals are *multicellular*, having a body made of numerous cells.

During the years 1838–1839, the *cell theory* was formulated by two eminent scientists of the day. Matthias Schleiden, a botanist, and Theodor Schwann, a zoologist, put together some of their fundamental ideas about the structure of plants and animals in what has developed into a basic concept of biological thought. The cell theory states (1) that cells are the basic units of life; (2) that all plants and animals are made of cells; and (3) that all cells arise from preexisting cells.

The Cell as a Basic Unit

UNIT OF STRUCTURE

Microscopic examination of plant and animal parts indicates that the bodies of living organisms are composed of cells. Cells provide structure and form to the body. They appear in a variety of shapes: round, concave, rectangular, elongate, tapered, spherical and other. Cell shape seems to be related to specialized function (Fig. 3.1).

Cells not only vary in shape, they also differ in size. Most plant and animal cells are quite small, ranging in size between 5 and 50 micrometers in diameter (Fig. 3.2). Cells are measured in units that are compatible with modern microscopes.

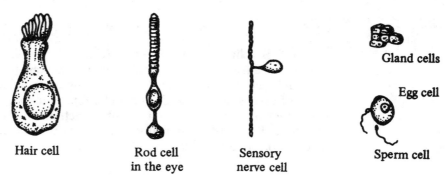

Hair cell

Rod cell
in the eye

Sensory
nerve cell

Gland cells

Egg cell

Sperm cell

Fig. 3.1 Types of human cells

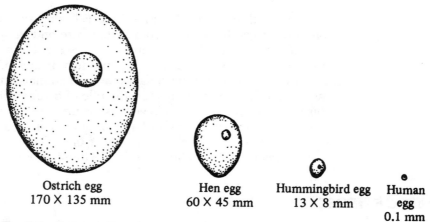

Ostrich egg
170 × 135 mm

Hen egg
60 × 45 mm

Hummingbird egg
13 × 8 mm

Human
egg
0.1 mm

Fig. 3.2 Comparative sizes of egg cells

TABLE 3.1.   Size Units Used In Microscopy

	Unit	Symbol	Equivalent	Measurement Use
Naked Eye Measurements	Meter	m	100 cm	Naked eye measurement; standard from which microscopic measurements are derived.
	Centimeter	cm	0.01 m; 0.4 in. 10 mm	Naked eye measurement of giant egg cells.
	Millimeter	mm	0.1 cm	Naked eye measurement of large cells.
Microscope Measurements	Micrometer **(Micron)	μm μ*	0.001 mm	Light microscope measurement of cells and larger organelles.
	Nanometer **(Millimicron)	nm mμ	0.001 μm	Electron microscopy measurement; cell fine structure; large macromolecules.
	**Angstrom unit	Å	0.1 nm	Electron microscopy measurement; molecules and atoms; X-ray methods.

* Pronounced *mew*.
** These units of measure are being phased out, but still appear in scientific literature.

UNIT OF FUNCTION

Each cell is a living unit. Whether living independently as a protist or confined in a tissue, a cell performs many metabolic functions to sustain life. Each cell is a biochemical factory using food molecules for energy, repair of tissues, growth and ultimately, reproduction. On the chemical level, the cell carries out all of the life functions that were discussed in Chapter 2. Living organisms function the way they do because their cells have the properties of life.

UNIT OF GROWTH

Each living thing begins life as a single cell. Protists remain unicellular. The protist cell body grows to a certain size and then divides into two new cells. A multicellular organism also begins life as a single cell. Its original cell grows to a certain size and then divides. Unlike the cells of protists, the cells of multicellular plants and animals hold together forming *tissues*. Later in this chapter, the role of tissues will be discussed. At this point, we are interested in cells as the unit of growth. As the number of cells increases in the body of a plant or animal, so does its size. Large organisms have more cells than smaller ones. Thus the size of a living thing depends upon the increase in the number of its cells.

UNIT OF HEREDITY

New cells arise only from preexisting cells. A cell grows to optimum size and then divides, producing two other cells *identical* to itself. Paramecia produce other paramecia; onion membrane cells produce new onion membrane cells identical in structure and function to themselves. From single cells, multicellular organisms produce other organisms like themselves. This is so because cells carry hereditary information from one generation to the next. The information is coded in molecules of DNA (deoxyribonucleic acid) and is usually transmitted accurately to new cells. The reproductive machinery in the nucleus of the cell serves as a carrier of hereditary instructions.

Parts of a Cell

Essentially, all cells have similar parts designed to contribute to the work of the whole cell. The cell may be described as a membrane-enclosed unit consisting of *cytoplasm* and a *nucleus*. Both nucleus and cytoplasm are packed with finer structures that can be resolved or distinguished clearly only through the electron microscope.

The parts of a cell are known as *organelles*, meaning "little organs." This term is appropriate because parts of the cell have special functions, somewhat like miniature body organs. Fig. 3.3 shows a typical animal cell as demonstrated through the light microscope. However, there are many organelles in the cell that cannot be seen by using the light mi-

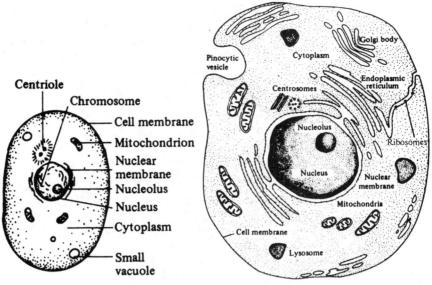

Fig. 3.3 Typical animal cell

Fig. 3.4 Electron micrograph of a cell

croscope. These structures can be resolved by the electron microscope and are known collectively as the *fine structure of the cell*. Fig. 3.4 presents an *electron micrograph* of a cell. You will find it helpful to refer to this diagram as you read about the structure and functions of parts of the cell.

CELL MEMBRANE

The outer boundary of the cell is called the *cell* or *plasma membrane*. Although it is no thicker than 10 nanometers, the cell membrane has an intricate molecular structure. It is composed of a bilayer of phospholipid molecules with proteins of various sizes embedded in the layers. *Pores in the cell membrane are lined with small protein molecules.*

The cell membrane controls the passage of materials into and out of the cell. It is often referred to as a living gatekeeper. The cell membrane is *semi-permeable* and highly selective: not every ion or molecule can cross its boundary. The movement of materials across the cell boundary and into or out of the cell is given the general term of *transport*. It is controlled by the globular proteins, the phospholipids and the pores of the membrane and by the electrochemical nature of *protoplasm*, the living substance of the cell.

Look at Fig. 3.5. This diagram shows the structure of the cell membrane as research biologists currently believe it to be. Using this *fluid mosaic model*, biologists can explain how some molecules move into and out of cells while others cannot penetrate the cell membrane boundary.

As you look at the fluid mosaic model, notice its structure. The core (center) of the membrane is made of compounds called *phospholipids*. Fig. 3.6 shows the structure of a phospholipid. The rounded head contains a phosphate group, the tails are fatty acids, and a variable organic group is attached to the phosphate head. Notice that the tails are posi-

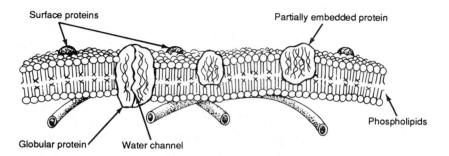

Fig. 3.5 Fluid mosaic model of cell membrane

Source: Biology the Science of Life. Wallace, King et al. 2nd Edition. Saunders p. 107

tioned inward, while the phosphate head points outward. The phosphate head is water-loving, but the lipid tails reject water. In other words, these fatty acid tails prevent water-soluble molecules from crossing the membrane.

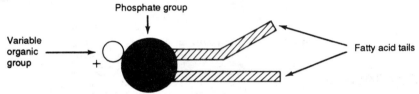

Fig. 3.6 Phospholipid

Nevertheless, water-soluble materials do enter the cell. They do so by way of the large, circular proteins that are set into the membrane. These proteins have channels through which water and its dissolved substances can pass. Other, smaller proteins lie on the surface of the cell membrane and admit water freely. Another group of small proteins, which have only one water-rejecting end, are partly embedded in the membrane.

Each of these three groups of proteins has special functions. One group interacts with hormones and helps special substances get into cells. Another group of proteins is enzymelike and carries out chemical reactions on the surface of the membrane. The large, globular proteins are free to pass back and forth in the membrane and function as carrier molecules. These proteins actually carry certain substances across the cell membrane into the cells. The cell membrane is not regarded as rigid—thus the concept of fluid. The scattered proteins indeed form a pattern or a mosaic. The fluid mosaic model of the cell membrane is a new concept about which much remains to be learned.

Overall, there are two major types of transport into and out of the cell: passive transport and active transport. *Passive transport* does not require the cell's chemical energy to move molecules. Passive transport does depend, however, on the heat energy within the cell to increase the frequency *with which molecules move.* One type of passive transport is *diffusion,* which is the process by which molecules move from an area of greater concentration to an area of lesser concentration. *Osmosis* is the movement of water across a semipermeable membrane. *Plasmolysis* is the shrinking of cytoplasm due to the movement of water out of the cell. Osmosis and plasmolysis are forms of diffusion and thus are examples of passive transport.

Active transport requires the expenditure (use) of chemical energy that is stored in ATP molecules in the cell. *Pinocytosis*, or "cell drinking," is a form of active transport by which fluid molecules are engulfed (taken in) by cells through the formation of vesicles (pockets) in the cell membrane. Solid particles are ingested by cells through a process known as *phagocytosis*; white blood cells ingest bacteria by means of phagocytosis. Collectively, pinocytosis and phagocytosis are known as *endocytosis*. At times molecules are forced out of cells by *exocytosis*, a means by which they are carried to the cell surface by vacuoles or vesicles.

The *sodium/potassium ion-exchange pump* is an important mechanism in active transport. Sodium/potassium ion exchange pumps push sodium ions out of the cell and force potassium ions into the cell. This process involves a protein carrier that is lodged in the cell membrane. Three sodium ions become attached to a special site (place) on the protein. Interaction with ATP causes the protein to change shape and release two of the sodium ions to the outside of the cell. Potassium ions take the place of the released sodium ions and are deposited within the cell. Sodium/potassium ion-exchange pumps are the means by which sodium ions are forced out of nerve cells, while potassium ions are pulled into the cells.

CYTOPLASM

All of the living material of the cell that lies outside the nucleus is the *cytoplasm*. It is packed full of organelles and is highly structured by a network of fine tubes and fibers that are spread throughout the cell. These supporting tubes (microtubules) and fibers (microfilaments) are known collectively as the *cytoskeleton*. *Microtubules* are long, thin, hollow tubules (little tubes), measuring 25 nanometers in diameter. They are composed of globular proteins and act as the framework of the submicroscopic cytoskeleton. Microtubules also seem to direct the flow of circulating materials within the cytoplasm and to create pathways for organelle movement. *Microfilaments* are long protein threads that measure about 6 nanometers in diameter and function in cell movement.

ENDOPLASMIC RETICULUM

Spreading throughout the cytoplasm, extending from the cell membrane to the membranes of the nucleus is a network of membranes that form channels, tubes and flattened sacs; this network is named the *endoplasmic reticulum*. One function of the endoplasmic reticulum is the movement of materials throughout the cytoplasm and to the plasma membrane. The endoplasmic reticulum has other important functions related to the synthesis of materials and their packaging and distribution to sites needed. Some membranes of the endoplasmic reticulum are dotted with thousands of organelles known as ribosomes; others are smooth.

RIBOSOMES

Ribosomes are small, circular organelles measuring 25 nanometers in diameter. They are by far the most numerous organelles in a cell; in one bacterial cell ribosomes may number upward of 15,000. Some ribosomes are attached to the membranes of the endoplasmic reticulum. These are engaged in the synthesis of proteins that will be exported from the cell to be used by various organs of the body. Digestive enzymes and hormones are examples of the kinds of protein molecules that are secreted by cells to be used elsewhere in the body. Ribosomes that lie free in the cytoplasm synthesize proteins to be used in the cell in activities such as cellular respiration. In general, ribosomes are the sites of protein synthesis.

GOLGI BODY

The *Golgi body*, sometimes referred to as the Golgi apparatus, consists of a series of membranes loosely applied to one another and forming *vesicles* (fluid-filled pouches) that are surrounded by microtubules. The Golgi body receives vesicles and their fluids from membranes of the endoplasmic reticulum. The vesicles are rewrapped in membranes by the Golgi body and transported to the cell membrane where they leave the cell. In general, the function of the Golgi body is to temporarily store, package and transport materials synthesized by other parts of the cell. A number of compounds formed in the endoplasmic reticulum are funneled into the Golgi apparatus, where they are modified and concentrated. These compounds are then rewrapped and transported to other parts of the cell. Among these compounds are many types of proteins, glycoproteins, proteins that become part of the plasma membrane, and glycolipids. Plant cells have several hundred Golgi bodies; animal cells, usually 10 to 20.

LYSOSOMES

The *lysosomes* are vacuoles bounded by double membranes which keep them isolated from other cellular organelles. The lysosomes contain hydrolytic enzymes that are capable of destroying the cell. By a well-controlled mechanism, lysosomes become attached to food vacuoles and release some of the digestive enzymes into the food vacuoles. Bacteria engulfed by white blood cells are digested in this way.

PEROXISOMES

Peroxisomes are membrane-bound sacs that resemble lysosomes. Unlike lysosomes, however, peroxisomes contain oxidizing enzymes. These enzymes make certain substances that are toxic to the cell harmless by adding oxygen to them. Membranes that form peroxisomes are derived from the endoplasmic reticulum.

GLYOXYSOMES

Glyoxysomes are membrane-bound sacs in the cytoplasm that resemble peroxisomes. However, glyoxysomes contain enzymes that convert fatty acids to carbohydrates. Glyoxysomes are usually found in plant seedlings.

The enzymes in glyoxysomes change the fats stored in the seed into carbohydrates that are used to build cell wall structures.

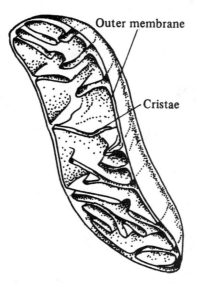

Fig. 3.7 Mitochondrion

MITOCHONDRION

Aside from the nucleus, the *mitochondrion* is the largest organelle in the cell. Because there are so many of these organelles, the plural mitochondria is used more frequently than the singular mitochondrion. In diameter, mitochondria measure between 0.5 and 1 micrometer; in length they measure up to 7 micrometers. Mitochondria, known as the "power houses of the cell," exhibit a variety of shapes: round, ovoid, elongated, and the like. Each mitochondrion is surrounded by a double membrane and has a membrane of many folds fitted into its internal structure. The many folds are covered with enzymes necessary for the chemical reactions that release energy. The degree of folding of the inner membrane is related to the energy requirements of the cell. The internal membranes, *cristae*, actually increase the surface area, permitting a great amount of biochemical activity to take place. In addition to increasing surface area, the cristae form compartments that provide additional work and storage areas for the complex task of *cellular respiration*. Depending on their energy needs, cells may have more than 2,500 mitochondria (Fig. 3.7).

NUCLEUS

The largest and most prominent organelle in the cell is the nucleus. A few primitive species such as bacteria and blue-green algae do not have an organized nucleus. These types of cells are known as *prokaryotes*. Most cells do, however, have a double membrane-bound nucleus and are classified as *eukaryotes*.

The double membrane of the nucleus fuses at certain places forming openings called *pores*. The pores measure about 65 nanometers in diameter. Inside the nuclear membrane a clear semi-solid material seems to fill up the nucleus. Embedded in this material are one or two small spherical bodies called *nucleoli* (singular, nucleolus). The nucleolus is the site of the synthesis and storage of the nucleic acid RNA (ribonucleic acid). In the nucleus of the nondividing cell is a tangle of very fine threads which absorb stain quite readily. In the granular stage these threads are known as *chromatin*. The chromatin threads come together, shorten and thicken forming *chromosomes* that can be seen quite prominently in the dividing cell.

Comparison of Plant and Animal Cells____

A typical plant cell is shown in Fig. 3.8; a typical animal cell is also shown for comparison. Notice that a plant cell has as its outer boundary a *cell wall* which surrounds the plasma membrane. The cell wall is a nonliving structure composed of *cellulose*, a complex starch molecule. Cellulose molecules are bound together in an intricate pattern and held in place by gluey carbohydrates known as *pectins*. The carbohydrate molecules that compose the cell wall are synthesized by the cytoplasm of the cell and secreted through the cell membrane to form a rigid boundary around the cell.

The cell wall serves to support the cell, to protect it from drying out and to inhibit bacterial invasion. If the cytoplasm inside the cell loses water and shrinks, the cell wall still retains its shape and remains fairly rigid. Animal cells do not have a structure that can be compared to the cellulose cell wall.

Fig. 3.8 shows a plant cell with a very large *vacuole*. A vacuole is a space in the cytoplasm filled with water and dissolved substances such as salts, sugars, minerals and other materials. The *cell sap* in the vacuoles of some cells contains pigments that color the flowers, fruits, leaves and stems of plants. Like other organelles, the vacuole is surrounded by a membrane which separates it from the cytoplasm of the cell.

Storage areas in animal cells are quite small, occupying very little space in the cytoplasm. These storage areas in animal cells are known as *vesicles*. They are neither fixed in number nor as lasting as organelles, but rather appear and disappear as they move materials from the endoplasmic reticulum to the Golgi body to the cell membrane.

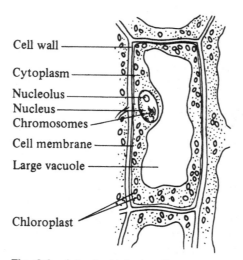

Fig. 3.8 A typical plant cell

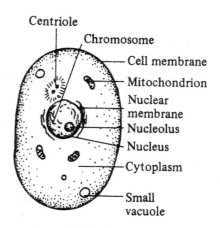

A typical animal cell

Chloroplasts belong to a group of structures which have the general name *plastids*. Plastids are membrane-bound organelles found only in plant cells. Usually plastids are spherical bodies that float freely in the cytoplasm, holding pigment molecules or starch. Storage plastids include leucoplasts and chromoplasts. Chloroplasts contain the green pigment *chlorophyll*, a substance that gives plants their green color. Chlorophyll is a special molecule that has the ability to trap light and to convert it to a form of energy that plants can use in carrying out the chemical steps of the food-making process known as *photosynthesis*.

Each chloroplast is surrounded by a double membrane. Inside the chloroplast are numerous flattened membranous sacs called *thylakoids* (formerly called *grana*). The thylakoids are the structures that contain the chlorophyll and it is within these sacs that photosynthesis takes place. *Stroma* is the name given to the dense ground substance that cushions the thylakoids. Animal cells do not have chloroplasts and therefore cannot make their own food. Fig. 3.9 shows the fine structure of a chloroplast.

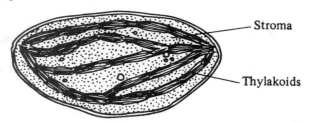

Stroma

Thylakoids

Fig. 3.9   A chloroplast

Centrioles are paired structures that lie just outside of the nucleus of nearly all animal cells and some cells of lower plants. They are absent in cells of higher plants. Under the light microscope, the centrioles look like two insignificant granules, but the electron microscope demonstrates that they have a very intricate structure (Fig. 3.10).

Flagella and *cilia* are fine threads of cytoplasm that extend from the surfaces of some cells. Both of these structures are involved in the locomotion of some protist species. Cilia are relatively short extensions but appear in great numbers, usually surrounding the body of the protist. Flagella are much longer than cilia and appear in fewer numbers.

Microtubules

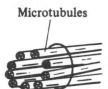

Fig. 3.10 The fine structure of a centriole

In addition to serving the locomotive needs of one-celled organisms, flagella and cilia help functions of other types of cells. Sperm cells of animals and plants are propelled through fluid media by the whip-like actions of their flagella. Tissue cells of the human windpipe are lined with cilia which wave back and forth catching dust particles and pushing them away from the lungs. The microstrucure of the flagella and cilia resembles that of the centrioles.

Organization of Cells and Tissues _____

Cells in the body of the multicellular organism are arranged in structural and functional groups called *tissues*. A tissue is a group of similar cells that work together to perform a particular function. Tissues that are grouped together and work for a common cause form *organs*. Groups of organs that contribute to a particular set of functions are called *systems*. The ability of cells to carry out special functions in addition to the usual work of cells exemplifies *specialization*. When different jobs are accomplished by the various tissues in an organ, we call this *division of labor*.

Tables 3.2 and 3.3 provide a summary of major animal and plant tissues.

TABLE 3.2.   Animal Tissues

Tissue Type	Function of Tissue
Epithelial	Made of closely packed cells specialized for covering and lining organs and protecting underlying tissues from drying out, mechanical injury and bacterial invasion (Fig. 3.11).
squamous cells	Flattened epithelial cells; specialized for lining body cavities such as the mouth, esophagus, eardrum and vagina.
cuboidal cells	Cube-like cells; specialized for gland tissue and the lining of the kidney tubules.
columnar cells	Tall and column-like cells; specialized for lining the alimentary canal in such organs as the stomach and intestines.
ciliated columnar cells	Ciliated columnar epithelial cells line the tubes that serve as air passageways in the respiratory system; the cilia push dust particles away from the lungs.
goblet cells	Specialized for secreting and storing fluid products of cells such as milk, hormones, enzymes and oils.

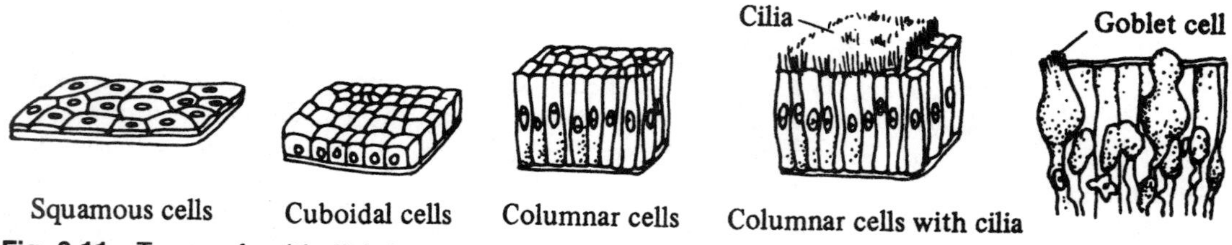

Squamous cells   Cuboidal cells   Columnar cells   Columnar cells with cilia

Fig. 3.11   Types of epithelial tissue

Nervous	Made of cells called *neurons* that are specialized for carrying impulses; the nervous system coordinates all of the body's activities (Fig. 3.12).
sensory neuron	Carries information from sense organs to other neurons in the brain and spinal cord.
motor neurons	Carries impulses to muscles and glands.
interneurons	Transmits impulses from sensory neurons to motor neurons.

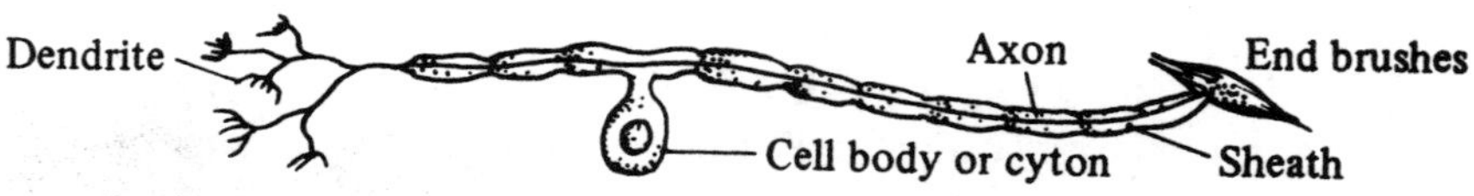

Fig. 3.12   Nerve cell

Tissue Type	Function of Tissue
Muscle	Specialized to respond to stimuli transmitted by motor neurons; characterized by electrical excitability and the ability to contract (Fig. 3.13).
smooth	Involuntary muscle; composes organs not under the control of the individual—e.g. small intestine.
cardiac	Specialized for the heart.
striated	Voluntary muscle; composes organs that are controlled by the will of the individual—e.g. arm and leg muscles.

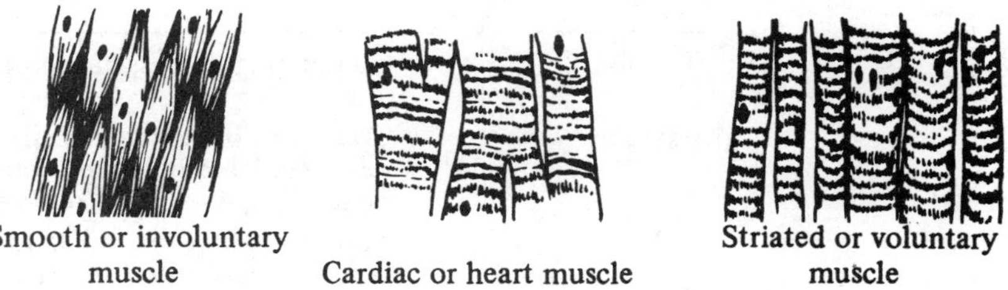

Fig. 3.13 Types of muscle tissue

Smooth or involuntary muscle

Cardiac or heart muscle

Striated or voluntary muscle

Blood	Composed of three types of cells suspended in plasma (Fig. 3.14).
erythrocytes	Red blood cells specialized for carrying oxygen.
leucocytes	White blood cells specialized for counteracting invasions by disease organisms.
platelets	Cell fragments that function in blood clotting.

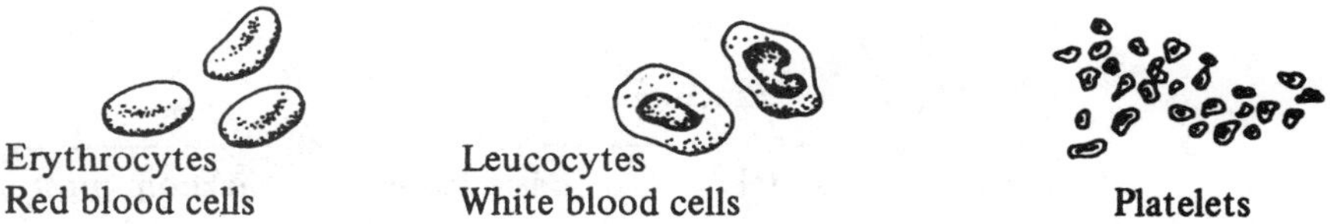

Fig. 3.14 Types of blood cells

Erythrocytes
Red blood cells

Leucocytes
White blood cells

Platelets

Connective	Characterized by large deposits of nonliving material that surrounds living cells; the nonliving *matrix* supports and binds tissues to the body skeleton (Fig. 3.15).
cartilage	Firm and elastic matrix found at tip of nose, end of long bones, ear lobes and wherever strength and flexibility are needed.
fibrous connective tissue	Parallel protein fibers give strength to the matrix; *ligaments*—elastic fibers—connect bone to bone. *Tendons*—nonelastic fibers—connect muscle to bone.

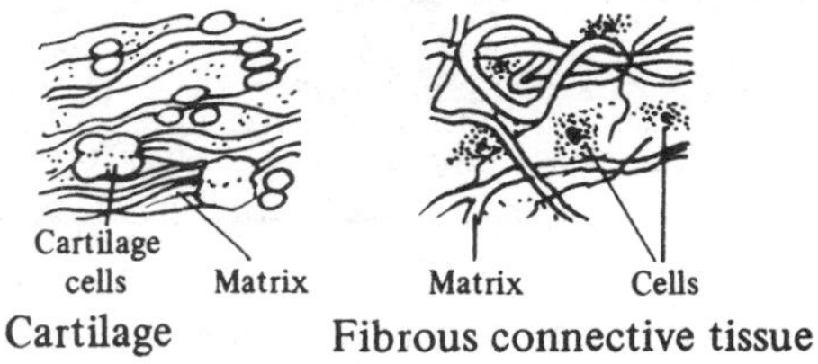

Fig. 3.15 Types of connective tissue

Cartilage
cells   Matrix
Cartilage

Matrix   Cells
Fibrous connective tissue

Bone	Calcium and phosphorus compounds make up the matrix; living cells arranged in rings around a canal through which nerve fibers and blood vessels extend. The skeleton of vertebrates is made of bone (Fig. 3.16).

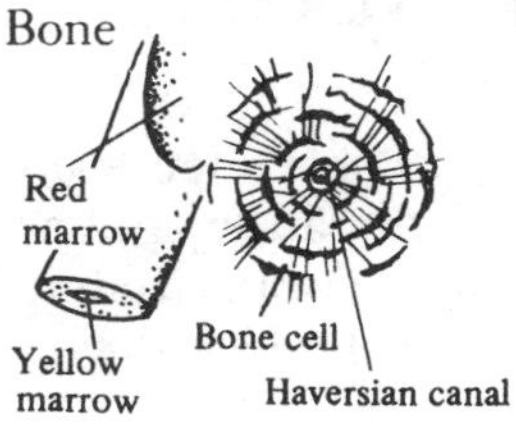

Fig. 3.16 Bone tissue

Red marrow
Yellow marrow
Bone cell
Haversian canal

TABLE 3.3. Plant Tissues

Tissue Type	Function of Tissue
Epidermis	One cell-layer thick; covering surfaces of leaves, stems and roots, protecting inner tissues (Fig. 3.17).

Leaf epidermis   Cork cells

Fig. 3.17 Epidermal plant tissue

Vascular	Conducting tissues; transport materials throughout plant (Fig. 3.18).
xylem	Composed of elongated cells known as *tracheids* and *vessels*, or conducting tubes; conducts water and its dissolved minerals upward from roots of plant.
phloem	Transports food materials to all parts of the plant; composed of *sieve tubes* and *companion cells*.

Xylem   Phloem

Fig. 3.18 Vascular plant tissues

Fundamental	Basic tissues that make up most of the plant body (Fig. 3.19).
parenchyma	Thin-walled cells containing chloroplasts and other plastids; tissues where photosynthesis takes place.
sclerenchyma	Tough supporting tissues composed of sturdy cell walls and dead cells; gives mechanical support to plant stems and forms tough coverings of seeds.
collenchyma	Living cells in stems and leaves that provide support.

Parenchyma   Sclerenchyma   Collenchyma

Fig. 3.19 Fundamental plant tissue

The Basic Chemistry of Cells

Most of the activities of the cell involve chemical changes. Small molecules may be joined together to form larger ones or complex substances may be broken down into their smaller units. Materials are changed from one form to another, used up or synthesized during biochemical events that take place in cells.

CHEMICAL COMPOUNDS IN LIVING CELLS

The cell is likened to a "chemical factory" that uses some of the elements present in the nonliving environment. Of the elements in the living material of the cell, carbon, hydrogen, oxygen, and nitrogen are present in the greatest amounts. Sulfur, phosphorus, magnesium, iodine, iron, calcium, sodium, chlorine and potassium are found in smaller quantities. These elements are present in inorganic and organic compounds that are utilized by the cell.

Inorganic compounds are compounds that do not have the elements carbon and hydrogen in chemical combination. The inorganic compounds in greatest percentages in cells are water, mineral salts, inorganic acids and bases. For example: hydrochloric acid (HCl), an inorganic acid, is secreted by the gastric glands in the stomach, acidifying stomach contents.

Organic compounds are compounds that contain the elements carbon and hydrogen in chemical combination. Organic compounds are produced by living plants and animals and can be synthesized in the laboratory, as well. The special bonding property of carbon permits it to form compounds that are structured as long chains of atoms or as rings of atoms. The types of organic compounds that are contained in the living material of the cell and used in its chemical activities are carbohydrates, lipids, proteins, and nucleic acids.

Carbohydrates

Carbohydrates are composed of the elements carbon, hydrogen, and oxygen. Hydrogen and oxygen atoms are usually present in carbohydrates in the ratio of 2:1. Glucose ($C_6H_{12}O_6$) represents the basic unit of carbohydrate structure, a *monosaccharide* molecule. A monosaccharide is a simple sugar from which larger carbohydrate molecules are built. When two monosaccharide molecules are joined together chemically, a double sugar or *disaccharide* is formed. As the two molecules join, a molecule of water is produced during the process in addition to the double sugar. A synthesis of this type is known as *dehydration synthesis*.

$$C_6H_{12}O_6 + C_6H_{12}O_6 \rightarrow C_{12}H_{22}O_{11} + H_2O$$

$$\text{glucose} + \text{glucose} \rightarrow \text{maltose} + \text{water}$$

Within living cells, many monosaccharides chemically combine by dehydration synthesis and form a *polysaccharide*, such as cellulose.

Carbohydrates are used by cells primarily as sources of energy. In plant cells, polysaccharides are utilized in cell structures such as cell walls. In some animal cells, glycoproteins (carbohydrate and protein compounds) form recognition sites on the membranes of certain cells.

Lipids

The lipids are a group of organic compounds that include the fats and fat-like substances. A lipid molecule contains the elements carbon, hydrogen and oxygen like that of a carbohydrate. Unlike the carbohydrates, however, in lipid molecules the ratio of hydrogen to oxygen is much greater than 2:1. A lipid molecule is made up of two basic units: an *alcohol* (usually glycerol) and a class of compounds called *fatty acids*.

Lipids are sources for biologically usable energy. Most lipid molecules provide twice as much energy per gram as do carbohydrate molecules. However, the energy in lipid molecules is not so easy to extract as that of glucose. Besides being useful for energy, lipids are essential to the structure of cells. The plasma membrane is composed of lipid molecules as is the myelin sheath of some nerve fibers.

Proteins

Proteins are organic compounds composed of the elements carbon, hydrogen, oxygen and nitrogen. Some proteins also contain sulfur. All proteins are built from small molecular units known as *amino acids*. The amino acid molecules link together in a particular way through *peptide* bonds. A *dipeptide* consists of two amino acids. A *polypeptide* contains many amino acid molecules. A *protein* is composed of one or more polypeptide chains.

About 20 amino acids are essential to living systems. From these a large number of different kinds of proteins are formed. The great variety of protein molecules is possible because of the many ways in which amino acid molecules can be arranged. Changing the sequence of just one amino acid in a chain will change the protein molecule.

Much of the work of the cell is concerned with the synthesizing of protein molecules. Some of these proteins such as hormones, enzymes and hemoglobin are used in complex biochemical activities. Other proteins contribute to the structure of cells such as those that make up the plasma membrane and other cellular membranes.

Nucleic Acids

Nucleic acids are a group of organic compounds that are essential to life. These are the compounds that pass hereditary information from one generation to another, making possible a remarkable continuity of life within the various species of living things. Deoxyribonucleic acid (DNA) molecules are the particular type of nucleic acid out of which genes are made. Genes are the bearers of hereditary traits from parent to offspring.

Another important characteristic of nucleic acids is their ability to carry information from genes in the cell nucleus to certain structures in the cytoplasm which direct major biochemical processes. For example: the building of proteins is controlled by the group of nucleic acids known in general as ribonucleic acid (RNA). The structures and functions of the three types of RNA molecules and the structure and function of DNA are presented in detailed discussion in Chapter 14.

THE ROLE OF ENZYMES IN LIVING CELLS

Cells are always engaged in chemical activity. The major difference between living things and nonliving matter is that living systems carry out vital chemical activities on a continuous and controlled basis. The control of chemical processes in cells requires the work of *enzymes*.

Enzymes are *organic catalysts*. A catalyst is a molecule that controls the rate of a chemical reaction but is itself *not* used up in the process. Enzymes control the rate of chemical reactions that take place in cells, tissues and organs. Each chemical reaction that occurs in a living system requires the assistance of a specific enzyme.

The Structure and Nature of Enzymes

Enzymes are large complex proteins made up of one or more polypeptide chains. Enzymes may be entirely protein or they may have nonprotein parts known as *coenzymes*. Frequently, vitamins function as coenzymes. For example: riboflavin, popularly called vitamin B_2, is important in a coenzyme function during the cellular process of respiration. Scientists affix the *ase* ending to names of enzymes.

Enzymes have several characteristics that are important to the chemistry of living cells. These characteristics are due to their protein nature. First, enzymes are made inactive by heat. In the human body enzymes work best at normal body temperature of 98.6°F, or 37°C. An increase in temperature of only a few degrees can render enzymes inactive. Second, the action of enzymes can be blocked by certain compounds. We call these enzyme blocks *poisons*. Hydrogen cyanide blocks one of the enzymes that is involved in cellular respiration. Third, enzymes are specific in their activity. Usually they catalyze one particular reaction. For example: the enzyme maltase will only catalyze the digestion of the sugar maltose into two glucose molecules.

Enzymes and pH

The strength of an acid or base (alkali) is measured by its *pH* (see Table 3.4). An *acid* is a substance that releases hydrogen (H^+) ions in solution. The greater the number of hydrogen ions released, the stronger the acid. Thus pH is the measure of H^+ concentration. On the pH scale, the measure of acidity ranges from 0 to 6; the stronger the acid, the lower the pH. Therefore gastric juice, measuring 1 on the pH scale, is far more acidic than saliva, which has a pH of 6. Neutral substances, such as pure water, measure 7 on the pH scale. A *base*, or alkali, is a substance that releases hydroxyl (OH^-) ions in solution. The strength of the base depends on the number of hydroxyl ions released in solution. The numbers between 8 and 14 on the pH scale represent bases of increasing strength; the higher the pH reading, the stronger the base. Therefore molar sodium hydroxide, which has a pH of 14, is a much stronger base than seawater, which measures 8 on the pH scale.

TABLE 3.4.   pH Values Scale

pH	Example
0	1 Molar nitric acid
1	Gastric juice
2	Lemon juice
3	Vinegar
4	Tomato juice
5	Sour milk
6	Saliva
7	Pure water
8	Seawater
9	Baking soda
10	Great Salt Lake
11	Liquid soap
12	Washing soda
13	Oven cleaner
14	1 Molar sodium hydroxide

The pH influences enzyme activity. The stomach enzyme, pepsin, works most effectively in a strongly acidic environment. (The pH of gastric juice is 1.) Most enzymes, however, cannot function to catalyze reactions if the pH of their environmental solutions is not neutral. It seems as though the presence of too many hydrogen or hydroxyl ions interferes with the conforming shape of the enzymes.

How Enzymes Work

Biochemical evidence provides insight into how enzymes work. The surface configuration of an enzyme is known as its *active site*. The *substrate* is the substance upon which the enzyme works. An enzyme affects the rate of reaction of the substrate molecule that fits the enzyme's activity site. In order for this to happen, a close physical association must take place between enzyme and substrate. This association is called the *enzyme-substrate complex* (Fig. 3.20). While the enzyme-substrate complex is formed, the internal energy state of the substrate molecule is changed, bringing about a chemical reaction. As soon as the reaction is completed, the enzyme and the products separate. Each enzyme works at its own maximum rate. The rate at which an enzyme can catalyze all of the substrate molecules that it can in a given time is called the *turnover number*. Different enzymes have different turnover numbers. However, the general rate of enzyme action is influenced by certain environmental conditions such as temperature and relative amounts of enzyme and substrate.

Traditionally, a simple analogy has been used to explain the *specificity* of enzymes. Specificity refers to the characteristic of enzymes that permits a particular enzyme to form a complex with a specific substrate molecule only. The "lock and key" analogy explains enzyme specificity: the substrate is viewed as a padlock, and the enzyme as the key able to unlock it. When unlocked (in the analogy) or acted upon by the key, the padlock comes completely apart. The key remains unchanged and ready to work again on another padlock of the same type (Fig. 3.21).

Newer biochemical evidence, however, supports the *induced fit* hypothesis as an explanation of the specificity of enzymes. Biochemists now tell us that the active site on the enzyme is not rigid, as suggested in the

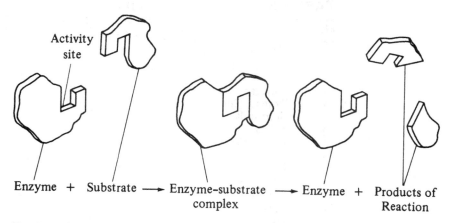

Enzyme + Substrate ⟶ Enzyme–substrate ⟶ Enzyme + Products of
complex   Reaction

Fig. 3.20   Enzyme-substrate complex

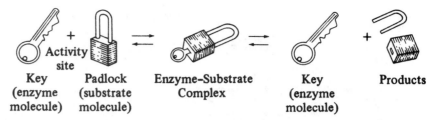

Fig. 3.21   Lock and key analogy of an enzyme and its substrate

lock and key analogy. As the substrate attaches to the enzyme's active site, the site changes shape to fit the substrate. The substrate, now joined to the enzyme's active site, becomes stressed, and the stress weakens certain chemical bonds in the substrate. The weakened bonds give way, and the substrate becomes broken into parts. At the end of this reaction, the substrate and the enzyme separate. The enzyme is unaltered, but the substrate has been chemically changed.

CELLULAR RESPIRATION

Living cells require a constant supply of energy to fuel the chemical activities that sustain life. Glucose is the major supplier of the cell's energy. The cell is able to extract energy from glucose in small packets. The energy release is then stored in two molecules: *adenosine diphosphate* (ADP) and *adenosine triphosphate* (ATP).

Types of Cellular Respiration

Energy is produced by the cell during two major pathways: *anaerobic respiration*, also known as *glycolysis*, and aerobic respiration.

Anaerobic Respiration

As is implied by the name, oxygen is not used during anaerobic respiration. Fig. 3.22 provides a summary of the events that take place during this phase of cellular respiration.

Anaerobic respiration takes place in the cytoplasm of cells. It begins with a molecule of glucose. It becomes activated by energy supplied by ATP. With the assistance of enzymes, glucose is converted through a series of steps to *pyruvic acid*. At each step either a hydrogen atom is given up or a molecule of water is formed. Every time a hydrogen bond is broken, energy within the molecule becomes a bit more concentrated. This energy is ultimately released to ADP molecules and stored between phosphate bonds. Whenever ADP accepts energy, inorganic phosphate from the cellular fluid is attached to the ADP molecule upgrading it to ATP. There is a net gain of two ATP molecules.

Aerobic Respiration

The *aerobic phase* of cellular respiration takes place inside the mitochondria. As the name implies, oxygen is used, serving the important function of the final hydrogen carrier. Fig. 3.23 provides a summary of the events that take place during aerobic respiration, also known as the Krebs cycle.

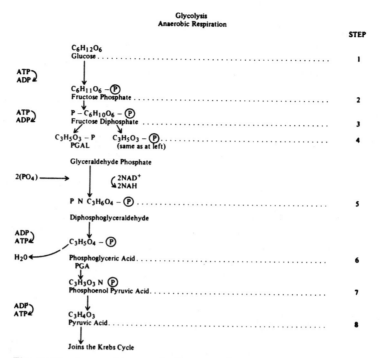

Fig. 3.22   Anaerobic respiration, or glycolysis

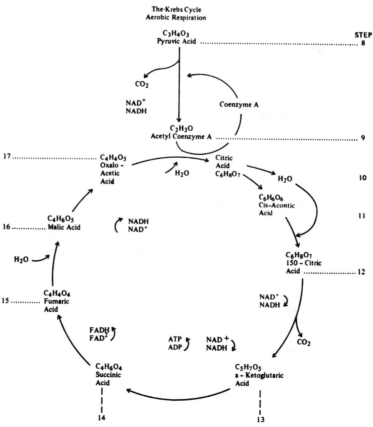

Fig. 3.23   Krebs citric acid cycle

Aerobic respiration begins with pyruvic acid which is quickly converted to acetyl coenzyme A. Through a cycle of chemical changes, fuel molecules are broken apart bit-by-bit, releasing a great deal of energy which is stored in ATP molecules. Hydrogen atoms from compounds formed during glycolysis and aerobic respiration then enter the next phase—oxidative phosphorylation.

Oxidative Phosphorylation

The coenzyme nicotinamide adenine dinucleotide (NAD) receives the hydrogen atoms. These hydrogen atoms are passed along an electron transport chain which is embedded in a mitochondrial membrane. Hydrogen electrons are passed down the chain of acceptors (Fig. 3.24). Meanwhile, hydrogen protons are pushed to the outside of the membrane. This establishes an electrochemical gradient across the membrane. Energy resulting from the difference in potential across the membrane is used to form ATP molecules. At the end of the transport chain, the hydrogen electron joins the proton. These hydrogen ions combine with oxygen to form water. During the electron transport process, inorganic phosphate is joined chemically to ADP forming ATP. Therefore, this phase of aerobic respiration is called *oxidative phosphorylation*.

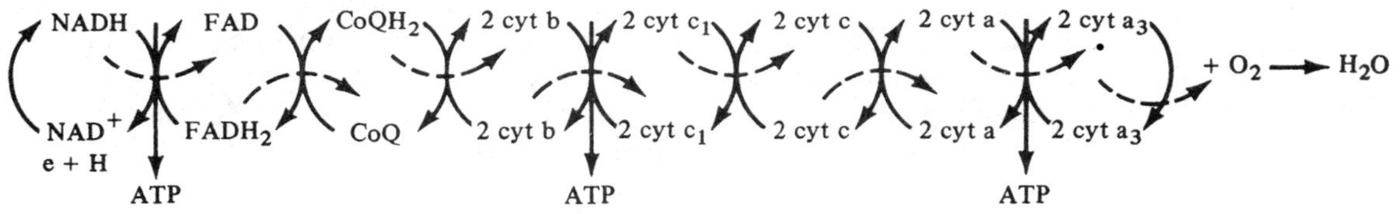

Fig. 3.24 Steps of oxidative phosphorylation including an electron, or cytochrome (cyt), transport system, leading to the formation of ATP

THE CELL CYCLE

Most cells go through an endless cycle of growth, replication of chromosomes, mitosis, and cytokinesis. You will learn about each of these cell events in the sections that follow.

The *cell cycle* consists of four repeating phases known as the *M phase*, *G-1 phase*, *S phase*, and *G-2 phase* (M stands for mitosis, G for gap, and S for synthesis). Mitosis and cell division take place during the M phase. During the G-1 phase, chromosomes are single stranded. Proteins necessary for the growth of the cell are synthesized in this phase. During the S phase DNA molecules are replicated. In the G-2 stage, the cell is getting ready for mitosis. The G-1, S, and G-2 phases are known collectively as *interphase*.

Cells that do not cycle have a much shortened life span. Human red blood cells do not have nuclei and die without undergoing cell division. The nuclei in the red blood cells of fish, birds, and reptiles are not active and cannot function in the process of cell division. These cells also die without reproducing themselves. Most cells, however, cycle continuously. Red blood cells are examples of the exceptions.

Cell Division _____

When a cell reaches a certain size, it divides into two new cells, identical to each other and very similar to the original *parent* cell. The new cells are known as *daughter* cells. The events marking cell division differ in prokaryotes and eukaryotes.

As you recall, a prokaryotic cell does not have an organized nucleus. The nuclear material is in the form of a single circular chromosome attached to the *mesosome* on the cell membrane. Before cell division, the chromosome replicates, forming another chromosome exactly like itself. The new chromosome is attached to its own mesosome on the cell membrane. As the cell elongates, the chromosomes move further apart. Just as the cell doubles in length, the cell membrane pinches inward. New cell wall material surrounds the pinched in membrane and separates the two new cells (Fig. 3.25).

Cell division in the eukaryotic cell includes two separate events: events in the nucleus through which the chromosomes are distributed to the daughter cells and *cytokinesis*, the division of the cytoplasm and its organelles.

Each species has a characteristic number of chromosomes in the nuclei of its cells. It is referred to as the *species number*, or simply as the *chromosome number*. As indicated in Table 3.5, the number of chromosomes in the cells of a given species is not an indicator of size or complexity. The somatic, or body, cells of an organism contain the full complement of chromosomes, referred to as the *diploid number* and designated by the symbol 2N. The nucleus of a somatic cell divides through *mitosis*, and the chromosomes are distributed equally to the daughter cells.

The organism's sex cells, or gametes—eggs and sperm, contain half the species number of chromosomes. This number is called the *haploid number* and symbolized as N. *Meiosis* is the kind of nuclear division that leads to the formation of sperm and egg cells.

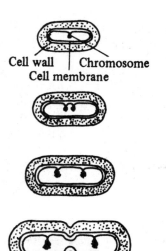

Cell wall | Chromosome
Cell membrane

Fig. 3.25 Division of a bacterium

TABLE 3.5. Chromosome Numbers of Some Common Species

Organism	Haploid No.	Diploid No.
mosquito	3	6
fruit fly	4	8
gall midge	20*	8*
evening primrose	7	14
onion	8	16
corn	10	20
grasshopper (female)	11	22
grasshopper (male)	10	21**
frog	13	26
sunflower	17	34
cat	19	38
human	23	46
plum	24	48
dog	39	78
sugar cane	40	80
goldfish	47	94

* In the fertilized egg of the gall midge, 32 chromosomes become nonfunctional leaving 8 functional chromosomes.
** The male grasshopper has only one sex chromosome.

MITOSIS

Mitosis (also known as karyokinesis) concerns the cell nucleus and its chromosomes. Before the onset of mitosis, the cell is in a stage known as *interphase*. During interphase, the chromosomes are exceptionally long and very thin, appearing as fine granules through the light microscope. It is during this stage that DNA molecules in the nucleus *replicate*. The result of replication is that each chromosome now has an exact copy of itself. When interphase comes to an end, the cell has enough nuclear material for two cells. The orderly process that divides the chromosomes equally between the two daughter cells is known as *mitosis*.

There are four stages of mitosis: prophase, metaphase, anaphase, and telophase. The significant events which mark each of these stages and interphase are outlined below and shown in Fig. 3.26.

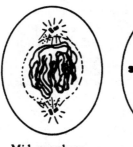   

Mid-prophase   Metaphase   Late anaphase   Late telophase

Fig. 3.26   Stages of mitosis

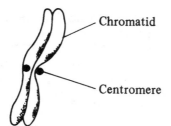
Chromatid

Centromere

Fig. 3.27   Chromatids hold together at centromere

Prophase

1. Chromosomes become shorter and thicker. Each chromosome consists of two *chromatids* attached at the centromere (Fig. 3.27).
2. The nuclear membrane begins to disintegrate.
3. Spindle fibers form extending from the centromeres to the poles.
4. The centrioles in animals cells, fungi, algae and some ferns replicate and a pair migrates toward each pole.
5. Chromosomes begin to move toward the equator of the cell.

Metaphase

1. The centrioles have migrated to the poles.
2. The chromosomes are lined up at the equator of the spindle.
3. Spindle fibers are attached to the centromeres connecting them to the poles of the spindle.
4. Both the nuclear membrane and the nucleolus have disappeared.

Anaphase

1. The centromeres split apart.
2. The chromatid pairs of each chromosome separate from each other. They move quickly in opposite directions, one toward each pole.

Telophase

1. The recently separated chromosomes reach the poles. A pole is the place where the new nucleus of each daughter cell will be located.
2. The spindle fibers extending from the poles to the centromeres disappear. Those fibers that lie in the plane between the opposing rows of chromosomes remain for a longer time.
3. A nuclear membrane reforms around each bundle of chromosomes at the poles. At this time, all remnants of the spindle fibers have disappeared.
4. At the equator of animal cells, the cytoplasm turns inward, pinching the old cell into two new ones. In plant cells, a cell plate of rigid cellulose separates the two new cells.

Interphase

Two new daughter cells are formed identical in genetic material to the parent and to each other. In a relatively short span of time, the daughter cells will accumulate more cytoplasm and grow to the size of the original parent cell. The cell nucleus controls the biochemical activities of the cell during interphase. The only work that the "resting" cell is not doing is dividing.

CYTOKINESIS

Cytokinesis is the separation of the cytoplasm following nuclear division. At first there is a doubling of the molecules that make up the cytoplasm. Cell structures that are composed of protein subunits, such as microtubules, microfilaments, and ribosomes, are synthesized from molecules within the cytoplasm. The membranous organelles, such as the Golgi apparatus, lysosomes, vacuoles and vesicles, are assembled by the membranes of the endoplasmic reticulum, which is restored and enlarged by self-assembling molecules. Chloroplasts and mitochondria replicate themselves from existing chloroplasts and mitochondria, respectively; it is theorized that these organelles have their own chromosomes much like that in a prokaryotic cell. (Some scientists believe that the chloroplast and the mitochondrion were once independently living organisms that now live symbiotically in cells.)

The cytokinetic activities discussed above cannot be seen. The visible part of cytokinesis commences in late anaphase and reaches completion in telophase. The first sign of cytokinesis is the formation of a *cleavage furrow* which cuts across the equator of the cell through the spindle. In animal cells, microfilaments appear at the furrow which separates the daughter cells. In plants a cell plate separates the daughter cells. The cell plate is produced from a series of vesicles that are provided by the Golgi apparatus.

In some algae and fungi, mitosis without cytokinesis is common. When cytokinesis does not occur, cells with many nuclei but without separating membranes or walls result. Plant bodies made up of multinucleated cells without membrane or cell wall separations are described as being *coenocytic*.

MEIOSIS

Meiosis, or reduction division, is a kind of cell division that occurs in the primary sex cells leading to the formation of viable egg and sperm cells. The purpose of meiosis is to reduce the number of chromosomes to one half in each gamete so that upon fertilization (the fusing of sperm and egg nuclei) the species chromosome number is kept constant. If the gametes contained the full complement of chromosomes, on fertilization the number would be doubled. Instead, a gamete contains the haploid or monoploid number of chromosomes, one of each type of chromosome.

Meiosis occurs in primary sex cells—oocytes in the female, spermatocytes in the male—which have the diploid chromosome number. The process includes two cell division events—Meiosis I and Meiosis II—coming one after the other (Fig. 3.28).

Meiosis I

The stages of meiosis I result in the reduction of the number of chromosomes.

Prophase I   1. The chromosomes become shorter and thicker.
                           2. The nucleolus disappears.

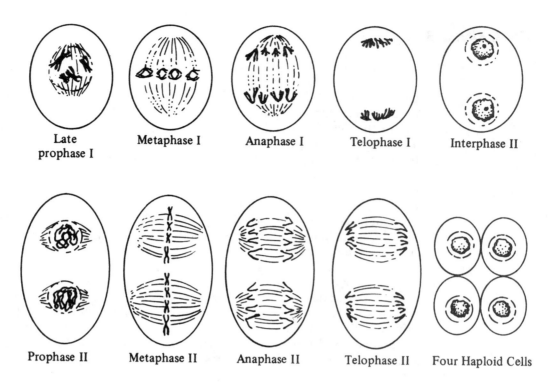

| Late prophase I | Metaphase I | Anaphase I | Telophase I | Interphase II |

| Prophase II | Metaphase II | Anaphase II | Telophase II | Four Haploid Cells |

Fig. 3.28   Stages of meiosis

3. Chromosomes pair with their homologues (mates) forming a group of four chromatids referred to as a *tetrad*.
4. The tetrads wrap around each other (synapse) and may exchange like parts.
5. The centrioles migrate and the spindle fibers appear.
6. The nuclear membrane disappears.

Metaphase I
1. The tetrads move as a unit to the equator.
2. The centromeres (kinetochores) of each of the homologous pairs of chromosomes become attached to spindle fibers extending from opposite poles.

Anaphase I
1. Each pair of double-stranded chromosomes (a set of sister chromatids) is pulled away from its homologue toward opposite poles.
2. The centromeres do not uncouple and the sister chromatids remain attached.

Telophase I
1. The chromosomes are double-stranded.
2. In some organisms the nuclear membrane reappears; in others, it does not and metaphase II starts immediately.

Interkinesis
1. The chromosomes disappear.
2. There is no replication of DNA.
3. Two haploid nuclei are present.
4. Interkinesis lasts for a very short time.

Meiosis II

This second stage separates the chromatids, terminating in four haploid cells.

Prophase II	1. The chromosomes reappear as do the spindle fibers.
	2. The centrioles migrate to opposite poles.
Metaphase II	1. Spindles form.
	2. The double-stranded chromosomes migrate to the equator. Their centromeres become attached to the spindle fibers.
	3. The centromeres uncouple as they did in mitosis.
Anaphase II	The chromosomes pull apart to opposite poles.
Telophase II	1. Four haploid nuclei are formed. Each nucleus has one member of each pair of chromosomes that began the original meiosis.
	2. The nuclear membrane reforms.
	3. Cytokinesis comes to completion.

Mechanically, meiosis II is primarily a mitotic division. As in mitosis, the chromosomes do not synapse. Since the nucleus is now haploid, there are no homologous chromosome pairs. Each double-stranded chromosome moves to the equator independently, not being attached to the spindle with a homologue. Each haploid cell produced during meiosis I divides again during meiosis II, producing four new haploid cells.

CLONING

By definition a *clone* is a population of cells (or whole organisms) that has descended from an original parent cell, which was stimulated to reproduce by asexual means. This means that a clone is genetically identical to the original cell. Over the past 40 years, tissue culture experiments have been carried out with plants in which single, nonembryonic cells have been induced to develop along a pathway of a fertilized egg. The results with such species as carrot, African violet, Boston fern and Cape sundew indicate that a whole plant can be propagated from a single nonreproductive cell. Cloning techniques have been tried with animal cells with varying degrees of success. Geneticists have been able to clone frogs, mice, and rats for laboratory research.

Chronology of Famous Names in Biology

1665   **Robert Hooke** (English)—was the first to view the pores in cork through the microscope and applied the word "cell" to these box-like structures.

1702   **Anton von Leeuwenhoek** (Dutch)—described red blood cells and protists.

1828	**Robert Brown** (English)—discovered and described the nucleus in orchid plants.
1830	**Johann Evangelista Purkinje** (Czech)—described the nucleus of the hen's egg.
1835	**Felix Dujardin** (French)—described cell sap as the essential substance of life.
1838	**Matthias Schleiden** and **Theodor Schwann** (German)—formulated the cell theory: all living organisms are made of cells.
1839	**Johann Evangelista Purkinje** (Czech)—invented the term protoplasm.
1858	**Rudolf Virchow** (German)—founded the study of cellular pathology and discovered that new cells could arise only from existing cells.
1869	**Fredrick Miescher** (Swiss)—was the first to isolate nucleic acid which he did from pus cells.
1880	**Walther Flemming** (Austrian)—discovered and described mitosis in the cells of amphibious larvae. He called mitosis "the dance of the chromosomes."
1890	**Emil Fischer** (German)—described the "lock and key" relationship between enzyme and substrate.
1898	**Camillio Golgi** (Italian)—discovered the membranes known as the Golgi apparatus by staining cells with silver nitrate and osmium tetroxide.
1937	**Sir Hans A. Krebs** (English)—was the first to describe the citric acid cycle of cellular respiration. He was awarded the Nobel Prize for this work in 1953.
1938	**James Danielli** (American)—accurately described the phospholipid layers of the cell membrane.
1945	**Albert Claude** (Belgian) and **Keith Porter** (American)—discovered the endoplasmic reticulum.
1948	**Albert Lehninger** and **Eugene Kennedy** (American)—discovered that synthesis of ATP occurs in the mitochondria.
1949	**Christian de Duve** (Belgian)—discovered the lysosomes.
1951	**Maurice Wilkins** and **Rosalind Franklin** (English)—used X-ray diffraction to study DNA.
1952	**Hugh Huxley** (English)—elucidated the biochemical processes involved in muscle contraction.
1952	**Fritiof Sjostrand** (Swedish) and **George Palade** (American)—discovered cristae in the mitochondria.
1953	**Francis Crick** (English) and **James Watson** (American)—elucidated the double helix structure of DNA.
1953	**Keith Porter** (American)—gave the name endoplasmic reticulum to the network of membranes in the cytoplasm.

1954   **George Palade** (American)—discovered ribosomes. In 1956 he discovered the function of rough endoplasmic reticulum.

1954   **F. C. Steward** (American)—produced the first carrot clone from cells of the mature carrot root.

1956   **Philip Siekevitz** and **George Palade** (American)—isolated the ribosomes.

1956   **Joe Hin Tjio** (American) and **Albert Levan** (Swedish)—established the number 46 for the human chromosome count.

1960   **Hans Moor** (Swiss)—developed the freeze-fracture technique used to study the interior of membranes.

1962   **Earl W. Sutherland** (American)—discovered the role of cyclic AMP in the transport of certain hormones across the cell membranes. He was awarded the Nobel Prize for this work in 1971.

1962   **Marshall Nirenberg, Severo Ochoa,** and **Har Gobind Khorana** (American all)—deciphered the genetic code.

1962   **M. F. Perutz** and **J. C. Kendrew** (English)—awarded the Nobel Prize for discovering the structure of the muscle protein myoglobin.

1966   **S. J. Singer** (American)—proposed the "fluid mosaic" model of the cell membrane.

1967   **Edwin Taylor** (American)—discovered the role of microtubules in mitosis.

1976   **Har Gobind Khorana** (American)—produced an artificial gene.

1978   **Gunther Blobel** (American)—discovered a "signal sequence" for RNA.

1978   **Masayaso Nomura** (American)—took apart RNA proteins in subunits of bacterial ribosomes and put them back together again.

Words for Study

acid	aerobic respiration	catalyst
active site	amino acid	cell
active transport	anaerobic respiration	cell cycle
adenosine diphosphate (ADP)	anaphase	cell membrane
	base	cell plate
adenosine triphosphate (ATP)	carbohydrate	cell theory
	carboxyl group	cellulose

centriole	interphase	pinocytosis
centromere	Krebs citric acid	plasma membrane
chlorophyll	cycle	plasmolysis
chloroplast	lipid	plastid
chromatin	lysosome	polypeptide
chromosome	meiosis	polysaccharide
cilia	mesosome	prokaryote
cleavage furrow	metaphase	prophase
clone	microfilament	protein
cristae	microtubule	protoplasm
cytokinesis	mitochondrion	pyruvic acid
cytoplasm	mitosis	replicate
cytoskeleton	monosaccharide	resolving power
diffusion	multicellular	ribosome
diploid number	nanometer	RNA
disaccharide	nucleic acid	semi-permeable
DNA	nucleolus	specificity
endocytosis	nucleus	sodium potassium ion-
endoplasmic	organ	exchange pump
reticulum	organelle	stroma
enzyme	organic	substrate
eukaryote	osmosis	system
exocytosis	oxidative	telophase
flagella	phosphorylation	thylakoid
fluid mosaic model	passive transport	tissue
Golgi body	pectin	turnover number
glycolysis	peptide bond	unicellular
glyoxysome	peroxisome	vacuole
haploid number	pH	vesicle
induced fit	phagocytosis	
inorganic	phospholipid	

Questions for Review

PART A.   Completion. Write in the word that correctly completes each statement.

1. Hereditary instructions are carried in the cell in the organelle named the ..1..

2. The general term given to the movement of materials into and out of cells is ..2..

3. The movement of water across a cell membrane is known as ..3..
4. Solid particles are ingested by cells through a process known as ..4..
5. The flow of circulating materials inside of a cell is directed by protein structures known as ..5..
6. The network of membranous canals that are spread through the cytoplasm are known collectively as the ..6..
7. Globular proteins are embedded in the ..7.. bilayer in the cell membrane.
8. Small storage areas in the cytoplasm of animal cells are known as ..8..
9. Proteins for export are synthesized by ..9.. that are attached to membranes of the endoplasmic reticulum.
10. The granular stage of the fine threads that are contained in the nucleus during interphase is known as ..10..
11. Cell walls of green plants are for the most part made of the compound ..11..
12. Plant cells are colored green by the pigment named ..12..
13. Protists use for locomotion fine cytoplasmic hairs known as flagella and ..13..
14. The type of tissue that covers and lines organs is ..14.. tissue.
15. Cells specialized for the carrying of nervous impulses are nerve cells, also known as ..15..
16. Amino acids are the building molecules of the nutrient molecules ..16..
17. Types of compounds that control chemical reactions are organic catalysts, also known as ..17..
18. The letters ATP are an abbreviation for ..18..
19. The symbol 2N refers to the ..19.. number of chromosomes.
20. Meiosis takes place in egg and sperm cells which are known collectively as ..20..
21. Mitosis is a process in cells that involves the ..21.. and its chromosomes.
22. Mitosis takes place during the ..22.. phase of the cell cycle.
23. The G-1, S, and G-2 phases of the cell cycle are known collectively as ..23..
24. As a direct result of mitosis, the parent and daughter cells have ..24.. genetic material.
25. The formation of the cleavage furrow signals the onset of ..25..

PART B. **Multiple Choice.** Circle the letter of the item that correctly completes each statement.

1. Multicellular organisms are produced from
   (a) one cell   (c) three cells
   (b) two cells   (d) four cells

2. The fine structures of the cell can be seen with the
   (a) naked eye
   (b) light microscope
   (c) phase contrast microscope
   (d) electron microscope

3. The cell's energy is used in the process of
   (a) passive transport
   (b) active transport
   (c) diffusion
   (d) osmosis

4. Because the cell membrane is highly selective in regard to the materials that can cross its boundary, it is described as being
   (a) semi-porous
   (b) semi-permeable
   (c) semi-fluid
   (d) semi-solid

5. The shrinking of cytoplasm due to the loss of water molecules is known as
   (a) evaporation
   (b) osmosis
   (c) diffusion
   (d) plasmolysis

6. Molecules carried to the cell surface by vesicles are forced out of the cell by the process of
   (a) exocytosis
   (b) endocytosis
   (c) facilitated transport
   (d) adsorption

7. The most numerous organelles in the cell are the
   (a) mitochondria
   (b) lysosomes
   (c) ribosomes
   (d) microtubules

8. Fluid substances for export outside of the cell are stored temporarily in the membranes of the
   (a) ribosomes
   (b) lysosomes
   (c) endoplasmic reticulum
   (d) Golgi apparatus

9. RNA is synthesized and temporarily stored in the
   (a) ribosomes
   (b) mitochondria
   (c) nucleoli
   (d) lysosomes

10. The internal membranes of the mitochondria are known collectively as
    (a) grana
    (b) thylakoids
    (c) mesosoma
    (d) cristae

11. Cell sap is stored in areas of plant cells known as
    (a) lysosomes
    (b) centrosomes
    (c) vesicles
    (d) vacuoles

12. Thylakoids are the same as
    (a) cristae
    (b) nucleoli
    (c) grana
    (d) stroma

13. The type of tissue that is specialized to respond to stimuli transmitted by motor nerve cells is
    (a) blood
    (b) muscle
    (c) connective
    (d) bone

14. A nonliving matrix is most correctly associated with the type of tissue classified as
    (a) epithelial
    (b) muscle
    (c) connective
    (d) nerve

15. A type of plant tissue specialized for conducting water and its dissolved materials is
    (a) vascular   (c) fundamental
    (b) epidermal   (d) collenchyma

16. Substrate molecules are acted upon by
    (a) enzymes   (c) carbohydrates
    (b) hormones   (d) lipids

17. Enzymes are synthesized from organic molecules known as
    (a) minerals   (c) proteins
    (b) carbohydrates   (d) polysaccharides

18. Anaerobic respiration is correctly known as
    (a) hydrolysis   (c) glycolysis
    (b) osmosis   (d) plasmolysis

19. The aerobic phase of respiration takes place inside of the
    (a) cytoplasm   (c) lysosome
    (b) nucleus   (d) mitochondrion

20. The circular chromosome of a prokaryotic cell attaches itself to a site on the cell membrane named the
    (a) centrosome   (c) mesosome
    (b) kinetosome   (d) centriole

21. The two strands that make up a chromosome are the
    (a) centromeres   (c) centrioles
    (b) chromatids   (d) spindle fibers

22. The red blood cells of birds
    (a) lack nuclei   (c) divide asexually
    (b) carry out mitosis (d) do not cycle

23. Passive transport requires
    (a) a semipermeable membrane (c) heat energy
    (b) chemical energy from ATP (d) an ion transfer pump

24. Examples of passive transport are
    (a) diffusion and osmosis   (c) diffusion and exocytosis
    (b) osmosis and endocytosis   (d) endocytosis and exocytosis

25. Solid particles are engulfed by cells during a process called
    (a) phagocytosis   (c) exocytosis
    (b) plasmolysis   (d) pinocytosis

PART C.   Modified True-False. If a statement is correct, write "true" for your answer. If a statement is incorrect, change the underlined word to a word that will make the statement true.

1. Scientists believe that cell shape seems to be related to its <u>age</u>.

2. Groups of similar cells that work together to do a particular job are designated as <u>systems</u>.

3. Large organisms have <u>fewer</u> cells than small organisms.

4. Hereditary instructions are carried in molecules of CBA.

5. The parts of a cell are known as organs.

6. Passage of materials into and out of the cell is controlled by the cell wall.

7. Another name for "cell drinking" is imbibing.

8. All of the material that is outside of the nucleus and inside of the cell membrane is called protoplasm.

9. Digestive enzymes and hormones are examples of lipid molecules that are exported from cells.

10. Aside from the nucleus, the centriole is the largest organelle in the cell.

11. Areas in plant cells that store starch and pigment molecules are given the general name of plastics.

12. The food-making process of green plants is known as phototropism.

13. The names Schleiden and Schwann are correctly associated with the chromosome theory.

14. A type of white blood cell specialized to counteract disease organisms that enter the body is the erythrocyte.

15. Another name for involuntary muscle is striated muscle.

16. Tracheids and vessels are best associated with phloem tissues.

17. The Krebs' citric acid cycle is the anaerobic phase of respiration.

18. As a result of mitosis, the daughter cells receive the haploid number of chromosomes.

19. The number of chromosomes in a human skin cell is 23.

20. Chromosome replication takes place during prophase.

21. The induced fit hypothesis explains the specificity of chromosomes.

22. Emil Fisher used the lock and key analogy to explain the relationship between substance and hormones.

23. Plasmolysis is the shrinking of peroxisomes due to water loss.

24. The sodium/potassium pump forces sodium ions out of the cell.

25. The enzyme-laden sacs within plant cells that change fatty acids to carbohydrates are the ribosomes.

Think and Discuss

1. Why is the cell the unit of life for all living things?

2. Why is the selectivity of the cell membrane important to the life of the cell?

3. What is the relationship between the endoplasmic reticulum and transport?

4. What would be the outcome of a lack of enzymes in a cell?
5. How does the cell cycle differ from cell division?

Answers to Questions for Review

PART A

1. nucleus
2. transport
3. osmosis
4. phagocytosis
5. microtubules
6. endoplasmic reticulum
7. phospholipid
8. vesicles
9. ribosomes
10. chromatin
11. cellulose
12. chlorophyll
13. cilia
14. epithelial
15. neurons
16. proteins
17. enzymes
18. adenosine triphosphate
19. diploid
20. gametes or sex cells
21. nucleus
22. M
23. interphase
24. identical
25. cytokinesis

PART B

1. a
2. d
3. b
4. b
5. d
6. a
7. c
8. d
9. c
10. d
11. d
12. c
13. b
14. c
15. a
16. a
17. c
18. c
19. d
20. c
21. b
22. d
23. c
24. a
25. a

PART C

1. function
2. tissues
3. more
4. DNA
5. organelles
6. cell membrane
7. pinocytosis
8. cytoplasm
9. protein
10. mitochondrion
11. plastids
12. photosynthesis
13. cell
14. leucocyte
15. smooth
16. xylem
17. aerobic
18. diploid
19. 46
20. interphase
21. enzymes
22. substrate and enzyme
23. cytoplasm
24. true
25. glyoxysomes

CHAPTER 4

BACTERIA AND VIRUSES

A bacterium (singular of *bacteria*) is an organism composed of a single cell. The bacterial cell is different from the cells of higher organisms in basic structure and in chemical composition. A bacterium does not have a nucleus bound by a membrane and is therefore classified as a prokaryote. (Derived from the Greek word *karyon*, prokaryote means "before nucleus.")

Until recently, bacteria and cyanobacteria (blue-green algae) were classified in the kingdom Monera. Newer evidence obtained from biochemical research has prompted investigators to divide the kingdom Monera into two new kingdoms, the Archaeobacteria and the Eubacteria.

Virus particles are not living cells and do not belong in the classification system of living things. Nevertheless, viruses play an important part in our lives. They are useful as experimental materials in gene studies and significant as causes of diseases in humans, other animals, and plants. For lack of a better place to include a discussion of viruses, the author has placed it here, where these particles can be easily compared with living cells.

Before proceeding, let us review some terms. A *prokaryote* is a cell that does not have a nucleus or other membrane-bound organelles; it lacks mitochondria, an endoplasmic reticulum, Golgi bodies, and lysosomes. A *eukaryote* is a cell that contains an organized nucleus and other membrane-bound organelles. *Archaeobacteria* is the kingdom of primitive bacteria. The species of Archaeobacteria show marked chemical differences from the true bacteria, which are classified in the kingdom *Eubacteria*.

Bacteria

Since all bacteria are prokaryotes, it is important to understand the characteristics of these cells.

GENERAL CHARACTERISTICS OF PROKARYOTES

Cell Wall

Present in most prokaryotic cells, the cell wall is an important structure providing shape to the cell and protecting it against osmotic damage. (However, *mycoplasmas*, the smallest living cells, do not have a cell wall.) Ranging from 5 to 80 nanometers in diameter, the walls of prokaryotic cells get their tensile strength from *murein*. Murein is composed of an enormous number of polysaccharide molecules held together by short chains of amino acids. Muramic acid, one of the major molecules in murein, never occurs in the cell walls of eukaryotic cells. Penicillin is an effective drug against bacteria because it inhibits the synthesis of murein and thus prevents the reproduction of bacteria. Penicillin does not act, however, against eukaryotic cells.

Cell Membrane

The cell membrane of the prokaryotic cell is very much like that of the eukaryotic cell except that it lacks cholesterol and other steroids. In the Archaeobacteria, the cell membrane is composed of modified branched fatty acids. Straight-chain fatty acids comprise the structure of the plasma membrane in the Eubacteria. The surface area of the cell membrane in some prokaryotes is greatly increased by *convolutions* (folds and loops). The convoluted cell membranes have incorporated in their structure electron transport systems and enzymes necessary for the chemical events taking place during respiration. The circular DNA molecule is attached to a site on the cell membrane called the *mesosome*.

Cytoplasm

The cytoplasm of the prokaryotes does not have the complex fine structure characteristic of the eukaryotes. Bacterial cytoplasm appears to be highly granulated due to the presence of a large number of ribosomes which are smaller in size than those found in nucleated cells. The ribosomes—the only cellular organelles in most prokaryotes—function just the same as their larger counterparts in the eukaryotic cells by being the sites of protein synthesis. A peripheral membrane system is present in cells of the cyanobacteria; these membranes house chlorophyll and accessory pigments.

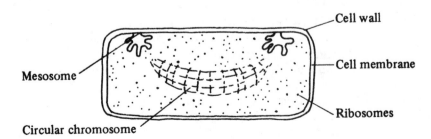

Fig. 4.1   A prokaryotic cell

Genetic Material

Prokaryotic cells contain a circular molecule of naked DNA. Prokaryotic DNA is unique in that it is associated with a small amount of RNA and nonhistone protein. The DNA chromosome is replicated before cell division. Then the cell wall and the plasma membrane grow inward, dividing the cell in two. This type of cell division is known as *binary fission*. The mesosome is formed by a loop in the plasma membrane and seems to function in the separation of the circular chromosomes during cell division.

Other Characteristics of Prokaryotes

Other differences separate nonnucleated cells from cells with a nucleus. Most prokaryotic cells are quite small, ranging in size from 0.5 to 10 micrometers. The average eukaryotic cell is much larger, although some nucleated cells measure only 7 micrometers in diameter. Some bacteria produce a slimy capsule made of carbohydrate that surrounds the cell wall and protects the bacterial cell from being engulfed by phagocytic white blood cells. Some prokaryotic cells move by means of a stiff, inflexible, rodlike structure called a *flagellum*.

CLASSIFICATION OF BACTERIA

As stated previously, because of distinct differences in the chemical composition of their organelles bacteria are now divided into two separate kingdoms: the Archaeobacteria and the Eubacteria. The Archaeobacteria are the "ancient or first bacteria"; the Eubacteria, the "true bacteria." All the species contained in both of these kingdoms are prokaryotes.

Groups of Bacteria

Many biologists refer to "divisions" or "groups" of bacteria, because the concept of species is difficult to apply to these organisms. By accepted definition, a species is a group of organisms so closely related that they can mate and produce viable offspring. Since bacteria usually reproduce asexually, and since the evolutionary history is difficult to determine, many "species" of bacteria do not easily meet the requirements of the definition. However, since "species" is a convenient word to show close structure and functional relationship, we shall continue to use it when necessary.

Gram Staining

In 1884 the Danish microbiologist Hans Christian Gram developed a technique for staining bacteria that is useful in distinguishing groups of organisms, particularly among the Eubacteria. The procedure of *gram staining* identifies cells based on their ability to absorb certain dyes. Bacterial cells walls that are made of peptidoglycans (amino sugars) absorb the purple dye gentian violet with such tenacity that washing the cells with alcohol does not remove the dye. Such purple-staining cells are designated as *gram-positive*. Cell walls composed of lipopolysaccharides, however, do not hold the purple dye and are easily bleached with alcohol. These cells stain red with dyes such as safranin or carbol fuchsin. Cells that take up the red counterstain are known as *gram-negative*.

Gram staining is one means by which bacteria are identified and classified. As a matter of fact, the composition of the cell wall affects the behavior of bacteria. Gram-positive bacteria are more susceptible to the effects of antibiotics than are gram-negative organisms. Gram-positive organisms are also more susceptible to the lysing effects of the enzyme lysozyme, which is found in human secretions such as tears, saliva, and nasal discharges.

KINGDOM ARCHAEOBACTERIA

The species of bacteria classified in the kingdom Archaeobacteria exhibit unusual characteristics. All are *anaerobes*, and all are unable to survive in an environment of oxygen. Rather, they live in harsh environments—stagnant marsh water, boiling hot springs, the salt water of lakes and ponds, and the anaerobic sludge at the bottom of lakes.

Biochemical Uniqueness of Archaeobacteria

The archaeobacteria differ biochemically from all other organisms living on earth today. First, their cell membranes are composed of branched lipid molecules unlike the straight-chain fatty acids of other cells. Second, in their metabolic processes, enzymes different from those in other cells are used in the synthesis of lipid and in RNA synthesis activities. Biologists use these and other differences as reasons to place the archaeobacteria in a separate kingdom.

Major Groups of Archaeobacteria

The largest group of archaeobacteria are the *methanogens*. As the name implies, these bacteria produce methane gas. They live where anaerobic conditions exist: in airless marshes, anaerobic lake bottoms, sewage treatment ponds, and even in the lower bowels of cattle and sheep, where anaerobic conditions prevail. The methanogens use hydrogen gas to reduce carbon dioxide in their energy-making process, producing methane gas as a byproduct. The reaction can be summarized as follows:

$$4H_2 + CO_2 \rightarrow CH_4 + 2H_2O$$

The methanogens must live where the inorganic gases carbon dioxide and hydrogen are available for their use and where oxygen is not present. In addition, they require three organic compounds for nutrition: methanol (H_3COH), formate ($HCOO-$), and acetate ($H_3C—COO-$). Evidence indicates that these five simple molecules, two inorganic and three organic, were present on early Earth at the time when oxygen was absent from the atmosphere. Thus it is believed that the methanogens are descendants from an ancient line of bacteria unrelated to the Eubacteria.

The *salt-loving bacteria* belong to the genus *Halobacterium*. The members of this genus are able to produce their own energy by carrying out a simple type of chemosynthesis. Embedded in the plasma membranes of the halophiles (salt-loving bacteria) are patches of a purple pigment, bacteriorhodopsin. In an atmosphere free of oxygen, this pigment is able to capture light energy and use it to form ATP molecules, the energy-storing molecules of cells.

Other species included in the kingdom Archaeobacteria include the *thermophiles* and the *thermoacidophiles*. As the name implies, thermophiles thrive in extremes of heat. They live in hot springs and can survive in temperatures near the boiling point of water, 90°C (194°F). Thermoacidophiles live in harsh environments of extreme heat and extreme acid. Of interest to scientists is the fact that the internal pH (acid-base measure) of these bacteria remains neutral.

KINGDOM EUBACTERIA

The eubacteria, referred to as the "true bacteria," comprise a large number of subgroups. Members of the kingdom Eubacteria live under less harsh conditions than do the archaeobacteria. A large number of prokaryotes are classified in this kingdom because of strong biochemical similarities in their cell walls, which are composed of amino sugars known as peptidoglycans. Their cell (plasma) membranes are similar in that they are built from straight-chain fatty acids rather than the short, branched chains found in the archaeobacteria.

Characteristics of the Eubacteria

All of these bacteria have thick, rigid cell walls. Some species are *non-motile* (nonmoving), while others are *motile*, using flagella (Fig. 4.2) or a sling motion to move from place to place.

Shape is one means of identifying eubacteria. The rod-shaped bacteria are known as *bacilli* (bacillus, sing.), the round bacteria as *cocci* (coccus, sing.) and the spiral-shaped as *spirilla* (spirillum, sing.) (Fig. 4.3). Some species typically remain attached. The *diplococci* occur in pairs, the *streptococci* in chains, and the *staphylococci* in clusters.

Reproduction in the eubacteria takes place by *binary fission*. There is no mitosis as in the eukaryotic cells. Replication of the circular chromosome is followed by equal splitting of the cytoplasm. The reproductive potential of bacteria is enormous because cell division is possible every 20 minutes. In this short span of time, a parent cell divides into two identical daughter cells. It has been estimated if the rate of reproduction were not slowed by environmental changes, one bacterium at the end of 6 hours could give rise to 500,000 descendants.

Eubacteria are able to survive unfavorable environmental conditions in the rather unique way of forming *endospores*. The spore is a vegetative cell containing DNA and little else. It becomes surrounded by a thick

Fig. 4.2 A flagellated bacterium

Rods

Cocci

Spirals

Fig. 4.3 Three shapes of bacteria

and almost indestructible cell wall secreted by the cytoplasm. In this condition, the endospore is able to withstand boiling, freezing, drying, treatment with disinfectants and other extremes of environment.

Capsule formation occurs in certain disease-producing species of eubacteria. The cytoplasm of the bacterium secretes a mucoid coat of polysaccharide materials which surrounds the cell wall. The capsule is resistant to phagocytosis (engulfing by white blood cells) and may even, itself, release toxins into the tissues of the host.

Most eubacteria are *heterotrophs* that have to depend on sources outside of their own bodies for organic food materials. Many species obtain food by absorbing nutrients from dead organic matter; organisms that feed in this way are known as *saprophytes*, or, in new terminology, *saprobes*. Bacteria that live on or inside of the bodies of animals or plants and cause disease are *parasites*.

A number of groups in the eubacteria are *autotrophs* and are able to change inorganic materials into organic compounds. Among these are the *photosynthetic bacteria*, which, like green plants, use light energy to produce food, and the *chemosynthetic bacteria*, which are capable of oxidizing the inorganic compounds of ammonia, nitrites, sulfur, or hydrogen gas into high-energy organic compounds without the need of light energy.

Nitrifying bacteria are excellent examples of the chemosynthetic bacteria. These bacteria live in little *nodules* (bumps) on the roots of leguminous plants such as peas, beans, peanuts, alfalfa and clover (Fig. 4.4). In two separate steps, the nitrifying bacteria convert ammonia, released into the soil by the breakdown of proteins from the bodies of dead plants and animals, to nitrites and then they change the nitrites to nitrates:

$$(1) \quad 2NH_4^+ + 3O_2 \rightarrow 2NO_2^- + 4H^+ + \text{energy}$$

$$(2) \quad 2NO_2^- + O_2 \rightarrow 2NO_3^- + \text{energy}$$

Rod-shaped bacteria belonging to the genus *Nitrobacter* are examples of nitrifying bacteria. Bacteria of decay, responsible for the breakdown of proteins in dead organic matter, belong to the genera *Pseudomonas* and *Thiobacillus*. Species belonging to the nitrifying and denitrifying genera aid in the recycling of nitrogen (Fig. 4.5).

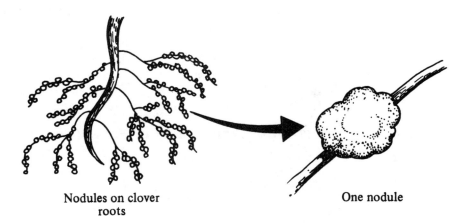

Nodules on clover
roots

One nodule

Fig. 4.4 Nodules on the roots of leguminous plants

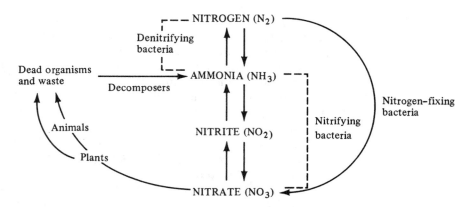

Fig. 4.5 Nitrogen cycle

Sulfur bacteria are also chemosynthetic bacteria. Sulfur, an element necessary for life, is present in soil, free and combined, in inorganic and organic compounds. The decomposition of rock, the breakdown of organic products, and rainwater are the major sources of soil sulfur. Living in the soil, the "sulfur" bacteria oxidize free sulfur in sulfates, producing energy for themselves and providing compounds of sulfur that can be used by plants.

$$2S + 3O_2 + 2H_2O \rightarrow 2SO_2 + 4H^+ + \text{energy}$$

Photosynthetic bacteria (Endothiobacteria) use pigment systems which are contained in the cell membrane to capture light energy and use it in the synthesis of organic molecules. Unlike green plants, however, photosynthetic bacteria do not have the pigment chlorophyll *a* and they do not produce molecular oxygen. In fact, most photosynthetic bacteria do not need or use oxygen.

Most groups of the eubacteria are *aerobic*, using molecular oxygen in the process of breaking down carbohydrates to carbon dioxide and water. *Obligate aerobes* are organisms that can live only in an environment that provides free or atmospheric oxygen. An example of an obligate aerobe is *Bacillus subtilis*.

Some bacteria are *obligate anaerobes* and derive their energy by fermentation. These organisms cannot live in an environment of free oxygen. Usually, the obligate anaerobes are disease producers; included in this group are *Clostridium tetani*, the causative organism of tetanus, and *Clostridium botulinum*, the bacterium that induces food poisoning. Still other bacteria are *facultative anaerobes*. These are basically aerobic bacteria, but they can live and grow in an environment that lacks free oxygen.

Four Important Groups of Eubacteria

Myxobacteria—Slime Bacteria The Myxobacteria are also known as the slime bacteria. The cells develop as a colony and move together in a flowing mass of slime resembling the movements of amoebae. The individual cells are long, thin rods. Myxobacteria live in the soil and obtain nutrients by absorbing dead organic matter (saprobes). At times the slime mass, or *pseudoplasmodium*, forms reproductive structures called *fruiting bodies* which look like small mushrooms. The fruiting bodies produce *cysts* which are much like the endospores of other bacteria in regard to their ability to withstand adverse environmental conditions. When environmental conditions improve, the encysted cells become metabolically active and reproduce by binary fission.

Spirochetes Most bacteria belonging to this group are anaerobic; many are disease producers. These bacteria are long, thin and curved, moving with a wriggling, corkscrew-like motion, made possible by an axial filament (Fig. 4.6). In some ways, the spirochetes resemble protozoa, but they are nonnucleated. They do not form spores or branches. They reproduce by transverse fission. The spirochete *Treponema pallidum* causes syphilis.

Rickettsiae The rickettsiae are very small gram-negative intracellular parasites (Fig. 4.7). They were first described in 1909 by Harold Taylor Ricketts, who found them in the blood of patients suffering from Rocky Mountain spotted fever. The rickettsiae are nonmotile, non-spore-forming, nonencapsulated organisms. In length, these organisms range from 0.3 to 1 micrometer. They live in the cells of ticks and mites and are transmitted to humans through insect bites. The rickettsiae are responsible for several febrile (fever-producing) diseases in humans, such as typhus fever, trench fever and Q fever.

Actinomycetes The Actinomycetes (also known as the Actinomycota) have characteristics of both the eubacteria and the fungi. Like the latter, their body structure takes the form of branching multicellular filaments. They live in soil where they obtain nutrition saprophytically. Some species are anaerobes. These organisms are mostly nonmotile and break down waxes and lipids in dead plants and animals. Some species cause diseases in animals and humans. *Actinomyces bovis* causes "lumpy" jaw in cattle. *Mycobacterium tuberculosis* causes tuberculosis in humans and *Mycobacterium leprae* causes leprosy. Some other actinomycetes have important medical and commercial value—primarily as sources of *antibiotics*. Antibiotics in common use that are produced by the actinomycetes are tetracycline, erythromycin, neomycin, nystatin and chloramphenicol.

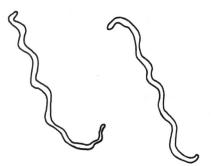

Fig. 4.6 Spirochetes are highly motile but they lack polarity

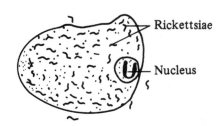

Fig. 4.7 Rickettsiae in tissue cells

Cyanobacteria

A controversial phylum in kingdom Eubacteria is Cyanobacteria. Some biologists classified these one-celled photosynthetic organisms as blue-green algae—thus the former name Cyanophyta. Now some biologists regard these organisms as bacteria and call them *cyanobacteria*. Others avoid the issue and refer to this group simply as the blue-greens.

In structure, the cyanobacteria are prokaryotes. As in the bacteria, the cytoplasm of the blue-greens is crowded with ribosomes. In addition, the cytoplasm of most of these organisms stores particles of protein materials and special carbohydrate molecules known as *polyglucans*. The cells of the blue-greens are larger than those of bacteria but smaller than eukaryotic cells. The cell wall is composed of murein, as in the bacteria, but is surrounded by a gelatinous material composed of pectin and glycoproteins. The cyanobacteria have no visible means of locomotion (no flagella), but some species move with a gliding motion, as do the myxobacteria.

The blue-greens derive their name from the pigment molecules that are contained in flattened membranous vesicles called *thylakoids*. Several types of pigments have been isolated from the cyanobacteria. As in the higher plants, chlorophyll *a* is the pigment used for photosynthesis. Present also in these organisms is the blue pigment *phycocyanin*, the red pigment *phycoerythrin*, and several types of carotenoids (yellow pigments). Not all of the cyanobacteria are blue-green. Some are black, brown, yellow, red, and bright green. At times the Red Sea takes on a reddish hue due to a species of cyanobacteria that inhabits these waters in great numbers and contains large amounts of phycoerythrin.

The cyanobacteria may live singly or occur in filaments or colonies. Although the cells of some species live in close association with one another, they usually carry on independent life functions. Interestingly enough, cytoplasmic bridges called *plasmadesmata* join the cytoplasm of some of the cells of the filamentous groups. At certain times the plasmadesmata may serve as the passageways for special materials.

Some of the filamentous blue-greens have the ability to fix atmospheric nitrogen. Under anaerobic conditions, special cells called *heterocysts* produce the enzyme *nitrogenase*, which has the ability to convert atmospheric nitrogen into usable nitrates.

Fresh water is the habitat of most of the cyanobacteria, but a few species are marine, and some species live in tropical soils where the nitrogen content is poor. Other habitats include microfissures in desert rocks, where light penetrates and a small amount of water is trapped. Cyanobacteria can also be found on tree bark, on damp rocks, and on flower pots. They reproduce by binary fission or by fragmentation of filaments.

The cyanobacteria are an important part of every environment. In the process of photosynthesis they produce oxygen as a by-product which helps to replenish the atmosphere. They also produce food for invertebrate and vertebrate species that feed on *phytoplankton*, the floating plants that live on the surfaces of oceans and lakes.

Cyanobacteria can also be harmful to humans and fish in their role as water pollutants. The rapid overpopulation of blue-greens such as *Oscillatoria* coats lakes with a slimy, smelly mass that is toxic to fish. The "blooms," aided by excessive nitrate compounds in lakes, have contributed to a serious decline in the freshwater fish populations in some areas.

Prochloron

The Prochloron were discovered in 1976 by R. A. Levin. These cells are prokaryotic and seem to be an evolutionary link between the cyanobacteria and the eukaryotic algae. Their pigment system seems to resemble that of the green algae and higher green plants. The only known representatives are those that live in association with a group of marine invertebrates known as the tunicates.

IMPORTANCE OF BACTERIA

Bacteria and Disease

As we have already mentioned, some species of bacteria cause disease in humans, animals and plants; these bacteria are termed *pathogenic*. A disease is any condition that interrupts the normal functioning of body cells preventing completion of a particular biochemical task. When pathogenic bacteria invade body tissues, they introduce a particular set of circumstances that change the environment of the cells.

There are three significant ways in which the cellular environment can be altered resulting in a disease condition. One way is by sheer numbers of organisms: enormous numbers of bacteria will affect adversely the functioning of cells. For example, *Escherichia coli*, gram-negative rods that normally live in the human intestine will cause disease if present in increased numbers. Another way is by the destruction of cells and tissues. A third change brought about by pathogenic bacteria is directly related to their production of *toxins*. Toxins are poisonous substances that inhibit the metabolic activities of the host cells.

Helpful Bacteria

Contrary to popular belief, most species of bacteria are helpful to humans. The nitrogen-fixing bacteria enable plants to obtain the nitrates necessary for protein synthesis. The bacteria of decay release ammonia and nitrates from dead organic matter into the soil. Bacteria that live in the human intestines synthesize several vitamins and contribute to the synthesis of a digestive enzyme. Manufacturers of vinegar, acetone, butanol, lactic acid and certain vitamins depend upon the action of bacteria in the production processes of these products. The retting of flax and hemp is a process in which bacteria are used to digest the pectin compounds that hold together the cellulose fibers. Once these fibers are free they can be used to make linen, textiles and rope. Bacteria are useful in the preparation of skins for leather and in the curing of tobacco. Manufacturers of dairy products use bacteria to ripen cheese and to improve the flavor of special items such as Swiss cheese. Farmers depend on bacteria in their fermenting of silage that is used for cattle feed. The pharmaceutical industry produces antibiotics such as aureomycin, teramycin and streptomycin from bacteria.

Viruses

Viruses are not living. Virus particles are not cells and do not exhibit the characteristics of life, as do cells. Lacking the enzymes and other cell machinery for carrying out metabolic processes, they remain inert and cannot function as living systems. Reproduction is the only function of viruses. To reproduce they must invade (enter) living cells. Viruses are parasites.

STRUCTURE OF VIRUSES

The virus particle is known as a *viron*. It consists merely of a protein coat called a *capsid* and a nucleic acid core.

In shape, most virus particles are either helical (spiral) or polyhedral (many-sided). The capsid of the helical viruses is composed of subunits called *capsomeres*. The tobacco mosaic virus (Fig. 4.9), the first virus to be described, and the virus that causes influenza are helical viruses. A few viruses are shaped like a cube, and a few are brick-shaped. The T-even bacteriophage (a virus that enters a bacterium) is polyhedral and has a tail with extending fibers. The smallpox virus is brick-shaped.

In some viruses, including those that cause influenza and herpes, a cytoplasmic membrane surrounds the protein coat. This surrounding *envelope* may come from the plasma membrane of the host cell or may be synthesized by the host's cytoplasm. Virologists have found that the envelope contains proteins that are virus-specific.

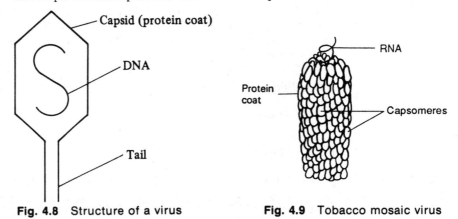

Fig. 4.8   Structure of a virus   Fig. 4.9   Tobacco mosaic virus

VIRAL NUCLEIC ACID

Viral nucleic acid may be a single molecule consisting of as few as five genes or may be multimolecular and have as many as several hundred genes. Viral nucleic acid may be single-stranded or double stranded; it may be circular or linear. Some viral nucleic acid is made of DNA (deoxyribonucleic acid); some has only an RNA (ribonucleic acid) core. Viruses never contain both DNA and RNA. For example, the polio virus contains only RNA, while the herpes viruses contain only DNA.

REPRODUCTION OF VIRUSES

A general description of the way in which viruses reproduce follows. A virus infects (enters) a host cell. The nucleic acid of the virus captures the nucleic acid of the host cell and uses it to make more virus nucleic acid and virus proteins. New virus particles, complete with the nucleic acid core and the protein coat, are assembled inside of the host cell until no more can fit. The host then bursts, releasing the newly assembled virions. This generalized pattern of virus reproduction varies with different viruses that typically infect bacteria, plant, or animal cells. A virus that infects a bacterial cell is known as a *bacteriophage*, or simply as a *phage*.

IMPORTANCE OF VIRUSES

Viruses are notorious for their disease-producing potential in animals and plants. In human beings, viruses cause such diseases as measles, the common cold, smallpox, rabies, chicken pox, yellow fever, influenza, viral pneumonia, encephalitis, infectious mononucleosis, AIDS, and several types of hepatitis, as well as fever blisters and venereal herpes. It is believed that some cancers are caused by viruses. Virus diseases of plants include tobacco mosaic disease, tobacco necrosis and rice dwarf.

VIROIDS

In the 1960's, it was discovered that several plant diseases are caused by *viroids*, particles smaller than viruses. The viroid contains a short RNA chain and lacks the protective protein coat. New viroids are produced in the nucleus of the host cell. Diseases such as the stunting of chrysanthemums, spindle tuber of potatoes and disease of citrus trees are caused by viroids.

Chronology of Famous Names in Biology

1684   **Anton von Leeuwenhoek** (Dutch)—was the first person to see bacteria.

1796   **Edward Jenner** (English)—demonstrated the use of cowpox as an immunizing vaccine against smallpox.

1850   **Ignaz Semmelweis** (German)—discovered that attending physicians carried the organisms that cause childbed fever from patient to patient on their unwashed hands.

1861   **Louis Pasteur** (French)—made major contributions to the understanding of bacteria; helped to prove the germ theory of disease; discovered a method for the prevention of rabies.

1867   **Joseph Lister** (English)—developed antiseptic principles in the practice of surgery.

1884   **Robert Koch** (German)—discovered the bacteria that cause tuberculosis and cholera. He formulated Koch's postulates, the steps to determine if a bacterium causes a disease.

1884	**Hans Christian Gram** (Danish)—developed differential staining techniques for bacteria which have become known as the Gram stain.
1892	**Dimitri Iwanowski** (Russian)—discovered the tobacco mosaic virus.
1929	**Alexander Fleming** (English)—discovered lysozyme in human secretions.
1935	**Wendell Stanley** (American)—isolated the tobacco mosaic virus.
1946	**Joshua Lederberg** and **Edward Tatum** (American)—were the first to demonstrate genetic recombination in bacteria.
1975	**Carl R. Woese** (American)—developed the idea that the prokaryotes include two distinct, unrelated kingdoms.

Words for Study

aerobic	eukaryote	pathogenic
anaerobe	facultative anaerobe	peptidoglycans
Archaeobacteria	flagellum	phage
autotroph	fruiting body	photosynthetic bacteria
bacillus	gram stain	phytoplankton
bacteriophage	heterocyst	plasmadesmata
binary fission	heterotroph	Prochloron
capsid	mesosome	prokaryote
capsomere	methanogens	spirillum
chemosynthetic	motile	spirochete
bacteria	murein	saprobe
coccus	mycoplasma	thermoacidophile
convolution	nodule	thermophile
cyanobacteria	obligate aerobe	toxin
endospore	obligate anaerobe	viron
Eubacteria	parasite	viroid

Questions for Review

PART A.   **Completion.** Write in the word that correctly completes each statement.

1. The prokaryotes are cells that lack ..1.. organelles. (2 words)
2. The cell membrane of the prokaryotes lacks the steroid ..2..

3. The DNA molecule is attached to a place on the cell membrane of bacteria called a ..3..

4. Mycoplasmas, unlike other prokaryotic cells, do not have a ..4.. (2 words)

5. Peptidoglycans are necessary for ..5.. formation in bacteria.

6. The only cellular organelle in bacteria is the ..6.. ⁎.

7. A chromosome makes an identical copy of itself in a process called ..7..

8. The methanogens belong to the kingdom ..8..

9. A group of interbreeding organisms is called a ..9..

10. The ..10.. are referred to as the true bacteria.

11. Bacteria reproduce by the method known as ..11..

12. In newer terminology, a saprophyte is referred to as a ..12..

13. Rod-shaped bacteria are known as ..13..

14. *Nitrobacter* are examples of ..14.. bacteria.

15. Bacteria that require molecular oxygen for respiration are known as ..15..

16. Bacteria that are indifferent to oxygen are called ..16..

17. Gram-positive bacteria absorb a ..17.. colored dye.

18. Toxins inhibit the ..18.. activities of cells.

19. True bacteria can survive unfavorable environmental conditions by forming ..19..

20. The ..20.. are evolutionary links between prokaryotic and eukaryotic cells.

21. In some cyanobacteria cells known as ..21.. can fix atmospheric nitrogen.

22. The habitat of most blue-greens is ..22.. water.

23. A virus particle is known as a ..23..

24. A virus that infects bacteria is called a ..24..

25. Floating green plants that inhabit surfaces of lakes and oceans are known collectively as ..25..

PART B.   Multiple Choice. Circle the letter of the item that correctly completes each statement.

1. Bacteria are best described as
   (a) eukaryotic cells   (c) prokaryotic cells
   (b) photosynthetic cells (d) filamentous cells

2. A chain of cells is known as a
   (a) filament   (c) chord
   (b) mass   (d) convolution

3. The DNA of prokaryotes is structured into
   (a) scattered chromatin granules (c) several nucleoids
   (b) histone chains   (d) a single chromosome

4. Muramic acid is isolated from the bacterial structure known as
   (a) mesosome   (c) cell wall
   (b) cell membrane   (d) chromosome

5. The function of the ribosomes is to
   (a) synthesize cell walls
   (b) carry hereditary traits
   (c) direct replication of the mesosome
   (d) translate mRNA into protein

6. Prokaryotic cells contain
   (a) DNA only
   (b) RNA only
   (c) both DNA and RNA
   (d) no nucleic acids

7. An incorrect statement concerning bacteria is that they
   (a) build ATP molecules
   (b) lack multienzyme systems
   (c) lack Golgi bodies
   (d) assemble proteins

8. Survival of the eubacteria is increased by the ability to
   (a) replicate DNA
   (b) synthesize proteins
   (c) generate cells walls
   (d) form endospores

9. Encapsulated bacteria resist
   (a) reproduction
   (b) endospore formation
   (c) phagocytosis
   (d) DNA replication

10. Nitrifying bacteria
    (a) convert ammonia into nitrates
    (b) convert ammonia into nitrogen gas
    (c) release ammonia from decaying bodies
    (d) synthesize legumes

11. The bacteria that live as parasites in the cells of ticks and mites are
    (a) spirochetes
    (b) actinomycetes
    (c) mycoplasmas
    (d) rickettsiae

12. It is true that viruses
    (a) reproduce
    (b) ingest
    (c) synthesize ATP
    (d) reduce H

13. Tunicates are
    (a) blue-green algae
    (b) fruiting bodies
    (c) a type of virus
    (d) marine invertebrates

14. Bacteriorhodopsin is correctly associated with the
    (a) chlorophyta
    (b) prochlorons
    (c) halophiles
    (d) viroids

15. Capsomeres are subunits of
    (a) bacteria cell walls
    (b) viral coats
    (c) clover nodules
    (d) slime mold

16. Thermophiles live best
    (a) on the bottom of bogs
    (b) in hot springs
    (c) in salt pools
    (d) in stagnant marshes

17. Lysozyme is a (an)
    (a) human secretion
    (b) small vacuole
    (c) enzyme
    (d) hydrogen carrier molecule

18. Fruiting bodies are best associated with
    (a) spirochetes
    (b) Eubacteria
    (c) Prochlorophyta
    (d) Myxobacteria

19. The organism that causes syphilis is a
    (a) spirochete   (c) prochlorophyte
    (b) eubacteria   (d) myxobacterium

20. Rickettsiae are best classified as
    (a) molds   (c) fungi
    (b) bacteria   (d) protists

21. Plasmadesmata are
    (a) shrinking cell walls
    (b) oversized vacuoles
    (c) cytoplasmic bridges between cells
    (d) several attached cell membranes

22. Heterocysts are most closely associated with
    (a) nitrogen fixation   (c) water storage
    (b) cell protection   (d) binary fission

23. Actinomycetes are a type of
    (a) virus   (c) fungus
    (b) rock   (d) bacteria

24. The protein coat of a virus is known as a
    (a) desmid   (c) capsid
    (b) plastid   (d) capsule

25. A virus
    (a) can reproduce itself independently
    (b) can reproduce only within living cells
    (c) does not contain either DNA or RNA
    (d) can be considered a type of cell

PART C.   Modified True-False. If a statement is true, write "true" for your answer. If a statement is incorrect, change the underlined word to one that will make the statement true.

1. At one time bacteria were classified as one-celled <u>plants</u>.

2. Archaeobacteria are <u>visible</u> to the naked eye.

3. The smallest living cells are <u>bacteria</u>.

4. In addition to the circular chromosome the only other organelle found in bacteria is the <u>Golgi body</u>.

5. A micrometer is $\frac{1}{100}$ of a millimeter.

6. Mycoplasmas have <u>twice</u> as much DNA as the true bacteria.

7. <u>Photosynthetic</u> bacteria do not need light energy to synthesize food molecules.

8. During favorable conditions, bacteria can reproduce every <u>2 hours</u>.

9. Disease-producing bacteria may be surrounded by a polysaccharide <u>cell wall</u>.

10. Small bumps on roots of clover plants that house bacteria are known as <u>tumors</u>.

11. Round bacteria are called <u>spirilla</u>.

12. *Escherichia coli* live normally in the human <u>heart</u>.

13. When oxygen is not available, facultative anaerobes obtain energy by way of <u>the Krebs cycle</u>.

14. Obligate anaerobes must live in an environment free of <u>carbon dioxide</u>.

15. Bacterial cells <u>can</u> build their own ATP molecules.

16. <u>Gram-positive</u> bacteria are more susceptible to the effects of antibiotics.

17. Most species of bacteria are <u>harmful</u>.

18. Retting is a process in which <u>viruses</u> are used to digest the pectin in flax plants.

19. Hemp fiber is used to make <u>linen</u>.

20. Polyglucans are molecules of <u>protein</u>.

21. The enzyme nitrogenase is produced under <u>aerobic</u> conditions.

22. The envelope surrounding the coat of a virus comes from the <u>host</u>.

23. Hepatitis is a <u>viroid</u> disease.

24. Viroids are <u>smaller</u> than viruses.

25. Viroids are known to cause several <u>animal</u> diseases.

Think and Discuss

1. What are the distinguishing characteristics of prokaryotic cells?
2. Why is the classification of species not appropriate for bacteria?
3. Why are virus particles not considered to be cells?
4. In what ways are bacteria harmful to cells?

Answers to Questions for Review

PART A

1. membrane-bound
2. cholesterol
3. mesosome
4. cell wall
5. cell wall
6. ribosome
7. replication
8. Archaeobacteria
9. species
10. eubacteria
11. binary fission
12. saprobe
13. bacilli
14. nitrifying
15. aerobes
16. facultative anaerobes

17. purple
18. metabolic
19. endospores
20. prochlorons
21. heterocysts

22. fresh
23. virion
24. bacteriophage
25. phytoplankton

PART B

1. c
2. a
3. d
4. c
5. d
6. c
7. b
8. d
9. c

10. a
11. d
12. a
13. d
14. c
15. b
16. b
17. c
18. d

19. a
20. b
21. c
22. a
23. d
24. c
25. b

PART C

1. true
2. invisible
3. mycoplasmas
4. ribosome
5. $\frac{1}{1000}$
6. one half
7. Chemosynthetic
8. 20 minutes
9. capsule
10. nodules
11. cocci
12. intestine
13. fermentation

14. oxygen
15. true
16. true
17. helpful
18. bacteria
19. rope
20. carbohydrate
21. anaerobic
22. true
23. virus
24. true
25. plant

CHAPTER 5

THE PROTIST KINGDOM: PROTOZOA, FUNGUS-LIKE PROTISTS, AND PLANT-LIKE PROTISTS

All of the species assigned to the kingdom Protista are eukaryotic. Most protists carry out their lives within a single cell as free-living organisms. However, some protist species are organized into *colonies* where each cell carries out its own life functions and where, also, there may be some simple division of labor among the cells in the grouping. An impressive variety of species are classified as protists, and they probably descended from diverse evolutionary lines. The protists themselves represent evolutionary modification and are probably the ancestors of the modern fungi, plants and animals.

Major Groups of Protists

Most modern classification schemes divide the Protista into three major groups: the protozoa, or animal-like protists; the fungus-like protists; and the plant-like protists.

PROTOZOA, THE ANIMAL-LIKE PROTISTS

Protozoa, meaning "first animals," are one-celled heterotrophs. Species of protozoa number in the thousands. They live in fresh water, salt water, dry sand and moist soil. Some species live as parasites on or inside of the bodies of other organisms. Reproduction in the protozoans is usually described as being asexual by means of mitosis, but recent research has revealed that many protozoa augment asexual reproduction with a sexual cycle. Usually, the sexual cycle occurs during periods of adverse environmental conditions, and the cell arising from the fusion of gametes (*zygote*) can resist unfavorable conditions. The thick wall and the decreased metabolic rate of the *cyst* permits survival during periods of cold, drought or famine.

The protozoa are divided into four phyla, based primarily on the methods of locomotion.

Mastigophora

The Mastigophora are protozoa that have one or more flagella. This phylum, also known as the Zoomastigina or zooflagellates, includes a rather heterogeneous group of organisms. Some species are free-living and inhabit fresh or salt water; of these, some are free-swimming; others glide over the surface of rocks; and some are sessile, attached to available submerged surfaces. Other Mastigophora species live in a *symbiotic* relationship with organisms of other species, each helping the other with a particular life function. For example, several species live in the intestines of termites, cockroaches and woodroaches, where they digest cellulose for these insects. The genus *Trypanosoma* includes parasites that cause debilitating diseases in human beings. *Trypanosoma gambiense* (Fig. 5.1) is the zooflagellate that causes African sleeping sickness. Humans are infected with the trypanosome by the bite of an infected tsetse fly.

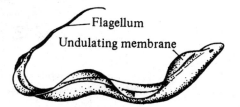

Flagellum
Undulating membrane

Fig. 5.1 The causative organism of African sleeping sickness, *Trypanosoma gambiense*

Sarcodina

The members of the phylum Sarcodina are described as being amoeboid. *Amoeba proteus* (Fig. 5.2) is the type species. Species included in the Sarcodina move by means of *pseudopods*, flowing extensions of the flexible and amorphous body. The pseudopods also serve in food-catching. Most of the sarcodines live in fresh water. A *contractile vacuole*, an organelle designed to expel excess water from the protist cell body, plays an important role in maintaining water balance. Food is temporarily stored in a food vacuole where it is digested by the action of enzymes.

The Foraminifera and the Radiolaria are groups of marine sarcodines that secrete hard shells of mineral compounds around themselves. When they die, their shells become an important constituent of the bottom mud of the ocean floor. The Foraminifera have built the limestone and chalk deposits that date back to the Cambrian period; the Radiolarians, the siliceous rocks dating back to the Precambrian period.

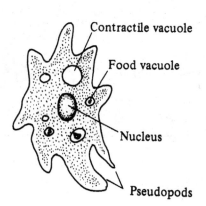

Contractile vacuole

Food vacuole

Nucleus

Pseudopods   **Fig. 5.2** *Amoeba proteus*

Sporozoa

The Sporozoa are parasitic spore-formers. The adult forms are incapable of locomotion, although immature organisms may move by means of pseudopodia. Some species of sporozoa go through a complicated life cycle requiring different hosts during different life stages. For example, the species *Plasmodium vivax*—the agent that causes malaria—requires two hosts: the *Anopheles* mosquito and a human (Fig. 5.3).

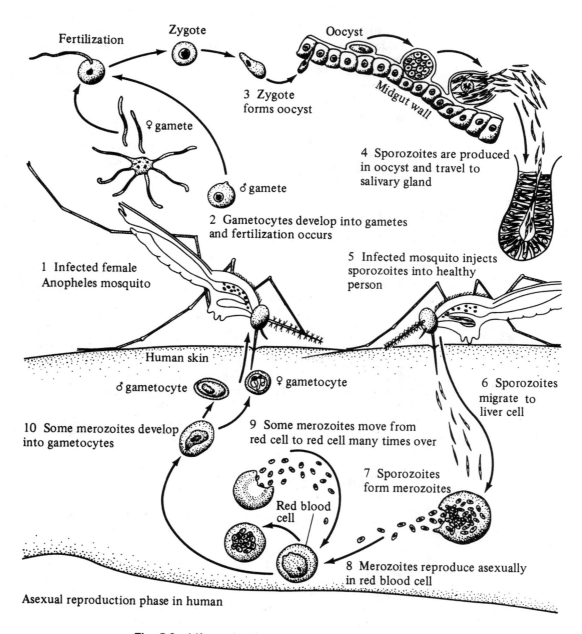

Fig. 5.3   Life cycle of *Plasmodium vivax*

Ciliata

Of the four protozoan divisions, the phylum Ciliata has the greatest number of species. Species belonging to this phylum have *cilia*, short cytoplasmic strands which are used for locomotion and in some cases to sweep food particles into an opening called the *oral groove*. Protozoans included in this phylum inhabit fresh water and salt water. Some are free-swimming; some creep; others are sessile; and some are parasitic in other animals.

The cytoplasm in ciliates is differentiated into rigid outer ectoplasm and a more fluid inner endoplasm. A *pellicle* lies just inside of the cell membrane. Some species respond to adverse environmental stimuli by discharging elongated threads called *trichocysts* which serve as defense mechanisms or a means of anchoring the protist to floating pond material while feeding. Characteristic of the ciliata is the presence of two kinds of nuclei. The *macronucleus* controls metabolic activities, while the smaller *micronucleus* directs cell division. Fig. 5.4 shows the structure of the *Paramecium*, a typical representative of the ciliata. Fig. 5.5 shows some other Ciliata.

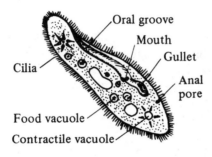

Fig. 5.4   Paramecium

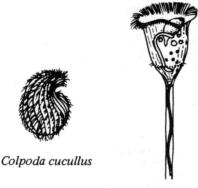

Colpoda cucullus

Vorticella

Fig. 5.5   Some examples of ciliates

Conjugation is a form of sexual reproduction which is occasionally demonstrated by the ciliates. Two organisms will join together at the oral groove. The micronucleus of each will undergo meiosis, producing several cells. All but two of these in each organism disintegrate. One of these haploid micronuclei remains in each cell, while the other migrates into the other cell, fusing with the stationary gamete. The new nucleus—which is now diploid and contains a new genetic combination—goes through cell division producing a new macronucleus and a new micronucleus (Fig. 5.6).

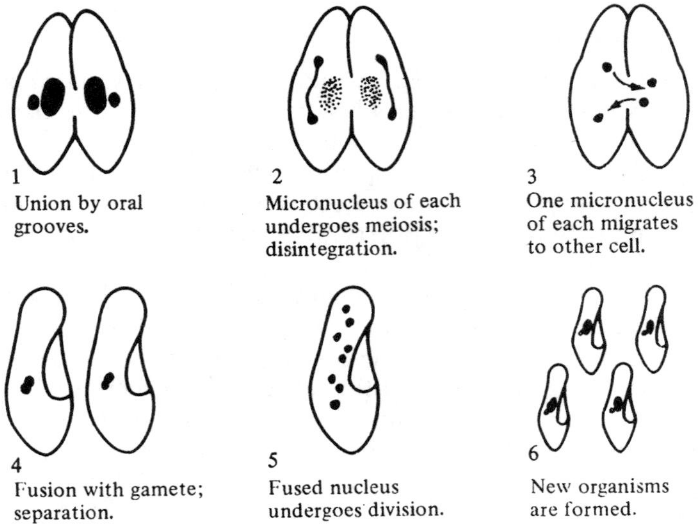

1
Union by oral grooves.

2
Micronucleus of each undergoes meiosis; disintegration.

3
One micronucleus of each migrates to other cell.

4
Fusion with gamete; separation.

5
Fused nucleus undergoes division.

6
New organisms are formed.

Fig. 5.6 Conjugation in paramecia

THE FUNGUS-LIKE PROTISTS

A *fungus* is an organism that obtains food by *absorbing* it from dead organic matter or from the body of a living host. There are two groups of organisms that are considered protists and yet their way of life and mode of nutrition are like those of the true fungi. These funguslike protists are the *Protomycota* and the *Gymnomycota*.

Protomycota

The Protomycota are small, unicellular organisms not readily visible. Many of them live as parasites or saprophytes on water-dwelling plants. A few species live in the soil. Others live as parasites inside of invertebrates such as liver fluke, nematodes or mosquito larvae.

The *chytrids* and the *hypochytrids*—two representative groups—live in association with a host organism or cell. One form of chytrid is a haploid cell that lives within another cell. At a particular time the nucleus goes through a series of mitotic divisions without division of the cytoplasm. When the nuclear divisions are completed, a little of the cytoplasm surrounds each nucleus. Each new cell develops a flagellum and these motile *zoospores* are released into the surrounding water (Fig. 5.7). Some may fuse and go through a sexual reproduction. Others merely develop into another vegetative cell and then the process repeats. Chytrids show various adaptations for absorbing nutrients. Some have cytoplasmic extensions known as *rhizoids* specialized for absorption.

Gymnomycota—Slime Molds

Vegetative thallus   Zoospore

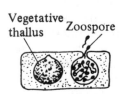

Fig. 5.7 Vegetative chytrid cells

There are two major groups of slime molds: the Myxomycota, or the true *slime molds*, and the Acrasiomycota, the *cellular slime molds*.

True Slime Molds

Slime molds live on the forest floor where they grow in damp soil, on or around rotting logs and on decaying vegetation. They appear to be shapeless globs of slime of varying colors: white, yellow or red.

The life cycle of the true slime molds begins with a multinucleate mass known as a *plasmodium*. This plasmodium glides about in amoeboid fashion, engulfing bacteria and small bits of organic material. The nuclei that make up the plasmodium are diploid. At some time in its life cycle the plasmodium stops moving and undergoes change, developing stalk-like structures with rounded knobs on top. These are *fruiting bodies*, and they support the structures known as *sporangia* (*sporangium*, sing.).

A sporangium is a structure that contains *spores*. The spores go through meiosis, producing flagellated gametes. The gametes fuse and form a *zygote* which is not flagellated, but instead, resembles an amoeba. This amoeba-like organism glides along the soil, engulfing food materials in phagocytic fashion. Its diploid nucleus goes through a series of mitotic divisions not accompanied by division of the cytoplasm. In this way the mutlinucleate plasmodium develops. The life cycle then repeats (Fig. 5.8).

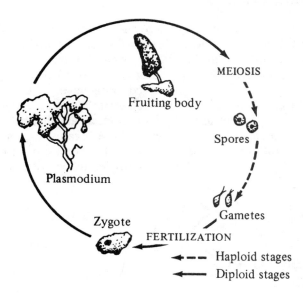

MEIOSIS

Fruiting body

Spores

Plasmodium

Gametes

Zygote

FERTILIZATION

- - - Haploid stages

⟵ Diploid stages

Fig. 5.8 Life cycle of a true slime mold

Cellular Slime Molds

The life cycle of the cellular slime molds, often called the *social amoebae*, is much different from that of the true slime molds. The fruiting bodies of the social amoeba are called *sorocarps*. The spores are carried by wind, water, insects or other means to new environments. Those spores placed in favorable environments will germinate. As the spore wall disintegrates, an amoeba, using pseudopods, pushes its way out onto the soil. These amoebae resemble other amoebae in structure. Each has a cell membrane, a nucleus, nucleolus, contractile vacuoles, food vacuoles, mitochondria and endoplasmic reticulum.

The free-living haploid amoebae feed on organic matter from the soil and grow to an optimum size. They then divide by mitosis and cytokinesis, producing daughter cells which, like the parent, have one haploid nucleus. Then their behavior changes. They stop feeding and begin to move in a directed manner toward definite centers or collecting points where they form closely packed groups and take on the formation of a compact mass of cells.

Next they become grouped together forming a *pseudo-plasmodium* or slug. A very thin sheath of polysaccharide material surrounds the mass, but the cells maintain their individuality. The finger-shaped pseudo-plasmodium moves across the soil substrate slowly but in a directed and coordinated way. Eventually, the front end of the pseudoplasmodium stops moving and the rear segment moves underneath the front end to form a mound of cells. At this time, differentiation of cells begins to take place. Stalk cells which give rise to a fruiting body are formed, and the life cycle repeats (Fig. 5.9).

THE PLANT-LIKE PROTISTS

There are three major groups of plant-like protists.

The Euglenophyta

The euglenoids are represented by the organism *Euglena* (Fig. 5.10). This unicellular organism has both plant-like and animal-like characteristics. It has chlorophyll *a* and *b* and some carotenoids, and it is able to carry on photosynthesis. However, *Euglena* lacks a cell wall, swims by using a flagellum, has a light-sensitive red-orange eyespot known as the *stigma* and a large contractile vacuole. *Euglena* also has a *pyrenoid body* that functions in the synthesis of *paramylum*, a carbohydrate storage product that is peculiar to the euglenoids. These organisms reproduce asexually by longitudinal mitotic cell division. However, during mitosis, the cell membrane remains intact and the nucleolus persists.

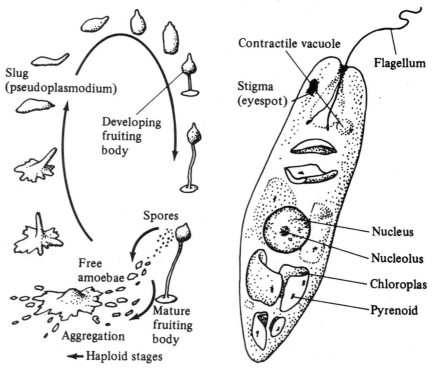

Fig. 5.9 Life cycle of a cellular slime mold

Fig. 5.10 Euglena

The Chrysophyta

Fig. 5.11 A chrysophyte

The yellow-green algae, the golden-brown algae and the diatoms are included in this group. Most species in the Chrysophyta group are single-celled and reproduce asexually. There are a few simple multicellular forms. The members of this group share certain characteristics. They are pigmented with chlorophyll *a* and *c* (lacking *b*) and contain the carotenoid *fucoxanthin*, which gives them the golden color. These protists live in salt water, fresh water, and in damp places between rocks. They use a polysaccharide called chrysolaminaran instead of starch. Many of the Chrysophyta have two flagella of unequal length at the anterior end of the body (Fig. 5.11); but some have only one flagellum and others, none.

The *diatoms* are diploid cells that lack flagella. Their cell walls, composed of two pieces, one fitting over the other, are impregnated with silica and pectin and often have intricate patterns in the forms of pits and ridges. When the organisms die, the shells fall to the bottom of the water where they disintegrate and form *diatomaceous earth*, a substance used as an abrasive in silver polish and detergents, as the packing in air and water filters, and in paint removers, deodorizing oils and fertilizers.

The Pyrrophyta

The *dinoflagellates* are small protists and usually unicellular. Most of these organisms have two unequal flagella, one extending longitudinally from the posterior end of the cell, the other encircling the central part of the cell. Some dinoflagellates extend trichocysts like the *Paramecium*; others have *nematocysts*, *stinging cells* common in the coelenterates.

The dinoflagellates have a most unusual nucleus. The chromosomes do not have centromeres and even during interphase remain in evidence as short, thickened rods. During mitosis, the nuclear membrane and the nucleolus remain, and no spindle is formed. Most species reproduce asexually by cell division and make up one of the two main groups of phytoplankton. Some species—*Noctiluca*, for example—are bioluminescent, giving off light like a firefly.

Importance to Humans

The quality of human life is in part determined by the ability to be free of disease. The protist disease-producers affect the health of large populations of people throughout the world. One such protist-initiated disease is amebic dysentery, a painful condition of bleeding ulcers caused by a type of amoeba that parasitizes human intestines. This protist-pathogen is passed to people through contaminated food and water. We have already mentioned two pathogenic protists carried by insects—*Trypanosoma gambiense* and *Plasmodium vivax*.

The red pigmented dinoflagellate *Gonyaulax* live in the waters of the Gulf of Mexico off the Florida coast. At times this organism reproduces explosively and uncontrollably coloring the waters red. This so-called "red tide" poisons millions of fish and does harm to people who eat these fish.

Many other protist species are helpful to humans. Slime molds help to keep an ecological balance by feeding on decayed plant and animal matter. These organisms also have a great deal of value as research specimens for investigators who are trying to ferret out the secrets of cell specialization and differentiation.

Plankton is the mass of green that floats on rivers, lakes and oceans. In reality, protists—microscopic "plant" life (phytoplankton) and microscopic "animal" life (zooplankton)—intermingle with mutual benefit in this floating mass which serves vital roles in the food chains of aquatic species.

Chronology of Famous Names in Biology

1667   **Anton Leeuwenhoek** (Dutch)—was the first person to see and describe protozoa.

1786   **Otto Frederick Muller** (Danish)—wrote the first treatise on protozoa titled *Animalcula Infusoria.*

1836   **Christian G. Ehrenberg** (German)—wrote a beautifully illustrated book describing 69 protozoa accurately.

1841   **Felix Dujardin** (French)—discovered protoplasm in protozoa.

1845   **Friedrich Stein** (Austrian)—named the orders and suborders of protozoa.

1880   **Richard Hertwig** (German)—discovered chromatin in the protozoan nucleus.

1959   **John T. Bonner** (American)—made a number of investigative studies which elucidated basic facts about the behavior and structure of cellular slime mold.

1962   **John T. Bonner** (American)—discovered that cAMP directs the movement of the social amoeba toward a collection center.

Words for Study

chytrid	contractile vacuole	dinoflagellate
cilia	cyst	fungus
colony	diatom	hypochytrid
conjugation	diatomaceous earth	macronucleus

micronucleus	pseudoplasmodium	stigma
nematocyst	pseudopod	symbiosis
oral groove	pyrenoid body	trichocysts
paramylum	rhizoid	tsetse
pellicle	slime mold	zooflagellate
plankton	social amoeba	zooplankton
plasmodium	sorocarp	zoospore
protist	sporangium	zygote
protozoa		

Questions for Review

PART A.   Completion. Write in the word that correctly completes the statement.

1. One-celled protists that resemble animal cells are the ..1..
2. Another name for a "self-feeder" is a(an) ..2..
3. *Amoeba* move by means of false feet known as ..3..
4. The usual mode of reproduction in protozoa is ..4..
5. The relationship in which two organisms of different species live together and neither is harmed by the association is known as ..5..
6. Digestion in *Amoeba proteus* takes place in the ..6..
7. The Sporozoa are harmful to organisms of other species and are therefore classified as ..7..
8. *Anopheles* is the genus name of a ..8..
9. The organelle that expels excess water from the protist is the ..9..
10. In the *Paramecium*, metabolic activity is controlled by which nucleus?
11. The body of the paramecium is prevented from being totally flexible by the ..11..
12. The form of reproduction in which like gametes fuse is called ..12..
13. The plasmodium is a stage in the life cycle of ..13..
14. A spore-producing structure is known as a ..14..
15. Sorocarps are correctly associated with the ..15.. slime molds.
16. In *Euglena*, the stigma is sensitive to ..16..
17. The cell walls of the diatoms are impregnated with pectin and ..17..
18. Diatomaceous earth forms from the ..18.. of diatoms.
19. The number of flagella usually found in dinoflagellates is ..19..
20. The number of flagella usually present in diatoms is ..20..

PART B. Multiple Choice. Circle the letter of the item that correctly completes each statement.

1. Classification systems are best described as being
   (a) unchanging   (c) unreliable
   (b) fixed   (d) artificial

2. The evolution of protists involved the development of
   (a) membrane bound organelles (c) functioning mitochondria
   (b) a discrete nucleus   (d) an active Golgi

3. Protists live
   (a) on land   (c) in fresh water
   (b) in the sea   (d) in all of these

4. The primary purpose of the protozoan cyst is to
   (a) produce new organisms
   (b) survive unfavorable conditions
   (c) increase the metabolic rate
   (d) aid in better nutrition

5. *Trypanosoma gambiense* is the organism that causes
   (a) botulism
   (b) cholera
   (c) African sleeping sickness
   (d) malaria

6. Limestone and chalk deposits are found in the shells of dead
   (a) diatoms   (c) Foraminifera
   (b) Sarcodina   (d) zooflagellates

7. *Plasmodium vivax* is the organism that causes
   (a) botulism   (c) African sleeping sickness
   (b) dysentery   (d) malaria

8. Trichocysts in ciliates serve mainly
   (a) as defense mechanisms (c) to expel water
   (b) to gather food   (d) for movement

9. Structures used for locomotion in the protists include all of the following except
   (a) pseudopodia   (c) cilia
   (b) legs   (d) flagella

10. The best way to prevent malaria is by destroying the breeding places of
    (a) *Plasmodium vivax*   (c) tsetse flies
    (b) *Trypanosoma vivax*   (d) *Anopheles* mosquitos

11. *Paramecia* move by means of
    (a) flagella   (c) trichocysts
    (b) cilia   (d) pseudopodia

12. Flagellated gametes produced by chytrids are the
    (a) zygotes   (c) zoospores
    (b) zooflagellates   (d) zymogens

13. Rhizoids are structures specialized for
    (a) reproduction   (c) locomotion
    (b) absorbing nutrients (d) providing rigidity

14. The plasmodium of true slime molds is best described as
    (a) multinucleate   (c) binucleate
    (b) uninucleate   (d) prokaryotic

15. In part of the life cycle of the true slime molds, sporangia are supported by structures known as
    (a) hyphae   (c) rhizoids
    (b) mycelia   (d) fruiting bodies

16. During mitosis in *Euglena*
    (a) the nucleolus disappears
    (b) chromosomes do not form
    (c) the nuclear membrane persists
    (d) transverse fission occurs

17. The yellow-green algae lack
    (a) chlorophyll *a*   (c) chlorophyll *c*
    (b) chlorophyll *b*   (d) caratenoids

18. An organism that lives inside of another and does harm to its host is known as a (an)
    (a) autotroph   (c) saprophyte
    (b) symbiont   (d) parasite

19. *Noctiluca* is noted for its
    (a) reproduction mode   (c) swimming ability
    (b) bioluminescense   (d) apparent immortality

20. Nematocysts are used for
    (a) swimming   (c) gliding
    (b) slinging   (d) stinging

PART C.   **Modified True-False.** If a statement is correct, write "true" for your answer. If a statement is incorrect, change the underlined word to one that will make it correct.

1. Protists are prokaryotes.

2. Chromosomes in the eukaryotes are best described as circular.

3. Protists descended from the same evolutionary lines.

4. Protozoa are unicellular autotrophs.

5. Protozoa are classified according to methods of nutrition.

6. *Trypanosoma gambiense* is carried by the insect known as the house fly.

7. *Trypanosoma gambiense* is best classified as a dinoflagellate.

8. *Plasmodium vivax* requires two hosts: fly and humans.

9. Adult sporozoans are motile.

10. The pseudoplasmodium of social amoebae moves in a directed way.

11. In *Paramecia*, reproduction is controlled by the macronucleus.

12. Chytrids are best classified as fungi.

13. The nuclei in the plasmodium of the true slime molds is haploid.

14. The pyrenoid body in *Euglena* stores <u>glycogen</u>.

15. The chloroplasts enable *Euglena* to carry out <u>heterotrophic</u> nutrition.

16. The pigment fucoxanthin causes the <u>green</u> color in the chrysophyta.

17. *Euglena* swims by means of a <u>flagellum</u>.

18. The diatoms have <u>no</u> commercial value.

19. Water balance in the amoeba is controlled by the <u>food</u> vacuole.

20. The chromosomes in the dinoflagellate nucleus lack structures called <u>chromatids</u>.

Think and Discuss

1. Why are protozoa classified as animal-like protists?

2. Study Fig. 5.8. Discuss each stage in the life cycle of a true slime mold.

3. What is the function of chlorophyll in *Euglena?*

4. Of what commercial value are the diatoms?

Answers to Questions for Review

PART A

1. protozoa
2. autotroph
3. pseudopodia
4. asexual
5. symbiosis
6. food vacuole
7. parasites
8. mosquito
9. contractile vacuole
10. macronucleus
11. pellicle
12. conjugation
13. slime molds
14. sporangium
15. cellular
16. light
17. silica
18. cell walls
19. two
20. zero

PART B

1. d
2. a
3. d
4. b
5. c
6. c
7. d
8. a
9. b
10. d
11. b
12. c
13. b
14. a
15. d
16. c
17. b
18. d
19. b
20. d

PART C

1. eukaryotes
2. linear
3. diverse
4. heterotrophs
5. locomotion
6. tsetse
7. zooflagellate
8. mosquito
9. nonmotile
10. true
11. micronucleus
12. protists
13. diploid
14. paramylum
15. autotrophic
16. yellow
17. true
18. great
19. contractile
20. centromeres

CHAPTER 6

THE FUNGI

Most fungi are eukaryotic, multicellular and multinucleate organisms. Yeasts are unicellular forms. The cells of fungi are different from those of other species because the boundaries separating the cells are either entirely missing or only partially formed. Thus fungi are primarily *coenocytic* organisms; this means that the cells have more than one nucleus in a single mass of cytoplasm. However, the characteristic that most distinguishes the fungi from other organisms is their mode of nutrition.

General Features

NUTRITION

Since fungi do not have chlorophyll, they cannot produce their food autotrophically by photosynthesis. Fungi are heterotrophs. However, they cannot engulf food phagocytically as do the amoeba, nor can they ingest food as do animals equipped with a mouth. Some fungi are restricted to a saprophytic way of life, obtaining nutrition by absorbing dead organic matter, while others are parasitic, absorbing their nutrition from living hosts. In either case, fungi must live very near their food supply in order to stay alive. Specialized fungal structures secrete digestive (hydrolytic) enzymes into the food substrate. The organic molecules of the substrate are made smaller by these enzymes and thus can be absorbed by the fungus.

BODY ORGANIZATION

Fig. 6.1 Spore

The fungus begins its life as a *spore* (Fig. 6.1). A spore is a microscopic cell that, under favorable conditions, will develop into a new individual. The cell wall of the spore is thick and tough, resistant to adverse environmental conditions. However, the spore is light in weight and is trans-

Fig. 6.2 Germinating spore

Fig. 6.3 Mycelium

Fig. 6.4 Internal view of a hypha

ported to new habitats by currents of air. When a spore settles on its substrate in favorable environmental conditions, its cell wall breaks open and the spore commences to sprout or *germinate* (Fig. 6.2). The germinating spore absorbs food and grows an elongated thread called a *hypha*. As growth continues, many hyphae develop until they appear as a tangled mass of threads. When many hyphae appear, the body of the fungus is now called a *mycelium* (Fig. 6.3). A hypha gives off a chemical that makes other hyphae grow away from it. In this way, competition for food is reduced among hyphae. In general, the function of the mycelium is to absorb food and to produce new fungus plants.

A microscopic study of a hypha provides interesting information about its internal structure. Some hyphae are multinucleate with many nuclei sharing the same mass of cytoplasm without dividing membranes (Fig. 6.4). Other hypha are divided into compartments by *septa*. These compartments contain one or two nuclei.

Parasitic fungi have special structures called *haustoria* that are specialized to penetrate the cell walls of plants. The haustoria grow into plant cell cytoplasm and absorb nutrients directly from the cells which they parasitize (Fig. 6.5).

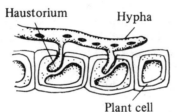

Fig. 6.5 Haustoria

METHODS OF REPRODUCTION

Reproduction in fungi usually occurs by the production of spores. Special reproductive hyphae called *sporangiophores* have growing at their tips *sporangia* (sing. sporangium), or spore cases. It is within the spore case that reproductive spores form.

Some fungi also reproduce vegetatively. *Vegetative reproduction* is the process by which a new individual is produced from a part of the parent's body without involving sex cells. New individuals produced vegetatively are identical to the parent. Broken fragments of mycelia produce new fungus individuals identical to themselves.

In most fungi asexual reproduction (vegetative and sporulation) is augmented by a sexual cycle. The sexual phase begins with the fusion of *gametangia* (gametangium, sing.), special gamete-producing structures. This is followed by a fusion of special nuclei or gametes resulting in the formation of a zygote.

Major Classes of Fungi

The phylum Mycota is the only phylum in the kingdom Fungi. It is divided into five classes: Oomycota, Ascomycota, Zygomycota, Basidiomycota, and Deuteromycota.

CLASS OOMYCOTA: WATER MOLDS

The Oomycota are primarily water molds, although some species live on land. This is the only fungus class in which the cell walls are made of cellulose, not of chitin, and in which the gametes are differentiated into male sperm and female egg cells. Another characteristic of the class is that the spores are flagellated and require free water for swimming.

Most of the Oomycota are *saprobes*, absorbing their nutrients from dead organic matter. Some species are parasitic and disease-producing. For example: *Albugo candida* causes white rust on cabbage and other leafy plants. *Saprolegnia*, a saprobe, grows as mold on the water-borne bodies of dead insects, fish and frogs.

CLASS ASCOMYCOTA: SAC FUNGI

Yeast is an example of a single-celled member of the Ascomycota. Yeast cells are small, oval structures that reproduce by budding. Most Ascomycota, however, are multicellular. This is the largest class of fungi, with about 30,000 species, including the powdery mildews, black and blue-green molds, and the truffles and morels.

The Ascomycota reproduce asexually by means of very fine spores known as *conidia*. The hypha of the Ascomycota are divided into compartments by septa. Each compartment contains its own nucleus, but pores in the septa allow the migration of cell structures from one compartment to another.

The Ascomycota are so named because during part of the life cycle reproductive cells are held in a little sac or *ascus*. At a certain time, two hyphae grow together. Although their cytoplasm intermingles, the nuclei remain separate and do not fuse. The new hyphae that grow from this fused structure have nuclei of different genetic strains. This is known as a *dikaryon*. The dikaryon fuses with nonreproductive hyphae to form a fruiting body in which the asci form. At this time, within the cell that is to become the ascus, the two nuclei of the dikaryon fuse. They now go through a series of meiotic and mitotic divisions, resulting in eight haploid nuclei, which soon become surrounded by their own walls to form eight *ascospores* that disperse when the ascus ruptures.

CLASS ZYGOMYCOTA

Most of the Zygomycota are land-dwelling organisms that inhabit the soil and carry out a saprobic way of life. Their hyphae are coenocytic and the cell walls are composed of *chitin*. Chitin is a tough, nitrogen-containing polysaccharide that is present in the exoskeletons of insects and in the cell walls of most species of fungi.

The type species of this class is *Rhizopus stolonifer* (Fig. 6.6), the black bread mold. The life cycle of *Rhizopus* begins with a germinating spore that is established on a favorable substrate such as a piece of bread in a moist, warm, dark environment. Hyphae grow from the spore, eventually forming a mycelium. Some of the hyphae become specialized into rhizoids, downward growing threads that secure the mycelium to the substrate. The rhizoids have additional functions, also. They send out

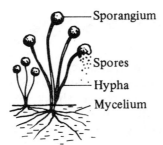

Sporangium

Spores

Hypha

Mycelium

Fig. 6.6 Bread mold and spores

digestive juices which break down the large organic molecules of the substrate into smaller ones in the process of digestion. The rhizoids then perform their third function by absorbing the digested material. Other specialized hyphae grow upward. These are known as *sporangiophores* because they bear at their tips sporangia which produce haploid spores.

During part of its life cycle, *Rhizopus* goes through sexual reproduction. Hyphae of opposite mating strains form a *gametangium*, a structure that produces special nuclei that function as gametes. The gametangia of opposite mating strains fuse. Some of their nuclei pair off and fuse forming diploid nuclei. A thick wall covers the fused gametangia and their zygote nuclei to form a *zygospore*. The zygospore remains dormant for a while. When the zygospore germinates, meiosis takes place in the zygotes. An aerial hypha grows from the activated zygospore. This hypha is a sporangiophore and supports a sporangium which produces many haploid spores. The life cycle repeats.

CLASS BASIDIOMYCOTA: CLUB FUNGI

The Basidiomycota include the mushrooms, bracket fungi and the smuts. The life cycle of the Basidiomycota begins when a haploid spore germinates, giving rise to hyphae. Hyphae of different mating types fuse forming mycelia that are dikaryotic. Most of the life history of the mushrooms and its relatives is spent in the dikaryotic stage.

The name of this class is derived from the *basidium* (club) that forms at the tip of each reproductive hypha. Inside each basidium, the two haploid nuclei fuse, forming a diploid nucleus. This nucleus undergoes meiosis, thereby producing four haploid nuclei. The haploid nuclei move to the outer edge of the basidium. Each nucleus becomes surrounded by an elongated cell wall in a structure known as the *basidiospore*. The basidiospore is maintained on a delicate stalk which separates it from the rest of the fungus. Dissemination of the spore takes place when the stalk breaks and is carried away by the wind.

CLASS DEUTEROMYCOTA: IMPERFECT FUNGI

This class contains the species known as the "Fungi Imperfecti." The species that puzzle taxonomists are put into this class. Most of these organisms have life cycles that are not typical of fungi. Since sexual reproduction has not been observed in these species, they are called imperfect fungi. The organisms that cause the human diseases of ringworm, athlete's foot and thrush are classed as Deuteromycota.

Special Nutritional Relationships _____

Fungi are often found associated with other species in special relationships. When two different species of organisms live together, the relationship is called *symbiosis*. If the relationship is of mutual benefit to both species, it is called *mutualism*. When one species benefits and the other does not but is not harmed by the association, the condition is known as *commensalism*. When one species lives at the expense of another, doing harm to its *host*, the relationship is *parasitism*. Disease-producing organisms are parasites.

LICHENS

Lichens are pioneer organisms that can inhabit bare rock and other uninviting substrates. They live on the barks of trees and even on stone walls. A lichen is a combination of two organisms—an alga and a fungus—that live together in a mutualistic relationship. The alga carries on photosynthesis, while the fungus absorbs water and mineral matter for its partner. The fungus also anchors the lichen to the substrate. Scientists have determined that the alga in this partnership can live alone. The fungus, which may belong to the Ascomycota, the Deuteromycota, or the Basidiomycota, cannot exist by itself and is therefore the dependent member of the team. New lichens are formed by the capture of an alga by a fungus. If the fungus kills the alga, the fungus also dies.

MYCORRHIZA

Several fungus species, including mushrooms, live in close association with plant roots. The mushroom absorbs minerals from the soil and passes them along to the plant on which it lives. This condition is known as *mycorrhiza*, meaning "fungus root." Scientists are not too sure how the plant helps the fungus, but it is known that most plant families grow better when in association with fungi.

Importance to Humans _____

From the viewpoint of environment and ecology, fungi help to keep the natural environment in balance. Species dependent on dead organic matter as their source of nutrition assist with the breaking down of fallen leaves, the dead bodies of plants and animals, and animal wastes. Fungi have the ability to absorb moisture from the air, a characteristic that permits them to live in environments that do not have adequate soil water as required by other species.

Most fungi are not pathogenic to humans. However, some nonparasitic fungi do work against human interests. There are fungus species that live quite well digesting such unlikely substrates as the insulation on telephone wires, leather, polyvinyl plastics, cork, hair and wax. There are those species that cause the mildew of clothing, wall paper and books. Then there are those wood-rotting fungi that destroy the wood construction in houses and ships.

Parasitic fungi invade plants more readily than animal bodies. Diseases of food crops can have disastrous effects on human populations. This occurred when the oomycete *Phytophthora infestans* caused the potato blight in Ireland, resulting in a devastating famine between the years 1845 and 1851, and when *Plasmopara viticola*, the cause of downy mildew of grapes, nearly ruined the wine-making industry of France. Ergot, a disease of rye, is caused by the ascomycete *Claviceps purpurea*. Ergotism, the disease that affects humans who eat the infected rye, induces gangrene, nervous spasms, convulsions, and psychotic delusions. Ergot is the source of the hallucinogen LSD. Other plant diseases caused by the Ascomycota are peach leaf curl, Dutch elm disease, chestnut blight and apple scab.

Some Ascomycota, however, are beneficial to humans. These include the truffles and the morels, edible fruiting bodies, that cost $400.00 a pound in New York. *Saccharomyces cerevisiae* is the species of yeast needed in the fermenting processes of malt, barley and hops to make beer. The fermenting of grapes produces wine. Another species of yeast is used in making dough.

The Fungi Imperfecti include *Penicillium notatum*, the source of the antibiotic penicillin. This group of fungi also includes those species that help to ripen Roquefort and Camenbert cheeses. Soy sauce is prepared by fermenting soybeans with the mold *Aspergillus oryzae*.

Chronology of Famous Names in Biology

1929   **Alexander Fleming** (English)—discovered the antibiotic properties of the fungus *Penicillium notatum*.

1943   **Albert Hoffman** (Swiss)—isolated LSD from ergot and discovered the powerful hallucinogenic properties.

Words for Study

ascospore	germinate	saphrophyte
ascus	haustoria	saprobe
basidiospore	host	septa
basidium	hypha	sporangiophores
chitin	lichen	sporangium
coenocytic	mutualism	spore
commensalism	mycelium	symbiosis
conidia	mycorrhiza	vegetative
dikaryon	parasitism	reproduction
gametangium	rhizoid	zygospore

Questions for Review

PART A.   Completion. Write in the word that correctly completes each statement.

1. A cell that has several nuclei in a single mass of protoplasm is called a ..1..
2. Fungi absorb dissolved food molecules from dead organic matter and thus are classified as ..2..
3. Hydrolytic enzymes are the same as ..3.. enyzymes.
4. Vegetative reproduction does not involve ..4.. cells.
5. Conidia are cells known more commonly as ..5..
6. An example of a unicellular ascomycete is ..6..
7. Rhizoids secrete substances that serve the purpose of ..7..
8. The combination of an alga and a fungus living in a mutualistic relationship is called a ..8..
9. Ringworm is a ..9.. infection and not one caused by worms.
10. A cell that has two nuclei originating from different genetic strains is known as a ..10..

PART B.   Multiple Choice. Circle the letter of the item that correctly completes each statement.

1. A saprobe is an organism that
   (a) absorbs material from living cells
   (b) absorbs material from dead cells
   (c) ingests dead organic matter
   (d) ingests living organic matter

2. Septa are
   (a) compartments
   (b) double nuclei
   (c) dividing walls
   (d) masses of protoplasm

3. Gametangia are assoiated with
   (a) sporulation
   (b) vegetative reproduction
   (c) germination
   (d) sexual reproduction

4. The only fungus class in which the gametes are differentiated into sperm and egg is the
   (a) Oomycota
   (b) Zygomycota
   (c) Ascomycota
   (d) Basidiomycota

5. A dikaryon is a cell that has
   (a) a single nucleus and lots of cytoplasm
   (b) perforated cell membranes
   (c) two nuclei of different genetic strains
   (d) a micronucleus and a macronucleus

6. Zygospores are characteristic of the fungus class
   - (a) Oomycota
   - (b) Fungi Imperfecti
   - (c) Basidiomycota
   - (d) Zygomycota

7. Black bread mold is the type species for the fungus class
   - (a) Oomycota
   - (b) Basidiomycota
   - (c) Deuteromycota
   - (d) Zygomycota

8. Each class in the kingdom Fungi uses special methods to produce
   - (a) hypha
   - (b) rhizoids
   - (c) mycelia
   - (d) spores

9. Fungi Imperfecti are so named because
   - (a) sexual reproduction has not been observed in them
   - (b) they cause human disease
   - (c) many of their hyphae are missing
   - (d) they do not reproduce by spores

10. The disease of rye which is devastating to humans is known as
   - (a) morel
   - (b) truffle
   - (c) ergot
   - (d) ringworm

PART C.   Modified True-False. If a statement is true, write "true" for your answer. If a statement is incorrect, change the underlined word to one that will make the statement true.

1. Most of the fungi begin life as a <u>zygote</u>.

2. To germinate means to <u>contaminate</u>.

3. Flagellated swimming spores are characteristic of the fungus class <u>Zygomycota</u>.

4. Ascospores are <u>diploid</u>.

5. A nitrogen-containing polysaccharide that composes the cell walls of most fungi is <u>cellulose</u>.

6. *Rhizopus stolonifer* is the scientific name for <u>blue-green mold</u>.

7. When two gametes fuse, a <u>zygospore</u> is formed.

8. An example of the class <u>Basidiomycota</u> is water mold.

9. The word basidium means <u>mold</u>.

10. Pioneer organisms that can inhabit bare rocks are <u>lichens</u>.

Think and Discuss _____

1. Why do fungi have to live near their food supply to stay alive?
2. What role is played by spores in the lives of fungi?
3. How do the water molds differ from other fungi?
4. Of what commercial value is yeast?

Answers to Questions for Review

PART A

1. coenocyte
2. saprobes
3. digestive
4. sex
5. spores
6. yeast
7. digestion
8. lichen
9. fungus
10. dikaryon

PART B

1. b
2. c
3. d
4. a
5. c
6. d
7. d
8. d
9. a
10. c

PART C

1. spore
2. sprout
3. Oomycota
4. haploid
5. chitin
6. black bread mold
7. zygote
8. Oomycota
9. club
10. true

CHAPTER 7

THE GREEN PLANTS

Taxonomists group all green plants in the kingdom Plantae. This kingdom consists of some single-celled species which include free-living cells, cells that live together in colonies and some cells that adhere together in long filaments. However, most members of the Plantae are multicellular. Although multicellular means "having many cells," the concept of multicellularity involves more than numbers of cells. It embraces several ideas about cell structure and function. One such idea concerns *cell specialization* in which cells are "programmed" to carry out special tasks. Cell specialization in multicellular plants also brings with it a *division of labor* in which groups of cells in tissue formation work together to perform some special life function of benefit to the entire plant organism. Another condition necessary to all specialization is the evolution of cell structures that permit specialization and difference in cell function.

As implied in the term "green plants," members of the kingdom Plantae contain the green pigment chlorophyll. Not only does chlorophyll color plant leaves and some stems green, it, more importantly, traps light energy which is used in the process of photosynthesis. As an outcome of photosynthesis nutrient molecules are made, serving as food for both plants and animals.

Most species of the kingdom Plantae are nonmotile, anchored to one place, and unable to move about, but a few of the lower plants are motile for at least part of the life cycle. However, the evolutionary trend exhibited in green plants is toward stationary organisms that carry out their life functions on land in locations where they remain for life. Lower plant species equipped to swim about live in salt and fresh water. Higher plants are *terrestrial* (land-dwelling).

The Plant Kingdom _____

LOWER PLANTS

The lower plants are not differentiated into roots, stems and leaves. The simple body of lower plants is either a single cell or a flat sheet of simple cells. Lower plants do not have specialized tissues to carry water, to anchor the plants, or to grow new cells. As a rule, the sex cells of the lower plants are produced in rather simple sex organs that are not protected by a surrounding wall of cells. The zygotes do not develop into embryos that are contained in a female reproductive organ. Most of these plants live in fresh water, although there are a few saltwater forms. Some species live in damp soil or on the bark of trees.

The green algae, the brown algae and the red algae compose the three divisions of lower plants.

Chlorophyta—Green Algae

Green algae belong to the Chlorophyta. The scientific name tells us that these algae contain the green pigment chlorophyll. Of interest to botanists is the fact that members of the Chlorophyta are considered to be the ancestors of the higher land plants based on three points of evidence. First of all, green algae have chlorophyll *a* and *b* in the same amounts as cells in higher plants. Second, green algae store food in the same form as do the higher plants. Third, green algae, like the higher plants, have well-defined cell walls made of cellulose.

The Chlorophyta includes forms such as *Chlamydomonas* that typically live as single cells, simple and complex colonial forms, and forms like sea lettuce that typically undergo a life cycle known as alternation of generation.

Chlamydomonas

Chlamydomonas is a unicellular autotroph that is a representative type of green algae. In some ways it resembles cells of the protist kingdom, because it lives in water, is motile, and has a pair of flagella (Fig. 7.1). However, biochemical analysis of its chlorophylls, carotenoids, and stored food reveals that these compounds are identical to those in the cells of green land plants. Scientists believe that *Chlamydomonas* represents the general type of unicellular algae from which multicellular green plants evolved.

Chlamydomonas is an oval-shaped cell with a haploid nucleus. Although its pigments and starch granules are identical with those of higher plants, its cell wall, unlike that of other green algae, is composed of glycoprotein, not cellulose. Sticking out from the anterior end of the organism are two flagella of equal length. The large cup-shaped chloroplast fills up nearly the entire volume of the cell. Inside the chloroplast there are membraneous sacs that contain the chlorophyll. At the bottom portion of the chloroplast is a rather large circular *pyrenoid body* that functions in the synthesis and storage of starch. The *stigma*, a light sen-

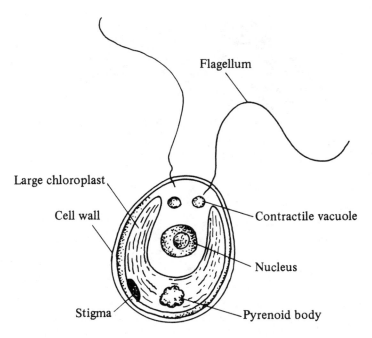

Fig. 7.1 *Chlamydomonas*

sitive eyespot, is located inside the chloroplast. Two contractile vacuoles which open and close alternately lie at the base of the flagella.

Before the onset of cell division, certain changes take place in the vegetative *Chlamydomonas* cell. The organism attaches itself to some object in the water and both of the flagella are resorbed into the cytoplasm. The once motile cell is now sessile. Mitosis and cytokinesis take place resulting in two identical daughter cells. These cells are retained within the membrane of the parent cell and quickly divide once more. Four identical flagellated *zoospores* are now released from the confines of the old cell wall. These cells grow and mature into vegetative *Chlamydomonas* organisms. This form of reproduction is asexual, and it is the usual mode of reproduction in *Chlamydomonas*.

Under unfavorable conditions (low nitrogen content in the water, for example), *Chlamydomonas* undergoes a primitive form of sexual reproduction. The vegetative *Chlamydomonas* cells go through several mitotic divisions, releasing several small flagellated cells into the water. These cells are known as *isogametes*, gametes that are indistinguishable from each other. The isogametes pair off, attached to each other end to end by the flagella. The cell wall slips away from each cell. Their cytoplasms fuse, forming a diploid zygote. A thick protective cell wall surrounds the zygote before it falls to the bottom of the pond. The zygote remains inactive and survives such an extreme condition as the drying of the pond. When environmental conditions improve, the zygote goes through a meiotic division forming zoospores that mature into haploid vegetative *Chlamydomonas* cells (Fig. 7.2).

Colonial Forms of Chlorophyta

The evolutionary trend from the unicellular *Chlamydomonas* to cells that can survive only when associated in a colony can be traced quite effi-

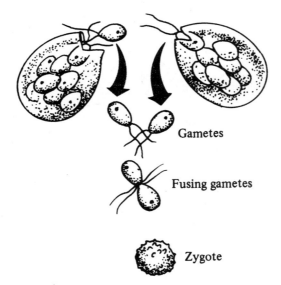

Gametes

Fusing gametes

Zygote

Fig. 7.2 Gamete formation in *Chlamydomonas*

ciently in the Chlorophyta. *Gonium* is a genus in the division Chlorophyta. Each *Gonium* species lives as a simple colony of cells. The cells of *Gonium* look very much like the single cell of *Chlamydomonas*. However, instead of living singly, they are held together in a gelatinous disk. The colony swims as a unit and the cells divide at the same time. Sexual reproduction in *Gonium* follows the pattern set by *Chlamydomonas* in that isogametes fuse to form a zygote.

Pandorina, another colonial form of green algae, shows a bit more complexity than *Gonium*. In *Pandorina*, the colony has an anterior end and a posterior end as shown by the orientation in swimming. Cells in the *Pandorina* colony cannot live alone and the colony dies if broken apart. The male gametes of *Pandorina* are smaller than the female gametes. Since the gametes can be distinguished, this type of sexual reproduction is known as *heterogamy*. Since their only distinguishing characteristic is size, they are known as *anisogametes* (Fig. 7.3).

Further colonial development is shown in the genus *Eudorina*. The cells of the anterior portion are more dominant than those of the posterior region, and the nonmotile female gametes are fertilized by the smaller,

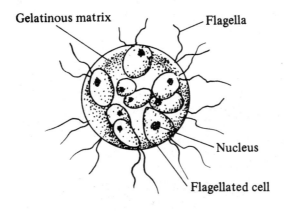

Gelatinous matrix

Flagella

Nucleus

Flagellated cell

Fig. 7.3 *Pandorina.* The cells are confined in an intercellular bed or matrix of gelatinous material.

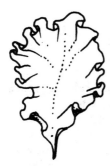

Fig. 7.4 Sea lettuce

free-swimming male gametes inside the colony. The condition in which the sperm cell is motile and the egg cell is nonflagellated and nonmotile is called *oogamy*.

Colony development in *Volvox* is significantly more advanced than in any of the other green algae. Colonies in the *Volvox* genus are large, consisting of 500 to 50,000 cells. Cytoplasmic strands between the cells permit communication. Most of the cells are vegetative and do not function in reproduction. Scattered in the posterior half of the colony are a few large cells which are specialized for reproduction. Among these colonial forms we see an increase in the number of cells that make up the colonies, in the communication among cells, and in the specialization and differentiation of cells.

Alternation of Generations

The sea lettuce *Ulva* is a green algae that lives in saltwater. Its body is in the form of a flat leafy thallus and is two cell layers thick (Fig. 7.4). The life cycle of *Ulva* is described as *alternation of generations* because one generation of *Ulva* is produced sexually by gametes while the next generation is produced asexually by zoospores. To the naked eye, there is no difference in the structure of the sea lettuce produced sexually or asexually.

The *gametophyte generation* is a haploid thallus from which small, flagellated gametes are released into the water. A thallus is a flat sheet of photosynthetic cells. They pair off and fuse. Each fused pair of gametes forms a zygote that, after a short time, becomes a diploid thallus of a new generation of *Ulva*.

The diploid thallus is the *sporophyte generation*. It produces haploid zoospores that develop and grow into a haploid thallus which is now the gametophyte (Fig. 7.5).

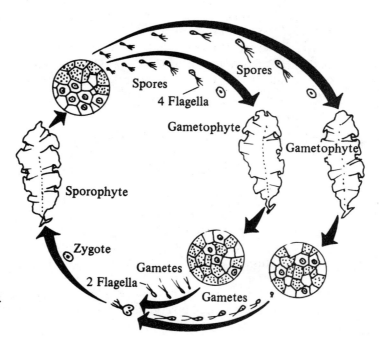

Fig. 7.5 Alternation of generations in sea lettuce (*Ulva*)

Phaeophyta—Brown Algae

Most species of brown algae are marine, inhabiting the cooler ocean waters. They grow attached to rocks along the shallow waters of seacoasts. The brown algae are commonly called seaweed.

Species of the Phaeophyta are the largest of the algae and may reach lengths of 45 meters or more. *Kelps* are massive brown algae found usually on the Pacific Coast. They contain chlorophylls *a* and *c*, but lack chlorophyll *b*. The brownish pigment *fucoxanthin* gives them their characteristic color. In structure, the brown algae are quite complex, showing considerable differentiation among the cells. Most species have *holdfasts* which secure them to rock substrates. Some species have stem-like and leaf-like parts. Many species have *air bladders* which give them buoyancy.

Fucus is a representative species of the Phaeophyta. Fig. 7.6 shows its general structure, including the very prominent air bladders. *Fucus*, also known as bladder wrack and rock weed, lives on the rocky sea shores of temperate seas. The thallus of *Fucus* is unusual in that it has repeated double branches, with enlarged tips commonly called *conceptacles*. These conceptacles hold the sex organs. The *antheridia* contain the sperm, while the *oogonia* hold the egg cells. In some species the male and female sex organs are on separate plants; in other species they are produced in the same conceptacle. Gametes are released into the water through pores at the tips of the conceptacles. An egg cell fuses with a sperm cell to form a zygote. The zygote grows into a diploid plant. The haploid gametophyte stage is completely missing in *Fucus*.

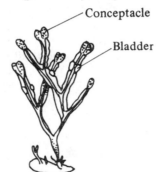

Fig. 7.6   *Fucus*, the rockweed

Rhodophyta—Red Algae

The red algae are seaweeds inhabiting the warmer waters of the oceans and living at considerably greater depths than the brown algae. Very few species grow in fresh water. The leaf-like structures of a red alga are branched and filamentous, appearing rather feathery. Some species resemble gelatinous ribbons.

The life cycles of most species of red algae are quite complex. In general, they go through some sort of alternation of generations. During the asexual phase, the red algae reproduce by nonmotile spores. The heterogametes involved in the sexual generation are nonmotile also. The male gametes are carried by currents of water to the female sex organs known as *carpogonia*. For a short while, the sperm cell sticks to the elongated tip of the carpogonium. Then the sperm cell nucleus passes through the wall of the elongated tip and into the inner region of the carpogonium. There, the sperm nucleus fuses with the nucleus of the egg cell.

Some representative species of red algae are *Chondrus* (also called Irish "moss"), *Polysiphonia* and *Nemalion*. The cell walls of the red algae are made of a combination of cellulose and a gelatinous material. The reserve carbohydrate stored in the cells of red algae is known as *floridean starch*; it is not a true starch.

The Rhodophyta contain both chlorophyll *a* and chlorophyll *d*. The latter is a type of chlorophyll not found in any other species of plants. The characteristic color of the red algae is given by the pigment *phycoerythrin*.

The value of accessory pigments to the photosynthetic process is demonstrated by the red algae. Chlorophyll *a* is the pigment actively involved in trapping light energy for use in photosynthesis. However, chlorophyll *a* cannot trap light energy at the depths at which the red algae grow. The pigments *phycocyanins* and phycoerythrins can and they pass the energy along to chlorophyll *a*.

Red algae have a great deal of commercial value. They are dried and ground up to be used as agar for bacterial media. They also form colloids which are used as suspending materials in ice cream and binders in puddings and chocolate milk.

HIGHER PLANTS

Higher plants have developed a number of adaptations for life on land—structures and biochemical methods for conserving water which is necessary for all of the biochemical activities of the cell and thus for the maintenance of life. Chief among these adaptations was the evolution of sex organs that are protected by a surrounding layer of nonreproductive cells. *Antheridia* are male sex organs where sperm cells are produced. Egg cells are contained in organs called *archegonia*. Gametes enclosed in these organs are protected against drying out. Another adaptation for the prevention of water loss in the embryophytes is the *cutin* covering, a waxy substance impregnated in cell walls, which provides waterproofing to epidermal tissue and prevents evaporation of water. The higher plants also have special *vascular* or water-carrying tissues designed to distribute water efficiently throughout the plant body.

Bryophyta—Mosses, Liverworts and Hornworts

The bryophytes are the first green land plants. They are primitive, small and inconspicuous. Although multicellular, the tissue differentiation is quite simple. Bryophyte species have no tissues that are specialized for water-carrying and no cambium specialized for growing new cells. Bryophyte species do not have true stems, leaves or roots. Simple rootlike structures called *rhizoids* anchor the plants to the ground and absorb moisture from the soil.

Alternation of generations occurs in all bryophytes. The larger and more noticeable generation is the gametophyte which usually supports a smaller (sometimes parasitic) sporophyte. Ciliated sperm cells, produced in the antheridia, swim to the archegonia and fertilize the egg cells held therein. The sporophyte generation begins with the zygote and is therefore diploid. Special cells in the sporophyte go through meiosis and produce haploid spores that begin the gametophyte generation.

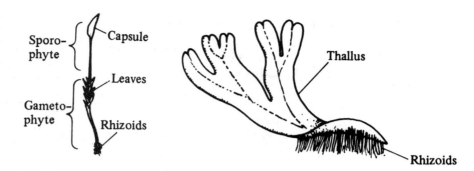

Fig. 7.7 Moss   **Fig. 7.8** Liverwort

There are 23,000 species of bryophytes among which are the mosses (Fig. 7.7), the liverworts (Fig. 7.8) and the hornworts, a group of plants structurally similar to the liverworts.

Tracheophyta—Vascular Plants

The *vascular* plants are truly land-dwelling plants. They have developed adaptations that permit them to live on land independent of bodies of water. The word "vascular" means that these plants have a water-carrying system. Water is conducted upward from the roots by *xylem* tubules. Fluid compounds are conducted downward from the leaves to lower plant organs by the *phloem* tubules.

The tracheophytes are divided into five subdivisions: psilopsids, club mosses, horsetails, ferns and seed plants.

Psilopsida

There is disagreement among botanists as to whether there are two living genera—*Psilotum* and *Tmesipteris*—of psilopsids or whether all members of this subdivision are extinct. The plants that may represent the psilopsids have rather simple bodies, with branched stems, but no roots. The leaves are absent or very small.

Lycopsida—Club Mosses

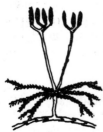

Fig. 7.9 Club moss—*Lycopodium*

There are about 900 living species of club mosses. Plants in this group are usually one meter or less in height. Many are ground creepers. The club mosses have water-carrying (vascular) tissues and true roots, stems and leaves. The leaves are small and spirally arranged on the stems. The popular name of the Lycopsida is derived from the arrangement of the sporangia which are clustered on leaves formed into *cones* or *strobili*. The cones are positioned on the tips of stems as shown in Fig. 7.9. Some club moss species bear one type of spore and one type of gametophyte; such species are described as being *homosporous*. Other species of club moss are *heterosporous* because they bear two types of spores and produce two types of gametophytes.

Sphenopsida—Horsetails

Only 25 living species of Sphenopsida remain. The horsetails are true land plants having a vascular system, true roots, stems and leaves. Although the leaves are small and scale-like, they carry on photosynthesis. Horsetails grow well in both tropical and temperate climates, along river banks and in moist tracts of land (Fig. 7.10).

The life cycle of the horsetails involves alternation of generations. The sporophyte generation is the conspicuous generation. Its haploid spores give rise to a small plant known as a *prothallus*. The prothallus represents the gametophyte generation. The zygote remains attached to the gametophyte and ultimately becomes the sporophyte.

The cell walls in the leaves and stems of horsetails contain silica (sand), making the dried, ground stems useful as scouring powders.

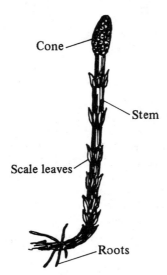

Cone

Stem

Scale leaves

Roots

Fig. 7.10   The horsetail—
Equisetum

Pteropsida—Ferns

The fern plant used in flower bouquets is the sporophyte generation. Remember that the sporophyte generation produces asexual spores. The mature fern has true roots, leaves and stem. Ferns growing in temperate climates have an underground stem called a *rhizome* which grows in a horizontal position. The rhizome not only stores food materials, but also gives rise to new fern plants which grow along its length. The stems of tropical species grow upright in a vertical position and serve as trunks of tree ferns.

Fern stems are composed of xylem and phloem vascular tissues but they do not have the growth layer of embryonic cells known as *cambium*. Growing from the lower side of the rhizome are fibrous roots which absorb water and dissolved minerals from the soil. The leaves grow out from the top of the rhizome and break through the ground. Ferns have compound leaves composed of many leaflets.

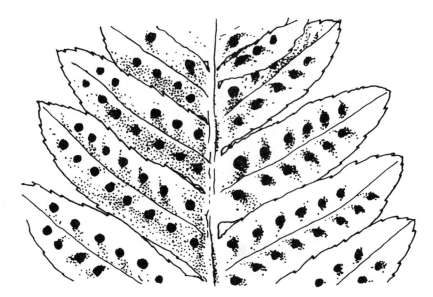

Fig. 7.11 Fern
leaf showing
spore cases

The leaves of ferns serve a dual purpose. First of all, the green pigment chlorophyll enables them to carry on photosynthesis to produce food for the plant. Secondly, the undersides of certain green leaves are covered with small structures that resemble brown dots. Each dot, called a *sorus*, is really a cluster of spore cases (*sporangia*) (Fig. 7.11). Inside of the sporangia haploid spores are produced. When the spores are ripe, they are discharged from the spore cases and carried by the wind to new soil habitats. A germinating spore grows into a small, inconspicuous heart-shaped gametophyte plant. The fertilized egg cell becomes a zygote that develops into a young sporophyte plant that is nourished by the gametophyte. Many botanists believe that seed plants evolved from the ferns.

Spermopsida—Seed Plants

The seed plants are divided into two groups: the *gymnosperms* and the *angiosperms*. The gymnosperms are known as the naked seed plants, while the angiosperms are referred to as the covered seed plants. Seed plants are the most successful land plants that have ever lived because they have reproductive mechanisms that do not require water. The plants that evolved before the seed plants produce sperm cells that can reach the egg cells only by swimming through a film of water.

Angiosperms are the most recently evolved major group of plants, dating from the early Cretaceous period, about 65 million years ago. Their success is measured in terms of the increase in number of species and their emergence as the dominant groups of plants, inhabiting varying environments throughout the world.

Gymnosperms Gymnosperms are cone-bearers. They are woody plants, chiefly evergreens, with needle-like or scale-like leaves. Cone-bearing plants grow in many parts of the world including tropical climates. However, most species are found in the cooler parts of temperate regions. Examples of gymnosperm species are pines, spruces, firs, cedars, yews, California redwoods, bald cypresses and Douglas firs. Most biologists do not think that the gymnosperms evolved directly from ferns.

Reproduction in gymnosperms takes place on special structures called *cones*. The cones are specialized nongreen leaves where seeds are produced. Most gymnosperm species bear two different kinds of cones: seed cones, called *megasporophylls*, and pollen cones, called *microsporophylls*.

Seed cones are large and their woody "leaves" (sporophylls) bear on them an *ovule* (called also a *megasporangium*). Inside of the ovule is a *megaspore mother cell*. This mother cell goes through a meiotic division producing both an egg cell and a food storage cell.

Pollen cones are smaller in size than the seed cones. Their leaves are known as *microsporophylls*. These microsporophylls are really *stamens*, reproductive structures that produce pollen. Inside of each pollen grain is a *microspore* cell. When the pollen grain germinates, the microspore goes through a reduction division producing two sperm cells.

Pollen grains are carried by the wind from the pollen cones to the seed cones. When a pollen grain lands on an ovule, it begins to germinate (sprout). The germinating pollen grain commences to grow a pollen tube during its first summer and the following spring. As the pollen tube grows, changes take place in the "leaves" of the seed cone where the ovules are pollinated. These leaves grow together holding the ovules tightly inside of the cone. The pollen tube enters a pore (*micropyle*) in the ovule. The two sperm nuclei from the pollen grain travel through the pollen tube into the ovule. One sperm cell fertilizes the egg cell. The other sperm cell and the pollen tube cells disintegrate. The fertilized egg becomes a zygote and then develops into the embryo plant. Changes occur in the outer walls of the ovule where a tough seed coat develops.

The ripened ovule is now a *seed* containing both an embryo plant and food for the embryo plant. The stored food known as *endosperm* is derived from the female gametophyte. The embryo plant represents the new sporophyte generation. Seeds are carried to new locations by wind, water or animals. The embryo plant inside of a seed can live for long periods in a resting state. The outer coat of the seed protects the embryo from extremes of temperature, drying, chemical corrosion and even burning. The germinating pollen grain and the developing ovule represent the gametophyte generation. The large conspicuous plant is the sporophyte generation.

Angiosperms—Flowering Plants The reproductive structure of the angiosperm is the *flower* which encloses the male and female sex organs. Fig. 7.12 shows the parts of the flower. On the outside of the flower are the green leaflike *sepals*. The sepals protect the flower when in the bud stage. Collectively, sepals are known as the *calyx*. Just inside of the calyx are colored *petals*, showy and conspicuous in flowers that are pollinated by insects or birds. All of the petals in a flower are known as the *corolla*. Look at Fig. 7.12 and locate the stamens. Notice that the top portion is called the *anther* and the stalk, the *filament*. The stamens are the male reproductive structures; pollen grains are produced in the anther. Locate the *pistil* which is in the center of the flower. The top portion of the pistil is the *stigma*. The *style* is the long stalk that leads to the rounded portion of the pistil called the *ovary*. Inside of the ovary are many *ovules*. The pistil and its many parts compose the female portion of the flower. The pistil is really a modified sporophyll which in the flowering plant is called the *carpel*.

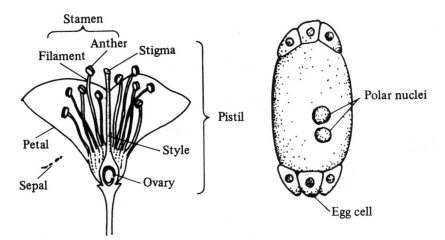

Fig. 7.12   Parts of a flower

Fig. 7.13   Maturation of the ovule

When an ovule is ready for fertilization, its megaspore mother cell goes through a reduction division and two mitotic divisions yielding eight nuclei. One of these nuclei forms the *embryo sac*. A nucleus near the ovule micropyle (pore) becomes the egg cell (Fig. 7.13).

Each pollen mother cell in the anther divides by meiosis and produces four pollen grains, each with a haploid nucleus. When a pollen grain lands on the stigma, its cell divides by mitosis to form two nuclei. One, the tube nucleus, directs the growth of the pollen tube down to the micropyle in an ovule. The other nucleus (*generative nucleus*) divides and forms two sperm nuclei (Fig. 7.14).

The sperm nuclei enter the ovule through the micropyle. One sperm fertilizes the egg cell and forms a diploid zygote. The other sperm cell fertilizes two polar nuclei in the embryo sac to form a triploid endosperm nucleus (Fig. 7.15). The behavior of both sperm cells is described as a *double fertilization*. The zygote then undergoes a number of changes and develops into the embryo plant. The ovule coats increase in number and harden, changing into seed coats. After fertilization in the ovules, the ovary enlarges and usually becomes a *fruit*. The ovules enlarge, change shape and form *seeds*, each containing an embryo plant and stored food. All accessory nuclei disintegrate.

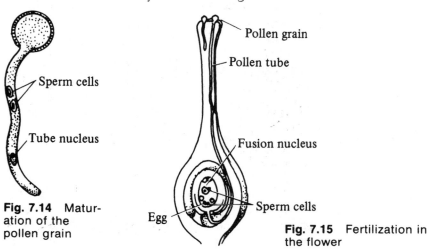

Fig. 7.14   Maturation of the pollen grain

Fig. 7.15   Fertilization in the flower

Parts of Higher Plants _____

The bodies of the higher plants are differentiated into roots, stems, and leaves with specialized tissues to carry water, transport nutrients, anchor the plants, grow new cells, and carry out other functions of the plant.

ROOT

Root Function

Roots anchor plants to the soil and absorb water and dissolved minerals from the ground. The absorbed materials enter the root by way of root hairs which are one-cell extensions of the epidermis. From the root hairs, dissolved materials pass through the cortex, endodermis, and pericycle into xylem cells. The xylem cells conduct the dissolved materials upward. The root cortex serves in the storage of food and water. A small amount of storage occurs in the parenchyma cells of the stele. Cells in the tips of the roots are responsible for the growth in length of the root. Growth in diameter of the root is controlled by the cambium between the xylem tissue and the phloem tissue. Asexual reproduction (vegetative propagation) is brought about in some plants by adventitious buds present on roots.

Gross Structure

There are two types of root systems: fibrous roots and tap roots. The *fibrous root* has numerous slender main roots of equal size with many branch roots smaller in size. Examples of plants with fibrous roots are corn, wheat, grasses (Fig. 7.16). The *taproot* is the main root of the plant. It is longer and thicker than the smaller branch roots. Examples of taproots are carrots, beets, dandelions (Fig. 7.17).

Microscopic Structure

A longitudinal section of a young root shows four zones distinguished by the cell types within each region (Fig. 7.18).

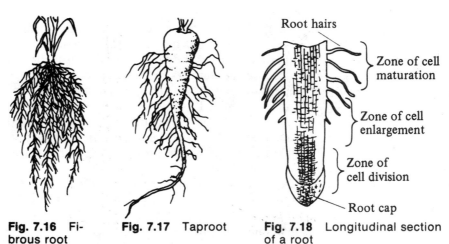

Fig. 7.16 Fibrous root

Fig. 7.17 Taproot

Fig. 7.18 Longitudinal section of a root

1. *Root cap.* A semicircular cap of cells forming the tip of the root and protecting the dividing cells just above it. Cells in the root cap are of moderate size and thick-walled. They protect the thinner-walled cells just above.
2. *Zone of cell division (meristemic region).* Here the cells are small, thin-walled with dense cytoplasm. These cells reproduce rapidly by mitosis and contribute to increased length of the root.
3. *Zone of cell enlargement or elongation.* The cells in this zone were recently formed in the zone of cell division. These cells become elongated, produce new cytoplasm and develop larger vacuoles.
4. *Zone of maturation.* In this zone the enlarged cells become *differentiated* into xylem, phloem, cambium, cortex and other tissues. Root hairs grow from the lower portion of the maturation zone.

A cross section of the root taken through the zone of maturation reveals the following tissues (Fig. 7.19):

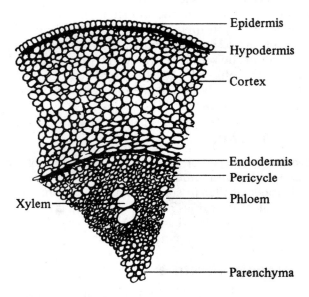

Fig. 7.19   Cross section of a root

1. The *epidermis* is the outer layer of cells from which the root hairs develop. Epidermal tissue is specialized for the absorption of water and minerals from the soil and for the protection of underlying tissues.
2. The *cortex* contains rather large, irregularly shaped parenchyma cells applied loosely to one another with a good deal of intercellular space. The function of this region is to store water and food. Water leakage from the cortex into the inner tissues is prevented by a band of thick-walled cells known as the *Casparian strip.* A waxy material, *suberin,* makes these cells waterproof and prevents leakage of water into the inner root tissues.
3. The *pericycle* is a layer of cells just inside the Casparian strip (endodermis) from which branch roots are produced. The cells of the branch roots grow through the cortex and through the epidermis and extend outside of the root into the soil.

4. The *xylem* is composed of conducting cells called *tracheids* and thin tubules known as *vessels*. The function of xylem tissue is to conduct water and dissolved substances upward through the root into the stem.
5. The *phloem* is composed of companion cells and sieve tubes. The function of the phloem is to transport water with dissolved food downward from the leaves through the stem into the root.
6. The *parenchyma* stores food and water and lends support to other tissues.

STEM

Stems have three major functions. First, they conduct water upward from the roots to the leaves and conduct dissolved food materials downward from the leaves to the roots. Second, stems produce and support leaves and flowers. Third, they provide the mechanisms for the storage of food.

Specialized stems have additional functions. The *tendrils* of grape vines function as climbing organs. *Thorns* on rose bushes offer protection against animal invaders. The desert cacti have stems specialized for the storage of water and food. *Runners* of strawberry plants and spider plants serve as organs of *vegetative propagation* in which new plants are produced at their nodes. *Rhizomes* are underground stems of ferns which serve the purpose of producing new plants. The underground stem of the white potato is called a *tuber* and its function is to store carbohydrate in the form of starch. The *corms* of the crocus and gladiolus are underground storage stems consisting of fleshy leaves.

Botanists call a stem with its leaves a *shoot*. The *shoot system* is the total of all of the stems, branches and leaves of a plant.

Fig. 7.20 shows a longitudinal section of a young stem.

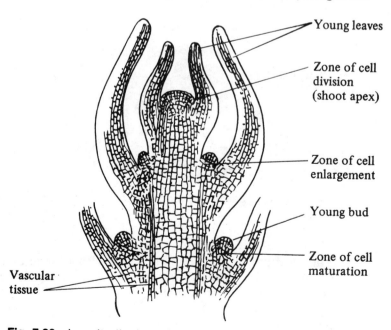

Young leaves

Zone of cell division (shoot apex)

Zone of cell enlargement

Young bud

Zone of cell maturation

Vascular tissue

Fig. 7.20 Longitudinal section of a young stem

TABLE 7.1. Types of Aerial Stems

Characteristic	Herbaceous Stems	Woody Stems
Texture	soft	tough
Color	green	nongreen
Growth	little in diameter	much growth in diameter
Tissues	primary	secondary
Life cycle	annual	perennial
Protective covering	epidermis	bark
Buds	naked	covered with scales
Examples	monocotyledons	all gymnosperms
	dicotyledons	dicotyledons

There are two types of above ground (aerial) stems: woody stems and herbaceous stems. Major characteristics of each are shown in Table 7.1.

Herbaceous stems of dicotyledons have the following tissues: epidermis, schlerenchyma, cortical parenchyma, pericycle, phloem, cambium, xylem, and pith. Herbaceous stems of monocotyledons do not have cambium. Since no secondary growth takes place, the stems do not increase in size appreciably. Microscopic study of the stem cross section shows that the xylem and phloem are organized into vascular bundles. These are scattered throughout the parenchyma tissue which fills the stem.

Woody stems are composed of primary and secondary tissues. Primary tissues are those that develop from the meristems (embryonic tissue) of the buds on twigs during the first year of growth. After the first year, growth in the the woody stem takes place in the secondary tissues. These are tissues that arise from the *cambium*.

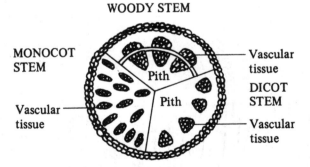

Fig. 7.21 Comparison of monocot, herbaceous, and woody stems

LEAF

The most important function of green leaves is to carry out photosynthesis, the food-making process in which inorganic raw materials are changed into organic nutrients.

A leaf consists of two parts: a stalk or *petiole* and the *blade*. The petiole attaches the blade to the stem. The blade is the place where photosynthesis takes place. Leaves vary greatly in shape (Fig. 7.22).

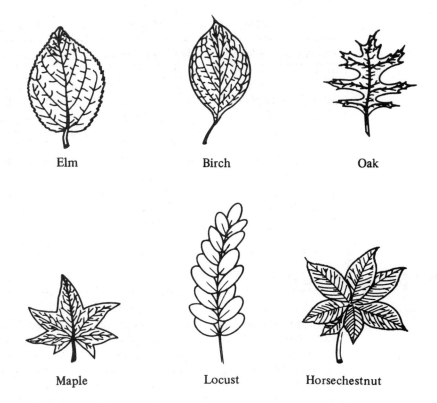

Fig. 7.22 Some common leaf shapes

Study of a leaf cross section under the microscope reveals **three types of tissue**: upper and lower epidermis, mesophyll, and the vascular bundles (Fig. 7.23).

The *epidermis* is a single layer of cells at the upper and lower surfaces of the leaf. The cells have thick walls made of cutin and lack chloroplasts. Their main function is to protect the underlying or overlying tissues from drying, bacterial invasion and from mechanical injury. On the underside of the leaf, the lower epidermis has pores known as *stomates*, the size openings of which are regulated by a pair of *guard cells*. The stomates serve as passageways for oxygen and carbon dioxide.

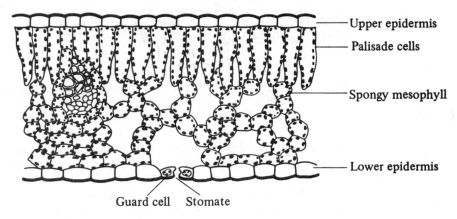

Fig. 7.23 Leaf cross section

The *mesophyll* is positioned between the two epidermal layers. It is composed of two distinct tissues: the *palisade cells* and the *spongy cells*. Packed with chloroplasts, the cells of both types of mesophyll tissue are the sites for photosynthesis in the leaf.

The *vascular bundles* or veins consist of xylem cells (vessels and tracheids) and phloem cells (sieve tubes and companion cells). The xylem cells transport water and dissolved minerals upward into the mesophyll. The phloem conducts dissolved food materials downward to the lower parts of the plant.

Loss of water vapor through plant leaves is termed *transpiration*, a process which is responsible for the rise of sap in trees. *Guttation* is the loss of liquid water through the leaves of plants with short stems. Guttation is caused by the effect of root pressure on water flow and has hardly any physiological advantage to the plant.

Photosynthesis

Photosynthesis takes place in the chloroplasts of green leaves and stems. Photosynthesis is the food-making process of green plants made possible by the light-trapping pigment chlorophyll. An important event of photosynthesis is the changing of light energy into chemical energy, which is ultimately stored in sugar molecules. The raw materials of photosynthesis are carbon dioxide and water. The overall equation for photosynthesis is shown below:

$$CO_2 + 2H_2O + light \xrightarrow{\text{chlorophyll}} O_2 + (CH_2O) + H_2O$$

REDUCTION-OXIDATION REACTIONS (REDOX)

Reduction is the addition of an electron (e) to an acceptor molecule. Oxidation is the removal of an electron from a molecule. The addition of an electron (reduction) stores energy in a compound. The removal of an electron (oxidation) releases energy. Whenever one substance is reduced, another substance is oxidized.

A^e	+	B	→	A	+	B^e
electron donor		electron acceptor		oxidized (loss of energy)		reduced (gain of energy)

In biological systems, removal or addition of an electron derived from hydrogen is the most frequent mechanism of reduction-oxidation reactions. Redox reactions play a major role in photosynthesis. For example: the synthesis of sugar from CO_2 is the reduction of CO_2. Hydrogen, obtained by splitting H_2O molecules, is added to CO_2 to form CH_2O units.

THE PROCESS OF PHOTOSYNTHESIS

Photosynthesis takes place inside of chloroplasts, membranous structures within the cells of the leaf mesophyll. The chloroplasts have fine structures within—flattened membranous sacs named *thylakoids*. On the membranes of the thylakoids, chlorophyll and the accessory pigments are organized into functional groups known as *photosystems*. Each of these photosystems contains about 300 pigment molecules which are involved directly or indirectly in the process of photosynthesis.

Each of these photosystems has a *reaction center* or a *light trap* where a special chlorophyll *a* molecule traps light energy. There are two types of photosystems: Photosystem I and Photosystem II. In Photosystem I, the chlorophyll *a* molecule is named P 700 because it absorbs light energy from the 700 nanometer wavelength. The chlorophyll molecule of Photosystem II is designated as P 680 because this pigment molecule (chlorophyll *a* absorbs light at the wavelength of 680 nanometers.

Photosynthesis involves four sets of biochemical events: photochemical reactions, electron transport, chemiosmosis and carbon fixation. The photochemical reactions and electron transport activities take place on the membranes of the thylakoids. The oval membranes of a thylakoid surround a vacuole or reservoir in which hydrogen ions are stored until needed in the Calvin cycle, or carbon fixation. Each thylakoid rests in the *stroma* or ground substance of the chloroplast. The stroma is the place where carbon fixation occurs.

The Events of Photosystem I

Cyclic Phosphorylation—Electron Transport

Light energy strikes a photosystem. Pigment molecules absorb this energy and pass it on to the reaction center molecule. The energy level of an electron in P700 is raised to a higher level. This increased energy in the electron causes it to escape the confines of the P700 (chlorophyll *a*) molecule and to become temporarily attached to an acceptor molecule called X. In accepting the electron, molecule X is reduced. Molecule X passes the electron on to another acceptor molecule and becomes oxidized in the process. In a series of redox reactions, the electron is passed from one acceptor molecule to another. It finally returns to P700. Each step of these reduction-oxidation reactions is catalyzed by a specific enzyme.

The energy released as the electron is passed along the transport chain is used to synthesize ATP (adenosine triphosphate). Excess hydrogen ions released when ATP is formed are stored in the reservoir of the thylakoid. Inorganic phosphate from the fluid of the stroma is incorporated in the ATP molecule during photosynthetic phosphorylation. Photosynthesis requires the energy from ATP to synthesize carbohydrates.

Noncyclic Phosphorylation

During the events of Photosystem I, "excited" electrons may travel in another pathway different from that which builds ATP. Chlorophyll acts as an electron donor and later as an electron acceptor. It donates energy-rich (excited) electrons and accepts back energy-poor electrons.

Light energy strikes a chlorophyll *a* molecule. An electron in its reaction center molecule, P700, becomes elevated to a higher energy state. The electron passes from P700 to acceptor X. From acceptor X, the electron passes to ferridoxin (Fd), an iron containing compound. Fd passes the electron to an intermediate compound and then to nicotinamide adenine dinucleotide phosphate (NADP). Actually two P700 molecules release electrons along this pathway simultaneously. NADP accepts both electrons (2e) and becomes NADPH. NADPH keeps both electrons and does not pass them along. The energy from $NADPH_2$ will serve as an energy source when carbon dioxide is reduced to form sugar. By acquiring two extra electrons, the NADPH also attracts a H proton. Thus NADPH + H is written and $NADP_{re}$.

Review: Photosystem I

1. Photons of light strike a chlorophyll *a* molecule.
2. The reactioin center molecule (P700) absorbs the light.
3. One of its electrons is raised to a higher energy level.
4. The pathways followed by "excited" electrons are shown below.

    Cyclic   e from P700 to X to acceptors to ATP

    Noncyclic   2e from P700 to X to Fd to NADP $\rightarrow$ $NADP_{re}$

The Events of Photosystem II

Photosystem II involves about 200 molecules in the reaction center of chlorophyll *a*, the light trapping pigment of green plants. In the bluegreens and in the bryophytes, the light trapping pigment is chlorophyll *b*; in the brown algae chlorophyll *c* and in the red algae chlorophyll *d*.

When light strikes the chlorophyll in Photosystem II, an electron in the reaction center molecule, P680, becomes "excited". The energized electron is passed to an electron "acceptor" molecule designated *Q*. Molecule *Q* passes the electron through a chain of acceptor molecules which pass the electrons to the holes in Photosystem I formed during the noncyclic synthesis of $NADP_{re}$. As electrons move along the chain of transport, they lose energy step by step. Some of the energy forms ATP. It is believed that P680 pulls replacement electrons from water, leaving behind free protons and molecular oxygen:

$$2H_2O \rightarrow 4e + 4H^+ + O_2$$
$$\text{to P680}$$

The protons become associated with $NADP_{re}$.

A summary equation reviewing the electron pathways in Photosystem I and in Photosystem II follows:

$H_2O \rightarrow$ 2e to Photo I to *Q* to transport chain to Photo II to X to transport chain to $NADP_{re}$ to Calvin Cycle.

Fig. 7.24 shows the pathways of photosynthesis.

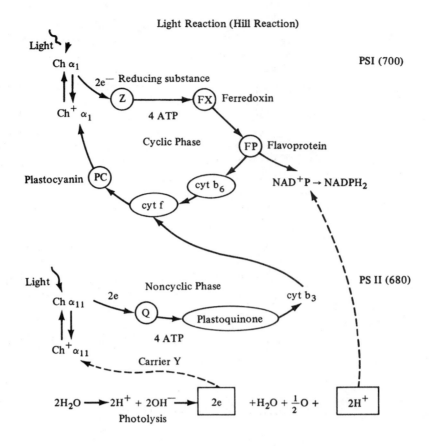

Fig. 7.24 The pathways of photosynthesis involve many steps and many intermediate products and catalysts, including flavoprotein and cytochrome (cyt).

The Calvin Cycle

The Calvin cycle is the series of events in photosynthesis during which carbon dioxide fixation takes place in the stroma of the chloroplast. $NADP_{re}$ and ATP that were produced during Photosystem I and Photosystem II are now used to attach CO_2 to a preexisting organic molecule. The enzymes that catalyze the events of the Calvin cycle are present in the stroma.

Carbon dioxide combines with the 5-carbon sugar, ribulose biphosphate (RuBp) forming an unstable 6-carbon compound. This compound breaks into two 3-carbon compounds, phosphoglyceric acid (PGA). The two PGA molecules are reduced to two molecules of phosphoglyceraldehyde (PGAL) in two successive steps. First, the PGA molecules receive a high energy phosphate from ATP. The high energy phosphate bond is broken, the phosphate is removed and replaced by a hydrogen atom from NADPH. Then, the joining together of two PGAL molecules results in the formation of a molecule of 6-carbon sugar. Some of the PGAL is used to replenish the store of ribulose biphosphate, the starting point of the Calvin cycle (Fig. 7.25).

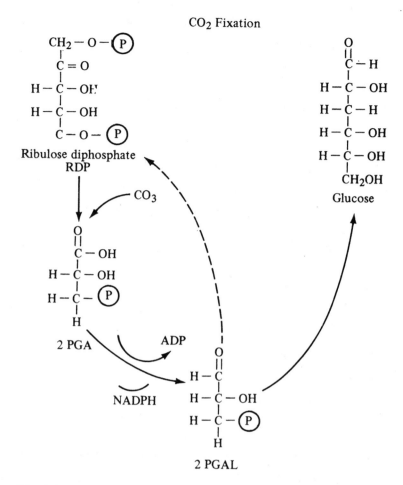

Fig. 7.25   The Calvin cycle showing the complex steps leading from ribulose diphosphate (biphosphate) to glucose, a 6-carbon sugar

PHOTORESPIRATION

Photorespiration is a rather strange series of events that occurs in the cells of green plants in the presence of sunlight. In the course of normal events, the enzyme ribulose biphosphate (RuBP) carboxylase joins a carboxyl group to ribulose biphosphate. The biochemical activity that follows was explained in the section entitled "The Calvin Cycle."

During photorespiration, oxygen, instead of carbon dioxide, binds to RuBP carboxylase. When RuBP carboxylase has oxygen bound to it, oxidation of ribulose biphosphate occurs. One molecule of PGA and a 2-carbon are released. The PGA remains in the C_3 cycle, but the 2-carbon molecule leaves the chloroplast and enters into chemical reactions in a peroxisome and a mitochondrion. Some of the CO_2 produced in these reactions is released; the rest is returned to the chloroplast to take part in photosynthesis.

Photorespiration oxidizes organic compounds using oxygen and results in the release of carbon dioxide. This process does not utilize the electron transport system and therefore does not produce energy. Rather, it uses up energy, thereby appearing to be wasteful. To date, scientists do not really know how photorespiration benefits the cell's work in photosynthesis.

THE C₄ OR HATCH-SLACK PHOTOSYNTHETIC PATHWAY

In the late 1960s three research botanists (Kortschak, Hatch, and Slack) discovered another photosynthetic pathway, which has become known as the C_4 or Hatch-Slack photosynthetic pathway. Essentially, this is what happens. Carbon dioxide combines with a compound known as PEP (phosphoenolpyruvate), forming a 4-carbon compound, malate. Malate is transferred to bundle sheath cells in the leaf. This 4-carbon compound gives up carbon dioxide, which enters the C_3 or Calvin cycle in the photosynthetic bundle sheath cells.

Plants that carry on C_4 photosynthesis exhibit special arrangement in their leaf tissues. In this special arrangement, termed *Kranz anatomy*, the *bundle sheath cells* are positioned in a circle around the vascular bundles (made up of phloem and xylem tubes). The mesophyl cells comprise the rest of the leaf's interior. The air spaces are very small. (Fig. 7.26). Plants in tropical and desert regions with especially high rates of photosynthesis are C_4 plants; among them are crabgrass, sugarcane, millet, and sorghum. Interestingly enough, corn, a temperate-region plant, also carries on C_4 photosynthesis.

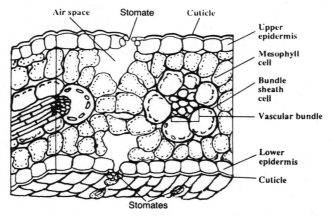

Fig. 7.26 Kranz anatomy. Bundle sheath cells surround vascular bundles.

Plant Hormones

Hormones, which are chemical messengers produced by tissues in small concentrations, effect change in the organism. In general, plant hormones affect plant growth and responses to the environment. Plant hormones differ from the hormones produced by animals in a number of ways. Animal hormones are produced by endocrine glands and affect target organs. Plant hormones, on the other hand, are produced by different tissues at different stages and affect virtually every tissue in the plant. Unlike animal hormones, plant hormones do not affect homeostasis (steady-state control of physiological activity). A summary of plant hormones and their effects follows.

Auxins (indoleacetic acid or IAA) cause plant stems to bend (grow) toward the light by promoting enlargement of the stem cells. Plant growth response to light is known as *phototropism*. *Apical dominance* is the control exercised by the growing tip of the stem over lower stem structures. Auxin accumulates in the growing stem tip and restrains the development of side stems or branches.

Gibberellins, isolated from the fungus *Gibberella fujikuroi*, belongs to a family of 57 closely related compounds. Gibberellic acid, produced in the young leaves of plants, controls stem growth, leaf growth, and root elongation. The gibberellins have another important function. The embryos of germinating cereal grains, such as corn and barley, release gibberellins, which, in turn, stimulate the release of food from the endosperm.

Cytokinins, produced in the roots, are distributed upward through the xylem. Together with auxins, the cytokinins control differentiation of plant tissue, stimulate cell division and the production of fruits and seeds, and slow the process of aging in picked leaves. Cytokinins also prevent the onset of dormancy in plants.

Abscisic acid, produced in mature leaves and transported to other plant tissues and organs, is a growth inhibitor, causing dormancy in trees. Roots grow downward in a positive response to gravity known as *geotropism*. It is believed that abscisic acid controls root growth response.

Ethylene, produced as a gas by plants, controls the ripening of fruit and lays an important role in the growth of seedlings up through the soil.

Oligosaccharines are sugar compounds that protect the cell wall against insect invasion by stimulating cells to produce antibiotics known as phytoalexins. Phytoalexins inhibit the synthesis of protein-digesting enzymes of insects and thus prevent predation.

Photoperiodicity

Each species of plant has its own *photoperiod* for flowering. Although plants flower in response to length of darkness, photoperiodicity is described in terms of *day length*. Some plants belong in the category of *long-day* plants because they flower in response to a shortening night. Plants that flower in response to lengthening nights are called *short-day* plants. *Day-neutral* plants seem not to be affected by the ratio of daylight to darkness.

Phytochrome has been identified as the pigment that controls the "dark clock" of flowering. Cocklebur, Maryland mammoth (a species of tobacco), and Biloxi soybeans are short-day plants. Spinach, radish, and barley are long-day plants requiring shorter hours of darkness. Tomatoes and cucumbers are day-neutral and flower without regard to length of night.

Chronology of Famous Names in Biology

1679   **Marcello Malpighi** (Italian)—determined the functions of xylem and phloem by conducting a series of girdling experiments.

1729   **Stephen Hales** (English)—demonstrated that transpiration can pull sap up through the xylem and phloem.

1772   **Joseph Priestley** (English)—demonstrated that the air is replenished by green plants.

1782   **Jean Senebier** (French)—showed that photosynthesis depends on "fixed air," now known as carbon dioxide.

1796   **Jan Ingen-Housz** (Dutch)—concluded that plants use carbon dioxide in photosynthesis.

1804   **Nicolas Theodore De Saussure** (French)—demonstrated that water is necessary for photosynthesis.

1879   **James Clerk Maxwell** (English)—discovered that light travels in waves.

1880   **Charles Darwin** and **Francis Darwin** (English)—were the first to propose the existence of a plant hormone.

1883   **T. W. Engelmann** (Germany)—provided evidence that chlorophyll plays a role in photosynthesis and demonstrated that red light was most effective in photosynthesis.

1904   **B. Haberland** (German)—discovered plants native to tropical climates have a different arrangement of the bundle sheath cells.

1905   **F. F. Blackman** (English)—was the first to present evidence that photosynthesis has a light-driven stage and a stage not requiring light.

1905   **Albert Einstein** (American)—proposed that light energy travels in packets called photons.

1910   **Max Planck** (German)—established that the energy of radiation is contained in packets called quanta.

1926   **Frits W. Went** (Dutch)—gave the name "auxin" to the substance in plant stems that controls their elongation.

1934   **C. B. van Niel** (American)—proposed that water is the source of the oxygen in photosynthesis.

1941   **Samuel Ruben, Merle Randall, Martin Kamen** and **James L. Hyde** (American)—reported that oxygen liberated in photosynthesis comes from water.

1950's **George Wald** (American)—greatest living expert on light and life.

1950	**Daniel Arnon** (American)—identified the reactions that take place in light reactions of photosynthesis.
1954	**Hugo Kortschak** (American)—discovered that Kranz anatomy plants begin carbon dioxide fixation with a 4-carbon compound.
1960	**Haraguro Yomo** (Japanese)—discovered that cereal grain embryos release gibberellin, which dissolves stored food in the endosperm.
1960	**H. P. Kortschak** (American), **M. D. Hatch** and **C. R. Slack** (both Australian)—elucidated photosynthetic pathways in C_4 plants such as sugarcane, corn and sorghum.
1961	**Melvin Calvin** (American)—discovered the pathway of events by which green plants incorporate carbon dioxide into carbohydrate molecules.

Words for Study

abscisic acid	flower	photosystem
alternation of	fucoxanthin	phototropism
generations	gametophyte	phytochrome
angiosperm	gibberellin	pistil
anther	gymnosperm	Plantae
antheridia	heterogamy	polar bodies
apical dominance	holdfasts	pollen
archegonia	hormones	prothallus
auxin	isogametes	pyrenoid body
Calvin cycle	kelps	rhizoids
calyx	Kranz anatomy	rhizome
cambium	megasporophyll	sepal
carpel	micropyle	sorus
carpogonia	microspore	specialization
conceptacle	microsporophyll	sporophyte
cone	multicellular	stamen
corolla	oligosaccharine	stigma
cutin	oogamy	strobilus
cytokinins	oogonia	style
division of labor	ovary	taproot
embryo sac	ovule	thallophytes
ethylene	petal	thallus
endosperm	phloem	vascular
fibrous root	photoperiod	xylem
filament	photorespiration	

Questions for Review

PART A.   Completion. Write in the word that correctly completes each statement.

1. Single-cell plants include free-living forms, cells that live in colonies, and ..1.. forms.

2. The chromsome number of vegetative *Chlamydomonas* cells is best described as being ..2..

3. Primitive gametes that are similar in appearance are known as ..3..

4. *Gonium* species live as a simple ..4.. of cells.

5. The *Volvox* series of cells demonstrates an evolutionary trend toward ..5..

6. Chlorophyll ..6.. is found only in the red algae.

7. The water-carrying tissues of green plants are known collectively as ..7.. tissues.

8. The cycle of a sexual generation following an asexual generation is known as ..8..

9. Fluid compounds are conducted downward in plants by ..9.. tubules.

10. A sorous (sori, pl) can be found on the leaves of ..10.. plants.

11. The dominant generation in the ferns is the ..11.. generation.

12. The gymnosperms are best described as ..12.. bearers.

13. The prefix "mega" means ..13..

14. The microsporophylls are ..14.. cones.

15. In gymnosperms, the agent of pollination is ..15..

16. In gymnosperms, the megasporangium produces the ..16..

17. Petals are known collectively as the ..17..

18. The tube nucleus develops from the germinating ..18..

19. A fruit is a ripened ..19..

20. Growth in length of roots occurs at the ..20..

21. A stem with its leaves is called a ..21..

22. The Casparian strip is located in what part of the plant?

23. Secondary plant tissues arise from the ..23..

24. The most effective light-trapping pigment is ..24..

25. Carotene and phycoerythrin function as ..25.. pigments during the light driven stages of photosynthesis.

26. Carbon fixation takes place in the ..26.. of the chloroplast.

27. The response that plants make to light is called ..27..

28. Indoleacetic acid belongs to a group of hormones known by the general name ..28..

29. A plant's photoperiod is described in terms of ..29..

30. The group of hormones produced in plant roots and distributed upward are known as ..30..

PART B. **Multiple Choice.** Circle the letter of the item that correctly completes each statement.

1. *Chlamydomonas* resembles the protists because it lives in water and swims by means of
   (a) cilia
   (b) pseudopods
   (c) tentacles
   (d) flagella

2. The function of the pyrenoid body is to
   (a) synthesize glucose
   (b) store starch
   (c) resorb flagella
   (d) secrete cellulose

3. Sea lettuce is best classified as a (an)
   (a) protist
   (b) bryophyte
   (c) euglenoid
   (d) alga

4. A true statement about *Volvox* is
   (a) Most of its cells do not function in reproduction.
   (b) All of its cells are specialized for reproduction.
   (c) There is no communication between its cells.
   (d) The egg cells have paired flagella.

5. The sporophyte generation of sea lettuce is
   (a) haploid
   (b) monoploid
   (c) diploid
   (d) tetraploid

6. Fucus does not have
   (a) air bladders
   (b) an asexual cycle
   (c) a sexual cycle
   (d) a branching thallus

7. The greatest hazard to land-dwelling plants is
   (a) drying out
   (b) insect infestation
   (c) loss of sperm cells
   (d) nonmotile egg cells

8. Waterproofing of land cells is made possible by cell walls composed of
   (a) chitin
   (b) cutin
   (c) cellulose
   (d) glycoprotein

9. Examples of bryophytes are
   (a) moss, *Volvox*, *Fucus*
   (b) Ulva, hornwort, *Pandorina*
   (c) *Euglena*, moss, red algae
   (d) hornwort, liverwort, moss

10. Species that bear two types of spores are described as being
    (a) homosporous
    (b) heterosporous
    (c) homozygous
    (d) heterozygous

11. Dried, ground stems of the horsetails are used in scouring powder because their cell walls contain
    (a) cutin
    (b) cellulose
    (c) silica
    (d) suberin

12. An underground stem is called a
    (a) Prothallus
    (b) root
    (c) strobilus
    (d) rhizome

13. Ferns are successful land plants because they have well developed
    (a) vascular systems
    (b) digestive systems
    (c) reproductive systems
    (d) excretory systems

14. In the gymnosperms, the megasporophyll is a (an)
(a) large spore
(b) seed cone
(c) egg cell
(d) large ovule

15. The embryo plant is protected inside of the
(a) ovule
(b) egg
(c) pollen
(d) seed

16. The production of two types of gametes is known as
(a) heterospory
(b) heterogamy
(c) isogamy
(d) isospory

17. A true statement about the gymnosperms is that they do not produce
(a) flagellated sperm
(b) seed cones
(c) nonmotile eggs
(d) pollen cones

18. In the flowering plants, the male reproductive structures are the
(a) calyx
(b) corolla
(c) pistils
(d) stamens

19. The chromosome number of the fertilized endosperm nucleus is best described as
(a) haploid
(b) monoploid
(c) diploid
(d) triploid

20. The part of the young root that pushes its way through the soil is the
(a) root hair
(b) tap root
(c) root cap
(d) root zone

21. The word "meristem" refers to
(a) rapidly growing cells
(b) old cells
(c) cells specialized for reproduction
(d) differentiated cells.

22. The primary function of the root cortex is to
(a) absorb minerals from the soil
(b) differentiate into other cells
(c) store water and food
(d) give rise to sperm cells

23. Secondary tissues in roots arise from the
(a) meristem
(b) embryo
(c) cambium
(d) phloem

24. A tendril is a specialized
(a) stem
(b) root
(c) flower
(d) leaf

25. Sap rises in trees due to the force created by
(a) evaporation
(b) transportation
(c) transpiration
(d) guttation

26. Photosystems are functional pigment groups located on the
(a) proteins of the plasma membrane
(b) membranes of the thylakoids
(c) in the stroma of the chloroplasts
(d) in fluids of vacuoles

27. As an outcome of cyclic phosphorylation
    (a) P700 is destroyed
    (b) ATP is formed
    (c) NADPH is synthesized
    (d) carbon is fixed

28. Ribulose biphosphate
    (a) begins the Calvin cycle
    (b) functions as an enzyme
    (c) breaks into two equal parts
    (d) is a 6-carbon sugar

29. During photorespiration, O_2 binds to
    (a) phosphoglyceric acid
    (b) RuBP carboxylase
    (c) malate
    (d) carbon dioxide

30. A plant that has Kranz anatomy in the arrangement of its leaf tissues is
    (a) *Volvox*
    (b) white oak
    (c) corn
    (d) wheat

PART C. Modified True-False. If a statement is true, write "true" for your answer. If a statement is incorrect, change the underlined word to one that will make the statement true.

1. *Chlamydomonas* is classified as a protist.

2. Zoospores represent sexual reproduction.

3. If the original parent colony of *Gonium* contained 8 cells, each new colony produced will contain 32 cells.

4. Oogamy refers to conditions in which the egg cell is motile.

5. The sporophyte generation is produced by spores.

6. Brown algae are called seaweed.

7. Water is carried upward in plants by special tubules known as phloem.

8. A strobilus is a stem.

9. The fern plant is the gametophyte generation.

10. A germinating fern spore develops into a gametophyte plant.

11. The reproductive mechanisms of seed plants do not require gametes.

12. A megaspore mother cell is found inside of an egg.

13. Reproduction in the gymnosperms takes place on special structures called stamens.

14. In gymnosperms, the egg cell develops from the microspore cell.

15. The flowering plants are called gymnosperms.

16. Green leaves that protect flower buds are called petals.

17. The stigma and the style are best associated with the filament.

18. The ovules are found inside of the ovary.

19. A seed is a ripened ovary.

20. Growth in diameter of roots takes place in the epidermis.

21. Two types of root systems are fibrous roots and root hairs.

22. The phloem is composed of companion cells and <u>sieve</u> tubes.
23. The underground stem of the white potato is a <u>rhizome</u>.
24. Runners of strawberry plants are specialized for <u>climbing</u>.
25. All gymnosperms have <u>herbaceous</u> stems.
26. The two types of cells that compose the leaf mesophyll are the spongy cells and <u>meristemic</u> cells.
27. Oxidation occurs by <u>loss</u> of an electron.
28. An "excited" electron escapes to a <u>lower</u> energy level.
29. During photorespiration, a two-carbon molecule leaves the chloroplast and enters the reactions of the mitochondrion and the <u>lysosome</u>.
30. The compound PEP is actively involved during C_3 photosynthesis.

Think and Discuss

1. Distinguish between heterotroph and autotroph.
2. Why is chlorophyll important to the life process of green algae?
3. How do cells that live in colonies differ from cells that live independently?
4. Would life on Earth be possible without the existence of green plants? Explain.
5. For each of the following, list the functions that would be lost to green plants if their cells could not synthesize these products: auxins, hormones, phytochrome.

Answers to Questions for Review

PART A

1. filamentous
2. haploid
3. isogametes
4. colony
5. multicellularity (specialization)
6. d
7. vascular
8. alternation of generations
9. phloem
10. fern
11. sporophyte
12. cone
13. large
14. pollen
15. wind
16. ovule

17. corolla
18. pollen grains
19. ovary
20. tips
21. shoot
22. root
23. cambium

24. chlorophyll *a*
25. accessory
26. stroma
27. phototropism
28. auxins
29. daylength
30. cytokinins

PART B

1. d
2. b
3. d
4. a
5. c
6. b
7. a
8. b
9. d
10. b

11. c
12. d
13. a
14. b
15. d
16. b
17. a
18. d
19. d
20. c

21. a
22. c
23. c
24. a
25. c
26. b
27. b
28. a
29. b
30. c

PART C

1. green alga
2. asexual
3. 8
4. nonmotile
5. gametes
6. true
7. xylem
8. cone
9. sporophyte
10. true
11. water
12. ovule
13. cones
14. megaspore mother cell
15. angiosperms

16. sepals
17. pistil
18. true
19. ovule
20. cambium
21. taproots
22. true
23. tuber
24. reproduction
25. woody
26. palisade
27. true
28. higher
29. peroxisome
30. C_4

CHAPTER 8

INVERTEBRATES: SPONGES TO MOLLUSKS

All animals belong to the kingdom Animalia, a grouping of 29 phyla. Twenty-eight of these phyla include animals called invertebrates because they do not have a vertebral column, or true backbone; the 29th phylum includes the vertebrates, animals with a vertebral column. This chapter discusses the basic organization of the animal body, some general characteristics of invertebrates, and six phyla of invertebrates.

Basic Organization of the Animal Body ____

The organization of the animal body is different from that of the plant and requires a different set of terms for accurate descriptions.

SYMMETRY

The animal body form is often described in terms of *symmetry*, the relative position of parts on opposite sides of a dividing line. Symmetry helps to define the degree of similarity between two species, or between parts of the same animal. *Spherical* symmetry describes the symmetry of a ball. Any section through the center of a ball-shaped organism divides the animal into equal and symmetrical halves (Fig. 8.1). *Radial symmetry* describes the symmetry of a wheel in which the parts are arranged in a circle around a central hub or axis. A vertical cut through the central axis divides a wheel-shaped organism into equal (symmetrical) halves (Fig. 8.2). Radial symmetry is a characteristic of animals that are sessile. However, radially symmetrical animals that have some head development are capable of rather quick movements. *Bilateral symmetry* is characteristic of animals that have a head end, a tail end, and right and left sides. The body of a dog or a horse is a good example of bilateral symmetry. A line drawn through the center of the body lengthwise divides the body into halves that are mirror images of each other (Fig. 8.3).

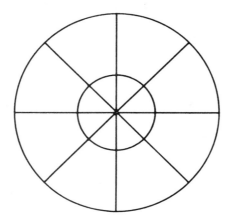

Fig. 8.1   Symmetry through a ball

Fig. 8.2   Radial symmetry

Fig. 8.3   Bilateral symmetry of the human body

EMBRYONIC CELL LAYERS

All of the tissues, organs, and systems of animals develop from two or three *embryonic cell layers*, often referred to as *primary germ layers*. These three cell layers are the *ectoderm* (meaning outside skin), the *mesoderm* (middle skin), and the *endoderm* (inside skin). Animals with simple body structures may develop from only ectoderm and endoderm because the middle tissue layer remains undifferentiated. The tissues of animals with more complex body form develop from the three primary germ layers. In these animals there is a body cavity or *coelom* (see-lom) which is lined with endoderm.

BODY DIRECTIONS

The kinds of directions used to locate body structures permit precise description. The forward end of an animal is the *anterior* end. The *posterior* end is opposite to the anterior region. Example: the head end of a dog is the anterior end; its tail represents the posterior end. *Dorsal* refers to the back of an animal. In all animals except humans (and some primates) the dorsal side faces upward. The underside or bellyside of an animal is the *ventral* side. The sides of an animal such as the right hand and left hand side of a human or the right fore and hind limbs or left fore and hind limbs of a horse are the *lateral* sides of the body. The point of attachment of a structure, such as the wing of an insect, is the *proximal* end. The free, unattached end is known as the *distal* end.

General Characteristics of Invertebrates

Invertebrates are animals without backbones. About ninety percent of all animal species are invertebrates. Just like all members of the kingdom Animalia, invertebrates are multicellular. They are composed of cells that lack walls. Animals are heterotrophs, dependent upon food supplied by the autotrophs. Animals ingest food; they take it in, digest it, and then by some means characteristic to the species, distribute nutrient molecules to cells that make up the body. Nutrients not utilized for energy or incorporated into the tissue-building processes are stored in cells as glycogen or fat.

Most invertebrates are capable of locomotion and have specialized cells with contractile proteins that facilitate movement. However, the adult forms of some lower invertebrate species are sessile, belonging to a group of *filter feeders*. These animals use cilia, flagella, tentacles or gills to sweep smaller organisms from the currents of water that flow over or through their bodies into the digestive cavities.

Some of the lower invertebrates reproduce vegetatively by *budding*. This means simply that a new organism grows from cells of the parent, breaks off, and then continues its own existence. Other invertebrate species reproduce sexually, utilizing sperm and egg. Still other species reproduce asexually by *parthenogenesis* in which an unfertilized egg develops into a complete individual. Some invertebrates have marvelous powers of *regeneration*, the growing back of lost parts or the production of a new individual from an aggregate of cells or from a piece broken off from the parent organism.

Phylum Porifera—Sponges

In a practical sense, you are familiar with the word *sponge*. You know that a sponge is able to soak up large amounts of water because of the many pores and spaces that form its structure. Living sponges are pore-bearing animals and, therefore, were given the scientific name of *Porifera*. These are the simplest of the multicellular animals. About 15,000 sponge species live attached to rocks bathed by ocean waters. Only a few species are freshwater dwellers.

STRUCTURE

Fig. 8.4 illustrates the structure of a simple sponge. Notice that the sponge has a cylindrical shape resembling a vase. At the anterior end is the opening of the central cavity which extends through the Porifera body. Sponges are composed of undifferentiated cells not organized into specialized tissues. The cells show a division of labor much like that of the cells in an algal colony. The cells of the sponge are arranged around the central cavity. In some species this cavity is divided into compartments; in others, it is a continuous space. The central cavity is brought into contact with the surrounding water by means of pores that lead into a system of canals.

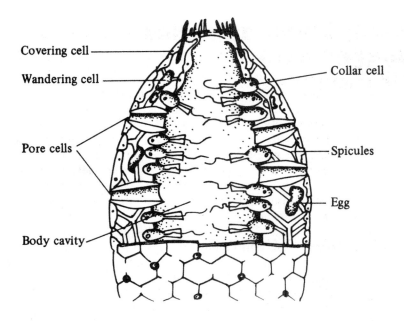

Covering cell

Wandering cell

Pore cells

Body cavity

Collar cell

Spicules

Egg

Fig. 8.4   Internal structure of the sponge

If you have ever held a natural sponge, you are aware that it is not flimsy. Sponges have an internal skeleton. In some species the skeleton is made of needle-like crystals called *spicules*, which are made of lime or silica (glass). In other species, a fibrous protein called *spongin* forms the skeleton, making the sponge firm, but pliable, wettable and absorbent.

Sponges are filter-feeders. The internal cavity is lined with *collar cells* equipped with flagella which capture microorganisms from water as it flows through the central cavity. Epithelial cells cover the outer body wall of the sponge. Between the outer wall and the cells that line the inner cavity is a jelly-like layer known as the *mesoglea*. Here amoeboid wandering cells known as *mesenchyme cells* help to transport digested molecules from the collar cells to cells in other parts of the body. The mesenchyme cells have the ability to change their form and function. They can change into collar cells, epithelial cells, or cells secreting the materials that make spicules for the skeleton.

REPRODUCTION AND REGENERATION

Sponges have exceptional powers of regeneration. If a sponge is sieved through a fine silk mesh and the cells left undisturbed, the cells will come together to reform a complete sponge. Porifera reproduce asexually by *budding*. A new individual grows from cells on the parent body, breaks off, and grows into an adult sponge. Sponges also reproduce sexually by means of nonmotile egg cells and motile flagellated sperm. Sponges are *hermaphrodites* because one sponge organism produces both egg and sperm. The gametes may be carried out to sea by flowing water currents or fertilization may take place in the central cavity. The zygote develops into an *amphiblastula* or early embryo. The embryo escapes from the cavity through a pore, swims about for a short while, and then attaches itself, settling down to grow into an adult sponge.

Phylum Cnidaria (Coelenterates)— Hydrozoa, Jellyfish, Sea Anemones, and Their Relatives

The next group on the evolutionary scale is the phylum Cnidaria. Like the sponges, the Cnidaria are aquatic animals. Species such as jellyfish, Portuguese-man-of-war, sea anemones, and corals live in ocean waters; *Hydra* is a freshwater genus. The cells that compose the body of the Cnidaria show more specialization than the body cells of sponges and are grouped together into simple tissues.

GENERAL CHARACTERISTICS

In general, the Cnidaria body is shaped like a hollow sac composed of two tissue layers. The *ectoderm* covers the outer surface of the body; *Endoderm* lines inner body surfaces. Between these two tissues is a gelatinous mass of undifferentiated material known as the *mesoglea*.

A unique feature of the Cnidaria is the *medusa* and *polyp* forms shown by representative classes in this phylum. Fig. 8.5 illustrates the structure of a jellyfish. Note that its body shows radial symmetry. Notice also that the outer surface of the body curves outward very much like a bowl or a bell. The under surface curves inward and is best described as being concave. This type of body shape is known as a *medusa*. Some cnidarians have the medusa shape all of their lives. Other species exhibit the medusa in part of the life cycle. Fig. 8.6 shows a simplified diagram of the medusa. Notice the *gastrovascular cavity*; extending downward from it is the centrally located *manubrium*, where the mouth is positioned. Tentacles hang from the outer edges of the jellylike bowl. The gonads are suspended under the radial canals and open inward in some species and outward in others.

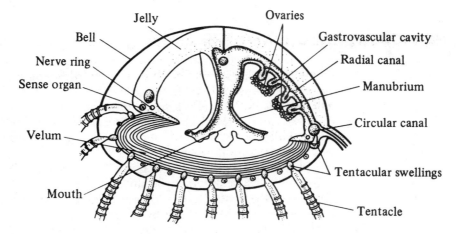

Fig. 8.5 The structure of a jellyfish

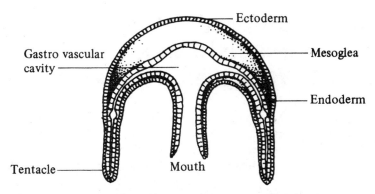

Fig. 8.6 Diagram of a medusa

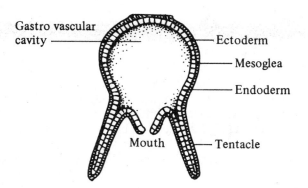

Fig. 8.7 A diagram of a polyp

A number of Cnidaria species have a body shape resembling a vase or cylinder. This type of body is known as a *polyp*. For some species the polyp is a stage in the life cycle; in others, the polyp is the adult body form. Fig. 8.7 provides a simple diagram of a polyp. Notice that the proximal end is attached. The mouth, surrounded by tentacles, is at the distal end of the animal.

The gastrovascular cavity, where digestion takes place, is a distinctive feature of the animals in this phylum. Enzymes are secreted into the cavity, where food is partially digested extracellularly (outside of cells). The partially digested food molecules are then engulfed by cells lining the cavity; these cells complete the digestion intracellularly (inside of cells). In an evolutionary sense, the cnidarian method of digestion is a signpost pointing toward increased specialization of body tissues and organs for digestion. Particles not digested are expelled through the single *mouth-anus*.

Biologists prefer to use the name Cnidaria for this phylum, instead of the former Coelenterata, because of the *cnidocytes* or stinging cells that are in the tentacles of Cnidaria. These animals feed on live prey. They capture smaller animals by immobilizing them with toxins secreted by the stinging cells. Each cnidocyte contains a thread capsule called a *nematocyst*, which discharges a thread in response to touch or chemical stimulation of the *cnidocil*, a trigger-like device. The thread may be barbed or coated with toxin. The nematocyst thread either hooks the prey, ensnares it, paralyzes it with toxin, or does all three of these. Once immobilized, the small animal is swept by the tentacles into the mouth-anus.

REPRESENTATIVE CLASSES

Class Hydrozoa

The Hydrozoa are the most primitive of the Cnidaria. Many of the Hydrozoa species are colonial, with the members of a colony showing a division of labor. Some of the hydrozoans go through an alternation of generations in which an asexual polyp generation alternates with a sexual medusa generation. In this Cnidaria class, the polyp form is dominant. However, the polyps of some species reproduce medusae by the asexual means of budding. The medusae produce gametes (eggs and sperm) and thus begin the sexual generation. Species in the class Hydrozoa include *Hydra*, *Obelia*, *Gonionemus* and the Portuguese man-of-war.

Hydra is a freshwater cnidarian that is representative of a genus of the same name. *Hydra* is a polyp and has no medusa form in its life history. In length, this hydrozoan is about 12 millimeters and has about eight tentacles that surround the mouth-anus (Fig. 8.8). Hydras move about by somersaulting, end over end. The animal's locomotion is made possible by cells that have contractile fibers called *myonemes*. These epitheliomuscular cells have locomotor and sensory functions. Reproduction in hydra is sexual and asexual. A single organism produces both egg and sperm which are discharged into the water where fertilization takes place. Asexual reproduction occurs by budding.

Nervous response in *Hydra* is controlled by a simple nervous system. Slim, pointed *sensory cells* scattered throughout the endoderm and ectoderm layers receive stimuli. From the sensory cells, the sensory im-

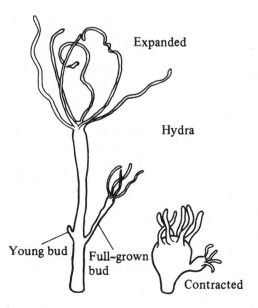

Fig. 8.8 Hydra expanded and contracted

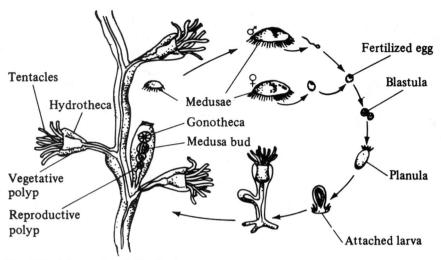

Fig. 8.9 Life cycle of *Obelia*

pulses are passed to nerve cells which form a *nerve net* spread throughout the ectoderm. The nerve net coordinates *Hydra's* activities, enabling the animal to respond to chemical and tactile stimuli in the environment. The nerve net is a very primitive nervous system in which nervous impulses travel in either direction.

Obelia is a marine hydroid—a colony of subindividuals. The size of the colony may range from 2.5 to 10 centimeters in height. It is sessile, attached to substrate rocks along the shoreline. Look at Fig. 8.9. You can see that a number of polyps are attached to a stalk. These polyps are specialized for carrying out specific life functions. Some polyps are vegetative and attend to the feeding needs of the group; others have a reproductive function. The *hydrotheca* is a transparent covering that surrounds the vegetative polyps; the *gonotheca* covers the reproductive polyps.

The reproductive polyps do not have tentacles and consequently cannot feed. They produce new polyp individuals by budding. The reproductive polyps produce another kind of bud that resembles a saucer. This bud escapes the polyp through an opening in the gonotheca, becomes free swimming and develops into a medusa, resembling a jellyfish. The medusa is the sexual stage producing egg and sperm. Fertilization takes place in the water. The zygote develops into a ball of cells which changes into a ciliated larva and then becomes a young polyp. The polyp attaches itself and settles down. On maturity, the polyp begins colony building by budding and the cycle repeats. Thus, there is an alternation of an asexual generation with a sexual generation.

Class Scyphozoa—Jellyfish

The Scyphozoa represent the true jellyfish. These cnidarians are free-swimming and inhabit marine waters. In size, a jellyfish may be as small as 2 centimeters in diameter or may be an enormous 4 meters across with trailing tentacles about 10 meters in length. As shown in Fig. 8.5, a jellyfish is the medusa body form with notches in the margin of the bell. Four *oral arms* extend from the mouth opening. Surrounding these arms are four *gastric pouches* where digestion occurs. A complex of canals

Fig. 8.10 Sea anenome

radiate through the medusa. Attached to the membranes of the canals and encircling the gastric pouches are the gonads.

In some species of jellyfish a polyp stage occurs, but it is subordinate and inconspicuous compared to the medusa. Reproduction by the polyp is asexual, accomplished by a kind of terminal budding known as *strobilation*. Reproduction by the medusa is sexual.

Jellyfish show greater cell specialization than *Hydra* or *Obelia*. Underlying the ectoderm are true muscle cells that propel the animal through water by regular contractions. Sensory nerves connect with the fibers of the nerve net and serve as the nerve supply for the tentacles, the muscle cells and the sense receptors. Now for the first time we see the emergence of true sense organs: *statocysts* and *ocelli*. The statocyst is a sense organ specialized for receiving and coordinating information that enables an organism to orient itself in respect to gravity. Statocysts are spaced around the margin of the bell. Each statocyst is composed of a circle of *hair cells* which surround a central hardened crystal of calcium carbonate known as the *statolith*. In response to movement of the statolith, the hair cells send impulses to the nerve fibers. The organism can adjust to an up or down position in the water as indicated by the statocyst. The ocelli are groups of light receptor cells located at the base of the tentacles.

Class Anthozoa—Sea Anemones

This class of Cnidaria includes the sea anemones, sometimes called "animal-flowers," because of their brightly colored tentacles (Fig. 8.10), and the coral-building animals. Both of these groups are polyps and have no medusa stage. They are sessile, attached to substrate rocks at the edge of the sea. Anthozoans have numerous tentacles that surround an elongated mouth which opens into a tube (stomodaeum) that extends into the gastrovascular cavity. Reproduction may be asexual by budding or sexual involving eggs and sperm.

The epithelial cells of the coral-building cnidarians secrete calcium carbonate walls in which the living polyps hide themselves. The compounds secreted by these polyps build limestone coral reefs.

Phylum Platyhelminthes—Flatworms _____

The tissues and the organs of the flatworm are developed from the three primary germ layers: ectoderm, mesoderm and endoderm. The simplest of the flatworms demonstrate bilateral symmetry. This phylum represents a step up the evolutionary scale showing definite development of excretory, nervous and reproductive systems. Most of the flatworms are hermaphrodites, and most are parasites, several serious parasites of humans and other animals. Among the flatworms are the planaria, flukes, and tapeworms.

THE PARASITIC WAY OF LIFE

By definition, a parasite is an organism that lives on or inside of the body of a plant or animal of another species and does harm to the host. Parasites offer physical discomfort to the host and tend to kill slowly, meanwhile having had time to reproduce themselves for several generations. *Ectoparasites* live on the host's body: body lice, dog fleas, ticks. *Endoparasites* live within the hosts's body and exhibit several adaptations for life in an intestine or in muscle or in the blood.

REPRESENTATIVE CLASSES

Class Turbellaria—Planaria

Planaria is a genus of small, freshwater, free-living flatworms that are studied extensively in biology classes. Fig. 8.11 illustrates the general external structure of a planarian. The digestive system consists of a ventral mouth-anus, a pharynx that can be pushed outward, a digestive cavity and an intestine. A true body cavity (coelom) is not present in *Planaria* as indicated by lack of an anus separate from the mouth. The excretory system of *Planaria* consists of a network of branching tubes. The outer ends of these tubes open to the outside through an excretory pore. The inner portion of the tube connects to ciliated cells called *flame cells*, which remove excess water from spaces around the cells.

A consequence of bilateral symmetry is the refinement of body systems. The nervous system of *Planaria* represents an evolutionary advancement. The nerve net of *Hydra* is replaced by two lateral nerve cords extending longitudinally from the anterior end to the posterior end of the body. The presence of anterior *ganglia*, or groups of nerve cells, to sort and coordinate nerve impulses is a signpost to *cephalization*, the development of the head region. The eyespots or *ocelli* can distinguish light from dark and can also discern the direction from which the light comes. The head region has many *chemoreceptors* which aid in the locating of food.

Of special interest to students of biology is the regenerative powers of *Planaria*. Fig. 8.12 illustrates the regeneration story. Some turbellarians are parthenogenic, producing individuals from nonfertilized eggs.

Class Trematoda—Flukes

Opistorchis sinensis is the species name of the Chinese liver *fluke*, a parasite for which the life cycle involves three hosts: snails, fish, and humans. Aquatic (water-dwelling) snails eat fluke eggs that have been discharged into the water in contaminated human feces. Once inside the snail, the eggs reproduce asexually, resulting in hundreds of offspring. Three generations are nonmotile. The fourth generation of offspring, free-swimming larvae, is discharged into the water. These *cercaria* larvae

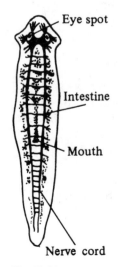

Fig. 8.11
Planaria

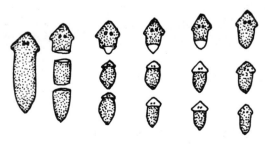

Fig. 8.12 Regeneration in planaria

burrow under the scales of fish, where they encyst in the muscles. When the infected fish are eaten by humans, the encysted cercariae become activated, mature into adult flukes, and move into the bile ducts and livers of their human hosts.

The body plan of the fluke is adapted for parasitism (Fig. 8.13). Most flukes range in length from 1 to 5 centimeters. A sucker at the head end and one on the underside enable them to adhere to host organs. The internal body consists of a muscular throat (pharynx) that sucks body fluids in through the mouth and passes them into a two-branched intestine. The greater portion of the internal body is filled with reproductive organs. Flukes are *hermaphroditic*, possessing the reproductive machinery for producing both eggs and sperm.

Blood flukes of the genus *Shistosoma* infect more than 200 million people in 70 nations of the tropics. The human host is robbed of nutrition by the fluke invaders, which cause a wasting of the limbs, distended abdomen, intestinal malfunction, and urinary disorders. At this time the newer, more effective drugs now available seem to be the best means of diminishing the effects of fluke infection.

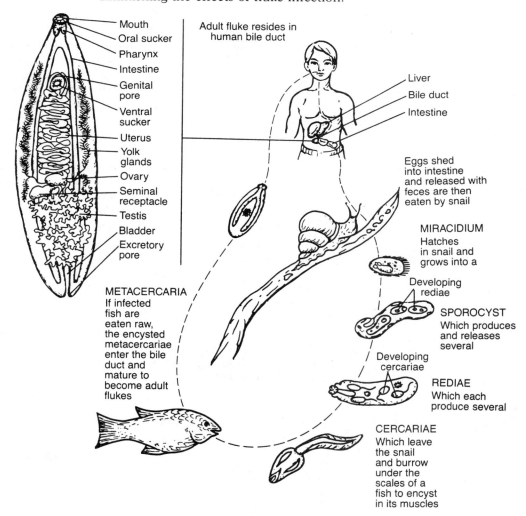

ADULT FLUKE

Mouth
Oral sucker
Pharynx
Intestine
Genital pore
Ventral sucker
Uterus
Yolk glands
Ovary
Seminal receptacle
Testis
Bladder
Excretory pore

Adult fluke resides in human bile duct

Liver
Bile duct
Intestine

Eggs shed into intestine and released with feces are then eaten by snail

MIRACIDIUM
Hatches in snail and grows into a

Developing rediae

SPOROCYST
Which produces and releases several

Developing cercariae

REDIAE
Which each produce several

METACERCARIA
If infected fish are eaten raw, the encysted metacercariae enter the bile duct and mature to become adult flukes

CERCARIAE
Which leave the snail and burrow under the scales of a fish to encyst in its muscles

Fig. 8.13 Life history of the Chinese Liver fluke

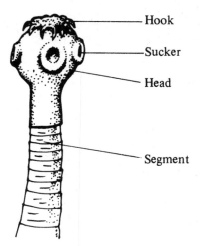

Hook

Sucker

Head

Segment

Fig. 8.14   Tapeworm

Class Cestoda—Tapeworm

Taenia solium is the species name of the tapeworm that infects pigs and people. Having no mouth or digestive system, the adult worm obtains nourishment by absorption through its body wall from the human intestine where it lives. The body of *Taenia* consists of a head, called the *scolex*, attached to a neck which is followed by a series of body segments called *proglottids*. The scolex, which is about 2 millimeters in diameter, is fitted with hooks and four suckers which enable the worm to attach itself to the intestine wall. The proglottids bud from the neck and become progressively more mature and larger as the chain of segments moves toward the posterior end of the animal. A chain of proglottids may be two to three meters long. Each proglottid has a complete reproductive system producing both egg and sperm. The nervous and excretory systems extend through the chain of proglottids (Fig. 8.14).

The life cycle of the tapeworm involves two hosts. A mature proglottid contains a sac filled with hundreds of fertilized eggs. When the proglottid walls rupture, the ground becomes infected with fertilized eggs. If these eggs are ingested by a pig, the protective walls surrounding each egg are digested, releasing developing embryos of the tapeworm into the digestive system of the pig. These embryos bore into the pig's capillaries and are carried by the blood to the muscles where the scolex forms a cyst. The worm now remains encysted in the muscles (meat) of the pig. If the butchered pig (now called pork) is improperly cooked and eaten by a human, the encysted worm becomes activated. Its head begins to bud proglottids and the cycle of infection repeats.

Phylum Nematoda—Roundworms

The roundworms, also called thread worms, are widely distributed, living in the mud of salt and fresh water and in soil. Roundworms are small, tapered at both ends; the elongated body is covered with an enzyme-resistant cuticle. The free-living species have well-developed sense organs, including eyespots and complex mouthparts; parasitic forms are simpler in structure. Anterior ganglia connect with dorsal and ventral nerve cords, articulating with smaller branching nerve fibers. Male and female gametes are produced by separate sexes.

Ascaris lumbricoides is a roundworm parasite that lives in the human intestinal tract. It ranges in length from 15 to 40 centimeters. The eggs escape through the feces of an infected individual and contaminate soil or water. People become infected with *Ascaris* by eating contaminated food.

Hookworms are small nematodes, measuring about seven centimeters in length. Well-developed hooks surround the male genitalia and are present in the mouths of both sexes. Embryos in contaminated soil bore through the skin of the feet and travel through the blood vessels to the lungs. Then they bore through the lung tissue into the bronchi and the windpipe into the throat. They are swallowed and pass into the intestines where they become attached to the intestinal wall. Hookworms drain blood from the host and often cause a severe anemia.

Phylum Annelida—Segmented Worms____

An important indicator of evolutionary advancement is the development of the body cavity known in technical language as the *coelom*. Species in the phylum Annelida possess a true coelom. The coelom is a body cavity that has two openings, beginning with an anterior *mouth* and terminating in a posterior *anus*. It is lined with endoderm and separates the internal body organ systems from the muscles of the body wall. The coelom allows space for the development of complex and specialized systems: circulatory, digestive, reproductive and excretory.

GENERAL CHARACTERISTICS

The annelids are segmented worms that live in soil, fresh water, or the sea. Most of the annelids are free-living, although some of the marine forms burrow in tubes and some species (class Myzostoma) are parasites on echinoderms.

The body of an annelid is divided into a series of similar segments and is said to be *metamerically segmented*. Most annelids have a closed *circulatory system* where the blood is contained in vessels. Enlarged muscular blood vessels function as hearts and pump the blood through the system of vessels. Annelids may be *dioecious* (have separate sexes) or hermaphroditic. Most annelid species go through a ciliated larval stage known as the *trochophore* larva. This is a larva of evolutionary importance because the same type appears in several phyla.

Table 8.1 summarizes the important characteristics of three classes of annelids. The first two classes are discussed in more detail.

REPRESENTATIVE CLASSES

Class Polychaeta—Sandworms

Nereis inhabits burrows in sand and rocks at the edge of the sea (Fig. 8.15). The body of the worm is markedly segmented. The head end is

TABLE 8.1.   Phylum Annelida

Class	Examples	Characteristics
Polychaeta	sandworm, sea mouse, fanworm, lugworm	Marine; segmented worms; setae, parapodia; tube-dwelling, free-living; sexes separate
Oligochaeta	earthworm, giant Australian worm	Freshwater and land (*Tubifex*) dwelling; segmented worms; setae, no parapodia; hermaphroditic
Hirudinea	leeches	Mostly freshwater, some marine, few terrestrial; body flattened; reduced segmentation and body cavity; no circulatory system or setae; ectoparasites, predators, scavengers

Fig. 8.15   Nereis, the sandworm

well-developed. It has a muscular pharynx that can be extended outward to capture food. The pharynx is equipped with a pair of hard curved jaws which are designed for grasping. The head bears four simple eyes, four short tentacles, and two longer ones. Each body segment following the head has fleshy protruding appendages called *parapodia* from which grow bristles, or *setae* (*seta*, sing.).

The digestive system of *Nereis* is a straight tube consisting of a pharynx, esophagus and a stomach-intestine, where the major part of digestion takes place. Undigested food is eliminated through the anus. Oxygen is taken into the body through the skin covering the parapodia, and it diffuses through the thin walls of the blood vessels. The blood of *Nereis* contains the red pigment hemoglobin and is enclosed in muscular vessels. A pair of coiled *nephridia* in each segment filter out waste materials. Well developed ganglia in the dorsal part of the head relay nerve signals from the sense organs to the ventral nerve cord. Lateral branching nerves from the ventral nerve innervate the body organs. The sexes are separate. Mature gametes are discharged into the water where fertilization takes place. The zygote develops into a trochophore larva.

Class Oligochaeta—Earthworm

Lumbricus terrestis is representative of the earthworms and is studied extensively in biology classrooms. The earthworm burrows in moist soil, feeding on organic materials in the earth (Fig. 8.16). The body is seg-

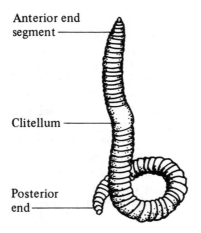

Anterior end
segment

Clitellum

Posterior
end

Fig. 8.16 Earthworm

mented and may have over one hundred metameres. Its under-developed head without eyes and tentacles is well suited to a burrowing way of life.

The digestive system consists of a mouth, a pharynx, a long narrow esophagus, a thin-walled crop, a muscular gizzard, an intestine and an anus. The intestine has a dorsal infolding called the *typhlosole*, an adaptation for increasing surface area necessary for absorptive purposes. The respiratory system of *Lumbricus* is similar to that of *Nereis* except in the earthworm oxygen diffuses through moist body skin and not through parapodia. Paired nephridia are present in each body segment and are used to filter out wastes from the coelomic fluid. Excretion of these wastes takes place through the *nephrostome*, a ventral excretory pore. In the skin, there are sense organs that function as light and touch receptors. The ventral nerve cord with its branching nerve fibers and the anterior ganglia are not unlike that of the sandworm.

Lumbricus is a hermaphrodite. However, earthworms do not self-fertilize; they exchange sperm during copulation. The *clitellum*, a smooth circle of tissue on the outside of the body, is the place where copulating worms attach to each other.

Phylum Mollusca—Clams and Their Relatives

The mollusks include the chitons, snails, clams, scallops, squids and octapuses. This is one of the largest animal phyla and includes about 1,000 species.

GENERAL CHARACTERISTICS

Mollusks are generally soft-bodied, nonsegmented, and usually enclosed within a calcium carbonate shell. They are most abundant in marine waters, although some species inhabit fresh water and others live on land. All mollusks have a *mantle*, a flattened piece of tissue which covers the body and which may secrete the calcareous shell. The body of the mollusk is described as being a *head-foot*, a muscular mass having different shapes and functions in the various classes. Between the body and the mantle is the *mantle cavity*, which functions in respiration. A large *visceral mass* contains most of the body organs. Water enters through an *incurrent siphon* and is expelled through an *excurrent siphon*.

The mollusks are the most highly developed of the nonsegmented animals and are considered to be the most advanced invertebrates. They have well-developed digestive, respiratory, excretory and reproductive systems. The circulatory system includes a two-chambered heart equipped with an auricle and a ventricle. Oxygen-laden blood is pumped both anteriorly and posteriorly through two arteries to parts of the body. Blood carrying respiratory wastes is collected through a vein that is closely applied to the nephridia. The deoxygenated blood is transported back to the gills. The excretory system has a pair of nephridia that excrete filtered wastes through an excretory pore into the mantle cavity. The excurrent siphon removes the dissolved wastes out of the shell.

The nervous system contains three pairs of ganglia positioned near the esophagus, in the foot, and at the end of the visceral mass. These ganglia are connected by transverse nerve fibers. Simple sense receptors in the form of sensory cells sensitive to light and touch are positioned in the margin of the mantle.

In most mollusks, the sexes are separate. The reproductive system is a mass of gland tissue that lies in the muscular foot near the coiled intestine. In some species sperm are conducted through the excurrent siphon of the male into the mantle cavity of the female by way of the incurrent siphon, and fertilization takes place in the mantle cavity of the female. In other species, eggs and sperm are released into the water where

TABLE 8.2. Phylum Mollusca

Class	Examples	Characteristics
Amphineura	chiton	Marine; bilaterally symmetrical; ventral foot; shell of eight calcareous plates; gills in mantle cavity.
Pelecypoda (Bivalvia)	clam, mussel, oyster, scallop, shipworm	Marine and freshwater; body enclosed in right and left shells; head reduced; filter feeders; compressed foot; sexes separate or hermaphroditic.
Gastropods	snail, whelk, limpet, slug	Marine, freshwater, land; most with spiral, single shells; well-developed head having tentacles and eyes; foot for locomotion; shell absent in slugs; trochophore larva.
Cephalopods	squid, octopus nautilus	Marine; well-developed head with large eyes; shell absent or present; tentacles; siphon used for locomotion; ink gland; sexes separate.

fertilization takes place. Some mollusks—oysters, scallops, and primitive worm-like forms—are hermaphroditic.

The fertilized eggs of clams and other bivalves develop into a larval stage known as a *glochidium*. The glochidium is discharged into the water and attaches as a parasite to the gills of fish, where it stays until reaching maturity.

Table 8.2 summarizes the important characteristics of the major classes of mollusks. One class—that of the bivalves—is discussed in more detail.

A REPRESENTATIVE CLASS—PELECYPODA (BIVALVIA)

The clam is an excellent representative of this class, usually known as the bivalves because of the presence of two valves or shells. The right and left valves are hinged on the dorsal side and held tightly together by muscles attached to the inner surfaces of the shells. Lining these inner surfaces is the membranous mantle. The cavity inside of the shells is the mantle cavity.

The clam has a large muscular foot which can be thrust between the opening of the shells and which contracts and expands during locomotion due to the action of protractor and retractor muscles attached to the inner faces of the shells. There is a large visceral mass at the base of the foot containing most of the animal's organs.

Four parallel gill plates are adjacent to the visceral mass. These gills function in respiration. Oxygen from the water in the mantle cavity diffuses into the gills, which are surrounded by capillaries. Oxygen diffuses into the capillaries and then into the bloodstream. Carbon dioxide diffuses out into the surrounding water by way of the capillaries and the gills.

The digestive system of the clam is complete. A mouth occupies the anterior end of the visceral mass. Leading from the mouth is a short esophagus followed by a stomach and a coiled intestine extending partly into the foot. At the end of the intestine is an anus. A bilobed digestive gland is on either side of the stomach (Fig. 8.17).

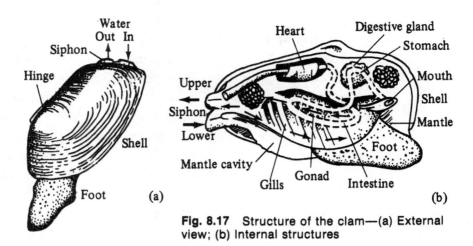

Fig. 8.17   Structure of the clam—(a) External view; (b) Internal structures

Chronology of Famous Names in Biology

1907	**H. V. Wilson** (American)—discovered that individual cells of the sponge, if left undisturbed in a culture dish, will reform sponge aggregates.
1929	**Charles M. Yonge** (Scots)—wrote a treatise on the biology of coral reefs in which he elucidated the patterns of behavior of coral reef invertebrates.
1939	**Ernest E. Just** (American)—developed basic research techniques to study eggs of marine animals.
1940	**Libbie Hyman** (American)—noted authority on the invertebrates, demonstrated metabolic gradients in *Planaria*.
1965	**T. L. Lentz** and **R. J. Barrnett** (American)—elucidated the fine structure of the *Hydra* nervous system; demonstrated how the feeding response is stimulated in *Hydra*.
1967	**Donald Kennedy** (American)—found the effects of the abdominal ganglia on reflex behavior in the slug.
1971	**A.O.D. Williams** (American)—discovered the function of brain cells in mollusks as they relate to behavior.
1975	**Charles M. Yonge** (Scots)—elucidated the life cycle of the giant clam *Tridocna gigas*.

Words for Study

amphiblastula	distal	larva
anterior	dorsal	lateral
antimeres	ectoderm	mantle
blastula	ectoparasite	manubrium
budding	endoderm	medusa
cephalization	endoparasite	mesenchyme cells
cercaria	filter feeder	mesoderm
chemoreceptors	flame cells	mesoglea
clitellum	fluke	metameres
cnidoblast	ganglia	myoneme
coelom	gastrovascular cavity	nematocyst
cnidarian	glochidium	nephridia
cnidocil	gonotheca	nephrostome
cnidocyte	hair cells	ocellus
collar cells	hermaphrodite	parapodia
dioecious	hydrotheca	parasite

parthenogenesis	regeneration	statolith
polyp	scolex	strobilation
posterior	setae	symmetry
primary germ layer	spicules	trochophore larva
proglottids	spongin	typhlosole
proximal	statocyst	ventral

Questions for Review

PART A. **Completion.** Write in the word that correctly completes each statement.

1. Relative positions of body organs on opposite sides of a dividing line are described in terms of ..1..
2. The portion of the fish fin that is attached to the body is the ..2.. end.
3. Invertebrates are animals without ..3..
4. Budding is a form of ..4.. reproduction.
5. The name Porifera means ..5.. bearing.
6. Needle-like crystals that make up the skeleton of a sponge are called ..6..
7. An organism that produces both egg and sperm is known as an ..7..
8. Cnidocytes are ..8.. cells.
9. A thread capsule found in Cnidaria tentacles is the ..9..
10. *Hydra* lives in ..10.. water.
11. The polyps in a colony of *Obelia* reproduce asexually by ..11..
12. Orientation to gravity is controlled by jellyfish sense organs known as ..12..
13. *Taenia* is the genus of the ..13..
14. *Ascaris* is a parasite belonging to phylum ..14..
15. Infection by hookworm can be avoided by wearing ..15..
16. The true coelom first appears in the ..16..
17. Animals in which the sexes are separate are said to be ..17..
18. The body of all mollusks is covered by a membranous ..18..
19. The incurrent and excurrent siphons are characteristic of species in the phylum ..19..
20. The head-foot body is characteristic of the phylum ..20..

PART B. **Multiple Choice.** Circle the letter of the item that correctly completes each statement.

1. Bilateral symmetry is characteristic of animals that have a
    (a) shell and plates
    (b) head and tail
    (c) foot and coelom
    (d) spines and tube feet

2. Distal is a body directional term that refers to
   (a) a forward end   (c) an attached end
   (b) a backward end   (d) a free end

3. In animal cells, excess nutrients are stored as
   (a) glucose and glycogen   (c) glycogen and fat
   (b) glucagon and fatty acid (d) glycogen and protein

4. The growing back of a lost part is known as
   (a) regeneration   (c) vegetative propagation
   (b) budding   (d) parthenogenesis

5. A true statement about sponges is
   (a) The cells are organized into tissues.
   (b) They move quite rapidly.
   (c) The digestive system is well developed.
   (d) The cells show a division of labor.

6. Types of sponge cells that can change their form and function are
   (a) collar cells   (c) mesenchyme cells
   (b) flame cells   (d) epithelial cells

7. In the Cnidaria the body is composed of
   (a) only undifferentiated cells (c) two tissue layers
   (b) one tissue layer   (d) three tissue layers

8. In jellyfish, digestion takes place in the
   (a) velum   (c) manubrium
   (b) tentacle   (d) gastrovascular cavity

9. Medusae produce by
   (a) budding   (c) regeneration
   (b) eggs and sperm   (d) parthenogenesis

10. Alternation of generations is demonstrated in the life cycle of
    (a) *Hydra*   (c) pelecypods
    (b) sea lily   (d) *Obelia*

11. Coral reefs are formed by
    (a) lime-secreting polyps
    (b) the bodies of sea anemomes
    (c) the shells of foraminifera
    (d) jellyfish statocysts

12. Flatworm embryos differentiate into
    (a) ectoderm, mesoglea, mesoderm
    (b) ectoderm, mesoglea, endoderm
    (c) ectoderm, mesoderm, endoderm
    (d) ectoderm, mesenchyme, endoderm

13. Bilateral symmetry brings with it
    (a) refinement of body systems
    (b) the development of the nerve net
    (c) increased reproduction
    (d) alternation of generations

14. Each tapeworm segment has a complete
    (a) digestive system   (c) reproductive system
    (b) nervous system   (d) locomotor organ

15. A true statement about *Planaria* is
    (a) *Planaria* are parasites.
    (b) *Planaria* lack powers of regeneration.
    (c) *Planaria* show the first signs of cephalization.
    (d) *Planaria* have segmented bodies.

16. The circulatory system of the annelid is
    (a) open   (c) partly closed
    (b) closed   (d) varying

17. Metamerism refers to
    (a) male and female gonads in the same animal
    (b) a parasitic infestation of humans
    (c) an evolutionary trend in roundworms
    (d) a series of identical segments

18. The trochophore larva
    (a) is a fossil remnant
    (b) shows relationships between phyla
    (c) is part of the roundworm life cycle
    (d) has free flowing pseudopods

19. A true statement about the earthworm is
    (a) Earthworms are self-fertilized.
    (b) The earthworm sexes are separate.
    (c) Earthworms copulate and exchange sperm.
    (d) Earthworms copulate and exchange eggs.

20. Cercaria larvae are best associated with
    (a) hydras   (c) flukes
    (b) earthworms   (d) jellyfish

PART C. Modified True-False. If a statement is true, write "true" for your answer. If a statement is incorrect, change the under-lined word to one that will make the statement true.

1. A true body cavity is lined with mesoderm.

2. The dorsal region of the body refers to the under surface.

3. Contractile proteins are associated with feeding.

4. The Cnidaria are the first group to show tissue organization.

5. Sponges have an external skeleton.

6. Spongin is a fibrous carbohydrate.

7. The bowl shape of a jellyfish is known as an umbrella.

8. The body form of hydra is a polyp.

9. The nerve net is associated with sponges.

10. Hair cells are part of the ocelli.

11. "Animal-flowers" refer to jellyfish.

12. The function of flame cells is related to digestion.

13. Proglottids are segments in Planaria.

14. The head of a tapeworm is known as the <u>proglottid</u>.

15. *Taenia* lives in two hosts: human and the <u>snail</u>.

16. Annelids are <u>round</u> worms.

17. *Nereis* is the genus name of the <u>tapeworm</u>.

18. The sandworm has a well developed <u>head</u>.

19. The most advanced phylum of invertebrates is <u>Annelida</u>.

20. The life cycle of the Chinese liver fluke includes <u>four</u> host organisms.

Think and Discuss

1. What purpose is served by describing the animal body in terms of symmetry?
2. How do powers of regeneration affect the mortality of the sponge?
3. How has the body of the tapeworm been adapted for a life of parsitism?
4. Why are the mollusks considered to be the most advanced of the invertebrates?

Answers to Questions for Review

PART A

1. symmetry
2. proximal
3. backbones
4. asexual
5. pore
6. spicules
7. hermaphrodite
8. stinging
9. nematocyst
10. fresh
11. budding
12. statocysts
13. tapeworm
14. Nematoda
15. shoes
16. Annelids
17. dioecius
18. mantle
19. Mollusca
20. Mollusca

PART B

1. b
2. d
3. c
4. a
5. d
6. c
7. c
8. d
9. b
10. d
11. a
12. c
13. a
14. c
15. c
16. b
17. d
18. b
19. c
20. c

PART C

1. endoderm
2. back
3. movement
4. true
5. internal
6. protein
7. medusa
8. true
9. *Hydra*
10. statolith

11. sea anenomes
12. excretion
13. tapeworms
14. scolex
15. pig
16. segmented
17. sandworm
18. true
19. Mollusca
20. three

CHAPTER 9

INVERTEBRATES: THE ARTHROPODA

The Arthropoda is the largest animal phylum in numbers of individuals, encompassing approximately 800,000 species, more than all of the other animal species combined. It includes spiders, ticks, mites, lobsters, crabs, insects, centipedes and millipedes (Fig. 9.1). Arthropods are widely distributed, occupying habitats in marine, freshwater and terrestrial environments.

The name *arthropod* in literal translation means "jointed foot," a distinctive characteristic of this group, expressed traditionally as "*jointed appendages.*" The arthropods are segmented animals protected by an exoskeleton made of protein and the flexible but tough carbohydrate chitin. The chitinous exoskeleton is fashioned in articulating plates held together by hinges covering both the body and the appendages, and attached to muscles that make possible quick and unencumbered movements.

The body cavity (coelom) is reduced in size and in its place is a *hemocoel* (blood cavity), a part of the *open circulatory system* in which a pulsating section functions as a dorsal heart. *Hemolymph* bathes the body organs directly because it is not confined to vessels. Respiration in many arthropod species makes use of *tracheal tubes* which communicate to the external environment by *spiracles*, pores that appear one pair to a segment. Some arthropods use gills as organs of respiration. Nervous impulses are carried on a ventral nerve cord occupying a mid-position in the body. The sexes are separate. Some of the insects reproduce parthenogenetically, a process in which eggs develop without fertilization.

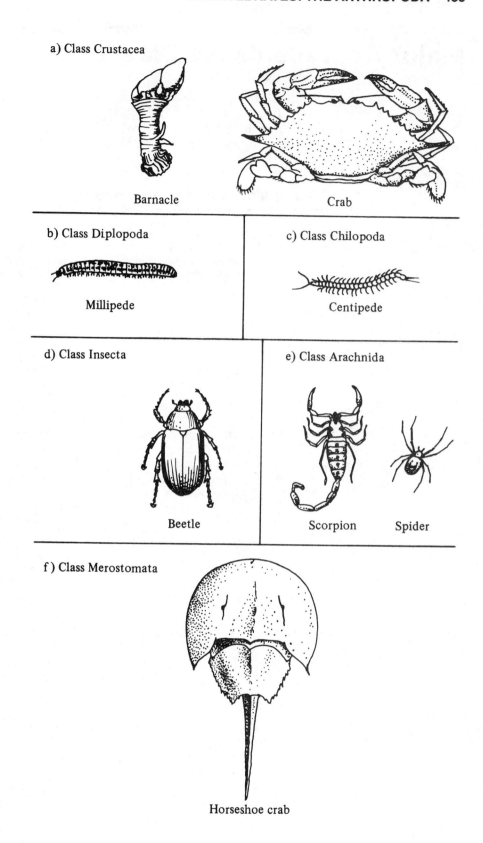

a) Class Crustacea

Barnacle

Crab

b) Class Diplopoda

Millipede

c) Class Chilopoda

Centipede

d) Class Insecta

Beetle

e) Class Arachnida

Scorpion   Spider

f) Class Merostomata

Horseshoe crab

Fig. 9.1   Some representative classes of arthropods

Major Representative Classes

The phylum Arthropoda is divided into several classes. The Arachnida, Crustacea, Insecta, Diploda, and Chilopoda are the most important. Table 9.1 summarizes the characteristics of these classes, while a more detailed discussion of certain classes follows.

TABLE 9.1.   Phylum Arthropoda

Class	Examples	Characteristics
Arachnida	spider, tick, mite, scorpion, horseshoe crab, harvestmen	Mostly terrestrial; small-to-moderate size; body divided into cephalothorax and abdomen; 6 pairs of appendages; 4 pairs of walking legs; no antennae; simple eyes; book lungs or trachae.
Crustacea	crab, shrimp, lobster, barnacle	Mostly aquatic; marine; gill breathers; exoskeleton of chitin; chewing mouthparts; 2 pairs of antennae; 3 or more pairs of legs.
	crayfish, water flea sow bug	freshwater terrestrial
Insecta	grasshopper, dragonfly, cockroach, termite, louse, flea, fly, butterfly, bee, beetle	Mostly terrestrial, freshwater, no marine forms; body divided into head, thorax, abdomen; mouthparts for biting, sucking, lapping; 2 pairs of wings; 3 pairs of legs; breathing by tracheae; excretion by Malpighian tubules; one pair of antennae; 2 compound eyes.
Diplopoda	millipede	Terrestrial; herbivores; head with antennae and chewing mouthparts; body segmented; 2 pairs of walking legs ventral on each segment; 2 eyes; breathing by trachea; 2 pairs of spiracles on each segment.
Chilopoda	centipede	Terrestrial; carnivorous; body segmented; one pair walking legs lateral on each segment; one pair long antennae; poison fangs on first body segment; chewing mouthparts.

CLASS ARACHNIDA—SPIDERS

By careful examination of a spider, you would be able to note the major external characteristics. The body is divided into two regions: the *cephalothorax* and the *abdomen*. The cephalothorax, an arthropod characteristic, is the fusion of the head and the thorax. The cephalothorax of the spider supports six pairs of jointed appendages. The first appendage has been modified into jaws called *chelicerae*. The second are the *palps*, sense receptors and grasping organs, leg-like in females but bulbous in males. The remaining appendages are four pairs of walking legs characteristic of arachnids.

In spiders in the posterior end of the abdomen lying underneath (ventral to) the anus are several pairs of rounded projections. These are called the *spinnerets*. They contain a group of flexible tubules through which silk secreted by the silk glands leaves the body when the spider spins a web.

On the anterior dorsal surface of the cephalothorax are eight simple eyes. These are the external parts of specialized systems that control the physiological processes of the spider. Nerve fibers connect the eight eyes with a nerve mass that surrounds the esophagus. This is where the ventral and dorsal ganglia unite. Branching nerve fibers service the various parts of the body.

The food of the spider consists of body juices of other animals. These juices are drawn through the mouth and the esophagus by action of a sucking stomach, which then passes the food on to another stomach where digestion takes place. Digestion is aided by five pairs of pouched glands. The digestive tract ends in a sac-like rectum and an anus.

Most spiders breathe by means of *book lungs*. These are chambers or cavities containing many thin, hollow membranous plates. The hollow spaces are connected to the outside by fine tracheal tubes. Oxygen is distributed by the blood circulating around the hollow spaces (sinuses). The heart is in the dorsal part of the abdomen fitted into the *pericardial cavity*. Blood enters the heart through a pair of valve-like openings and is pumped out through vessels which empty into the body spaces.

Excretion in the spider is controlled by a pair of *Malpighian tubules*. These are long slender tubules attached at one end to the digestive tract. Nitrogen-containing wastes in the body fluid are changed into uric acid which is then moved through the Malpighian tubule to the end of the digestive tract where it is ultimately excreted as dry crystals (Fig. 9.2).

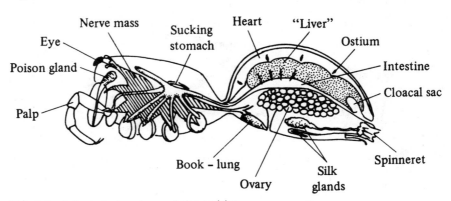

Fig. 9.2 Internal structure of the spider

The sexes in the spider are separate. The sex organs are positioned in the ventral part of the abdomen and communicate with the outside through a ventral opening. During copulation, the male uses modified palps to transfer sperm to the female, effecting internal fertilization. The fertilized eggs are laid in a silk cocoon by some species of spiders. In other species, the eggs are carried around by the female until they are hatched.

CLASS CRUSTACEA—LOBSTERS AND THEIR RELATIVES

Derived from the Latin *cursta*, meaning "crust," the name Crustacea describes the lobsters and their relatives aptly. The crustacean body is covered by a tough exoskeleton arrangement in the form of arched plates which thin out at the joints to permit maximum movement. The lobster is representative of this class.

External Characteristics

The segmented body is divided into a cephalothorax and an abdomen. Six segments of the head and eight segments of the thorax are fused into a single cephalothorax; seven segments compose the abdomen. Pairs of jointed appendages are outgrowths of each of the body segments except the first and the last (Fig. 9.3).

The first body segment of the cephalothorax has a pair of *compound eyes* (Fig. 9.4) positioned at the ends of long flexible stalks that can be extended or retracted. The compound eye is an arthropod characteristic that represents an evolutionary development in light receptors. The compound eye is a collection of thousands of light-gathering units, the *ommatidia*, each with its own lens and light-sensitive cells. Each unit makes its own picture of part of an object. These separate pictures are put together in a pattern resembling a collage so that the compound eye perceives thousands of pictures of the same object but at slightly different angles. The compound eye does not render a clear picture of an object but is a very efficient device for spotting movement of prey.

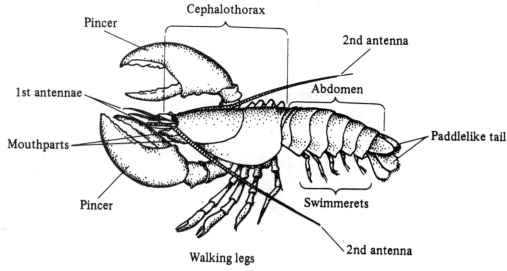

Fig. 9.3   External structure of the lobster

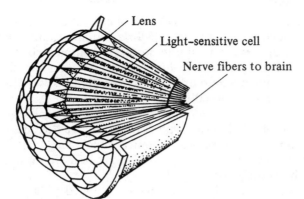

Lens
Light–sensitive cell
Nerve fibers to brain

Fig. 9.4   A diagram of the arthropod compound eye

Segments 2 and 3 on the cephalothorax bear pairs of antennae, sense receptors for touch and chemical stimuli. On segments 4–9 are appendages that have been modified into biting and chewing mouthparts. Segment 10 bears two large pincers (claws) used to fight off enemies and to capture food. Segments 11–14 have four pairs of walking legs. Segments 15–19 are abdominal segments; they bear appendages called *swimmerets* that function mainly in circulating water over the gills. There are two paddle-like appendages on segment 20 which constitute the tail, an appendage modified for swimming.

Internal Structure

The respiratory system consists of feathery *gills*, outgrowths of the body wall that lie on both sides of the body. The gills, surrounded by blood vessels, are bathed by water in the *gill chamber*, a space under the anterior part of the exoskeleton. The gills absorb oxygen from the water and release carbon dioxide into the water.

The heart is located dorsally. Blood enters the heart through three pairs of valve-like openings. It is pumped out through several arteries which branch to all parts of the body, where it delivers oxygen and food to the cells. Blood is emptied into the body sinuses which drain into the capillaries of the gills. Here blood exchanges carbon dioxide, a waste product of respiration, for oxygen.

The excretory system consists of a pair of *green glands* that are located in the head. These glands filter out the nitrogenous wastes produced during protein metabolism and pass them into a small *bladder*. The bladder empties out through a small pore located at the base of the antennae (Fig. 9.5).

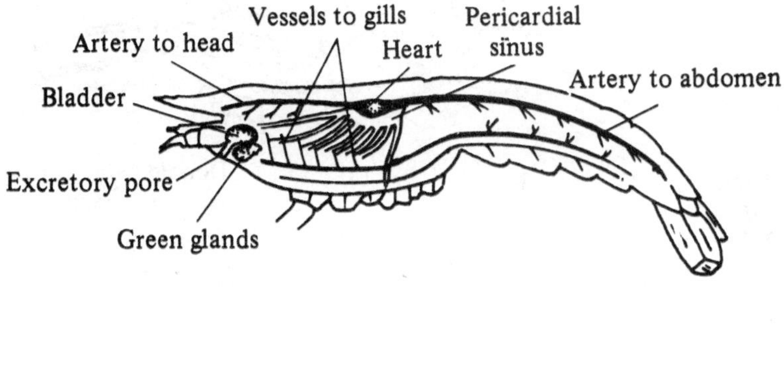

Vessels to gills   Pericardial
Artery to head   Heart   sinus
                                    Artery to abdomen
Bladder

Excretory pore

Green glands

Fig. 9.5   Excretory system of the lobster

The digestive system consists of a mouth, a short esophagus, a stomach divided into cardiac and pyloric chambers, an intestine and an anus. The cardiac portion of the stomach contains special grinding organs known as the *gastric mill*. Finely ground food enters the pyloric chamber of the stomach where most of the digestion takes place and where useful nutrients are absorbed into the blood.

The nervous system includes a dorsal "brain" which is really a large ganglion situated in the cephalothorax above the digestive system. These rings of tissue connect with a ventral chain of ganglia from which a double nerve cord extends into the tail. Nerves branch out from the dorsal brain and the ventral cord (Fig. 9.6).

Sexes in the lobster are separate. Bilobed gonads are located ventrally in the cephalothorax. Sperm cells leave the testis by way of *sperm ducts* that open into pores at the base of the fifth pair of walking legs. *Oviducts* in the female conduct eggs from the ovary to openings at the base of the third pair of walking legs. The early embryos develop while they are attached to the swimmerets of the female.

CLASS INSECTA

The Insecta represent the most advanced class of the modern arthropods. Scientists estimate that the numbers of species in this class are probably as high as several million. To date, more than one half million have been described in scientific literature. It is known that there are more kinds of insects than any of the other animals combined. The insects are by far the dominant form of terrestrial life. The diversity of the insects is unmatched.

There are many reasons for the success of the insects. The modern insects are relatively small organisms ranging in size from 1.5 to 50 millimeters. Their small body size and consequently small food requirement has made them quite successful in the struggle for survival. Another reason for their survival is the fact that they have wings and can fly away from predators or toward food sources. (In fact the insects are the only invertebrates that can fly.) That insects have developed adaptations for survival can be attested to by their wide distribution in terrestrial and aquatic environments. Insect species occupy habitats from the equator to the arctic and from sea level to the snow fields of the highest mountains. Many live in fresh water during the larval stages and several species spend their adult lives near the water.

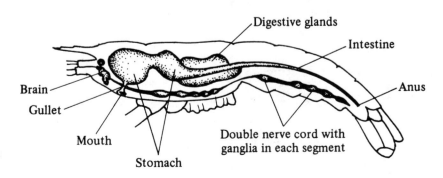

Fig. 9.6 Digestive and nervous systems of the lobster

General Characteristics

There are several characteristics that all insects have in common. The insect body is divided into a well-defined head, thorax and abdomen. Segments 1 to 6 are fused to form the head on which there are a pair of compound eyes and one or more simple eyes. Segment 2 carries a single pair of antennae. The mouthparts vary according to the species of insect. The grasshopper, for example, has biting mouthparts, whereas in the butterfly the mouth is structured for sucking and in the mosquito the mouth serves as a piercing device. Segments 7–9 make up the thorax. Each of these segments bears a pair of walking legs. Most insects bear a pair of wings on the last two segments of the thorax.

The composition and structure of wings differ among orders of insects. For example: the front wings of the beetle serve as protective armor and are generously impregnated with chitin, while the wings of the dragonfly are delicate and membranous. The second wing pair in mosquitoes and grasshoppers is greatly reduced in size and these stubs are used as balancing organs rather than for flight purposes.

The number of segments that make up the insect abdomen varies from a maximum of 11 downward. Although the abdomen bears no appendages, it does have a number of breathing pores (spiracles). The spiracles open into air tubes called trachea that conduct oxygen to all of the body cells.

The life history of an insect may involve several stages in which the young form of the insect does not resemble the parent. Such a life cycle in which there is change of body form is known as *metamorphosis*.

Metamorphosis

Metamorphosis usually involves distinct stages. The life cycle of the butterfly provides an excellent example of four stages: (1) the egg or embryonic stage (2) the *larva* or feeding stage (3) the *pupa* or cocoon stage and (4) the adult stage. The larva is an active stage of life when the feeding organism is an entirely different organism in body form from the parent. The caterpillar, for example, does not resemble the adult butterfly. The pupa is a quiescent nonfeeding stage in the life cycle (Fig. 9.7).

The process of metamorphosis in insects is under control of *hormones* secreted by cells in the brain. Hormones are protein molecules that are secreted by endocrine (ductless) gland cells into the blood for transport to the sites of action. Hormones are specific and stimulate certain *target* cells or organs to grow or to carry out general metabolic activities. *Brain hormone*, secreted by certain cells in the insect's brain, stimulates the *prothoracic gland*, which, in turn, secretes *ecdysone*, a growth and differentiation hormone. Ecdysone controls the molting of the larva and the change into the pupa stage. Another endocrine gland, the *corpus*

Egg   Larva   Pupa   Adult

Fig. 9.7 Complete metamorphosis in the butterfly

allatum, lies near the brain in the larva (caterpillar) and secretes a hormone called *juvenile hormone*. This hormone encourages larval growth and molting (shedding of skin), but prevents the larva from changing body form. After the *corpus allatum* stops secreting juvenile hormone, the metamorphosis to the adult body form occurs.

Representative Orders

The large number of insect species are classified into more than 15 orders. The members of these orders vary in size, feeding habits, habitat, life cycle, and behavior, with some showing complex patterns of social behavior. The grasshopper illustrates the structure of a typical insect, the honeybee the life style of a social insect.

Order Orthoptera—Grasshopper

Fig. 9.8 shows parts of the grasshopper. The head of the grasshopper is clearly defined. There are two lateral compound eyes separated by three simple eyes or ocelli. The single pair of antennae have sensory functions. The grasshopper has biting mouthparts which include a pair of chewing jaws, a pair of flap-like structures for manipulating food, a lower lip-like structure and a tongue-like organ used for the mastication and manipulation of food. The forewings are hard, leathery and opaque. The hind wings are thin, membranous and transparent.

The thorax of the adult grasshopper is divided into three clearly seen parts: the prothorax (with a saddle-like covering called the *pronotum*), mesothorax, and metathorax. The end of the abdomen is modified to form the sex organs. In males, the posterior tip of the abdomen is rounded and the projection used in copulation is present. The posterior end of the female's abdomen is a forked structure, the *ovipositor*, used to dig holes into which the eggs are deposited.

The grasshopper has three pairs of legs; one pair each is on the

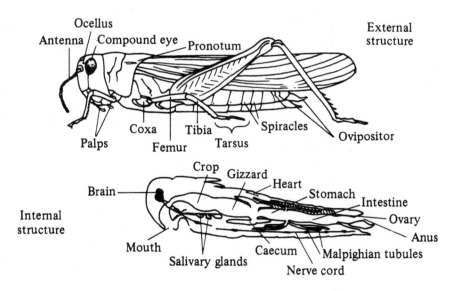

Fig. 9.8 Parts of the grasshopper

prothorax, mesothorax, and metathorax. The third pair of legs are strong and elongated, adapted for jumping.

The digestive system consists of a mouth that leads into a narrow esophagus which connects with a thick-walled grinding organ, the *gizzard*, which is followed by a thin-walled stomach. The narrow intestine ends with the anus. The cavity of the mouth gets secretions from two salivary glands. Six double-blind sacs surround the gizzard and stomach and empty enzymes into the stomach.

On the first segment of the abdomen, one on each side of the animal, are two *tympanic* membranes. These structures respond to vibrations and sound waves. On two thoracic segments and on eight segments of the abdomen are the spiracles, the external openings of the tracheal tubes and air sacs that branch (ramify) throughout the body. The small branches of the tracheal tubes are in contact with the body cells conducting oxygen to them and removing carbon dioxide wastes from them.

The circulatory system includes a dorsal heart. Blood enters the heart through five pairs of valve-like openings called *ostia* and is pumped into the body sinuses. The hemocoel receives blood from all parts of the body and returns it to the heart. The blood of the grasshopper is merely a circulating medium for food and wastes.

The excretory system consists of numerous fine Malpighian tubules which reabsorb water and excrete protein wastes in the form of dry uric acid crystals.

The sexes in the grasshopper are separate. The male reproductive system consists of two *testes* located near the intestine, a sperm duct, and a copulatory organ situated at the tip of the abdomen. In the female, eggs are formed in two ovaries and travel through an oviduct to the vagina. During copulation, the seminal receptacle of the female receives and stores sperm. When eggs pass through the vagina, they are fertilized by the stored sperm. The female deposits the eggs in the soil by digging a hole with the ovipositor.

The zygote develops quite rapidly into an embryo, but then growth and development stop for a while. This rest period in embryonic growth is called *diapause*, which, in the case of the grasshopper, lasts over the winter. Development resumes in the spring and by early summer young grasshoppers, called *nymphs*, emerge. The nymphs are wingless but, after a series of molts, grow to adult size.

Order Hymenoptera—Honeybee

The honeybee exhibits a specialized social structure called *polymorphism*, a condition in which individuals of the same species are specialized for different functions. In a honeybee colony, three classes of individuals arise: fertile males called *drones*; fertile females, or *queens*; and sterile females, or *workers*. The workers have a special concave surface on the second pair of walking legs called a *pollen basket* used to carry pollen.

The queen bee receives sperm from the drone once during her lifetime. The sperm are stored in a special organ called the *spermatotheca* in which they may live for years. Fertilized eggs give rise to females, most of which remain workers. A special female may be selected by the colony and fed a diet of "royal jelly" which causes her to grow larger than the others and become fertile. This fertile female will become a queen, and either take over the existing colony or start a colony of her own. Drones develop from unfertilized eggs by the process of parthenogenesis.

Importance to Humans

The arthropods have very definite effects on human life. The arachnids influence the quality of human life in diverse ways. Spiders are predators on insects that may offer discomfort or harm to humans, and some feed on decomposed organic matter and serve as decomposers. On the other hand, ticks cause Rocky Mountain spotted fever and mites are spoilers of grain. The crustaceans for the most part have a great deal of value. Crabs, lobsters and shrimp serve the cause of human nutrition deliciously.

Insects cannot be avoided. Some insects are helpful and others are harmful. Honeybees provide honey and pollinate flowers. Ladybird beetles destroy other insects that are crop destroyers. Many insects such as flies, fleas, and mosquitoes are carriers of disease; others such as Japanese beetles, tent caterpillars and grasshoppers cause serious damage to foliage and food crops. A greater-than-billion-dollar industry has been built for exterminating insects that infest our homes and gardens. Insects help in the balance of nature and also in certain situations help to unbalance natural communities. This very successful species is human's greatest competitor for food on the earth.

Chronology of Famous Names in Biology

1669   **Marcello Malpighi** (Italian)—first to describe metamorphosis in the silkworm; discovered the excretory tubules of insects.

1737   **Jan Swammerdam** (Dutch)—prepared a detailed monograph on insect structure including many fine illustrations.

1824   **Straus-Durckheim** (German)—produced an illustrated work on insect anatomy: a detailed study of the European beetle, the cockchafer.

1834   **Leon Dufour** (French)—published several monographs describing the anatomy of several insect families.

1841   **George Newport** (English)—elucidated the embryology of insect phyla.

1864   **Franz Leydig** (German)—first to study tissues of insects under the microscope.

1959   **V. B. Wigglesworth** (English)—elucidated the physiological process that controls metamorphosis in insects.

1963   **G. G. Johnson** (English)—determined the means by which insect flight carries them over large distances.

1964   **Wolfgang Beerman** and **Ulrich Clever** (German)—discovered that chromosome puffs seen in insect species are active genes.

1965   **Miriam Rothschild** (English)—found the role of hormones in the life cycle of the rabbit flea.

1967	**Suzanne Batra** and **Lekh Batra** (American)—studied the mutualistic relationship between some insects and fungi.
1977	**G. Adrian Horridge** (English)—explained the workings of the ommatidia of the insect compound eye.
1978	**Lorus Milne** and **Margery Milne** (American)—presented a research study on the four types of insects that live on the surface of quiet waters.
1980	**William G. Eberhard** (American)—discovered the function of the horns in the horned beetle as organs of lifting.
1982	**Thomas D. Seeley** (American)—discovered how honeybees locate a site for a hive.

Words for Study

abdomen	green glands	palp
arthropod	hemocoel	pericardial cavity
book lungs	hemolymph	polymorphism
cephalothorax	hormone	pronotum
chelicerae	jointed appendages	prothoracic gland
compound eyes	juvenile hormone	pupa
corpus allatum	larva	queens
crustacean	Malpighian tubules	spermatotheca
diapause	metamorphosis	sperm duct
drones	nymphs	spinneret
ecdysone	ommatidia	spiracles
gastric mill	open circulatory system	swimmerets
gill	ostia	tracheal tubes
gizzard	oviduct	tympanic
	ovipositor	workers

Questions for Review

PART A.   **Completion.** Write in the word that correctly completes each statement.

1. The animal phylum that is largest in number of species and number of individuals is the ..1..
2. The development of eggs without fertilization is known as ..2..
3. The body of a spider is divided into ..3.. and an abdomen.
4. Simple eyes are known as ..4..
5. Green glands have a (an) ..5.. function.

6. The most advanced class of the modern anthropods are the ..6..
7. The grasshopper has ..7.. mouthparts.
8. Wing stubs in mosquitoes are used for the purpose of ..8..
9. A hormone is secreted by ..9.. or ductless glands.
10. A membrane specialized for gathering sound vibrations is the ..10.. membrane.

PART B.   Multiple Choice. Circle the letter of the item that correctly completes each statement.

1. The most outstanding characteristic of the anthropods is
   (a) membranous wings   (c) simple eye
   (b) bilobed antennae   (d) jointed appendages

2. Book lungs are the breathing mechanisms of
   (a) arachnids   (c) lobsters
   (b) termites   (d) grasshoppers

3. Examples of crustacea are
   (a) barnacles and beetles   (c) horseshoe crab and shrimp
   (b) lobsters and sow bugs   (d) water flea and mite

4. Ommatidia are structural and functional units of
   (a) thorax   (c) compound eye
   (b) flame cell   (d) brain

5. Lobster embryos
   (a) are free-swimming
   (b) fall to the bottom of the sea
   (c) are maintained in male the green gland
   (d) attach to the female swimmerets

6. The group that represents the dominant form of terrestrial life is the
   (a) insect   (c) scorpions
   (b) human   (d) spiders

7. The insect body is divided into the
   (a) cephalothorax and abdomen (c) head, tail, wing
   (b) wing, abdomen, head   (d) head, thorax, abdomen

8. The breathing holes on the abdomen of the grasshopper connect to
   (a) green glands   (c) Malpighian tubules
   (b) nephridia   (d) tracheal tubules

9. A nymph refers to a young
   (a) maggot   (c) honey bee
   (b) grasshopper   (d) blow fly

10. The growth and differentiation that takes place in insect larva is controlled by the hormone known as
   (a) juvenile   (c) ecdysone
   (b) brain   (d) prothoracic

PART C.   Modified True-False. If a statement is true, write "true" for your answer. If a statement is incorrect, change the underlined word to one that will make the statement true.

1. The body cavity of the arthropods is replaced by the <u>coelom</u>.

2. Arachnids are <u>ants</u>.

3. Malpighian tubules change nitrogenous wastes into dry crystals of <u>bile</u>.

4. <u>Spinnerets</u> circulate water over the gills of the lobster.

5. The lobster has a (an) <u>closed</u> circulatory system.

6. The only flying invertebrates are the <u>waterfleas</u>.

7. Butterfly mouth parts are structured for <u>biting</u>.

8. The feeding stage of the developing insect is the <u>pupa</u>.

9. Change in body form of the butterfly is known as <u>cocoon</u>.

10. Juvenile hormone is secreted by the <u>corpus allatum</u>.

Think and Discuss

1. There are 80,000 species of Arthropoda. Why is this phylum so successful?

2. How does metamorphosis extend the life of the butterfly?

3. How does the compound eye of the lobster work?

4. Describe polymorphism in the honeybee.

Answers to Questions for Review

PART A
1. Arthropoda
2. parthenogenesis
3. cephalothorax
4. ocelli
5. excretory
6. insects
7. biting
8. balancing
9. endocrine
10. tympanic

PART B
1. d	5. d	8. d
2. a	6. a	9. b
3. b	7. d	10. c
4. c		

PART C
1. hemocoel
2. spiders
3. uric acid
4. swimmerets
5. open
6. insects
7. sucking
8. larva
9. metamorphosis
10. true

CHAPTER 10

CHORDATES: PREVERTEBRATES TO MAMMALS

Characteristics of the Chordates _____

The chordates are animals that have a notochord at some time in their life cycle. The notochord is a living, internal skeletal axis in the form of a compact cellular rod-like structure that extends the length of the body. It lies dorsal to the digestive tract and is not to be confused with the backbone. In lower chordates (invertebrate chordates), the notochord prevents the body from shortening when muscles contract. During the embryonic stages of the vertebrate chordates, the notochord is replaced by a column of bones, the *vertebrae*, which form the backbone.

Chordates are set apart from lower animals by several distinguishing characteristics in addition to having a notochord. First, all chordate embryos have the three primary germ layers from which all specialized tissues and organs develop. Secondly, chordates are bilaterally symmetrical animals with anterior-posterior differentiation. Thirdly, the body has a true coelom and a digestive tract that begins with a mouth and ends with an anus. Other characteristics which differentiate the chordates from other animals are the presence of *pharyngeal gill slits* and the *dorsal hollow nerve cord*. The pharyngeal gill slits are paired vertical slits in the wall of the *pharynx* (throat) positioned directly behind the mouth and connecting to the outside of the body. The gill slits may be present only in the embryo stage, or they may persist and be used in breathing. The dorsal hollow nerve cord is a hollow cylindrical tube, usually expanded into a brain at the anterior end. It lies dorsal to the notochord. The hollow space is called the *neurocoel*.

The phylum Chordata is divided into five subphyla: Echinodermata, Hemichordata, Urochordata, Chephalochordata, and Vertebrata. The first four phyla represent the prevertebrates, also referred to as the invertebrate chordates.

The Prevertebrates _____

PHYLUM ECHINODERMATA—SEA STARS AND THEIR RELATIVES

Echinoderms are spiny-skinned invertebrates that include the sea stars brittle stars, sand dollars, sea urchins and sea cucumbers. Although they do not look very much like vertebrate animals, the development of the echinoderm embryo strongly resembles that of the chordates in the early stages. The larval stage is free-swimming and shows bilateral symmetry.

General Characteristics

All echinoderms live in the sea. Most species are capable of a very slow, creeping locomotion. The only group of sessile echinoderms is the sea lilies.

The name echinoderm means spiny-skin, a distinctive feature of the phylum members. Just under the skin, calcareous spines and plates form a skeleton. Another distinctive characteristic of the echinoderms is pentaradial symmetry: the body is built on a plan of five *antimeres* radiating from a central disc in which the mouth is in the middle. The digestive system is complete, although the anus does not function. The echinoderms have no head and no excretory and respiratory systems. They do, however, have a *water vascular system* composed of a series of fluid-filled tubes that are used in locomotion. Changes of pressure in this system enable an echinoderm to extend and retract *tube feet*. The tube feet are

TABLE 10.1. Echinoderms

Class	Examples	Characteristics
Crinoidea	sea lily, feather	Sessile, attached by a stalk; branched arms; ciliated tube feet used for feeding; some species are free swimming; more abundant during Paleozoic era.
Asteroidea	sea star	Free moving by means of tube feet; arms branching from a central disc.
Ophiuroidea	brittle stars, serpent stars, basket stars	Free moving; thin flexible arms marked off from disc; tube feet used as sensory organs and for feeding.
Echinoidea	sand dollar, sea biscuit, sea urchin	Free moving; body fused plates or flattened disc, without free rays, covered with calcareous plates; some species covered with spines.
Holothurioidea	sea cucumber	Free moving; elongated flexible body with mouth at one end; sometimes with tentacles; skeletal elements of the skin reduced.

used in locomotion and in some species they are used to capture prey. In the echinoderms, the sexes are separate.

Table 10.1 summarizes the important characteristics of the classes of echinoderms. One class—that of the sea star—is discussed in more detail.

A Representative Class—Asteroidea (Sea Stars)

The *sea star* is an excellent representative of the echinoderms. It has all of the distinguishing characteristics: pentaradial symmetry, spiny skin, tube feet controlled by a water vascular system, no head, and no excretory or respiratory system. Protruding from the wall of the coelom and extending out between the calcareous plates into the sea water, are the *papulae*, sac-like structures that function as respiratory and excretory organs.

The mouth is located in the center of the disc on the underside of the body. The mouth side of the body is called the oral side. The aboral side is the upper surface without the mouth. A short esophagus leads from the mouth to the cardiac portion of the stomach. A constriction in the stomach wall separates the cardiac portion of the stomach from the pyloric part. The cardiac stomach is turned inside out and pushed through the mouth when the starfish is eating. The stomach engulfs the food, usually molluscs or crustaceans, and digests it before pulling the stomach back to the inside. The intestine and the anus of the starfish are practically nonfunctional.

On the aboral side of the sea star is a colored plate known as the *madreporite*. Water enters the sea star through minute openings in this plate. Water is drawn by ciliary action down into the stone canal (made rigid by calcareous rings) to the ring canal that encircles the central disc. The ring canal has five radiating canals which extend into the arms of the sea star. Short side branches connect the radial canals with many pairs of tube feet, which contract and expand in response to the water pressure in the *ampulla*, a muscular sac at the upper end of the tube feet (Fig. 10.1).

The nervous system is composed of a nerve ring located in the disc from which a ventral and radial nerve branch into each arm. The radial nerves have finer branches which extend throughout the body. At the tip of each arm is a light sensitive eyespot that is innervated by the radial nerve.

Sea star sexes are separate. Paired gonads are to be found in each ray. The eggs of the female and sperm of the male escape through pores

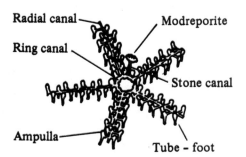

Radial canal
Ring canal
Modreporite
Stone canal
Ampulla
Tube - foot

Fig. 10.1   Water vascular system of the sea star

on the aboral surface of the sea star. Fertilization takes place in the water. During embryonic development, sea stars pass through several larval stages.

Sea stars have remarkable powers of regeneration. If an arm breaks off, the arm grows back. Should a piece of the central disc be attached to the amputated arm, a new individual will grow from the dismembered part. Sea stars prey on oysters. At one time, oyster "farmers" would clear the oyster beds of sea stars, cut them up, and throw the cut pieces back into the water. What they accomplished was an increase in the numbers of sea stars. In effect, they aided the process of sea star regeneration.

SUBPHYLUM HEMICHORDATA (ENTEROPNEUSTA)

The hemichordates, wormlike animals, are entirely marine. Some species live near the shore, while others burrow in sediments at the bottom of shallow seas. Their bodies are divided into three regions: the *proboscis*, covered with cilia; a short *collar*; and a long *trunk*. The proboscis, used for burrowing, resembles an acorn—thus the name "acorn worms." The cilia that cover the proboscis sweep water, food, and mucus into the mouth. The flattened body contains gill slits, through which excess water is pushed out. A digestive tract extends through the entire length of the body. Food is carried into the pharynx and filtered into the digestive tract.

At one time biologists classified the hemichordates as chordates, believing that the connective tissue in the collar was hollow notochord. Newer evidence shows, however, that this is not so. Hemichordates are considered to be the evolutionary link between the echinoderms and the chordates. The ciliated larva of the hemichords resembles the larval stage of some echinoderms. The pharangeal gill slits are a major chordate characteristic found in all chordates but nowhere else in the kingdom Animalia.

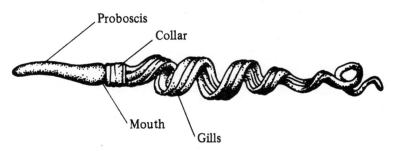

Fig. 10.2 *Balanoglossus*, the acorn worm

SUBPHYLUM UROCHORDATA—TUNICATA

The adult tunicates are sessile, sac-like animals, often referred to as *sea squirts*. They are filter-feeders, taking in water which passes through the gill slits into the atrial chamber and out by way of the *atriopore*. The tunicates reproduce asexually by budding and also sexually by eggs and sperm. The larval forms have a notochord, a nerve cord, a pharynx with gill slits, and an *endostyle*. An endostyle is aciliated or glandular outpocketing from the wall of the throat of the urochords. Cilia and mucus from the gland sweep food backwards into the gullet. The adult forms lose the chordate characteristics. Examples of the tunicates are *Cynthia*, *Salpa* and Sea Pork.

SUBPHYLUM CEPHALOCHORDATA—LANCELETS

The representative organism for the group is *Amphioxus*, the lancelet. It burrows in the sand at the shoreline of tropical or temperate waters. The body is small, elongated, and shaped like a fish without paired fins. It has gill slits, a well-developed notochord and a dorsal hollow nerve cord (Fig. 10.3).

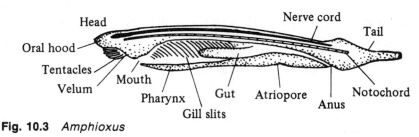

Fig. 10.3 *Amphioxus*

The Vertebrates

Animals that have a true backbone composed of segmented parts called *vertebrae* belong to the chordate subphylum Vertebrata. The vertebrae may be made of cartilage or bone: if made of the latter, cartilage cushions prevent the bones from rubbing together. The backbone is built around the notochord and usually obliterates it. Vertebrates vary in size from large to small, but all have a living endoskeleton usually made of bone. A limited number of water-dwelling species exhibit an endoskeleton made of cartilage. All vertebrate species have marked development of the head where a brain is enclosed in a *cranium*.

Blood is pumped through a *closed circulatory system* by means of a ventral heart, having at least two chambers: an *atrium* and a *ventricle*. The *hepatic portal system* carries blood laden with food from the intestines to the liver before it reaches the body cells. Vertebrate red blood cells contain the iron-bearing pigment hemoglobin which is specialized to carry oxygen. Such a system of closed blood vessels prevents blood from entering the body cavity.

Most vertebrates (except human) have a post-anal tail which is a continuation of the vertebral column. Although there are never more than two sets of paired appendages, some adult vertebrates show only one such set or none at all, the appendages having been lost over evolutionary time. Evidence of lost appendages may be seen in embryonic forms or may be demonstrated by *vestigial* structures. The coccyx bone in humans is a remnant (vestigial structure) of a post-anal tail. Other characteristics of vertebrates include a mouth that is closed by a movable lower jaw and a thyroid gland derived from the ventral wall of the pharynx. In the invertebrate chordates the endostyle is an evolutionary signpost pointing to the development of the thyroid gland.

The structural differences between the invertebrates and the vertebrates are summarized in Table 10.2.

TABLE 10.2.   Differences in Structure Between the Invertebrates and the Vertebrates

	Invertebrate	Vertebrate
skeleton	nonliving exoskeleton	living endoskeleton
nerve cord	ventral, double and solid; formed by delamination* from ectoderm.	dorsal, single and hollow; formed by invagination of** ectoderm.
heart	dorsal	ventral
hemoglobin	in plasma, when present	in blood cells
circulatory system	open with hemocoel and sinuses	closed, contained in vessels

* Delamination means splitting off from ectoderm
** Invagination refers to an inpocketing

CLASS AGNATHA—CYCLOSTOMES

The Agnatha are the most primitive of the vertebrates. These are the jawless fishes—lampreys and hagfish—characterized by a round mouth that has earned them the general name *cyclostome*. The agnathans have only a single nasal opening at the tip of the snout and no paired fins or paired limbs of any kind. The notochord is present throughout life and the skeletal structures are made of cartilage. The skin is smooth and slimy, lacking scales. The eyes are rudimentary.

The lampreys (Fig. 10.4) live in freshwater where they are parasite-predators on bony fish. The mouth of the lamprey is a sucker disk used by the fish to attach itself to rocks or other organisms. After attaching itself to a fish, the lamprey uses a rasping tongue-like structure to break the skin and suck the blood of the host. The lampreys produce an *am-*

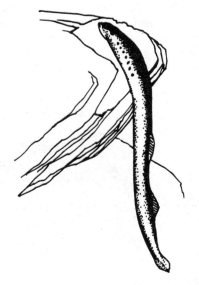

Fig. 10.4 Lamprey

nocoete larva, a filter-feeder living buried in river mud, which lasts for about seven years before changing to adult form. The adult life span is short, the adults dying after going upstream to spawn.

Hagfish live in temperate and tropical marine waters where they are scavengers, feeding on dead, disabled, and diseased fish.

CLASS CHONDRICHTHYES — CARTILAGINOUS FISH

The sharks (including the dogfish), rays and skates are examples of the Chondrichthyes. Most species in this class live in the ocean. The Chondrichthyes have paired fins, a lower jaw, and gill arches. The body of the shark is covered with *placoid scales* which arise from the ectoderm, also forming the teeth in the jaws and on the roof of the mouth. A distinctive feature of this group of fish is that the skeleton is made of cartilage, not bone as in the Osteichthyes, or bony fish. The sharks and their relatives differ from the bony fish in other ways as well: they have no swim bladder, no true scales and the gill slits are uncovered.

Sharks are predators and feed upon bony fish. Most of the digestion takes place in the stomach while absorption of digested food occurs in the intestine. A *spiral valve* that extends the length of the short, fat intestine increases surface area for absorption. The mouth is located ventrally (Fig. 10.5).

Of particular interest is the sensory system of the shark. The eye of the shark, except for shape, size and absence of eyelids, is very much like the human eye. The paired nostrils are pits on the ventral surface of the head that open externally only and do not empty into the throat. The nostrils are lined with an olfactory membrane which is connected by nerve fibers to the olfactory lobes in the brain. Along the sides of the body are pressure receptors known as *lateral line systems*. In the head pits of the lateral line organs have become modified into long canals filled with mucous. These canals, known as *ampullae of Lorenzini*, are sensitive electrochemical receptors.

In all cartilaginous fish, fertilization is internal with the male using *claspers*, modified pelvic fins, to place the sperm in the female's body. In some species the embryos are nourished *ovoviviparously*, obtaining food from the egg. In other species, the embryo is maintained inside the

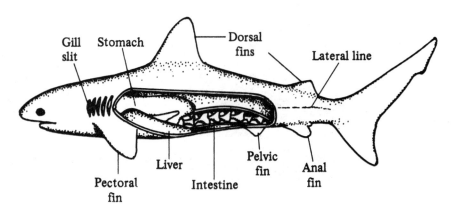

Fig. 10.5 Structure of the shark

mother's body and nourishment is passed from the blood vessels of the mother into the blood system of the developing embryo; this method of embryo nutrition is known as *viviparous*.

SUPERCLASS PISCES

Superclass Pisces includes all of the bony fish. Like the cartilaginous fish, the Pisces are water-dwelling vertebrates that breathe by means of gills, have paired eyes, and a two-chambered heart with blood flowing from the heart through the gills and then to the other parts of the body. The Pisces differ from cartilaginous fish in having *dermal* scales and bone, not cartilage, in the skeleton.

Major Groups

The superclass Pisces is divided into four subgroups: classes Dipnoi, Crossopterygii, Ganoidei, and Osteichthyes.

The Dipnoi are lungfish that live in seasonally dry estuaries in Africa, Australia and South America. These fish breathe by means of lungs and by gills. The lung allows these animals to obtain oxygen from the air and permits them to live in water that is too foul for gill breathing. During the dry season, they aestivate in dry mud; they become activated when the river bed fills up with water during the rainy season.

Some biologists consider the lungfish to be intermediate between the fish and the amphibia. Lungfish have several body structures that look like those found in primitive land dwellers, and the early part of the life cycle of the lungfish and the frog are almost identical. However, the shape

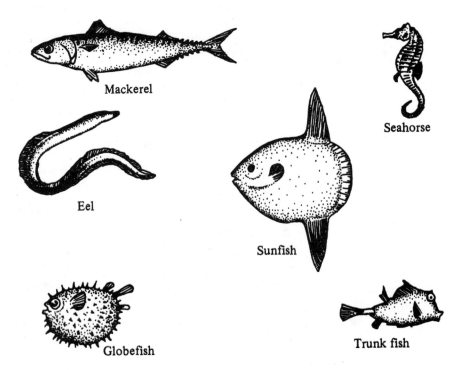

Mackerel

Seahorse

Eel

Sunfish

Globefish

Trunk fish

Fig. 10.6 A diversity of bony fishes

of the teeth and jaws are not froglike and indicate that direct line descendency of the amphibians may not have taken place.

The Crossopterygii, for the most part, represent fish that have become extinct. This group is thought to be the ancestral line from which modern fish and amphibia descended. A living member of this group called a *coelocanth* was caught off the coast of South Africa. It was a large, lobe-finned fish with bluish scales.

Members of the Ganoidei are the most primitive of the bony fish. Most species in this class live in fresh water and their distinctive feature is the heavy armor of *ganoid* scales which are flat and form heavy plates. Once the armored fish were the dominant form of fish, but now only a few scattered species remain. Living examples of the Ganoidei are sturgeon, gar, pike and freshwater dogfish. Some species have partly cartilaginous skeletons and a degenerating spiral valve indicating possible relationship to the sharks.

The Osteichthyes, or true Bony Fish, are the most highly organized of all the fish. They belong to the order Teleostei and are referred to as *teleosts*. They are a very successful and widely distributed group, differing widely in aquatic habitat, size, feeding patterns, and shape. Fig. 10.6 illustrates the physical diversity among the teleosts. However, there are some characteristics that are typical of all species of bony fish.

Distinctive Features

The distinctive features of the teleosts that make them different from the cartilaginous fish are primarily the skeleton, the scales and the gills. The internal skeleton is made of bone. The scales are either *cycloid* (smooth) or *ctenoid* (rough) and are constructed as thin bony plates rather than like the thick plates of the ganoid type. The gills are reduced in number to four pairs, instead of the seven pairs in sharks, and are covered with a flap of bone called the *operculum*. The fins are paired. Fig. 10.7 shows the location of the dorsal, anal, pelvic and pectoral fins. The *swim bladder* is an outgrowth of the pharynx, an oxygen-filled structure that enables the fish to float.

Some sense organs in bony fish are well developed; others are not. The lateral line receptors are sensitive to movements of current. The olfactory pits are not connected to the mouth and therefore have no respiratory function, but they do enable fish to respond to chemical stim-

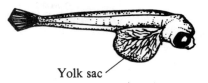

Fig. 10.8 An immature fish

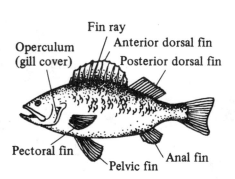

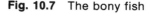

Fig. 10.7 The bony fish

uli. Eyes of fish vary in size according to species. The taste buds located in the mucosa of the mouth are poorly developed. Fish probably cannot hear in air. There are three semicircular canals that control equilibrium, but auditory vibrations reach the ear through the skull bones.

Reproductive Behavior

Patterns of reproduction vary among the bony fish, but in most species fertilization is external. During certain breeding seasons the female lays thousands of eggs known as *roe*. The male deposits sperm cells known as *milt* close to the eggs. The sperm, attracted by some chemical substance given off by the eggs, swim to them. One sperm enters an egg and fertilizes it. The fertilized egg (zygote) then goes through a series of mitotic divisions (cleavage), resulting in a new immature fish called a *fry*. The fry has attached to it a sac appropriately called the *yolk sac* because it contains yolk which nourishes the young fish until it can feed independently (Fig. 10.8).

Fertilization in fish is chancy: many sperm never reach the eggs and many fertilized eggs die before development. Hence there is an overproduction of gametes to ensure the survival of the species.

In a few fish species fertilization is internal and in a few parental care is given to the fertilized eggs. The stickleback male, for example, takes care of the fertilized eggs in nests and the male seahorse carries them around in a brood pouch.

CLASS AMPHIBIA—FROGS AND THEIR RELATIVES

From an evolutionary perspective, amphibia are transitional animals, living first in water and then on land during specific times in the life cycle. They are equipped for this double life, so to speak, by body structures and organs that change to meet their needs.

An amphibian must spend part of its life cycle in the water where its eggs are laid and fertilized. The eggs develop into a larval stage, or tadpole, that has fish-like characteristics. In tadpoles breathing is by means of gills, blood is pumped by a two-chambered heart, and swimming is by means of a tail and body movements made possible by muscles in the body wall. The change to adult form is known as *metamorphosis*, a process controlled by the thyroid gland. The adult amphibian loses the gills, lateral line senses, tail, unpaired fins and muscles controlling them—the fish characteristics—and develops structures adapted for life on land. An adult amphibian breathes by means of lungs and has a three-chambered heart which is more efficient at pumping blood between the lungs, the heart, and the rest of the body. In most species the adult also has limbs for movement, but no tail (Fig. 10.9).

There are three general types of amphibia. The Apoda are wormlike, legless, ground-burrowing forms found in tropical and semitropical regions; *Caecilia* (Fig. 10.10) is a representative genus. The Urodela are amphibians that do not lose their tadpole-like tail in metamorphosis; salamanders, newts and mud puppies (Fig. 10.11) are examples. The Anura, represented by frogs and toads, lose their tails on becoming adults. Frogs and toads are the first vertebrates to become vocal.

In frogs fertilization is external. Sperm leave the testes through tubules called *vasa efferentia* which communicate with the kidney. The

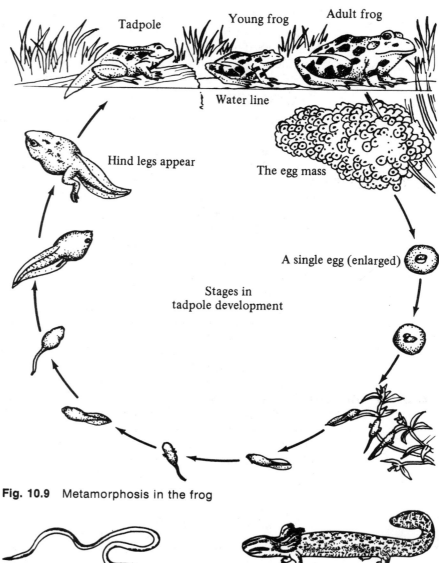

Tadpole Young frog Adult frog

Water line

Hind legs appear

The egg mass

A single egg (enlarged)

Stages in
tadpole development

Fig. 10.9 Metamorphosis in the frog

Fig. 10.10 *Caecilia*, a wormlike
amphibian

Fig. 10.11 *Necturus*, the
mud puppy

sperm cells then pass into the *Wolffian duct* which leads to the *cloaca*,
a passageway that opens to the outside of the body. In the female large
egg masses are released into the body cavity from two ovaries, located at
the anterior end of each kidney. Beating cilia sweep the eggs into coiled
tubules known as *oviducts* where they are propelled to the cloaca and
then out of the body. As the eggs pass through the oviducts they are
coated with a thin layer of jelly-like material. At the time when the female
is depositing eggs in the shallow waters of a pond or brook, the male
deposits sperm over them. The sperm swim to the eggs and as each sperm
reaches an egg, it digests its way through the jelly and into the egg,
effecting fertilization. After fertilization the jelly coating on the eggs
swells due to the absorption of large amounts of water. The swelling of

the black jelly causes the eggs to adhere together and protects them from predation by fish and other animals. The fertilized egg, or zygote, undergoes cleavage, forming a tadpole.

Amphibians demonstrate some interesting mechanisms for prolonging life. Many have powers of regeneration to the extent that entire organs can grow back after having been lost. The axolotyl and the urodele (*Triton cristatus*) are examples of animals in which regeneration occurs. Many amphibians are able to hibernate or aestivate and survive unfavorable conditions.

CLASS REPTILIA

Reptiles are the first true land vertebrates relieved of the necessity of returning to the water to reproduce. Reproducing on land was made possible by the development of special embryonic membranes that preserve on land the protection offered by the aquatic environment to the embryos of lower vertebrates. One such membrane is the *amnion*, a membranous sac surrounding the embryo and filled with water. This sac of water prevents the delicate, rapidly dividing embryonic cells from drying out and protects them from shock and mechanical injury. Animals having the amnion are called *amniotes*; those animals without an amnion are known as *anamniotes*. Reptiles, birds and mammals are amniotes and therefore adapted to life on land.

The eggs of reptiles are fertilized internally, and the female lays fertilized eggs. Reptile embryos develop encased in an egg surrounded by a leathery shell. Lining the shell is the *chorion*, an embryonic membrane that mediates the two-way exchange of gases between the outside air and the embryo. Serving as a temporary organ of respiration and excretion is a third embryonic membrane called the *allantois*. The allantois, bearing a capillary network, grows into the space between the chorion and the amnion.

General Characteristics

Reptiles have a dry leathery skin covered with epidermal scales. A somewhat flattened skull contains a brain having a cerebrum much larger than that of the fish or amphibians. The eyes have secreting glands which keep the surface moist. Some species of reptiles have an external pore (*meatus*) positioned on the side of the head connected to the auditory ossicle or bone in the middle ear. Reptiles are air-breathers and have rather well-developed lungs. The heart is composed of two atria and a ventricle; in some species the ventricle is almost divided into two compartments, an evolutionary signpost pointing to the four-chambered heart. The body temperature of reptiles is not constant, changing with the external environment. In popular speech, such animals are called cold-blooded; in technical language, *poikilotherms*.

Major Groups

Extinct Forms

Reptiles flourished during the Mesozoic Era which lasted for about 130 million years and came to an end about 65 million years ago. At that time there were more than twelve orders of reptiles. Today only four orders of reptiles remain. Among the ancient reptiles there was a great diversity of form and adaptations that permitted life in a variety of habitats: dry land, water, swamps, and air. The *stem reptiles* had many features in common with primitive amphibians and resembled modern-day lizards externally. The *therapsids;* however, were more advanced, having teeth which looked much like those of the mammals. The dinosaurs are remembered for being very large and yet some were quite small. It is believed that many of the dinosaur species were able to walk on two legs, using the enormous tail for balance. The shorter front legs were probably used when walking slowly or resting. Some of the larger dinosaurs returned to the sea. The *icthyosaurs* were best adapted for aquatic life. A few of the ancient reptiles were able to fly; the *pterosaurs* had wings but they probably did more gliding than true flying.

Living Forms

The living reptiles are classified in four orders: Rhynchocephalia, Chelonia, Squamata and Crocodilia.

Order Rhynchocephalia—Tuatara   The only living representative of this order is the New Zealand "tuatara" of the genus *Sphenodon*. This lizard is about one and a half meters long having a median eyestalk at the top of the head. *Sphenodon* has a number of primitive reptilian characteristics, including no tear glands, teeth in the roof of the mouth, a lung that resembles a cluster of toad's lungs, abdominal ribs, unfused frontal skull bones, and a vertebral column with a fishlike structure.

Order Chelonia—Turtles and Tortoises   The chelonians are the turtles and tortoises. The skeleton is modified to form a box-like covering, the upper curved portion of which is called the *carapace*, the lower part, the *plastron*. The head and the tail are the only movable parts of the animal. The jaws are horny and toothless. Chelonians live on land, in freshwater, and in the sea. More turtles live on the American continent than anywhere else (Fig. 10.12).

Order Squamata—Lizards and Snakes   Lizards and snakes are squamates. Their bodies are covered with a great number of small flexible scales that cannot be removed easily like the scales of bony fish. Lizards have movable eyelids, visible earpits and usually have legs. Snakes, on the other hand, are legless, and do not have eyelids or earpits.

Lizards are typically land dwellers requiring the warmth of the sunshine. The iguana is a large tree-living lizard native to Mexico. The chameleon, native to Africa, has a flexible grasping tail and grasping feet; its coat is capable of changing colors to match the background. The gila monster *Heloderna* is the only lizard whose bite is poisonous.

Fig. 10.12   Turtle

Fig. 10.13   Crocodile

Although snakes are cold to the touch, they are not slimy. The body is heavily muscled and strongly ribbed. Snakes use the ribs in walking. Of the 110 species of snakes in the United States only 20 species are poisonous. Among the poisonous snakes are the copperhead, rattlesnake, water moccasin, and the coral snake. The harmless varieties include the puff adder, the garter snake, the black snake and the milk snake.

Order Crocodilia—Crocodiles and Alligators This order includes the crocodiles and the alligators, large thick-bodied reptiles. These are considered to be the most advanced of the reptiles. The heart has a ventricle that is almost completely divided into two compartments. The lungs are very well developed and the brain has a rather large cerebrum. They live in shallow salt or fresh water where they are better able to locomote than on land (Fig. 10.13).

CLASS AVES—BIRDS

Birds are terrestrial vertebrates with feathers. Feathers are the distinctive feature of birds: all birds have them and no other animals are so covered. The forelimb is modified into wings for flight, leaving the hindlimbs for walking (bipedal locomotion). Birds are built for flight; special adaptations in body structure effect lightness in weight, efficiency and strength. Not only are the feathers light in weight and easily moved and lifted by wind, but they also create warmth next to the body. Body heat warms the air that is in contact with the bird's body. Warm air becomes lighter and rises. Other adaptations for flight are the compact, but hollow, bones, numerous air sacs occupying all available body spaces, reduced rectum, loss of teeth, and tail feathers replacing a bony tail (Fig. 10.14).

In most bird species, the wings are organs of flight. However, in some species, the wings are modified for other purposes. For example: the wings of penguins serve as flippers for swimming.

According to fossil evidence, the evolutionary link between the reptiles and the birds was *Archaeopteryx*. Biologists consider *Archaeopteryx* to have been a bird because it had feathers, but it also had some very distinctive reptilian features, such as teeth and a long, bony tail.

Modern birds have lost reptilian characteristics. The feet of birds of various species are modified for all kinds of uses: running, scratching, grasping, swimming and wading. The horny bill is a characteristic of all birds. It is toothless and modified in various shapes to carry out special functions. The colors of bird feathers are due to pigments or irridescence and indicate *sexual dimorphism*, a state in which the male is often brightly and elaborately colored while the female is drab.

The brain of the bird shows greater development than the brain of the reptile, and the heart has four chambers—two atria and two ventricles, permitting complete separation of oxygenated and deoxygenated blood. Unlike the reptiles, birds maintain constant body temperature: they are *homiotherms*, or warm-blooded. The digestive system consists of a crop, a stomach, a gizzard, an intestine and a much reduced rectum. Birds do not retain their feces, a condition which is an adaptation for flight. There is a well developed kidney that empties by way of the ureter. Birds have remarkable eyesight, a feature that makes hawks and eagles such successful predators. On the other hand, the sense of smell is very

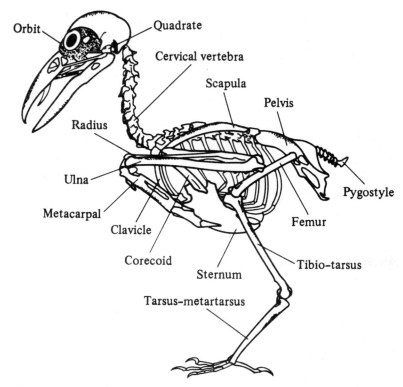

Fig. 10.14   Bone structure of the bird

poorly developed in birds. The *nictitating membrane* (third eyelid) is a translucent membrane that covers the eye crosswise serving as protection during flight.

Birds show another advancement over reptiles in behavior. Song is used to establish territorial rights, an important aspect of bird social behavior. The song of birds is produced by air passing over the *syrinx*, a secondary larynx located at the lower end of the windpipe (trachea) at its junction with the bronchi. The pattern of reproduction in birds also differs from that of reptiles and amphibians. Birds practice monogamy, the mating of one male with one female either for life or for the duration of the breeding season. Before mating birds go through courtship, a special type of behavior engaged in by males trying to gain the attention of reproductive females. In birds, fertilization is *internal*; the development of the young, external. In most species the female lays the eggs in a characteristic-type nest and both the male and the female take turns *setting*, a process of incubation by which the eggs are kept warm, and later caring for the young.

The testes of birds are positioned in the back just above the kidneys. Mature sperm leave the testes through tubules called the *ductus deferens* which lead into the cloaca. The gonads of females are the ovaries, glands where eggs are produced. In female birds, there is one functioning ovary; the other having degenerated early in the bird's life. A bird's egg cell consists of a nucleus and a little protoplasm. The cell is surrounded by yolk. A mature cell and its yolk is drawn into the upper end of the oviduct which is funnel-shaped and contains waving cilia. As the cell with its yolk

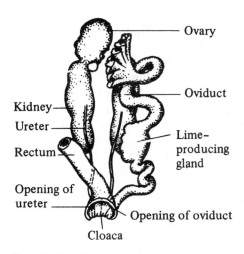

Ovary

Oviduct

Kidney

Ureter

Lime-producing gland

Rectum

Opening of ureter

Opening of oviduct

Cloaca

Fig. 10.15 Reproductive system of the female bird

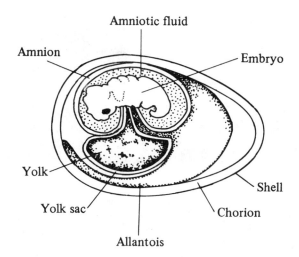

Amniotic fluid

Amnion

Embryo

Yolk

Shell

Yolk sac

Chorion

Allantois

Fig. 10.16 Development of the bird embryo

travels through the oviduct, it is surrounded with layers of the protein *albumen*. The albumen is the white of the egg. The albumen is surrounded by a thin membrane which then is covered by a calcareous shell secreted by lime-producing glands that line the lower end of the oviduct (Fig. 10.15).

Fertilization is accomplished during mating at which time the male and female place their cloacas close together. Sperm swim from the cloaca of the male into the female cloaca and up into the oviduct. Fertilization takes place high up in the oviduct before the albumen and the other surrounding membranes are secreted by the oviduct cells. Most birds lay a *clutch* of fewer than six eggs. However, ducks may lay as many as 15 eggs at one time (Fig. 10.16).

CLASS MAMMALIA

The first mammals appeared on earth about 65 million years ago. The early mammals were small and insignificant, but had developed adaptations enabling them to survive at the same time that the giant reptiles roamed the earth. It is believed that the early mammals were no larger than field mice, feeding on insects and scraps of vegetation. The large eye sockets of the fossil forms indicate that these primitive mammals were nocturnal and probably lived in trees.

Scientists present various reasons for the success of the mammals. First among these is the fact that they were fast movers and their food requirements and behavioral patterns kept them out of the way of carnivorous (flesh-eating) reptiles. Second, the early mammals have been accused of contributing to the demise of the dinosaurs by eating the unprotected eggs of these reptiles. Third, the change in form of the skeleton which allowed the positioning of relatively thin legs underneath the animal led to a narrow walking track and contributed to faster movement. (Remember how the thick legs of a crocodile are spread out from the body on a wide track.) Fourth, the refinement of the lower jaw—movable and composed of a single bone—coupled with the development of different kinds of teeth for biting, tearing and chewing made the mammals successful carnivores.

Modern mammals show a great deal of diversity. They vary in size from a shrew hardly more than 2.5 centimeters in length to the enormous blue whale which may measure more than 30 meters in length. Mammals are widely distributed over the earth and are adapted to live in diverse habitats on land (wolves), in water (sea lions), in underground burrows (gophers), in desert burrows (kangaroo rat), in open seas (whales), in the air (bats) and in forests (baboons).

General Characteristics of Mammals

The characteristics that set mammals apart from other animals and made them adaptable to a wide range of habitats are as follows:

1. Mammals have *mammary glands* (from whence the name mammal) that supply the young with milk directly after birth. Newborn mammals are usually quite helpless and depend upon the mother for nourishment.
2. At some time during the life cycle, all mammals have *hair*. Hair is as typical to mammals as feathers are to birds and scales to bony fish and reptiles.
3. Mammals are warm-blooded. Constant body temperature is due, in part, to the four-chambered heart, a device which prevents the mixing of oxygenated and deoxygenated blood. The four-chambered heart first appears in birds. However, in mammals there is another characteristic of blood that contributes mightily to stable body temperature. The red blood cells on maturity lose their nuclei and mitochondria, restricting them to function as oxygen carriers, not as users of oxygen.
4. Most species of mammals have *sweat glands* which provide a secondary means of excreting water and salts.
5. Mammalian teeth have evolved into three different types: incisors for tearing; canines for biting; molars and premolars for grinding.
6. All but a few species have seven vertebrae in the neck. These neck bones are known as *cervical vertebrae*.
7. A muscular diaphragm separates the thoracic cavity (containing the lungs and the heart) from the abdominal cavity (housing part of the digestive system, the reproductive organs and the excretory system).

Major Groups

The class Mammalia is divided into three major groups: Protheria, Metatheria, and Eutheria.

Subclass Prototheria—Monotremes

Fig. 10.17 Duckbill or platypus

The subclass Prototheria consists of a single order, Monotremata. The monotremes are primitive egg-laying mammals. The eggs, large and full of yolk, house the developing monotreme embryos. Examples of the monotremes are the "duckbill" or platypus (*Ornithorhynchus*) indigenous to Australia and Tasmania (Fig. 10.17); the spiny anteater (*Echidna*), also an inhabitant of Australia; and a long snouted anteater (*Proechidna*) indigenous to New Guinea. Modified sweat glands of the anteater secrete a milk substitute which the young lick up from tufts of hair on the mother's belly.

Subclass Metatheria—Marsupials

The subclass Metatheria consists of a single order, *Marsupialia*. The marsupials are primitive mammals that do not have a placenta. The young are about 5 centimeters long at birth and are in an extremely immature condition. At birth they crawl into the mother's pouch or *marsupium*. The rounded mouth is attached to a nipple and the mother expresses milk down the throat of the helpless fetus. As development occurs, the young marsupial is then able to obtain milk by sucking. There are 29 living genera of marsupials, 28 of which live in Australia. The oppossum *Didelphys* is indigenous to North, South and Central Americas and *Caenolestes* inhabits regions of Central America only. Besides the oppossum, other marsupials are the kangaroo, koala bear, Tasmanian wolf, wombat, wallaby and native cat. It is believed that at one time a land bridge connected South America to Australia. With the disappearance of this land link, Australia became geographically isolated, permitting the existence of marsupials that do not have to compete with more advanced mammals.

Subclass Eutheria—Placental Mammals

The subclass Eutheria includes all of the modern mammals that have a placenta. You will recall that in the reptile and the bird, a membranous sac called the allantois serves as an organ of respiration and excretion for the developing embryo. The reptile and bird embryos develop outside the mother's body and are nourished on stored yolk in the egg. The eutherians develop inside of the mother's body in a muscular sac, the *uterus*. The region where the allantois comes in contact with the uterine wall is heavily supplied with capillaries, the smallest blood vessels in the body. At this particular site the *placenta* is formed. The embryo is attached to the placenta by an *umbilical cord* which contains a fetal artery and a fetal vein. By diffusion from the capillaries of the mother, food nutrients and oxygen travel into the capillaries in the embryonic placenta. The fetal artery collects blood from the capillaries of the placenta and conducts it to the capillaries of the embryo. The fetal vein collects the blood of the embryo laden with the wastes of fetal respiration and conducts it away from the embryo. It is important to emphasize that the blood of the mother and the blood of the embryo do not mix. Gasses and nutrients from the maternal circulation pass to the fetal circulation by diffusion across capillary walls.

Table 10.3 provides a quick summary of some of the placental mammals.

TABLE 10.3. Some Placental Mammals

Orders	Characteristics	Examples
Insectivora	Small; nocturnal; burrowing or tree-living; feed on insects; sharp-snouted.	Hedgehog, moles, shrews

TABLE 10.3. Some Placental Mammals (cont.)

Orders	Characteristics	Examples
Dermoptera	Link between the insectivores and the bats.	Flying lemur
Chiroptera	Winged, only mammals able to fly; nocturnal; identify objects by echolocation; according to genus feed on insects, blood, fruit.	Bats
Carnivora	Predators; swift of foot, collar bones reduced, teeth specialized for meat-eating; cerebral development.	Dogs, wolves, coyotes, bears, cats, lions, cheetah, foxes, raccoons, weasels, skunks, etc.
Rodentia	Gnawing mammals; sharp incisor teeth; plant eaters; most numerous of all living mammals.	Hares, squirrels, guinea pigs, rats and mice, beavers, muskrats, porcupines, prairie dogs, woodchucks
Primates	Evolved from tree-dwelling Insectivora ancestors; most species arboreal; teeth unspecialized; marked development of eyes and brain, especially the cerebrum; quadrupeds, but upright sitting posture.	
Lemuroidea	Tree-dwelling; primitive; small, pointed ears; long snout; big toes and thumbs set apart from other digits.	Lemurs
Tarsioidea	Binocular vision; reduced sense of smell.	Tarsiers
Anthropoidea	Enlarged, convoluted cerebrum; well-developed eyes; prehensile tails; broad, flat noses; thumbs reduced; external nostrils close together.	Monkeys, baboons, macaques, gibbons, orangutans, gorillas, chimpanzees, humans

Chronology of Famous Names in Biology

1750   **Rene Antoino deReaumur** (French)—wrote a monumental work in six volumes on the anatomical structure of insects.

1757   **Pierre Lyonet** (Dutch)—presented a brilliant piece of research on the life cycle of the goat-moth caterpillar.

1760   **Petrus Camper** (Dutch)—published anatomical studies on elephant, rhinoceros and the reindeer. Published an excellent paper on the anatomy of the orangutan.

1763   **John Hunter** (English)—introduced "modern" methods in the study of comparative anatomy of vertebrates.

1804   **Alexander Brongniart** (French)—first to classify the amphibia separately from the reptiles.

1830   **Johannes Peter Muller** (Germany)—devoted himself to marine research and made valuable contributions to the study of the evolution of marine forms.

1837   **Karl Ernst von Baer** (Germany)—discovered that animal tissues arise from three primary germ layers.

1876   **Max Furbringer** (Germany)—produced valuable information on the comparative anatomy of the breast, wing, and shoulder of birds.

1957   **James Gray** (England)—carried out research which elucidated the mechanisms by which small fish gain swimming speed.

1965   **Archie Carr** (English)—studied the navigational habits of the green turtle.

1967   **Neal Griffith Smith** (England)—discovered that gulls recognize species mates by visual signals.

1968   **Kjell Johansen** (Swedish)—presented a remarkable study on the evolution of air-breathing fishes.

1969   **Crawford H. Greenewalt** (American)—elucidated the mechanical means by which birds produce song.

1971   **Knut Schmidt-Nielsen** (American)—discovered the pathway that air takes through the lungs of birds.

1977   **T. J. Dauson** (English)—studied the adaptive strategies of kangaroos that enable them to maintain their reproductive potential.

Words for Study

albumen	atriopore	chorion
allantois	atrium	cloaca
amnion	auricle	clutch
ampulla	bipedal locomotion	cranium
ampullae of Lorenzini	carapace	ctenoid
antimeres	chordate	cyclostome

dorsal hollow nerve cord
ductus deferens
endostyle
fry
ganoid
hepatic portal system
homiotherm
lateral line
madreporite
mammary gland
marsupial
metamorphosis
milt
monotreme
neurocoel

nictitating membrane
notochord
operculum
oviduct
oviparous
ovoviviparous
papulae
pharyngeal gill slits
pharynx
placenta
placoid scales
plastron
poikilotherm
proboscis
roe
sea star

sexual dimorphism
spiral valve
swim bladder
syrinx
teleost
tube feet
umbilical cord
uterus
vasa efferentia
ventricle
vertebrae
vestigial
viviparous
Wolffian duct
yolk sac

Questions for Review

PART A. **Completion.** Write in the word that correctly completes each statement.

1. The ..2.. stage of the echinoderms resembles that of the chordates.
2. Spines and plates made of calcium form the echinoderm ..2..
3. Antimeres are the radiating ..3.. of the sea stars.
4. All hemichords live in a ..4.. environment
5. The hemichord proboscis is used to ..5..
6. The most primitive of the vertebrates are the ..6.. fish because of their feeding patterns.
7. Hagfish are best described as ..7..
8. The lamprey has ..8.. paired fins.
9. The amnocoete lives buried in ..9..
10. The type of scales that cover shark's body are ..10..
11. The pressure receptors along the sides of fish are known as ..11.. systems.
12. A living "fossil" fish is the ..12..
13. The teleosts are the ..13.. fish.
14. The larval stage of the frog is the ..14..
15. The first vertebrates to become vocal are the ..15..
16. The adaptation that protects the embryos of terrestrial animals against drying and dessication is the ..16..

17. Dinosaurs are best classified as ..17..
18. The carapace and the plastron are best associated with the ..18..
19. All birds have ..19..
20. An example of a monotreme is the ..20..
21. An immature fish is called a ..21..
22. Egg masses of frogs leave the body cavity through tubes known as ..22..
23. Before mating birds go through ..23.. behavior.
24. The sperm of fish are known as ..24..
25. The marsupial young are born immature because there has been no development of the ..25..

PART B.   Multiple Choice. Circle the letter of the item that correctly completes each statement.

1. A distinctive feature of the phylum Echinodermata is the
   (a) excretory system
   (b) spiny skin
   (c) antimeres
   (d) notochord

2. Sea stars have a functional
   (a) excretory system
   (b) respiratory system
   (c) water vascular system
   (d) head

3. The nonfunctioning organs of the sea star are the
   (a) stomach and anus
   (b) mouth and intestine
   (c) tube feet and stomach
   (d) anus and intestine

4. Water enters the sea star by way of the
   (a) madreporite
   (b) stone canals
   (c) ring canals
   (d) cardiac stomach

5. In the sea star a light-sensitive eyespot is located
   (a) near the madreporite
   (b) adjacent to the stomach
   (c) next to the ampulla
   (d) at the tip of each arm

6. The function of the spiral valve is to
   (a) increase surface area
   (b) digest food
   (c) circulate blood
   (d) store wastes

7. The process in which embryos obtain nourishment from the yolk in the egg is described as
   (a) viviparous
   (b) ovoviviparous
   (c) amnocoete
   (d) osmosis

8. The dogfish is a (an)
   (a) amphibian
   (b) lamprey
   (c) bony fish
   (d) shark

9. Dipnoi is the class of the
   (a) sharks
   (b) skates
   (c) lungfish
   (d) cyclostomes

10. Teleosts are the
    (a) sharks
    (b) lung fish
    (c) lamprey
    (d) bony fish

11. The gill covering of the bony fish is the
    (a) spiral valve
    (b) operculum
    (c) cranium
    (d) chorion

12. It is true that fish
    (a) have well-developed sense organs
    (b) usually are farsighted
    (c) have three semicircular canals
    (d) have mouth-connected olfactory pits

13. The urodeles
    (a) give birth to live young
    (b) have tails
    (c) are usually legless
    (d) represent the reptiles

14. The fetal membrane that functions as an organ of respiration and excretion is the
    (a) amnion
    (b) allantois
    (c) chorion
    (d) shell

15. When an organ or structure "bears a capillary network," it
    (a) has hair
    (b) is divided
    (c) becomes septate
    (d) has blood vessels

16. The reptiles are true land animals because they
    (a) breathe free air
    (b) have strong walking legs
    (c) reproduce on land
    (d) feed in vegetation

17. Bipedal locomotion is a characteristic of
    (a) toads
    (b) turtles
    (c) crocodiles
    (d) birds

18. Hollow bones, numerous air sacs and loss of teeth are adaptations for
    (a) floating
    (b) swimming
    (c) diving
    (d) flying

19. The four chambered heart first appears in
    (a) mammals
    (b) birds
    (c) reptiles
    (d) amphibians

20. Metamorphosis is best associated with
    (a) fish
    (b) rabbits
    (c) frogs
    (d) strawberries

21. Roe and milt are best associated with
    (a) fish
    (b) frogs
    (c) starfish
    (d) jellyfish

22. The number of sperm cells necessary to fertilize an egg is (are)
    (a) one
    (b) two
    (c) three
    (d) four

23. The number of ovaries in a female bird is
    (a) one
    (b) two
    (c) three
    (d) four

24. Egg-laying mammals are known as
    (a) marsupials
    (b) tadpoles
    (c) monotremes
    (d) placentals

25. The production of song in birds is associated with the
    (a) nictitating membrane   (c) gizzard
    (b) syrinx   (d) bill

PART C   Modified True-False. If a statement is true, write "true" for
your answer. If a statement is incorrect, change the underlined word to one that will make the statement true.

1. An important chordate characteristic of the hemichords is the presence of a hollow notochord.
2. The trunk of the hemichord sweeps water into the mouth.
3. The function of the gill slits is to remove excess food from the body.
4. In sea stars, the sexes are separate.
5. Acorn worms belong to the echinoderms.
6. All vertebrates have an endoskeleton made of bone.
7. Vertebrates never have more than 4 sets of paired appendages.
8. A cyclostome refers to a type of organ.
9. Lamprey eels are scavengers.
10. The shark's skeleton is made of bone.
11. The words placoid, dermal and ganoid describe types of fins.
12. The most highly organized of all the fish are the bony fish.
13. Most of the teleosts are ovoviviparous.
14. Metamorphosis of the tadpole is controlled by the gastric gland.
15. Sphenodon is the most advanced of the reptiles.
16. In mammals the allantois evolved into the chorion.
17. Fertilization in most fish takes place internally.
18. A cloaca is a (an) tube.
19. Egg white is the protein gelatin.
20. A group of eggs laid at one time by a bird is a catch.
21. A tadpole has a three-chambered heart.
22. Frog eggs are coated with a cellulose-like substance.
23. In mammals the abdominal cavity houses the lungs and heart.
24. Mammals are warm-blooded animals.
25. Only reptiles and mammals have a four-chambered heart.

Think and Discuss _____

1. What distinguishing characteristics set the chordates apart from lower
   animals?
2. Why is "sea star" a better name than "starfish"?

3. Why is a closed circulatory system more efficient than an open circulatory system?
4. How is the body of a bird structured for flight?
5. Why is the blue whale correctly classified as a mammal, not as a fish?

Answers to Questions for Review

PART A

1. larval
2. exoskeleton
3. arms
4. marine
5. burrow
6. jawless
7. scavengers
8. no
9. mud
10. placoid
11. lateral line
12. coelacanth
13. bony
14. tadpole
15. frogs and toads
16. amnion
17. reptiles
18. turtle
19. feathers
20. duckbill platypus
21. fry
22. oviducts
23. courtship
24. milt
25. placenta

PART B

1. b
2. c
3. d
4. a
5. d
6. a
7. b
8. d
9. c
10. d
11. b
12. c
13. b
14. b
15. d
16. c
17. d
18. d
19. b
20. c
21. a
22. a
23. a
24. c
25. b

PART C

1. gill slits
2. cilia
3. water
4. true
5. hemichords
6. Most
7. two
8. fish
9. predators
10. cartilage
11. scales
12. true
13. oviparous
14. thyroid
15. primitive
16. placenta
17. externally
18. opening
19. albumen
20. clutch
21. two
22. jelly
23. thoracic
24. true
25. birds

CHAPTER 11

HOMO SAPIENS: A SPECIAL VERTEBRATE

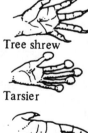

Tree shrew

Tarsier

Orangutan

Human

Fig. 11.1 A comparison of primate hands

Humans are *primates*. They share this mammalian order with gorillas, chimpanzees, monkeys, lemurs, tarsiers, slow lorises and the tailed bush babies. What are the special features that distinguish a primate from other mammals?

Scientists believe that at one time all primates lived in trees. The successful species were those that developed adaptations for arboreal life. Over evolutionary time, species such as chimpanzees, baboons and humans have left the trees and have adapted to life on land. However, certain characteristics persist in the ground-living species that show close relationship to their tree-living "cousins".

Fig. 11.1 illustrates the comparative shapes of hands of some primate species. These hands were adapted for grasping objects such as tree limbs, enabling the animals to locomote by swinging from tree limb to tree limb. Even the feet and tails of some primate species are *prehensile*, adapted for grasping. A second primate characteristic is the well-developed sense of sight. The eyes of most other mammals are located at the sides of the head. The eyes of a primate are directed forward, a structural arrangement that permits *stereoscopic vision*. This means that primates can see in three dimensions (length, width and breadth), a characteristic that enables them to see ahead with clarity the branches that they grasp. Another primate adaptation is the development of a larger and more *convoluted* brain. Convolutions are folds in the brain which increase surface area and allow for a greater number of nerve cells. Mention should also be made of two other primate characteristics. Primates have teeth that are less specialized for tearing and more useful for the grinding and chewing of a varied diet. Primates also have a small number of offspring and provide extended parental care for the young.

A SPECIAL PRIMATE

Though humans are primates, they are set apart from other primates by some unusual and important adaptations. *Bipedalism*, the ability to walk on two legs (instead of four), has freed the forelimbs for doing work. The human can walk great distances and carry things from region to region— things that aid in the causes of hunting, or gathering or building.

Humans can live in the most forbidding of environments and are the most ecologically versatile of all of the primates. The human adjustment to various climates and land surfaces is probably due to the greater brain development. In all probability, brain development is the single greatest factor contributing to human's ability to form words and to speak. Speech is tied to a myriad of intricate brain functions: learning words, associating ideas, remembering. The formulation of the spoken word leads to the creation of the written word. When ideas are written, they are not lost in time. The human being is the only animal that can act in terms of history. Human civilizations function in terms of learned behavior called *culture*.

BASIC HUMAN STRUCTURE

The human body is divided into a *head*, *neck*, *trunk* and two pairs of appendages—namely, the *arms* and *legs*. Hair is present on the head, under the arms, sparsely on the arms and legs, and around the pubic area; in males, there is a greater distribution of body hair, often on the chest and heavily on the back of the hands, on the arms and the legs and on the face. The face is directed forward in a vertical position, the main axis of the body is set vertically, also.

In humans, as in other mammals (birds, too), the coelomic cavity is divided into three different areas. The *pericardial* cavity encloses the heart; *two pleural cavities* contain the lungs; and the *peritoneal* cavity holds the major part of the digestive system, the reproductive system and the urinary system.

The Skeletal System _____

The human skeleton, like that of all vertebrates, is a living endoskeleton that grows with the body. At birth, the human baby has a body that is made up of 270 bones. Due to the fusion of separate bones, the mature skeleton is composed of 206 bones. Fig. 11.2 shows some of the bones that make up the skeleton. Table 11.1 provides a more complete summary.

The human skeleton is a magnificent feat of engineering. The primary purpose of the skeleton is to carry the weight of the body and to support and protect the internal organs. The skeleton must be strong and able to absorb reasonable amounts of shock without fracturing. At the same time, the body framework must be flexible and light enough in weight to permit movement. Skeletal bones move in response to muscles that work like levers, allowing a variety of movements such as walking, running, hopping, sitting, bending, lifting and stooping.

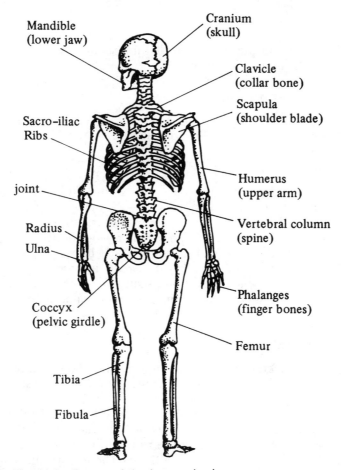

Mandible (lower jaw)

Cranium (skull)

Clavicle (collar bone)

Scapula (shoulder blade)

Sacro-iliac
Ribs

joint

Radius

Ulna

Humerus (upper arm)

Vertebral column (spine)

Phalanges (finger bones)

Coccyx (pelvic girdle)

Femur

Tibia

Fibula

Fig. 11.2 Bones of the human body

TABLE 11.1. The Human Skeleton

Skeleton	Region		Names of Bones
Axial Skeleton	Skull	Cranium	Frontal, parietal (2), temporal (2), occipital, sphenoid, ethmoid
		Face	Nasal, vomer, inferior turbinals, lacrimals, malars, palatines, maxillae, mandible, hyoid
	Vertebral Column		Vertebrae: cervical, thoracic, lumbar, sacral, coccyx
	Thorax		Ribs, sternum
Appendicular Skeleton	Pectoral Girdle		Clavicle, scapula
	Pectoral Appendages		Humerus, radius, ulna, carpals, metacarpals, phalanges
	Pelvic Girdle		Ilium, ischium, pubis (compose the os innominatum)
	Pelvic Appendages		Femur, patella, tibia, fibula, tarsals, metatarsals, phalanges

PARTS OF THE SKELETON

The human skeleton is divided into two major parts: the *axial skeleton* and the *appendicular skeleton.*

Axial Skeleton

The skull, the thorax (rib cage) and the vertebral, or spinal column are the three regions of the axial skeleton.

Skull

All the bones of the head compose the skull. The two regions of the skull—the *cranium* and the face—are made up of 22 flat and irregularly shaped bones. Eight bones form the cranium, which functions in the protection of the brain. The facial region, designed to protect the eyes, nose, mouth and ears, is composed of 14 bones. The *sinuses* are air spaces in the facial bones which aid in reducing the weight of the skull. The bones of the middle ear that function in transmitting sound to the inner ear are the smallest bones in the body—namely, the *hammer, anvil* and *stirrups.*

Vertebral Column

The vertebral column is composed of 26 bones known as vertebrae. At birth the vertebral column consists of 33 bones: seven cervical (neck) vertebrae, twelve thoracic vertebrae, five lumbar vertebrae, five sacral vertebrae, and four coccygeal (tail) vertebrae, but the five sacral bones fuse into one large triangular bone—the *sacrum*—at the back of the pelvis and the four coccygeal bones fuse into a single *coccyx.*

The vertebral column is the backbone made flexible by the cartilage and ligaments that join the individual vertebrae. Such a flexible backbone permits movement of the head and the bending of the trunk. Of major importance is the backbone's function in the protection of the spinal cord, which extends downward from the brain through the opening in each vertebra. Nerves branching from the spinal cord radiate to all parts of the body through openings in the sides of the vertebrae. Fig. 11.3 shows the discs of cartilage that separate the individual vertebrae. These discs prevent friction by the rubbing of the bones and serve as shock absorbers.

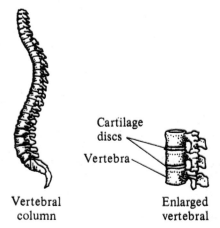

Cartilage discs

Vertebra

Vertebral column

Enlarged vertebral

Fig. 11.3 The human vertebral column

Thorax

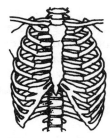

Fig. 11.4 Human rib cage

Just below the neck are 12 pairs of ribs that are attached to the vertebral column. The general shape of the *thoracic basket*, or rib cage, is shown in Fig. 11.4. The first ten pairs of ribs are attached to the breastbone, known also as the *sternum*, by cartilage strips, forming a structure that is smaller on top. The loose connections of the ribs to the vertebral column and the flexible cartilage connections at the sternum allow the ribs to move when the lungs are inflated. The 11th and 12th pair of ribs are often referred to as "floating ribs" because they are attached to the vertebral column but not to the breastbone.

The Appendicular Skeleton

The arms and hands, the legs and feet, and the bones of the shoulder and the pelvis make up the *appendicular skeleton*. You probably realize that "appendicular" is the adjective of the word *appendage*. An appendage is an attachment to a main body or structure. Legs and arms are attachments to the axial skeleton. The sites where arms and legs are attached to the axial skeleton are bones referred to as *girdles*. The *pectoral* (shoulder) girdle where the arms are attached is composed of the *scapula* (shoulder blade), a large triangular bone, and the *clavicle* (collarbone), a smaller curved bone. The legs are attached to the *pelvic girdle*, which is formed by the fusion of three bones: the ilium, the ischium, and the pubis on each side of the midline of the body (Fig. 11.5).

The bones of the legs and arms are appropriately called the *long bones*. The *femur*, the long bone between the hip and the knee is the longest and strongest bone in the body, supporting the weight of the body. The long bones in the lower leg are the thinner *fibula* and the thicker *tibia*.

Between the shoulder blade and the elbow is the bone of the upper arm, the *humerus*, a thinner version of the femur. The two long bones of the lower arm are the *radius* and the *ulna*. Finger and toe bones are known as *phalanges*; the bones of the foot, the *metatarsals*, while *metacarpals* are the bones of the hand.

In general, long bones are shaped like tubes with rounded processes at the ends which are designed to fit into other bones to form *joints*. The ends of the long bones are filled with *spongy bone*, a structural device which makes the bones light in weight but strong. The open spaces in spongy bone are filled with red marrow; while yellow marrow fills the shaft of the long bone.

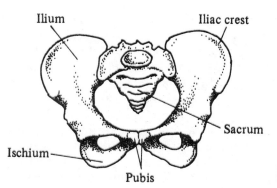

Fig. 11.5 The pelvic girdle

Bone marrow is an important substance. *Red marrow* in the spongy areas of the long bones and in the ribs and vertebrae are the sites ,where red blood cells are produced at the rates of millions per minute. The yellow marrow in the shaft of the long bones contains mostly fat. However, when the blood-making capability of the red marrow is low, yellow marrow is somehow converted into red marrow.

COMPOSITION OF BONE

About twenty percent of living bone is water; the remaining eighty percent consists of mineral matter and protein. Mineral matter deposited in bone usually forms the compound tricalcium phosphate ($Ca_3(PO_4)_2$). Magnesium and other elements may also contribute to the mineral composition. The protein portion of the matrix is made of *collagen fibers* that are found in tendons, skin and connective tissue. The protein fibers and the mineral matter form the nonliving matrix of bone. However, bone also contains a variety of living cells and blood vessels that provide the pathways for nourishment to the cells and permit the removal of respiratory wastes from the cells.

Living bone cells are nestled in small spaces in the mineral matrix of bone. These cells are of three types, each adapted to carry out a specific function that has to do with the building and maintaining of bone. New bone material and the repair of broken bones is the work of *osteoblasts*, which are responsible for secreting the mineral and protein compounds that form the matrix. A second type of bone cell is the *osteoclast*, a bone breaker, able to dissolve bits of bone that are in the way of the efficient design of the skeleton. The destructive work of the osteoclasts is often followed by constructive work of the osteoblasts in the rebuilding of bone. The third type of bone cell, the *osteocyte*, functions as the caretaker of the bone tissue nearby.

The surface of nearly all parts of bone is covered by a tough membrane, the *periosteum*. The periosteum is perforated by microscopic blood vessels that supply the bone cells with nourishment. Although bone matrix appears to be solid, it is pierced by a network of *Haversian canals* through which blood vessels pass. Nerve fibers also extend into the bone interior through Haversian canals. The larger blood vessels pass directly into spongy bone and into the yellow marrow areas of the long bones.

The Muscular System _____

Muscles represent 40 percent of the total weight of the human body. Muscle tissue is characterized by contractility and electrical excitability, two distinctive properties that enable it to effect movement of the body and its parts. Two common disorders of locomotion and/or other movement are described in Table 11.2.

There are three types of muscle tissue: smooth, striated, and cardiac. The movements of smooth and striated muscle tissue are controlled by contractile proteins and innervation from the nervous system; these muscles respond to the electrical stimulation of a nerve impulse. Cardiac muscle, on the other hand, functions to a great degree because of its own inherent ability to generate and conduct electrical impulses.

TABLE 11.2. Two Common Disorders of Locomotion and/or Other Movement

Disorder	Description
Arthritis	Inflammation of the joints and their supporting structures
Tendonitis	Inflammation of the tendon at the point of attachment to the bone

TYPES OF MUSCLE

Smooth Muscle

Smooth muscle is present in the walls of the internal organs, including the digestive tract, reproductive organs, bladder, arteries and veins. Because smooth muscle is contained in organs that do not respond to the will of a person, these muscles are called *involuntary muscles*. The most common function of smooth muscle is to squeeze, exerting pressure on the space inside the tube or organ it surrounds. Food is moved down the esophagus by the squeezing action of smooth muscle. The action of smooth muscles causes urine to be expelled from the bladder, semen to be discharged from the seminal vesicles and blood to be pumped through arteries. The opening and closing of the iris of the eye in response to light is accomplished by the action of smooth muscles.

Smooth muscle is made up of cells packed with contractile proteins, the cells forming sheets of tissue. Smooth muscle tissue is *innervated* by nerve cells and fibers from the sympathetic nervous system, that part of the nervous system that controls the activities of the internal organs. The contraction of smooth muscle is in response to stimulation by nerve cells, neurohumors or hormones.

Striated Muscle

Striated muscle is variously referred to as striped muscle, *voluntary muscle* or *skeletal muscle*—terms describing its structure and function. Located in the legs, arms, back and torso, striated muscles attach to and move the skeleton; since they are moved by the will of the person, they are often termed *voluntary muscles*. A striated or skeletal muscle is made up of a great number of *muscle fibers*, each of which extends the entire length of the muscle. There are probably around six billion fibers in more than 600 muscles scattered throughout the body.

If you look at a bit of striated muscle through the microscope, you will note that the muscle fibers contain many nuclei that seem not to be separated from each other by a plasma membrane. Such an arrangement in which plasma membranes are missing is called a *syncytium*. Each muscle fiber is innervated by at least one motor neuron. A *neuromuscular junction* is the space (synapse) between a nerve cell and a muscle.

Cardiac Muscle

Cardiac muscle is present only in the heart, where the cells form long rows of fibers. Unlike other muscle tissue, cardiac muscle contracts independently of nerve supply since reflex activity and electrical stimuli are contained within the cardiac muscle cells themselves. *Purkinje fibers*, part of the mechanism that controls heartbeat, are so specialized for conducting electrical impulses that they do not have contractile proteins. Each heartbeat is started by self-activating electrical activity of the heart's *pacemaker* known as the *sinoatrial node* (S-A node), positioned in the wall of the right atrium. From the S-A node, the impulse spreads throughout the atrium to the *atrioventricular node* (A-V node), a specialized bundle of cardiac muscle located on the atrium near the ventricles. The impulse spreads from the A-V node to all parts of the ventricles, causing simultaneous contractions in the ventricles.

Cardiac muscle is innervated by the tenth cranial nerve, the *vagus* nerve. The vagus is a mixed nerve containing some nerves that speed up heart beat and others that retard it. However, cardiac muscle contracts independently of nerve supply, and the effect of nerves on heartbeat is not too well understood.

HOW MUSCLES CONTRACT

The fine structure of a muscle fiber controls muscle contraction. A muscle fiber is made up of a bundle of finer fibers called *myofibrils*. A single myofibril is composed of smaller units named *sarcomeres* which

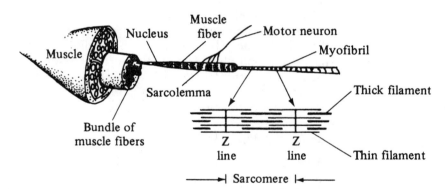

Fig. 11.6   The fine structure of skeletal muscle

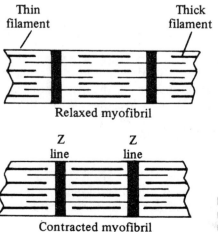

Fig. 11.7   Sliding filament mechanism of muscle contraction

are arranged in single file along the length of the myofibril. Within the sarcomere are alternating rows of thin and thick filaments. The thin filaments are attached to two vertical bands of thick protein called the Z *lines*. Fig. 11.6 shows the fine structure of a muscle myofibril.

Muscles contract due to a sliding filament mechanism. When the thick and thin filaments slide past each other, the Z lines of the sarcomeres are pulled closer together: in effect, contracting. When the sarcomeres contract, the myofibrils contract, which causes the contraction of muscle fibers. Fig. 11.7 illustrates the sliding filament mechanism of muscle contraction.

The Nervous System

The nervous system in humans is made up of two major parts: the central nervous system and the peripheral nervous system. Nervous tissue is specialized to receive stimuli from the outside environment and to conduct impulses to other body tissues. The development of the nervous system and particularly of the brain is what makes humans significantly different from other animals.

The basic unit of function of the nervous system is the *neuron*, or nerve cell. An understanding of the structure and function of this cell and how it transmits nerve impulses is important before dealing with the parts of the nervous system.

NERVE CELLS

The parts of the nerve cell are the *cyton*, or cell body; the *dendrites* and the *axon*. The dendrites receive signals from sense organs or from other nerve cells and transmit them to the cyton. The cell body passes signals to the axon, which then conducts the signals away from the dendrites and cell body. Notice that a nerve cell has only one axon and many dendrites. The axon terminating in *end brushes* (known also as *terminal branches*) is popularly called a *nerve fiber*. Many of the axons in the ver-

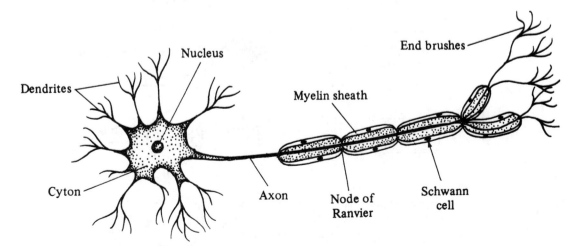

Fig. 11.8 Motor neurons

tebrate body are covered by a fatty *myelin sheath* made of *Schwann cells.* The space between each Schwann cell is known as a *node of Ranvier.* Axons with the myelin sheath transmit impulses more quickly than those without the fatty coverings.

The nervous system has three types of neurons. *Sensory* or *afferent* neurons receive impulses from the sense organs and transmit them to the brain or spinal cord. *Associative* or *interneurons* are located within the brain or spinal cord. These transmit signals from sensory neurons and pass them along to motor neurons. *Motor* or *efferent* (Fig. 11.8) neurons conduct signals away from the brain or spinal cord to muscles or glands, so-called *effector* organs.

THE NATURE OF THE NERVE IMPULSE

The primary function of the nervous system is to permit communication between the external and internal environments. Communication in the nervous system is made possible by signals or impulses carried in a one-way direction along nerve cells. These impulses are electrical and chemical in nature.

When a neuron is not carrying an impulse, it is said to be at *resting potential.* When a nerve cell is stimulated to carry an impulse, its electrical charge changes and it is said to have an *action potential.* Action potentials (nerve impulses) from any one nerve cell are always the same. All impulses are of the same size, there being no graded responses. This is known as the *"all or none response,"* meaning that a nerve cell will transmit an impulse totally or not at all (Fig. 11.9).

The electrical changes that occur in a nerve cell are due to differences in the distribution of certain ions, or charged particles on either side of the nerve membrane. Three conditions are responsible for the distribution of ions. First of these factors is the nature of the cell membrane itself. Remember, it is highly selective. The cell membrane is im-

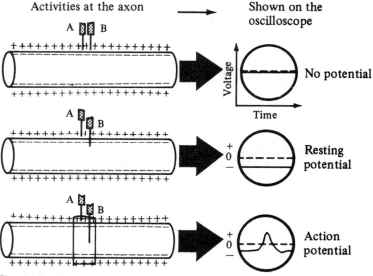

Fig. 11.9 Resting and action potentials

permeable to the sodium ion (Na^+); at the same time, the cell membrane is highly permeable to the potassium ion (K^+). A second factor involves diffusion, the process by which molecules move from an area of greater concentration to one of lesser concentration. A third factor has to do with the attraction of ions of opposite charge and the repulsion of ions of like charge.

Let us begin with the situation in which there is a large concentration of K^+ inside of the cell and a large concentration of Na^+ on the outside of the nerve cell membrane. By diffusion, K^+ ions will cross the plasma membrane to the outside of the cell. The Na^+ cannot follow because the cell membrane is impermeable to them. K^+ will move to the outside until an equilibrium is established. This, in effect, causes the outside of the nerve cell membrane to be more positive than the inside of the membrane. The cell now can be described as being at its resting potential. The cell is also *polarized*.

As an impulse travels through the nerve cell, several events take place. As the impulse touches a given point along the length of the plasma membrane, that site becomes permeable to Na^+ ions and Na^+ crosses the membrane and enters the cell. At the same time, the cell membrane becomes even more permeable to K^+ which then leaks out of the cell at a greater rate. These events lead to the *depolarization* of the cell in which there is a wave-like reversal of electrical charge along the length of the cell membrane. After the impulse is transmitted, the cell uses energy in mechanisms known as the *sodium-potassium pump* and carrier-facilitated transport to bring the cell back to its resting potential.

Axons with myelinated sheaths can conduct impulses at the rate of 200 meters per second. Naked axons may conduct impulses at the rate of a few millimeters per second.

Impulses travel from one neuron to another crossing a specialized gap called the *synapse*. The synapse is a space between nerve cells that measures about 20 nanometers in width—just enough distance to prevent the touching of nerve cells. The terminal branches of the axons have synaptic knobs at their ends. As impulses travel along an axon, the synaptic knobs release chemicals called *neurotransmitters* or *neurohumors*. The neurotransmitters carry the impulse across the synapse onto the dendrites of the receiving nerve cell. The two main neurohumors are *acetylcholine* and *norepinephrine*, each with inhibitory or excitatory capabilities. After an impulse has crossed a synapse, the neurotransmitter is destroyed by an enzyme such as *cholinesterase*. (Many drugs of abuse, including LSD and cocaine, interfere with nerve impulse transmission.)

The endbrushes of the motor neurons are buried in a muscle or a gland. The junction between the nerve fibers and the muscle is known as the *neuromuscular junction*. The neurotransmitter *acetylcholine* carries the impulse from the nerve fiber to the muscle or gland. Nerve transmission across the neuromuscular junction is always excitatory, never inhibitory.

THE CENTRAL NERVOUS SYSTEM

The brain and the spinal cord compose the central nervous system. In the vertebrate body, the organs of the central nervous system are well protected by being wrapped in connective tissue and enclosed in bone.

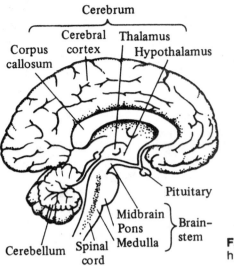

Fig. 11.10. The human brain

The brain, covered by the membranous *meninges*, rests in the skull cavity where it is enclosed by the cranium. The spinal cord, also covered by connective tissue, is circled by the vertebral column.

Brain

The human brain is divided into many parts, each with special functions. Among the most important parts are the *cerebrum, cerebellum,* and *medulla* (Fig. 11.10).

Cerebrum

In mammals the largest part of the brain is the cerebrum. The cerebrum is the seat of intelligence. It controls the voluntary, conscious activities of the human body. All voluntary muscle movements, all speech, thinking and memory are controlled by the cerebrum. It is also the center for control of interpretation of all stimuli that enter the brain through the sense organs.

The cerebrum folds back on itself in many places forming wrinkles known as convolutions. These convolutions increase the surface area of the brain. The cerebrum is divided into two equal halves—the right and left cerebral hemispheres. Each of these hemispheres is subdivided into distinct regions of nervous control by fissures and convolutions. These regions control sensory areas for different parts of the body and are not haphazardly arranged.

The outer layer of the cerebrum is called the *cerebral cortex* and is composed of *gray matter*—unsheathed nerve cell bodies. Under the cortex is the *white matter* made up of sheathed axons. These fibers connect the various parts of the cerebrum and the cerebrum with other parts of the brain.

Cerebellum

The second largest part of the brain is the cerebellum. It is a bulbous mass of nerve tissue, which in human beings, lies underneath the back portion of the cerebrum. In general, the cerebellum controls balance. Its front and back areas regulate muscle tone. A region within the back lobe controls equilibrium; other sections coordinate voluntary movements. The cerebellum controls the precision and coordination of voluntary movements such as walking, running, dancing, skating, writing and typing.

Medulla

The medulla, or medulla oblongata, lies below the cerebrum and connects with the spinal cord. It contains a great number of *ganglia* (cytons) that receive sensory impulses and send out motor signals. Through the medulla pass many of the sensory and all of the motor nerves on their way to or from the higher centers in the brain. The medulla controls automatic, involuntary activities such as the contraction of smooth muscles, reflex movements, dilation and constriction of blood vessels, swallowing, breathing, and the like.

Other Parts

Other important parts of the brain include the thalamus and hypothalamus. The *thalamus* is the region of the brain where integration of sensory information occurs. The *hypothalamus* controls body temperature, osmoregulatory activities, maturity, thirst, hunger and sex drive. The hypothalamus is also the region where the nervous and hormonal systems interact.

Spinal Cord

The spinal cord is an elongated tubular structure containing masses of nerve cells and fibers and lying within the vertebral column. It is composed of a central H-shaped core of gray matter surrounded by white matter. The spinal cord conducts impulses to and from the brain; the impulses enter and leave the spinal cord through spinal nerves which extend from the spinal cord to the other organs of the body. The spinal cord is also the center for simple reflex activity.

In a simple *reflex*, often known as a *reflex arc*, only sensory nerves, the spinal cord, and motor nerves are involved. It allows instantaneous response without involving transmission to and from the brain. An example of a reflex is pulling your hand from a hot stove. When you touch a hot stove sensory nerves pick up the stimulus from receptors in the skin, transmit it to the spinal cord, which signals motor nerves to signal muscles for you to pull your hand away—all instantaneously.

THE PERIPHERAL NERVOUS SYSTEM

The peripheral nervous system connects the central nervous system—the brain and spinal cord—with the other organs of the body. It has two parts—somatic and automatic.

Somatic Peripheral System

The somatic peripheral system is composed of cranial nerves and spinal nerves. The fibers of both sensory and motor neurons are bundled together to form the cranial nerves. Twelve cranial nerves extend between the brain and the sense organs (eyes, ears, nose, etc.), heart, and other internal organs. Thirty-one pairs of mixed sensory and motor nerves extend from the spinal cord to the muscles and organs of the body. Each of the spinal nerves separates into sensory fibers and motor fibers as they join to form the spinal cord. The sensory fibers lead into the dorsal side of the spinal cord. Some of the sensory fibers synapse with the associative (interneuron) neurons; others lead to the brain. The cell bodies of the sensory nerves are in the dorsal root ganglia outside of the spinal cord. The motor nerves lead out from the spinal cord on the ventral side. Their cell bodies are in the spinal cord where they synapse with the associative neurons in the spinal cord.

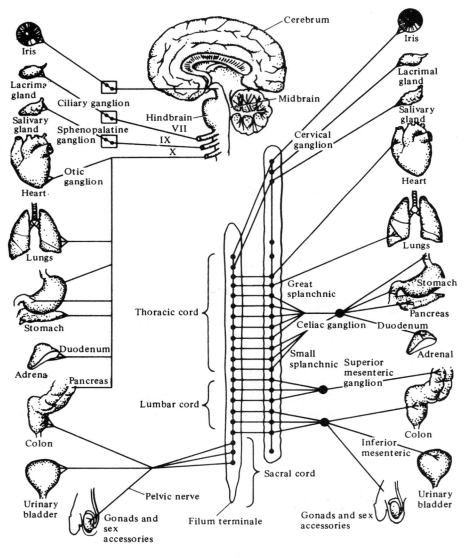

Fig. 11.11 The autonomic nervous system

Parasympathetic system
(craniosacral)

Sympathetic system
(thoracolumbar)

Autonomic Nervous System

A network of nerves known as the autonomic nervous system controls the body's involuntary activities and the smooth muscles of the internal organs, glands and heart muscle. It is composed of motor (efferent) neurons leaving the brain and spinal cord and also of peripheral efferent neurons (Fig. 11.11).

The autonomic nervous system is divided into the *sympathetic system* and the *parasympathetic system*. These subsystems are antagonists. When one set of nerves activates the smooth muscles of the body, the other set inhibits the action. For example: the parasympathetic nerves dilate the blood vessels and slow the heartbeat; the sympathetic nerves constrict the blood vessels and quicken heartbeat.

Fig. 11.12 summarizes the relationships of the parts of the nervous system.

DISORDERS OF THE NERVOUS SYSTEM

Table 11.3 lists and describes four disorders of the nervous system.

TABLE 11.3. Some Disorders of the Nervous System

Disorder	Description
Cerebral Palsy	A form of paralysis denoted by jerky, spastic, writhing movements resulting from damage to the portion of the brain that controls muscles. Cerebral palsy is a group of syndromes with a common result.
Meningitis	Inflammation of the meninges, the membranes that surround the brain and spinal cord
Stroke	A disorder resulting from a hemorrhage in the brain or a blood clot in a cerebral blood vessel which may cause brain damage
Polio	A viral disease that affects the central nervous system and often results in muscle damage to the legs and/or arms

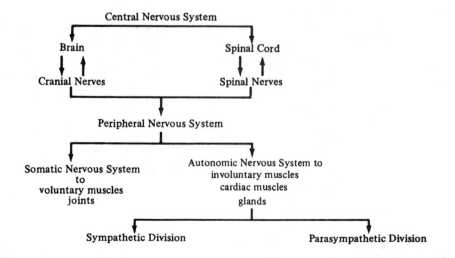

Fig. 11.12 The nervous system the easy way

The Endocrine System

Within the mammalian body there is a constellation of *ductless glands* known as the *endocrine system.* Fig. 11.13 shows the location of these glands in the human body. You will notice that these glands are not grouped together but are distributed throughout the body. Although these glands are not grouped together, they are considered to be a system because of similarities in structure and function.

As the name implies ductless glands do not have ducts and therefore do not discharge their secretions directly into another organ. This is in contrast to most of the glands in the body which are duct glands, delivering their secretions directly into a contiguous or nearby organ. For example, the salivary gland has a duct that delivers saliva directly into the mouth and sweat glands have ducts that conduct perspiration to the skin. *Endocrine glands,* also known as *glands of internal secretion,* deliver their secretions—*hormones*—into the bloodstream, which then carries

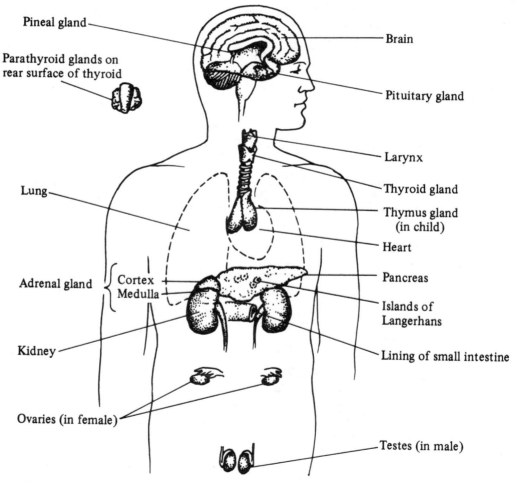

Fig. 11.13 Diagram showing the location of the major endocrine glands: pineal, pituitary, thyroid, parathyroid, thymus, pancreas (part), lining of the small intestine, adrenal glands, and sex glands

them to their target organs. Hormones regulate many of the important metabolic activities of cells and organs.

The endocrine system is made up of the pituitary gland, thyroid gland, parathyroid glands, the adrenal gland, the isles of Langerhans in the pancreas, the thymus gland, the pineal gland, and the gonads—testes in the male and ovaries in the female. Certain secretions of the stomach and small intestine are also hormones and thus part of the endocrine system.

Through their secretions the endocrine glands regulate growth, rate of metabolism, response to stress, blood pressure, muscle contraction, digestion, immune responses, and the development and functioning of the reproductive system. Hormones exert their influence by becoming involved with the genetic machinery of cells and by affecting the metabolic activities of cells working through the cellular respiration pathways.

Prostaglandins are a recently discovered group of hormones that are not produced in any particular gland. Prostaglandins of one kind or another are produced by most tissues in response to other hormones or to irritation of the tissues. One group of prostaglandins are responsible for the pain brought on by inflammatory responses. One specific prostaglandin causes the blood to clot; another type enhances circulation. It seems that some prostaglandins can cause adverse effects in the body, while others prevent them.

The pituitary gland, sometimes called the hypophysis, hangs from the base of the brain and is thought to exert control over much of the functioning of the other endocrine glands. It does this through its *trophic hormones*—hormones that stimulate the activity of other glands. Trophic hormones from the pituitary are known to stimulate secretions of the thyroid gland and of the adrenal gland and to regulate functioning of the sex organs. The pituitary, and in turn the rest of the endocrine system, is itself thought to be controlled by the hypothalamus region of the brain. The hypothalamus releases *neurosecretions*, known as *releasing factors*, which are transmitted to the pituitary where they, in turn, regulate the release of trophic hormones (Fig. 11.14).

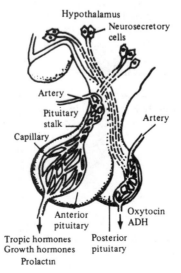

Fig. 11.14 The functional relationship between the pituitary and the hypothalamus

The level of hormones in the blood and the release of hormones from endocrine glands is under a feedback control mechanism, with the blood "feeding" back to the brain information on how much hormone is circulating in the bloodstream. If, for example, the blood level of thyroxin, the thyroid hormone, falls too low, the hypothalamus secretes thyroid-releasing factor which stimulates the pituitary to release thyroid-stimulating hormone (TSH) and in turn the thyroid is stimulated to secrete more thyroxin. As the blood levels of thyroxin rise, the release of TSH and thyroxin is slowed.

Table 11.4 summarizes the endocrine system, listing the major endocrine glands, their hormones and their functions, and the problems associated with excess or diminished secretion.

TABLE 11.4. Endocrines and Their Hormones

Name of Gland	Location	Hormone	Normal Function	Excess Secretion	Diminished Secretion
Anterior pituitary	Base of brain forward portion	Growth hormone (STH)	Affect skeletal growth, protein synthesis, blood glucose concentration	Gigantism acromegaly	Dwarfism
		Trophic hormones TSH ACTH FSH LH	Stimulate target glands thyroid adrenal cortex ovarian follicles; testes gonads	Oversecretion of glands	Undersecretion of glands
Posterior pituitary	Hind portion	Vasopressin	Control of blood pressure; reabsorption of water by kidney tubules	Increased blood pressure; glycogen converted to sugar	Decreased blood pressure; excess sugar changed to fat; kidney tubules not reabsorbing water
		Oxytocin	Contractions of uterus		
Thyroid	2 lobes on either side of larynx	Thyroxin (65% iodine)	Controls rate of oxidation in cells	Increased oxidation; nervous exophthalmic goiter	Lowered oxidation; in a child-cretinism; in an adult myxedemic goiter due to lack of iodine in drinking water
Parathyroid	Four glands above thyroid	Parathyroxin	Regulates amount of calcium in blood	Trembling due to lack of muscular control	Contraction of muscles (tetany); death.

TABLE 11.4.   Endocrines and Their Hormones (Continued)

Name of Gland	Location	Hormone	Normal Function	Excess Secretion	Diminished Secretion
Stomach	Mucous lining (mucoos)	Gastrin	Stimulates secretion of gastric juice	Promotes ulceration of stomach wall	Inhibits gastric digestion
Small intestine	Mucous lining	Secretin	Activates the liver and pancreas to secrete and release their secretions	Excessive pancreatic and liver secretions	Diminished pancreatic and liver secretion
Adrenal medulla	Two glands above kidney	Adrenalin	Controls release of sugar from liver; contraction of arteries; clotting	Increases blood pressure; promotes clotting; releases glycogen; strengthens heart beat	
Adrenal Cortex		Glucocorticoids	Affects normal functioning of gonads; helps maintain normal blood sugar levels		Addison's disease: muscular weakness, darkening of skin, low blood pressure; death
		Mineralo-corticoids	Stimulates kidney tubules to reabsorb sodium		
Pancreas Isles of Langerhans	Embedded in pancreas	Insulin	Regulates storage of glycogen in liver; accelerates oxidation of sugar in cells		Diabetes; unused sugar remains in blood and is excreted with urine
Gonads	Abdominal region	Testosterone (Males) Estrogen Progesterone (Females)	Regulates normal growth and development of sex glands; regulates reproduction; controls sex characteristics	Premature development of gonads; effects on secondary sex characteristics	Interference with normal reproductive functions; diminished growth of sex characteristics
Thymus	Chest region	Thymosin	Stimulates immunological activity of lymphoid tissue		Breakdown of immune system
Pineal	Base of brain	Melatonin	Regulates gonadotropins by anterior pituitary		

The Respiratory System

The process of respiration consists of external breathing and internal respiration. Breathing concerns the intake of air (*inhaling*) and the letting out of carbon dioxide and water vapor (exhaling). Internal respiration (cellular respiration) takes place in cells and is the series of biochemical events by which energy is released from food molecules.

Air is taken in through the nose. The nasal cavity is divided into two pathways by a septum made of cartilage and bone, and the bony *turbinates* increase the tissue surface along the dividing wall. The surfaces of the septum and the walls of the nasal cavities are covered with mucous membranes, some of which are lined with fine hairs. The nasal cavities have several small openings that lead to spaces in the facial bones called *sinuses*. These eight sinuses help to equalize the air pressure in the nasal cavity, reduce the weight of the skull, and contribute to the sound of the voice.

As air passes through the nasal cavity, it is warmed, humidified and filtered for dust particles. Incoming air passes through the nasal cavity into the *pharynx* (throat) through the *larynx* (voice box) and then into the *trachea* (windpipe). The trachea extends from the back of the throat down into the chest. It is held open by a series of C-shaped cartilage rings.

Just behind the middle of the breastbone, the trachea divides into two branches: the left and right *bronchi* (bronchus, sing.). Each bronchus divides and subdivides into smaller tubules called *bronchioles* that ramify throughout the lungs. Each bronchiole ends in a tiny air sac called an *alveolus*. Each alveolus is surrounded by blood capillaries. Oxygen diffuses from the lungs into the bloodstream and is transported to all parts of the body by way of the red blood cells. Conversely, carbon dioxide diffuses out of the blood into the air sacs and makes the reverse trip through the respiratory tubes, finally leaving the body through the nose.

The human body has two lungs. Each of these is enclosed in a double membranous sac known as the *pleural sac*. Not only is this sac airtight, but it also contains a lubricating fluid. The pleural sac and the lubricating fluid prevent friction that might be caused by the rubbing of the lungs against the chest wall.

The respiratory centers in the medulla of the brain and in other brain regions control breathing. The size of the chest cavity is regulated by the *diaphragm*, a flat sheet of muscle that separates the chest cavity from the abdominal cavity. The diaphragm is attached to the breastbone at the front, to the spinal column at the back and to the lower ribs on the sides. When the diaphragm muscle contracts, it is drawn downward creating a partial vacuum in the chest cavity. This causes air to flow through the respiratory tubes into the lungs. When the diaphragm is relaxed, the chest cavity becomes smaller, forcing the air out. The average rate of respiration in humans is about 18 breaths per minute.

Table 11.5 describes five common disorders of the respiratory system.

TABLE 11.5.   Some Disorders of the Respiratory System

Disorder	Description
Pleurisy	Inflammation of the pleural linings of the lungs, caused by an accumulation of fluid between the pleural layers
Bronchitis	Inflammation of the membranes that line the bronchial tubes
Asthma	An allergic response resulting in the constriction of the bronchial tubes
Emphysema	A condition characterized by the loss of elasticity in the muscle fibers of the air sacs in the lungs, causing their enlargement and degeneration. The result is difficulty in breathing, overwork of the heart, and very often death.
Hiccough also spelled "hiccup"	Irritation of the nerves that control the diaphragm, resulting in an irregular intake of air which makes a peculiar noise as the glottis closes down

The Circulatory System

The human circulatory system consists of the heart and the system of blood vessels that transport blood throughout the body.

THE HEART

The human heart lies in the chest cavity behind the breastbone and slightly to the left. The heart is a bundle of cardiac muscles specialized for rhythmic contractions and relaxations known as *heartbeat*. The rate of average heartbeat is 72 times per minute.

Fig. 11.15 shows the external structure of the heart. In size, the heart is about as large as a person's clenched fist. The walls are thicker on one side than on the other. The surface is covered with a number of small arteries and veins; these small arteries are the *coronary arteries* which carry blood laden with oxygen and nutrients to the muscle fibers of the

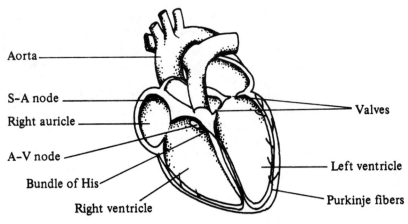

Aorta

S-A node

Right auricle

A-V node

Bundle of His

Right ventricle

Valves

Left ventricle

Purkinje fibers

Fig. 11.15 Structure of the heart

heart. A number of large arteries and veins lead into the top of the heart. These carry blood to and from the other parts of the body.

The inside of the heart is divided into four chambers. The two chambers at the top are the receiving chambers, or the *atria*. The lower chambers, the *ventricles*, are pumping chambers. Each atrium is separated from the ventricle below by a valve. The atrium and the ventricle on the right are separated from the left atrium and ventricle by a thick wall of muscle called the *septum*.

THE PATH OF THE BLOOD

The heart is a double pump. Blood flows from the right atrium into the right ventricle which pumps the blood through the pulmonary artery to the lungs where it receives oxygen and gives up waste products. The right ventricle represents the first pump. Blood returns from the lungs to the heart by way of the *pulmonary vein* and empties into the left atrium. The left atrium then sends the oxygenated blood into the left ventricle which then pumps it out to all parts of the body. The left ventricle represents the second pump.

Blood is oxygenated in the lungs. Deoxygenated blood leaves the heart (pumped by the right ventricle) by way of the pulmonary artery. Oxygenated blood is returned to the left atrium by way of the pulmonary vein. Arteries always carry blood away from the heart; veins carry it to the heart.

Heart valves prevent the backflow of blood. Separating the right atrium from the right ventricle is the *tricuspid valve*, so named because it has three flaps of tissue or cusps. The opening and closing of these cusps is controlled by papillary muscles. A valve with two cusps, the *bicuspid valve*, separates the left atrium from the left ventricle. (The bicuspid valve is also known as the *mitral* valve). Other valves are located where the aorta and the pulmonary arteries join the ventricles.

Several effects of the heart's pumping action can provide information about the condition of the heart and circulation. When a doctor listens to the heart through a stethoscope, he or she hears a series of sounds like "lubb-dub". The "lubb" is caused by the closing shut of the valves between the atria and the ventricles. At the same time the ventricles contract. The "dub" sound occurs when the semilunar valves of the aorta and the pulmonary artery close. The pause between the "dub" sound and the next "lubb" sound is the short time in which the heart rests.

The pulse is caused by the force of the blood on the arteries as the heart beats. The pulse rate, like that of heartbeat, is 72 beats per minute under normal conditions. Strenuous exercise increases the pulse rate.

Blood pressure measurement is a very valuable diagnostic procedure. Each time the ventricle contracts, blood is forced through an artery, increasing blood flow. The contraction phase is known as *systole*: relaxation, *diastole*. Normal systolic pressure is 120; diastolic, 80. This information is written as 120/80.

WORK OF THE BLOOD

Blood consists of a liquid medium called *plasma* and three kinds of blood cells: red blood cells (*erythrocytes*), white blood cells (*leucocytes*) and platelets.

The human body contains about 25 trillion erythrocytes, each one lasting about 120 days. New red cells are produced by the bone marrow at the rate of one million per second. Erythrocytes contain hundreds of molecules of the iron-protein compound *hemoglobin*. In the lungs, oxygen binds loosely to hemoglobin forming the compound oxyhemoglobin. As erythrocytes pass body cells with low oxygen content, oxygen is released from hemoglobin and diffuses into tissue cells. Carbon dioxide combines with another portion of the hemoglobin molecule and is transported to the lungs where it is exhaled.

For each 600 red blood cells there is one white blood cell, numbering in the billions in the blood. There are five types of white blood cells, functioning to protect the body against invading foreign proteins. The amoeboid-like *neutrophils* and *monocytes* behave as phagocytes, engulfing bacteria and other foreign proteins. *Eosinophils* detoxify histamine-like secretions. *Lymphocytes* participate in immune responses, and *basophils* produce anticoagulants.

BLOOD CLOTTING

Platelets are the smallest blood cells. When a capillary is cut, the platelets collect at the site of the injury. There they break into smaller fragments and initiate the complicated chemical process of blood clotting in which more than 15 factors, including thromboplastin, calcium (Ca), and fibrinogen, are involved in the formation of a clot containing blood cells in a fibrin meshwork (Fig. 11.16).

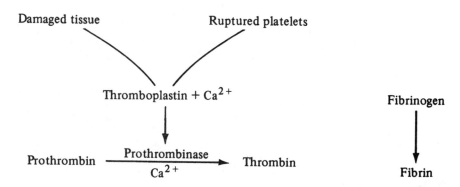

Fig. 11.16   Some steps of the blood clotting process

BLOOD TYPES

The main types of blood are A, B, AB, and O. Transfusions of blood are possible only when the blood types are compatible. If the blood types are not compatible proteins in the plasma will recognize foreign antigens on red blood cells and respond by causing the cells to *agglutinate*, or clump, a condition that causes blockage in small blood vessels. Table 11.6 summarizes the blood proteins involved in blood types.

TABLE 11.6. Proteins of Blood Types

Blood Type	Cell Antigen	Plasma Antibody
A	A	b
B	B	a
AB	AB	none
O	none	a and b

NOTE: Type AB—universal recipient
Type O—universal donor

DISORDERS OF THE CIRCULATORY SYSTEM

Table 11.7 describes some common disorders of the blood circulatory system

TABLE 11.7. Some Disorders of the Blood Circulatory System

Disorder	Description
Cardiovascular disorders	Cardiovascular disorders involve the heart and blood vessels.
Hypertension	High blood pressure indicates that pressure in the arteries is high. High arterial pressure may be caused by hardening of the arterial walls, stress, heredity, improper diet, cigarette smoking, and aging.
Coronary thrombosis	A form of heart attack in which a blockage in the coronary artery or one of its branches results in oxygen debt in the heart muscle
Angina pectoris	A form of heart attack in which narrowing of the coronary arteries deprives the heart muscle of oxygen
Blood disorders	Abnormalities in blood cells
Anemia	Inability of the blood to carry adequate amounts of oxygen because of faulty red blood cells
Leukemia	A disease of the bone marrow in which there is uncontrolled production of nonfunctional white blood cells

The Lymphatic System

Homo sapiens actually has two circulatory systems. One is the blood circulatory system, the other is the lymphatic system. The body cells are bathed with tissue fluid called *lymph.* Lymph comes from the blood plasma, diffusing out of the capillaries into the tissue spaces in the body. Lymph differs from plasma in that it has 50 percent fewer proteins and does not contain red blood cells. Lymph has the important function of bringing nutrients and oxygen to cells and removing from them the waste products of respiration.

Although there is a constant flow of lymph from the blood plasma, neither the blood volume nor its protein content is diminished. This is so because as fast as lymph is drained from the blood plasma, it is returned to the blood. Radiating throughout the body are tiny lymph capillaries which join together to form larger lymph vessels and ducts. Tissue fluids are propelled through the body by differences in capillary pressure, muscle action, intestinal movements and respiratory movements. These movements squeeze the lymph vessels and push the fluid along. Lymph moves only in one direction toward the heart. Lymphatic valves prevent its backflow.

Lymph is returned to the circulating blood through a large lymphatic vessel called the *thoracic duct*, which discharges its contents into the left subclavian vein. The right lymphatic duct empties its contents into the right subclavian vein. These veins then merge into other veins that empty into the heart.

In addition to vessels, the lymphatic system has lumpy masses of cells known as *lymph nodes* distributed throughout the body. These lymph nodes or glands are filtering organs that clear the tissue fluids of bacteria and other foreign particles. Lymph nodes are in the head, face, neck, thoracic region, armpits, groin and pelvic and abdominal regions. These nodes help the body in defense against disease. In addition to filtering out bacteria, they produce lymphocytes and antibodies.

Edema is the swelling that results from inadequate drainage of lymph from the body tissue spaces. Edema is brought about by heart and kidney disorders. malnutrition, injury, or other causes.

The Digestive System

The human digestive system begins with a mouth and ends with an anus, and is often described as a "tube within a tube". Variously called the *gut*, *alimentary canal* or the *gastrointestinal tract*, the digestive system extends from the lower part of the head region through the entire torso (Fig. 11.17).

Essentially, this system carries out five separate jobs that have to do with the processing and distribution of nutrients. First of all, it governs ingestion, or food intake. Secondly, it transports food to organs for temporary storage. Thirdly, it controls the mechanical breakdown of food and its chemical digestion. A fourth function is the absorption of nutrient molecules. Its final piece of work is the temporary storage and then elimination of waste products.

Digestion begins in the mouth. Teeth grind the food while three pairs of *salivary glands* pour salivary juice (saliva) into the mouth. Saliva contains the enzyme *salivary amylase* (ptyalin), which begins the digestion of starch. The moistened, chewed food is swallowed and moves through the throat into the food tube, or *esophagus*. The esophagus has no digestive function but moves the food into the stomach by waves of muscle contractions called *peristalsis*.

Chemical digestion is also known as *hydrolysis*. As the name indicates, hydrolysis is the splitting of large, insoluble molecules into small molecules that are able to dissolve in water. In the digestive system, hydrolysis is regulated by digestive enzymes, as shown in the examples that follow:

$$\text{maltose} + \text{water} \xrightarrow{\text{(maltase)}} \text{glucose} + \text{glucose}$$

$$\text{proteins} + \text{water} \xrightarrow{\text{(protease)}} \text{amino acids}$$

$$\text{lipids} + \text{water} \xrightarrow{\text{(lipase)}} \text{3 fatty acids and 1 glycerol}$$

The *stomach* is the widest organ in the alimentary canal. It stores food while it churns and squeezes it, turning it into the consistency of a thick pea soup. In this semi-liquid form food can be worked upon by enzymes. *Gastric glands* embedded in the walls of the stomach secrete *gastric juice*, a combination of hydrochloric acid, and two enzymes, rennin and pepsin. Rennin is specialized for digesting the protein in milk, pepsin for hydrolyzing several plant and animal proteins. The semi-liquid food, often referred to as a *bolus* or *chyme*, is released a little at a time into the upper part of the small intestine.

Between the stomach and the small intestine is a ring of muscle called the *pyloric sphincter* that closes the stomach off from the *duodenum*, the upper part of the small intestine. In effect, the sphincter muscle regulates the flow of chyme from the stomach into the intestine.

The major work of digestion occurs in the small intestine. Lying outside of the alimentary canal are two important glands that are necessary for the many processes of digestion. The largest of these is the liver. It synthesizes *bile* and stores it in a pouch known as the *gall bladder*. Through bile ducts, bile is released into the small intestine where it serves as an emulsifier of fat, enabling it to be acted upon by the fat-digesting enzyme, lipase. The other accessory gland is the *pancreas*, a dual gland that synthesizes both hormones and enzymes. The pancreas releases pan-

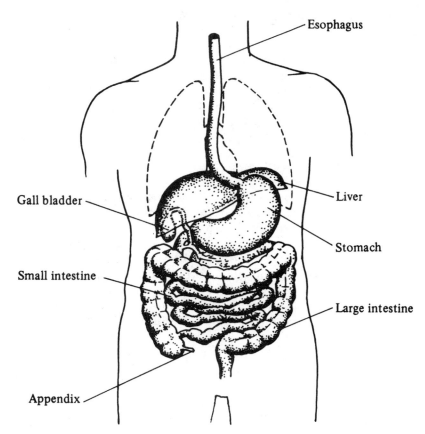

Fig. 11.17 The human digestive system

creatic juice into the small intestine. This digestive juice is a combination of water and several digestive enzymes, each of which is specific for the digestion of fat, carbohydrate or protein.

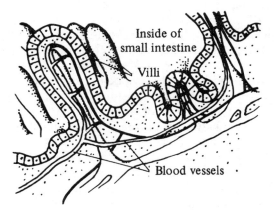

Fig. 11.18 A villus

In the walls of the small intestine are *intestinal glands* which manufacture and secrete intestinal juice, a combination of enzymes that digest starches, sugars and proteins. The outcome of all digestion is that nutrient molecules are reduced to soluble forms that enable them to cross cell membranes. Carbohydrates are digested into glucose or fructose. Proteins are broken down into amino acids. Fats are hydrolyzed into fatty acids and glycerol. These nutrients are absorbed by finger-like *villi* (villus, sing.), an adaptation in the small intestine for increasing surface area (Fig. 11.18).

Digested food diffuses into the capillaries of the villi. Blood then carries the food molecules to the liver through the *portal vein*. In the liver, sugar is removed from the blood and stored as *glycogen*. Digested fat molecules are absorbed into the *lacteals* (lymph vessels) and then enter the bloodstream through the *thoracic duct* which is in the chest cavity.

Food that is not digested passes into the large intestine, also called the *colon*. The large intestine absorbs a great deal of water and dissolved minerals. Undigested food or *feces* is pushed into the rectum where it is stored temporarily until eliminated through the anus.

Some common disorders of the digestive system are described in Table 11.8.

TABLE 11.8.   Some Disorders of the Digestive System

Disorder	Description
Ulcer	A punched-out defect in the wall of the alimentary canal, caused by the digestive action of gastric juice
Constipation	Difficult passage of stools from the large intestine, caused by excessive absorption of water
Diarrhea	Passage of frequent and watery stools, associated with decreased water absorption and possibly resulting in dehydration
Appendicitis	Acute inflammation of the appendix
Diverticulosis	Grapelike outpocketings on the colon wall, which may become infected and obstruct the bowel
Gallstones	An aggregation of hardened bile salts, cholesterol, and calcium in the gall bladder

The Excretory System

In human beings, the lungs, the skin and the urinary system work to expel the wastes produced in metabolic activities. The lungs excrete carbon dioxide and water. The skin expels water and salts from the sweat glands and a small amount of oil from the sebaceous glands. The urinary system handles the major work of excretion.

The human urinary system is located dorsally in the abdomen. Fig. 11.19 shows the organs that make up the urinary system and their locations. This system consists of two *kidneys*, tubes known as *ureters* extending from each kidney to a *urinary bladder* and a single *urethra*, a tube that leads out of the bladder.

The unit of structure and function of the kidney is the *nephron*. There are about one million of these microscopic units in each kidney. They actively remove waste products from the blood and return water, glucose, sodium ions and chloride ions to the blood. Fig. 11.20 shows the structure of a nephron.

The nephron is made up of several structures. The first of these is a knot of capillaries called the *glomerulus*. The glomerulus fits into a second portion—the *Bowman's capsule*, a cup-shaped cellular structure that leads into the third part, the kidney tubule. There are four main parts of each kidney tubule: the *proximal convoluted tubule*, the *loop of Henle*, the *distal convoluted tubule* and the *collecting duct*.

The *renal artery* transports blood into the kidney. The artery divides into smaller vessels called *arterioles* which divide into smaller blood vessels called capillaries. It is this network of capillaries that is called the glomerulus. The pressure of the blood in the glomerulus is quite high and forces fluid from the blood through the walls of the capillaries into

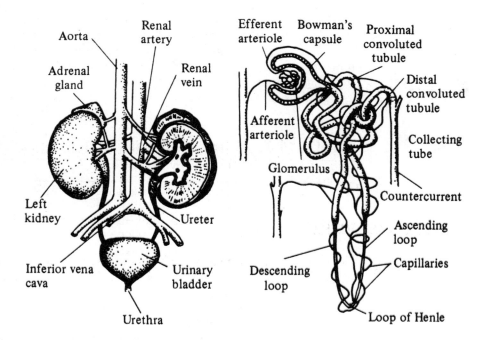

Fig. 11.19 The urinary system

Fig. 11.20 The kidney nephron

the hollow cup of the Bowman's capsule. This process is known as *pressure filtration*. The fluid entering the nephron has the same composition of the blood except that it lacks blood cells, large protein molecules and lipids, which are not able to cross the membranes of the cells that compose the capillary walls.

As the filtrate (the water filtered out of the blood) moves into the proximal convoluted tubule, much of the water is reabsorbed back into the blood. Glucose and sodium are reabsorbed by active transport. As the water that is left in the tubule moves down into the *loop of Henle*, chloride ions and sodium are reabsorbed into the bloodstream by active transport. The remaining water and urea move into the distal convoluted tubule where additional sodium ions are reabsorbed into the bloodstream by active transport. The remaining water and urea waste pass into the collecting duct. From there it passes through the ureters to the bladder. This *urine* is now stored in the bladder until released from the body through the urethra.

The actions of the posterior pituitary hormone vasopressin and the steroid hormone aldosterone from the adrenal cortex regulate absorption in the kidney.

Table 11.9 describes three disorders of the urinary system.

TABLE 11.9. Some Disorders of the Urinary System

Disorder	Description
Kidney stones	An accumulation of mineral salts that precipitate out from the urine and form stones in the kidneys and other parts of the urinary tract
Nephritis	Inflammation of the kidney, often caused by toxins from bacteria that infect other parts of the body
Gout	A disease denoted by the accumulation of the chalky salts of uric acid in joints and at the ends of bones, and caused by malfunction of the kidney in secreting uric acid

Sense Organs

The human body has five major senses—sight, hearing, taste, smell, and touch—that provide information about the external environment and transmit the stimuli to sensory nerves and ultimately to the brain for processing.

THE EYE AND VISION

The human *eyeball* measures about 2.5 centimeters in diameter. Most of the eyeball rests in the bony eyesocket of the skull. Only about one sixth of the eye is exposed. External structures associated with the eye are eyelids and lashes and eyebrows. A delicate protective membrane, the *conjunctiva*, covers the eye. Three pairs of small muscles attach the eye to the eyesocket. Secretions from tear glands help to keep the eye moist.

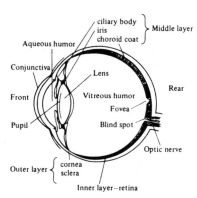

Fig. 11.21 Structure of the human eye

Fig. 11.21 shows the structure of the human eye. Notice that the eyeball is divided into two chambers that are separated by a *lens*. The front chamber contains a clear fluid called the *aqueous humor*. The back chamber contains a transparent jelly-like material called the *vitreous humor*.

The lens is transparent and is made of a great many layers of protein fibers. It measures about 8 millimeters in diameter. The function of the lens is to focus light on the *retina* at the back of the eyeball. The shape of the lens changes; it flattens when focusing on distant objects and thickens when focusing on near objects. The ability to bring objects into focus although they are located at different distances is called *accommodation*.

The colored portion of the front of the eye is the *iris*; in its center is a hole called the *pupil*. Light enters the eye through the pupil and passes through the cornea, the aqueous humor, the lens and the vitreous humor. Light reaches the *retina* where a barrage of signals are set up and are conducted to the *optic nerve* which carries these signals to the visual portions of the brain. The lens turns the image upside down and reverses it from left to right. The visual centers in the brain correct the inversions and reversals of the lens to make the image right side up.

Buried in the retina are cells called *rods* and *cones*. There are about 7 million cones and 120 million rods. The rods help the eye to accommodate in dim light and aid in night vision. The cones are responsible for color vision.

THE EAR AND HEARING

The human ear is made up of three divisions: the *outer ear*, the *middle ear* and the *inner ear*. Fig. 11.22 shows the structure of the ear. The outer ear catches sound waves and transports them to the *eardrum*, a membrane that stretches across the outer canal separating it from the middle ear. Sound waves cause the eardrum to vibrate.

The middle ear is a small cavity that is filled with air. It lies inside of the skull bone between the outer ear and the inner ear. At the bottom of the middle ear is an opening that leads into a canal. This canal is called the *Eustachian tube*, a passageway that connects the middle ear to the throat. This tube equalizes the air pressure in the ear with that of the throat. This is accomplished by yawning or swallowing.

The middle ear contains three very small bones called the *hammer*, the *anvil* and the *stirrup*. These are the smallest bones in the body. These

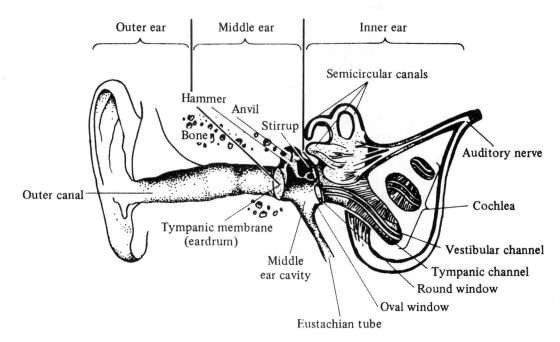

Fig. 11.22 Structure of the human ear

bones accept the vibrations from the eardrum and transmit them to the oval window, one of two small membrane-covered openings between the middle ear and the inner ear.

The inner ear, which is entirely encased in bone, has a fluid-filled structure called the *cochlea*, so named because it resembles a snail in shape. The cochlea has numerous canals that are lined with hair cells. The vibrations from the oval window are transmitted to the hair cells in the cochlea and thence on to the *auditory nerve*, which conducts the vibrations to the brain. In the brain, these signals are interpreted into sounds.

THE OTHER MAJOR SENSES

The organs of smell are located in the mucous membranes of the upper part of the nasal cavities. Special *olfactory cells* respond to odors and pass the impuse along the *olfactory nerve* to the brain. The sense of smell is far more important in lower animals than it is in humans.

The organs of taste are found chiefly on the tongue. These *taste buds*, as they are called, distinguish basic qualities such as bitter, sweet, sour, and salty.

The organs of touch are located on the skin surface and they respond to temperature, pain and pressure.

Reproduction

In this section we will discuss the human reproductive system and, very briefly, the development of the human young.

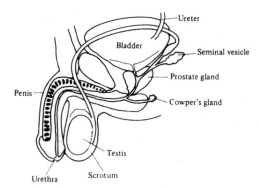

Fig. 11.23   Male reproductive system

MALE REPRODUCTIVE SYSTEM

In the male reproductive system some organs are located outside of the body and others are positioned internally. Look at Fig. 11.23 and locate the scrotum. It is a sac-like organ that is located outside of the body. The *scrotum* contains the testes, glands that produce sperm and the male hormone *testosterone*. Also positioned outside of the body is the *penis*, the organ that delivers the sperm into the body of the female.

Each testis contains thousands of very small tubes called *seminiferous tubules*. Within these tubules, the sperm cells are manufactured. At the top of and behind each testis are coiled tubules called the *epididymis*. Each epididymis tubule functions as a storage place for sperm and also serves as a pathway which carries the sperm to a duct called a *vas deferens*. In its travels to the vas deferens, the sperm pass the *seminal vesicles* where they obtain nutrients. From the vas deferens, the sperm are conducted to the urethra, a single tube that extends from the bladder through the penis. Sperm cells leave the body through the penis. Erection of the penis and ejaculation of *semen* (a mixture of sperm and secretions from the seminal vesicle, the prostate gland and Cowper's gland) are processes necessary for the placement of sperm in the female reproductive tract.

FEMALE REPRODUCTIVE SYSTEM

The female reproductive system serves three important functions: the production of egg cells, the disintegration of nonfertilized egg cells, and the protection of the developing embryo. The reproductive system has specialized organs to carry out these functions.

Two oval-shaped ovaries lie one on each side of the midline of the body in the lower region of the abdomen. On a monthly alternating basis each ovary produces a mature egg. Eggs are located in spaces in the ovary called *follicles*. As an egg matures, it bursts out of the ovarian follicle and is released into the appropriate branch of the *fallopian tube*, a tube that leads from the region of the ovary to the uterus (Fig. 11.24).

If the egg is not fertilized, it is discharged from the body in a process called *menstruation*. The vascularized lining of the uterus, known as the *endometrium*, distintegrates in response to decreased levels of estrogen and progesterone in the blood. Menstrual bleeding lasts four to seven days.

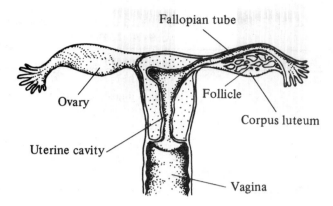

Fig. 11.24 Female reproductive system

If the egg is fertilized, it becomes implanted in the uterus where it goes through a series of cell divisions known as *cleavage*.

CLEAVAGE

The fertilized egg goes through a series of cell divisions in which there is no growth in size of the zygote nor separation of the cells. Fig. 11.25 shows the stages of cleavage.

The first division of cleavage results in the formation of two cells; the second, four. Succeeding divisions result in eight cells, then 16, thirty two, and so forth, until a solid ball of cells called a *morula* is formed. Cells in the morula migrate to the periphery and the solid ball of cells changes to a hollow ball of cells called a *blastula*.

Within a very short time, one side of the blastula pushes inward, forming what resembles a double-walled cup. This stage of cleavage is known as the *gastrula*. During the gastrula stage three distinct layers of cells—the *ectoderm* (the outer layer), the mesoderm (the middle layer), and the endoderm (the inner layer)—are formed. These layers, known

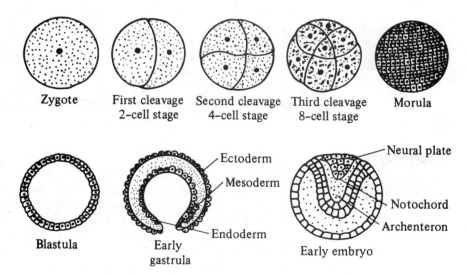

Fig. 11.25 Stages of cleavage

TABLE 11.10. Differentiation of the Three Primary Germ Layers

Ectoderm	Endoderm	Mesoderm
skin	lining of lungs	muscles
nervous system	lining of digestive system	skeleton
sense organs	pancreas	heart
	liver	blood vessels
	respiratory system	blood
		ovaries, testes
		kidneys

as *primary germ layers*, develop into the tissues and organs of the body through a process known as *differentiation*. An *embryo* is now forming. Table 11.10 summarizes the development of the primary germ layers.

DEVELOPMENT OF THE HUMAN EMBRYO

As a result of cleavage of the fertilized egg and its implantation in the uterus many changes occur in the body. The follicle from which the egg cell bursts becomes filled with some yellowish glandular material and is now known as the *corpus luteum*. The *corpus luteum* acts as an endocrine gland, secreting *progesterone* which prevents any other eggs in the ovary from developing further. The menstrual cycle is halted. No more eggs are discharged for the duration of the pregnancy.

The embryo produces several membranes that do not form any part of the new baby but which are necessary to the development and well being of the embryo. One of these membranes is the *amnion*, a water-filled sac that completely surrounds and protects the embryo. The water absorbs shocks and prevents friction that might damage the embryo.

The implanted embryo is attached to the uterus by means of the *umbilical cord*, a structure that contains blood vessels that function in carrying nutrients and oxygen to the embryo and transporting wastes away from the embryo. The umbilical cord connects with the *placenta*, a vascularized organ made up of tissues of the mother and tissues of the embryo. The blood of the embryo that circulates in the capillaries of the placenta is separated from the blood of the mother by layers of cells thin enough to allow diffusion between the two circulatory systems. There is no mixing of the blood of the mother with the blood of the embryo. (The term *fetus* is used when the embryo takes human form).

PARTURITION

The birth process is known as *parturition*. In humans the period of gestation (period of embryo development) is about 9 months or 40 weeks. At the end of that time, the uterus begins to contract in a process called *labor* to expel the baby. The onset of uterine contractions is probably caused by the release of oxytocin into the bloodstream by the posterior pituitary. The human newborn passes through the neck of the uterus (cervix) head first and then through the vagina to the outside.

MULTIPLE BIRTHS

Although humans usually produce only one offspring at a time, sometimes two, three, or even more young may be born at the same time. Of these multiple births, twins are the most common. There are two types of twins: identical and fraternal. Identical twins result from the fertilization of one egg and have the same genetic makeup. They are of the same sex and are almost identical in appearance. They develop in a common chorionic sac and share a common placenta. However, the umbilical cords are separate.

Fraternal twins develop from two separate fertilized eggs. They do not share a common genetic makeup and are no more alike than siblings born at separate times. The sexes may be different. Each fraternal twin has its own chorionic membrane and its own placenta.

Human Beings and Race

All humans belong to the species *Homo sapiens*. This means that the genetic material of all people is so similar that all humans can interbreed and produce fertile offspring. The human species is really a group of interbreeding populations. Populations that have adapted to certain environments become genetically different based on the frequency with which certain genes appear. Skin color, hair texture, body build and facial bone structure are a few of the characteristics that identify human population groups known as *races*.

Although we can make broad generalizations about the identifying characteristics of racial groups, not every member of each group fits these specifications. A set of physical characteristics can be drawn up that will fit individuals of several different races. Therefore, it is difficult for biologists and anthropologists to agree on the number of human races. Modern anthropologists divide *Homo sapiens* into three major stocks: Caucasoid, Mongoloid and African. Each of these groupings is subdivided into several human populations distinguishable by certain pronounced characteristics. The Caucasoid stock is composed of four white races: Nordic, Alpine, Mediterranean, and Hindu. The Asian stock is divided into the Malaysian, American Indian, and the Mongolian populations. The African stock is separated into Black, Melanesian, Pygmy Black, and Bushman populations. Of doubtful grouping are the Polynesian and the Australoid peoples. Many anthropologists classify these groups as Asian; others disagree. No matter the grouping, the differences among human races are very small.

Chronology of Famous Names in Biology

1637 **William Harvey** (English)—discovered how blood circulates in the body.

1759 **Philibert Gueneau de Montbeillard** (French)—made the first systematic measurements of the growth of a child.

1771   **John Hunter** (English)—dissected the body of the Irish giant, Charles Byrne, and discovered a much enlarged pituitary gland.

1790   **Luigi Galvani** (Italian)—discovered that electrical stimuli can cause muscles to contract.

1830   **Thomas Addison** (English)—discovered the effect of damaged adrenals on health, causing mottled skin, weight loss, anorexia, irritability.

1830   **Jan Evangelista Purkinje** (Czech)—discovered the sweat glands in the skin and the fiber network in cardiac muscle; demonstrated the importance of fingerprints.

1849   **Arnold Adolph Gerthold** (Germany)—discovered the endocrine function of testes.

1853   **Thomas Curling** (English)—first reported myxedema.

1855   **Claude Bernard** (French)—discovered the glycogenic function of the liver.

1860   **Anders Retzius** (Swedish)—devised and named the cephalic index: maximum length to maximum breadth of the skull.

1885   **Paul Langerhans** (German)—discovered the islets in the pancreas.

1886   **Pierre Marie** (French)—discovered that an oversecretion of growth hormone causes acromegaly.

1873   **Camillo Golgi** (Italian)—devised a method of impregnating metallic salts into nerve cells to better determine their structure. He discovered the dendrites and axons of nerve cells.

1882   **Richard Owen** (English)—discovered the existence of parathyroid glands by dissecting an Indian rhinoceros.

1904   **Ernest Starling** and **William Bayliss** (English)—devised the term *hormone*.

1914   **Edward C. Kendall** (American)—isolated thyroxin from cattle thyroid.

1915   **David Marine** (American)—discovered the link between iodine and goiter.

1921   **Frederick Barting** and **Charles Best** (Candian)—isolated insulin.

1925   **David Marine** (American)—associated iodine deficiency with the high incidence of goiter in Cleveland.

1929   **Edward Doisy** (American)—isolated the ovarian hormone estrone.

1930   **Karl Landsteiner** (American)—discovered human blood groups.

1942   **Charles Drew** (American)—devised a more effective method for preserving blood for transfusions.

1944	Herbert Evans (American)—discovered growth hormone.
1944	Choh Hao Li (American)—isolated growth hormone and other pituitary secretions.
1948	Otto Loewi (French)—discovered the presence and actions of neurohumors.
1950	Andrew F. Huxley and R. Niedegarde (English)—found that two lines of skeletal muscle fibers move close together when a muscle contracts.
1950	Gregory Pincus (American)—developed a steroid to suppress ovulation from the roots of a wild Mexican yam.
1955	Alan L. Hodgkin and Andrew F. Huxley (English)—won Nobel Prize for showing how a difference of potential contributes to the functioning of a neuron. They studied the giant axons of the squid.
1958	Hans Selye (Canadian)—discovered that stress affects the endocrine system.
1959	Morris Goodman (American)—prepared animal serums with antibodies which measure the degree of relationship among the various species of primates.
1965	Hugh E. Huxley and Jean Hanson (English)—developed the sliding filament theory of muscle contraction.

Words for Study

accommodation	cardiac muscle	diastole
action potential	cerebellum	differentiation
afferent neuron	cerebrum	duodenum
agglutinate	chyme	eardrum
alimentary canal	clavicle	edema
alveolus	cleavage	effector organ
antibody	coccyx	endocrine gland
appendage	cochlea	eosinophil
atrioventricular node	cones	epididymis
atrium	convolution	erythrocyte
axon	coronary artery	Eustachian tube
basophil	corpus luteum	femur
biscuspid valve	cranium	fetus
bipedalism	culture	fibula
blastula	cyton	follicle
bronchiole	dendrite	gastrula
bronchus	diaphragm	girdle

glomerulus	optic nerve	sinoatrial node
gray matter	osteoblast	sinus
Haversian canals	osteoclast	skull
hormone	osteocyte	smooth muscle
humerus	oval window	sodium-potassium pump
hydrolysis	parturition	stereoscopic vision
hypophysis	pericardial	sternum
hypothalamus	periosteum	striated muscle
interneuron	peritoneal	synapse
involuntary muscle	phalanges	syncytium
lacteals	platelet	systole
lens	pleural sac	thalamus
leucocyte	prehensile	thoracic duct
lymph	primate	thorax
lymphocyte	prostaglandins	tibia
medulla	Purkinje fibers	tricuspid valve
meninges	pyloric sphincter	trophic hormone
menstruation	race	ulna
metacarpal	radius	umbilical cord
metatarsal	reflex	ureter
mitral valve	releasing factor	urethra
monocyte	resting potential	vagus nerve
morula	retina	vas deferens
myofibril	rods	ventricle
nephron	round window	vertebrae
neuromuscular junction	sacrum	vertebral column
neuron	sarcomere	villus
neurotransmitter	scapula	voluntary muscle
neutrophil	Schwann cell	white matter
node of Ranvier	scrotum	
olfactory nerve	semen	

Questions for Review

PART A.   **Completion.** Write in the word that correctly completes each statement.

1. Humans belong to the species named ..1..
2. The hands of primates are adapted for ..2..
3. The ability to walk on two legs is known as ..3..
4. Human patterns of learned behavior are known collectively as ..4..

5. The human pericardial cavity encloses the ..5..

6. One function of the facial sinuses is to reduce the ..6.. of the skull.

7. A common name for the thoracic basket is the ..7..

8. Bones that attach arms and legs to the axial skeleton are known as ..8..

9. Cells that build bone matrix are called ..9..

10. Cells that dissolve the mineral matrix of bone are the ..10..

11. The function of smooth muscle is to ..11.. organs it surrounds.

12. The type of muscle that moves the skeleton is called voluntary or ..12.. muscle.

13. The space between a nerve cell and a muscle is the ..13.. junction.

14. Two distinctive properties of muscle are electrical excitability and ..14..

15. The S-A node acts as the ..15.. of the heart.

16. Myofibrils are correctly associated with ..16.. cells.

17. Sensory information is integrated in that brain region known as the ..17..

18. The nature of a nerve impulse is both electrical and ..18..

19. Balance and coordination are controlled by the part of the brain known as the ..19..

20. The peripheral nervous system is divided into the somatic and ..20.. systems.

21. A distinctive structural characteristic of endocrine glands is that they do not have ..21..

22. The homeostatic mechanism that regulates the pituitary gland and its target organs is known as a ..22.. system.

23. The endocrine glands that cap the kidneys are the ..23..

24. The contraction of a ventricle in the heart is known as ..24..

25. The force of the blood on the arteries is known as ..25..

26. The largest lymphatic in the body is the ..26..

27. The unit of structure and function in the kidney is the ..27..

28. The part of the eye that bends light rays is the ..28..

29. The ..29.. in the eye are responsible for color vision.

30. Pressure in the middle ear is regulated by the ..30.. tube.

31. The solid ball of cells formed during cleavage is known as the ..31..

32. The corpus luteum behaves as a ..32.. gland.

33. The birth process is known as ..33..

34. Food moves through the digestive tract by waves of muscle contraction known as ..34..

35. The ..35.. closes the stomach from the duodenum.

36. Bile is stored in the ..36..

37. The large intestine is also called the ..37..

38. Tubes leading from the kidney to the urinary bladder are called ..38..

39. The ..39.. nerve carries signals from the eye to the brain.

40. Semen is a mixture of sperm and secretions from the seminal vesicle, Cowper's gland and the ..40.. gland.

PART B. **Multiple Choice.** Circle the *letter* of the item that correctly completes each statement.

1. The eutheria are
   (a) flying birds
   (b) infertile frogs
   (c) female kangaroos
   (d) placental mammals

2. A prehensile tail is one that is adapted for
   (a) swimming
   (b) grasping
   (c) gliding
   (d) balancing

3. Convolutions are best associated with the
   (a) pharynx
   (b) lung
   (c) brain
   (d) nerve cord

4. The number of bones in the mature human skeleton is
   (a) 206
   (b) 270
   (c) 290
   (d) 305

5. The major function of the backbone is to
   (a) protect the spinal cord
   (b) bend the trunk
   (c) to hold the neck and head
   (d) to support the legs

6. The strongest bone in the body is the
   (a) clavicle
   (b) sternum
   (c) femur
   (d) metatarsal

7. Which is a true statement about bone?
   (a) The matrix is composed of living material.
   (b) Yellow marrow is in the spongy areas of the long bones.
   (c) Bone has a variety of living cells.
   (d) The ends of long bones are made of solid bone.

8. Bone is covered by a membrane known as the
   (a) peritoneum
   (b) pericardium
   (c) perithecium
   (d) periosteum

9. Which of the following are three names for voluntary muscle:
   (a) striped, cardiac, smooth
   (b) striated, skeletal, smooth
   (c) skeletal, cardiac, striated
   (d) striated, striped, skeletal

10. A syncytium refers to
    (a) a group of cells without plasma membrane boundaries
    (b) a segment of smooth muscle that rings the trachea
    (c) a group of nerve cells that stimulates muscle
    (d) a single muscle fiber when stimulated by a neuron

11. Which of the following is a true statement about Purkinje fibers?
    (a) They are found in smooth muscle.
    (b) They lack contractile proteins.
    (c) They form a syncytium with skeletal muscle.
    (d) They cannot conduct electrical impulses.

12. If nerves to the heart are severed, the heart will
    (a) cease functioning
    (b) beat faster
    (c) beat regularly
    (d) beat erratically

13. Impulses enter nerve cells by way of the
    (a) dendrites
    (b) cytons
    (c) axons
    (d) terminal branches

14. Myelin is a nerve cell covering composed of
    (a) $CaCO_3$ (c) fat
    (b) collagen   (d) carbohydrate

15. The fact that a nerve cell transmits an impulse totally or not at all
    is a type of response known as the
    (a) take it or leave it   (c) make it and take it
    (b) all or none   (d) all or some

16. The width of a synapse is approximately
    (a) 80 nm   (c) 40 nm
    (b) 60 nm   (d) 20 nm

17. The part of the brain that is necessary for life is the
    (a) cerebrum   (c) fissure of Rolando
    (b) cerebellum   (d) medulla

18. The sympathetic and parasympathetic systems function
    (a) antagonistically to each other
    (b) to reinforce the activities of each other
    (c) without relationship to the autonomic system
    (d) to counteract the control of the cerebrum

19. The hypothalamus is a part of the brain where
    (a) integration of sensory information occurs
    (b) integration of memory and higher thought processes occur
    (c) follicle stimulating hormone is released into the blood stream
    (d) the nervous and hormonal systems interact

20. Blood in the pulmonary artery
    (a) lacks oxygen
    (b) lacks carbon dioxide
    (c) contains nitrogen
    (d) contains all three gases

21. The most important function of erythrocytes is to
    (a) carry nutrients from cell to cell
    (b) protect the body against disease
    (c) carry oxygen to all cells
    (d) remove carbon from all cells

22. The lymph nodes are glands that
    (a) secrete hormones and neurohumors
    (b) propel tissue fluids through the body
    (c) control the production of red blood cells
    (d) filter bacteria from the tissue fluids

23. Lymphocytes are white blood cells that
    (a) phagocytize bacteria
    (b) detoxify histamines
    (c) produce anticoagulants
    (d) participate in immune responses

24. Breathing is controlled by the
    (a) diaphragm
    (b) respiratory centers in the brain
    (c) level of carbon dioxide in the blood
    (d) all three of the above

25. The lungs are enclosed in a set of double membranes known as the
    (a) pericardium
    (b) periosteum
    (c) pleural sac
    (d) peritoneum

26. In human beings the organs of excretion are the
    (a) kidneys, lungs, rectum
    (b) large intestine, sweat glands, lungs
    (c) kidneys, lungs, sweat glands
    (d) sweat glands, kidneys, anus

27. The function of the urinary bladder is to
    (a) store urine
    (b) detoxify urea
    (c) add CO_2 to ammonia
    (d) filter out glucose

28. Blood is transported to the kidney from the dorsal aorta by the
    (a) renal vein
    (b) renal artery
    (c) arterioles
    (d) glomerulus

29. The cup-shaped portion of the nephron is the
    (a) loop of Henle
    (b) glomerulus
    (c) Bowman's capsule
    (d) convoluted tubules

30. The widest organ in the alimentary canal is the
    (a) stomach
    (b) large intestine
    (c) colon
    (d) gall bladder

31. Two glands lying outside the alimentary canal but important to digestion are the
    (a) liver and kidney
    (b) pancreas and thoracic duct
    (c) liver and pancreas
    (d) liver and colon

32. The major work of digestion occurs in the
    (a) stomach
    (b) small intestine
    (c) large intestine
    (d) esophagus

33. Digestion begins in the
    (a) stomach
    (b) small intestine
    (c) esophagus
    (d) mouth

34. In the human male, sperm is stored in tubules known as
    (a) sperm ducts
    (b) Cowper's gland
    (c) vas deferens
    (d) epididymis

35. Follicles are sites where
    (a) fertilization occurs
    (b) chromosomes are halved
    (c) ova are produced
    (d) sperm receive nutrients

36. The primary germ layer that gives rise to the blood vessels is the
    (a) endoderm
    (b) mesoderm
    (c) ectoderm
    (d) protoderm

37. The membranous sac of water that surrounds the developing human fetus is the
    (a) amnion
    (b) chorion
    (c) allantois
    (d) placenta

38. Identical twins are produced from
    (a) two eggs fertilized by two sperm
    (b) two eggs fertilized by one sperm
    (c) one egg fertilized by one sperm
    (d) one egg fertilized by two sperm

39. The implanted embryo is attached to the uterus by means of the
(a) placenta   (c) yolk stalk
(b) umbilical cord   (d) allantois

40. Modern anthropologists divide *Homo sapiens* into three major stocks:
(a) Caucasoid, Mongoloid, African
(b) Caucasoid, Bushman, Melanesian
(c) Caucasoid, Hindu, Mongolian
(d) Caucasoid, Nordic, Pygmy Black

PART C.   Modified True-False. If a statement is true, write "true" for your answer. If a statement is incorrect, change the *underlined* word to one that will make the statement true.

  1. Scientists believe that at one time all primates lived in savannas.

  2. Due to stereoscopic vision, primates can see in two dimensions.

  3. The primate with the best developed brain is the chimpanzee.

  4. The hammer, anvil, and stirrups are sinuses in the middle ear.

  5. The humerus is the long bone in the lower leg.

  6. Red blood cells are produced in the red marrow of long bones.

  7. "Bone breakers" are cells known as osteoblasts.

  8. Striated muscle is called involuntary muscle.

  9. The sympathetic nervous system innervates striated muscle.

 10. The type of muscle that makes up the heart is smooth muscle.

 11. Z lines are bands of fat.

 12. The unit of function and structure of the nervous system is the cyton.

 13. Schwann cells are best associated with the myofibril sheath.

 14. Nervous impulses are conducted by efferent neurons to the brain or spinal cord.

 15. Motor neurons conduct impulses to the effectors.

 16. Interneurons are located outside of the brain and spinal cord.

 17. The reversal of polarity on a nerve cell is known as the resting potential.

 18. The brain and the vertebral column make up the central nervous system.

 19. The membranes that cover the brain are known as the phalanges.

 20. The parasympathetic and sympathetic systems are divisions of the somatic nervous system.

 21. Glands of internal secretion release chemicals collectively known as enzymes.

 22. Chemicals known as releasing factors are secreted by the anterior pituitary.

 23. The presence of estrogen in the blood signals the uterus to secrete luteinizing hormone.

24. The coronary arteries are positioned <u>inside</u> of the heart.
25. The liquid part of the blood is called <u>serum</u>.
26. The life of a red blood cell lasts for <u>30</u> days.
27. White blood cells function to protect the body against invasion by foreign <u>worms</u>.
28. Platelets initiate the <u>clumping</u> of blood.
29. Another name for the voice box is the <u>pharynx</u>.
30. Each bronchiole ends in an air sac known as a <u>syrinx</u>.
31. Rennin digests protein in <u>milk</u>.
32. The <u>gall bladder</u> synthesizes bile.
33. The gall bladder <u>stores</u> bile.
34. Digested fat molecules are absorbed into <u>villi</u> and then enter the bloodstream through the thoracic duct.
35. Vibrations are carried to the brain by the <u>optic</u> nerve.
36. The back chamber of the eye is filled with <u>aqueous</u> humor.
37. The three small bones in the middle ear are the anvil, <u>club</u>, and stirrup.
38. Each <u>identical</u> twin has its own chorionic membrane and its own placenta.
39. The human female has <u>one</u> ovary (ovaries).
40. In the <u>male</u> reproductive system some organs are located outside the body and some are positioned internally.

Think and Discuss _____

1. Why are human beings classified as primates?
2. How does the skeletal structure enable the human being to stand upright?
3. Explain what is meant by the "all or none response" of the nerve cell.
4. Why is the pituitary gland considered to be the master gland of the body?

Answers to Questions for Review

PART A

1. *Homo sapiens*
2. grasping
3. bipedalism
4. culture
5. heart
6. weight
7. rib cage
8. girdles
9. osteoblasts
10. osteoclasts

11. squeeze
12. striated or striped
13. neuromuscular
14. contractility
15. pacemaker
16. muscle
17. thalamus
18. chemical
19. cerebellum
20. autonomic
21. ducts
22. feedback
23. adrenals
24. systole
25. pulse

26. thoracic duct
27. nephron
28. lens
29. cones
30. Eustachian
31. morula
32. ductless or endocrine
33. parturition
34. peristalsis
35. pyloric sphincter
36. gall bladder
37. colon
38. ureters
39. optic
40. prostate

PART B

1. d
2. b
3. c
4. a
5. a
6. c
7. c
8. d
9. d
10. a
11. b
12. c
13. a
14. c

15. b
16. d
17. d
18. a
19. d
20. a
21. c
22. d
23. d
24. d
25. c
26. c
27. a

28. b
29. c
30. a
31. c
32. b
33. d
34. d
35. c
36. b
37. a
38. c
39. b
40. a

PART C

1. trees
2. three
3. human
4. bones
5. upper arm
6. true
7. osteoclasts
8. voluntary
9. smooth
10. cardiac
11. protein
12. neuron
13. myelin
14. afferent
15. true
16. inside
17. action
18. spinal cord
19. meninges
20. autonomic

21. hormones
22. hypothalamus
23. anterior pituitary
24. outside
25. plasma
26. 120
27. proteins
28. clotting
29. larynx
30. alveolus
31. true
32. liver
33. true
34. lacteals
35. auditory
36. vitreous
37. hammer
38. fraternal
39. two
40. true

CHAPTER 12

NUTRITION

Nutrition is the totality of methods by which an organism satisfies the energy, fuel and regulatory needs of its body cells. Those substances that contribute to the nutritional needs of cells are the *nutrients*. Animals take these nutrients into the body by the *ingestion* of food. Food, therefore, refers to edible materials that supply the body nutrients. Nutrients needed in large amounts are classified as *macronutrients*: carbohydrates, proteins and fats. *Micronutrients*—vitamins and minerals—are needed in smaller amounts. Vitamins are organic compounds; minerals are inorganic. *Malnutrition* results from the improper intake of nutrients. This may be due to a person's eating too little food or to the intake of too much of one nutrient and not enough of others.

Nutritionists urge the eating of a *balanced diet*. A balanced diet is a good mixed diet that includes choices from the four major groups of food: the milk group, the meat group, the vegetable and fruit group and the breads and cereals group. Three or four servings from each of these groups each day will ensure a nutritionally useful diet.

Macronutrients

CARBOHYDRATES

Carbohydrates, which include all starches and sugars, contain the elements carbon, hydrogen, and oxygen. Usually, for every two atoms of hydrogen there is one atom of oxygen; thus hydrogen and oxygen are present in the ratio of 2:1. All starches and sugars are built from the basic unit $C_6H_{12}O_6$, that is, glucose, a *monosaccharide* or single sugar. Examples of other monosaccharides are fructose (fruit sugar) and ribose (found in nucleic acids). Fig. 12.1 shows the structural formula for glucose.

When two molecules of single sugars are combined chemically, a double sugar or *disaccharide* is formed. Examples of disaccharides are maltose and sucrose (table sugar). When many single sugar units are joined chemically, *polysaccharides* or complex sugars are formed. Examples of complex sugars are starches, cellulose and glycogen.

When two molecules of a monosaccharide are joined together, a molecule of water is given off. When four molecules of single sugars are

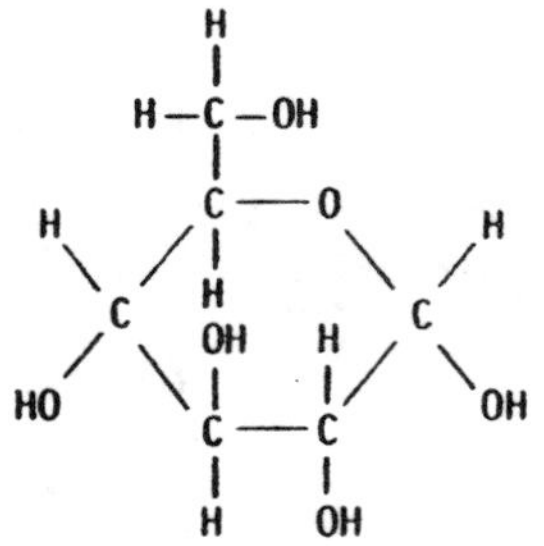

Fig. 12.1 Structural formula for the monosaccharide glucose

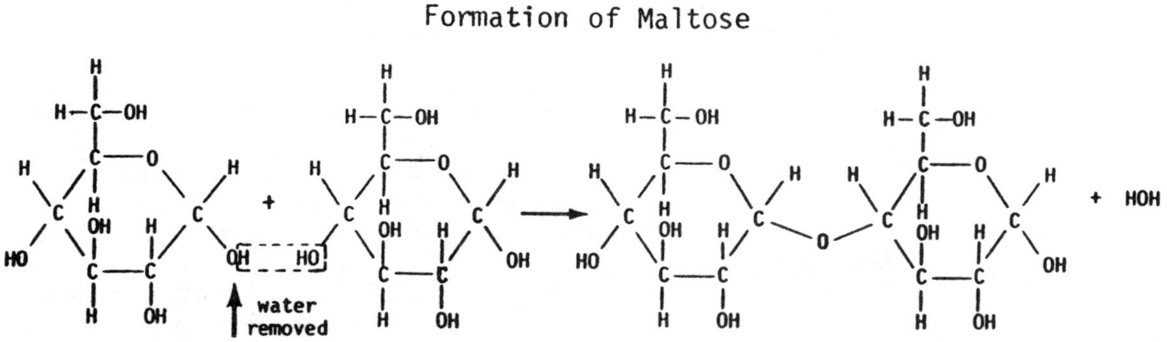

Formation of Maltose

Fig. 12.2 Dehydration synthesis, in which two single sugars combine to form a double sugar

joined chemically, two molecules of water are formed. The formation of water during synthesis (building up of compounds) is called *dehydration synthesis*. Dehydration synthesis makes possible the close "packaging" of complex molecules. Fig. 12.2 shows the formation of maltose.

The primary function of carbohydrates in the diet is to serve as fuel for the body cells. All carbohydrates must be broken down into glucose or fructose by digestion before they can be used by cells. If body cells receive more simple sugar than they can use as energy, some of the excess sugar is stored in the liver and muscles as *glycogen*, commonly called *animal starch*. However, if carbohydrates are taken into the body in much larger quantities than the body needs, they are converted into fat and stored under the skin and around body organs.

Carbohydrates help form the structures of some important biological compounds, including parts of the cell membrane, and they assist the body in the manufacture of biotin and other of the B-complex vitamins.

Food sources of carbohydrates include potatoes, fruits, vegetables, cereal grains, beans, peas, sugar cane, beets, milk, baked goods and pasta.

PROTEINS

All proteins contain the elements carbon, hydrogen, oxygen, and nitrogen. In addition to these elements, some proteins contain sulfur. A protein molecule is constructed from building block units known as *amino acids*. Fig. 12.3 shows a generalized structural formula of an amino acid. Notice the parts of an amino acid. The *amino group* at the left end of

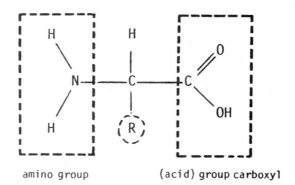

Fig. 12.3 Generalized structure of an amino acid. ® represents a variable group and is the basis for the variety of amino acids.

the molecule contains the elements hydrogen and nitrogen in the ratio of 2:1. At the right end of the amino acid molecule is a carboxyl or acid group, which contains the elements carbon and oxygen and an hydroxyl (OH) group. The center of the molecule contains carbon, hydrogen, and an "R" group, that is, a variable group, thereby allowing for variety among the amino acids.

Twenty amino acids are found in living cells; these are called essential amino acids. However, some proteins are composed of special amino acids, usually formed by a change in a common amino acid, that supplement the basic set of twenty essential amino acids.

Like complex carbohydrates, proteins are formed by dehydration synthesis when two or more amino acids are joined chemically. *Dipeptides* are formed when two amino acids are joined together (Fig. 12.4). *Polypeptides* are formed by the dehydration synthesis of a number of amino acid groups. A protein is usually made up of one or more polypeptides.

The number of different proteins is enormous. Variety in proteins is made possible by variability in structure. Although proteins have a basic structural similarity, each kind is different because of the number, types, and order of amino acids that compose it.

Proteins are the most abundant of the organic compounds in body cells. They compose all of the fibrous structures in the body, including hair, nails, ligaments, the microfilaments in cells, and the myofibrils of muscles. They also form part of hemoglobin, certain hormones such as insulin, and thousands of enzymes that control biochemical processes of cells. Proteins are assimilated into protoplasm and are vital to the formation of DNA molecules. Proteins are also necessary in forming antibodies, molecules that constitute an important part of the immune system

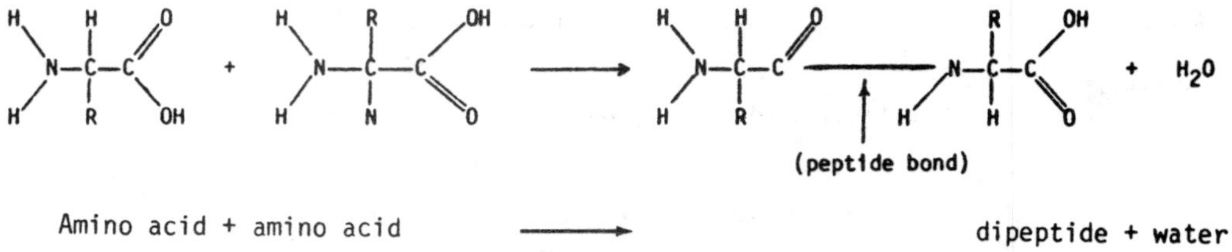

Amino acid + amino acid $\longrightarrow$ dipeptide + water

Fig. 12.4 Formation of a dipeptide

which functions to ward off disease, and in regulating the water balance and acid-base balance in the body.

Food sources of protein include meat, fish, eggs, milk, cheese, and soy, lima, and pea beans. A serious protein *deficiency disease* is *kwashiorkor*. This disease, which threatens the lives of many children in Africa, causes misshapen heads, barrel chests, bloated stomachs, spindly legs and arms, decreased mental abilities, and poor vision.

LIPIDS (FATS)

Lipids include the fats and oils and are a major group of biological compounds. Like the carbohydrates, lipids contain the elements carbon, hydrogen, and oxygen. (Some lipids also contain phosphorus and nitrogen.) However, in lipids the ratio of hydrogen to oxygen is much greater than 2:1 and varies from one lipid to another.

Some lipids are formed during the process of dehydration synthesis. One molecule of glycerol is chemically joined to three molecules of *fatty acid* with the accompanying loss of water (Fig. 12.5).

Fats, like carbohydrates, are fuel foods, supplying the cells with energy. Certain fats are essential to the structure and function of body cells, to the building of cell membranes, and to the synthesis of certain hormones. Fats also aid in the transport of fat-soluble vitamins. Foods rich in fats include butter, bacon, egg yolk, cream, and certain cheeses.

The energy potential of food is measured in Calories. The Calorie is the unit of heat necessary to raise 1 liter of water 1 degree Celsius. (The large Calorie is always written with a capital C.) Fats are concentrated sources of energy. One gram of fat provides nine Calories, while one gram of protein or carbohydrate provides only four Calories. Therefore, foods rich in fats add to the caloric content of the human diet. Some body fat is necessary to cushion body organs and to prevent heat loss through the body's surface. Excessive fat intake causes overweight.

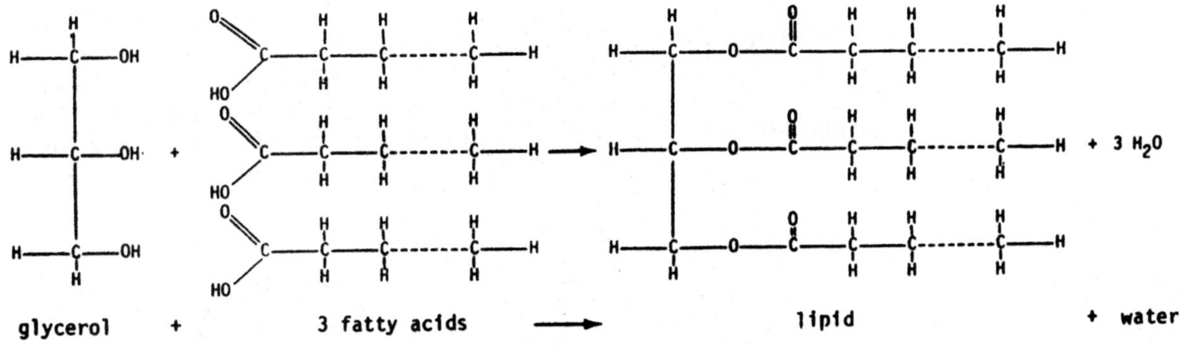

glycerol + 3 fatty acids ⟶ lipid + water

Fig. 12.5 Formation of a lipid

FIBER

The *fiber* in the human diet comes only from plant sources. Fiber is not a nutrient, but it is important in the diet to stimulate the normal action of the intestines in the elimination of wastes. Fiber absorbs many times its weight in water and aids in the formation of softer stools. It also provides bulk, which promotes regularity and more frequent elimination.

Currently, it is suggested that dietary fiber may contribute protection against many noninfectious diseases of the large intestine, such as cancer of the colon, hemorrhoids, appendicitis, colitis and diverticulosis. Incidences of these diseases seem to be much lower in countries where the diets are high in fiber. It is also believed that increased dietary fiber reduces blood cholesterol levels and helps to prevent the formation of fatty deposits on the inner walls of the arteries.

Raw fruits and vegetables, whole cereals and bread, and fruits with seeds (strawberries, figs, raspberries) are excellent sources of fiber.

Micronutrients

VITAMINS

The study of micronutrients and their effects on the body was begun in 1906 by Dr. Frederick Gowland Hopkins, a physiology professor at Cambridge University, England. Dr. Hopkins did not isolate these microfactors in food, but he was able to demonstrate serious effects of deficiency diets on white laboratory rats. These illnesses had been noted many times over in human populations. In 1921, Casimir Funk, a Polish scientist, attached the name "vitamine" to these micro food substances which, when missing, caused human illness and body disorders.

Vitamins are organic compounds. They are classified as *water soluble* or *fat soluble*. In general, the water soluble vitamins are coenzymes necessary to the proper sequence of biochemical events that occur during cellular respiration. It is interesting to note that the primates (*Homo sapiens* included) and guinea pigs are the only vertebrate animals that cannot synthesize their own vitamin C from carbohydrates. Therefore, the daily requirements of ascorbic acid must be met through food intake. The functions of the fat soluble vitamins are not clearly understood.

Table 12.1 reviews the major vitamins, their functions and food sources, and the symptoms that commonly result from a deficiency.

MINERALS

Minerals are inorganic compounds. Some such as calcium and sodium are needed in relatively large amounts. Calcium, together with phosphorus, is used in building bones and teeth. Calcium is also a regulator of muscle activity. Nerve cells could not carry impulses nor could muscles contract without the assistance of sodium and potassium. Sodium also functions in the regulation of body temperature, since large amounts of its salts are excreted by the sweat glands. Other minerals are needed by the body in only small amounts. These are known as the *trace* minerals.

TABLE 12.1. Vitamins and Their Uses

Vitamin	Necessary For	Deficiency Symptoms	Food Sources
A-Retinol (fat soluble)	Healthy visual pigments in eye; healthy skin membranes	Dryness of membranes; poor growth; night blindness; inflamed eyelids	Fish liver oil, butter, cream, milk, margarine, brightly colored fruits and vegetables, leafy vegetables
C-Ascorbic Acid[1] (water soluble)	Intercellular cement for teeth and bones; healthy capillary walls; resistance to infection	Sore gums, tendency to bruise easily, painful joints, loss of weight—all these symptoms are associated with scurvy.	Citrus fruits, tomatoes, cabbage, green leafy vegetables, green peppers
D-Calciferol[2] (fat soluble)	Regulates calcium and phosphorus metabolism and growth, building strong bones and teeth	Soft bones, poor tooth development and dental decay—rickets	Fish liver oil, irradiated feed, egg yolk, salmon
E-Tocopherol	Prevention of oxidation by red blood cells; muscle tone	Hemolysis of red blood cells	Wheat germ, green leafy vegetables
K-Menadione	Synthesis of prothrombin, clotting of blood	Prolonged bleeding from wound	Green vegetables, tomatoes
All the vitamins below belong to vitamin B complex:			
Thiamin and Niacin[3]	Growth; healthy digestion; normal nerve function; good appetite; carbohydrate metabolism	Poor digestion; depression; nerve disorders; loss of appetite	Yeast, wheat germ, liver, enriched foods, bread, green vegetables
Riboflavin (water soluble)	Growth; health of skin and mouth; functioning of eyes; carbohydrate metabolism	Retarded growth; sores at corner of mouth; disturbances of vision; inflammation of the tongue	Same as for thiamin and niacin; meat

1. Vitamin C is not stored by body, oxidizes rapidly and is readily destroyed by exposure to air.
2. Humans make own vitamin D when skin is exposed to sunlight.
3. Prolonged deficiency of thiamin results in beri-beri, a nerve disease that may result in paralysis. Prolonged deficiency of niacin results in pellagra, a disease characterized by a rash, graying and falling out of hair, depression, loss of weight and digestive disturbance.

TABLE 12.2.   Minerals and Their Uses

Mineral	Necessary For	Deficiency Symptoms	Food Sources
Magnesium	Healthy bones and teeth; involved in protein metabolism	Weakening of bones and teeth; faulty metabolism	Green leafy vegetables
Sodium	Functioning of sodium-potassium pump; regulates water balance in cells; regulates nerve impulse; maintains acid-base balance of tissue fluids and blood	Leads to cardiovascular diseases and disorders of the nervous system	Salt
Iron	Synthesis of hemoglobin, myoglobin and the cytochromes	Anemia, difficulties in cellular respiration	Liver, red meats, egg yolk, whole grain cereals
Iodine	Synthesis of thyroxin	Goiter, sluggish metabolism	Marine fish, iodized salts
Fluorine	Aids in resistance to tooth decay	Breakdown of tooth enamel	Water treatment
Calcium	Building of bones and teeth; muscle contraction; nerve impulse transmission; permeability of cell membrane; activation of ATP enzymes	Loss of minerals from bone, anemia, nerve and muscle disorders	Milk and dairy products, eggs, whole grain cereals, green leafy vegetables
Potassium	Functioning of sodium-potassium pump; regulation of nerve impulse; muscle function; glycogen formation; protein synthesis	Nerve and muscle disorders; irregular heart beat	Beans and peas, fruits, vegetables
Phosphorus	Building of bones and teeth; phosphorylation of glucose; building of ATP molecules; functions in cellular respiration; present in nucleic acids	Malfunctions of basic cell processes	Milk and dairy products

In general, the functions of trace minerals are regulatory in that they enable the enzymes of metabolism to work.

Table 12.2 reviews the minerals, their function and major food sources, and the symptoms that commonly result from their lack.

Chronology of Famous Names in Biology

1890	**Theodore Palm** (English)—wrote a treatise on the absence of bone deformity (rickets) in poor Japanese children.
1897	**Christian Eijkman** (Indian)—discovered that chickens and people kept on a diet of polished rice developed the disease beriberi.
1898	**J. Lind** (English)—discovered that a diet containing citrus fruits prevents scurvy.
1906	**Frederick G. Hopkins** (English)—applied research methods in inducing and curing deficiency diseases in rats.
1919	**Kurt Huldschinsky** (German)—discovered that ultraviolet radiation and the hormone calciferol can cure rickets.
1920	**E. V. McCollum** and **Lafayette Mendel** (American)—identified the first vitamin and named it A.
1920	**Joseph Goldberger** (American)—discovered that pellagra is caused by a vitamin deficiency.
1921	**Casimir Funk** (Polish)—invented the name "vitamine" to describe a special group of nutrients.
1922	**E. V. McCollum** (American)—discovered vitamin D.
1924	**Harry Steenbock** and **A. F. Hess** (American)—discovered that irradiation of milk increases the vitamin D content.
1950	**D. M. Hadjimarkos** (American)—researched the roles of micronutrients such as selenium, vanadium and barium in the formation of dental caries.
1963	**Marcel El Conrad** (American)—completed original research which elucidated the pathways of iron metabolism in the human body.
1964	**Reginald R. W. Townley** (Tasmanian)—researched carbohydrate malabsorptive disorders in children.
1965	**David Paige** (American)—published a major study on lactose intolerant populations showing lactose enzyme deficiency as the genetic cause for milk intolerance.
1968	**Jana Parizkova** (Czech)—completed research on problems of obesity which included studies of fat deposition and metabolism.

Words for Study

amino acid
balanced diet
deficiency disease
dehydration synthesis
dipeptide
disaccharide
fatty acid

fiber
glycogen
kwashiorkor
lipid
macronutrient
malnutrition
micronutrient

monosaccharide
nutrient
nutrition
polypeptide
polysaccharide
roughage
vitamin

Questions for Review

PART A.   Completion. Write in the word that correctly completes each statement.

1. The molecules in food that are used by cells for energy, growth and repair of tissue are ..1..

2. The three major types of macronutrients are carbohydrates, proteins and ..2..

3. Fibrous structures such as hair and nails, enzymes, and hemoglobin all contain the macronutrient ..3..

4. The energy potential of food is measured in ..4..

5. Fiber stimulates the action of the ..5..

6. Some fatty acids are used to synthesize chemical messengers known as ..6..

7. Vitamins and minerals belong to a class of substances known as ..7..

8. Muscle activity is regulated by the mineral ..8..

9. In general the water soluble vitamins are necessary for the events of cellular ..9..

10. Minerals necessary only in small amounts are known as ..10.. minerals.

11. Carbohydrates include all sugars and ..11..

12. A disaccharide is a ..12.. sugar.

13. Huge starch molecules composed of many units of simple sugar are known as ..13..

14. Glucose is composed of the elements carbon, oxygen, and ..14..

15. The number of carbon atoms in a molecule of glucose is ..15..

PART B. Multiple Choice. Circle the letter of the item that correctly completes each statement.

1. Of the following pairs, the one in which both items are micronutrients is
   (a) calcium and ascorbic acid
   (b) protein and sucrose
   (c) maltose and calcium
   (d) glycerol and copper

2. Another name for animal starch is
   (a) glycerol
   (b) glycogen
   (c) glucose
   (d) glucagon

3. Two protein molecules are
   (a) levulose and vitamin E
   (b) copper and mannose
   (c) nitrogen and starch
   (d) insulin and antibodies

4. The role of fats in the body is
   (a) functional but not structural
   (b) structural but not functional
   (c) functional and structural
   (d) only structural

5. Fiber in the diet
   (a) causes diverticulosis
   (b) prevents infectious diseases
   (c) increases blood cholesterol
   (d) absorbs large amounts of water

6. The name Casimir Funk is most correctly associated with
   (a) the discovery of rickets
   (b) inventing the name "vitamine"
   (c) finding the cause of beriberi
   (d) inventing the term "roughage"

7. The mineral sodium plays a role in all of the following except
   (a) regulation of body temperature
   (b) regulation of water balance
   (c) transmission of nerve impulses
   (d) synthesis of hemoglobin

8. Prolonged bleeding from a wound can result from a deficiency of
   (a) vitamin A
   (b) vitamin D
   (c) vitamin K
   (d) vitamin E

9. Vitamins belonging to the B complex group include
   (a) tocopherol and niacin
   (b) thiamine and niacin
   (c) ascorbic acid and menadione
   (d) calciferol and thiamine

10. Goiter is caused by a deficiency of
    (a) fluorine
    (b) zinc
    (c) iodine
    (d) magnesium

11. All proteins contain the elements carbon, hydrogen, oxygen, and
    (a) copper
    (b) phosphorus
    (c) sulfur
    (d) nitrogen

12. The number of amino acids essential to all cells is
   (a) 11   (c) 33
   (b) 22   (d) 44

13. The carboxyl or acid group present in amino acids is correctly represented as
   (a) CHOO   (c) COOH
   (b) COHO   (d) COHH

14. When a large number of amino acids are joined together chemically, the result is the formation of a
   (a) polypeptide   (c) polysaccharide
   (b) dipeptide   (d) disaccharide

15. In the structural formula of an amino acid, "R" stands for
   (a) an amino group   (c) a variable group
   (b) a carboxyl group   (d) an hydroxyl group

PART C.   Modified True-False. If a statement is true, write "true" for your answer. If a statement is incorrect, change the *underlined* word to one that will make the statement true.

1. Kwashiorkor is a <u>carbohydrate</u> deficiency disease.

2. Nightblindness is caused by lack of vitamin <u>D</u>.

3. The mineral essential to the formation of hemoglobin is <u>zinc</u>.

4. Guinea pigs and *Homo sapiens* <u>can</u> synthesize vitamin C.

5. <u>Fats</u> are the most abundant of the organic compounds in body cells.

6. The major fuel foods are carbohydrates and <u>proteins</u>.

7. Raw fruits and vegetables are excellent sources of <u>fiber</u>.

8. Rickets is associated with a lack of vitamin <u>C</u>.

9. The mineral <u>phosphorus</u> is necessary for the synthesis of the hormone thyroxin.

10. <u>Fluorine</u> aids in resistance to tooth decay.

11. During dehydration synthesis one or more molecules of <u>hydrogen</u> form.

12. Sucrose is an example of a <u>monosaccharide</u>.

13. Enzymes are composed of <u>fats</u>.

14. Lipids are composed of glycerol and <u>amino</u> acids.

15. The three principal elements from which lipids are composed are carbon, <u>nitrogen</u>, and oxygen.

Think and Discuss

1. Why do body cells require carbohydrate foods?
2. What are the functions of proteins in the body?

3. If fats were not available to body cells, what processes or functions would be lost?
4. Why is fiber necessary in the human diet?

Answers to Questions for Review

PART A

1. nutrients
2. fats
3. protein
4. Calories
5. intestine
6. hormones
7. micronutrients
8. calcium
9. respiration
10. trace
11. starches
12. double
13. polysaccharides
14. hydrogen
15. six

PART B

1. a
2. b
3. d
4. c
5. d
6. b
7. d
8. c
9. b
10. c
11. d
12. b
13. c
14. a
15. c

PART C

1. protein
2. A
3. iron
4. cannot
5. Proteins
6. fats
7. true
8. D
9. iodine
10. true
11. water
12. disaccharide
13. proteins
14. fatty
15. hydrogen

CHAPTER 13

DISEASES OF *HOMO SAPIENS*, AND THE IMMUNE SYSTEM

A disease is a disorder that prevents the body organs from working as they should. In general, diseases can be classified as being *infectious* or *noninfectious*. Infectious diseases are caused by organisms that invade the body and do harm to the cells, tissues and organs. As a rule, there is disease specificity whereby a specific disease-producing organism causes a particular disease. Disease-producing organisms are said to be *pathogens* and are described as being *pathogenic*. Usually, pathogens— or germs, as they are often called—are microorganisms, organisms microscopic in size. Pathogenic microorganisms include certain bacteria, protozoans, spirochetes, richettsias, mycoplasmas, and fungi. Parasitic worms and viruses also often produce disease in humans. Most infectious diseases are *contagious*—capable of being passed from one person to another by means of body contact or by droplet infection.

Noninfectious diseases are caused by factors other than pathogenic organisms. Among the factors that are responsible for noninfectious diseases are genetic causes, malnutrition, exposure to radiation, emotional disturbances, organ failure, poisoning, endocrine malfunctioning and immunological disorders. Whatever the cause, a disease works counter to the well being of a human.

Infectious Diseases

METHODS OF SPREADING

Contagious or *communicable* diseases are spread from one person to another in a number of ways. One such way is *direct contact* whereby the infected person touches a well person. This may be accomplished by handshake, kissing, or through sexual intercourse (*sexually transmitted diseases*). Microorganisms are also spread *indirectly* when a noninfected person handles objects that have been in contact with the infected person. Eating from the same plate, drinking from the same glass, handling bed

TABLE 13.1.   Sexually Transmitted Diseases

Disease	Causative Organism	Symptoms
Syphilis	T. pallidum (spirochete)	Body lesions, tumors, dementia
Gonorrhea	N. gonorrhoeae (coccus)	Infection in genital/reproductive system
Venereal Herpes	Herpes simplex-2 (virus)	Sores at infection site, pains in joints
AIDS	HIV (retrovirus)	Syndrome of killer infections
Trichomoniasis	T. vaginalis (protozoan)	Venereal warts
Crab Lice	P. pubis (insect)	External parasite in pubic hair
Monilial Vaginitis	C. albicans (fungus)	Vaginal penile infections
Chancroid	H. ducreyi (bacterium)	Genital system infection
Hepatitis A,B	Virus	Infection of liver

clothes or towels or any number of items are means by which germs are passed. *Droplet infection* is a common method of passing germs along. Disease germs are present in droplets of water that escape from the nose and mouth when sneezing, coughing and talking. If these infected droplets are inhaled or taken in by mouth, the germs then enter the body of another person. Animals may be vectors, or carriers, of diseases. Biting insects or mammals may carry disease germs in their salivary glands and pass them on to a human who is bitten. Animal hair carries insects which may be disease carriers. Contaminated food and water spread diseases throughout human populations. Finally, human carriers of disease organisms (who remain unaffected by the germs that they carry) can spread pathogenic organisms to noncarriers.

COMMON INFECTIOUS DISEASES AND THEIR CAUSES

Bacteria are responsible for many human diseases, as outlined in Table 13.2.

TABLE 13.2.   Diseases Caused by Bacteria

Diseases Caused by Bacilli (Rods)	Diseases Caused by Cocci (Spheres)	Diseases Caused by Spirellae (Spirals)
Tuberculosis	Pneumonia (some forms)	Syphilis
Diphtheria	Gonorrhea	Asiatic Cholera
Tetanus	Scarlet Fever	Yaws
Typhoid Fever	Rheumatic Fever	Lyme Disease
Bubonic Plague	Streptococcus sore throat	
Whooping Cough	Meningitis (some forms)	
Tuleremia	Childbed fever	
Leprosy		

TABLE 13.3. Rickettsia Diseases

Human Disease	Disease Vector
Typhus	Lice
Trench Fever	Lice
Rocky Mountain Spotted Fever	Ticks
Rickettsial Pox	Mites

TABLE 13.4. Worm-Caused Diseases

Diseases Caused by Flatworms	Diseases Caused by Roundworms
Tapeworm Infection	Hookworm
Sheep Liver Fluke Infection	Trichinosis
Chinese Liver Fluke Infection	Pinworm Infection
	Ascaris Infection
	Filariasis

TABLE 13.5. Common Virus Diseases

Virus Disease	Organ Affected
Poliomyelitis	Nervous system and muscles
Rabies	Nervous system
Encephalitis	Nervous system
Viral Pneumonia	Lungs
Common Cold	Respiratory system
Influenza	Lungs and respiratory system
Fever blisters	Skin around the lips
Genital Herpes	Genital organs
Mumps	Salivary glands
Viral Hepatitis	Liver
Trachoma	Eyes
Measles	
Smallpox	
Chicken pox	
Yellow fever	spread throughout the body
Dengue fever	
Psittacosis (parrot fever)	
Warts	Hands and feet
AIDS	Immune system

Certain rickettsia can also produce disease in humans when they enter the body through the bites of their hosts—mites, ticks, lice, and fleas. Rickettsia diseases are often serious, characterized by high fever and rash, and often lead to death (Table 13.3). Some protozoa and some worms can also produce disease in humans. Most parasitic worms are ingested in contaminated food, usually encysted in the muscles of cows, pigs, sheep, and snails (Table 13.4).

A few species of yeasts and molds are pathogenic for humans. They attack the skin, mucous membranes and the lungs. Ringworm and athlete's foot are two such diseases. Although irritating and inconvenient, these diseases are not generally serious and can usually be readily treated.

Viruses—nonliving particles that come "alive" when they invade a cell—are major disease-producers in humans. Some viruses cause disease that affects the entire body; others cause disorders that affect a particular organ. Table 13.5 lists some common viral diseases.

HOW PATHOGENS DAMAGE THE BODY

Once pathogenic organisms enter the body there is interaction between the body and the germs that results in disease. The type of disease is determined by the type of pathogen that invades the body. The severity of the disease depends on the ability of the body to ward off infection (*resistance*) and the strength or *virulence* of the infecting germ.

Pathogens may affect body tissues and functions in a number of ways. Some pathogens produce enzymes that dissolve the materials that hold cells together, creating pathways for germs to enter tissues. Other germs produce substances that kill certain body cells. Pathogens frequently damage only certain cells and tissues. The rickettsia of Rocky Mountain spotted fever damages the liver as does the Chinese liver fluke. The polio virus destroys nerve cells. The spirochete of syphilis often destroys brain tissue. The typhoid bacillus attacks the lymph tissue of the intestinal wall.

Some disease-producing microorganisms secrete *toxins* that interfere with the metabolic activities of cells. Certain germs block vital passageways of the body and thus prevent normal functioning; for example, the organisms that cause diphtheria seal the throat with membranes and prevent breathing. Worm parasites frequently compete with the host for nutrients and produce malnutrition in the host.

NONSPECIFIC DEFENSES AGAINST DISEASE

When the body is attacked by pathogenic organisms it puts up a series of defenses, designed to destroy the enemy and maintain health.

Skin

The first line of defense against invasion of the body by germs is the skin. The clean, unbroken skin is thick enough and tough enough to prevent most germs from penetrating. As a rule, germs that land on the skin do not live long enough to cause trouble because the skin itself has a germicidal quality that inhibits the growth of germs on its surface.

Other Defenses

The eyes, nose and mouth are in effect breaks in the skin that could permit the penetration of germs into the body. Most germs entering the eye do not live long enough to cause distress. They are dissolved by *lysozyme*, an enzyme in tears. Nevertheless, some virulent germs survive and produce eye infections such as conjunctivitis (pink eye) or *tracoma*. Tracoma, a virus disease, is especially dangerous because it often causes blindness.

Germs numbering in the thousands enter the mouth daily with food and drink. Few of these survive to reach the intestines. The saliva in the mouth is able to kill many of the invaders. Those that reach the stomach face the killing action of hydrochloric acid and the digesting power of pepsin. However, some germs manage to survive. The germs of Asiatic cholera, typhoid, paratyphoid and other serious intestinal diseases are able to resist these body defenses and cause illness.

Uncountable numbers of disease germs are breathed in through the nose from the surrounding air. However, few of these ever reach the lungs. The nasal passages present a complicated maze of filters guarded by hairs that trap many germs. In addition, the mucus membranes lining the air passages secrete sticky mucus which traps disease germs, rendering them inactive. Sneezing forces germs to the outside, also.

Germs that manage to reach the breathing tubes are for the most part trapped in mucus secretions from the cells that compose these tubes. In addition, the cilia of the cells that line the air tubes sweep the mucus-trapped germs back to the throat where they are swallowed and then destroyed in the stomach by hydrochloric acid and pepsin. Special "dust cells" in the air sacs of the lungs pick up some germs and carry them out. Despite these active defenses, some germs survive and cause respiratory diseases such as colds, influenza and pneumonia.

The Immune System

The purpose of the *immune system* is to protect the body against infection. When microorganisms (germs) infect the body, the immune system is called into action. Its function is to destroy the invading germs or foreign proteins.

The organs and tissues of the immune system are distributed throughout the body. The immune system is made up of lymphoid tissue, fluid called lymph, and white blood cells which act against foreign matter, germs and proteins, that enter the body. The immune system is closely involved with the blood circulatory system.

The cells of the immune system gather in the lymhoid tissues. Lymphoid tissues include the adenoids and tonsils in the head region, the thymus gland in the chest cavity, bone marrow in the center of long bones, the spleen just below the heart, lymph nodes positioned under the arms and in the groin, and Peyer's patches in the small intestine (Fig. 11.1).

CELLS OF THE IMMUNE SYSTEM

The cells of the immune system are able to recognize and act upon microorganisms and foreign proteins that enter the body. Any foreign substance or organism that causes the immune system to react is called an *antigen*. Antigens are usually proteins, glycoproteins (carbohydrate-protein molecules), or carbohydrates. These molecules may be carried on the cell membranes of invading microorganisms. Fats are usually not antigenic.

Phygocytes

Macrophages and *neutrophils* are two types of white blood cells that engulf (phagocitize) invading microorganisms. These cells provide the human body with a very effective first line of defense.

Huge, amoeboid macrophages and the smaller neutrophils are specialized for attacking invading microorganisms. Some macrophages stay within the spleen and lymph nodes, where they engulf any microorganism invaders that pass their way. Other macrophages and the neutrophils travel through the body searching for invaders.

Chemicals released from damaged blood platelets attract the traveling macrophages and neutrophils. These cells gather at the site of infection and ingest foreign bacteria—hence the name *phagocytes*. The phagocytes ingest large numbers of bacteria and are themselves killed by the bacterial toxins (poisons). The accumulated dead bodies of macrophages and neutrophils forms *pus*.

Lymphocytes

The second line of defense provided by the immune system is made possible by a type of white blood cell known as a *lymphocyte*. There are two kinds of lymphocytes: the *B-cell lymphocyte* and the *T-cell lymphocyte*. Working together, these cells carry out a complex series of events known as the *immune response*.

The human immune response embraces two major biochemical events: the *humoral immune response* and the *cell-mediated response*. The humoral immune response involves the production of protein molecules called *antibodies*, and the cell-mediated response demands direct cellular action. Each of these responses is driven by a specific type of lymphocyte: either the B-cell or the T-cell.

Lymphocytes are produced by the stem cells which reside in the marrow of long bones. Stem cells produce all of the blood cells. In some way still unknown, changes (differentiation) take place in the primary blood cells, enabling them to carry out specific functions. The B-cell lymphocytes mature in the bone marrow (thus the name B-cell); the T-cell lymphocytes, in the thymus gland (hence the name T-cell).

The B-cells and T-cells look alike in the inactive state. When activated, however, they differ significantly. Activated B-cells have a very rough endoplasmic reticulum caused by the enormous number of ribosomes that are attached to these membranes. Activated T-cells have large concentrations of free ribosomes in the cytoplasm.

Primary Immune Response

On the outer surface of their cell membranes, B-cell lymphocytes carry specific antigen-recognition proteins. Each cell is a specialist, carrying only one kind of recognition protein. When a newly produced B-cell meets with a matching antigen, the B-cell is activated and the antigen attaches to the recognition site on the membrane of the B-cell. This B-cell with its attached antigen becomes much larger and undergoes mitosis and differentiation, producing two new kinds of cells—*plasma cells* and *memory cells*.

The activated plasma cells produce an antibody that matches the captured antigen. Antibody production by the plasma cells is enormous—about 2000 molecules of antibody per minute! The antibodies attach to the foreign antigen and immobilize it. Now the macrophages and neutrophils are called into action. They ingest the antigen-antibody complexes. This early reaction of the immune system is called the *primary immune response*.

Secondary Immune Response

The memory cells store all the information needed to build the same kind of antibodies that the plasma cells make during the primary immune response. Plasma cells have very short lives. In contrast, memory cells have very long lives and can remain in existence for many years. Should a second attack by the same antigen occur, the memory cells go into action, producing huge numbers of antibodies through very rapid reproduction of plasma cells. The plasma cells are now produced in greater numbers than during the primary immune response. Thus the *secondary immune response* is much more effective than the primary immune response.

Subpopulations of the T-cell Lymphocyte

Antibodies can inactivate viruses and prevent them from entering cells. Once a virus has entered a cell, however, it is safe from antibody. Therefore, for virus-infected cells, the immune system makes use of the cytotoxic T-cell.

Usually, a cell that is infected with a virus has viral antigens on its cell surface. The cytotoxic T-cell recognizes these foreign antigens on the surface of the cell membrane and kills that cell as it comes in contact with it. The ability of cytotoxic T-cells to destroy antigens without the aid of antibodies is known as *cell-mediated immunity*. Another type of T-cell forms a capsule around foreign cells, thereby localizing infection and preventing its spread.

A different subpopulation of T-cell lymphocytes is called *regulatory T-cells*. Of these, there are helper T-cells and suppressor T-cells. The helper T-cells assist other T-cells and B-cells in responding to antigens. The helper T-cells also activate some macrophages.

AIDS AND THE IMMUNE SYSTEM

Acquired immune deficiency syndrome (*AIDS*) is a killer disease that has reached epidemic proportions. Infection by the human immunodeficiency virus (HIV) causes the immune system to collapse, leaving the body open to devastating infections. HIV viruses attack the T-4 lymphocytes, which function as helper cells. When made inactive, the T-helper cells cannot stimulate reproduction of B-cell lymphocytes. The B-cells give rise to the antibody-producing plasma and memory cells. The result is that AIDS shuts down the patient's entire immune system. The infected person becomes host to a number of infectious organisms that cause diseases such as Kaposi's sarcoma, pneumocystic pneumonia, and yeast infections.

AIDS is not a single disease but is a syndrome of symptoms caused by the various invading microorganisms that take advantage of an immune system that cannot function. Chapter 14 presents a discussion of the way in which the HIV retrovirus works.

PROTECTION AGAINST DISEASE

Immunization

Immunity is the ability to resist the attack of a particular disease-producing organism. Immunity to one kind of disease germ does not automatically make a person immune to other types of disease germs. *Active immunity* is brought about by antibody production by a person's own body cells. Active immunity can be stimulated in either of two ways; by getting the disease and recovering from it or by being immunized against the disease. Immunization that produces active immunity involves the injection of weakened disease agents that stimulate antibody production but produce only mild symptoms or none at all. Active immunity is long-lasting because the body cells continue to produce the antibodies.

An injection of gamma globulins can give a person temporary immunity against certain specific diseases. This means that a person has borrowed antibodies in the blood and not those made by his (her) own cells. This kind of immunity is called *passive immunity*. It lasts only as long as the antibodies last; when they are used up, the immunity ceases.

Smallpox is a disease that has been almost entirely eradicated from even remote corners of the earth. The fight against this disease began in the 18th century when Edward Jenner vaccinated people with cowpox. Vaccination with cowpox stimulates the body to build its own antibodies and give lasting immunity to the dread disease of smallpox. Cowpox is a mild disease in humans and stimulates the cells to produce antibodies which happen to be effective against smallpox.

Successful methods of immunization have been developed against diphtheria, polio, whooping cough, lockjaw, plague, typhoid, yellow fever, cholera, measles, mumps, rubella, and typhus. In every community in the United States children are automatically immunized against diphtheria, tetanus, whooping cough, polio and smallpox. Many of the states have laws that prevent children from attending public school without the prescribed immunizations.

Safe Drinking Water

At one time diseases such as typhoid fever and cholera devastated the population of New York. Today not one death can be attributed to either of these diseases. The reason for the marked decrease in the incidences of these diseases is attributed to the development of a safe water supply.

Sanitary water engineers are employed to protect the community against the dangers of contaminated water. They use many water purification techniques. *Settling* is a process in which water is held in large tanks until suspended solids settle out. *Filtering* is accomplished by allowing water to trickle through sand beds several feet deep. This removes 90–95 percent of all bacteria as well as fine particles of solid matter. *Aeration* is the process in which water is sprayed up into the air. This technique kills some bacteria and allows more air to dissolve into the water. *Chlorination* is the chemical purification of water; chlorine gas or hypochlorite is added to the water in small quantities to kill any remaining pathogens.

Sanitary engineers and bacteriologists watch the community water supply very closely. They test the water supply constantly for *E. coli*. The presence of this bacterium is known as the *index of fecal contamination*. Since *E. coli* are normally present in the human intestines, their presence in drinking water indicates that the water supply must be contaminated with sewage. When *E. coli* is not present, the water is free of human wastes and is probably free of organisms that cause typhoid, cholera, and other diseases of the human intestines.

Safe Milk and Food

At one time diseases such as tuberculosis, Q fever, septic sore throat, brucellosis and stomach upsets were transmitted to people from contaminated milk. To prevent the spread of disease through milk most local laws require that milk be *pasteurized* before sales. Commercial dairies chill the milk as quickly as possible after collection. This temporarily inhibits the growth of bacteria which may thrive at the body temperature of the cow. Then the milk is heated for a certain time and at a certain temperature that will kill the toughest disease germs present. This process is called *pasteurization*. After pasteurization, milk still contains thousands of living bacteria, but the disease-producing germs have been destroyed.

Methods of food preservation are used to prevent food from being spoiled and contaminated by bacteria. Canning sterilizes food and seals it up so that no bacteria can get in. Other methods are used that prevent the growth of bacteria in food products and prevent rapid spoilage. Some of these methods were devised by the ancients; some are relatively new. They include drying, salting, sugaring, pickling, fermenting, smoking, refrigeration, fast feeezing, sterilization by heat or radiation, and the addition of chemical preservatives.

Noninfectious Diseases _____

The noninfectious diseases are not communicable because they are not caused by infectious organisms. These diseases have various causes other than germs. Table 13.6 provides a listing of some of the noncommunicable diseases.

Not all of the diseases listed in the table will be discussed in this chapter. Diseases resulting from endocrine insufficiency or oversecretion were summarized in Chapter 11, and some disorders that result from nutritional deficiency were summarized in Chapter 12.

TABLE 13.6. Some Noninfectious Diseases

Disease Type	Malfunction
Cardiovascular disease	Heart and blood vessels
Allergy	Hypersensitivity to certain substances
Genetic	Inborn defects
Emotional including drug-related psychoses	Psychological problems and mental stress
Occupational	Physical problems caused by work conditions
Poisoning	Illness caused by tissue toxins
Nutritional	Nutrient deficiencies in diet
Hormonal	Imbalance in endocrine secretions
Cancer	Wild reproduction of cells

CARDIOVASCULAR DISEASES

Problems of the heart and blood vessels are known as *cardiovascular diseases*. Because diseases of the heart and the circulatory system are leading causes of death in the United States, a great deal of research effort has been put forth for many years to determine the causes of heart defects and to find ways to prevent and/or cure these disorders. Heart defects trouble people of all ages.

Many infants are born with *congenital* heart defects. Sometimes the heart defect is a hole in the heart wall which divides the left and right ventricles. In some cases, the large arteries leaving the heart are in the wrong place. It also happens that a blood vessel that should have closed at birth failed to do so. Today many of these defects in childrens' hearts are corrected through procedures of *open heart surgery*. During the operation, the blood is sent through a machine that serves as a mechanical heart and lungs. After the real heart is repaired, the blood is returned to its normal pathways through the body.

In young and old alike, defective heart valves may cause trouble. Valves that are beyond self repair are replaced by plastic ones.

Many things may go wrong with the adult heart. Sometimes the Purkinje fibers in the heart lose their ability to contract rhythmically and thus the pacemaking ability of the heart is impaired. This defect is helped by implanting an artificial *pacemaker* in the patient's chest which makes possible appropriate heart stimulation.

Coronary artery disease prevents the proper blood supply from reaching the heart. The cells in heart muscle become damaged when not enough oxygen and nutrients reach them. Damaged heart cells cause heart attacks. If the damage to the heart is not too severe, a person will recover. Severe heart attacks cause death. High blood pressure (*hypertension*) is another major cause of heart attacks and strokes.

In recent years, some highly skilled heart surgeons have attempted exotic methods for dealing wih patients who have badly diseased hearts. One of these methods has been heart transplant in which a diseased heart is exchanged for a healthy one. This technique has met with limited success in which the lives of transplant patients have been prolonged for a few months to a year. Recently, a patient was kept alive for about three months by means of an artificial heart. Neither of these methods has proven practical for large numbers of cases.

ALLERGY

Some people are very sensitive to substances that are quite harmless to most other people. The cells of these sensitive people produce antibodies to ward off whatever substance affects them. The antibodies become attached to the tissue cells, rendering the person *sensitized*.

Whenever that particular substance enters the body again, it reacts with the attached antibodies and damages the cells. These damaged cells prompt certain symptoms such as itching, sneezing, tearing eyes, red welts, large hives, fever, and a general feeling of not being well. The injured cells may release chemicals called *histamines* which are responsible for the symptoms. *Anti-histamines* are substances that may neutralize the histamines and relieve the symptoms. People may be sensitive to things that they inhale such as pollen, dust or powders. Other people may be sensitive to things that they eat such as wheat, eggs, milk, fish, nuts, chocolate, strawberries or bananas. There are those people who get violent reactions from drugs such as penicillin, aspirin, or streptomycin. A person may become *allergic* to a given substance at any time in life.

OCCUPATIONAL DISEASES

The jobs of many people put them in contact with substances that present hazards to health. Fumes, dust, gases, vapors, fibers and chemicals may have devastating effects on the health of a large number of people. Not too long ago it was discovered that asbestos causes lung cancer in those who have regular exposure to its fibers. Similarly, people who inhale the fibers of cotton or sugar cane may develop debilitating lung conditions and breathing problems. *Silicosis* is a type of lung disease occurring in miners and sandblasters who are in constant contact with rock dust. *Asbestosis* is a progressive inflammation of the lungs that results from breathing in fine fibers of asbestos. The disease occurs in workers in the construction trades and in miners.

Some occupations put workers in contact with poisons. Painters used to be subjected to lead poisoning from lead-base paints. (Today, most household paints are made without lead.) Lead is toxic to the body, settles in the brain and damages brain tissue. It is not unusual for farmers and exterminators to become poisoned by pesticides which they may absorb into the body by contact or by breathing. Recently, the pesticide Chlordane was shown to have adverse effects on the health of persons whose homes were contaminated by misuse of the product.

People in the medical professions are subject to occupational hazards. The ionizing radiation produced by X-ray machines can cause a wide variety of unfavorable conditions.

CANCER

Cancer is not a single disease. It is a whole constellation of different diseases that share a common characteristic. In every form of cancer there is an abnormal, uncontrolled growth of body cells. The cells that grow so wildly are not foreign cells that have invaded the body, but regular body cells that somehow have gone wrong. They grow in a disorganized fashion and compete with normal cells for space and nutrition.

These wildly growing cells form a malignant mass called a *neoplasm* or a *cancer*. The malignant mass sends out fingers of cancerous cells that burrow into the normal tissues around it. As the invasion progresses, the normal tissue is gradually destroyed.

In the beginning cancer is localized in a particular part of the body. As time passes, fragments of malignant tissue separate away from the original malignant mass. They are carried off in the blood and lymph, reaching many new sites in the body. Wherever they remain, a new colony is established and grows like the original mass. This process in which cancer spreads is called *metastasis*.

All cancerous cells do not grow at the same rate. Some types of cancer grow quickly; others grow slowly. All cancer ultimately destroys normal tissue and kills the host. Every living cell has the potential to turn cancerous. Every variety of vertebrate animal is subject to the disease. So are humans of every race, nation or habitat. Cancer cannot be blamed on modern living. Cancers have been found in Egyptian mummies and in fossils of ancient dinosaurs.

Many causes for cancer have been suggested. Some of these suggestions are more than guesses or hunches having been based on statistical studies of the incidences of disease. A case in point is cigarette smoking. Rather extensive research work has shown a very definite link between smoking and lung cancer (heart disease, too). Smokers also show a high incidence of cancer of the lips, mouth and throat.

Research has also indicated that some forms of cancer seem to run in families. For example: the daughters and sisters of women with breast cancer, and the children of patients with rectal cancer should be informed of their increased risk. There seems to be familial patterns of occurrence in these diseases. Genes that cause cancer, *oncogenes*, have been isolated.

There is strong evidence that many types of chemicals are *carcinogens*, or cancer producers. Contact over a period of time with certain defoliants, insecticides and "buried" chemical wastes seems to have caused cancers of multiple varieties in large segments of the population. It is known that radium workers tend to get bone cancer. Workers in factories that produce aniline dyes often get cancer of the bladder.

Homo sapiens is constantly bombarded by various kinds of radiation: ultraviolet rays, X rays, gamma rays, cosmic rays, radioactive fallout and the like. Any form of radiation, natural or human-made, can produce leukemia, bone cancer, skin cancer or other types of cancer, such as lung, breast or thyroid cancers. The effect of the atomic bomb explosions on Hiroshima and Nagasaki in Japan leave no doubt that a single exposure to a high dose of radiation can produce leukemia.

There is some indication that certain forms of cancer may be communicable, transmitted from one person to another by way of a virus. For example: cancer of the cervix may be initiated by infection with herpes virus. These virus particles are found commonly in cancerous cervix cells. It is also known that the second wife of a man whose first wife died from cancer of the cervix sometimes develops cancer of the cervix also. The spread of Kaposi's sarcoma (a rare form of cancer associated with AIDS) indicates viral transmission. The electron microscope has revealed that some leukemia cells contain virus particles.

The best protection against any type of disease is to avoid contact with causative agents or chemicals whenever possible.

Chronology of Famous Names in Biology

460 BC   **Hippocrates** (Greek)—"father of medicine" because he was the first of the ancients to attempt scientific explanations of disease.

1527   **Jacques de Bothencourt** (French)—gave the name "venereal" to those diseases transmitted by sexual intercourse.

1530   **Girolamo Fracastoro** (Italian)—gave the name "syphilis" to the heretofore unnamed venereal disease.

1796   **Edward Jenner** (English)—discovered that vaccination with cowpox renders immunity against smallpox.

1854   **Louis Pasteur** (French)—proved that microorganisms caused fermentation. In 1885 he developed the treatment for rabies.

1860   **Philippe Ricord** (American)—determined that gonorrhea and syphilis are two separate diseases.

1872   **Moritz Kaposi** (Russian)—described the symptoms of Kaposi's sarcoma.

1879   **Albert Neisser** (German)—isolated the organism that causes gonorrhea.

1880   **Charles Laveran** (French)—described the protozoan parasite of malaria.

1882   **Elie Metchnikoff** (Russian)—discovered the phagocytic activities of white blood cells.

1883   **Robert Koch** (German)—discovered the bacterium that causes tuberculosis.

1892   **Walter Reed** (American)—discovered that yellow fever was transmitted by Aedes mosquito.

1896   **Joseph Lister** (English)—developed aseptic techniques to prevent infection during and after surgery.

1906   **Howard T. Ricketts** (American)—discovered that ticks infected with microorganisms (Rickettsia) cause Rocky Mountain spotted fever.

1908   **Paul Ehrlich** (German)—developed the first chemical substance—salvarsan—to inhibit the growth of the syphilis spirochete.

1911   **Emil von Behring** (German)—discovered that immunity to diphtheria could be given by injecting a person with antitoxin.

1929   **Alexander Fleming** (English)—discovered the growth inhibitions effect of penicillin on staphylococcus.

1932   **Gerhard Domagk** (German)—discovered that *prontosil* is effective against Streptococcus infections.

1933   **George Dick** (American)—devised a test to determine susceptibility to scarlet fever and discovered the germ that causes that disease.

1933   **Bela Schick** (American)—developed the Schick test for susceptibility to diphtheria.

1935   **Wendell Stanley** (American)—isolated the virus that causes tobacco mosaic disease.

1943   **Howard Florey** (American)—developed an efficient method of mass producing penicillin from the mold *Penicillium notatum.*

1944   **Peter Medawar** (English)—did the basic research which brought to light the problems of tissue and organ transplant techniques and immunology. **1967**—discovered acquired immune tolerance factors that prevent tissue transplants.

1948   **Selman Waksman** (American)—first to isolate streptomycin from *Streptomyces griseus*

1954   **Jonas Salk** (American)—developed a vaccine effective against poliomyelitis.

1963   **Albert Sabin** (American)—developed oral polio vaccine.

1965   **K. Ishizaka** (American)—discovered gamma globulin E (IgE) which is present in normal serum in small amounts.

1985   **Flossie Wong-Staal** (American)—did pioneering research on the structure of the AIDS virus.

Words for Study

active immunity
aeration
AIDS
allergy
antibody
antigen
anti-histamine
carcinogen
cardiovascular diseases
contagious
cytotoxic T-cell
droplet infection
histamine
hypertension

immune system
immunization
index of fecal
　contamination
infections
lymph
lymphocyte
macrophage
memory cells
metastasis
neoplasm
neutrophil
oncogene

passive immunity
pasteurization
phagocyte
plasma cells
regulatory T-cell
resistance
settling
tracoma
tubercle
vaccine
virulence
virus
venereal disease

Questions for Review

PART A.   Completion. Write in the word that correctly completes each statement.

1. Diseases that are caused by viruses, bacteria or other pathogens are known collectively as ..1.. diseases.
2. A disease that is spread from one person to another is said to be ..2..
3. Rocky Mountain spotted fever is caused by the organism called a ..3..
4. Polio is caused by ..4.. infection.
5. Most parasitic worms enter the body by way of contaminated ..5..
6. The disease-producing ability of a pathogen is summed up by the term ..6..
7. Tracoma is a disease of the eye caused by a ..7..
8. Phagocytosis is best associated with ..8.. blood cells.
9. Fluid that bathes the body spaces is called ..9..
10. Immune blood proteins are the ..10.. globulins.
11. The ability to resist disease is known as ..11..
12. Cholera is spread through unclean ..12..
13. Milk is ..13.. to prevent the spread of tuberculosis and Q fever.
14. The work of the Purkinje fibers can be taken over by an artificial ..14..
15. During allergic attacks, cells release chemicals known as ..15..
16. Lymphoid tissues in the head region include the tonsils and the ..16..
17. The fluid of the immune system is known as ..17..
18. Foreign proteins that cause an immune system to react are known as ..18..
19. The purpose of the immune system is to protect the body against ..19..
20. Lymphatic tissue known as Peyer's patches is located in the ..20..

PART B.   Multiple Choice. Circle the letter of the item that correctly completes each statement.

1. The pathogens that are not microscopic in size are
   (a) virus particles   (c) worms
   (b) bacteria   (d) mycoplasmas

2. Transmission of germs by direct contact may be accomplished by
   (a) droplet infection   (c) handling clothing
   (b) handshake   (d) sneezing

3. An example of a disease caused by a bacillus is
   (a) meningitis   (c) syphilis
   (b) yaws   (d) tetanus

4. Viruses reproduce
   (a) in living cells
   (b) on dead organic matter
   (c) in quiet waters
   (d) in blood plasma

5. Ringworm is a disease of the skin that is caused by infection with
   (a) protozoa
   (b) fungus
   (c) hookworm
   (d) bacteria

6. When germs break through the skin, certain chemicals are released from body cells that
   (a) kill the germs immediately
   (b) seal up the wound
   (c) cause the capillaries to expand
   (d) prevent pain and tenderness

7. Microorganisms that are enclosed in capsules are usually
   (a) phagocytic
   (b) anaerobic
   (c) harmless
   (d) pathogenic

8. By the time lymph leaves the lymph vessels and is returned to the blood, the lymph is
   (a) absolutely sterile
   (b) able to destroy bacteria
   (c) almost free of bacteria
   (d) crowded with bacteria

9. An injection of gamma globulins can give a person the type of immunity best described as
   (a) partial
   (b) temporary
   (c) lasting
   (d) active

10. The purpose of the aeration of water is to
    (a) remove debris
    (b) improve the color
    (c) prevent tooth decay
    (d) kill anaerobes

11. Allergy is caused by sensitizing
    (a) lymphocytes
    (b) antibodies
    (c) leucocytes
    (d) antitoxins

12. A malignant mass of cells is known as a (an)
    (a) metastasis
    (b) protoplasm
    (c) ectoplasm
    (d) neoplasm

13. Carcinogens are
    (a) virulent pathogens
    (b) immune proteins
    (c) contact poisons
    (d) cancer producers

14. A disease vector
    (a) carries the disease
    (b) causes the disease
    (c) cures the disease
    (d) controls the disease

15. The phagocytic activities of white blood cells were discovered by
    (a) Jenner
    (b) Metchnikoff
    (c) Salk
    (d) Sabin

16. Human body cells that ingest dead bacteria are the
    (a) pathogens
    (b) phagocytes
    (c) bacteriophages
    (d) antigens

17. Huge, amoeboid cells that engulf microorganisms that invade the body are the
    (a) eosinophils
    (b) stem cells
    (c) erythrocytes
    (d) macrophages

18. B-cells and T-cells are best classified as
   (a) lymphocytes   (c) neutrophils
   (b) scavenger cells   (d) macrophages

19. The humoral immune response produces
   (a) auxins   (c) antibodies
   (b) antigens   (d) autosomes

20. Lymphocytes are produced by
   (a) T-cells   (c) plasma cells
   (b) stem cells   (d) neutrophils

PART C.   **Modified True-False.** If a statement is true, write "true" for your answer. If a statement is incorrect, change the underlined word to one that will make the statement true.

1. Malnutrition is an example of a <u>contagious</u> disease.

2. Human carriers <u>are</u> affected by the germs they carry.

3. Leprosy is a disease that is caused by a <u>coccus</u>.

4. Rickettsia are microorganisms that live in the body of <u>protozoa</u>.

5. Smallpox is a <u>bacterial</u> disease that affects the whole body.

6. Hepatitis is an infection of the <u>liver</u>.

7. The ability of the body to ward off infection by disease organisms is known as <u>virulence</u>.

8. The virus of polio destroys <u>lung</u> tissue.

9. Lysozyme is <u>a hormone</u> in tears.

10. Phagocytic white cells <u>egest</u> bacteria.

11. A boil is the result of a <u>spreading</u> infection.

12. Macrophages are very <u>small</u> white blood cells.

13. The name Edward Jenner is best associated with the disease <u>polio</u>.

14. Cowpox provides <u>passive</u> immunity.

15. Methods of food preservation include salting, freezing and <u>drying</u>.

16. The organism that causes syphilis is classified as a <u>rickettsia</u>.

17. Antibodies are produced in enormous quantities by the <u>B-cells</u>.

18. Lymphocytes that have very long life spans are the <u>macrophages</u>.

19. Infection by the AIDS virus causes a shutdown of the <u>circulatory</u> system.

20. Cytotoxic T-cells kill cells that have foreign <u>antibodies</u> on their cell membranes.

Think and Discuss _____

1. Why is Lyme disease classified as an infectious disease?
2. Thousands of germs enter the mouth daily, but few survive. Explain.
3. Describe the functions of the memory and plasma cells of the immune system.
4. When the T-cell lymphocytes fail to function, why does the entire immune system break down?

Answers to Questions for Review

PART A

1. infectious
2. communicable or contagious
3. rickettsia
4. virus (viral)
5. food
6. virulence
7. virus
8. white
9. lymph
10. gamma
11. immunity (resistance)
12. water
13. pasteurized
14. pacemaker
15. histamines
16. adenoids
17. lymph
18. antigens
19. infection
20. small intestine

PART B

1. c
2. b
3. d
4. a
5. b
6. c
7. d
8. c
9. b
10. d
11. b
12. d
13. d
14. a
15. b
16. b
17. d
18. a
19. c
20. b

PART C

1. noninfectious
2. are not
3. bacillus
4. ticks, mites, lice, fleas
5. virus
6. true
7. resistance
8. nerve or muscle
9. enzyme
10. engulf or ingest
11. local (contained)
12. large
13. smallpox
14. active
15. true
16. spirochete
17. plasma cells
18. memory cells
19. immune
20. antigens

CHAPTER 14

HEREDITY AND GENETICS

The science of genetics has made tremendous advances since the 1960's but the foundations lay at the beginning of the 20th century. Before discussing the modern findings of genetics we should review the classical principles upon which the science is founded.

Classical Principles of Heredity

MENDEL'S LAWS OF HEREDITY

Gregor Mendel, an Austrian monk, began the first organized and mathematical study of how traits are inherited. Using the garden pea as the test organism, Mendel identified seven different traits which were easily recognizable in this self-pollinating plant. He called each of these traits *unit characters*. Mendel began his work in 1856 when little was known about chromosomes and their functions in cell division and the concept of the gene had not yet been developed.

Mendel not only identified characteristics that seemed to be inherited, but for each unit character, he identified an opposite trait. For example: if the unit character was height, the opposite traits were short and tall. If the unit character was seed coat color, the opposite traits were yellow and green.

Based on his observations of crosses that he made in pea plants by means of hand pollination, Mendel formulated three major laws or principles of inheritance. It should be noted that he kept very careful records of his experiments and converted his results into mathematical ratios.

The Law of Dominance

If two organisms that exhibit contrasting traits are crossed, the trait that shows up in the first filial generation (F_1) is the dominant trait. For example: when a pure-bred tall pea plant is crossed with a short pea plant, all of the offspring will be tall. The offspring will not be pure tall, however, and are therefore known as *hybrids*. The factor for shortness is hidden. We say today that the *phenotype* of the F_1 plants is tall. A phenotype refers to the traits that we can see. The *genotypes* or genetic makeup of these plants is said to be hybrid or *heterozygous*, meaning mixed.

When organisms with contrasting traits are crossed, the trait that shows up in the F_1 generation is called the *dominant* trait. The trait that is hidden is called the *recessive* trait. Since Mendel did not have the concept of the gene, he called the conditions that make for dominance and recessiveness factors. From the results of crosses made with the garden pea, he concluded that (a) two factors determine a characteristic and that (b) two recessive factors are needed in order for the recessive characteristic to show up and (c) that one dominant factor and one recessive factor produce offspring that have the dominant trait.

The Law of Segregation

When hybrids are crossed, the recessive trait segregates out at a ratio of three individuals with the dominant trait to one individual with the recessive trait. The 3:1 ratio is known as the *phenotypic* ratio, because it refers to the traits that can be seen and not those factors hidden in the germplasm. The hybrid cross is also known as the F_1 cross and the offspring produced by this cross are known as the second filial generation, or F_2. In terms of modern knowledge, the F_2 generation also produces another type of ratio called the *genotypic ratio* which refers to *gene* makeup. The genotypic ratio is 1:2:1, translated into 1 *homozygous* dominant (pure) individual: 2 *heterozygous* (hybrid) individuals: 1 *homozygous* recessive. Only the homozygous recessive shows the recessive trait.

The Law of Independent Assortment

Mendel believed that each trait is inherited independently of others and remains unaltered throughout all generations. We now know that Mendel's "factors" are genes that are linked together on chromosomes and that if genes are on the same chromosome, they are inherited together.

THE CONCEPT OF THE GENE

Around 1911, Thomas Hunt Morgan introduced the tiny fruit fly *Drosophila melanogaster* as the new experimental organism for work in the field of heredity. The experimental work of Morgan resulted in the discovery that the chromosome is the means by which hereditary traits are transmitted from one generation to another. Morgan's chromosome theory of inheritance includes the concept that chromosomes are composed of discrete units called *genes*. Genes are the actual carriers of specific traits and move with the chromosomes in mitotic and meiotic cell divisions. Morgan further proposed that genes control the development of traits in each organism. When genes change, or *mutate*, the traits they control change.

SOME GENETIC SHORTHAND

A combination of the theories of Mendel and Morgan led to the development of a system of genetic "shorthand" which is used to represent the genetic makeup of a trait and to show rather simply what happens when organisms with specific traits are crossed. In this shorthand capital letters are used to indicate dominant genes, lowercase letters for recessive genes. A Punnett square is a diagrammatic device used to predict the genotypic and phenotypic ratios that will result when certain gametes fuse. Remember that as a result of meiosis each gamete has only one half the number of chromosomes that are in the somatic cells.

Problem: In fruitflies, long wing (L) is dominant over vestigial wing (*l*). What is the result of a cross between two flies that are heterozygous (*Ll*) for wing length?

Solution:

Parents: Male   ×   Female
                    Ll   *Ll*

Gametes: Ⓛ Ⓛ   Ⓛ Ⓛ

Punnett square

	L	l
L	LL	Ll
l	Ll	ll

F_2 *LL* = 1 homozygous dominant long winged fly
     Ll = 2 heterozygous dominant long winged flies
     ll = 1 homozygous recessive short winged fly

INTERMEDIATE INHERITANCE

Geneticists have discovered that in many cases a trait is not controlled by a single gene, but rather by the cooperative action of two or more genes. There are many instances in which Mendel's "law of dominance" does not hold true. A case in point is what was once called *blending inheritance,* or *incomplete dominance.* It is now known as *codominance.* When red-flowered evening primroses are crossed with white-flowered primroses, the hybrids are pink. Neither red nor white color is dominant and therefore the result is a blend. In sweet peas, the expression of red or white flowers is dependent upon two genes: a (C) gene for color and an (R) gene for enzyme. If C and R are inherited together, the flower color is red. If the dominant C is missing, the flower is white and if the dominant R is missing, the color is white. Therefore white flowers are the result of several different genotypes: *ccrr, ccRR, CCrr, Ccrr.*

CROSSING OVER

Genes are linked on chromosomes and are inherited in a group on a particular chromosome. However, linkage groups are broken by *crossing over,* a phenomenon which may occur during meiosis when homologous chromosomes are intertwined during synapsis. It is at this time that chromosomes may exchange homologous parts and thus assort linkage groups.

MUTATIONS

It was stated before that genes can change and that changes in genes are known as *mutations*. As a rule, mutations are usually recessive and they are usually harmful. Mutations usually occur at random and spontaneously. However, mutations may be induced by radiation or by chemical contamination.

There are several types of mutations. A loss of a piece of a chromosome is known as a *deletion*. The genes on the broken off piece of chromosome are lost. Sometimes a broken piece of chromosome sticks on to another chromosome, thus adding too many genes; this type of mutation is known as *duplication*. Sometimes a piece of chromosome becomes rearranged in the chromosome where it belongs, thus changing the sequence of the genes on that chromosome; this is known as an *inversion*, and it prevents gene for gene matching when chromosomes line up during meiosis. *Point mutations* are changes in individual genes.

Polyploidy is a condition in which cells develop extra sets of chromosomes: 3N, 4N. These extra sets of chromosomes change the characteristics of organisms. Usually polyploidy occurs in plants. Breeders may treat special plants with a chemical such as colchicine which prevents the division of the cell after the nucleus has divided. Polyploid fruits and flowers are quite large.

SEX DETERMINATION

In human beings, there are 22 pairs of *autosomes*, chromosomes that affect all characteristics except sex determination. One pair of chromosomes determines the sex of an individual. In females, the sex chromosomes are designated as XX. In males, the sex chromosomes are XY. Certain disorders are sex-linked, usually passed from mother to son by a defective gene on the X chromosome; red-green color blindness is one such sex-linked trait that is found more frequently in males and hardly at all in females. Hemophilia is another sex-linked trait that affects males with greater frequency than females.

MULTIPLE ALLELES

Alleles are two or more genes that have the same positions on homologous chromosomes. Alleles are separated from each other during meiosis and come together again at fertilization when homologous alleles are paired, one from the sperm cell and one from the egg cell. Two or more alleles determine a trait.

The inheritance of some characteristics cannot be explained by the action of a single pair of alleles. It has been shown experimentally that multiple alleles determine certain traits. However, within each cell no more than two alleles are present.

The inheritance of the ABO blood group in humans is an example of the existence of multiple alleles: I^A, I^B, and i. In this example, alleles I^A and I^B are codominant with each other and i is recessive to both I^A and I^B. Table 14.1 shows blood types and possible genotype.

TABLE 14.1.   Blood Genotypes

Blood Type	Genotype
A	$I^A I^A$ or $I^A i$
B	$I^B I^B$ or $I^B i$
AB	$I^A I^B$
O	ii

Modern Genetics

THE ROLE OF NUCLEIC ACIDS

DNA is an important part of the chromosome structure of all cells. DNA is a nucleic acid as is RNA. The unit of structure and function in the nucleic acid is called a *nucleotide*. A nucleotide is composed of a *phosphate group*, a five-carbon sugar, and a protein base. If the five-carbon sugar is *ribose*, the nucleic acid is *ribonucleic acid* (RNA). If the five-carbon sugar is deoxyribose, then the nucleic acid is *deoxyribose nucleic acid* (DNA). Fig. 14.1 is a diagrammatic presentation of a nucleotide.

Key:
P = phosphate
S = sugar
A = adenine
G = guanine
C = cytosine
T = thymine

Fig. 14.1 Nucleotides are the units on which DNA molecules are built.

The protein bases in nucleic acids are ring compounds. Those bases with single rings are *pyrimidines*. Bases with double rings are *purines*. The pyrimidines in nucleic acid are *thymine, cytosine,* and *uracil*. The purines are *adenine* and *guanine*. The four bases that make up the DNA molecule are adenine (A), guanine (G), thymine (T) and cytosine (C). The four bases that make up the RNA molecule are adenine (A), guanine (G), cytosine (C) and uracil (U).

In the early 1950's, James Watson, an American, and Francis Crick, an English investigator, unraveled the structure of DNA. The Watson-Crick model of DNA, as it has come to be known, indicates that the DNA molecule is shaped like a double helix. It consists of two long chains of nucleotides turned around each other in the shape of a double spiral. Fig. 14.2 shows the DNA molecule as a double helix. Notice that this diagram resembles a step ladder that has curved sides and straight rungs. The sides of the DNA molecule consist of alternate phosphate-sugar groups. The rungs of the "ladder" consist of protein bases joined by weak hydrogen bonds. It is known that adenine and thymine link together while guanine and cytosine link together (Fig. 14.3).

The RNA molecule is usually single stranded and much smaller than DNA. RNA is synthesized in the nucleus and functions in the cytoplasm where it controls the synthesis of proteins by the ribosomes.

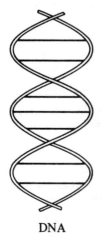

DNA

Fig. 14.2 DNA: double helix

HOW DNA FUNCTIONS

Deoxyribonucleic acid is able to function as genetic material because it has unusual characteristics. It is stable, can make more of itself, and can control the production of enzymes. DNA is stable as demonstrated by its ability to pass hereditary traits from one generation to another unchanged. However, DNA does change occasionally. This accounts for mutations.

DNA Replication

DNA can make more of itself in a process called *replication*. Replication refers to a duplication of molecules. Prior to the onset of cell division (mitosis and meiosis) DNA molecules replicate in a way that is at the same time both simple and precise. The double-stranded helix unwinds, forming a structure that resembles a straight-sided ladder with horizontal rungs. The rungs of the ladder are formed by the joining of a pyrimidine molecule with a purine through a weak hydrogen bond. The bond breaks and the two strands "unzip" between the base pairs. Complementary nucleotide chains are joined to the free base ends in the unzipped strands. Fig. 14.4 illustrates replication of DNA. As the result of the incorporation of free nucleotides into the unzipped strands, two new molecules of DNA are formed which are identical to each other and to the original molecule.

Genetic Code

The DNA molecule carries coded instructions for controlling all functions of the cell. At the present time, scientists know most about the functions of the *genetic code* that controls protein synthesis. They have determined that triplet combinations of bases code for each of 20 amino acids. The coded sequence of amino acids determines the formation of

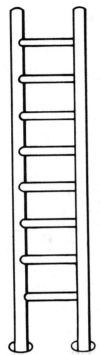

Fig. 14.3
Straight-sided
DNA

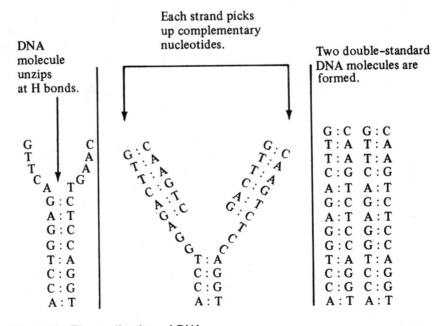

Fig. 14.4   The replication of DNA

different types of proteins. The code for proteins is present in messenger RNA molecules which are complementary to DNA molecules. For example: let us suppose that a portion of a DNA molecule carries a code such as AAC GGC AAA TTT. Its mRNA complement would be as follows: UUG CCG UUU AAA.

The mRNA, carrying this code, now moves from the nucleus to the cytoplasm. The mRNA attaches itself to several ribosomes, each having its own ribosomal RNA. Specific transfer RNA (tRNA) molecules bring to the ribosomes their own kind of activated amino acids. Transfer RNA molecules that fit the active sites of mRNA's on the ribosomes temporarily attach to them. As a result, amino acids are lined up in the proper sequence (Fig. 14.5). Note that the RNA code is a *triplet code* with one triplet, or *codon*, made up of three bases coding for a specific amino acid.

One Gene–One Polypeptide Hypothesis

In 1941 Beadle and Tatum used the red bread mold *Neurospora crassa* to find out how genes influence the synthesis of enzymes. As a result of their work the "one gene—one enzyme" hypothesis was formed. This hypothesis suggested that the synthesis of each enzyme in a cell was controlled by the action of a single gene. At present, it is known that a single enzyme may be composed of several polypeptides. The synthesis of each polypeptide is governed by a different gene. Hence, the new hypothesis "one gene—one polypeptide" is considered to be more accurate.

POPULATION GENETICS

A population includes all members of a species that live in a given location. Modern geneticists are concerned about the factors in populations that affect gene frequencies. All of the genes that can be inherited (heritable genes) in a population are known collectively as the *gene pool*. The Hardy-Weinberg Principle uses an algebraic equation to compute the gene frequencies in human populations. The conditions set by the Hardy-Weinberg Principle for determining the stability of a gene pool are as follows: large populations, random mating, no migration, and no mutation.

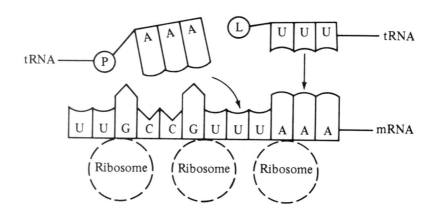

Fig. 14.5 Protein synthesis

Some New Directions in Genetics _____

Through the biotechnology of genetic engineering, scientists have found it possible to transfer genetic information from one organism to another. In a series of technical steps known as *gene splicing*, geneticists are able to remove a specific gene from the cells of a human and splice it within the circular DNA strand from a bacterium.

In a process called *transduction*, this spliced strand of DNA can be introduced into a bacterial cell, using a virus as a carrier. The bacterium will then reproduce *clones*, organisms bearing identical genes. The spliced ring of DNA is now *recombinant DNA*, bearing a human gene and bacterial genes. Then if, for example, the human gene carries the genetic code for insulin, the bacterial clones bearing this gene will synthesize insulin, the hormone that regulates sugar metabolism.

Genetic engineering has accomplished a great deal. The techniques of genetic engineering, which include gene splicing and transduction, have made possible medicinal drug substances for treating human ailments. *Interferons* are proteins that prevent the replication of RNA viruses. After certain cells have been infected by viruses, they produce interferons. These immune proteins are then used medically to prevent and treat certain diseases. Interferons are produced abundantly and inexpensively by means of recombinant DNA.

To date, the technology of producing recombinant DNA has proved very useful in the:

(1) production of insulin, using bacterial cells;
(2) production of interferons;
(3) production of human growth hormone;
(4) production of enzymes for the cheese industry;
(5) development of processes for safer and more efficient cleanup of organic wastes generated in the food-processing industries;
(6) treatment of hepatitis.

The Retrovirus of AIDS—Genetic Coding in RNA _____

The human immunodeficiency virus (HIV), the virus that causes AIDS, is a *retrovirus*, that is, a virus that carries its genetic information coded in ribonucleic acid (RNA) instead of in deoxyribonucleic acid (DNA). It is usual for RNA to carry only the genetic codes that build proteins. Retroviruses, however, produce the enzyme reverse transcriptase, which enables the virus to carry out its *replication* processes. Through replication, HIV makes more of itself.

When HIV infects (enters) the type of white blood cell known as a T-4 lymphocyte, the RNA of the virus captures the DNA of the invaded cell. Now reverse transcriptase goes to work, forcing the DNA of the T-4 cell to follow the coded instructions contained in the viral RNA. This results in the production of HIV viruses within the T-4 cell. When the T-4 cell can no longer hold the number of viruses produced, the cell membrane bursts, releasing numerous HIV viruses into the blood—and so the process continues. These viruses infect other T-4 lymphocytes, ultimately destroying them.

To summarize, under usual circumstances, the information of heredity is coded in DNA molecules. The genetic codes for protein synthesis are contained in DNA and then transferred to RNA. Through a series of steps involving three kinds of RNA, proteins useful to the cell are synthesized. The virus of AIDS, however, reverses the process. Genetic codes are transferred from RNA to DNA, forcing the captured cell to make HIV viruses instead of proteins.

Research workers have found that most retroviruses that infect animal cells have three genes, which have been named *pol*, *env*, and *gag*. *Env* contains the genetic blueprint that determines the structure of the viral outer coat. *Gag* holds the code for the internal structure of the virus. *Pol* provides instructions for the code for the characteristc enzymes of the virus.

Unlike other retroviruses, the virus of AIDS contains *eight* genes. The functions of four of these genes are unknown. However, a gene that research workers have named *tat* codes for the enormous production rate of the HIV virus, which accounts for the enormous infecting power of the AIDS virus. At present virologists are hard at work trying to unravel the mysteries of the other four HIV genes.

Importance to Humans

The concept of heredity has always been important to humans. The ancients living five thousand years ago in Babylonia and Assyria carried out the process of pollination to make their date palms produce fruit. Since they did not understand the mechanisms of fertilization and genetics, they turned to superstitious practices to solve their problems.

PLANT AND ANIMAL GENETICS

Plant and animal breeders throughout history have engaged in practices to improve their crops and their domesticated animals. Using the method of *selection* they mated organisms based on observed characteristics that were desirable to the breeder. But selection has its limits. Species cannot be improved beyond their genetic capabilities. Species improvements have been made through hybridization, in which varieties of organisms are crossed (mated) in the hope that the favorable characteristics of each will show up in the offspring. For example: a cross between the Texas longhorned cattle and the Indian Brahman bull has resulted in hybrids that are resistant to infections and produce good quality meat. Many varieties of hybrid sweet corn have been produced by selective breeding and carefully planned genetic crosses between strains having desired characteristics. These new strains of corn have improved texture, delicious taste and resistance to certain fungus diseases that infect corn plants.

TABLE 14.2.   Some Human Genetic Disorders

Disorder	Description
Phenylketonuria	Results from the inability of a gene to synthesize a single enzyme needed for the normal metabolism of the amino acid phenylalanine. Phenylketonuria renders a child mentally retarded. Testing the urine of newborns allows for corrective dietary treatment to prevent the disease.
Sickle-Cell Anemia	Results from the inheritance of two mutant genes for abnormal hemoglobin, which causes the sickling of red blood cells. The afflicted person experiences severe pain caused by obstructed blood vessels and anemia caused by the decreased hemoglobin content of the sickle-shaped cells.
Tay-Sachs Disease	Results from the inheritance of two recessive genes that cause a malfunctioning of the nervous system whereby nerve cells are destroyed. The deterioration in nerve tissue is caused by the accumulation of fatty material due to inability to synthesize a specific enzyme. The disorder greatly shortens the life span; it is rare for afflicted children to live beyond the age of 6 years.

HUMAN GENETICS

In 1956 the work of Joe Hin Tjio and Albert Levan provided a method of counting human chromosomes accurately. They developed the technique of producing a *karyotype*, which is a photograph of matched chromosome pairs. Shortly after this work was done, geneticists were able to point to the cause of Down's syndrome, a condition of mental retardation and physical handicap caused by a tripling of chromosome number 21. Table 14.2 describes three other human genetic disorders.

The study of human disease caused by genetic disorders has resulted in some remarkable findings. For example: a large number of human disorders (club foot, cleft lip and palate, spina bifida, and water on the brain—to name a few) are caused by the interaction of several genes. One mutated gene can cause a series of biochemical defects which are then translated into human abnormalities. The procedure of *amniocentesis* in which a small amount of amniotic fluid is removed from a pregnant woman is used to study cells of the embryo. In this way certain chromosomal defects can be determined before birth.

Chronology of Famous Names in Biology

1863   **Gregor Mendel** (Austrian)—initiated the first mathematical study of inheritance.

1869   **Fredrich Meischer** (German)—extracted nucleic acid from cells and named the substance.

1900   **Hugo deVries** (Dutch)—discovered mutations in the evening primrose.

1903   **Walter S. Sutton** (American)—discovered the mechanism of meiosis while studying the sperm cells of grasshoppers.

1908   **Sir Archibald Garron** (English)—discovered "inborn errors of metabolism".

1911   **Thomas Hunt Morgan** (American)—developed the theory of the gene. He was the first investigator to use the fruitfly for genetic research.

1914   **Robert Feulgen** (German)—found that fuchsin-red dye is specific for DNA.

1928   **Frederick Griffith** (English)—accidentally discovered that harmless diplococci could be transformed into harmful bacteria that cause pneumonia.

1941   **George Beadle** and **Edward L. Tatum** (American)—developed the "one gene-one enzyme" theory in work with *Neurospora crassa*.

1944   **Oswald T. Avery** (American)—found the transforming factor in DNA.

1949   **P. A. Levene** (American)—showed that nucleic acid could be broken down into four nitrogenous bases.

1950   **Edwin Chargaff** (American)—found that nitrogenous bases in DNA do not occur in equal proportions.

1952   **Alfred Hershey** and **Martha Chase** (English)—performed experiments with labeled viral DNA showing that it carries the complete hereditary message.

1952   **Alfred Mirsky** (American)—showed that tissue cells contain equal amounts of DNA.

1952   **James Watson** (American)—and **Francis Crick** (England)—were the first to demonstrate the double helix structure of DNA.

1952   **Norton Zinder** and **Joshua Lederberg** (American)—discovered that DNA is the genetic material.

1958   **Linus Pauling** (American)—found a difference in electrophoresis patterns between sickle cell hemoglobin and normal hemoglobin.

1959   **Vernon Ingram** (American)—showed that the difference between normal and sickle cell hemoglobin is one amino acid in 300.

1960   **Hans Gruneberg** (Indian)—discovered that a single mutant gene in rats causes a complex of disorders.

1960   **Arthur Kornberg** (American)—first to synthesize DNA *in vitro*.

1961   **Francis Crick** (English)—provided experimental evidence supporting the triplet code while working with the T_4 virus.

1985   **William A. Hazeltine** (American)—discovered the function of the *tat* gene in the HIV virus.

Words for Study

adenine	gene pool	polyploidy
alleles	gene splicing	purine
aminocentesis	genetic code	pyrimidine
autosome	genotype	recessive
clone	guanine	recombinant DNA
codominance	Hardy-Weinberg Principle	replication
codon	hybrid	retrovirus
crossing over	interferon	RNA
cytosine	intermediate inheritance	thymine
deletion	karyotype	transduction
dominant	mutation	translocation
DNA	nucleotide	triplet code
duplication	phenotype	uracil
gene	point mutation	

Questions for Review

PART A.   Completion. Write in the word that correctly completes each statement.

1. The first organized study of heredity was made by ..1..
2. Characteristics that can be observed are called ..2..
3. Genes on the same chromosome are said to be ..3..
4. Short wing in fruitflies is known as ..4.. wing.
5. Linkage groups are broken by ..5..
6. The chemical ..6.. induces polyploidy.
7. Two or more genes that have the same positions on homologous chromosomes are said to be ..7..
8. Pyrimidines are nucleic acids with ..8.. rings.
9. The complete name for DNA is ..9..
10. Messenger RNA is synthesized in the ..10..
11. DNA that has been changed by adding a gene from another species is known as ..11.. DNA.
12. Organisms arising from the same parent cell and bearing identical genes are called ..12..
13. Bacterial cells bearing the human gene for insulin will synthesize ..13..
14. The technique of adding a human gene to a circular strand of bacterial DNA is termed gene ..14..

15. The overall technology of transferring genetic information from one organism to another has become known as genetic ..15..

16. The shortened name of the virus of AIDS is ..16..

17. The nucleic acid ..17.. codes the genetic information of the virus that causes AIDS.

18. Genetic information in cells is coded in the nucleic acid ..18..

19. The process in which a virus makes more of itself is known as ..19..

20. The genetic material of a T-4 lymphocyte is captured by a (an) ..20.. virus.

PART B.   Multiple Choice. Circle the letter of the item that correctly completes each statement.

1. If two organisms that exhibit contrasting traits are crossed, the trait that shows up in the F_1 generation is called
   (a) codominant   (c) recessive
   (b) dominant   (d) allelic

2. The trait that remains hidden in the F_1 generation is the
   (a) dominant   (c) recessive
   (b) codominant   (d) phenotype

3. The specimen used by Mendel in his work was the
   (a) fruitfly   (c) firefly
   (b) snapdragon   (d) garden pea

4. A 3:1 ratio is characteristic of an
   (a) F_1 cross   (c) F_3 cross
   (b) F_2 cross   (d) F_4 cross

5. The genotypic ratio of an F_1 cross is
   (a) 3:1   (c) 1:2:1
   (b) 8:2   (d) 1:3:1

6. *Drosophila melanogaster* is a
   (a) fruit fly   (c) bread mold
   (b) gene   (d) cross over

7. Red cattle crossed with white cattle produce a red and white hybrid. This type of inheritance is known as
   (a) mutation   (c) recessive
   (b) codominance   (d) transformation

8. A loss of a piece of chromosome is known as a
   (a) translocation   (c) deletion
   (b) transfiguration   (d) duplication

9. Homologous chromosomes intertwine during
   (a) meiosis   (c) transduction
   (b) mitosis   (d) translocation

10. Human blood groups are formed by
    (a) a single pair of genes only
    (b) multiple alleles
    (c) several pairs of genes
    (d) no genes

11. Introducing recombinant DNA into a bacterial cell by means of a virus carrier is accomplished by
    (a) transference
    (b) translocation
    (c) transduction
    (d) translation

12. Certain cells produce interferons after having been infected by
    (a) viruses
    (b) fungi
    (c) bacteria
    (d) molds

13. Interferons are best classified as
    (a) hormones
    (b) toxins
    (c) catalytic enzymes
    (d) immune proteins

14. Cells of human embryos can be studied by the technique of
    (a) genetic engineering
    (b) amniocentesis
    (c) translocation
    (d) gene splicing

15. A photograph of matched chromosome pairs is a
    (a) micrograph
    (b) karyotype
    (c) daguerrotype
    (d) pictograph

16. Reverse transcriptase is best associated with
    (a) a retrovirus
    (b) AIDS infection
    (c) DNA transcription
    (d) T-4 lymphocytes

17. The human immunodeficiency virus carries its genetic code in molecules of
    (a) HIV
    (b) DNA
    (c) ATP
    (d) RNA

18. When a T-cell is infected with an HIV virus, the T-cell ultimately
    (a) makes more of its kind
    (b) makes more protein
    (c) increases its DNA content
    (d) is destroyed

19. The usual number of genes in an animal virus is
    (a) one
    (b) two
    (c) three
    (d) four

20. The function of RNA is to code for
    (a) carbohydrates
    (b) proteins
    (c) fats
    (d) minerals

PART C.   Modified True-False. If a statement is true, write "true" for your answer. If a statement is incorrect, change the underlined word to one that will make the statement true.

1. Point mutations are changes in single <u>cells</u>.

2. When a piece of broken chromosome sticks to another complete chromosome, the defect is known as <u>linkage</u>.

3. The five-carbon sugar in DNA is named <u>ribose</u>.

4. The sex chromosomes of human males are <u>XX</u>.

5. A person with genotype <u>ii</u> has blood type <u>AB</u>.

6. Thymine is always joined to <u>guanine</u>.

7. The DNA molecule is shaped like a double <u>circle</u>.

8. DNA molecules are composed of units called <u>nucleic acids</u>.

9. RNA does not contain the base <u>uracil</u>.

10. Pyrimidine molecules are joined to purines through weak <u>oxygen</u> bonds.

11. Down's syndrome is a genetic disease caused by the <u>doubling</u> of chromosome number 21.

12. Recombinant DNA is carried into a bacterial cell by means of a <u>catalyst</u>.

13. Recombinant DNA is now used to produce two human <u>enzymes</u>.

14. Interferons prevent the replication of <u>DNA</u>.

15. The names Tjio and Levan are correctly associated with counting <u>genes</u>.

16. The viral gene *env* codes for the <u>environment</u> of the virus.

17. Thus far, the number of genes identified in the AIDS virus is <u>six</u>.

18. A virus that passes its genetic code from RNA to DNA is a <u>reverse virus</u>.

19. Reverse transcriptase helps the HIV virus to carry out its <u>metabolic</u> processes.

20. A cell that is invaded by a virus is called a <u>home</u> cell.

Think and Discuss

1. Although Gregor Mendel did not know about genes, he formulated three basic principles of heredity. Explain.

2. Cite an example in which Mendel's law of dominance does not hold true.

3. What does the "one gene-one enzyme" hypothesis suggest?

4. How does a retrovirus differ from a regular virus?

5. What ethical questions might be raised against the use of genetic engineering?

Answers to Questions for Review

PART A

1. Mendel
2. phenotypes
3. linked
4. vestigial
5. crossing over
6. colchicine
7. alleles
8. single
9. deoxyribonucleic acid
10. nucleus

11. recombinant
12. clones
13. insulin
14. splicing
15. engineering
16. HIV
17. RNA
18. DNA
19. replication
20. invading

PART B

1. b	8. c	15. b
2. c	9. a	16. a
3. d	10. b	17. d
4. b	11. c	18. d
5. c	12. a	19. c
6. a	13. d	20. b
7. b	14. b	

PART C

1. genes
2. duplication
3. deoxyribose
4. XY
5. 0
6. adenine
7. helix
8. nucleotides
9. thymine
10. hydrogen
11. tripling
12. virus
13. hormones
14. RNA
15. chromosomes
16. outer coat
17. eight
18. retrovirus
19. replication
20. host

CHAPTER 15

PRINCIPLES OF EVOLUTION

Evolution concerns the orderly changes that have shaped the earth and that have modified the living species that inhabit the earth. Evolution is a fusion of biological and physical sciences that have provided supporting data which confirm the fact that over periods of time major changes have occurred in the interior of the earth and on its surface, accompanied by modifications in climate. All of the changes in the earth are classified as nonbiological or *inorganic* evolution. Changes that have taken place in living organisms are known as biological or *organic evolution*.

Evidence of Evolution

Evidence that evolution—gradual change over a period of time—has occurred in living things is provided by many sciences and includes facts from the geologic record, the study of fossils, and evidence from cell studies, biochemistry, comparative anatomy and comparative embryology.

THE GEOLOGIC RECORD

Look at Table 15.1—The Geologic Time Scale. This scale is to be read from the bottom upward because it represents the age of the earth as determined by the rock layers of the earth. Geologists believe that the earth is between 4.5 and 5 billion years old. The age of the earth is measured by a process called *radioactive dating*.

TABLE 15.1. The Geologic Time Scale

Era	Period	Epoch	Millions of Years Ago	Climate and Life
CENOZOIC	Quarternary	Recent	0.01	4 ice ages; *Homo sapiens*.
		Pleistocene	2.5	Increase in herb population; first *Homo*.
	Tertiary	Pliocene	7.0	Cool; hominoid apes; first humans.
		Miocene	26.0	Forests decrease; dominance of angiosperms.
		Oligocene	38	Dominance on land by mammals, (anthropoid apes, ungulates, whales); birds; insects.
		Eocene	54	Mild to tropical weather; small horses.
		Paleocene	65	First primates and carnivores.
MESOZOIC	Cretaceous		136	Rise of angiosperms; decline of gymnosperms; extinction of dinosaurs; second great radiation of insects.
	Jurassic		190	Europe covered by ocean; last of the seed ferns; gymnosperms dominant; reptiles dominant; origin of birds; dinosaurs abundant.
	Triassic		225	Extensive arid and mountainous areas; dominance of land by gymnosperms; first dinosaurs; first mammals.
PALEOZOIC	Permian		280	Glaciers in southern hemispheres; Appalachians rising; first conifers, cycads, ginkos; expansion of reptile; decline of amphibians; extinction of trilobites.
	Carboniferous	Pennsylvanian Mississippian	345	Subtropical climate; swamps; great coal forests; lycopsids, sphenopsids, ferns, gymnosperms; reptiles evolve; amphibians dominant; first insects; fungi.
	Devonian		395	U.S. covered by oceans, Europe mountainous; primitive tracheophytes; origin of first seed plants; first liverworts; age of fishes, sharks.

TABLE 15.1. The Geologic Time Scale (*Continued*)

Era	Period	Epoch	Millions of Years Ago	Climate and Life
PALEOZOIC (*cont'd*)	Silurian		430	Rise of mountains in Europe; continents flat; first vascular plants; arthropods appear on land.
	Ordovician		500	Mild climate; seas cover continents; plants invade land; marine algae abundant; jawless fishes evolve.
	Cambrian		570	Primitive marine algae; marine invertebrates in great numbers.
PRECAMBRIAN				Earth cooling; shallow seas; eukaryotes evolve; cyanobacteria; bacteria.

Scientists have determined that certain elements disintegrate by giving off radiations spontaneously and at a regular rate. Such elements are said to be *radioactive*. In the process of emitting radiations, the radioactive substance changes to something else. For example: uranium-238 changes to lead. The *half-life* of U-238, the rate at which one half of the uranium in a rock sample will change to lead, is 4.5 billion years. Uranium's rate of decay is not affected by any chemical or physical conditions. Therefore measuring the uranium-lead ratio in a sample of rock is a very reliable method for estimating the age of the rock.

Dating of the oldest rocks found on earth indicates that they are about 3 billion years old. To allow time for the original formation of the rocks, geologists add another 2 billion years to this figure, thus arriving at the 4.5 to 5 billion-year estimate of the age of the earth.

FOSSIL EVIDENCE

Fossils are the preserved remains of plants and animals. They are often found in sedimentary rock, which is formed by the gradual settling of *sediments*. The age of fossils is estimated by the use of *carbon dating*. The sample plant or animal fossil is tested for the ratio of radioactive carbon (carbon 14) to nonradioactive carbon (carbon 12). By using the rate of decay of carbon 14 to carbon 12, the age of a fossil can be determined.

The fossil records contained in the layers of sedimentary rock provide reliable evidences of change in plant and animal species. The lower down the rock layer, the older the fossil. Top layers contain more recent fossil remains of more complex species. The hard parts of animals, such as a shell or a skeleton, become fossilized in the hardened sediments of rock. Imprints, casts, or molds are other types of fossil remains, produced by an organism leaving a footprint, a track or form in the sediment.

Other types of fossil remains have also provided evidence of ancient species. *Amber* is the hardened resin of trees. Insects trapped in the sticky gum remain preserved as the resin hardens into amber. Ice has preserved some rather huge animals, such as the woolly mammoth, which was probably caught in the glaciers of Siberia. Leaf imprints have been preserved in coal while it was being formed. *Petrification*, the absorbing of mineral matter by dead plant and animals, preserves species in stone. Bogs have been the sources for preserved wood fossils. Since bacteria of decay cannot thrive in the acid environment of bogs, wood samples have been kept intact. Saber-toothed tiger fossils have been found in the La Brea tar pits in Los Angeles.

EVIDENCE FROM CELL STUDIES AND BIOCHEMISTRY

The cells of all living organisms have comparable structures that function in similar ways. All eukaryotic cells have a cell membrane, a nucleus (except mature red blood cells), cytoplasm with energy-producing mitochondria, and ribosomes where proteins are synthesized. The fact that the cells of all living things have similar structures that perform the same tasks indicates that there is evolutionary unity in all living things.

On the molecular level, there is similarity in genetic material of cells. Similar genes direct the formation of similar cell structures and similar proteins. For example: insulin produced in the hog pancreas is so similar to human insulin that hog pancreas is the source of insulin that is prescribed for diabetics. This means that the human and the hog have some very similar DNA molecules. It is not unusual for animals of various species to synthesize some proteins of similar natures. Interestingly enough, the primates (including humans and apes) and the guinea pigs show an unusual kind of biochemical relationship. These are the only vertebrate animals that cannot synthesize vitamin C from carbohydrates.

EVIDENCE FROM COMPARATIVE ANATOMY

A comparative study of the bone structures and body systems of animals from the various phyla reveals a great deal of similarity. A comparative study of the skeletal systems of vertebrates shows that many of the bones are very much alike. Much of our evidence for evolution comes from a study of *homologous* structures. Homologous structures are bones that look alike and have the same evolutionary origin although they may be used for different purposes. Fig. 15.1 shows the bones in the forelimbs of vertebrates. Notice the similarity of structure of these bones. The flipper of a whale, the arm of a human and the wing of a bird are *homologous* structures having the same evolutionary origin and maintaining similarity of structure.

A comparison of the digestive, nervous and circulatory systems of vertebrates indicates similar evolutionary origin. Biologists believe that there must have been a common ancestor from which the vertebrate line descended. Fig. 15.2 shows the brains of some common vertebrate species. Notice their similarities and differences.

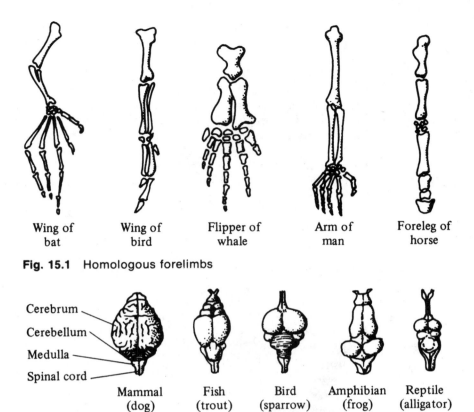

| Wing of bat | Wing of bird | Flipper of whale | Arm of man | Foreleg of horse |

Fig. 15.1   Homologous forelimbs

Cerebrum
Cerebellum
Medulla
Spinal cord

| Mammal (dog) | Fish (trout) | Bird (sparrow) | Amphibian (frog) | Reptile (alligator) |

Fig. 15.2   A comparison of vertebrate brains

EVIDENCE FROM COMPARATIVE EMBRYOLOGY

Embryology is the study of developing forms or embryos. A comparative study of the embryos of seemingly unrelated vertebrates indicates similar evolutionary origins. Fig. 15.3 shows some of these embryos. Notice the marked similarity in structure. As the embryonic development continues, the distinctive traits of each species begin to take form.

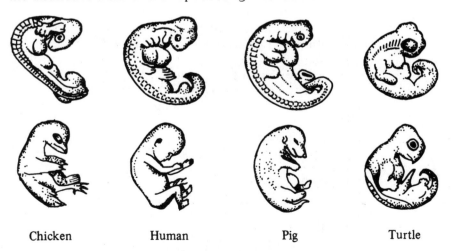

| Chicken | Human | Pig | Turtle |

Fig. 15.3   A comparison of vertebrate embryos

Theories of Evolution

Since the 18th century several theories have been proposed to explain evolution. Among these are the use and disuse theory of Lamarck and the theory of *natural selection* formulated by Charles Darwin. Recently advances in genetics and biochemistry have led to the formulation of a modern theory of evolution.

USE AND DISUSE

The French biologist Jean Baptiste Lamarck (1744–1829) presented the first well organized theory of evolution. Lamarck recognized that living animals and plants throughout the world were different. He proposed that these differences were due to changes which caused significant variations among species. He also believed that changes are going on all of the time, because evolution is a continuing process.

Lamarck believed that organisms evolved in straight line fashion from the "less perfect to the more perfect". We know today that branching evolution better explains the diversity of the many species. Lamarck also believed that the changes that take place in living organisms are stimulated by needs in the environment. He believed that systems and body structures developed in response to use and that those structures in "disuse" eventually disappeared. Lamarck also believed that acquired characteristics could be passed on from parent to offspring.

German biologist August Weismann (1834–1914) used a rather dramatic experiment to disprove Lamarck's theory of inheritance of acquired characteristics. Weismann cut off the tails of mice and mated them. He carried out this process for 25 generations of mice. In each mating he found that mice grew tails just as long as the tails of those mice in the first generation. Weismann proved that cutting off the tails of mice (acquired characteristic) did not alter the "germplasm" of future generations.

NATURAL SELECTION

Charles Darwin (1809–1882) was an English naturalist who together with his cousin, Alfred Russel Wallace (1823–1915), developed a theory of evolution which laid the groundwork for modern biological thinking. At the age of 22 in the year of 1831, Charles Darwin commenced on a trip around the world as the official naturalist of the ship *H.M.S. Beagle*. Darwin was commissioned to collect as many specimens of animal and plant life as possible and to record observations of the natural phenomena that he saw. Darwin's trip on the *Beagle* extended over a five-year period. The ship traveled around the coast of South America to the Galápagos Islands. From there the ship pursued a westerly course passing the southern coast of Australia, then going north to the southern coast of Asia and then south along the southern coast of Africa back to England.

In all of the places where the ship stopped, Darwin collected specimens of living organisms and recorded what he observed. For the next twenty-five years Darwin studied and organized his notes. In 1859 he put together his thoughts in a published work titled *On the Origin of Species by Means of Natural Selection, or the Preservation of Favored Races in the Struggle for Life.* In this book, Darwin's theory of natural selection was clearly set forth. Today this work is referred to by a shorter title, *The Origin of Species.*

Darwin's theory of natural selection can be summed up thusly: large numbers of new plants and animals are produced by nature. Many of these do not survive because nature "weeds out" weak and feeble organisms by killing off those that cannot adapt to changing environmental conditions. Only the strongest and most efficient survive and produce progeny. Specific tenets of the Darwin-Wallace theory of evolution follow.

Overproduction

The theory of natural selection cites the fact that every organism produces more gametes and/or organisms that can possibly survive. If every gamete produced by a given species united in fertilization and developed into offspring, the world would become so overcrowded in a short period of time that there would be no room for successive generations. This does not happen. There is a balance that is maintained in the reproduction of all species and therefore natural populations remain fairly stable, unless upset by a change in conditions.

Competition

As stated above, not all the offspring (and gametes) survive. There is competition for life among organisms: competition for food, room and space. Therefore there is a *struggle for existence* in which some organisms die and the more hardy survive.

Survival of the Fittest

Some organisms are better able to compete for survival than others. The differences that exist between organisms of the same species making one more fit to survive than another can be explained in terms of *variations.* Variations exist in every species and in every trait in members of the species. Therefore some organisms can compete more successfully for the available food or space in which to grow or can elude their enemies better. These variations are said to add survival value to an organism. Survival value traits are passed on to the offspring by those individuals that live long enough to reproduce. As weaker individuals are weeded out of the species, those individuals that remain do so because they are best adapted to live in their environment. As time goes on the special adaptations for survival are perpetuated and new species evolve from a common ancestral species.

The environment is the selecting agent in natural selection because it determines what variations are satisfactory for survival and which are not. Any change in an environment can affect the usefulness of a given variation. A change in the environment can render a once useful variation useless. In this way the direction of evolutionary development can change.

MODERN EVOLUTIONARY THEORY

The major weakness in Darwin's theory of natural selection is that he did not explain the source, or genetic basis, for variations. He did not distinguish between variations that are hereditary and those that are non-hereditary. He made the assumption that all variations that have survival value are passed on to the progeny. Like Lamarck, Darwin believed in the inheritance of acquired characteristics.

We now know that there are several ways in which variation is produced within a species.

Causes of Variation

Hugo De Vries (1848–1935), a Dutch botanist, explained variations in terms of *mutations*. His study of 50,000 plants belonging to the evening primrose species enabled him to identify changes in leaf shape and texture and in plant height that were passed on from parent plant to offspring. In 1901 De Vries offered his mutation theory to explain organic evolution. He did not know how mutations come about or where they occur. Today, we know that mutations are changes in genes that can come about spontaneously or can be induced by some mutagenic agent.

Spontaneous mutation rates are very low. In *E. coli* one cell in every 10^9 will mutate from streptomycin sensitivity to streptomycin resistance. In human beings the rate of mutation in Huntington's disease occurs in 5 out of every 10^6 gametes. It is a known fact that different genes have different mutation rates. This is probably due to the different chemical composition of genes or perhaps to the different places that they occupy on chromosomes. Mutations alone do not effect major changes in the frequencies of alleles.

An important cause of variation within species is genetic recombination that results from sexual reproduction wherein the genes of two individuals are sorted out and recombined into a new combination, producing new traits—and thus variation.

Gene flow is another agent of evolution, responsible for the development of variations. It is the movement of new genes into a population. Gene flow often acts against the effects of natural selection. *Genetic drift* is a change in a gene pool that takes place in a population as a result of chance. Let us suppose that a mutation takes place in a gene of one person. If that person does not reproduce, the gene is lost to the population. Sometimes a small population breaks off from a larger one. Within that small population is a mutant gene. Because now that the mating within the small population is very close, the frequencies of the mutant gene will increase. In the Amish population where there is very little (or no) outbreeding, there is an increase in the homozygosity of the genes in the gene pool. This is seen in the high frequencies of genetic dwarfism and polydactyly (six fingers). The isolated smaller population has a different gene frequency than the larger population from which it came. This is known as the *founder principle*.

Genetic drift and the random mutations that increase or decrease as the result of genetic drift are known as non-Darwinian evolution.

Speciation is the forming of one or more new species from a species already in existence. This can happen when a population becomes divided and part of the original species continues life in a new habitat. Islands cut off from the mainland by surrounding water, isolated mountain peaks, migration of a part of the species to a new area—all these are examples of geographic isolation from the main population. The separated populations cannot interbreed. Over evolutionary time, different environments present different selective pressures, and the change in gene pools will eventually produce new species. For example: Darwin identified 14 different species of finches living on the Galapagos Islands that had decended from a single species of finch that lived on mainland Peru.

Time Frame

Currently, it is believed that there is a time frame for evolution. The concept of *gradualism* supports the idea that evolutionary change is slow, gradual and continuous. The concept of *punctuated equilibrium* sets forth the idea that species have long periods of stability, lasting for four or five million years, and then suddenly change as the result of some geological or other environmental change.

The Origin of Life

Theories of evolution are concerned not only with how living organisms have changed through time but also with how life began—how living organisms first evolved on earth. In the 1920's the Russian scientist A. I. Oparin began investigations into how life could have evolved from inorganic compounds under the conditions of early earth. He found that he could produce coacervate droplets that had the ability to incorporate simple enzymes in their structure. This initial work opened the door to many studies of how life could have begun.

In the 1950's Stanley Miller set up an experiment in which he duplicated the chemical conditions and the temperature of the early seas. Into sterile water he put some methane gas, hydrogen and ammonia. He sealed off the system so that nothing could leave or enter it. The water was heated to a temperature similar to that of the early seas. The water and gas mixture were subjected to electric sparks. Miller let the experiment run for about a week. At the end of that time, he tested the water and found present in it organic compounds that were not there before. Miller's experiments help to explain how organic matter appeared in the early seas.

In 1963, Carl Sagan, duplicating Miller's experiment, was able to produce ATP (adenosine triphosphate) from inorganic matter. ATP molecules are essential to all living cells and function as energy storage molecules.

Following the lead of Stanley Miller, other investigators demonstrated how organic molecules could have been formed in the early seas. Melvin Calvin demonstrated the polymerization of complex molecules. Sidney Fox produced microspheres, long chain peptides surrounded by something resembling a membrane. All of these experiments have contributed to the theory of the origin of life.

Scientists believe that the first organisms were heterotrophs that obtained nutrition from the organic molecules in the "hot, thin soup", an expression used to describe the early oceans. As oxygen began to increase in the atmosphere, conditions changed for the heterotrophs. Those heterotrophs that could not adjust to the changing atmospheric conditions were destroyed. Eventually, organisms that could make their own organic compounds from inorganic materials in the presence of sunlight began to evolve. These early autotrophs were the blue-green algae.

The blue-green algae increased the oxygen content of the air—remember oxygen is a byproduct of photosynthesis—and this further threatened the continued existence of the heterotrophs. As a result of the high concentration of oxygen in the atmosphere, an ozone layer developed which further diminished the organic compounds available to the groups of existing heterotrophs. From this group there evolved heterotrophs that could utilize oxygen in cellular respiration during which energy was released from organic molecules taken in as nutrients. Thus the carbohydrates produced by the autotrophs and the oxygen in the air supplied the newly evolved heterotrophs with the nutrients necessary for survival. This description is known as the *heterotroph hypothesis*. It is an attempt to explain the origin of autotrophic and heterotrophic cells.

Evolution of Humans _____

The evolution of *Homo sapiens* has always been of interest to modern humans. People like to know their origins. In 1924 the first specimen of *Australopithecus africanus* was discovered in South Africa. It is estimated that this prehuman form lived about 2 or 3 million years ago. Studies of the several australopithecine fossils found indicate that this species may well have been the forerunner of humans. The brain case indicates that the brain was within the size range of modern apes. The teeth were more human-like than those of modern apes. They were arranged in rows without gaps between the canines and the premolars, and the canine teeth did not stick out beyond the adjacent teeth. The bones of the pelvis show that the australophithecines were bipedal, walking on two feet. Because not all of the australopithecine fossils are alike, anthropologists believe that two species of man-ape lived at about the same time: *Australopithecus africanus* and *Australopithecus robustus*. Very little is known about the life patterns of these two species.

Fossil finds indicate that another species of prehuman lived on earth probably for as long as 500,000 to 1,000,000 years. *Homo erectus*, as this species is called, lived on various continents of the earth. The brain case is smaller than that of modern man, but the teeth are larger. The fossil teeth show traces of having had an enamel collar, a primitive trait in apes and in man. The species became extinct (Fig. 15.4).

In 1856 the fossil remnants of a human species called *Homo sapiens neanderthales* were found in a cave in Germany. It is believed that Neanderthal man appeared on earth about 125,000 years ago. He lived through

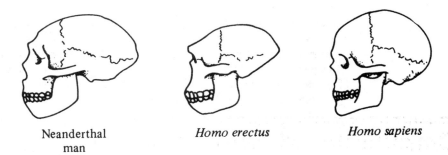

Neanderthal man

Homo erectus

Homo sapiens

Fig. 15.4   A comparison of skulls of three human species

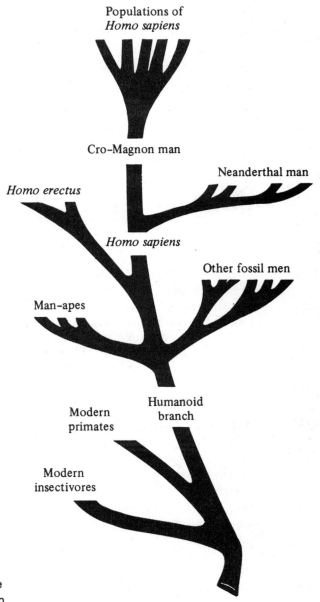

Populations of
Homo sapiens

Cro-Magnon man

Neanderthal man

Homo erectus

Homo sapiens

Other fossil men

Man–apes

Humanoid
branch

Modern
primates

Modern
insectivores

Fig. 15.5   The tree
of human evolution

two glacial periods and then became extinct between 50,000 and 15,000 years ago. It is commonly accepted theory that over thousands of years, modern humans gradually replaced the Neanderthal grade of man, until there remained only one species of human being—*Homo sapiens* (Fig. 15.5).

Chronology of Famous Names in Biology

1747	**Comte de Buffon** (French)—was the first scientist to state that living things can change.
1771	**Erasmus Darwin** (English)—was an early advocate of the idea of evolution.
1784	**Jean Baptiste Lamarck** (French)—presented the first organized theory of evolution, although his ideas are not acceptable today. He proposed the theory of "use and disuse" which suggested that organs that were well developed from use would be passed on to offspring in the well developed stage.
1800	**Thomas Malthus** (English)—wrote the notable "Essay on Population" in which he said that food supplies cannot keep pace with rapid increases in human population.
1830	**Charles Darwin** (English)—developed the theory of natural selection on which the modern ideas of evolution are based.
1830	**Alfred Wallace** (English)—proposed a theory of evolution similar to that of Charles Darwin's.
1874	**August Weismann** (German)—proposed that germplasm carries the hereditary material and this affects the successive generations of progeny.
1891	**Eugene Dubois** (Javanese)—found the braincase of the first known fossil of *Homo erectus*.
1924	**Raymond Dart** (South Africa)—identified the first recovered specimen of *Australopithecus*.
1935	**E. B. Ford** (English)—introduced the concept of industrial melanism to explain change in color of species of moths.
1940	**Trofim D. Lysenko** (Russian)—tried to change genes and inheritance by subjecting plants to periods of heat and cold. His methods in agriculture failed. Lysenko is remembered for his fallacious proposals concerning the "vernalization of wheat"
1947	**Robert Broom** (South African)—found fossil bones of *Australopithecus africanus*.
1950	**Bernard Kettlewell** (English)—provided an excellent investigative study of evolutionary selection in the peppered moth.
1960	**Louis Leakey** and **Mary Leakey** (Tanzanian)—found fossils that predate humans.

Words for Study

amber
evolution
fossil
founder principle
gene flow
genetic drift
gradualism

heterotroph hypothesis
homologous
natural selection
petrification
punctuated equilibrium
radioactive dating
speciation

Questions for Review

PART A. **Completion.** Write in the word that correctly completes each statement.

1. Changes that have taken place in the structure and energy of the earth are classified as ..1.. evolution.
2. Elements that emit atomic particles are said to be ..2..
3. The kind of rock in which fossils are found is ..3.. rock.
4. Cells of all living things have ..4.. structures.
5. The flipper of a whale and the wing of a bird are ..5.. structures having the same evolutionary origin.
6. The theory of "use and disuse" was proposed by ..6..
7. The theory of natural selection was proposed by ..7..
8. Hugo De Vries explained variations in terms of ..8..
9. Loss of genes from a population is known as ..9.. (2 words).
10. The experiments of ..10.. helped to explain how organic molecules appeared in the early seas.

PART B. **Multiple Choice.** Circle the letter of the item that correctly completes each statement.

1. *Australopithecus africanus* is considered to have been a
   (a) man-ape
   (b) man
   (c) ape
   (d) old world monkey
2. A species of early man, the fossils of which have been found all over the world, is
   (a) *Australopithecus*
   (b) Neanderthal
   (c) *Homo erectus*
   (d) *Homo habilis*
3. The early autotrophs were probably
   (a) bacteria
   (b) viruses
   (c) amoeba
   (d) blue-green algae

4. The first living cells were probably
   (a) autotrophs   (c) parasites
   (b) heterotrophs   (d) symbionts

5. The modern theory of the origin of life was initially developed by
   (a) Miller   (c) Sagan
   (b) Fox   (d) Oparin

6. The founder principle explains
   (a) the origin of life
   (b) abnormal gene frequencies in a small population
   (c) the methods by which genes leave a population
   (d) finding and securing of lost genes

7. Genes are sorted out and recombined by the process of
   (a) genetic drift   (c) sexual reproduction
   (b) mutation   (d) vegetative reproduction

8. The evening primrose was the experimental specimen used in the study of
   (a) genetic drift   (c) gene flow
   (b) mutation   (d) vegetative propagation

9. Darwin's theory of evolution is known as
   (a) natural selection   (c) survival of the fittest
   (b) struggle for existence   (d) competition

10. The "continuity of the germplasm" was demonstrated in an experiment using mice by the investigator named
    (a) Miller   (c) Fox
    (b) Lamarck   (d) Weismann

PART C. **Modified True False.** If a statement is true, write "true" for your answer. If a statement is incorrect, change the underlined word to one that will make the statement true.

1. The age of the earth is measured by a process called <u>reactive</u> dating.

2. Uranium-238 ultimately turns into <u>boron</u>.

3. It is estimated that the oldest rocks on earth are about <u>5</u> billion years old.

4. Fossil plants and animals are dated by measuring the <u>hydrogen</u> content.

5. Fossil insects are most likely to be found in <u>ambergris</u>.

6. The absorbing of mineral matter by dead plant and animal bodies which turns them into stone is known as <u>petrification</u>.

7. Human beings <u>can</u> synthesize vitamin C from carbohydrates.

8. The study of developing forms is known as <u>anatomy</u>.

9. The H.M.S. Beagle is most closely associated with <u>Wallace</u>.

10. The concept of punctuated equilibrium helps to explain the time frame of <u>reproduction</u>.

Think and Discuss

1. Explain how radioactive dating is used to determine the age of the earth.

2. How do you know that dinosaurs did not live when humans appeared on earth?

3. How does fossil evidence support the theory of evolution?

4. How can you prove that mutations are inherited?

Answers to Questions for Review

PART A

1. inorganic
2. radioactive
3. sedimentary
4. similar
5. homologous

6. Lamarck
7. Darwin
8. mutations
9. genetic drift
10. Miller

PART B

1. a
2. c
3. d
4. b

5. d
6. b
7. c

8. b
9. a
10. d

PART C

1. radioactive
2. lead
3. 3
4. carbon
5. amber

6. true
7. cannot
8. embryology
9. Darwin
10. evolution

CHAPTER 16

ECOLOGY

*E*cology is the science that studies the interrelationships between living species and their physical environment. The word "ecology" was coined in 1869 by the German zoologist Ernst Haeckel to emphasize the importance of the environment in which living things function. The environment includes living or *biotic* factors and nonliving factors referred to as *abiotic* factors.

The abiotic factors consist of physical and chemical conditions that affect the ability of a given species to live and reproduce in a particular place. Included in the abiotic factors are temperature, light, water, oxygen, pH (acid-base balance) of soil, type of substrate, and the availability of minerals. Certain kinds of plants and animals will flourish in a natural community if the conditions are present that permit their survival. Species interact to influence the survival of one another. One important principle of ecology is that no living organism is independent of other organisms or of the physical environment, if they share the same community.

The Concept of the Ecosystem _____

Certain terms are used in ecology to provide a consistent description of conditions and events. A *population* refers to all of the members of a given species that live in a particular location. For example: a beech-maple forest will contain a population of maple trees, a population of beech trees, a population of deer, and populations of other species of plants and animals. All of the plant and animal populations living and interacting in a given environment are known as a *community*.

The living community and the nonliving environment work together in a cooperative ecological system known as an *ecosystem*. An ecosystem has no size requirement or set boundaries. A forest, a pond and a field are examples of ecosystems. So is an unused city lot, a small aquarium, the lawn in front of a residential dwelling, or a crack in a sidewalk. All of these examples reflect areas where interaction is taking place between living organisms and the nonliving environment.

COMPONENTS OF THE ECOSYSTEM

Modern ecologists think of the ecosystem in terms of its interacting forces or components. One such component is the physical environment. This includes the air, which is made up of 21 percent oxygen, 78 percent nitrogen, 0.03 percent carbon dioxide, and the remainder inert gases. The soil is the source of minerals that supply plants with compounds of nitrogen, zinc, calcium, phosphorus, and other minerals.

The green plants in the ecosystem are another component. These are the *producers*, so-called because they are able to make their own food using the inorganic materials of carbon dioxide and water and minerals from the soil. Directly dependent upon the producers are the *primary consumers—herbivores*, or plant-eaters. Herbivores come in all sizes: crickets, leaf cutters, deer and cattle. The *carnivores*, or flesh-eaters, such as snakes, frogs, hawks, and coyotes are *secondary consumers* because they feed on the herbivores. The *tertiary consumers* are those that feed on the smaller carnivores and herbivores as well. There are also scavengers in the ecosystem. Earthworms and ants feed on particles of dead organic matter that have decayed in the soil. Vultures eat the bodies of dead animals.

The *decomposers* form another important part of ecosystems. Bacteria and fungi are organisms that break down dead organic matter and release from it organic compounds and minerals that are returned to the soil. Many of the materials returned to the soil are used by the producers in the process of food-making. Without the work of the decomposers the remains of dead plants and animals would pile up, not only occupying space needed by living organisms, but also keeping trapped within their dead bodies valuable minerals and compounds.

The structure of the ecosystem remains the same whether its location is on land or in water. A marine ecosystem illustrates this fact. The autotrophs are microscopic plants called *phytoplankton* which float on the top of the water. The phytoplankton is composed of billions of single-celled green algae which produce a vast quantity of food by means of photosynthesis just as the land plants do. Living in close association with the phytoplankton are heterotrophs such as protozoa and brine shrimp. Since these species feed on the algae, they are also herbivores. Other types of consumers are dependent upon the algae also. Among these are the secondary consumers, small fish that constitute the carnivores that feed on the herbivores. Then there are the larger fish that feed on the smaller species of fish. Dwelling on the bottom sediments are the scavenger worms and snails that feed on the dead plant and animal bodies that fall to the bottom. Living among the sediments are the decomposers—the fungi and bacteria—that break apart plant and animal remains.

THE ECOLOGICAL NICHE

An important concept of ecology is that of the *niche*. An ecological niche is a feeding pattern exhibited by species that compose a community. A niche is a feeding way-of-life in relationship to other organisms. For example: small woodpeckers and nuthatches feed on grubs that are present in the crevices of trees. Although the woodpecker and the nuthatch feed on grubs present in the same tree, these species are not in competition with each other. The woodpecker searches for its food in the crevices at the bottom of the tree and works its way upward. The nuthatch functions best at the top of the tree and works its way downward. Both of these species may live close together, may feed on similar organisms on the same tree, but select the grubs from different locations on the tree. Therefore, it can be said that the woodpecker and the nuthatch occupy different ecological niches.

When two species live in the same place and feed in the same way, using the same food at the same time, *competition* results. The species that has special adaptations for reaching the food first or has greater reproductive potential will be the survivor. The other species will be eliminated. The process by which elimination establishes one species per niche in a particular habitat is known as the *competition-exclusion principle*.

Energy Flow in an Ecosystem

The source of all energy in an ecosystem is the sun. Autotrophs are able to capture just a small portion of the sun's energy and use it to make food enough for all of the living organisms in the ecosystem. The green plant is able to store energy temporarily in ATP molecules and in the nutrients that it makes. Energy is then transferred from green plants into the animal body where it is used to power the vital functions necessary for life. Once energy is used to do work it is converted into heat which then escapes the body and radiates into the atmosphere. The cycles of photosynthesis (energy trapping and conversion) and respiration (energy release and use) must be repeated *ad infinitum* if the ecosystem is to continue.

ENERGY AND FOOD CHAINS

The flow of energy through an ecosystem can be studied by way of *food chains*, which show how energy is transferred from one organism to another through feeding patterns. An example of a food chain on a cultivated field might be as follows:

$$Lettuce \rightarrow Rabbit \rightarrow Snake \rightarrow Hawk$$

Each stage in the food chain represents a feeding or *trophic level*. Lettuce is the producer, the green plant that provides the food which supports the ecosystem. The rabbit is the primary consumer and represents the second trophic level. The carnivorous snake feeds upon the rabbit and thus represents the third trophic level, a secondary consumer. This food chain ends with the hawk, a tertiary consumer that occupies the smallest trophic level in terms of energy. Every food chain begins with an autotroph and ends with a carnivore that is not eaten by a larger animal. When carnivores die, their bodies usually serve as food for scavengers. The bacteria and fungi of decay decompose the remains not devoured by the scavengers.

The flow of energy in a food chain is in a straight line pattern. Most of the energy is concentrated in the level of the producer. At each succeeding level the energy is decreased. However, the feeding relationships among organisms in an ecosystem are not usually this simple, and, in actuality, are more complex.

ENERGY AND FOOD WEBS

Let us look again at a simple food chain.

Lettuce → Rabbit → Snake → Hawk

Suppose all of the rabbits disappear from the cultivated field. Will the snakes die of starvation? The answer to this question can be seen in Fig. 16.1, which illustrates a food web. Notice that the snake can feed upon a shrew, mouse, or owl. In other words, other primary consumers (herbivores) serve well as food for snakes. Fig. 16.1 shows that there are several alternative energy pathways in a *food web*. It is the alternative pathways that enable an ecosystem to keep its stability. One species does not eradicate another in the quest for food.

The relationships between predators and prey help to keep an ecological community stable. The predators help to keep the populations upon which they feed in check. For example: if rabbits were allowed to reproduce without their numbers being thinned by natural predators, the population of rabbits would become overwhelming. The burgeoning rabbit population would devour the available vegetation and then the species would experience starvation and death. Such imbalances in nature do occur when humans upset the natural stability of an ecosystem. Past experience has shown that ecosystems are upset when new species are introduced into an area where there are no predators or natural enemies of the species. Fig. 16.2 shows a *food pyramid*, another way of illustrating energy flow in an ecosystem. Notice that the autotrophs at the base of the pyramid support all of the heterotrophs (consumers) that exist at each nutritional level and that there is a decrease of available energy at each nutritional level.

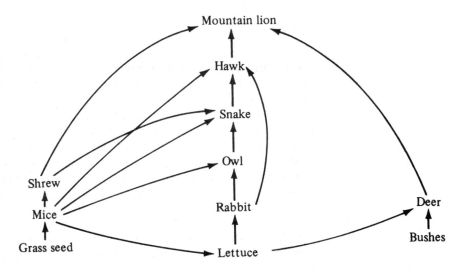

Fig. 16.1 A food web

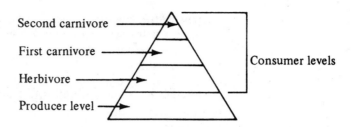

Fig. 16.2 Food pyramid

Biogeochemical Cycles

Certain compounds cycle through the abiotic portion and the biotic communities of ecosystems. These compounds contain elements that are necessary to the biochemical processes that are carried out in living cells. Among these elements are carbon, hydrogen, oxygen and nitrogen. In elemental form, they are useless to cells and must be combined in chemical compounds. Let us trace the pathways of some of the vital compounds from the earth to living organisms to the atmosphere and back to earth. This cycle of events is best described by the term *biogeochemical cycle* (Fig. 16.3).

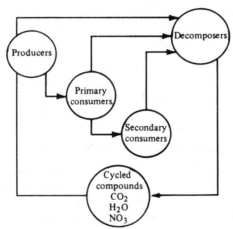

Fig. 16.3 Biogeochemical cycles

THE WATER CYCLE

Water is the source of hydrogen, one of the elements necessary for the synthesizing of carbohydrate molecules by green plants. Water is necessary as the dissolving medium for substances that cross cell membranes and enter cells. Without water, there can be no life.

There are three ways in which water vapor enters the atmosphere. Water evaporates from land surfaces and from the surfaces of all bodies of water. Water vapor enters the air as a waste product of respiration of animals and plants. For example: every time you exhale water vapor is released into the air. Great amounts of water are lost from plants through the openings in the leaves; this water loss due to evaporation is called *transpiration*.

Water vapor in the air is carrried to high altitudes where it is cooled and forms clouds by condensation. Eventually, clouds fall to the earth in the form of precipitation: rain, snow, or sleet. Most of the precipitation returns to the oceans, lakes or streams and less than 1 percent of it falls on land. Of the water that does fall on land, about 25 percent of it will evaporate from the various land surfaces before it can be absorbed by plants or used by animals. Water that does not evaporate enters the soil and becomes available to plant roots and soil organisms.

Soil water that is not absorbed by plants seeps down into the ground until it reaches an impervious layer of rock. The water moves along this rock as *groundwater* until it reaches an outlet into a larger body of water such as a lake or an ocean. The water cycle repeats. Fig. 16.4 illustrates the water cycle.

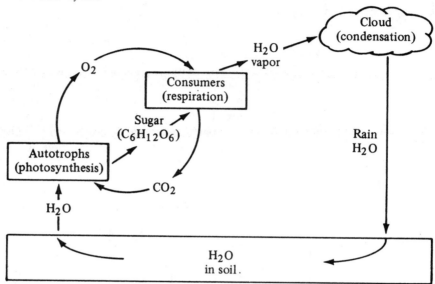

Fig. 16.4 Water cycle

CARBON DIOXIDE/OXYGEN CYCLE

Carbon dioxide and oxygen cycle through the abiotic and biotic components of ecosystems. Oxygen is released from plants into the atmosphere during photosynthesis. Oxygen from the air is taken in by animals and used during cellular respiration. Carbon dioxide, a waste product of respiration, is released into the air by animals and plants. Plants take in the carbon dioxide and use it during photosynthesis. The cycle repeats.

NITROGEN CYCLE

About 80 percent of the volume of the air is elemental nitrogen. Although it is an important component of proteins, nitrogen must be in a combined form before it can be used by the cells of living things.

Nitrogen becomes available to plants through the action of nitrogen-fixing bacteria that change atmospheric nitrogen into nitrates which are used by the plants for protein synthesis.

Nitrogen is also fixed by lightning. The energy produced by a discharge of lightning joins nitrogen with oxygen to form nitric oxides. These oxides combine with water vapor to form nitrous and nitric acids. When these acids fall on soil with rainwater, they are changed into nitrites and nitrates.

Denitrifying bacteria in the soil change nitrates back to atmospheric nitrogen. The cycle then repeats.

The Limiting Factor Concept _____

There are environmental factors such as light, temperature, and amount of rainfall that set the limit of conditions for species survival. Every species can live within a range of values for each factor in its physical environment but cannot survive beyond the low and high values called the *limits of tolerance*. For example: the eggs of the frog *Rana pipiens* exhibit a limit of tolerance for the temperature range between 0° and 30°C. The eggs of *Rana pipiens* will die in temperature ranges below or above these figures. However, at 22°C more *Rana pipiens* eggs hatch than at any other temperature. This means that 22°C is *optimum temperature* for the survival of this species. Among factors in the physical environment that set limits on growth and other physiological processes are light, temperature and amount of rainfall.

The effect of light as a *limiting factor* was demonstrated in 1920 when W. W. Garner and H. A. Allard showed that day length was an important factor in the flowering of plants. The physiological response made by plants to changes in daylength is known as *photoperiodism*. Some plants (short-day) flower only if they are exposed to light for *less than* a certain amount of time each day; other plants (long-day) must have a certain minimum length of photoperiod. Researchers explain the ability of plants to measure time by the action of phytochrome, a light absorbing pigment that is associated with the cell membrane and with some of the cell's internal membranes.

Between 0°C and 40°C is the range of temperature between which most species are active. In temperatures below 0°C there is a slowing down of biochemical processes. Freezing causes spicules of ice to form in cells which causes the lysing of cells and the disruption of cell processes. Above 40°C protein of most species denatures, rendering enzymes, hormones and the structural proteins of cells useless. However, some forms of life are able to exist outside the range of 0° to 40°. The seeds of most deciduous trees must go through a period of cold, very often in temperatures far below the freezing point of water, before they can germinate. It is also a fact that certain microorganisms such as bacteria and certain species of algae can live in hot springs and geysers where the temperature may exceed 90°C.

Warm-blooded mammals that live in cold climates have developed physiological mechanisms for preserving body heat. The combination of large amounts of body fat and the ability to hibernate during the most forbidding of the cold months serves to maintain the lives of these organisms. On the other hand, cold-blooded animals have body temperatures that are quite similar to the temperature of the external environment. They too become inactive in cold weather and then require protective shade during the hot season.

Temperature is a factor that regulates the growth of plants and the activities of animals. There is no doubt that plant and animal hormones evoke physiological changes in response to the rise and fall of temperature.

Rainfall is also a limiting factor. Land-dwelling plants and animals depend upon rainfall as the source of water. Species that live successfully in deserts or in areas where there is seasonal drought have developed adaptations for water conservation. The kangaroo rat lives in humid burrows under the desert sands where, huddled together with others of its kind, it licks the condensed drops of water vapor from the hair of its burrow mates. The lungfish (*Dipnoi*) estivates under the mud in dried ponds, awaiting the seasonal downpour of rain that will fill up its habitat and restore the species to active life. The seeds of several flowering plant species of the desert will germinate only when a heavy rain wets the substrate around them; the rainfall seems to remove from the seed coat a chemical compound that inhibits germination. Similarly, species in areas of heavy rainfall have special adaptations. Some plant species in the rain forests of Brazil have developed adaptations such as aerial roots that can absorb moisture from the air rather than becoming waterlogged in too wet soil.

Ecological Succession

It was stated before that an ecosystem consists of a living community and its nonliving environment. As physical conditions change in an ecosystem, so do the plants and animals that inhabit that system. The orderly change of the biotic community in an ecosystem is known as *ecological succession*. There are two levels of succession: primary and secondary.

As the name implies, primary succession is the introduction of living species into a region that was formerly barren. Let us suppose that a rocky cliff is devoid of soil. By the processes of erosion and weathering, the rock is pulverized. Lichens that inhabit the bare rock contribute to its chemical erosion. Dead lichens contribute organic matter to the pulverized rock material. Now the physical conditions have changed just enough to allow the growth of mosses. These too add organic matter to the substrate and also contribute some water-holding ability to the newly formed soil. The moss population is followed by the grasses and then the ferns. In orderly succession there follow the low-growing bushes, the higher-growing shrubs and then trees. Each population inhabiting the region makes it better for another species.

Primary succession culminates in the existence of a stable community, which is known as a *climax community* where one or two large trees predominate. Over a period of years secondary succession may occur in which one species of tree is gradually replaced by another. For example: white pine trees may replace the gray birch trees. Limiting factors such as temperature, light, rainfall and mineral content of the soil influence the type of climax community. A climax community in a broad geographical area having one type of climate is known as a *biome*.

World Biomes

The earth is divided into several biomes.

THE TUNDRA

Vast stretches of treeless plains surrounding the Arctic Ocean where cold is the limiting factor (60°F in the summer to −130°F in the winter) are known as the arctic tundra. Here the ground is permanently frozen a few feet below the surface and is responsible for the many lakes and bogs that characterize the region. The kinds of organisms that inhabit the tundra are those that have developed adaptations for survival in extremely cold temperatures. The plant species is composed of lichens, mosses, grasses and sedges. During the summer, the flowering herbs bloom in brilliant color for a very short time along with the dwarf willows. This seemingly meager plant life is able to support the food chains of the tundra. It feeds thousands of migrating birds and insect swarms that appear during the brief summer. Many mammals such as the musk ox, caribou, polar bears, wolves, foxes and marine mammals remain active for most of the year.

THE TAIGA

The coniferous forests of Canada survive well in the long severe winters. The spruce and fir trees predominate. The kinds of mammals that inhabit the taiga are the black bear, the wolf, lynx and squirrel.

THE DECIDUOUS FOREST

The forests of the temperate regions are dominated by broad-leaved trees that lose their leaves in the winter. Examples of the kinds of trees that compose these hardwood forests are oak and hickory, oak and chestnut, beech and maples and willows, cottonwood and sycamore. The types of animals inhabiting these forests are deer, fox, squirrel, skunk, woodchuck and raccoon.

THE DESERT

Deserts form in regions where the annual rainfall is less than 6.5 centimeters, or where rain occurs unevenly during the year and the rate of evaporation is high. The temperature changes drastically from hot days to cold nights. Plants that survive in the desert have specific adaptations for low moisture and high temperature. Shrubs, such as creosote and sagebrush, shed their short, thick leaves during dry spells and become dormant as protection against wilting. Fleshy desert plants, such as the cacti of American deserts and the euphorbias of African deserts, store water in their tissues. Cheat grass and wild flowers are annuals which grow quickly after a desert downpour, bloom, produce seed and die. Mosses, lichen and blue-green algae lie dormant on the sand and become active when moisture is present. The animals of the desert include lizards, insects, kangaroo rats and arachnids.

THE GRASSLANDS

Grasslands occur where the annual rainfall is low and irregular. Grasslands usually occupy large areas of interior continents that are sheltered from moisture-laden winds. In the United States the grasslands are known as the Great Plains; in Russia they are called the *steppes*, in South Africa, the *veldt* and in South America as the *pampas*. Grasses have adaptations for living in soil where the rainfall is low and erratic. Grazing animals are suited for life on the grasslands. At one time the American grasslands supported huge herds of antelope and bison. The settlers replaced these natural herds with cattle and sheep. Besides the domesticated animals, a number of predator species inhabit the grasslands, including coyotes, bobcats, badgers, hawks, kit foxes and owls. These feed primarily on burrowing rodents.

THE TROPICAL RAIN FOREST

The tropical rain forest is characterized by high temperatures and constant rainfall. This type of biome is found in Central and South America, in Southeast Asia and in West Africa. The trees are tall and the vegetation is so thick that the forest floor is shaded from light. The animals of the rain forest include monkeys, lizards, snakes and birds.

THE SEA

Ocean waters cover almost three fourths of the earth's surface and support the greatest abundance and diversity of organisms in the world. Averaging 3.5 to 4.5 kilometers in depth, a marine biome constitutes the thickest layer of living things in the biosphere. The dominating physical factors determine the type of living organisms that compose its communities. Temperature, light intensity, salinity, waves, tide currents and pressures are the limiting factors that set the conditions for life.

Light in the ocean extends for a depth of 180 meters; beyond this there is total darkness. The upper layers of water are known collectively as the continental shelf, a region that supports a vast array of living things. Phytoplankton live at the surface and serve as food for primary consumers such as sardines and anchovies. These herbivores serve as food for larger fish such as salmon, mackerel and tuna. Mud-burrowing animals such as clams, snails, worms and shrimp are eaten by crabs, lobsters, starfish, and a number of different kinds of fish such as cod, halibut, haddock, rays, and flounder. In the dark region of the sea, phytoplankton are absent. The small animals that live at the bottom of the sea feed on organic debris that falls from above. The temperature of the sea is fairly constant. Its salinity lowers the freezing point of water.

Humans and the Biosphere _____

People influence the physical and the living environments in which they live.

NEGATIVE EFFECTS OF HUMAN ACTIVITIES

Unfortunately, the changes that humans have produced have, for the most part, been negative. Table 16.1 summarizes these changes.

EFFORTS TO CORRECT HUMAN ERRORS

Increased awareness about the ecology of our environment has prompted attempts to stop the destruction. People have become aware of the life-threatening dangers of chemical pollution brought about by irresponsible dumping of chemical wastes. Powerful interest groups have been able to have laws enacted to prevent the illegal hunting of wildlife, and wild life

TABLE 16.1.   Effects of Humans on the Environment

Effect	Description
Population Growth	The human population of the world has grown so explosively that ecosystems cannot produce food to serve human needs.
Overhunting	Indiscriminate and uncontrolled hunting of wildlife has resulted in the extinction of such species as the great auk, dodo bird, passenger pigeon, Carolina parakeet, and heath hen. A number of other species are threatened with extinction: the blue whale, wild turkey, bison, grizzly bear, and American bald eagle.
Importation of Organisms	By accident or with definite intent, people have transported organisms into ecosystems where they have no natural enemies. Every year the Japanese beetle destroys innumerable species of green plants in the United States, at a cost of millions of dollars. The gypsy moth and the fungus that has destroyed elm trees are other examples of imported organisms that cause damage because of lack of natural enemies.
Exploitation	Humans have exploited wildlife for their own profit. Illegal hunting of the African elephant and the Pacific walrus for ivory, capture of the Colombian parrot for the pet trade, and the cutting down of trees in the tropical rain forests for plywood have upset ecosystems.
Thoughtless Land Use Management	The increased use of agricultural lands for the construction of houses, parking lots, huge shopping malls, and the like has destroyed the natural habitats of wildlife. In addition, poor use of agricultural land in the forms of overcropping and overgrazing has resulted in the loss of valuable nutrients from the soil.

preserves have been established to protect endangered species. Methods to control the human birth rate have been developed and shared with underdeveloped nations. Conservation measures are currently being applied to conserve arable land.

Conservation _____

The *conservation* of natural resources is of primary importance to humans if the land on which they live is to be preserved for future generations. Natural resources are the air we breathe, the water we drink, the soil that supports plant and animal life, and the minerals under the ground. All of these non-human-made resources are vital to the well being of nations. Ecologists think of natural resources as *renewable* and *nonrenewable*. Forests, grasslands and the wild species that inhabit them can be restored after their numbers have been somewhat depleted. These are the renewable resources. Once used up, minerals, including the fossil fuels of oil and coal, cannot be restored and therefore are classified as nonrenewable. Desirable conservation practices must be the concern of every human. Understanding principles of ecology and of conservation is paramount to the making of wise decisions.

SOIL CONSERVATION

The soil covering the earth is really a very thin layer measuring about 0.9 meter in depth. Soil is not made quickly. It takes thousands of years, much erosion and weathering of rock, and the gradual addition of plant and animal remains to make a fertile top soil. Top soil can be quite fragile and easily destroyed by *sheet erosion*, washed away from sloping hills by rain. Soil may also become exhausted or *leached*, having been over-cultivated so that the mineral components are lost.

Conservation measures protect top soil. Erosion can be prevented by the wise use of ground cover, rooted plants that hold the soil in place. *Terracing*, the planting of trees, bushes and low-growing plants in steps cut into the hillsides prevents loss of soil by sheet erosion. *Contour plowing*, the circular plowing of hillsides, is another method of preventing the erosion of soil. The leaching of minerals and nitrates from the soil is prevented by *strip cropping* (planting corn or cotton with alternate rows of legumes) and *crop rotation*, planting a field on alternate years.

CONSERVATION OF WATER

Water is rendered useless for drinking, for bathing, for irrigation and as a habitat for fish when it is polluted by the chemical wastes from industry and by human sewage. Sewage treatment plants clean up sewage before it is dumped into waterways. Special treatment must be given to chemical wastes to detoxify them before disposal.

Water is a renewable resource. However, people on the earth are using more water than even before in industry, refrigeration, agriculture and the like. Humans are dependent on rainfall to maintain an adequate *water table* (level of groundwater) and to replenish water stores in reservoirs. The wasting of water through careless use can have serious consequences for human life.

CONSERVATION OF FORESTS

Forests are important in the conserving of water, soil and habitats for wild life. Forests also supply us with lumber and natural resins that are used in the paint industry and for other industrial processes. Forests are destroyed by over-lumbering through which more trees are cut down than can be replaced by natural processes. Trees grow slowly, requiring about thirty or forty years to reach maturity. We need wood to make paper and to build houses, telephone poles and fences. However, if forests are lumbered imprudently, trees will not last. Most of the virgin forests in the United States have been destroyed, and now in their places are second growth forests.

Chronology of Famous Names in Biology

1840	**Justus von Liebig** (German)—developed the ecological concept of the Law of the Minimum in which he showed the limiting effect on growth of environmental factors.
1869	**Ernst Haeckel** (German)—coined the word "ecology" as the science that studies environment.

1920	**W. W. Garner** and **H. A. Allard** (American)—discovered that daylength affects the flowering of the Biloxi soybean and also the flowering of a mutant form of tobacco called "Maryland Mammoth".
1938	**K. C. Hamner** and **J. Bonner** (American)—discovered how the cockleburr responds to the photoperiod.
1939	**George Washington Carver** (American)—discovered that legumes such as peanuts returned nitrates and minerals to the soil. He developed 300 new industrial uses for peanuts and 118 byproducts from sweet potatoes.
1942	**Elton Charles** (American)—made a detailed study of irruption and death of lemmings.
1948	**Ralph Buchsbaum** (American)—wrote a very important work on basic ecology.
1952	**Rachel Carson** (American)—wrote vivid descriptions of the oceans as ecosystems and about the organisms that live therein.
1955	**F. W. Went** (American)—a modern botanist, who has elucidated the various roles of plants in ecosystems.
1955	**Herbert Zim** (American)—presented important information about the living organisms that inhabit seashores.
1957	**G. S. Avery** (American)—elucidated reasons for the death of oak trees in U.S. cities.
1958	**Eugene Odum** (American)—a modern researcher who has been a motivator in the field of ecology.
1959	**W. H. Amos** (American)—researched the living organisms of sand dunes.
1960	**Howard T. Odum** (American)—developed techniques for measuring the energy flow in forests and marine environments.
1960	**Marston Bates** (American)—a leading authority on the ecology of world biomes.
1964	**Rene Catala** (New Caledonian)—an authority on fluorescent corals.
1967	**George M. Woodwell** (American)—leading authority on the effects of radiation and toxic substances on ecosystems.

Words for Study

abiotic	biosphere	climax
autotroph	biotic	community
biogeochemical cycle	carnivore	conservation

crop rotation	food web	producer
cycling	herbivore	scavenger
deciduous	heterotroph	secondary consumer
decomposers	limiting factor	strip cropping
ecological succession	natural resources	succession
ecology	niche	taiga
ecosystem	nitrogen fixation	tertiary consumer
erosion	photoperiodism	tolerance
food chain	population	trophic level
food pyramid	primary consumer	tundra

Questions for Review

PART A. **Completion.** Write in the word that correctly completes each statement.

1. The living factors in the environment are known as ..1.. factors.
2. The living community and the nonliving environment are known collectively as the ..2..
3. Without the work of ..3.. the remains of dead organisms would pile up.
4. A nuthatch is a species of ..4..
5. The number of species that can successfully occupy a niche is ..5..
6. The flow of energy through an ecosystem can be studied by identifying simple ..6.. (two words).
7. An herbivore is also classified as a ..7.. consumer.
8. A great deal of water is lost from plants by an evaporation process known as ..8..
9. About 80 percent of the volume of air is the elemental gas ..9..
10. Among the limiting factors in the environment are light, temperature and ..10..
11. Phytochrome is a plant hormone that controls ..11..
12. The seeds of most deciduous plants must go through a period of ..12.. before they can germinate.
13. The compound that all living things need for the dissolving of chemical substances is ..13..
14. The orderly change of a biotic community is known as ..14..
15. A geographical area having one type of climate is a ..15..

PART B. **Multiple Choice.** Circle the letter of the item that correctly completes each statement.

1. The nonliving factors in an ecosystem are referred to as
   (a) adiabatic
   (b) abomasum
   (c) abiotic
   (d) abscissant

2. The percent of the air that is made up of oxygen is
   (a) 21
   (b) 50
   (c) 78
   (d) 90

3. The habitat of phytoplankton is
   (a) deciduous forest
   (b) ocean surface
   (c) Nairobi desert
   (d) Canadian taiga

4. An ecological niche is a (an)
   (a) organism
   (b) geographical region
   (c) habitat
   (d) feeding pattern

5. The lives of organisms in an ecosystem depend upon the
   (a) amount of light in the ecosystem
   (b) quality of environment
   (c) availability of pH in the ecosystem
   (d) flow of energy through the ecosystem

6. In a food chain, lettuce is best classified as a
   (a) producer
   (b) deciduous plant
   (c) vegetable
   (d) detritus

7. The arrow in a food web diagram stands for
   (a) an alternative energy pathway
   (b) increased energy
   (c) "is eaten by"
   (d) "eats up"

8. In a food web, hawks are best classified as
   (a) scavengers
   (b) predators
   (c) prey
   (d) primary consumers

9. In plant cells, phytochrome is associated with
   (a) nuclei
   (b) cytoplasm
   (c) lysosomes
   (d) plasma membranes

10. The washing away of top soil from the slopes of hills is known as
    (a) weathering
    (b) erosion
    (c) irrigation
    (d) leaching

11. An example of an adaptation that permits the survival of some plant species in the tropical rain forest is
    (a) degenerate stomates
    (b) needlelike leaves
    (c) aerial roots
    (d) cutinized stems

12. Chemical changes on the surfaces of rock may be brought about by
    (a) lichen colonies
    (b) heavy rainfall
    (c) tree roots
    (d) germinating bryophytes

13. Musk ox, caribou and foxes are part of the mammal population of the
    (a) temperate forest
    (b) taiga
    (c) grasslands
    (d) tundra

14. In desert biomes, the average rainfall is no greater than
    (a) 20 cm
    (b) 6.5 cm
    (c) 15 cm
    (d) 0 cm

15. Examples of primary consumers in the marine biome are
    (a) halibut and cod
    (b) rays and haddock
    (c) sardines and anchovies
    (d) cod and haddock

PART C. **Modified True-False.** If a statement is true, write "true" for your answer. If a statement is incorrect, change the underlined word to one that will make the statement true.

1. The cultivation of a field in alternate years is known as <u>crop rotation</u>.

2. Oil, coal and natural gas are <u>renewable</u> resources.

3. <u>Straight row</u> plowing prevents the erosion of hillsides.

4. All of the plant and animal species interacting in a given environment are known as a <u>population</u>.

5. Tertiary consumers are those that feed on smaller <u>herbivores</u>.

6. Vultures are best classified as <u>predators</u>.

7. In ecosystems, fungi and bacteria serve in the roles of <u>scavengers</u>.

8. Competition results when two species share the same <u>habitat</u>.

9. Once energy is used, it is converted into <u>fat</u>.

10. A process in which energy is released is <u>photosynthesis</u>.

11. Most of the energy in a food chain is concentrated at the <u>carnivore</u> level.

12. The alternative pathways of energy flow in an ecosystem are best shown in a food <u>pyramid</u>.

13. The relationships between predators and prey help to <u>disrupt</u> an ecological community.

14. The source of hydrogen utilized by green plants for photosynthesis is <u>sugar</u>.

15. A <u>changing</u> community is known as a climax community.

Think and Discuss

1. What is an ecosystem?
2. How does an ecological niche differ from an ecosystem?
3. What is the difference between a food chain and a food web?
4. Why is it important to preserve our forests?

Answers to Questions for Review

PART A

1. biotic
2. ecosystem
3. decomposers
4. bird
5. one
6. food chains

7. primary
8. transpiration
9. nitrogen
10. rainfall
11. flowering

12. cold
13. water
14. succession
15. biome

PART B

1. c
2. a
3. b
4. d
5. d

6. a
7. c
8. b
9. d
10. b

11. c
12. a
13. d
14. b
15. c

PART C

1. true
2. nonrenewable
3. Contour
4. community
5. carnivores
6. scavengers
7. decomposers
8. niche

9. heat
10. respiration
11. autotroph
12. web
13. stabilize
14. water
15. stable

INDEX

1,001 Statistics Practice Problems

FOR DUMMIES

A Wiley Brand

1,001 Statistics Practice Problems For Dummies®

Published by: John Wiley & Sons, Inc., 111 River Street, Hoboken, NJ 07030-5774, www.wiley.com

Copyright © 2014 by John Wiley & Sons, Inc., Hoboken, New Jersey

Published simultaneously in Canada

For general information on our other products and services, please contact our Customer Care Department within the U.S. at 877-762-2974, outside the U.S. at 317-572-3993, or fax 317-572-4002. For technical support, please visit www.wiley.com/techsupport.

Wiley publishes in a variety of print and electronic formats and by print-on-demand. Some material included with standard print versions of this book may not be included in e-books or in print-on-demand. If this book refers to media such as a CD or DVD that is not included in the version you purchased, you may download this material at http://booksupport.wiley.com. For more information about Wiley products, visit www.wiley.com.

Library of Congress Control Number: 2013954236

ISBN 978-1-118-77604-9 (pbk); ISBN 978-1-118-77605-6 (ebk); ISBN 978-1-118-77

Manufactured in the United States of America

V10007140_122818

Contents at a Glance

Contents at a Glance

Table of Contents

Introduction

● ●

One thousand and one practice problems for statistics! That's probably more than a professor would assign you in one semester (we hope!). And it's more than you'd ever want to tackle in one sitting (and we don't recommend you try). So why so many practice problems, and why this book?

Many textbooks are pretty thin on exercises, and even those that do contain a fair number of problems can't focus on all aspects of each topic. With so many problems available in this book, you get to choose how many problems you want to work on. And the way these problems are organized helps you quickly find and dig into problems on particular topics you need to study at the time. Whether you're into the normal distribution, hypothesis tests, the slope of a regression line, or histograms, it's all here and easy to find.

Then there's the entertainment factor. What better way to draw a crowd than to invite people over for a statistics practice problems marathon!

What You'll Find

This book contains 1,001 statistics problems divided into 17 chapters, organized by the major statistical topics in a first-semester introductory course. The problems basically take on three levels:

  ✔ **Statistical literacy:** Understanding the basic concepts of the topic, including terms and notation

  ✔ **Reasoning:** Applying the ideas within a context

  ✔ **Thinking:** Putting ideas and concepts together to solve more difficult problems

In addition to providing plenty of problems to work on in each chapter, this book also provides worked-out solutions with detailed explanations, so you aren't left high and dry if you get a wrong answer. So you can rest assured that when you work for 30 minutes on a problem, get an answer of 1.25, and go to the back of the book to see that the correct answer is actually 1,218.31, you'll find a detailed explanation to help you figure out what went wrong with your calculations.

How This Workbook Is Organized

This book is divided into two main parts: the questions and the answers.

Part 1: The Questions

The questions in this book center on the following areas:

- **Descriptive statistics and graphs:** After you collect and review data, your first job is to make sense of it. The way to do that is two-fold: (1) Organize the data in a visual manner so you can see it, and (2) crank out some numbers that describe it in a basic way.

- **Random variables:** A *random variable* is a characteristic of interest that varies in a random fashion. Each type of random variable has its own pattern in which the data falls (or is expected to fall), along with its own mean and standard deviation for the data. The pattern of a random variable is called its *distribution*.

  The random variables in this book include the binomial, the normal (or *Z*), and the *t*. For each random variable, you practice identifying its characteristics, seeing what its pattern (distribution) looks like, determining its mean and standard deviation, and, most commonly, finding probabilities and percentiles for it.

- **Inference:** This term can seem complex (and word on the street says it is), but inference basically just means taking the information from your data (your sample) and using it to draw conclusions about the group you're interested in (your population).

  The two basic types of statistical inferences are confidence intervals and hypothesis testing:

  - You use *confidence intervals* when you want to make an estimate regarding the population — for example, "What percentage of all kindergartners in the United States are obese?"

  - You use a *hypothesis test* when someone has a supposed value regarding the population, and you're putting it to the test. For example, a researcher claims that 14 percent of today's kindergartners are obese, but you question whether it's really that high.

  The underpinnings needed for both types of inference are margin of error, standard error, sampling distributions, and the central limit theorem. All of them play a major role in statistics and can be somewhat complex, so make sure you spend time on these elements as a backdrop for confidence intervals and hypothesis tests.

- **Relationships:** One of the most important and common uses of statistics is to look for relationships between two random variables. If variables are categorical (such as gender), you explore relationships by using two-way tables containing rows and columns, and you examine relationships by looking at and comparing percentages among and within groups. If both variables are numerical, you explore relationships graphically by using scatter plots, quantify them by using correlation, and use them to make predictions (one variable predicting the other) by using regression. Studying relationships helps you get at the essence of how statistics is applied in the real world.

- **Surveys:** Before you analyze data in all the ways mentioned in this list, you have to collect the data. Surveys are one of the most common means of data collection; the main ideas of surveys to address with practice are planning a survey, selecting a representative sample of individuals to survey, and carrying out the survey properly. The main goal in all of these areas is to avoid *bias* (systematic favoritism). Many types of bias exist, and in this book, you practice identifying and seeing ways to minimize them.

Part II: The Answers

This part provides detailed answers to every question in this book. You see how to set up and work through each problem and how to interpret the answer.

Beyond the Book

This book of 1,001 practice problems will keep you busy with pencil and paper for a while, but like the infomercials say, "Wait! There's more!" Your book purchase also comes with a free, one-year subscription to all 1,001 practice problems online. Track your progress and view personalized reports that show where you need to study the most and what you're pretty comfortable with.

What you'll find online

The online practice that comes free with this book offers you the same 1,001 questions and answers that are available here, presented in a multiple-choice format. Multiple-choice questions force you to zoom in on the details that can make or break your correct solution to the problem. Sometimes one of the possible wrong answers will catch you in the act of making a certain error. But that's great because after you identify a particular error (often a common error that many others make as well), you'll know not to fall into that trap again.

The beauty of the online problems is that you can customize your online practice — that is, you can select the types of problems and the number of problems you want to work on. The online program keeps track of your performance so you can focus on the areas where you need the biggest boost.

You can access this online tool by using a PIN code, as described in the next section. Keep in mind that you can create only one login with your PIN. Once the PIN is used, it's no longer valid and is nontransferable. So you can't share your PIN with others after you've established your login credentials. In other words, the problems are yours and only yours!

How to register

To gain access to additional tests and practice online, all you have to do is register. Just follow these simple steps:

1. **Find your PIN access code:**

   - **Print-book users:** If you purchased a print copy of this book, turn to the inside front cover of the book to find your access code.

   - **E-book users:** If you purchased this book as an e-book, you can get your access code by registering your e-book at www.dummies.com/go/getaccess. Go to this website, find your book and click it, and answer the security questions to verify your purchase. You'll receive an email with your access code.

2. Go to Dummies.com and click Activate Now.

3. **Find your product (Statistics: 1,001 Practice Problems For Dummies (+ Free Online Practice)) and then follow the on-screen prompts to activate your PIN.**

Now you're ready to go! You can come back to the program as often as you want — simply log on with the username and password you created during your initial login. No need to enter the access code a second time.

For Technical Support, please visit http://wiley.custhelp.com or call Wiley at 1-800-762-2974 (U.S.), +1-317-572-3994 (international).

Your registration is good for one year from the day you activate your PIN. After that time period has passed, you can renew your registration for a fee. The website tells you how.

Where to Go for Additional Help

The written solutions for the problems in this book are designed to show you what you need to do to get the correct answer to those particular problems. Although a bit of background information is injected at times, the solutions aren't meant to teach the material outright. Solutions to the problems on a given topic contain the normal statistical language, symbols, and formulas that are inherent to the topic, with the assumption that you're familiar with them.

If you're ever confused about why a problem is done a certain way, or you want more info to fill in between the lines, or you just feel like you need to go back and refresh your memory on some of the topics, several *For Dummies* books are available as a reference, including *Statistics For Dummies*, *Statistics Essentials For Dummies*, and *Statistics Workbook For Dummies*, all written by Deborah J. Rumsey, PhD, and published by Wiley.

Part I
The Questions

1,001 Questions

 web extras Visit www.dummies.com for great (and free!) Dummies content online.

In this part . . .

Statistics can give anyone problems. Terms, notation, formulas — where do you start? You start by practicing problems that hone the right skills. This book gives you practice — 1,001 problems worth of practice, to be exact. Working problems like these helps you figure out what you do and don't understand about setting up, working out, and interpreting your answers to statistics problems. Here's the breakdown in a nutshell:

- ✔ Warm up with statistical vocabulary, descriptive statistics, and graphs (Chapters 1 through 3).

- ✔ Work with random variables, including the binomial, normal, and *t*-distributions (Chapters 4 through 6).

- ✔ Decipher sampling distributions and margin of error and build confidence intervals for one- and two-population means and proportions (Chapters 7 through 10).

- ✔ Master the general concepts of hypothesis testing and perform tests for one- and two-population means and proportions (Chapters 11 through 13).

- ✔ Get behind the scenes on collecting good data and spotting bad data in surveys (Chapter 14).

- ✔ Explore relationships between two quantitative variables, using correlation and simple linear regression (Chapters 15 and 16).

- ✔ Look for relationships between two categorical variables, using two-way tables and independence (Chapter 17).

Chapter 1

Basic Vocabulary

• •

*E*verything's got its own lingo, and statistics is no exception. The trick is to get a handle on the lingo right from the get-go, so when it comes time to work problems, you'll pick up on cues from the wording and get going in the right direction. You can also use the terms to search quickly in the table of contents or the index of this book to find the problems you need to dive into in a flash. It's like with anything else: As soon as you understand what the language means, you immediately start feeling more comfortable.

The Problems You'll Work On

In this chapter, you get a bird's-eye view of some of the most common terms used in statistics and, perhaps more importantly, the context in which they're used. Here's an overview:

✔ The big four: population, sample, parameter, and statistic

✔ The statistics terms you'll calculate, such as the mean, median, standard deviation, *z*-score, and percentile

✔ Types of data, graphs, and distributions

✔ Data analysis terms, such as confidence intervals, margin of error, and hypothesis tests

What to Watch Out For

Pay particular attention to the following:

✔ Pick out the big four in every situation; they'll follow you wherever you go.

✔ Really get the idea of a distribution; it's one of the most confusing ideas in statistics, yet it's used over and over — so nail it now to avoid getting hammered later.

✔ Focus not only on the terms for the statistics and analyses you'll calculate but also on their interpretation, especially in the context of a problem.

Picking Out the Population, Sample, Parameter, and Statistic

1–4 You're interested in knowing what percent of all households in a large city have a single woman as the head of the household. To estimate this percentage, you conduct a survey with 200 households and determine how many of these 200 are headed by a single woman.

1. In this example, what is the population?

2. In this example, what is the sample?

3. In this example, what is the parameter?

4. In this example, what is the statistic?

Distinguishing Quantitative and Categorical Variables

5–6 Answer the problems about quantitative and categorical variables.

5. Which of the following is an example of a quantitative variable (also known as a numerical variable)?

(A) the color of an automobile

(B) a person's state of residence

(C) a person's zip code

(D) a person's height, recorded in inches

(E) Choices (C) and (D)

6. Which of the following is an example of a categorical variable (also known as a qualitative variable)?

(A) years of schooling completed

(B) college major

(C) high-school graduate or not

(D) annual income (in dollars)

(E) Choices (B) and (C)

Getting a Handle on Bias, Variables, and the Mean

7–11 You're interested in the percentage of female versus male shoppers at a department store. So one Saturday morning, you place data collectors at each of the store's four entrances for three hours, and you have them record how many men and women enter the store during that time.

7. Why can collecting data at the store on one Saturday morning for three hours cause bias in the data?

   (A) It assumes that Saturday shoppers represent the whole population of people who shop at the store during the week.

   (B) It assumes that the same percentage of female shoppers shop on Saturday mornings as any other time or day of the week.

   (C) Perhaps couples are more likely to shop together on Saturday mornings than during the rest of the week, bringing the percentage of males and females closer than during other times of the week.

   (D) The subjects in the study weren't selected at random.

   (E) All of these choices are true.

8. Because a variable is a characteristic of each individual on which data is collected, which of the following are variables in this study?

   (A) the day you chose to collect data

   (B) the store you chose to observe

   (C) the gender of each shopper who comes in during the time period

   (D) the number of men entering the store during the time period

   (E) Choices (C) and (D)

9. In this study, _____ is a categorical variable, and _____ is a quantitative variable.

10. Which chart or graph would be appropriate to display the proportion of males versus females among the shoppers?

   (A) a bar graph

   (B) a time plot

   (C) a pie chart

   (D) Choices (A) and (C)

   (E) Choices (A), (B), and (C)

11. How would you calculate the mean number of shoppers per hour?

Understanding Different Statistics and Data Analysis Terms

12–17 Answer the problems about different statistics and data analysis terms.

12. Which of the following data sets has a median of 3?

   (A) 3, 3, 3, 3, 3

   (B) 2, 5, 3, 1, 1

   (C) 1, 2, 3, 4, 5

   (D) 1, 2, 4, 4, 4

   (E) Choices (A) and (C)

13. Susan scores at the 90th percentile on a math exam. What does this mean?

14. You took a survey of 100 people and found that 60% of them like chocolate and 40% don't. Which of the following gives the distribution of the "chocolate versus no chocolate" variable?

(A) a table of the results

(B) a pie chart of the results

(C) a bar graph of the results

(D) a sentence describing the results

(E) all of the above

15. Suppose that the results of an exam tell you your z-score is 0.70. What does this tell you about how well you did on the exam?

16. A national poll reports that 65% of Americans sampled approve of the president, with a margin of error of 6 percentage points. What does this mean?

17. If you want to estimate the percentage of all Americans who plan to vacation for two weeks or more this summer, what statistical technique should you use to find a range of plausible values for the true percentage?

Using Statistical Techniques

18–19 You read a report that 60% of high-school graduates participated in sports during their high-school years.

18. You believe that the percentage of high-school graduates who played sports is higher than what was reported. What type of statistical technique do you use to see whether you're right?

19. You believe that the percentage of high-school graduates who played sports in high school is higher than what's in the report. If you do a hypothesis test to challenge the report, which of these p-values would you be happiest to get?

(A) $p = 0.95$

(B) $p = 0.50$

(C) $p = 1$

(D) $p = 0.05$

(E) $p = 0.001$

Working with the Standard Deviation

20 Solve the problem about standard deviation.

20. Which data set has the highest standard deviation (without doing calculations)?

(A) 1, 2, 3, 4

(B) 1, 1, 1, 4

(C) 1, 1, 4, 4

(D) 4, 4, 4, 4

(E) 1, 2, 2, 4

Chapter 2

Descriptive Statistics

● ●

Descriptive statistics are statistics that describe data. You've got the staple ingredients, such as the mean, median, and standard deviation, and then the concepts and graphs that build on them, such as percentiles, the five-number summary, and the box plot. Your first job in analyzing data is to identify, understand, and calculate these descriptive statistics. Then you need to interpret the results, which means to see and describe their importance in the context of the problem.

The Problems You'll Work On

The problems in this chapter focus on the following big ideas:

✔ Calculating, interpreting, and comparing basic statistics, such as mean and median, and standard deviation and variance

✔ Using the mean and standard deviation to give ranges for bell-shaped data

✔ Measuring where a certain value stands in a data set by using percentiles

✔ Creating a set of five numbers (using percentiles) that can reveal some aspects of the shape, center, and variation in a data set

What to Watch Out For

Pay particular attention to the following:

✔ Be sure you identify which descriptive statistic or set of descriptive statistics is needed for a particular problem.

✔ After you understand the terminology and calculations for these descriptive statistics, step back and look at the results — make comparisons, see if they make sense, and find the story they tell.

✔ Remember that a percentile isn't a percent, even though they sound the same! When used together, remember that a percentile is a cutoff value in the data set, while a percentage is the amount of data that lies below that cutoff value.

✔ Be aware of the units of any descriptive statistic you calculate (for example, dollars, feet, or miles per gallon). Some descriptive statistics are in the same units as the data, and some aren't.

Understanding the Mean and the Median

21–32 Solve the following problems about means and medians.

21. To the nearest tenth, what is the mean of the following data set? 14, 14, 15, 16, 28, 28, 32, 35, 37, 38

22. To the nearest tenth, what is the mean of the following data set? 15, 25, 35, 45, 50, 60, 70, 72, 100

23. To the nearest tenth, what is the mean of the following data set? 0.8, 1.8, 2.3, 4.5, 4.8, 16.1, 22.3

24. To the nearest thousandth, what is the mean of the following data set? 0.003, 0.045, 0.58, 0.687, 1.25, 10.38, 11.252, 12.001

25. To the nearest tenth, what is the median of the following data set? 6, 12, 22, 18, 16, 4, 20, 5, 15

26. To the nearest tenth, what is the median of the following data set? 18, 21, 17, 18, 16, 15.5, 12, 17, 10, 21, 17

27. To the nearest tenth, what is the median of the following data set? 14, 2, 21, 7, 30, 10, 1, 15, 6, 8

28. To the nearest hundredth, what is the median of the following data set? 25.2, 0.25, 8.2, 1.22, 0.001, 0.1, 6.85, 13.2

29. Compare the mean and median of a data set that has a distribution that is skewed right.

30. Compare the mean and the median of a data set that has a distribution that is skewed left.

31. Compare the mean and the median of a data set that has a symmetrical distribution.

32. Which measure of center is most resistant to (or least affected by) outliers?

Surveying Standard Deviation and Variance

33–48 Solve the following problems about standard deviation and variance.

33. What does the standard deviation measure?

34. According to the 68-95-99.7 rule, or the empirical rule, if a data set has a normal distribution, approximately what percentage of data will be within one standard deviation of the mean?

35. A realtor tells you that the average cost of houses in a town is $176,000. You want to know how much the prices of the houses may vary from this average. What measurement do you need?

    (A) standard deviation

    (B) interquartile range

    (C) variance

    (D) percentile

    (E) Choice (A) or (C)

36. What measure(s) of variation is/are sensitive to outliers?

    (A) margin of error

    (B) interquartile range

    (C) standard deviation

    (D) Choices (A) and (B)

    (E) Choices (A) and (C)

37. You take a random sample of ten car owners and ask them, "To the nearest year, how old is your current car?" Their responses are as follows: 0 years, 1 year, 2 years, 4 years, 8 years, 3 years, 10 years, 17 years, 2 years, 7 years. To the nearest year, what is the standard deviation of this sample?

38. A sample is taken of the ages in years of 12 people who attend a movie. The results are as follows: 12 years, 10 years, 16 years, 22 years, 24 years, 18 years, 30 years, 32 years, 19 years, 20 years, 35 years, 26 years. To the nearest year, what is the standard deviation for this sample?

39. A large math class takes a midterm exam worth a total of 100 points. Following is a random sample of 20 students' scores from the class:

    Score of 98 points: 2 students

    Score of 95 points: 1 student

    Score of 92 points: 3 students

    Score of 88 points: 4 students

    Score of 87 points: 2 students

    Score of 85 points: 2 students

    Score of 81 points: 1 student

    Score of 78 points: 2 students

    Score of 73 points: 1 student

    Score of 72 points: 1 student

    Score of 65 points: 1 student

To the nearest tenth of a point, what is the standard deviation of the exam scores for the students in this sample?

40. A manufacturer of jet engines measures a turbine part to the nearest 0.001 centimeters. A sample of parts has the following data set: 5.001, 5.002, 5.005, 5.000, 5.010, 5.009, 5.003, 5.002, 5.001, 5.000. What is the standard deviation for this sample?

41. Two companies pay their employees the same average salary of $42,000 per year. The salary data in Ace Corp. has a standard deviation of $10,000, whereas Magna Company salary data has a standard deviation of $30,000. What, if anything, does this mean?

42. In which of the following situations would having a small standard deviation be most important?

(A) determining the variation in the wealth of retired people

(B) measuring the variation in circuitry components when manufacturing computer chips

(C) comparing the population of cities in different areas of the country

(D) comparing the amount of time it takes to complete education courses on the Internet

(E) measuring the variation in the production of different varieties of apple trees

43. Suppose that you're comparing the means and standard deviations for the daily high temperatures for two cities during the months of November through March.

Sunshine City: $\mu = 46°F; \sigma = 18°F$

Lake Town: $\mu = 42°F; \sigma = 8°F$

What's the best analysis for comparing the temperatures in the two cities?

44. Everyone at a company is given a year-end bonus of $2,000. How will this affect the standard deviation of the annual salaries in the company that year?

45. Calculate the sample variance and the standard deviation for the following measurements of weights of apples: 7 oz, 6 oz, 5 oz, 6 oz, 9 oz. Express your answers in the proper units of measurement and round to the nearest tenth.

46. Calculate the sample variance and the standard deviation for the following measurements of assembly time required to build an MP3 player: 15 min, 16 min, 18 min, 10 min, 9 min. Express your answers in the proper units of measurement and round to the nearest whole number.

47. Calculate the standard deviation for these speeds of city traffic: 10 km/hr, 15 km/hr, 35 km/hr, 40 km/hr, 30 km/hr. Express your answers in the proper units of measurement and round to the nearest whole number.

48. Which of the following data sets has the same standard deviation as the data set with the numbers 1, 2, 3, 4, 5? (Do this problem without any calculations!)

(A) Data Set 1: 6, 7, 8, 9, 10

(B) Data Set 2: –2, –1, 0, 1, 2

(C) Data Set 3: 0.1, 0.2, 0.3, 0.4, 0.5

(D) Choices (A) and (B)

(E) None of the data sets gives the same standard deviation as the data set 1, 2, 3, 4, 5.

Employing the Empirical Rule

49–56 Use the empirical rule to solve the following problems.

49. According to the empirical rule (or the 68-95-99.7 rule), if a population has a normal distribution, approximately what percentage of values is within one standard deviation of the mean?

50. According to the empirical rule (or the 68-95-99.7 rule), if a population has a normal distribution, approximately what percentage of values is within two standard deviations of the mean?

51. If the average age of retirement for the entire population in a country is 64 years and the distribution is normal with a standard deviation of 3.5 years, what is the approximate age range in which 95% of people retire?

52. Last year's graduates from an engineering college, who entered jobs as engineers, had a mean first-year income of $48,000 with a standard deviation of $7,000. The distribution of salary levels is normal. What is the approximate percentage of first-year engineers that made more than $55,000?

53. What is a necessary condition for using the empirical rule (or 68-95-99.7 rule)?

54. What measures of data need to be known to use the empirical (68-95-99.7) rule?

55. The quality control specialists of a microscope manufacturing company test the lens for every microscope to make sure the dimensions are correct. In one month, 600 lenses are tested. The mean thickness is 2 millimeters. The standard deviation is 0.000025 millimeters. The distribution is normal. The company rejects any lens that is more than two standard deviations from the mean. Approximately how many lenses out of the 600 would be rejected?

56. Biologists gather data on a sample of fish in a large lake. They capture, measure the length of, and release 1,000 fish. They find that the standard deviation is 5 centimeters, and the mean is 25 centimeters. They also notice that the shape of the distribution (according to a histogram) is very much skewed to the left (which means that some fish are smaller than most of the others). Approximately what percentage of fish in the lake is likely to have a length within one standard deviation of the mean?

Measuring Relative Standing with Percentiles

57–64 Solve the following problems about percentiles.

57. What statistic reports the relative standing of a value in a set of data?

58. What is the statistical name for the 50th percentile?

59. Your score on a test is at the 85th percentile. What does this mean?

60. Suppose that in a class of 60 students, the final exam scores have an approximately normal distribution, with a mean of 70 points and a standard deviation of 5 points. Bob's score places him in the 90th percentile among students on this exam. What must be true about Bob's score?

61. On a multiple-choice test, your actual score was 82%, which was reported to be at the 70th percentile. What is the meaning of your test results?

62. Seven students got the following exam scores (percent correct) on a science exam: 0%, 40%, 50%, 65%, 75%, 90%, 100%. Which of these exam scores is at the 50th percentile?

63. Students scored the following grades on a statistics test: 80, 80, 82, 84, 85, 86, 88, 90, 91, 92, 92, 94, 96, 98, 100. Calculate the score that represents the 80th percentile.

64. Some of the students in a class are comparing their grades on a recent test. Mary says she almost scored in the 95th percentile. Lisa says she scored at the 84th percentile. Jose says he scored at the 88th percentile. Paul says he almost scored in the 70th percentile. Bill says he scored at the 95th percentile. Rank the five students from highest to lowest in their grades.

Delving into Data Sets and Descriptive Statistics

65–80 Solve the following problems about data sets and descriptive statistics.

65. Which of the following descriptive statistics is least affected by adding an outlier to a data set?

(A) the mean

(B) the median

(C) the range

(D) the standard deviation

(E) all of the above

66. Which of the following statements is incorrect?

(A) The median and the 1st quartile can be the same.

(B) The maximum and minimum value can be the same.

(C) The 1st and 3rd quartiles can be the same.

(D) The range and the IQR can be the same.

(E) None of the above.

67. Test scores for an English class are recorded as follows: 72, 74, 75, 77, 79, 82, 83, 87, 88, 90, 91, 91, 91, 92, 96, 97, 97, 98, 100. Find the 1st quartile, median, and 3rd quartile for the data set.

68. The average annual returns over the past ten years for 20 utility stocks have the following statistics:

1st quartile = 7

Median = 8

3rd quartile = 9

Mean = 8.5

Standard deviation = 2

Range = 5

Give the five numbers that make up the five-number summary for this data set.

69. Bob attempts to calculate the five-number summary for a set of exam scores. His results are as follows:

Minimum = 30

Maximum = 90

1st quartile = 50

3rd quartile = 80

Median = 85

What is wrong with Bob's five-number summary?

70. Which of the following data sets has a mean of 15 and standard deviation of 0?

(A) 0, 15, 30

(B) 15, 15, 15

(C) 0, 0, 0

(D) There is no data set with a standard deviation of 0.

(E) Choices (B) and (C)

71. The starting salaries (in dollars) of a random sample of 125 university graduates were analyzed. The following descriptive statistics were calculated and typed into a report:

Mean: 24,329

Median: 20,461

Variance: 4,683,459

Minimum: 18,958

Q_1: 22,663

Q_3: 29,155

Maximum: 31,123

What is the error in these descriptive statistics?

72. Which of the following statements is true?

(A) Fifty percent of the values in a data set lie between the 1st and 3rd quartiles.

(B) Fifty percent of the values in a data set lie between the median and the maximum value.

(C) Fifty percent of the values in a data set lie between the median and the minimum value.

(D) Fifty percent of the values in a data set lie at or below the median.

(E) All of the above.

73. Which of the following relationships holds true?

(A) The mean is always greater than the median.

(B) The variance is always larger than the standard deviation.

(C) The range is always less than the IQR.

(D) The IQR is always less than the standard deviation.

(E) None of the above.

74. Suppose that a data set contains the weights of a random sample of 100 newborn babies, in units of pounds. Which of the following descriptive statistics isn't measured in pounds?

(A) the mean of the weights

(B) the standard deviation of the weights

(C) the variance of the weights

(D) the median of the weights

(E) the range of the weights

75. Which of the following is not a measure of the spread (variability) in a data set?

(A) the range

(B) the standard deviation

(C) the IQR

(D) the variance

(E) none of the above

76. A data set contains five numbers with a mean of 3 and a standard deviation of 1. Which of the following data sets matches those criteria?

(A) 1, 2, 3, 4, 5

(B) 3, 3, 3, 3, 3

(C) 2, 2, 3, 4, 4

(D) 1, 1, 1, 1, 1

(E) 0, 0, 3, 6, 6

77. A supermarket surveyed customers one week to see how often each customer shopped at the store every month. The data is shown in the following graph. What are the best measures of spread and center for this distribution?

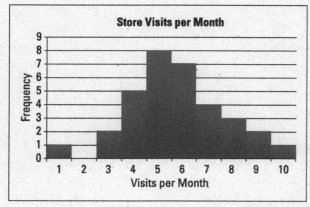

78. Students took a test that had 20 questions. The following graph shows the distribution of the scores. What are the best measures of spread and center for the data?

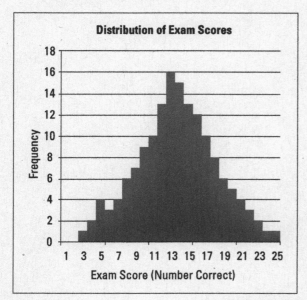

79. An Internet company sells computer parts and accessories. The annual salaries for all the employees have the following parameters:

  Mean: $78,000

  Median: $45,000

  Standard deviation: $40,800

  IQR (interquartile range): $12,000

  Range: $24,000 to $2 million

What are the best measures of spread and center for the data?

80. The distribution of scores for a final exam in math had the following parameters:

  Mean: 83%

  Median: 94%

  Standard deviation: 7%

  IQR (interquartile range): 9%

  Range: 65% to 100%

What are the best measures of spread and center for the data?

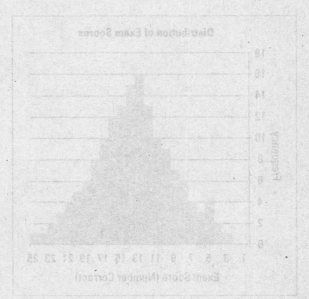

Chapter 3

Graphing

• •

Graphs should be able to stand alone and give all the information needed to identify the main point quickly and easily. The media gives the impression that making and interpreting graphs is no big deal. However, in statistics, you work with more complicated data, consequently taking your graphs up a notch.

The Problems You'll Work On

A good graph displays data in a way that's fair, makes sense, and makes a point. Not all graphs possess these qualities. When working the problems in this chapter, you get experience with the following:

- Identifying the graph that's needed for the particular situation at hand
- Graphing both categorical (qualitative) data and numerical (quantitative) data
- Putting together and correctly interpreting histograms
- Highlighting data collected over time, using a time plot
- Spotting and identifying problems with misleading graphs

What to Watch Out For

Some graphs are easy to make and interpret, some are hard to make but easy to interpret, and some graphs are tricky to make and even trickier to interpret. Be ready to handle the latter.

- Be sure to understand the circumstances under which each type of graph is to be used and how to construct it. (Rarely will someone actually tell you what type of graph to make!)
- Pay special attention to how a histogram shows the variability in a data set. Flat histograms can have a lot of variability in the data, but flat time plots have none — that's one eye-opener.
- Box plots are a huge issue. Making a box plot itself is one thing; understanding the do's and (especially) the don'ts of interpreting box plots is a whole other story.

Interpreting Pie Charts

81–86 The following pie chart shows the proportion of students enrolled in different colleges within a university.

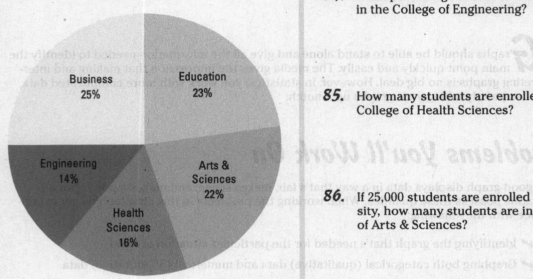

Enrollment by College

Business 25%
Education 23%
Engineering 14%
Arts & Sciences 22%
Health Sciences 16%

Illustration by Ryan Sneed

81. Which college has the largest enrollment?

82. If some students were enrolled in more than one college, what type of graph would be appropriate to show the percentage in each college?

   (A) the same pie chart

   (B) a separate pie chart for each college showing what percentage are enrolled and what percentage aren't

   (C) a bar graph where each bar represents a college and the height shows what percentage of students are enrolled

   (D) Choices (B) and (C)

   (E) none of the above

83. What percentage of students is enrolled in either the College of Education or the College of Health Sciences?

84. What percentage of students is not enrolled in the College of Engineering?

85. How many students are enrolled in the College of Health Sciences?

86. If 25,000 students are enrolled in the university, how many students are in the College of Arts & Sciences?

Considering Three-Dimensional Pie Charts

87 Answer the following problem about three-dimensional pie charts.

87. What characteristic of three-dimensional pie charts (also known as "exploding" pie charts) makes them misleading?

Interpreting Bar Charts

88–94 *The following bar chart represents the post-graduation plans of the graduating seniors from one high school. Assume that every student chose one of these five options. (**Note:** A gap year means that the student is taking a year off before deciding what to do.)*

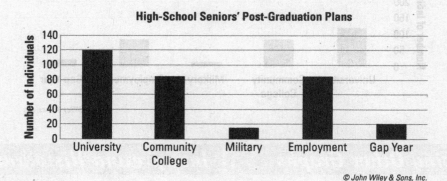

High-School Seniors' Post-Graduation Plans

© John Wiley & Sons, Inc.

88. What is the most common post-graduation plan for these seniors?

89. What is the least common post-graduation plan for these seniors?

90. Assuming that each student has chosen only one of the five possibilities, about how many students plan to either take a gap year or attend a university?

91. How many total students are represented in this chart?

92. What percentage of the graduating class is planning on attending a community college?

93. What percentage of the graduating class is not planning to attend a university?

94. This bar chart displays the same information but is more difficult to interpret. Why is this the case?

High-School Seniors' Post-Graduation Plans

© John Wiley & Sons, Inc.

Introducing Other Graphs

95–96 Solve the following problems about different types of graphs.

95. What type of graph would be the best choice to display data representing the height in centimeters of 1,000 high-school football players?

96. Is the order of bars significant in a histogram?

Interpreting Histograms

97–105 The following histogram represents the body mass index (BMI) of a sample of 101 U.S. adults.

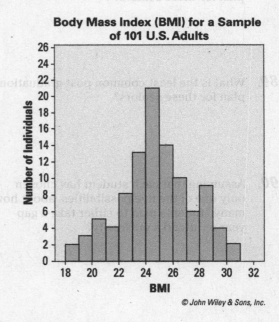

Body Mass Index (BMI) for a Sample of 101 U.S. Adults

© John Wiley & Sons, Inc.

97. Why are there no gaps between the bars of this histogram?

98. What does the x-axis of this histogram represent?

99. What do the widths of the bars represent?

100. What does the y-axis represent?

101. How would you describe the basic shape of this distribution?

102. What is the range of the data in this histogram?

103. Judging by this histogram, which bar contains the average value for this data (considering "average value" as similar to a balancing point)?

104. How many adults in this sample have a BMI in the range of 22 to 24?

105. What percentage of adults in this sample have a BMI of 28 or higher?

Digging Deeper into Histograms

106–112 The following histogram represents the reported income from a sample of 110 U.S. adults.

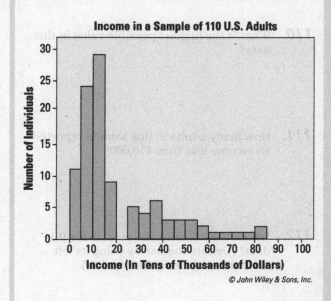

Income in a Sample of 110 U.S. Adults

Number of Individuals

Income (In Tens of Thousands of Dollars)

© John Wiley & Sons, Inc.

106. How would you describe the shape of this distribution?

107. What would be the most appropriate measure of the center for this data?

108. Which value will be higher in this distribution, the mean or the median?

109. What is the lowest possible value in this data?

110. What is the highest possible value in this data?

111. How many adults in this sample reported an income less than $10,000?

112. Which bar contains the median for this data? (Denote the bar by using its left endpoint and its right endpoint.)

Comparing Histograms

113–119 The following histograms represent the grades on a common final exam from two different sections of the same university calculus class.

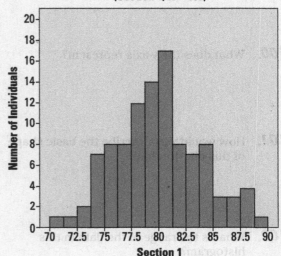

Grades in a University Course
(Section 1, N = 100)

Section 1

Illustration by Ryan Sneed

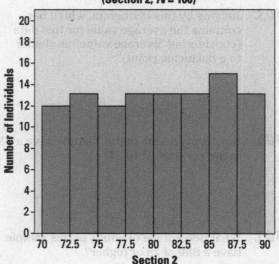

Grades in a University Course
(Section 2, N = 100)

Section 2

Illustration by Ryan Sneed

113. How would you describe the distributions of grades in these two sections?

114. Which section's grade distribution has the greater range?

115. How do you expect the mean and median of the grades in Section 1 to compare to each other?

116. How do you expect the mean and median of the grades in Section 2 to compare to each other?

117. Judging by the histogram, what is the best estimate for the median of Section 1's grades?

118. Judging by the histogram, which interval most likely contains the median of Section 2's grades?

    (A)   below 75

    (B)   75 to 77.5

    (C)   77.5 to 82.5

    (D)   85 to 90

    (E)   above 90

119. Which section's grade distribution do you expect to have a greater standard deviation, and why?

Describing the Center of a Distribution

120 Solve the following problem about the center of a distribution.

120. For the 2013 to 2014 season, salaries for the 450 players in the NBA ranged from slightly less than $1 million to more than $30 million, with 19 players making more than $15 million and about half making $2 million or less. Which would be the best statistic to describe the center of this distribution?

Interpreting Box Plots

121–128 The following box plot represents data on the GPA of 500 students at a high school.

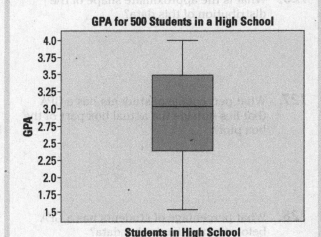

GPA for 500 Students in a High School

Illustration by Ryan Sneed

121. What is the range of GPAs in this data?

122. What is the median of the GPAs?

123. What is the IQR for this data?

124. What does the scale of the numerical axis signify in this box plot?

125. Where is the mean of this data set?

126. What is the approximate shape of the distribution of this data?

127. What percentage of students has a GPA that lies outside the actual box part of the box plot?

128. What percentage of students has a GPA below the median in this data?

Comparing Two Box Plots

129–133 The following box plots represent GPAs of students from two different colleges, call them College 1 and College 2.

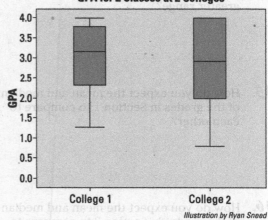

Illustration by Ryan Sneed

129. What information is missing on this graph and on the box plots?

   (A)   the total sample size

   (B)   the number of students in each college

   (C)   the mean of each data set

   (D)   Choices (A) and (B)

   (E)   Choices (A), (B), and (C)

130. Which data set has a greater median, College 1 or College 2?

131. Which data set has the greater IQR, College 1 or College 2?

132. Which data set has a larger sample size?

133. Which data set has a higher percentage of GPAs above its median?

Comparing Three Box Plots

134–139 These side-by-side box plots represent home sale prices (in thousands of dollars) in three cities in 2012.

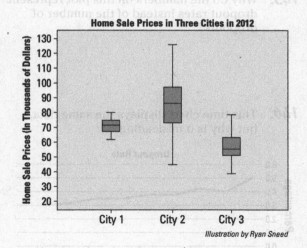

Home Sale Prices in Three Cities in 2012

Illustration by Ryan Sneed

134. From high to low, what is the order of the cities' median home sale prices?

135. If the number of homes sold in each city is the same, which city has the most homes that sold for more than $72,000?

136. Assuming 100 homes sold in each city in 2012, which city has the most homes that sold for more than $72,000?

137. Which city has the smallest range in home prices?

138. Which of the following statements is true?

(A) More than half of the homes in City 1 sold for more than $50,000.

(B) More than half of the homes in City 2 sold for more than $75,000.

(C) More than half of the homes in City 3 sold for more than $75,000.

(D) Choices (A) and (B).

(E) Choices (B) and (C).

139. Which of the following statements is true?

(A) About 25% of homes in City 1 sold for $75,000 or more.

(B) About 25% of homes in City 2 sold for $75,000 or more.

(C) About 25% of homes in City 2 sold for $98,000 or more.

(D) About 25% of homes in City 3 sold for $75,000 or more.

(E) Choices (A) and (C).

Interpreting Time Charts

140–146 The data in the following time chart shows the annual high-school dropout rate for a school system for the years 2001 to 2011.

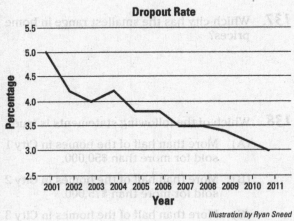

Dropout Rate

Illustration by Ryan Sneed

140. What is the general pattern in the dropout rate from 2001 to 2011?

141. What was the approximate dropout rate in 2005?

142. What was the approximate change in the dropout rate from 2001 to 2011?

143. What was the approximate change in the dropout rate from 2003 to 2004?

144. This time chart displays the same data, but why is it misleading?

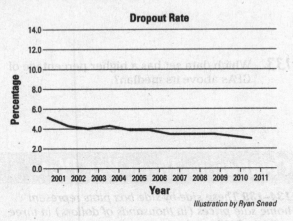

Dropout Rate

Illustration by Ryan Sneed

145. Why do the numbers on this plot represent dropout rates instead of the number of dropouts?

146. This time chart displays the same data, but why is it misleading?

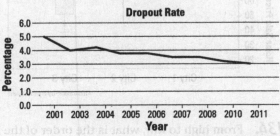

Dropout Rate

Illustration by Ryan Sneed

Getting More Practice with Histograms

147–148 The following three histograms represent reported annual incomes, in thousands of dollars, from samples of 100 individuals from three professions; call the different incomes Income 1, Income 2, and Income 3.

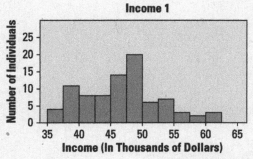

Income 1

Illustration by Ryan Sneed

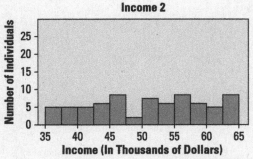

Income 2

Illustration by Ryan Sneed

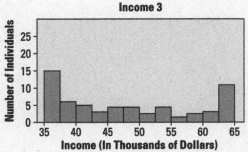

Income 3

Illustration by Ryan Sneed

147. How would you describe the approximate shape of these distributions (Income 1, Income 2, and Income 3)?

148. All three samples have the same range, from $35,000 to $65,000, but they differ in variability. Put the incomes of these three professions in order in terms of their variability, from largest to smallest, using their graphs.

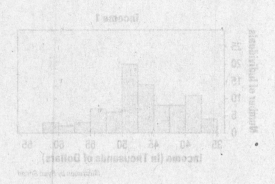

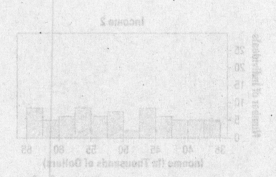

Chapter 4

Random Variables and the Binomial Distribution

● ●

Random variables represent quantities or qualities that randomly change within a population. For example, if you ask random people what their stress level is on a scale from 0 to 10, you don't know what they're going to say. But you do know what the possible values are, and you may have an idea of which numbers are likely to be reported more often (like 9 or 10) and less often (like 0 or 1). In this chapter, you focus on random variables: their types, their possible values and probabilities, their means, their standard deviations, and other characteristics.

The Problems You'll Work On

In this chapter, you see random variables in action and how you can use them to think about a population. Here are some items on the menu:

- ✔ Distinguishing discrete versus continuous random variables

- ✔ Finding probabilities for a random variable

- ✔ Calculating and interpreting the mean, variance, and standard deviation of a random variable

- ✔ Finding probabilities, mean, and standard deviation for a specific random variable, the binomial

What to Watch Out For

The problems in this chapter involve notation, formulas, and calculations. Paying attention to the details will make a difference.

- ✔ Understand the notation really well; several symbols are floating around in this chapter.

- ✔ Be able to interpret your results, not just do the calculations, including the proper use of units.

- ✔ Know the ways to find binomial probabilities; pay special attention to the normal approximation.

Comparing Discrete and Continuous Random Variables

149–154 Solve the following problems about discrete and continuous random variables.

149. Which of the following random variables is discrete?

(A) the length of time a battery lasts

(B) the weight of an adult

(C) the percentage of children in a population who have been vaccinated against measles

(D) the number of books purchased by a student in a year

(E) the distance between a pair of cities

150. Which of the following random variables isn't discrete?

(A) the number of children in a family

(B) the annual rainfall in a city

(C) the attendance at a football game

(D) the number of patients treated at an emergency room in a day

(E) the number of classes taken in one semester by a student

151. Which of the following random variables is discrete?

(A) the proportion of a population that voted in the last election

(B) the height of a college student

(C) the number of cars registered in a state

(D) the weight of flour in a sack advertised as containing ten pounds

(E) the length of a phone call

152. Which of the following random variables is continuous?

(A) the number of heads resulting from flipping a coin 30 times

(B) the number of deaths from plane crashes in a year

(C) the proportion of the American population that believes in ghosts

(D) the number of films produced in Canada in a year

(E) the number of people arrested for auto theft in a year

153. Which of the following random variables is continuous?

(A) the number of seniors in a college

(B) the number of gold medals won at the 2012 Summer Olympics by athletes from Germany

(C) the number of schools in a city

(D) the number of registered physicians in the United States

(E) the amount of gasoline used in the Unites States in 2012

154. Which of the following random variables isn't continuous?

(A) the proportion of adults on probation in a state

(B) the population growth rate for a city

(C) the amount of money spent by a household for food over a year

(D) the number of bird species observed in an area

(E) the length of time it takes to walk ten miles

Understanding the Probability Distribution of a Random Variable

155–157 The following table represents the probability distribution for X, the employment status of adults in a city.

X	P(X)
Employed full-time	0.65
Employed part-time	0.10
Unemployed	0.07
Retired	0.18

© John Wiley & Sons, Inc.

155. If you select one adult at random from this community, what is the probability that the individual is employed part-time?

156. If you select one adult at random from this community, what is the probability that the individual isn't retired?

157. If you select one adult at random from this community, what is the probability that the individual is working either part-time or full-time?

Determining the Mean of a Discrete Random Variable

158–159 Let X be the number of classes taken by a college student in a semester. Use the formula for the mean of a discrete random variable X to answer the following problems:

$$\mu_x = \sum x_i p_i$$

158. If 40% of all the students are taking four classes, and 60% of all the students are taking three classes, what is the mean (average) number of classes taken for this group of students?

159. If half of the students in a class are age 18, one-quarter are age 19, and one-quarter are age 20, what is the average age of the students in the class?

Digging Deeper into the Mean of a Discrete Random Variable

160–163 In the following table, X represents the number of automobiles owned by families in a neighborhood.

X	P(X)
0	0.25
1	0.60
2	
3	0.05

© John Wiley & Sons, Inc.

160. What is the missing value in this table (representing the number of automobiles owned by two families in a neighborhood)?

161. What is the mean number of automobiles owned?

162. If every family currently not owning a car bought one car, what would be the mean number of automobiles owned?

163. If all the families currently owning three cars bought a fourth car, what would be the mean number of automobiles owned?

Working with the Variance of a Discrete Random Variable

164–165 Use the following formula for the variance of a discrete random variable X as needed to answer the following problems (round each answer to two decimal places):

$$\sigma_x^2 = \sum (x_i - \mu_x)^2 p_i$$

164. If the variance of a discrete random variable X is 3, what is the standard deviation of X?

165. If the standard deviation of X is 0.65, what is the variance of X?

Putting Together the Mean, Variance, and Standard Deviation of a Random Variable

166–169 In the following table, X represents the number of siblings for the 29 students in a first-grade class.

X	P(X)
0	0.34
1	0.52
2	0.14

© John Wiley & Sons, Inc.

166. What is the mean number of siblings for these students?

167. What is the variance for the number of siblings for these students?

168. What is the standard deviation for the number of siblings for these students? Round your answer to two decimal places.

169. How would the variance and standard deviation change if the number of siblings, X, doubled in each case but the probabilities, $p(x)$, values stayed the same?

Digging Deeper into the Mean, Variance, and Standard Deviation of a Random Variable

170–173 In the following table, X represents the number of books required for classes at a university.

X	P(X)
0	0.30
1	0.25
2	0.25
3	0.10
4	0.10

© John Wiley & Sons, Inc.

170. What is the mean number of books required?

171. What is the variance of the number of books required? Round your answer to two decimal places.

172. What is the standard deviation of the number of books required? Round your answer to two decimal places.

173. How would the standard deviation and variance change if only 20% of the students required two books, but now 5% of the students require five books (with all other categories unchanged)?

Introducing Binomial Random Variables

174–178 Solve the following problems about the basics of binomial random variables.

174. What condition(s) must a random variable meet to be considered binomial?

(A) fixed number of trials

(B) exactly two possible outcomes on each trial: success and failure

(C) constant probability of success for all trials

(D) independent trials

(E) all of the above

175. You flip a coin 25 times and record the number of heads. What is the binomial random variable (X) in this experiment?

176. You roll a six-faced die ten times and record which face comes up each time (X). Why is X not a binomial random variable?

177. You interview a number of employees selected at random and ask them whether they've graduated from high school. You continue the interviews until you have 30 employees who say they graduated from high school. If X is the number of people you had to ask until you got 30 "yes" responses, why isn't X a binomial random variable?

178. You recruit 30 pairs of siblings and test each individual to see whether he or she is carrying a particular genetic mutation, and then you add up the total number of people (not pairs) who have the mutation. Why is this not a binomial experiment?

Figuring Out the Mean, Variance, and Standard Deviation of a Binomial Random Variable

179–183 Solve the following problems about the mean, standard deviation, and variance of binomial random variables.

179. What is the mean of a binomial random variable with $n = 18$ and $p = 0.4$?

180. What is the mean of a binomial random variable with $n = 25$ and $p = 0.35$?

181. What is the standard deviation of a binomial distribution with $n = 18$ and $p = 0.4$? Round your answer to two decimal places.

182. What is the variance of a binomial distribution with $n = 25$ and $p = 0.35$? Round your answer to two decimal places.

183. A binomial distribution with $p = 0.14$ has a mean of 18.2. What is n?

Finding Binomial Probabilities with a Formula

184–188 X is a binomial random variable with $p = 0.55$. Use the following formulas for the binomial distribution for the following problems.

$$P(X = x) = \binom{n}{x} p^x (1-p)^{n-x}$$

$$\text{where } \binom{n}{x} = \frac{n!}{x!\,(n-x)!} \text{ and}$$

$$n! = (n-1)(n-2)(n-3) \ldots (3)(2)(1)$$

184. What is the value of $\binom{n}{x}$ if $n = 8$ and $x = 1$?

185. What is the probability of exactly one success in eight trials? Round your answer to four decimal places.

186. What is the probability of exactly two successes in eight trials? Round your answer to four decimal places.

187. What is the value of $\begin{pmatrix} 8 \\ 0 \end{pmatrix}$?

188. What is the probability of getting at least one success in eight trials? Round your answer to four decimal places.

Digging Deeper into Binomial Probabilities Using a Formula

189–195 Suppose that X has a binomial distribution with p = 0.50.

189. What is the probability of exactly eight successes in ten trials? Round your answer to four decimal places.

190. What is the probability of exactly three successes in five trials?

191. What is the probability of exactly four successes in five trials? Round your answer to four decimal places.

192. What is the value of $\begin{pmatrix} 5 \\ 5 \end{pmatrix}$?

193. What is the probability of three or four successes in five trials? Round your answer to four decimal places.

194. What is the probability of at least three successes in five trials? Round your answer to four decimal places.

195. What is the probability of no more than two successes in five trials? Round your answer to four decimal places.

Finding Binomial Probabilities with the Binomial Table

196–200 X is a random variable with a binomial distribution with n = 15 and p = 0.7. Use the binomial table (Table A-3 in the appendix) to answer the following problems.

196. What is $P(X = 6)$?

197. What is $P(X = 11)$?

198. What is $P(X < 15)$?

199. What is $P(4 < X < 7)$?

200. Find $P(4 \leq X \leq 7)$.

Digging Deeper into Binomial Probabilities Using the Binomial Table

201–205 X is a random variable with a binomial distribution with n = 11 and p = 0.4. Use the binomial table (Table A-3 in the appendix) to answer the following problems.

201. What is $P(X = 5)$?

202. What is $P(X > 0)$?

203. What is $P(X \leq 2)$?

204. What is $P(X > 9)$?

205. What is $P(3 \leq X \leq 5)$?

Using the Normal Approximation to the Binomial

206–208 Solve the following problems about the normal approximation to the binomial.

206. Which of the following binomial distributions would allow you to use the normal approximation to the binomial?

(A) $n = 10, p = 0.3$

(B) $n = 10, p = 0.4$

(C) $n = 20, p = 0.9$

(D) $n = 25, p = 0.3$

(E) $n = 30, p = 0.4$

207. You're conducting a binomial experiment by flipping a fair coin ($p = 0.5$) and recording how many times it comes up heads. What is the minimum size for *n* that will allow you to use the normal approximation to the binomial?

208. You're conducting a binomial experiment by selecting marbles from a large jar and recording how many times you draw a green marble. Overall, 30% of the marbles in the jar are green. What is the minimum size for *n* that will allow you to use the normal approximation to the binomial?

Digging Deeper into the Normal Approximation to the Binomial

209–214 You conduct a binomial experiment by flipping a fair coin (p = 0.5) 80 times and recording how often it comes up heads (X). Use the normal approximation to the binomial to calculate any probabilities in this problem set. You can do so because the two required conditions are met: np and n(1 – p) are both at least 10.

209. How would you formally write the probability that you get at least 50 heads?

210. How would you formally write the probability that you get no more than 30 heads?

211. To use the normal approximation to the binomial, what value will you use for μ?

212. To use the normal approximation to the binomial, what value will you use for σ? Round your answer to two decimal places.

213. What is the z-value corresponding to the x value in $P(X \geq 45)$? Round your answer to two decimal places.

214. What is the probability that you will get at least 45 heads in 80 flips? Round your answer to four decimal places.

Getting More Practice with Binomial Variables

215–223 X is a binomial variable with p = 0.45 and n = 100.

215. What are the mean and standard deviation for the distribution of *X*? Round your answer to two decimal places.

216. What is the z-value corresponding to the x value in $P(x \leq 40)$? Round your answer to two decimal places.

217. What is $P(X \leq 40)$? Round your answer to four decimal places.

218. What is $P(X \geq 40)$? Round your answer to four decimal places.

219. Suppose you want to find $P(X = 40)$. You can try to solve this by using a normal approximation and converting $x = 40$ to its z-value of $z = -1.00$. What is $P(Z = -1.00)$? Round your answer to four decimal places.

220. What is the z-value corresponding to $x = 56$? Round your answer to two decimal places.

221. What is $P(X \geq 56)$? Round your answer to four decimal places.

222. What is $P(X \leq 56)$? Round your answer to four decimal places.

223. What is $P(56 \leq x \leq 60)$? Round your answer to four decimal places.

Chapter 5

The Normal Distribution

The normal distribution is the most common distribution of all. Its values take on that familiar bell shape, with more values near the center and fewer as you move away. Because of its nice qualities and practical nature, you see many applications of the normal distribution not only in this chapter but also in many chapters that follow. Understanding the normal distribution from the ground up is critical.

The Problems You'll Work On

In this chapter, you practice in detail everything you need to know about the normal distribution and then some. Here's an overview of what you'll work on:

- Finding probabilities for a normal distribution (less than, greater than, or in between)
- Understanding the standard normal (Z-) distribution, its properties, and how its values are interpreted and used
- Using the Z-table to find probabilities
- Determining percentiles for a normal distribution

What to Watch Out For

The normal distribution seems easy, but many important nuances exist as well, so stay on top of them. Here are some things to keep in mind:

- Be very clear about what a z-score is and what it means. You'll use it throughout this chapter.
- Make sure you understand how to use the Z-table to find the probabilities you want.
- Know when you're asked for a percent (probability) and when you're asked for a percentile (a value of the variable affiliated with a certain probability). Then know what to do about it.

Defining and Describing the Normal Distribution

224–234 Solve the following problems about the definition of the normal distribution and what it looks like.

224. What are properties of the normal distribution?

    (A)   It's symmetrical.

    (B)   Mean and median are the same.

    (C)   Most common values are near the mean; less common values are farther from it.

    (D)   Standard deviation marks the distance from the mean to the inflection point.

    (E)   All of the above.

225. In a normal distribution, about what percent of the values lie within one standard deviation of the mean?

226. In a normal distribution, about what percent of the values lie within two standard deviations of the mean?

227. In a normal distribution, about what percent of the values lie within three standard deviations of the mean?

228. What two parameters (pieces of information about the population) are needed to describe a normal distribution?

229. Which of the following normal distributions will have the greatest spread when graphed?

    (A) $\mu = 5, \sigma = 1.5$

    (B) $\mu = 10, \sigma = 1.0$

    (C) $\mu = 5, \sigma = 1.75$

    (D) $\mu = 5, \sigma = 1.2$

    (E) $\mu = 10, \sigma = 1.6$

230. For a normal distribution with $\mu = 5$ and $\sigma = 1.2$, 34% of the values lie between 5 and what number? (Assume that the number is above the mean.)

231. For a normal distribution with $\mu = 5$ and $\sigma = 1.2$, about 2.5% of the values lie above what value? (Assume that the number is above the mean.)

232. For a normal distribution with $\mu = 5$ and $\sigma = 1.2$, about 16% of the values lie below what value?

233. A normal distribution with $\mu = 8$ has 99.7% of its values between 3.5 and 12.5. What is the standard deviation for this distribution?

234. What are the mean and standard deviation of the Z-distribution?

Working with z-Scores and Values of X

235–239 *A random variable X has a normal distribution, with a mean of 17 and a standard deviation of 3.5.*

235. What is the z-score for a value of 21.2?

236. What is the z-score for a value of 13.5?

237. What is the z-score for a value of 25.75?

238. What value of X corresponds to a z-score of –0.4?

239. What value of X corresponds to a z-score of 2.2?

Digging Deeper into z-Scores and Values of X

240–245 *All the scores on a national exam have a normal distribution and a range of 0 to 100, but when split up, Form A of the exam has $\mu = 70$ and $\sigma = 10$, while Form B has $\mu = 74$ and $\sigma = 8$.*

240. What are the scores on each exam corresponding to a z-score of 1.5?

241. What are the scores on each exam corresponding to a z-score of –2.0?

242. In terms of standard deviations above or below the mean, what score on Form B corresponds to a score of 80 on Form A?

243. In terms of standard deviations above or below the mean, what score on Form B corresponds to a score of 85 on Form A?

244. In terms of standard deviations above or below the mean, what score on Form A corresponds to a score of 78 on Form B?

245. In terms of standard deviations above or below the mean, what score on Form A corresponds to a score of 68 on Form B?

Writing Probability Notations

246–248 Write probability notations for the following problems.

246. Write the probability notation for the area shaded in this Z-distribution (where $\mu = 0$ and $\sigma = 1$).

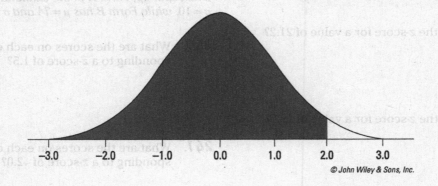

© John Wiley & Sons, Inc.

247. Write the probability notation for the area shaded in this Z-distribution (where $\mu = 0$ and $\sigma = 1$).

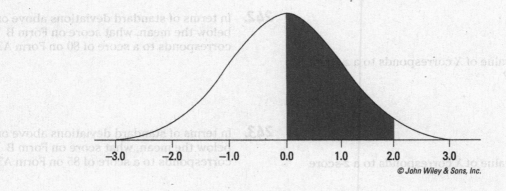

© John Wiley & Sons, Inc.

248. Write the probability notation for the area shaded in this Z-distribution (where $\mu = 0$ and $\sigma = 1$).

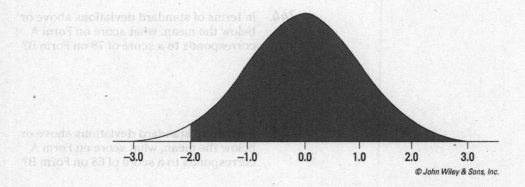

© John Wiley & Sons, Inc.

Introducing the Z-Table

249–253 Use the Z-table (Table A-1 in the appendix) as needed to answer the following problems.

249. What is $P(Z \leq 1.5)$?

250. What is $P(Z \geq 1.5)$?

251. What is $P(Z \geq -0.75)$?

252. What is $P(-0.5 \leq Z \leq 1.0)$?

253. What is $P(-1.0 \leq Z \leq 1.0)$?

Finding Probabilities for a Normal Distribution

254–258 The diameter of a machine part produced by a factory is normally distributed, with a mean of 10 centimeters and a standard deviation of 2 centimeters.

254. What is the z-score for a part with a diameter of 13 centimeters?

255. What is the probability of a part having a diameter of at least 13 centimeters?

256. What is the probability of a part having a diameter no greater than 13 centimeters?

257. What is the probability of a part having a diameter between 10 and 13 centimeters?

258. What is the probability of a part having a diameter between 7 and 10 centimeters?

Digging Deeper into z-Scores and Probabilities

259–270 The weight of adult men in a population is normally distributed, with a mean of 160 pounds and a standard deviation of 20 pounds.

259. What is the z-score for a weight of 135 pounds?

260. What is the z-score for a weight of 170 pounds?

261. What is the z-score for a weight of 115 pounds?

262. What is the z-score for a weight of 220 pounds?

263. What is the z-score for a weight of 205 pounds?

264. What is the probability of a weight greater than 220 pounds?

265. What is the probability of a weight less than 220 pounds?

266. What is the probability of a weight between 135 and 160 pounds?

267. What is the probability of a weight between 205 and 220 pounds?

268. What is the probability of a weight between 115 and 135 pounds?

269. What is the probability of a weight between 135 and 170 pounds?

270. What is the probability of a weight between 170 and 220 pounds?

Figuring Out Percentiles for a Normal Distribution

271–278 In a population of adults ages 18 to 65, BMI (body mass index) is normally distributed with a mean of 27 and a standard deviation of 5.

271. What is the BMI score for which half of the population has a lower value?

272. What BMI marks the bottom 25% of the distribution for this population?

273. What BMI marks the bottom 5% of the distribution for this population?

274. What BMI marks the bottom 10% of the distribution for this population?

275. What BMI value marks the upper 10% of the distribution for this population?

276. What BMI value marks the upper 5% of the distribution for this population?

277. What BMI value marks the upper 30% of the distribution for this population?

278. What two BMI values mark the 1st and 3rd quartiles of the distribution?

280. What is the value where only 5% of the students scored below it?

281. What is the value where only 10% of the students scored above it?

282. What is the value where only 1% of the students scored above it?

283. What is the value where only 2.5% of students scored above it?

284. What is the value where only 5% of the students scored above it?

Digging Deeper into Percentiles for a Normal Distribution

279–284 Scores on an exam are normally distributed, with a mean of 75 and a standard deviation of 6.

279. What is the value where only 20% of the students scored below it?

Getting More Practice with Percentiles

285–290 In a large population of military recruits, the time required to run a mile has a normal distribution with a mean of 360 seconds and a standard deviation of 30 seconds.

285. What time represents the cutoff for the 5% of fastest times?

286. What time represents the cutoff for the fastest 50% of times?

287. What time represents the cutoff for the slowest 10% of times?

288. What time represents the cutoff for the fastest 10% of times?

289. What time represents the cutoff for the fastest 25% of times?

290. What time represents the cutoff for the slowest 25% of times?

Chapter 6

The *t*-Distribution

The *t*-distribution is a relative of the normal distribution. It has a bell shape with values more spread out around the middle. That is, it's not as sharply curved as the normal distribution, which reflects its ability to work with problems that may not be exactly normal but are close.

The Problems You'll Work On

In this chapter, you'll work on all the ins and outs of the *t*-distribution and its relationship to the standard normal (*Z*-) distribution. Here's what you'll focus on:

- Understanding the characteristics that make the *t*-distribution similar to and different from the standard normal (*Z*-) distribution
- Finding probabilities for various *t*-distributions, using the *t*-table
- Finding critical values on the *t*-distribution that are used when calculating confidence intervals

What to Watch Out For

As you work through this chapter, keep the following in mind:

- Pick out the big four — population, parameter, sample, and statistic — in every situation; they'll follow you wherever you go.
- Know the circumstances for when you should use the *t*-distribution.
- The *t*-table (for the *t*-distribution) is different from the *Z*-table (for the *Z*-distribution); make sure you understand the values in the first and last rows.
- Pay special attention to the term *tail probability*, which appears a great deal in this chapter.

Understanding the t-Distribution and Comparing It to the Z-Distribution

291–300 Solve the following problems about the t-distribution, its traits, and how it compares to the Z-distribution.

291. Which of the following is true of the t-distribution, as compared to the Z-distribution? (Assume a low number of degrees of freedom.)

(A) The t-distribution has thicker tails than the Z-distribution.

(B) The t-distribution has a proportionately larger standard deviation than the Z-distribution.

(C) The t-distribution is bell-shaped but has a lower peak than the Z-distribution.

(D) Choices (A) and (C)

(E) Choices (A), (B), and (C)

292. Which t-distribution do you use for a study involving one population with a sample size of 30?

293. If a particular t-distribution from a study involving one population is t_{24}, what is the sample size?

294. If you graphed a standard normal distribution (Z-distribution) on the same number line as a t-distribution with 15 degrees of freedom, how would you expect them to differ?

295. If you graphed a standard normal distribution (Z-distribution) and a t-distribution with 20 degrees of freedom on the same number line, how would you expect them to differ?

296. Given different t-distributions with the following degrees of freedom, which one would you expect to most closely resemble the Z-distribution: 5, 10, 20, 30, or 100?

297. What minimal information do you need to recognize which particular t-distribution you have?

298. What statistical procedure would be most appropriate to test a claim about a population mean for a normal distribution when you don't know the population standard deviation?

299. What statistical procedure would be most appropriate for a study testing a claim about the mean of the difference when the subjects (and hence the data) are in pairs?

300. What is the particular t-distribution resulting from a paired design involving a total of 50 observations?

Using the t-Table

301–335 Use the t-table (Table A-2 in the appendix) as necessary to solve the following problems.

301. For a study involving one population and a sample size of 18 (assuming you have a *t*-distribution), what row of Table A-2 will you use to find the right-tail ("greater than") probability affiliated with the study results?

302. For a study involving a paired design with a total of 44 observations, with the results assuming a *t*-distribution, what row of Table A-2 will you use to find the probability affiliated with the study results?

303. What column of Table A-2 would you use to find an upper-tail ("greater than") probability of 0.025?

304. What column of Table A-2 would you use to find the *t*-value for a two-tailed test with $\alpha = 0.01$?

305. What column of Table A-2 would you use to find the *t*-value for an overall probability of 0.05 for a two-tailed test?

306. For a *t*-distribution with 10 degrees of freedom, what is $P(t \geq 1.81)$?

307. For a *t*-distribution with 25 degrees of freedom, what is $P(t \geq 2.49)$?

308. What is $P(t_{15} \geq 1.34)$?

309. A *t*-value of 1.80, from a *t*-distribution with 22 degrees of freedom, has an upper-tail ("greater than") probability between which two values on Table A-2?

310. A *t*-value of 2.35, from a *t*-distribution with 14 degrees of freedom, has an upper-tail ("greater than") probability between which two values on Table A-2?

311. What is $P(t_{15} \leq -2.60)$?

312. What is $P(t_{27} \leq -2.05)$?

313. What is $P(t_{27} \geq 2.05 \text{ or } t_{27} \leq -2.05)$?

314. What is $P(t_9 \geq 3.25 \text{ or } t_9 \leq -3.25)$?

315. What is the 95th percentile for a t-distribution with 10 degrees of freedom?

316. What is the 60th percentile for a t-distribution with 28 degrees of freedom?

317. What is the 10th percentile for a t-distribution with 20 degrees of freedom?

318. What is the 25th percentile for a t-distribution with 20 degrees of freedom?

319. What value is the 10th percentile for a t-distribution with 16 degrees of freedom?

320. What value is the 5th percentile for a t-distribution with 16 degrees of freedom?

321. Which of the following is a reason to use the t-distribution rather than the Z-distribution to calculate a confidence interval for the mean?

(A)　You have a small sample size.

(B)　You don't know the population standard deviation.

(C)　You don't know the population mean.

(D)　Choices (A) and (B)

(E)　Choices (A), (B), and (C)

322. Using the last row of Table A-2, what column would you use to find the t-value for a 99% confidence interval?

323. Using the first row (column headings) of Table A-2, what column would you use to find the t-value for a 99% confidence interval?

324. What is the t-value for a 95% confidence interval for a t-distribution with 15 degrees of freedom?

325. What is the t-value for a 99% confidence interval for a t-distribution with 23 degrees of freedom?

326. What is the t-value for a 90% confidence interval for a t-distribution with 30 degrees of freedom?

327. For a *t*-distribution with 19 degrees of freedom, what is the confidence level of a confidence interval with a *t*-value of 2.09?

328. For a *t*-distribution with 26 degrees of freedom, what is the confidence level of a confidence interval with a *t*-value of 2.78?

329. For a *t*-distribution with 12 degrees of freedom, what is the confidence level of a confidence interval with a *t*-value of 1.36?

330. Considering *t*-distributions with different degrees of freedom, which of the following would you expect to most closely resemble the *Z*-distribution: 30, 35, 40, 45, or 50?

331. Considering *t*-distributions with different degrees of freedom, which of the following would you expect to least resemble the *Z*-distribution: 10, 20, 30, 40, or 50?

332. Considering *t*-distributions with different degrees of freedom, which of the following would you expect to have the largest *t*-value for a right-tail probability of 0.05: 10, 20, 30, 40, or 50?

333. Considering *t*-distributions with different degrees of freedom, which of the following would you expect to have the smallest *t*-value for a right-tail probability of 0.10: 10, 20, 30, 40, or 50?

334. For a t_{40} distribution, which of the following confidence intervals would you expect to be narrowest: 80%, 90%, 95%, 98%, or 99%?

335. For a t_{50} distribution, which of the following confidence intervals would you expect to be widest: 80%, 90%, 95%, 98%, or 99%?

Using the t-Distribution to Calculate Confidence Intervals

336–340 Use Table A-2 in the appendix as needed and the following information to solve the following problems: The mean length for the population of all screws being produced by a certain factory is targeted to be $\mu_0 = 5$ centimeters. Assume that you don't know what the population standard deviation (σ) is. You draw a sample of 30 screws and calculate their mean length. The mean ($\bar{x}$) for your sample is 4.8, and the standard deviation of your sample (s) is 0.4 centimeters.

336. Calculate the appropriate test statistic for the following hypotheses:

$$H_0: \mu = 5 \text{ cm}$$
$$H_1: \mu \neq 5 \text{ cm}$$

337. Using Table A-2, what is the probability of getting a *t*-value of –2.74 or less?

338. What is the 95% confidence interval for the population mean? Round your answer to two decimal places.

339. What is the 90% confidence interval for the population mean? Round your answer to two decimal places.

340. What is the 99% confidence interval for the population mean? Round your answer to two decimal places.

Chapter 7

Sampling Distributions and the Central Limit Theorem

• •

Sample results vary — that's a major truth of statistics. You take a random sample of size 100, find the average, and repeat the process over and over with different samples of size 100. Those sample averages will differ, but the question is, by how much? And what affects the amount of difference? Understanding this concept of variability between all possible samples helps determine how typical or atypical your particular result may be. Sampling distributions provide a fundamental piece to answer these problems. One tool that's often needed in the process is the central limit theorem. In this chapter, you work on both.

The Problems You'll Work On

Here are the main areas of focus in this chapter:

✔ Working with the sample means as a random variable, with their own distribution, mean, and standard deviations (standard errors) and then doing the same with sample proportions

✔ Understanding the central limit theorem, how to use it, when to use it, and when it's not needed

✔ Calculating probabilities for $\bar{X}$ and $\hat{p}$

What to Watch Out For

Sampling distributions and the central limit theorem are difficult topics. Here are some main points to keep in mind:

✔ $\bar{x}$ is the mean of one sample, and $\bar{X}$ represents any possible sample mean from any sample.

✔ Everything builds on the understanding of what a sampling distribution is — work until you get it.

✔ The central limit theorem is used only in certain situations — know those situations.

✔ Finding probabilities for $\bar{X}$ and $\hat{p}$ involve dividing by a quantity involving the square root of n. Don't leave it out.

Introducing the Basics of Sampling Distributions

341–347 Solve the following problems that introduce the basics of sampling distributions.

341. A characteristic of interest that takes on certain values in a random manner is called a _____.

342. Suppose that 10,000 students took the AP statistics exam this year. If you take every possible sample of 100 students who took the AP exam and find the average exam score for each sample and then put all those average scores together, what would it represent?

343. A GPA is the grade point average of a single student. Suppose that you found the GPA for every student in a university and found that the mean of all those GPAs is 3.11. What statistical notation do you use to represent this value of 3.11?

344. If you roll two fair dice, look at the outcomes, and find the average value, you could get any number from 1 (where both your dice came up 1) to 6 (where both dice came up 6). However, in the long run, if you took the average of all possible pairs of dice, you'd get 3.5 (because that's the average value of the numbers 1 through 6). How do you represent 3.5 in this situation using statistical notation?

345. Suppose that you roll several ordinary six-sided dice, choose two of those dice at random, and average the two numbers. Suppose the average of these two dice is 3.5. How would you express the 3.5 in statistical notation?

346. Which of the following features is necessary for something to be considered a sampling distribution?

(A) Each value in the original population should be included in the distribution.

(B) The distribution must consist of proportions.

(C) The distribution can't consist of percentages.

(D) Each of the observations in the distribution must consist of a statistic that describes a collection of data points.

(E) The samples in the distribution must be drawn from a population of normally distributed values.

347. Which of the following would *not* ordinarily be considered a sampling distribution?

(A) a distribution showing the average weight per person in several hundred groups of three people picked at random at a state fair

(B) a distribution showing the average proportion of heads coming up in several thousand experiments in which ten coins were flipped each time

(C) a distribution showing the average percentage daily price change in Dow Jones Industrial Stocks for several hundred days chosen at random from the past 20 years

(D) a distribution showing the proportion of parts found to be deficient in each of several hundred shipments of parts, each of which has the same number of parts in it

(E) a distribution showing the weight of each individual football fan entering a stadium on game day

Checking Out Random Variables and Sample Means

348–352 You have a six-sided die with faces numbered 1 through 6, and each face is equally likely to come up on any given roll.

348. What is the meaning for X in this example?

349. What is the meaning of $\bar{X}$ in this example?

350. Suppose that you rolled the die five times and got the values of 3, 4, 2, 3, and 1. What is the value of $\bar{x}$?

351. Suppose that you rolled the die five times and got the values of 3, 4, 6, 3, and 5. What is the value of $\bar{x}$?

352. Which of the following is the formula for the standard error of a sample mean?

(A) $\sigma = \dfrac{\sigma_X}{\sqrt{n}}$

(B) $\sigma_{\bar{X}} = \dfrac{\sigma}{\sqrt{n}}$

(C) $\sigma_{\bar{X}} = \dfrac{\sigma_X}{\sqrt{n}}$

(D) $\sigma_{\bar{X}} = \dfrac{\sigma_X}{n}$

(E) $\sigma_X = \dfrac{\sigma_X}{\sqrt{n}}$

Examining Standard Error

353–357 Suppose that you have a population of 100,000 individuals and ask them 200 trivia questions; their score is the number of trivia questions they got right. For this population, the mean score is 158 and the population standard deviation of the scores is 26.

353. Suppose that you draw several samples of 50 people from this particular population, and for each sample you find $\bar{x}$, their average score. And you do this repeatedly until you have all possible average scores from all possible samples of 50 people. Fill in the blanks with the appropriate statistical terms: The _____ represents the amount of variability in the entire population of trivia test scores, and the _____ represents the amount of variability in all the average trivia scores for all groups of 50.

354. Suppose that you draw 100 samples of size 30 and calculate the mean and standard error. If you then draw 100 samples of size 50 from the same population, you would expect to see the mean _____ and the standard error _____.

(A) be approximately the same; be larger

(B) be approximately the same; be smaller

(C) be smaller; be smaller

(D) be smaller; be larger

(E) be smaller; be approximately the same

355. For this particular population, with its given mean and standard deviation, what is the standard error for a sample of size 50? Round your answer to two decimal places.

356. For this particular population, with its given mean and standard deviation, what is the standard error for a sample of size 60? Round your answer to two decimal places.

357. For this particular population, with its given mean and standard deviation, what is the standard error for a sample of size 30? Round your answer to two decimal places.

Surveying Notation and Symbols

358–364 Solve the following problems about notation and symbols.

358. What is the notation for the average of a set of scores sampled from a larger population of scores?

359. What is the notation for the average value of a random variable X in a population?

360. Which symbol represents the standard error of the sample means?

361. Which symbol represents the standard deviation of individual scores in a population?

362. Suppose that you assign a number value to each card in a standard deck of 52 playing cards (Ace = 1, 2 = 2, 3 = 3, and so on up to Jack = 11, Queen = 12, King = 13). Then, you draw a hand of five cards, compute the average face value of cards in that hand, return the cards to the deck, shuffle the deck, and do the same thing again. You do this over and over again — an infinite number of times — each time recording the average value of the cards in your hand. What symbol would best represent the standard deviation of all the averages you find by this method?

363. Every morning, the crew of a fishing boat pulls up a net full of fish, randomly chooses 50 of those fish, computes the average weight of the fish in the catch, and then records that number in a log. Over several years, the owner of the boat uses the standard deviation of those average weights to make financial projections to be sure to consider random fluctuations in the average size of the fish in a catch when estimating revenues. Which symbol best represents the value that the owner is trying to estimate to make financial projections?

364. A realtor looks at seven houses recently sold that are comparable to a house she'll be trying to sell. Which symbol best represents the sale price of any one of the homes?

Understanding What Affects Standard Error

365–368 Solve the following problems about items that affect standard error.

365. If everything else is held constant, which of the following is most likely to result in a larger standard error?

(A) a larger population mean

(B) a smaller population mean

(C) a larger sample size

(D) a smaller sample size

(E) a smaller population standard deviation

366. What effect would it have on standard error of the sample mean if you could raise the size of each of the samples without changing the population standard deviation?

367. Population A has a mean of 1,000 and a standard deviation of 70, while Population B has a mean of 1,200 and a standard deviation of 62. For samples of size 35, which will have a smaller standard error for the sample mean $\bar{X}$?

368. If a survey sampling researcher wants to cut the standard error of the mean by half, what should he do regarding the sample size next time?

Digging Deeper into Standard Error

369–374 A quality control expert is studying the length of bolts produced in his factory by drawing a sample of completed parts and studying their dimensions in centimeters (cm).

369. For a sample of size 9, if $\sigma_X = 20$, what is the standard error of the mean, $\sigma_{\bar{X}}$? Round your answer to four decimal places.

370. For a sample of size 16, if $\sigma_X = 20$, what is the standard error of the mean, $\sigma_{\bar{X}}$?

371. Which of the following would be expected to result in a larger standard error of the mean?

(A) a larger sample size

(B) a smaller sample size

(C) a smaller population standard deviation

(D) a larger population standard deviation

(E) Choices (B) and (D)

372. If the standard error of the sample mean is 5, for samples of size 25, what is the population standard deviation?

373. For a sample of size 40, what will be the units for the standard error?

374. To estimate the population mean, which of the following standard errors will give you the most precise estimate?

(A) 0.4856

(B) 0.6818

(C) 0.7241

(D) 1.3982

(E) 8.2158

Connecting Sample Means and Sampling Distributions

375–385 Solve the following problems about sample means and sampling distributions.

375. A researcher draws a series of samples of exam scores from a population of scores that has a normal distribution. What required sample size (if any) is necessary for the distribution of the sample means to also have a normal distribution?

376. Which of the following conditions is enough to ensure that the sampling distribution of the sample means has a normal distribution?

(A) Samples of size $n > 10$ are drawn.

(B) The population of all possible scores is very large.

(C) At least 30 samples are drawn, with replacement, from the distribution of possible scores.

(D) Individual scores x_i are normally distributed.

(E) None of the above.

377. The daily productivity of individual bees has a normal distribution, with each bee producing a slightly different amount of honey. If random samples of ten bees are taken, what is the shape of the sampling distribution of the sample means of honey production?

378. If the population distribution is _____ and the sample size is _____, you need to apply the central limit theorem to assume that the sampling distribution of the sample means is normal.

(A)   normal, 10

(B)   normal, 50

(C)   right-skewed, 60

(D)   Choices (B) and (C)

(E)   Choices (A), (B), and (C)

379. If individual scores are _____, the sampling distribution of sample means for scores from that population will be normal regardless of sample size *n*.

380. When individual scores are normally distributed, what can you conclude about the shape of the sampling distribution of the sample means?

381. You have a population of 10,000 test scores with a normal distribution. If you draw all possible samples, each of size 40, and graph their means, what shape do you expect this graph to take?

382. The sizes of adult shoes stocked in a store have a normal distribution. If a shoe store employee randomly pulls four pairs of shoes at a time out of the storeroom, finds the average size of the four pairs, returns them to the shelf, and repeats this process 30 times, how would the distribution of sample means most likely look?

383. Suppose that all possible samples of size 5 are repeatedly drawn from a normally distributed population of 10,000 scores, the average score in each sample is noted, and the distribution of sample averages is examined. What can you confidently say about the distribution of all the possible means in this case?

384. Suppose that samples of size 10 are repeatedly drawn from a population of stones, and the average weight of each sample is recorded. The distribution of weights in the population of stones is normally distributed. What shape do you expect the distribution of sample means to have?

385. A hardware store stocks nails in lengths from 20 to 100 millimeters, with sizes 5 millimeters apart (20, 25, 30, and so on). The store keeps the same number of nails of each length in stock. If a random sample of size 35 is drawn from the stock, the mean length calculated and recorded, the nails returned to the stock, and the experiment repeated 100 times, what shape do you expect the distribution of the sample means to assume?

Digging Deeper into Sampling Distributions of Sample Means

386–392 Use the descriptions of Populations A–D to answer the following problems. The distribution of values in Population A is strongly skewed right. The distribution of values in Population B is flat all the way across (this is called a uniform distribution). The distribution of values in Population C is strongly skewed left. The values in Population D have a normal distribution.

386. Imagine that samples are repeatedly taken from each population. If each sample has $n = 50$ observations in it, for which population will the sampling distribution of the sample means be approximately normal?

(A) Population A

(B) Population B

(C) Population C

(D) Population D

(E) all of the above

387. Suppose that samples of size 20 are repeatedly taken from the four populations. Which population is likely to have an approximately normal distribution for the sampling distribution of the sample mean?

388. Suppose that samples of size 40 are repeatedly taken from the four populations. What shape do you expect the sampling distribution of the sample means to take?

389. Suppose that each population has 100,000 units. If samples are repeatedly taken from each, what shape do you expect the sampling distributions of the sample means to have?

390. Suppose that you drew all possible samples of a particular size from Population C. How large would your sample size have to be before you would expect the sampling distribution of sample means to be normal?

391. Imagine that a researcher draws all possible samples of size 3 from a normal distribution. What general shape should the sampling distribution of sample means have?

392. Imagine that a researcher draws all possible samples of size 100 from a population that doesn't have a normal distribution. What general shape should the sampling distribution of sample means have?

Looking at the Central Limit Theorem

393–400 Solve the following problems that involve the central limit theorem.

393. Suppose that a researcher draws random samples of size 20 from an unknown distribution. Can the researcher claim that the sampling distribution of sample means is at least approximately normal?

394. As a *general* rule, approximately what is the smallest sample size that can be safely drawn from a non-normal distribution of observations if someone wants to produce a normal sampling distribution of sample means?

395. A researcher draws a sample of 500 values from a large population of scores that has a skewed distribution. The researcher then goes on to do statistical analyses that assume that the sampling distribution of sample means is at least approximately normal. What can you conclude about whether the researcher has violated a condition of his statistical procedure?

396. A researcher draws 150 samples of 10 apiece from a normally distributed population of individual observations. What can the researcher conclude in this case?

397. A researcher is going to draw a sample of 150 values from a population. The researcher wants to find probabilities for the sample mean. Which of the following statements is true?

    (A) The central limit theorem can't be used because you don't know the distribution of the population.

    (B) The central limit theorem can't be used because the researcher is taking only one sample.

    (C) The central limit theorem can be used because the sample size is large enough and the population distribution is unknown.

    (D) The central limit theorem isn't needed because you can assume that the population is normal if it isn't stated.

    (E) It can't be determined from the information given.

398. The distribution of the length of screws produced by a factory isn't normal. A researcher draws 150 samples of 10 apiece from the day's output of screws. Can the researcher use the central limit theorem in this case to make conclusions about the random variable $\bar{X}$?

399. A researcher draws 40 samples of 150 numbers each apiece from a non-normal skewed population of individual observations. What can the researcher conclude in this case?

400. A researcher draws a sample of size 32 from a population whose individual values are skewed right (not normal). The researcher wants to use the normal distribution to find probabilities regarding his results. Can he use the central limit theorem in this case?

Getting More Practice with Sample Mean Calculations

401–405 Assume that scores on a math exam are normally distributed for students in a particular university, with a mean of 10 and a standard deviation of 3. Samples of differing sizes are drawn from this population.

401. For a sample of 10 students, their scores are as follows: 7, 6, 7, 14, 9, 9, 11, 11, 11, and 11. What is the sample mean?

402. You draw a sample and calculate the sample mean. Which of the following is true?

(A) The sample mean will always be smaller than the population mean because the sample is only part of the population.

(B) The sample mean will always be the same as the population mean.

(C) The sample mean will be larger than the population mean because the sample is only part of the population.

(D) Larger samples tend to yield more precise estimates of the population mean.

(E) Not enough information to tell.

403. Which of the following is true about the relationship between the mean from a particular sample and the population mean?

(A) The sample mean is the same as the population mean.

(B) The sample mean is the best estimate of the population mean.

(C) Larger samples yield more precise estimates of the population mean.

(D) Smaller samples yield more precise estimates of the sample mean.

(E) Choices (B) and (C)

404. Assume that the true population mean is 10 and the population standard deviation is 3. What is the probability of getting a sample average at least as far from the mean as this sample gave, considering that the sample size is $n = 10$?

405. Suppose that in a sample of 500 students, the mean score on the test was found to be 9.7. What is the probability of finding a sample of size 500 whose mean is farther from the population mean than that of this particular sample?

Finding Probabilities for Sample Means

406–411 Find probabilities for sample means in the following problems.

406. If $\mu = 100$, $\sigma = 30$, and $n = 35$, what is the probability of finding a sample average 10 or more units from the population average?

407. If $\mu = 100$, $\sigma = 40$, and $n = 64$, what is the probability of finding a sample average 10 or more units from the population average?

408. If $\mu = 50$, $\sigma = 16$, and $n = 64$, what is the probability of finding a sample average 10 or more units from the population average?

409. If $\mu = 50$, $\sigma = 16$, and $n = 16$, and if the population is normally distributed, what is the probability of finding a sample average 10 or more units from the population average?

410. If $\mu = 0$, $\sigma = 160$, and $n = 100$, what is the probability of finding a sample average 10 or more units from the population average?

411. A medical technician is told that the mean white blood cell count in the population is 7,250 white blood cells per microliter of blood with a standard deviation of 1,375, but the technician doesn't know whether white blood cell counts are normally distributed in the population. The technician takes 40 tubes of blood from different randomly selected people and gets a mean white blood cell count of $\bar{x} = 7,616$ averaging across the 40 tubes. What is the probability of getting a sample mean this high or higher?

Digging Deeper into Probabilities for Sample Means

412–415 The owner of a cookie factory manufactures chocolate chip cookies to very precise specifications. The cookie population weight has a mean of 12 grams and a population standard deviation of 0.1 grams.

412. The statistician who works at the factory pulls a random sample of 36 cookies from the most recent batch one day and discovers that the average weight of a cookie in that sample is 12.011 grams. What is the probability of getting a value this close or closer to the mean?

413. The statistician who works at the factory pulls a random sample of 49 cookies from the most recent batch one day and discovers that the average weight of a cookie in that sample is 12.004 grams. What is the probability of getting a value this close or closer to the mean?

414. The statistician who works at the factory pulls a random sample of 36 cookies from the most recent batch one day and discovers that the average weight of a cookie in that sample is 12.02 grams. What is the probability of getting a value this close or closer to the mean?

415. The statistician who works at the factory pulls a random sample of 49 cookies from the most recent batch one day and discovers that the average weight of a cookie in that sample is 12.02 grams. What is the probability of getting a value this close or closer to the mean?

Adding Proportions to the Mix

416–419 Suppose that a researcher believes that gender bias exists in the presentation of zombies in movies. The researcher believes that exactly 50% of all clearly gendered zombies should be female if no bias is present. The researcher gets a list of all zombie movies ever made, randomly chooses 29 movies from the list, and then chooses a random spot in each movie to begin watching. The researcher codes the apparent gender of the first zombie that appears after the randomly chosen spot to begin watching. The results are that 20 of 29 zombies are apparently female.

416. Assuming that you're interested in the proportion of female zombies, what is $\hat{p}$? Round your answer to two decimal places.

417. What is p? Round your answer to four decimal places.

418. What is $\sigma_{\hat{p}}$? Round your answer to four decimal places.

419. In the research scenario presented, can the central limit theorem be used?

Figuring Out the Standard Error of the Sample Proportion

420–423 Calculate the standard error of the sample proportion in the following problems.

420. If $p = 0.9$ and $n = 100$ in a binomial distribution, what is the standard error of the sample proportion?

421. If $p = 0.1$ and $n = 100$ in a binomial distribution, what is $\sigma_{\hat{p}}$?

422. If $p = 0.5$ and $n = 100$ in a binomial distribution, what is $\sigma_{\hat{p}}$?

423. If $p = 0.67$ and $n = 60$ in a binomial distribution, what is $\sigma_{\hat{p}}$?

Using the Central Limit Theorem for Proportions

424–425 Determine how to use the central limit theorem for proportions in the following problems.

424. If a researcher wants to study a binomial population where $p = 0.1$, what is the minimum size n needed to make use of the central limit theorem?

425. If a researcher wants to study a population proportion where $p = 0.5$, what is the minimum size n needed to make use of the central limit theorem?

Matching z-Scores to Sample Proportions

426–428 Calculate z-scores for observed sample proportions in the following problems.

426. If $p = 0.5$, $n = 10$, and $\sigma_{\hat{p}} = 0.1581$, what z-score corresponds to an observed $\hat{p}$ of 0.25? Round your answer to two decimal places.

427. If $p = 0.5$, $n = 25$, and $\sigma_{\hat{p}} = 0.1$, what z-score corresponds to an observed $\hat{p}$ of 0.25?

428. If $p = 0.25$ and $n = 25$, what z-score corresponds to an observed $\hat{p}$ of 0.25?

Finding Approximate Probabilities

429–431 These problems all have a binomial distribution. In each case, find the approximate probability that $\hat{p} \leq 0.25$, using the central limit theorem if the appropriate conditions are met.

429. $p = 0.5$, $n = 10$

430. $p = 0.5$, $n = 25$

431. $p = 0.25$, $n = 40$

Getting More Practice with Probabilities

432–435 Imagine that a fair coin is tossed repeatedly, and the total number of heads across all tosses is noted.

432. If the coin is tossed 36 times, what is the probability of getting at least 21 heads?

433. If the coin is tossed 50 times, what is the probability of getting at least ten heads?

434. If the coin is tossed 100 times, what is the probability of getting more than 60 heads?

435. A large corporation buys computers in bulk. It knows that a certain manufacturer guarantees that no more than 1% of its machines are defective. The corporation has observed a defect rate of 1.5% in 1,000 machines it has bought from that company. If the manufacturer truly achieves the defect rate it claims, what is the probability of observing a defect rate at least as high as 1.5% in a random sample of 1,000 machines?

Digging Deeper into Approximate Probabilities

436–440 A researcher begins with the knowledge that p = 0.3. The researcher finds p̂ as stated, using n trials given in the question. Find the approximate probability in each case.

436. Given the information provided, what is the approximate probability of observing a proportion $\hat{p}$ outside of the range 0.2 to 0.4, with 36 trials?

437. Given the information provided, what is the probability of observing a proportion $\hat{p} < 2$ in 36 trials?

438. Given the information provided, what is the approximate probability of observing $\hat{p} < 0.4$ with 36 trials?

439. Using the information provided, what is the approximate probability of observing a sample proportion $\hat{p}$ outside the range 0.2 to 0.4 with 81 trials?

440. What is the probability of observing a sample proportion $\hat{p}$ in the range 0.2 to 0.4 with 144 trials?

Chapter 8

Finding Room for a Margin of Error

• •

A margin of error is the "plus or minus" part you have to add to your statistical results to tell everyone you acknowledge that sample results will vary from sample to sample. The margin of error helps you indicate how much you believe those results could vary, with a certain level of confidence. Anytime you're trying to estimate a number from a population (like the average gas price in the United States), you include a margin of error. And it's never a good idea to assume that a margin of error is small if it's not given.

The Problems You'll Work On

The set of problems in this chapter focus solely on the margin of error. It's related to the confidence interval, and it's being given its own spotlight because it's so important. Here's what you'll be working on:

- ✔ Understanding exactly what margin of error (MOE) means and when it's used
- ✔ Breaking down the components of margin of error and seeing the role of each component
- ✔ Examining the factors that increase or decrease the margin of error and the way they do so
- ✔ Working on margin of error for population means as well as a few population proportions
- ✔ Interpreting the results of a margin of error properly

What to Watch Out For

As you work through the problems in this chapter, pay special attention to the following:

- ✔ Knowing exactly what happens when certain values in a margin of error change
- ✔ Understanding the relationship between values of z and the confidence level needed for a margin of error
- ✔ Accurately calculating the sample size needed to get a particular margin of error you want
- ✔ Keeping your level of understanding high and not getting buried in the calculations

Defining and Calculating Margin of Error

441–446 Solve the following problems about margin of error basics.

441. An opinion survey reports that 60% of all voters who responded to a phone call support a bond measure to provide more money for schools, with a margin of error of ± 4%. What does the term *margin of error* add to your knowledge of the survey results?

442. A sports supplement maker claims that based on a random sample of 1,000 users of its product, 93% of all its users gain muscle and lose fat, with a margin of error of ± 1%. What critical information is missing from this claim?

443. A poll shows that Garcia is leading Smith by 54% to 46% with a margin of error of ± 5% at a 95% confidence level. What conclusion can you draw from this poll?

444. What is the margin of error for estimating a population mean, given the following information and a confidence level of 95%?

$$\sigma = 15$$
$$n = 100$$

445. What is the margin of error for estimating a population mean given the following information and a confidence level of 95%?

$$\sigma = 5$$
$$n = 500$$

446. What is the margin of error for estimating a population mean given the following information and a confidence level of 95%?

$$\sigma = \$10,000$$
$$n = 40$$

Using the Formula for Margin of Error When Estimating a Population Mean

447–449 A researcher conducted an Internet survey of 300 students at a particular college to estimate the average amount of money students spend on groceries per week. The researcher knows that the population standard deviation, σ, of weekly spending is $25. The mean of the sample is $85.

447. What is the margin of error if the researcher wants to be 99% confident of the result?

448. What is the margin of error if the researcher wants to be 95% confident in the result?

449. What is the lower limit of an 80% confidence interval for the population mean, based on this data?

Finding Appropriate z*-Values for Given Confidence Levels

450–452 Use Table A-4 to find the appropriate z-value for the confidence levels given except where noted.*

450. Table A-1 (the Z-table) and Table A-4 in the appendix are related but not the same. To see the connection, find the z*-value that you need for a 95% confidence interval by using Table A-1.

451. What is the z*-value for a 99% confidence level?

452. What is the z*-value for a confidence level of 80%?

Connecting Margin of Error to Sample Size

453–455 A sociologist is interested in the average age women get married. The sociologist knows that the population standard deviation of age at first marriage is three years for both men and women and hasn't changed for the last 60 years. The sociologist would like to demonstrate with 95% confidence that the mean age of the first marriage for women now is higher than the mean age in 1990, which was 24 years old.

453. If the sociologist can sample only 50 women, what will the margin of error be?

454. If the sociologist can sample 100 women, what will the margin of error be?

455. What is the smallest number of participants the sociologist can sample to estimate the average nuptial age for the population, with a margin of error of only two years?

Getting More Practice with the Formula for Margin of Error

456–460 Two hundred airplane engine screws are to be sampled from a factory run of many thousands of screws to estimate the average length of all the screws (in millimeters).

456. Assuming that $\sigma = 0.01$ millimeters and you're using a 99% level of confidence, find the margin of error for estimating the population mean.

457. Assuming that $\sigma = 0.05$ millimeters and you're using a 99% level of confidence, what is the margin of error for estimating the population mean?

458. Assuming that $\sigma = 0.10$ millimeters and you're using a 99% level of confidence, what is the margin of error for estimating the population mean?

459. Margin of error can help factories determine whether they're manufacturing parts within a satisfactory range of dimensions. Suppose that a factory discovers that the average weights of the parts it's making have a margin of error (with 99% confidence) that is twice as large as it should be, when they tried to estimate the average size of the parts. What step could the factory take regarding sample size, if it wants to reduce the margin of error by half?

460. An unscrupulous factory owner found that the margin of error for estimating the average size of a factory-produced part is twice as large as it should be. One way of fixing this is to make sure that the manufacturing processes are all very consistent, but doing so is expensive and the factory owner didn't want to go to that much cost and trouble. The factory owner was instantly able to report to the board of directors that he had cut the margin of error by a little more than half. Assuming no changes were made in the manufacturing process, what deceptive action might the factory owner have done to achieve this result?

Linking Margin of Error and Population Proportion

461–464 Solve the following problems related to margin of error and population proportion.

461. A market researcher samples 100 people to find a confidence interval for estimating the average age of their customers. She finds that the margin of error is three times larger than she wants it to be. How many people should the researcher add to the sample to bring the margin of error down to the desired size?

462. A survey of 10,000 randomly selected adults from across Europe finds that 53% are unhappy with the euro. What is the margin of error for estimating the proportion among all Europeans who are dissatisfied with the euro? (Use a confidence level of 95%.) Give your answer as a percentage.

463. A county election office wants to plan for having enough poll workers for election day, so it conducts a telephone survey to estimate what proportion of those who are eligible to vote intend to do so. The office calls 281 eligible people in the community and finds that 135 of them intend to vote. What is the margin of error for estimating the population proportion, using a 99% confidence level? Give your answer as a percentage.

464. A sample of 922 households finds that the proportion with one or more dogs is 0.46. The margin of error is reported as 2.7%. What confidence level is being used in this confidence interval for a proportion?

Chapter 9

Confidence Intervals: Basics for Single Population Means and Proportions

Suppose that you want to find the value of a certain population parameter (for example, the average gas price in Ohio). If the population is too large, you take a sample (such as 100 gas stations chosen at random) and use those results to *estimate* the population parameter. Knowing sample results vary, you attach a margin of error (plus or minus), to cover your bases. Put it all together and you get a *confidence interval* — a range of likely values for the population parameter.

The Problems You'll Work On

The problems in this chapter give you practice calculating and interpreting some basic confidence intervals. Here are the highlights:

- Calculating a confidence interval for a population mean when the population standard deviation is known (involves the *Z*-distribution) or unknown (involves a *t*-distribution)

- Calculating a confidence interval for a population proportion when the sample is large

- Finding and interpreting the margin of error of a confidence interval and what factors affect it

- Determining appropriate sample sizes required to achieve a certain margin of error

What to Watch Out For

When working through these problems, make sure you can do the following:

- Know when a confidence interval is needed — the word *estimate* is your cue.

- Know which conditions to check before deciding which formula to use.

- Understand each part of a confidence interval, what its role is, and how it can be affected.

- Focus particularly on how the results are *interpreted* (explained in real language) instead of focusing just on the calculations.

Introducing Confidence Intervals

465–468 A clothing store is interested in the mean amount spent by all of its customers during shopping trips, so it examines a random sample of 100 electronic cash-register records and discovers that, among those who made purchases, the average amount spent was $45 with a 95% confidence interval of $41 to $49.

465. Which of the following are true statements regarding the 95% confidence interval for this data?

    (A)   If the same study were repeated many times, about 95% of the time the confidence interval would include the average money spent from the sample, which is $45.

    (B)   If the same study were repeated many times, about 95% of the time, the confidence interval would contain the average money spent for all the customers.

    (C)   There is a 95% probability that the average money spent for all the customers is $45.

    (D)   Choices (A) and (B)

    (E)   Choices (A), (B), and (C)

466. Which of the following is a reason for reporting a confidence interval as well as a point estimate for this data?

    (A)   The store studied a sample of sales records rather than the entire population of sales records.

    (B)   The confidence interval is certain to contain the population parameter.

    (C)   Because sample results vary, the sample mean is not expected to correspond exactly to the population mean, so a range of likely values is required.

    (D)   Choices (A) and (B)

    (E)   Choices (A) and (C)

467. Which of the following statements is a valid argument for drawing a sample of size 500 rather than size 100?

    (A)   The larger sample will produce a less-biased estimate of the sample mean.

    (B)   The larger sample will produce a more precise estimate of the population mean.

    (C)   The 95% confidence interval calculated from the larger sample will be narrower.

    (D)   Choices (B) and (C)

    (E)   Choices (A), (B), and (C)

468. Which of the following statements is true regarding the sample mean of $45?

    (A)   It is the same as the population mean.

    (B)   It is a good number to use to estimate the population mean.

    (C)   If you drew another sample of 100 from the same population, you would expect the sample mean to be exactly $45.

    (D)   Choices (A) and (C)

    (E)   None of the above

Checking Out Components of Confidence Intervals

469–484 Solve the following problems about confidence interval components.

469. Consider the following samples of $n = 5$ from a population. Without doing any calculations, which would you expect to have the widest 95% confidence interval if you're using the sample to estimate the population mean?

    Sample A: 5, 5, 5, 5, 5

    Sample B: 5, 6, 6, 6, 7

    Sample C: 5, 6, 7, 8, 9, 10

    Sample D: 5, 6, 7, 8, 9, 20

470. When analyzing the same sample of data, which confidence interval would have the widest range of values?

(A) one with a confidence level of 80%

(B) one with a confidence level of 90%

(C) one with a confidence level of 95%

(D) one with a confidence level of 98%

(E) one with a confidence level of 99%

471. How will the confidence interval be affected if the confidence level increases from 95% to 98%?

472. With all factors remaining equal, why does increasing the confidence level increase the width of the confidence interval?

473. A survey of 100 Americans reports that 65% of them own a car. The 95% confidence interval for the percentage of all American households who own one car is 60% to 70%. What is the margin of error for the confidence interval in this example?

474. If all other factors stay the same, which sample size will create the widest confidence interval?

(A) $n = 100$

(B) $n = 200$

(C) $n = 300$

(D) $n = 500$

(E) $n = 1,000$

475. If all other factors are equal, which sample size results in the narrowest confidence interval?

(A) $n = 200$

(B) $n = 500$

(C) $n = 1,000$

(D) $n = 2,500$

(E) $n = 5,000$

476. A hospital is considering what size sample to draw in a study of the accuracy of its clinical records. If it decides to use a random sample of 500 records, rather than a random sample of 200 records, which of the following statements is true?

(A) The sample of 500 will have a wider 95% confidence interval.

(B) The sample of 500 will have a narrower 95% confidence interval.

(C) The width of the 95% confidence interval will not be affected by the sample size.

(D) The sample of 500 will produce a more precise estimate of the population mean.

(E) Choices (B) and (D)

477. What is the relationship between the confidence level and the width of the confidence interval?

478. Assume that Population A has substantially less variability than Population B. Comparing samples of the same size, and confidence intervals of the same confidence level, which of the following statements is true?

   (A) The confidence intervals related to Populations A and B are expected to be the same.

   (B) The confidence interval related to Population A is expected to be wider.

   (C) The confidence interval related to Population A is expected to be narrower.

   (D) It depends on how the data was collected.

   (E) Not enough information to tell.

479. A university is planning to study student satisfaction with technological services on campus, based on a survey of a random sample of students. Which of the following statements is true?

   (A) A confidence level of 80% will produce a wider confidence interval than a confidence level of 90%.

   (B) A confidence level of 80% will produce a narrower confidence interval than a confidence level of 90%.

   (C) A sample of 300 students will produce a narrower confidence interval than a sample of 150 students.

   (D) Choices (A) and (C)

   (E) Choices (B) and (C)

480. The following samples are drawn from different populations. Assuming that the samples are accurate reflections of the variability of the populations and the same confidence level is used for each, which sample will have the widest confidence interval?

      Sample A: 10, 20, 30, 40, 50, 60, 70, 80, 90, 100

      Sample B: 1, 2, 3, 40, 50, 60, 70, 800, 900, 1000

      Sample C: 41, 42, 43, 44, 45, 46, 47, 48, 49, 50

      Sample D: 10, 15, 20, 21, 22, 23, 24, 25, 30, 35

      Sample E: 510, 520, 530, 540, 550, 560, 570, 580, 590, 600

481. A random sample of 100 people taken from which of the following populations will yield the widest confidence interval for the mean income?

   (A) workers ages 22 to 30 who live in Denver

   (B) all U.S. workers ages 22 to 30

   (C) all U.S. workers ages 16 to 22

   (D) all workers ages 22 to 30 who live in North America (Canada, the U.S., and Mexico), adjusted in U.S. dollars

   (E) all workers ages 22 to 30 who live in U.S. cities with populations of fewer than 10,000 people

482. A random sample of 100 taken from which of the following populations will yield the narrowest confidence interval for average height?

   (A) children ages 1 to 5

   (B) children ages 5 to 10

   (C) children ages 10 to 16

   (D) teenagers ages 13 to 19

   (E) adults ages 55 to 65

483. Which of the following samples will yield the widest confidence interval for the same population mean?

(A) one with confidence level 95%, $n = 200$, and $\sigma = 8.5$

(B) one with confidence level 95%, $n = 200$, and $\sigma = 12.5$

(C) one with confidence level 95%, $n = 400$, and $\sigma = 8.5$

(D) one with confidence level 80%, $n = 200$, and $\sigma = 8.5$

(E) one with confidence level 80%, $n = 400$, and $\sigma = 8.5$

484. Which of the following will decrease the margin of error of a confidence interval?

(A) Increase the sample size from 200 to 1,000 subjects.

(B) Decrease the confidence level from 95% to 90%.

(C) Increase the confidence level from 95% to 98%.

(D) Choices (A) and (B)

(E) Choices (A) and (C)

Interpreting Confidence Intervals

485–489 Solve the following problems about interpreting confidence intervals.

485. A sample of the heights of boys in a school class shows the mean height is 5 feet 9 inches. The margin of error is ± 4 inches for a 95% confidence interval. Which of the following statements is true?

(A) The 95% confidence interval for the mean height of all the boys is between 5 feet 5 inches and 6 feet 1 inch.

(B) The mean of any one sample has a 95% chance of being between 5 feet 5 inches and 6 feet 1 inch.

(C) Based on the data, the sample mean height is 5 feet 9 inches, and 95% of all the boys will be between 5 feet 5 inches and 6 feet 1 inch.

(D) It means that based on the sample data, 95% of the heights are calculated to have a mean of 5 feet 9 inches, 5% of the boys are more than 4 inches shorter and 5% are at least 4 inches taller than 5 feet 9 inches.

(E) It means that a boy randomly selected from the class has a 95% chance of having a height between 5 feet 5 inches and 6 feet 1 inch.

486. A sample of college students showed that they earned a mean summer income of $4,500. The margin of error is ± $400 for a 95% confidence interval. What does this mean?

487. Which of the following correctly describes the margin of error?

(A)   The margin of error is the percentage of errors that were made in taking the sample.

(B)   The margin of error is an estimate that adjusts for the false reporting by the people surveyed.

(C)   The margin of error is used to calculate a range of likely values for a population parameter, based on a sample.

(D)   The margin of error identifies the quality of the sampling methods. A margin of error ± 5% indicates a well-designed study.

(E)   The margin of error shows the distance of the sample results from the population mean.

488. A sample of college students found that the mean amount that students spend on books, supplies, and lab fees was $450 per semester, with a margin of error of ± $50. The confidence level for these results is 99%. Based on these results, which of the following statements is true about the margin of error?

(A)   The margin of error means that 99% of the fees on books are within $50 of each other.

(B)   The margin of error measures the amount by which your sample results could change, with 99% confidence.

(C)   The margin of error means you have to adjust your results by $50 to account for inaccurate reporting by the people surveyed.

(D)   The margin of error identifies the quality of the sampling methods. A margin of error ± $50 indicates a poorly designed study.

(E)   The margin of error shows that the total range of all the purchases on average was $400 to $500, and the mean was $450.

489. A poll of 1,000 likely voters showed that Candidate Smith had 48% of the vote, and Candidate Jones had 52% of the vote. The margin of error was ± 3%, and the confidence level was 98%. Who is most likely to win the election?

Spotting Misleading Confidence Intervals

490–494 Solve the following problems about misleading confidence intervals.

490. In a survey, 6,500 of the first 10,000 fans at a football game chose chocolate ice cream as their favorite flavor. The ice-cream company running the survey then claims that 65% of all Americans prefer chocolate ice cream over other flavors, based on this survey. Which of the following choices best describes the conclusions of the ice-cream company?

(A)   The survey has a built-in bias.

(B)   The survey will have valid results because the sample size is high.

(C)   The results are biased because the confidence interval is too wide when only 10,000 people respond.

(D)   The survey isn't valuable because it doesn't list the choices of flavors made by the other fans.

(E)   The results of the survey are biased because it didn't account for people who don't like ice cream.

491. A company took a random sample of 30 first-year employees and asked them their level of satisfaction with their jobs. It found that 80% of those sampled were "very happy" with their employment, ± 3% at a confidence level of 95%. The company took this information and reported that 80% of all its employees were very happy with their jobs, ± 3%. There is at least one problem with the company's reported results. Choose the answer(s) that best describes the problem(s).

(A) The survey is accurate because it's based on a random sample.

(B) The survey is biased because it was based only on first-year employees, who may feel differently about their jobs than other employees.

(C) The survey is misleading because it doesn't report the results of the other first-year employees.

(D) The sample size is only 30. The margin of error must be higher than 3% based on the size of the sample and the confidence level.

(E) Choices (B) and (D)

492. A company conducted a random online survey of 1,000 visitors to its website during the past three months. The sample showed that the visitors had a mean of five online purchases at all Internet sites during the past 12 months. The margin of error was ± 0.6 purchases with a confidence level of 95%. So the company concludes that the average number of online purchases for all visitors to its website during the past three months is 5, plus or minus 0.6. Which of the following choices best describes the survey?

(A) The survey can be used by the company to help predict the spending habits of all its website visitors during the year.

(B) The survey can be used only by the company as part of its analysis of the Internet spending habits of all visitors to its website during the last three months.

(C) The survey can be used by the company as part of its analysis of the Internet spending habits of all of its customers (in-store and website) during the last three months.

(D) The sample is faulty and unusable because it surveys visitors to its website only, not necessarily those who make purchases.

(E) The survey is faulty because it's based on too small of a sample.

493. Over a three-month period, a random sample of teenagers visiting a local movie theater was asked how many times they went to a movie in the last year. The purpose of the survey is to estimate the average number of movies a teenager attended in the last year. The results of the survey were as follows:

$n = 1,000$ teenagers

$\bar{x} = 4.5$ visits

MOE $= 0.7$ visits

Confidence level $= 95\%$

Which of the following choices best describes the survey?

(A) The results are invalid because the survey was done at a movie theater.

(B) The 95% confidence interval for the average number of movies attended by any teenager is between 3.1 and 5.9 visits per year and is valid because that's what the formula tells you to calculate.

(C) This survey isn't a valid method for estimating how many times teenagers visited movie theaters in the past 12 months because the survey was taken only over a three-month period.

(D) This survey isn't valid because it was based only on a sample and not the population of all teenagers.

(E) Choices (A) and (C)

494. A Colorado fashion magazine that has 2 million readers found in a mail-in survey with 5,800 responses that 56% of respondents chose Colorado as their favorite place to live, ± 2% with a confidence level of 98%. Which of the following choices best describes the survey?

(A) The sample is probably based on a representative sample of the magazine's readers because it's so large.

(B) The sample isn't based on a representative sample of the magazine's readers.

(C) Because the magazine is based on Colorado, more readers are likely to buy the magazine who are from Colorado; therefore, they'd be more likely to vote it as the best place to live.

(D) The sample results are likely biased because the respondents had to make the effort to mail back the survey.

(E) Choices (B), (C), and (D)

Calculating a Confidence Interval for a Population Mean

495–522 Calculate confidence intervals for population means in the following problems.

495. In a random sample of 50 intramural basketball players at a large university, the average points per game was 8, with a standard deviation of 2.5 points and a 95% confidence level. Which of the following statements is correct?

(A) With 95% confidence, the average points scored by all intramural basketball players is between 7.3 and 8.7 points.

(B) With 95% confidence, the average points scored by all intramural basketball players is between 7.7 and 8.4 points.

(C) With 95% confidence, the average points scored by all intramural basketball players is between 5.5 and 10.5 points.

(D) With 95% confidence, the average points scored by all intramural basketball players is between 7.2 and 8.8 points.

(E) With 95% confidence, the average points scored by all intramural basketball players is between 7.6 and 8.4 points.

496. On the SAT Math test, a random sample of the scores of 100 students in a high school had a mean of 650. The standard deviation for the population is 100. What is the confidence interval if 99% is the confidence level?

497. An apple orchard harvested ten trees of apples. From a random sample of 50 apples, the mean weight of an apple was 7 ounces. The population standard deviation is 1.5 ounces. What is the confidence interval if 99% is the confidence level?

498. A random sample of 200 students at a university found that they spent an average of three hours a day on homework. The standard deviation for all the university's students is one hour. What is the confidence interval if 90% is the confidence level?

499. Analysis of a random sample of 200 people ages 18 to 22 showed that a person spent an average of $32.50 on a typical outing with a friend. The population standard deviation for this age group is $15.00. What is the confidence interval if 95% is the confidence level?

500. A random sample of 150 people over age 17 showed that a person spent an average of 30 minutes a day on vigorous exercise. The population standard deviation for exercise for this age group is 15 minutes. What is the confidence interval if 90% is the confidence level?

501. A random sample of 200 college graduates showed that a person made an average of $36,000 income in the first year after graduation. The standard deviation for income for all first-year college graduates is $8,000. What is the confidence interval if 95% is the confidence level?

502. A random sample of 300 trips by a city bus along a specific route showed that the average time to complete the bus route was 45 minutes. The standard deviation for all trips on this bus route is 3 minutes. What is the confidence interval if 95% is the confidence level?

503. A random sample of 1,100 travel itinerary requests on an airline ticket website showed that the average itinerary request took 4.5 seconds to be calculated and displayed to the traveler. The standard deviation for all itinerary requests is 2 seconds. What is the confidence interval if 98% is the confidence level?

504. A random sample of 200 MP3 players on an assembly line showed that the average amount of time to assemble an MP3 player was 12.25 minutes. The population standard deviation for assembly is 2.15 minutes. What is the confidence interval if 95% is the confidence level?

505. A random sample of 300 university students found that the average distance to the hometown of a student was 125 miles. The standard deviation for the distance for all students at the university is 40 miles. What is the confidence interval if 90% is the confidence level?

506. A random sample of 75 data entry specialists at a data center found that specialists made an average of 2.7 errors out of 10,000 data items. The standard deviation for the errors made by all specialists at the bank is 0.75 per 10,000 items. What is the confidence interval if 95% is the confidence level?

507. A random sample of 500 bats (of the type that has a standard length of 38 inches) made for major league baseball players found that the average bat had a length of 38.01 inches. The standard deviation for all 38-inch bats is 0.01 inches. What is the confidence interval if 99% is the confidence level?

508. A random sample of 2,000 special valve parts for an engine resulted in an average length of 3.2550 centimeters. The standard deviation for the population of special valve parts is 0.025 centimeters. What is the confidence interval if 99% is the confidence level?

509. A random sample of 40 purchases of medium-quality hardwood purchased over a 12-month period by a furniture maker from different suppliers showed a mean cost of $0.78 per board foot. The standard deviation for the year for all wood purchased was $0.12 per board foot. What is the confidence interval if 95% is the confidence level?

510. A standardized math ~~t~~ was given to a random sample of ~~c~~ore was 84% with students. Their m~~~~iation of 5%. What is a sample standa~~~~al if 95% is the confidence ~~~~dence level?

511. A ra~~~~th~~~~le of 25 households found ~~~~size household had tha~~~~ople. What is the confidence 3.~~9~~% is the confidence level?

~~~~om sample of 30 teenagers found ~~~~he average number of social network ~~~~ds each person had was 85, with a ~~~~ple standard deviation of 50 friends. ~~~~at is the confidence interval if 95% is ~~~~he confidence level?

513. A random sample of 24 first-year college students found that the students traveled an average of 400 miles on the longest trip they took last year, with a sample standard deviation of 300 miles. What is the confidence interval if 95% is the confidence level?

514. A random sample of 20 shoppers leaving a mall found that they spent an average of $78.50 that day, with a sample standard deviation of $50.75. What is the confidence interval if 95% is the confidence level?

515. A random sample of 20 visitors leaving a museum found that they spent an average of three hours in the museum, with a sample standard deviation of one hour. What is the confidence interval if 90% is the confidence level?

516. A random sample of 50 1-pound loaves of bread at a bakery found that the mean weight was 18 ounces, with a sample standard deviation of 1.5 ounces. What is the confidence interval if 90% is the confidence level?

517. A random sample of ten people shopping at a grocery store found that they visited the store an average of 2.8 times per month, with a sample standard deviation of 2 visits. You know from previous research that the number of shopping visits each month is approximately normally distributed. What is the confidence interval if 80% is the confidence level?

518. A random sample of 18 first-year students at a university found that they watched an average of five movies a month (whether in theaters, online, or on DVD), with a population standard deviation of three movies. What is the confidence interval if 90% is the confidence level?

519. A random sample of 25 visitors to an amusement park found that they spent an average of $32.00 that day while at the park. The population standard deviation is $6.00. What is the confidence interval if 99% is the confidence level?

520. In calculating the sample sizes needed for a particular margin of error, which of the following results will be rounded to 118 (the number of subjects required)?

(A) 117.2

(B) 117.6

(C) 118.1

(D) Choices (A) and (B)

(E) Choices (A), (B), and (C)

521. In calculating the sample sizes needed for a particular margin of error, how would the following results be rounded: 121.1, 121.5, 131.2, and 131.6?

522. Although in general terms researchers like to have larger rather than smaller sample sizes, to get more precise results, what are some limiting factors on the sample size used in a study?

(A) A larger sample often means greater costs.

(B) It may be difficult to recruit a larger sample (for example, if you're studying people with a rare disease).

(C) At some point, increasing the sample size may not significantly improve precision (for instance, increasing the sample size from 3,000 to 3,500).

(D) Choices (A) and (B)

(E) Choices (A), (B), and (C)

Determining the Sample Size Needed

523–529 Figure out the sample size needed in the following problems.

523. A physician wants to BMI (*body mass index,* combines information the average weight) for her adult pa that to draw a sample of clinic and retrieve this information frecides wants an estimate with a ma d 1.5 units of BMI, with 95% co e believes that the national pop of dard deviation of adult BMI of 4. applies to her patients. She know BMI is approximately normally dis for adults. How large a sample does need to draw?

524. A physician wants to estimate the height of 6-year-old boys in her community, using a random sample drawn from administrative records. She wants an estimate with a margin of error of 0.5 inches, with 95% confidence, and believes that the population standard deviation of 1.8 inches applies to her population. She also knows that height is approximately normally distributed for this population. How large a sample does she need to draw?

525. You want to estimate the average height of 10-year-old boys in your community. The population standard deviation is 3 inches. What size sample do you need for a margin of error of no more than ± 1 inch and a confidence level of 95% when constructing a confidence interval for the mean height of all 10-year-old boys?

526. You want to take a sample that measures the weekly job earnings of high-school students during the school year. The population standard deviation is $20. What size sample do you need for a margin of error of no more than ± $5 and a confidence level of 99% when constructing a confidence interval for the mean weekly earnings of all high-school students?

527. You want to take a sample that measures the weekly job earnings of university students during the school year. The population standard deviation is $55. What size sample do you need for a margin of error of no more than ± $10 and a confidence level of 90% when constructing a confidence interval for the mean weekly earnings of all university students?

528. You want to take a sample that measures the amount of sleep university students get each night. The population standard deviation is 1.2 hours. What size sample (number of students) do you need for a margin of error of no more than ± 0.25 hours and a confidence level of 95% when constructing a confidence interval for the mean amount of sleep of all university students?

529. You want to take a sample that measures the attendance at women's Division I university basketball games. The population standard deviation is 2,300. What size sample (number of games) do you need for a margin of error of no more than ± 800 and a confidence level of 95% when constructing a confidence interval for the mean attendance at all such games?

Introducing a Population Proportion

530–536 A random sample of 100 students at a university found that 38 students were thinking of changing their major.

530. You were asked to report both a point estimate (a single number) and a confidence interval (range of values) for your survey. Why would the confidence interval be requested?

(A) You may make a mistake with your calculations.

(B) You are using sample data to estimate a parameter.

(C) If you drew a different sample of the same size, you would expect the results to be slightly different.

(D) Choices (B) and (C)

(E) Choices (A), (B), and (C)

531. What does 0.38 represent in this example?

(A) the proportion of students at the university who are thinking of changing their major

(B) the number of students at the university who are thinking of changing their major

(C) an estimate of the proportion of all students at the university who are thinking of changing their major

(D) the proportion of students in the sample of 100 who are thinking of changing their major

(E) Choices (C) and (D)

532. Can you use the normal approximation to the binomial to calculate a confidence interval for this data? Why or why not?

533. What is the standard error of $\hat{p}$?

534. With all other relevant values being fixed, which of the following confidence levels will result in the widest confidence interval?

(A) 80%

(B) 90%

(C) 95%

(D) 98%

(E) 99%

535. At a confidence level of 90%, what is the confidence interval for the proportion of all students thinking of changing their major?

536. At a confidence level of 95%, what is the confidence interval for the proportion of all students thinking of changing their major?

Connecting a Population Proportion to a Survey

537–539 A website ran a random survey of 200 customers who purchased products online in the past 12 months. The survey found that 150 customers were "very satisfied."

537. What are the sample proportion and the standard error for the sample proportion, based on this data?

538. With a 95% confidence level, what is the margin of error for the estimate of the proportion of all customers who purchased products online in the past 12 months?

539. With a 99% confidence level, what is the confidence interval for the proportion of all customers who purchased products online in the past 12 months? Round to two decimal places.

Calculating a Confidence Interval for a Population Proportion

540–545 Assume that you have a random sample of 80 with a sample proportion of 0.15.

540. Can you use the normal approximation to the binomial for this data?

541. With an 80% level of confidence, what is the confidence interval for the population proportion? Round your answer to four decimal places.

542. With a 90% level of confidence, what is the confidence interval for the population proportion? Round your answer to four decimal places.

543. With a 95% level of confidence, what is the confidence interval for the population proportion? Round your answer to four decimal places.

544. Assuming a 98% confidence level, what is the confidence interval for the population proportion based on this data? Round your answer to four decimal places.

545. Assuming a 99% confidence level, what is the confidence interval for the population proportion based on this data? Round your answer to four decimal places.

Digging Deeper into Population Proportions

546–547 Solve the following problems about confidence intervals and population proportions.

546. Assume that the 95% confidence interval for a population proportion based on a certain data set is 0.20 to 0.30. Which of the following could be the 98% confidence interval for the population proportion using the same data?

(A)   0.15 to 0.35

(B)   0.21 to 0.29

(C)   0.22 to 0.38

(D)   0.23 to 0.27

(E)   0.24 to 0.26

547. Assume that the 95% confidence interval for a population proportion based on a certain data set is 0.20 to 0.30. Based on the same data and pertaining to the same population proportion, what level of confidence could the interval 0.22 to 0.28 represent?

(A)   80%

(B)   90%

(C)   99%

(D)   Choice (A) or (B)

(E)   Choice (A), (B), or (C)

Getting More Practice with Population Proportions

548–552 In a random sample of 160 adults in a large city, 88 were in favor of a new 0.5% sales tax. Assume that you can use the normal approximation to the binomial for this data.

548. If you had only one number to use to estimate the proportion of all the adults in the city who favor the new tax, what number would you use?

549. What is the standard error for $\hat{p}$?

550. With a 95% confidence level, what is the margin of error for estimating the proportion of all adults in the city who favor the new tax?

551. With a 99% confidence level, what is the margin of error for estimating the proportion of all adults in the city who favor the new tax?

552. With an 80% level of confidence, what is the margin of error for estimating the proportion of all adults in the city who favor the new tax?

Chapter 10

Confidence Intervals for Two Population Means and Proportions

• •

Many real-world scenarios are looking to compare two populations. For example, what is the difference in survival rates for cancer patients taking a new drug compared to cancer patients on the existing drug? What is the difference in the average salary for males versus females? What's the difference in the average price of gas this year compared to last year? All these questions are really asking you to compare two populations in terms of either their averages or their proportions to see how much difference exists (if any). The technique you use here is confidence intervals for two populations.

The Problems You'll Work On

The problems in this chapter give you practice with the following:

- ✔ Calculating and interpreting confidence intervals for the difference in two population means when the population standard deviations are known (involves the Z-distribution) or unknown (involves a t-distribution)

- ✔ Calculating and interpreting confidence intervals for the difference in two population proportions when the samples are large

What to Watch Out For

Keep the following in mind as you work through this chapter:

- ✔ This chapter is about the *difference in the means*, not *the mean of the differences*.

- ✔ When you look at the difference between two means (or two proportions), keep track of what populations you're calling Population 1 and Population 2. Subtracting two numbers in the opposite order changes the sign of the results!

- ✔ If a confidence interval for a difference in means (or proportions) contains negative values, positive values, and zero, you can't conclude that a difference exists.

Working with Confidence Intervals and Population Proportions

553–557 A random poll of 100 males and 100 females who were likely to vote in an upcoming election found that 55% of the males and 25% of the females supported candidate Johnson. Call the population of males "Population 1" and the population of females "Population 2" while working these problems.

553. If you could use only one number to estimate the difference in the proportions of all males and all females supporting candidate Johnson among all likely voters, what number would you use?

554. What is the standard error for the estimate of the difference in proportions in the male and female populations?

555. With a 95% confidence level, what is the confidence interval for the difference in the percentage of males and females favoring Johnson among all likely voters?

556. With a 90% confidence level, what is the confidence interval for the difference in the percentage of males and females favoring Johnson among all likely voters?

557. With a 99% confidence level, what is the confidence interval for the difference between the percentage of males and females favoring Johnson among all likely voters?

Digging Deeper into Confidence Intervals and Population Proportions

558–560 A random survey found that 220 out of 300 adults living in large cities (with populations of more than 1 million) wanted more state funding for public transportation. In small cities (with populations fewer than 100,000), 120 out 300 adults surveyed wanted more state funding. Call the population of adults living in large cities "Population 1" and the population of adults living in small cities "Population 2" while working these problems.

558. If you had only one number to estimate the difference between the proportion of adults in large cities and the proportion of adults in small cities favoring increased state funding for public transportation, what number would you use?

559. With a 90% confidence level, what is the confidence interval for the difference in the population proportions? Give your answer to four decimal places.

560. With an 80% confidence level, what is the confidence interval for the difference in the population proportions?

Working with Confidence Intervals and Population Means

561–565 A random sample of 70 12th-grade boys in a high school showed a mean height of 71 inches. (Assume that among all boys in the high school, their heights have a population standard deviation of 2 inches.) A random sample of 60 12th-grade girls in the same school showed a mean height of 67 inches. (Assume that among all girls in the high school, their heights have a population standard deviation of 1.8 inches.) Call the population of boys "Population 1" and the population of girls "Population 2" while working these problems.

561. Using a 95% confidence level, what is the confidence interval for the population difference in the mean heights of all 12th-grade boys compared to 12th-grade girls at this school? Round to the nearest hundredth.

562. Using an 80% confidence level, what is the confidence interval for the population difference in the mean heights of all 12th-grade boys compared to 12th-grade girls at this school? Round to the nearest hundredth.

563. Using a 99% confidence level, what is the confidence interval for the population difference in the mean heights of all 12th-grade boys compared to 12th-grade girls at this school? Round to the nearest hundredth.

564. Using a 98% confidence level, what is the confidence interval for the population difference in the mean heights of all 12th-grade boys compared to 12th-grade girls at this school? Round to the nearest hundredth.

565. Suppose that you want a confidence interval for the difference in the mean heights of all 12th-grade boys compared to all 12th-grade girls at this school. In one case, you treat the population of boys as Population 1 and the population of girls as Population 2. In another case, you switch the order and treat the population of girls as Population 1 and the population of boys as Population 2. How would the resulting confidence intervals differ?

Making Calculations When Population Standard Deviations Are Known

566–571 A random sample of 120 college students who were physics majors found that they spent an average of 25 hours a week on homework; the standard deviation for the population was 7 hours. A random sample of 130 college students who were English majors found that they spent an average of 18 hours a week on homework; the standard deviation for the population was 4 hours. Call the population of physics majors "Population 1" and the population of English majors "Population 2" while working these problems.

566. Using a 90% confidence level, what is the margin of error for the estimated difference in average time spent on homework for college physics majors versus college English majors? Round to one decimal place.

567. Using an 80% confidence level, what is the margin of error for the estimated difference in average time spent on homework, for college physics majors versus college English majors? Round to one decimal place.

568. Using a 95% confidence level, what is the confidence interval for the true difference in average time spent on homework for college physics majors versus college English majors? Round your answer to the nearest tenth.

569. Using a 99% confidence level, what is the confidence interval for the true difference in average time spent on homework for college physics majors versus college English majors? Round your answer to the nearest tenth.

570. If you did not know the population standard deviations, how would your calculation of confidence intervals differ?

(A) You would use t^* from a t-distribution rather than z^* from the standard normal distribution.

(B) You would use the sample standard deviations rather than the population standard deviations.

(C) You would combine the sample sizes and divide the sum of the standard deviations by $(n_1 + n_2)$.

(D) Choices (A) and (B)

(E) Choices (A), (B) and (C)

571. If your sample sizes were 35 English majors and 20 physics majors, how, if at all, would your calculation of confidence intervals differ?

Digging Deeper into Calculations When Population Standard Deviations Are Known

572–574 A random sample of 200 men in North America found that their average age at first marriage was 29 years old; the standard deviation for the population was 6 years. A random sample of 220 women in North America found that their average age at first marriage was 26 years old; the standard deviation for the population was 4 years. Call the group of men "Population 1" and the group of women "Population 2" while working these problems.

572. For an 80% confidence level, what is the margin of error for the estimate of the difference in average age at first marriage for men and women? Round your answer to the nearest tenth.

573. For a 90% confidence level, what is the margin of error for the estimate of the difference in average age at first marriage for men and women? Round your answer to the nearest tenth.

574. For a 95% confidence level, what is the confidence interval for the true difference in mean age at first marriage for men and women in North America? Round your answer to the nearest tenth.

Working with Unknown Population Standard Deviations and Small Sample Sizes

575–580 *A random sample of 20 12th-grade boys in a high school showed a mean weight of 170 pounds with a sample standard deviation of 18 pounds. A random sample of 20 9th-grade boys in the same school showed a mean weight of 140 pounds with a sample standard deviation of 12 pounds. Call the population of 12th graders "Population 1" and the population of 9th graders "Population 2" while working these problems.*

575. What degrees of freedom will you use to calculate the confidence interval for the difference in weights?

576. Using a 99% level of confidence, what is the margin of error for the estimated difference in mean weights between 12th-grade boys and 9th-grade boys at this school? Round to the nearest whole number of pounds.

577. Using an 80% level of confidence, what is the margin of error for the estimated difference in mean weights between 12th-grade boys and 9th-grade boys at this school? Round to the nearest whole number of pounds.

578. Using a 90% level of confidence, what is the margin of error for the estimated difference in mean weights between 12th-grade boys and 9th-grade boys at this school? Round to the nearest whole number of pounds.

579. Using a 95% confidence level, what is the confidence interval for the true difference in the mean weights of all the 12th-grade and 9th-grade boys at this school? Round the endpoints to the nearest whole number.

580. Using a 98% confidence level, what is the confidence interval for the true difference in the mean weights of the 12th-grade and 9th-grade boys at this school? Round the endpoints to the nearest whole number.

Digging Deeper into Unknown Population Standard Deviations and Small Sample Sizes

581–584 *A random sample of 20 men in North America found that their average annual income after five years of employment, in thousands of dollars, was 37, with a standard deviation for the sample of 3.5. A random sample of 25 women in North America found that their average income after five years of employment, in thousands of dollars, was 30, with a standard deviation for the sample of 3. Call the population of men "Population 1" and the population of women "Population 2" while working these problems.*

581. You want to calculate a confidence interval for the true difference in average salaries for all men and women after five years of employment (in thousands of dollars). What degrees of freedom will you use?

582. With a 90% level of confidence, what is the margin of error for the estimated difference between average male and female salaries after five years of employment (in thousands of dollars)? Round your answer to one decimal place.

583. With a 95% level of confidence, what is the margin of error for the estimated difference between average male and female salaries after five years of employment? Round your answer to one decimal place.

584. With a 99% level of confidence, what is the confidence interval for the true difference in average income between all North American men and women, after five years of employment? Round the confidence limits in your answer to one decimal place.

Chapter 11

Claims, Tests, and Conclusions

• •

Hypothesis testing is a scientific procedure for asking and answering questions. Hypothesis tests help people decide whether existing claims about a population are true, and they're also commonly used by researchers to see whether their ideas have enough evidence to be declared statistically significant. This chapter is about understanding the basics of hypothesis testing.

The Problems You'll Work On

In this chapter, you break down the basics elements of a hypothesis test. Here's an overview of the problems you'll work on:

✔ Setting up a pair of hypotheses: the current claim (null hypothesis) and the one challenging it (alternative hypothesis)

✔ Using data to form a test statistic, which determines how far apart the two hypotheses are

✔ Measuring the strength of the new evidence through a probability (*p*-value)

✔ Making your decision as to whether the current claim can be overturned

✔ Understanding that your decision can be wrong and what errors you can commit

What to Watch Out For

Hypothesis testing has two levels at each step: how to do it and figuring out what it means. Here are some things to pay special attention to as you work through this chapter:

✔ Realizing it's all about the null hypothesis and whether you have enough evidence to overturn it

✔ Knowing the role of the test statistic, beyond calculating it and looking it up on a table

✔ Understanding how to interpret a *p*-value properly and why a small *p*-value means you reject H_0

✔ Being clear on Type I and Type II errors in the real sense — as false alarms and missed opportunities

Knowing When to Use a Hypothesis Test

585–586 Solve the following problems about using a hypothesis test.

585. Which of the following statements as currently written could be tested using a hypothesis test?

(A) An automobile factory claims 99% of its parts meet stated specifications.

(B) An automobile factory claims that it produces the best-quality cars in the country.

(C) An automobile factory claims that it can assemble 500 automobiles an hour when the assembly line is fully staffed.

(D) Choices (A) and (B)

(E) Choices (A) and (C)

586. Which of the following scenarios as currently stated could *not* involve a hypothesis test without further clarification?

(A) A political party conducts a survey in an attempt to contradict published claims of the proportion of voter support for a proposed law.

(B) A commercial laboratory does sample tests on a hand sanitizer to see whether it kills the percentage of bacteria claimed by the manufacturer.

(C) A school gives its students standardized tests to measure levels of achievement compared to prior years.

(D) A laboratory takes samples of a yogurt to see whether the manufacturer has met its published standard of being 99% fat free.

(E) A university evaluation group gives random surveys to students to see whether university claims regarding the proportion of students who are satisfied with student life are valid.

Setting Up Null and Alternative Hypotheses

587–604 Determine null and alternative hypotheses in the following problems.

587. You decide to test the published claim that 75% of voters in your town favor a particular school bond issue. What will your null hypothesis be?

588. You decide to test the published claim that 75% of voters in your town favor a particular school bond issue. What will your alternative hypothesis be?

589. Given the null hypothesis $H_0: \mu = 132$, what is the correct alternative hypothesis?

590. A university claims that work-study students earn an average of $10.50 per hour. What is the null hypothesis for a hypothesis test of this statement?

591. The manufacturer of the new GVX Hybrid car claims that it gets an average of 52 miles per gallon of gas. What is the null hypothesis for this statement?

592. Suppose that μ is the average number of songs on an MP3 player owned by a college student. Write down the description of the null hypothesis $H_0: \mu = 228$.

593. A think tank announces that 78% of teenagers own cellphones. What is the null hypothesis for a hypothesis test of this statement?

594. A travel agency claims that people from States 1 and 2 are equally likely to have taken a vacation in Hawaii. What is the null hypothesis for this statement?

595. According to a newspaper report, seven out of ten Americans think that Congress is doing a good job. What alternative hypothesis would you use if you believe this stated proportion is too high?

596. Amtrak claims that a train trip from New York City to Washington, D.C., takes an average of 2.5 hours. What alternative hypothesis would you use if you think the average trip length is actually longer?

597. An airline company claims that its flights arrive early 92% of the time. What alternative hypothesis would you use if you think this statistic is too high?

598. A car manufacturer advertises that a new car averages 39 miles per gallon of gasoline. What alternative hypothesis would you use if you think this statistic is too high?

599. A company claims that only 1 out of every 200 computers it sells has a mechanical malfunction. What alternative hypothesis would you use if you think this statistic is too low?

600. A hospital claims that only 5% of its patients are unhappy with the care provided. What is the alternative hypothesis if you think this statistic is too low?

601. A health study states that American adults consume an average of 3,300 calories per day. What alternative hypothesis would you use if you think this statistic is incorrect?

602. A study claims that adults watch television an average of 1.8 hours per day. What alternative hypothesis would you use if you think this figure is too low?

603. An investment company claims that its clients make an average of 8% return on investments every year. What alternative hypothesis would you use if you think this figure is too high?

604. Someone claims that high-school students living in cities with populations more than 1 million (Population 1) are 25% more likely to attend college than high-school students living in cities with populations less than 1 million (Population 2). Write the alternative hypothesis if you think this statistic is incorrect.

Finding the Test Statistic and the p-Value

605–612 You're conducting an experiment with the following hypotheses:

$$H_0: \mu = 4$$

$$H_a: \mu \neq 4$$

The standard error is 0.5, and the alpha level is 0.05. The population of values is normally distributed.

605. If $\bar{x} = 3$ in your sample, what is the test statistic?

606. If $\bar{x} = 4.5$ in your sample, what is the test statistic?

607. If $\bar{x} = 5.2$ in your sample, what is the test statistic?

608. If $\bar{x} = 3.6$ in your sample, what is the test statistic?

609. Suppose that your test statistic is 1.42. What is the p-value for this result?

610. Suppose that your test statistic is –1.56. What is the p-value for this result?

611. Suppose that your test statistic is 0.75. What is the p-value for this result?

612. Suppose that your test statistic is –0.81. What is the p-value for this result?

Making Decisions Based on Alpha Levels and Test Statistics

613–618 Suppose that you're conducting a study with the following hypotheses:

$H_0: p = 0.45$

$H_a: p > 0.45$

613. If your alpha level (significance level) is 0.05 and your test statistic is 1.51, what will your decision be? Assume that n is large enough to use the central limit theorem.

614. If your alpha level is 0.10 and your test statistic is 1.51, what will your decision be? Assume that n is large enough to use the central limit theorem.

615. If your alpha level is 0.01 and your test statistic is 1.98, what will your decision be? Assume that n is large enough to use the central limit theorem.

616. If your alpha level is 0.05 and your test statistic is 1.98, what will your decision be? Assume that n is large enough to use the central limit theorem.

617. If your alpha level is 0.05 and your test statistic is –1.98, what will your decision be? Assume that n is large enough to use the central limit theorem.

618. If your alpha level is 0.01 and your test statistic is –3.0, what will your decision be? Assume that n is large enough to use the central limit theorem.

Making Conclusions

619–633 Make conclusions after reading the information in the following problems.

619. What does it mean if a test statistic has a p-value of 0.01?

620. You are conducting a statistical test with an alpha level of 0.10. Which of the following is true?

(A) There is a 10% chance that you will reject the null hypothesis when it is true.

(B) There is a 10% chance that you will fail to reject the null hypothesis when it is false.

(C) You should reject the null hypothesis if your test statistic has a p-value of 0.10 or less.

(D) Choices (A) and (C)

(E) Choices (B) and (C)

621. The alpha level of a test is 0.05. The p-value for your test statistic is 0.0515. What is your decision?

622. A test was done with a significance level (α level) of 0.05, and the p-value was 0.001. Write down the best description of this result.

623. A test is done to challenge the statistic that 70% of people spend their summer vacations at home. The significance level is $\alpha = 0.05$.

H_0: $p = 0.70$

H_a: $p < 0.70$

p-value = 0.03

What can you conclude about the results?

624. If the significance level α is 0.02, which p-value for a test statistic will result in a test conclusion to reject H_0?

(A) 0.03

(B) 0.01

(C) 0.05

(D) 0.97

(E) 0.98

625. If the significance level α is 0.05, which p-value for a test statistic will result in a test conclusion to reject H_0?

(A) 0.95

(B) 0.10

(C) 0.06

(D) 0.055

(E) 0.04

626. Based on the following information, what do you conclude?

H_0: $p = 0.03$

H_a: $p \neq 0.03$

$\alpha = 0.01$

p-value = 0.007

627. Based on the following information, what do you conclude?

H_0: $p = 0.65$

H_a: $p > 0.65$

$\alpha = 0.03$

p-value = 0.02

628. Based on the following information, what do you conclude?

H_0: $\mu = 220$

H_a: $\mu < 220$

$\alpha = 0.05$

p-value = 0.06

629. Based on the following information, what do you conclude?

H_0: $p = 0.42$

H_a: $p > 0.42$

$\alpha = 0.05$

p-value = 0.42

630. Based on the following information, what do you conclude?

H_0: $\mu = 0.2$

H_a: $\mu > 0.2$

$\alpha = 0.02$

p-value = 0.2

631. Based on the following information, what do you conclude?

$H_0: \mu = 10$

$H_a: \mu \neq 10$

$\alpha = 0.01$

p-value = 0.018

632. Based on the following information, what do you conclude?

$H_0: \mu = 9.65$

$H_a: \mu > 9.65$

$\alpha = 0.05$

Test statistic: -1.88

p-value = 0.03

633. Based on the following information, what is your conclusion?

$H_0: \mu = 348$

$H_a: \mu > 348$

$\alpha = 0.05$

p-value = 0.07

Understanding Type I and Type II Errors

634–640 Solve the following problems about Type I and Type II errors.

634. Which of the following describes a Type I error?

(A) accepting the null hypothesis when it is true

(B) failing to accept the alternative hypothesis when it is true

(C) rejecting the null hypothesis when it is true

(D) failing to reject the alternative hypothesis when it is false

(E) none of the above

635. Which of the following describes a Type II error?

(A) accepting the alternative hypothesis when it is true

(B) failing to accept the alternative hypothesis when it is true

(C) rejecting the null hypothesis when it is true

(D) failing to reject the null hypothesis when it is false

(E) none of the above

636. If the alpha level is 0.01, what is the probability of a Type I error?

637. If the alpha level is 0.05, what is the probability of a Type II error?

638. Which of the following is a description of the power of the test?

(A) the probability of accepting the alternative hypothesis when it is true

(B) the probability of failing to accept the alternative hypothesis when it is true

(C) the probability of rejecting the null hypothesis when it is true

(D) the probability of rejecting the null hypothesis when it is false

(E) none of the above

639. What is the key to avoiding a Type II (missed detection) error?

(A) having a low significance level

(B) having a random sample of data

(C) having a large sample size

(D) having a low *p*-value

(E) Choices (B) and (C)

640. What is the key to avoiding a Type I error?

(A) having a low significance level

(B) having a random sample of data

(C) having a large sample size

(D) having a low *p*-value

(E) Choices (A), (B), and (C)

Chapter 12

Hypothesis Testing Basics for a Single Population Mean: *z*- and *t*-Tests

Conducting a hypothesis test is somewhat like doing detective work. Every population has a mean, and it's usually unknown. Many people claim that they know what it is; others assume that it hasn't changed from a past value; and in many cases, the population mean is supposed to follow certain specifications. Your modus operandi is to challenge or test that value of the population mean that's already assumed, given, or specified and use data as your evidence. That's what hypothesis testing is all about.

The Problems You'll Work On

In this chapter, you'll work out the basic ideas of hypothesis testing in the context of the population mean, including the following areas:

- Setting up the original, or null, hypothesis (the assumed or specified value) and the alternative hypothesis (what you believe it to be)
- Working through the details of doing a hypothesis test
- Making conclusions and assessing the chance of being wrong

What to Watch Out For

Hypothesis testing can seem like a plug-and-chug operation, but that can take you only so far. To truly master the materials in this chapter, keep the following in mind:

- Pay careful attention to the problem to determine how to set up the alternative hypothesis.
- Make sure you can calculate a test statistic and, more importantly, know what it's telling you.
- Remember that a small *p*-value comes from a large test statistic, and both mean rejecting H_0.
- Get an intuitive feeling about what Type 1 and Type 2 errors are — don't simply memorize!

Knowing What You Need to Run a z-Test

641–642 Figure out what you need to know to run a z-test in the following problems.

641. Dr. Thompson, a health researcher, claims that a teenager in the United States drinks an average of 30 ounces of sugary carbonated soda per day. A high-school statistics class decides to test this claim and is open to soda consumption being, on average, either higher or lower than the claimed 30 ounces per day. They conduct a random survey of 15 of their classmates and find self-reported soda consumption is, on average, 25 ounces per day. What additional information would you need to run a z-test to determine whether the students in this school drink significantly more or less soda than Dr. Thompson claims is consumed by U.S. teens in general?

642. You want to run a z-test to determine whether the sample from which a sample mean is drawn differs significantly from a population mean. You have the sample mean and sample size ($n = 20$); what other information do you need to know?

(A) whether the characteristic of interest is normally distributed in the population

(B) the population size

(C) the population mean and standard deviation

(D) the sample standard deviation

(E) Choices (A) and (C)

Determining Null and Alternative Hypotheses

643–646 Figure out null and alternative hypotheses in the following problems.

643. A researcher believes that people who smoke have lower shyness scores than the population average on a shyness scale, which is 25. What is the null hypothesis in this case, given that μ is the mean shyness score for all smokers?

644. Suppose a researcher has heard that children watch an average of ten hours of TV per day. The researcher believes this is wrong but has no theory about whether it's an overestimate or an underestimate of the truth. If the researcher wants to do a z-test of one population mean, what will the researcher's alternative hypothesis be?

645. A computer store owner reads that its consumers buy five flash drives each year, on average. The owner feels that on average her customers actually buy more than that, so the owner does a random survey of the customers on her mailing list. What is the store owner's alternative hypothesis?

646. A man reads that the average cost of dry-cleaning a shirt is 3 dollars, but in his city, it seems cheaper. So he randomly chooses ten dry cleaners in town and asks them the price to dry-clean a shirt. What will the man's alternative hypothesis be?

Introducing p-Values

647–650 Calculate p-values in the following problems.

647. A researcher has a *less than* alternative hypothesis and wants to run a single sample mean *z*-test. The researcher calculates a test-statistic of $z = -1.5$ and then uses a *Z*-table (such as Table A-1 in the appendix) to find a corresponding area of 0.0668, which is the area under the curve to the left of that value of *z*. What is the *p*-value in this case?

648. Suppose that a researcher has a *not equal to* alternative hypothesis and calculates a test statistic that corresponds to $z = -1.5$ and then finds, using a *Z*-table (such as Table A-1 in the appendix), a corresponding area of 0.0668 (the area under the curve to the left of that value of *z*). What is the *p*-value in this case?

649. A researcher has a *not equal to* alternative hypothesis and calculates a test statistic that corresponds to $z = -2.0$. Using a *Z*-table (such as Table A-1 in the appendix), the researcher finds a corresponding area of 0.0228 to the left of –2.0. What is the *p*-value in this case?

650. A scientist with a *not equal to* alternative hypothesis calculates a test statistic that corresponds to $z = 1.1$. Using a *Z*-table (such as Table A-1 in the appendix), the scientist finds that this corresponds to a curve area of 0.8643 (to the left of the test statistic value). What is the *p*-value in this case?

Calculating the z-Test Statistic

651–652 Determine the z-test statistic in the following problems.

651. A coffee shop manager reads that the preferred temperature for coffee among the U.S. populous is 110 degrees Fahrenheit with a standard deviation of 10 degrees. However, the manager doesn't believe this is true of his customers. Through complex and extensive testing, the manager finds that a random sample of 50 of his customers prefer their coffee, on average, to be 115 degrees Fahrenheit. Calculate the *z*-test statistic for this case and give your answer to two decimal places.

652. A very mathematically oriented musician has read studies showing that the average piece of popular music has 186.39 chord changes with a standard deviation of 26.52. The musician examines a random sample of 40 of his favorite songs and finds a mean of 172.12 chord changes. Calculate the *z*-statistic for this case and give your answer to four decimal places.

Finding p-Values by Doing a Test of One Population Mean

653–654 Calculate p-values in the following problems.

653. A pen company surveys the market and finds that people, on average, hope to use a pen for 40 days before having to replace it, with a standard deviation of 9 days. The company director of research believes that customers are actually less ambitious and will hope to use a pen for fewer than 40 days. The research director conducts a customer survey of 25 customers and finds that its customers, on average, expect to replace a pen after 36 days. What is the *p*-value if the director uses a *Z*-distribution to do a test of one population mean?

654. A berry farmer decided to grow blackberries after reading that bushes give an average of 3 pounds of fruit per year, with a standard deviation of 1 pound. The farmer suspects she won't get quite so much fruit because growing conditions aren't entirely right. The farmer identifies a random sample of 100 bushes and keeps careful track of how much fruit each bush gives. At the end of the year, the farmer finds that the average yield for each of the bushes in her sample was 2.9 pounds of fruit. Assume that the farmer conducts a *z*-test of a single population mean. What *p*-value will the farmer get?

Drawing Conclusions about Hypotheses

655–657 Figure out the conclusions that can be drawn in the following problems.

655. A psychiatrist reads that the average age of someone who's first diagnosed with schizophrenia is 24 years, with a standard deviation of 2 years. The psychiatrist suspects that the schizophrenic patients in her clinic were diagnosed at a younger age than 24. She examines records for a random sample of clinic patients and finds that the age at first diagnosis in the sample is 23.5 years on average. If age at first diagnosis is normally distributed and the *p*-value found in this case is 0.02 and the psychiatrist wants to run a test with a 0.05 significance level, what conclusion can the psychiatrist draw?

656. Airplane passengers carry an average of 45 pounds of luggage on a flight, with a standard deviation of 10 pounds. A researcher suspects that business travelers carry less than that on average. The researcher randomly samples 250 business travelers and finds that they carry an average of 44.5 pounds of luggage when traveling. The researcher wants a significance level of 0.05. What conclusion can the researcher draw based on this data pattern if the researcher performs a *z*-test for one population mean?

657. In a particular city, a square foot of office space averages $2.00 per month in rent, with a standard deviation of $0.50. A shopkeeper hopes to open a store in a neighborhood with significantly cheaper rent than that, using a 0.05 level of significance. He samples 49 offices at random from the neighborhood he's most interested in. The shopkeeper finds that the average rent in his sample is $3.00 per month. He conducts a z-test and finds his p-value to be 0.10. What can the shopkeeper conclude about his average rent?

Digging Deeper into p-Values

658–659 Calculate p-values in the following problems.

658. Assume that the average temperature of a healthy human being is 98.6 degrees Fahrenheit with a standard deviation of 0.5 degrees. A doctor believes that her patients average a higher temperature than that, so she randomly selects 36 of her patients and finds that their temperature is 98.8 degrees on average. If the doctor carries out a z-test for one population mean, what is the p-value for these results? Give your answer to four decimal places.

659. The average satisfaction rating for a company's customers is a 5 (on a scale of 0 to 7), with a standard deviation of 0.5 points. A researcher suspects that the Northeastern division has customers who are more satisfied. After randomly sampling 60 customers from the Northeastern division and asking them about their satisfaction, the researcher finds an average satisfaction level of 5.1. What is the p-value in this case if the researcher does a z-test for a single sample mean?

Digging Deeper into Conclusions about Hypotheses

660–665 Figure out the conclusions that can be drawn in the following problems.

660. In a company, employees type an average of 20 words per minute. Typing rates are normally distributed with a standard deviation of 3. The manager of a large branch of the company believes that his employees do better than that. He randomly samples 30 employees from his branch and finds an average typing rate of 20.5 words per minute. If the manager wants a significance level of 0.05, what can he conclude?

661. A potter believes that her workshop assistants can cover a certain size of vase with only 2 ounces of glaze, which is the industry standard. She knows that the amount of glaze required to cover a vase of the specified size follows a normal distribution with a standard deviation of 0.8 ounces, and she believes that her workshop is still putting too much glaze on each vase. She samples 30 vases from a large production run and finds a sample mean of 2.3 ounces of glaze. Using a significance level of 0.01, what can she conclude?

662. A researcher believes that the tissue cultures in her lab are significantly denser than average. The researcher takes a random sample of 40 specimens of tissue from her lab and finds the average weight is 0.005 grams per cubic millimeter. Textbooks claim that such tissues should weigh 0.0047 grams per cubic millimeter with a standard deviation of 0.00047 grams. What can this researcher conclude if she uses a one-sample z-test and a significance level of 0.001?

663. On average, hens lay 15 eggs per month with a standard deviation of 5. A farmer tests this claim on his own hens. He samples 30 of his hens and finds that they lay an average of 16.5 eggs per month. Can he reject the null hypothesis that his hens on average lay the same number of eggs as the larger hen population? Use a one-sample z-test and a significance level of 0.05.

664. A nationally known moving company knows that the typical family uses 110 boxes in a long-distance move, with a standard deviation of 30 boxes. A company from Chicago wants to see how it compares. The company randomly samples 80 families from the Chicago area and finds that they used, on average, 103 boxes in their most recent move. Using a one-sample z-test and a significance level of 0.05, what is your decision?

665. A travel magazine claims that the American family who uses a car to go on vacation travels an average of 382 miles from home. A researcher believes that families with dogs who vacation by car on average travel a different distance. She draws a sample of 30 families with dogs who drive on their vacations and finds that they drive an average of 398 miles. (Assume that the population standard deviation is 150 miles.) At a 0.05 significance level and a one-sample z-test, what can you conclude from this data?

Digging Deeper into Null and Alternative Hypotheses

666–667 Figure out null and alternative hypotheses in the following problems.

666. A magazine reports that the average number of minutes that U.S. teenagers spend texting each day is 120. You believe it's less than that. What are your null and alternative hypotheses?

667. The average number of calories that a pizza slice contains is 250, with a standard deviation of 35 calories. A nutrition researcher suspects that slices of pizza on the college campus where he works contain a higher number of calories. That researcher randomly samples 35 pizza slices in the area of his campus and finds a mean of 265 calories. What are the null and alternative hypotheses?

Knowing When to Use a t-Test

668–670 Solve the following problems about knowing when to use a t-test.

668. Which of the following conditions indicate that you should use a *t*-test instead of using the *Z*-distribution to test a hypothesis about a single population mean? (Assume that the population has a normal distribution.)

    (A) The population standard deviation is unknown.

    (B) The population standard deviation is known.

    (C) The sample standard deviation is unknown.

    (D) The sample standard deviation is known.

    (E) None of the above conditions are related to a *t*-test.

669. A researcher is trying to decide whether to use a *Z*-distribution or a *t*-test to evaluate a hypothesis about a single population mean. Which of the following conditions would indicate that the researcher should use the *t*-test?

    (A) The population standard deviation isn't known.

    (B) The sample size is only $n = 50$.

    (C) The sample mean is less than the population mean.

    (D) The alternative hypothesis is a *not equal to* hypothesis.

    (E) The sampling distribution of sample means is normal.

670. A student finds that a sample of 50 of his friends reports, on average, that they spend 43 hours per week using social networking sites, with a sample standard deviation of 8 hours. The student believes that this is significantly less than the claims of journalists that students spend 50 hours per week using social networking sites. The student has a *less than* alternative hypothesis about a single population mean. How should the student test his hypothesis?

Connecting Hypotheses to t-Tests

671–674 Solve the following problems about hypotheses and t-tests.

671. A student believes that his friends spend less time on social networking sites, on average, than is claimed in the media. If μ_1 is the average amount of time spent by the student's friends and μ_0 is the amount of time claimed in the media, what is the null hypothesis the student would use to do a *t*-test on a single population mean?

672. A student believes that his friends spend less time on social networking sites, on average, than is claimed in the media. If μ_1 is the average amount of time spent by the student's friends and μ_0 is the amount of time claimed in the media, what is the alternative hypothesis the student would use to do a *t*-test on a single population mean?

673. A national retail chain says in its ad that the average price of a certain hair product it sells across the United States is $10 per bottle (the null hypothesis). A manager of one of the stores in the chain believes that it's more than that. She samples 30 bottles of the product at random from her own store and conducts a t-test. Her p-value is smaller than 0.05 (pre-specified significance level) so she rejects the null hypothesis. (Assume that the population standard deviation is unknown.) Based on the manager's data, can she conclude that the average price of this product across the United States is actually more than $10 per bottle?

674. Suppose that a dentist believes that her patients experience less pain than the average dental patient does. She samples 40 of her patients and receives pain ratings from them after she gives them a filling. The dentist then compares her sample mean and standard deviation with a population value she found in a medical journal. Her sample average pain rating was 3.2 (on a 10-point scale), and the medical journal reported an average pain rating of 3.5. What are the null and alternative hypotheses in this case?

Calculating Test Statistics

675–676 Figure out test statistics in the following problems.

675. Use the following information to calculate a t-value (test statistic).

Sample mean: 30

Claimed population mean: 35

Sample standard deviation: 10

Sample size: 16

What is the value for the test statistic, *t?*

676. It's believed that the average amount of sleep a person in the United States gets per night is 6.3 hours. A mom believes that mothers get far less sleep than that. She contacts a random sample of 20 other moms on a mom social networking site and finds that they get an average of 5.2 hours of sleep per night, with a standard deviation of 1.8 hours. Using this data, what is the value of the test statistic *t* from a t-test for a single population mean?

Working with Critical Values of t

677–680 Solve the following problems about the critical value of t.

677. A researcher hypothesizes that a population of interest has a mean greater than 6.1. The researcher uses a sample of 15. What is the critical value of *t* needed to reject the null hypothesis, using a significance level of 0.05?

678. Suppose that a researcher believes that a sampled population of interest has a mean that differs from a claimed value of 100 but is unsure of the direction of the difference. The researcher wants a 0.05 significance level with a sample size of *n* = 10. What critical value of *t* should the researcher use for a t-test involving a single population mean?

679. It's believed that the average amount that a person in the United States sleeps is 6.3 hours per night. A psychologist believes that mothers get far less sleep than that. She contacts a random sample of other moms on a social networking site for moms, and she finds among the sample of 20 moms an average of 5.2 hours of sleep per night with a standard deviation of 1.8 hours. If this psychologist wants a significance level of 0.05 and sets up a *less than* alternative hypothesis, what would she conclude based on the *t*-value of −2.733 in this case?

680. A researcher wants to use a 90% confidence interval to determine whether her sample of 17 ball bearings differs in either direction from an average target value of 0.0112 grams. In the sample, the standard deviation is 0.0019 grams, and the mean is 0.0123 grams. How does the critical value of *t* compare with the observed value of *t* in this case?

Linking p-Values and t-Tests

681–685 Solve the following problems about p-values and t-tests.

681. A study finds a test statistic *t*-value of 1.03 for a *t*-test on a single population mean. The sample size is 11, and the alternative hypothesis is $H_a: \mu \neq 5$. Using Table A-2 in the appendix, what range of values is sure to include the *p*-value for this value of *t*?

682. A researcher believes that her sample mean is smaller than 90 and finds a sample mean of 89.8 with a sample standard deviation of 1. Using the *t*-table (Table A-2 in the appendix), and given that the sample had 29 observations, what is the approximate *p*-value?

683. Imagine that it costs an average of $50,000 per year to incarcerate someone in the United States. A prison warden wants to know how the costs in her prison compare to the population average, so she randomly samples 12 of her inmates, reviews their records carefully, and finds an average cost of $58,660 per inmate, with a standard deviation of $10,000. If she performs a *t*-test on a single population mean, what is the approximate *p*-value for this data?

684. A researcher has a *greater than* alternative hypothesis and observes a sample mean greater than the claimed value. The test statistic *t* for a *t*-test for a single population mean is 2.5, with 14 degrees of freedom. What is the *p*-value associated with this test statistic?

685. A researcher with a *not equal to* alternative hypothesis observes a sample mean less than the claimed value. The test statistic is found to be −2.5 with 20 degrees of freedom. What is the *p*-value associated with this test statistic value of *t* if the researcher runs a *t*-test for a single population mean?

Drawing Conclusions from t-Tests

686–692 Make conclusions from t-tests in the following problems.

686. An instructor claims that her students take an average of 45 minutes to complete her exams. You think that the average time is higher than that. You decide to investigate this using a significance level of 0.01. You take a sample, find the test statistic, and find the *p*-value is 0.0001 (small by anyone's standards). What is your conclusion?

687. Suppose that it's claimed that people who play musical instruments have average skills in verbal ability. A scientist, however, has a hypothesis that people who play musical instruments are below average in verbal ability. On a verbal ability test with a population mean score of 100, a random sample of eight musicians yields an average score of 97.5, with a standard deviation of 5. Verbal ability, as measured by this test, is normally distributed in the population. What should the scientist conclude if the significance level for his test is 0.05?

688. The president of a large corporation believes that her employees donate less money to charity than the corporate target of $50 per year, per employee. She conducts a survey of ten randomly selected employees and finds an average annual donation of $43.40 with a standard deviation of $5.20. Donations are normally distributed in this population. If the president does a *t*-test for a single population mean, with a 0.05 level of significance, what should she conclude?

689. A coat maker promises that his heavy winter coats feel warm down to a temperature of –5 degrees Centigrade, but he believes that his coats actually protect people well at even colder temperatures. He surveys 15 of his customers, and they report, on average, that the coats feel warm down to a temperature of –6.5 degrees Centigrade, with a standard deviation of 1.0 degrees Centigrade. If the coat maker wants a 0.10 significance level ($\alpha = 0.10$), what should he conclude if he runs a one-sample *t*-test, assuming that temperatures are normally distributed?

690. A teacher believes that other teachers in her school get lower evaluations compared to teachers in other schools in the district. She knows that the average teaching evaluation in the district is a rating of 7.2 on a 10-point scale, with scores normally distributed. She surveys six randomly selected teachers in her school and finds an average teaching rating among them of 6.667 with a standard deviation of 2. What should she conclude from a single population *t*-test, using her data and a significance level of 0.05?

691. A popsicle company tries to keep its products at a temperature of –1.92 degrees Centigrade. The company president believes that the freezers are set too low, thus wasting money to keep products colder than necessary. She randomly samples five freezers and finds that they're running at an average temperature of –2.25 degrees Centigrade, with a sample standard deviation of 1.62 degrees. Overall, temperature is normally distributed for this brand of freezer. Using a significance level of 0.01, what should she conclude?

692. You're given the following information and asked to perform a t-test on a single population mean. The null hypothesis is that the mean weight of all parts of an object of a certain type is 50 grams per object: $H_0 : \mu = 50$. The alternative hypothesis is that the objects are heavier on average: $H_a : \mu > 50$. The sample mean is 54 grams, and the sample standard deviation is 8 grams. The sample size is 16. Weight is normally distributed in this population. Using a significance level of $\alpha = 0.05$, what is your conclusion?

Performing a t-Test for a Single Population Mean

693–694 Suppose you expect that the average number of beads in a 1-pound bag coming from the factory is 1,200. However, the retailer believes that the average number is larger. You draw a random sample of 30 bags and find that the mean of that sample is 1,350 beads, with a standard deviation of 500. Assuming that the population of values is normally distributed, you use this information to perform a t-test for a single population mean.

693. Using a significance level of 0.01, what is your decision?

694. Using a significance level of 0.05, what is your decision?

Drawing More Conclusions from t-Tests

695–700 Determine the conclusions that should be drawn in the following problems.

695. A popular financial advisor states that, on average, a person should spend no more than $100 per month on entertainment. A researcher believes that the average person spends more than that and conducts a survey of 25 randomly chosen people. In that survey, the average monthly spending on entertainment was found to be $118.44, with a standard deviation of $35.00. Assuming that spending follows a normal distribution in the population and using a significance level of $\alpha = 0.01$, what should the researcher conclude, based on a t-test of a single mean?

696. A Wisconsin cheese company read that, on average, a person in Europe consumes 25.83 kilograms of cheese per year. The company's sales director believes that Americans consume even more cheese. The company does a survey of a random sample of Americans and finds that they eat, on average, 27.86 kilograms of cheese per year with a standard deviation of 6.46 kilograms. Assume that cheese consumption among Americans is normally distributed. If there were 30 people in the sample and the significance level is 0.05, what can the cheese company managers conclude about their sales director's hypothesis?

697. A meditation teacher reads that the ideal amount of time to meditate each day is 20 minutes. He wants to know whether the practice of his students differs significantly from the ideal amount of meditation by running a *t*-test on a single mean, using a significance level of $\alpha = 0.10$. He surveys nine people randomly chosen from his meditation classes and finds that they meditate for an average of 24 minutes per day with a standard deviation of 5 minutes. Assuming that meditation time among all such students is normally distributed, what should the teacher conclude?

698. Suppose that a heavy-duty laser printer is claimed to produce an average of 20,000 pages of print before needing to be serviced. Also assume that the page output of such printers until service is needed is normally distributed. A company that uses a lot of heavy-duty laser printers randomly samples 16 of its printers to see how many pages it gets, per printer, before needing service. The study finds an average of 18,356 pages between servicing and a standard deviation of 2,741 pages. Using a significance level of 0.05 and a *not equal to* alternative hypothesis, what is your decision?

699. A doctoral student has heard that dissertations are, on average, 90 pages in length, but she believes that dissertations written by students in her program may have a different average length. She randomly selects ten dissertations completed by people who went through her program and finds a mean length of 85.2 pages with a standard deviation of 7.59 pages. If she assumes that dissertation page length is normally distributed and runs a *t*-test on a single mean with a significance level of 0.05, what should she conclude?

700. A logger walks through a forest with her husband. Her husband estimates that the average tree in the forest is 30 years old. Because all the trees in a certain area of the forest are to be cleared anyway, the logger randomly chooses five trees, cuts them down, and counts the rings. She's interested in knowing only whether these trees differ from her husband's estimated mean of 30 years on average, so she chooses a significance level of $\alpha = 0.50$. Her sample averaged 33 years of age with a standard deviation of 5.6 years. In this forest, tree age is approximately normally distributed. What is the logger's conclusion, based on a significance test for the true average age of all trees planned to be cleared?

Chapter 13

Hypothesis Tests for One Proportion, Two Proportions, or Two Population Means

● ●

*T*his chapter gives you practice doing hypothesis tests for three specific scenarios: testing one population proportion; testing for a difference between two population proportions; and testing for a difference between two population means. Most hypothesis tests use a similar framework, so patterns will develop, but each hypothesis test has its own special elements, and you work on those here.

The Problems You'll Work On

In this chapter, you practice setting up and carrying out three different hypothesis tests, focusing specifically on the following:

✔ Knowing what hypothesis test to use and when

✔ Setting up and understanding the null and alternative hypotheses correctly in all cases

✔ Working through the appropriate formulas and knowing where to get the needed numbers

✔ Calculating and interpreting the test statistics and *p*-values

What to Watch Out For

As you go through more hypothesis tests, knowing their similarities and differences is important. Here are a few notes about this chapter:

✔ Notation plays a big role — make a list for yourself to keep it all straight.

✔ The formulas go up a notch in this chapter. Be sure you label and understand the parts of the formula and what clues you need to set them up.

✔ If you do find a statistically significant difference, make it clear which population is the one with the larger proportion or mean.

Testing One Population Proportion

701–715 Solve the following problems about testing one population proportion.

701. A bank will open a new branch in a particular neighborhood if it can be reasonably sure that at least 10% of the residents will consider banking at the new branch. The bank will use a significance level of 0.05 to make its decision. The bank does a survey of residents of a particular neighborhood and finds that 19 out of 100 random people surveyed said they'd consider banking at the new branch. Run a z-test for a single proportion and determine whether the bank should open the new branch, considering its standard policy.

702. A corporate call center hopes to resolve 75% or more of customer calls through an automated computer voice recognition system. It randomly surveys 50 recent customers; 45 report that their issue was resolved. Can the management of the corporation conclude that the computer system is hitting its minimum target, using a significance level of 0.05? Use a z-test for a single proportion to provide an answer.

703. A used bookstore will buy a collection of books if it's reasonably sure that it can sell at least 50% of the books within six months. A customer comes in with 30 books; the clerk evaluates them and determines that 17 are likely to sell within six months. Using the threshold of a 0.05 significance level, should the clerk make an offer for the collection of books? Use the z-test for a single proportion to decide.

704. A factory owner hopes to maintain a standard of less than 1% defects. She randomly samples 1,000 ball bearings and finds that 6 are defective. Can the factory owner conclude that the process is producing a defect rate of 1% or less? Use a z-test for a single proportion and a level of significance of 0.05 to decide.

705. The two symbols $\hat{p}$ and p_0 appear throughout a hypothesis test for one proportion (for example, the formula for the test statistic is $\frac{\hat{p} - p_0}{SE}$). What is the difference between these two symbols?

706. A computer manufacturer is willing to buy components from a supplier only if it can be reasonably sure that the defect rate is less than 1%. If it inspects a shipment of 10,000 randomly selected components and finds that 90 are defective, should the computer manufacturer work with that supplier? Perform a hypothesis test using a z-test on a single proportion. The significance level is 0.01.

707. What is the difference between the symbol p and the term "p-value" in a hypothesis test for a proportion?

(A) p is the actual population proportion, and a p-value is a sample proportion.

(B)   A p-value is the actual population proportion, and p is a sample proportion.

(C)   A p-value is the claimed value for the population, and p is the actual value of the population proportion.

(D) p is the claimed value for the population proportion, and p-value is the actual value of the population proportion.

(E)   None of the above.

708. An antique dealer is willing to buy collections of antiques as long as no more than 5% of the items in the entire collection are found to be fakes. At a recent estate sale, the dealer randomly samples 10 items from 200 items for sale and finds that 2 are fakes. Should the dealer buy that entire collection, based on a z-test for a single proportion, using a significance level of 0.01?

709. A blood bank wants to ensure that none of its blood carries diseases to the recipients. It tests a sample of 1,000 specimens and finds that 2 of them have potentially deadly diseases. The director of research has asked that a z-test on a single proportion be done. With what confidence can the blood bank say that the true population proportion of illnesses in the blood specimens is greater than 0?

710. A European city council likes to ensure that no more than 3% of the pigeons that inhabit the city carry disease that humans can catch from droppings. As long as there is significant evidence that fewer than 3% of pigeons carry potential human diseases, the council leaves the pigeons alone. If there isn't enough evidence, it aggressively works to control the population of pigeons by trapping and moving them or shooting them.

The city council randomly samples 200 pigeons and finds that 6 have droppings with potential for human disease transmission. Should the council try to control the pigeon population or leave them alone, considering its guidelines and a threshold of $\alpha = 0.05$?

711. A fashion designer likes to use small defects to make his pieces more interesting. He sends designs off to manufacturing plants abroad and rejects shipments if the true defect rate is anything other than 25%. The designer doesn't want rates much lower or much greater than 25%. In a recent large shipment, a random sample of 50 pieces found a defect rate of 12%. Should the shipment be accepted or rejected, if a z-test for a single population proportion and a significance level of $\alpha = 0.05$ is used to make the decision?

712. Suppose that you flip a coin 100 times, and you get 55 heads and 45 tails. You want to know whether the coin is fair, so you conduct a hypothesis test to help you decide. What are the null and alternative hypotheses in this situation? (*Note: p* represents the overall probability of getting a heads, also known as the proportion of heads over an infinite number of coin flips.)

713. Joe claims that he has ESP, and he aims to prove it. He asks you to shuffle five regular decks of cards. One by one, you pick a card, note its suit (diamonds, spades, hearts, or clubs), and return it to the deck. As you select each card, Joe tells you what he thinks its suit is. You repeat this process 100 times and then look at the proportion of correct answers Joe has given (designated as $\hat{p}$). Suppose you want him to be at least 20 percentage points above the accuracy you would expect by chance. In this scenario, what are the null and alternative hypotheses?

714. A biologist wants to study cell lines where nearly 25% of the cells in a sample have a particular phenotype. As a result, the biologist wants to reject any sample of cells that differ from the target 25% phenotype present based on a significance level of 0.10. In a sample of 1,000,000 cells, 250,060 are found to have the phenotype. Should that sample be studied or rejected? Use a z-test for a single proportion to answer.

715. Suppose that Bob runs a grocery store, and he has been watching customers as they stand in the checkout aisle waiting to pay for their groceries. Based on his observations, he believes that more than 30% of the customers purchase at least one item in the checkout aisle. You believe the percentage is even higher than that. You conduct a hypothesis test to find out what the story is. Your null and alternative hypotheses are H_0: $p = 0.30$ and H_a: $p > 0.30$. You take a random sample of 100 customers and observe their behavior in the checkout aisle. Your sample finds that only 20% of the customers purchased items. What conclusion can you make regarding H_0 at this point (if any)?

Comparing Two Independent Population Means

716–720 *A manager of a large grocery store chain believes that happy employees are more productive than unhappy ones. He randomly samples 60 of his grocery checkout clerks and classifies them into one of two groups: those who smile a lot (Group 1) and those who don't (Group 2). After sorting the random sample, there just happens to be 30 in each group. He then examines their productivity scores (based on how quickly and accurately they're able to assist customers) and gets the following data for each group:*

> *Group 1 score mean: 33.3*

> *Group 2 score mean: 14.4*

The population of productivity scores is normally distributed with a standard deviation of 17.32, which is assumed to be the population standard deviation that applies both to Group 1 and Group 2.

Conduct an appropriate test to see whether a difference in productivity levels exists between the two groups among all employees, using $\alpha = 0.05$.

716. What are the appropriate null and alternative hypotheses for this test?

717. What is the critical value for a z-test for this hypothesis?

718. What is the standard error for this test?

719. What is the test statistic for this data?

720. What is your decision, given this data?

Digging Deeper into Two Independent Population Means

721–725 A psychologist read a claim that average intelligence differs between smokers and nonsmokers and decided to investigate. She sampled 30 smokers and 30 nonsmokers and gave them an IQ test. She used an alpha level of 0.10. The mean for the smokers is 51.9, and the mean for the nonsmokers is 52.6. The population variance for each group is 5.

721. What are the null and alternative hypotheses for this data?

722. What is the critical value for a z-test for this hypothesis?

723. What is the test statistic for this data?

724. What is your decision, given this data?

725. Suppose that you were using an alpha level of 0.05. What would your decision be regarding this data?

Getting More Practice with Two Independent Population Means

726–730 A sleep researcher investigates performance following a night with two different levels of sleep deprivation. In Group 1, 40 people were permitted to sleep for five hours. In Group 2, 35 people were permitted to sleep for three hours. The outcome measure was performance on a memory test with a population standard deviation of 6 in each group. The mean memory test score in the group that slept three hours was 58; in the group that slept five hours, it was 62.

726. If the researcher is interested in whether performance differs according to the amount of sleep, what are the null and alternative hypotheses?

727. If the researcher is interested in whether more sleep is associated with better performance, what are the null and alternative hypotheses?

728. Using a z-test for two population means, what is the test statistic for this data?

729. Using a z-test for two population means, a *not equal to* alternative hypothesis, and an alpha level of 0.01, what is the researcher's decision regarding this data?

730. Using a z-test for two population means, a *greater than* alternative hypothesis, and an alpha level of 0.05, what is the researcher's decision regarding this data?

Using the Paired t-Test

731–736 A city power company wants to encourage households to conserve energy. It decides to test two advertising campaigns. The first emphasizes the potential to save money by saving energy (the "wallet" appeal). The second emphasizes the humanitarian value of reducing carbon footprint (the "moral" appeal). The company randomly chooses ten households and presents one randomly chosen appeal of the two to them, observes their energy consumption for a month, and then presents the other appeal and observes their energy consumption for a month. Energy consumption is expressed in kilowatts.

With the "wallet" appeal as Group 1 and the "moral" appeal as Group 2, the results are as follows:

$$\bar{d} = -83.5$$
$$s_d = 46.39$$
$$n = 10$$
$$df = 9$$

Assume that power consumption is normally distributed at both time points. The research question is whether energy consumption differs following one appeal rather than the other.

731. What is the appropriate test to determine if one appeal is more successful than the other?

732. Given the information provided, which of the following is a true statement regarding this sample data?

(A) The wallet appeal was more successful.

(B) The moral appeal was more successful.

(C) There was greater variability in the wallet appeal data.

(D) There was greater variability in the moral appeal data.

(E) On average, there was no difference between the wallet and moral appeals.

733. What is the degrees of freedom for the appropriate test for this data?

734. What is the standard error for a paired t-test on this data? Round your answer to two decimal places.

735. What is the test statistic for this data? Round to two decimal places.

736. Using a *not equal to* alternative hypothesis and an alpha level of 0.10, what is your conclusion regarding this data?

Digging Deeper into the Paired t-Test

737–745 A precocious child wants to know which of two brands of batteries tends to last longer. She finds seven toys, each of which requires one battery. For each toy, she then randomly chooses one of the brands, puts a fresh battery of that brand in the toy, turns the toy on, and records the time before the battery dies. She repeats the experiment with a battery of the brand not used in the first trial for each toy.

Her data is listed in the following table (in terms of hours of battery life before failure):

	Brand 1	Brand 2	Difference
Toy 1	11.1	12.8	−1.7
Toy 2	10.2	12.4	−2.2
Toy 3	10.3	10.8	−0.5
Toy 4	7.9	8.7	−0.8
Toy 5	10.9	12.0	−1.1
Toy 6	14.2	13.5	0.7
Toy 7	7.1	8.0	−0.9

Illustration by Ryan Sneed

Assume that the difference scores are normally distributed. The sample standard deviation of difference scores (s_d) is 0.9214.

737. What is the appropriate test to determine whether the two brands have a different battery life?

738. If the child claims that one brand lasts longer than the other, what are the null and alternative hypotheses for this research question?

739. What is $\bar{d}$ for this data? Round your answer to four decimal places.

740. What is the standard error for this data? Round your answer to four decimal places.

741. At an alpha level of 0.05, given a *not equal to* alternative hypothesis, what is the critical value of t?

742. At an alpha level of 0.01, given a *not equal to* alternative hypothesis, what is the critical value of t?

743. What is the test statistic for this data? Round your answer to four decimal places.

744. At the $\alpha = 0.01$ level, what is your decision regarding this data?

745. At the $\alpha = 0.05$ level, what is your decision regarding this data?

Comparing Two Population Proportions

746–752 *A computer security consultant is interested in determining which of two password generation rules is more secure. One rule requires users to include at least one special character (*, @, !, %, or $); the other doesn't.*

The consultant creates 100,000 phantom accounts and observes them for six months, monitoring carefully for inappropriate logins. Among the 50,000 accounts with passwords that used the first rule (requiring special characters), 1,055 were breached by security threats. Among the 50,000 who followed the second rule, 2,572 security breaches occurred.

746. What is the appropriate statistical test to address this research question?

747. Assume that the researcher claims that the password systems differ in strength. What are the appropriate null and alternative hypotheses for this research question?

748. What are the values of $\hat{p}_1$ and $\hat{p}_2$ for this data?

749. What is the value of $\hat{p}$ for this data?

750. What is the standard error for this data? Round your answer to four decimal places.

751. What is the z-statistic for this data? Round your answer to two decimal places.

752. Using a significance level of 0.05 and a *not equal to* alternative hypothesis, what is your decision based on this data?

Digging Deeper into Two Population Proportions

753–760 *A prison warden wants to see whether it's beneficial or harmful to give prisoners access to the Internet. She randomly assigns prisoners in 100 cells to receive such access and 100 cells to be observed as controls. The outcome variable is whether each cell had a report of a behavioral problem during the week following introduction of Internet access.*

Of those in the Internet group (Group 1), 50 of the cells had a reported episode of behavioral disturbance, and of those in the non-Internet group (Group 2), 70 cells had such incidents. If appropriate, conduct a z-test for independent proportions and report your results. Use a 95% confidence level.

753. What are the values of $\hat{p}_1$ and $\hat{p}_2$ for this data?

754. What is the value of $\hat{p}$ for this data?

755. What is the standard error for this data? Round your answer to four decimal places.

756. For the following hypotheses and an alpha level of 0.01, what is the critical value for the test statistic?

$$H_0: p_1 = p_2$$
$$H_a: p_1 \neq p_2$$

757. For the following hypotheses and an alpha level of 0.01, what is the critical value for the test statistic?

$$H_0: p_1 = p_2$$
$$H_a: p_1 > p_2$$

758. What is the test statistic for this data, given the following hypotheses? Round your answer to four decimal places.

$$H_0: p_1 = p_2$$
$$H_a: p_1 \neq p_2$$

759. What is your decision regarding this data, given an alpha level of 0.05 and the following hypotheses?

$$H_0: p_1 = p_2$$
$$H_a: p_1 \neq p_2$$

760. What is your decision regarding this data, given an alpha level of 0.01 and the following hypotheses?

$$H_0: p_1 = p_2$$
$$H_a: p_1 \neq p_2$$

Chapter 14

Surveys

Surveys are everywhere, and their quality can range from good to bad to ugly. The ability to critically evaluate surveys is the focus of this chapter.

The Problems You'll Work On

The problems in this chapter focus on the following big ideas:

- Understanding the basics and nuances of good and bad surveys (issues involved in designing the survey, selecting a sample of participants, and collecting the data)
- Identifying problems that commonly occur in surveys and samples that create bias
- Using the proper terminology at the proper time to describe a problem that's been found with a survey or a sample

What to Watch Out For

Pay particular attention to the following:

- The way a sample is selected depends on the situation — know the differences.
- Specific terms are used to describe problems with surveys and samples. Make sure you pick up on subtle differences. Here are the most common terms:
  - **Sampling frame:** A list of all the members of the target population.
  - **Census:** Getting desired information from everyone in the target population.
  - **Random sample:** Each member of the population has an equal chance of being selected for the sample.
  - **Bias:** Systematic unfairness in sample selection or data collection.
  - **Convenience sample:** Chosen solely for convenience, not based on randomness.
  - **Volunteer/self-selected sample:** Sample where people determine on their own to be involved.
  - **Non-response bias:** Occurs when someone in the sample doesn't return or doesn't finish the survey.
  - **Response bias:** When the respondent takes the survey but doesn't give correct information.
  - **Undercoverage:** Sampling frame doesn't include adequate representation from certain groups within the target population.

Planning and Designing Surveys

761–766 *You're interested in the willingness of adult drivers (age 18 and over) in a metropolitan area to pay a toll to travel on less-congested roads. You draw a sample of 100 adult drivers and administer a survey on this topic to them.*

761. What is the target population for this study?

762. Suppose that you collect your data in a way that makes it likely that the survey respondents aren't representative of the target population. What is this called?

763. If you were to select your sample by drawing numbers at random from the published phone directory and calling during daytime hours on weekdays, how could these actions bias the results?

   (A)   Not everyone has a phone or a listed phone number.

   (B)   Not everyone is at home during the day on weekdays.

   (C)   Not everyone is willing to participate in telephone polls.

   (D)   Choices (A) and (B)

   (E)   Choices (A), (B), and (C)

764. What is the main problem with the survey question "Don't you agree that drivers should be willing to pay more for less-congested roads?"

765. Of the 100 people in your sample, 20 choose not to participate. You later discover that they're in a lower income bracket than those who did participate. What problem does this introduce into your study?

766. One of your questions asks whether the respondent voted in the last election. You find a much higher proportion of individuals claiming to have voted than is indicated by public records. What is this an example of?

Selecting Samples and Conducting Surveys

767–775 *Solve the following problems about selecting samples and conducting surveys.*

767. Why is bias particularly problematic in surveys?

768. Which of the following is an example of a good census of 2,000 students in a high school?

   (A)   calculating the mean age of all the students by using their official records

   (B)   asking the first 25 students who arrive at school on a given day their age and calculating the mean from this information

   (C)   sending an e-mail to all students asking them to respond with their age and calculating the mean from those who respond

   (D)   Choices (A) and (C)

   (E)   Choices (A), (B), and (C)

769. You want to survey students at a high school and calculate the mean age. Which of the following procedures will result in a simple random sample?

(A) classifying the students as male or female and drawing a random sample from each

(B) using an alphabetized student roster and selecting every 15th name, starting with the first one

(C) selecting three tables at random from the cafeteria during lunch hour and asking the students at those tables for their age

(D) selecting one student at random, asking him or her to suggest three friends to participate and continuing in this fashion until you have your sample size

(E) numbering the students by using the school's official roster and selecting the sample by using a random number generator

770. Suppose that you intend to conduct a survey among employees at a firm. You use a file supplied by the personnel department for your sampling file, not realizing that it excludes people hired in the past six months. What kind of bias is likely to result?

771. Suppose that you conduct a survey by showing an 800 number on the television screen during a popular program and inviting people to call in with their response to a posted question. What type of bias is likely to result?

772. Suppose that you conduct a survey by interviewing the first 50 people you see in a shopping mall. What type of bias is likely to result?

773. Sometimes questions are rephrases and scales are reversed during the course of a survey. For example, with Likert scale items, 1 may sometimes mean *strongly agree* and sometimes *strongly disagree*. What is one reason for this reversal?

774. What is the biggest problem that makes the question "Should everyone go to college and seek gainful employment?" less than ideal?

775. Why do some surveys include "don't know" as a response?

(A) because the respondent may be uninformed about the topic

(B) because the respondent may not remember enough information to answer the question

(C) because the respondent may find the question offensive

(D) Choices (A) and (B)

(E) Choices (A), (B), and (C)

Chapter 15

Correlation

• •

*I*n this chapter, you explore possible linear relationships between a pair of quantitative (numerical) variables, *X* and *Y*. Your basic question is this: As the *X* variable increases in value, does the *Y* variable increase with it, does it decrease in value, or does it just basically not react at all? Answering this question requires the use of graphs as well as the calculation and interpretation of a certain numerical measure of togetherness — correlation.

The Problems You'll Work On

Your job in this chapter is to look for, describe, and quantify possible linear relationships between two quantitative variables, using the following methods:

✔ Graphing pairs of data on a scatter plot and describing what you see

✔ Measuring the strength and direction of a linear relationship, using correlation

✔ Interpreting correlation properly and knowing its properties

✔ Understanding what elements can affect correlation

What to Watch Out For

Correlation is more than just a number; it's a way of describing relationships in a universal way.

✔ Understand that correlation applies to quantitative variables (like age and height), even though the "street definition" of correlation relates any variables (like gender and voting pattern).

✔ Know the many properties of correlation — some are counterintuitive (for example, you may think that switching the values of *X* and *Y* will change the correlation, but it doesn't).

✔ Always remember: "Man does not live by a correlation alone." You always need to look at a scatter plot of the data as well. (Neither is foolproof by itself.)

Interpreting Scatter Plots

776–781 This scatter plot represents the high-school and freshman college GPAs of 24 students.

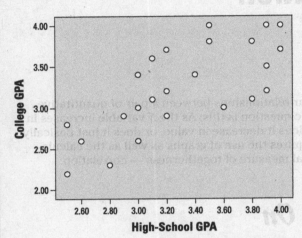

Illustration by Ryan Sneed

776. How would you describe the linear relationship between high-school GPA and college GPA?

(A) strong

(B) weak

(C) positive

(D) negative

(E) Choices (A) and (C)

777. Looking at the following high-school GPAs for five students, which one would you predict to have the highest college GPA?

(A) 2.5

(B) 2.8

(C) 3.1

(D) 3.4

(E) 4.0

778. How do you know this scatter plot displays a positive linear relationship between the two variables?

779. How do you know this scatter plot displays a relatively strong linear relationship between the two variables?

780. If these two quantitative variables had a correlation of 1, how would the scatter plot be different?

781. If these two quantitative variables had a correlation of –1, what would the scatter plot look like?

(A) The points would all lie on a straight line.

(B) All the points would have to be between –1 and 0.

(C) All the points would slope downward from left to right.

(D) Choices (A) and (C)

(E) Choices (A), (B), and (C)

Creating Scatter Plots

782–783 Solve the following problems about making scatter plots.

782. For a group of adults, suppose you want to create a scatter plot between height and one other variable. Which variable(s) would be the appropriate candidate?

(A) gender

(B) race/ethnicity

(C) weight

(D) zip code of residence

(E) Choices (C) and (D)

783. You have a data set containing four variables collected from a group of children ages 5 to 21 years. The variables are height, age, weight, and gender. Which pair(s) of variables is appropriate for making a scatter plot?

(A) height and age

(B) height and gender

(C) gender and age

(D) height and weight

(E) Choices (A) and (D)

Understanding What Correlations Indicate

784–787 Figure out what the correlations in the following problems indicate.

784. Which of the following correlations indicates a strong, negative linear relationship between two quantitative variables?

(A) –0.2

(B) –0.8

(C) 0

(D) 0.4

(E) 0.8

785. Which of the following correlations indicates a weak, positive linear relationship between two quantitative variables?

(A) –0.2

(B) –0.6

(C) 0.2

(D) 0.75

(E) 0.9

786. Which of the following correlations indicates a very strong, positive linear relationship between two quantitative variables?

(A) –0.7

(B) –0.1

(C) 0.2

(D) 0.4

(E) 0.9

787. Which of the following correlations indicates a weak, negative linear relationship between two quantitative variables?

(A) –0.2

(B) –0.8

(C) –1

(D) 0.4

(E) 0.8

Digging Deeper into Scatter Plots

788–790 In this scatter plot, X and Y both represent measurement variables.

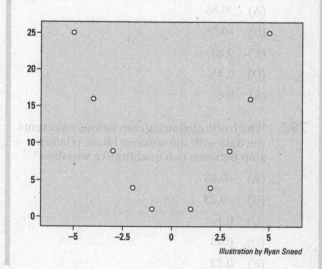

Illustration by Ryan Sneed

788. How do you describe the relationship between the two quantitative variables in this scatter plot?

789. What is your best guess for the correlation for these two quantitative variables?

790. Why is correlation a poor choice to describe the relationship between these two particular variables?

Digging Deeper into What Correlations Indicate

791–793 Figure out what the correlations in the following problems indicate.

791. Which of the following correlations represents the data with the strongest linear relationship between two quantitative variables?

(A) −0.85

(B) −0.56

(C) 0.23

(D) 0.45

(E) 0.6

792. Which of the following correlations represents the data with the weakest linear relationship between two quantitative variables?

(A) −0.63

(B) −0.23

(C) 0.1

(D) 0.45

(E) 0.73

793. Which of the following represents the strongest possible linear relationship between two quantitative variables?

(A) −1

(B) 0

(C) 1

(D) Choices (A) and (C)

(E) Choices (A), (B), and (C)

Calculating Correlations

794–799 Use the following data set to help with calculating the correlation in the following problems.

X	Y
1	2
2	2
3	4
4	3

Illustration by Ryan Sneed

794. What is the value of $\bar{x}$ for this data?

795. What is the value of $\bar{y}$ for this data?

796. What is the value of $n - 1$ in this case?

797. What is the value of s_x for this data?

798. What is the value of s_y for this data?

799. Suppose you have the following information about two variables, X and Y:

$$\sum_x \sum_y (x - \bar{x})(y - \bar{y}) = 2.5$$

Using the formula for correlation, calculate the correlation between these two variables.

Noting How Correlations Can Change

800–803 Use this information to answer the following problems: The following statistics describe two variables, X and Y:

$$\bar{x} = 8.00; \bar{y} = 8.53$$

$$s_x = 4.47; s_y = 5.36$$

$$n = 15$$

$$\sum_x \sum_y (x - \bar{x})(y - \bar{y}) = 274$$

800. What is the correlation of X and Y in this case?

801. If the sample size is changed to 20, with all the other given values the same, how will the correlation change?

802. If s_y is changed to 4.82, with all the other given values the same, how will the correlation of X and Y change?

803. If $\sum_x \sum_y (x - \bar{x})(y - \bar{y})$ is changed to 349, how does the correlation of X and Y change?

Looking at the Properties of Correlations

804–806 Solve the following problems about correlation properties.

804. Suppose that you calculate the correlation between the heights of fathers and sons, measured in inches, and then you convert the data to centimeters; how does the correlation change?

805. Which of the following values isn't possible for a correlation?

(A) −2.64

(B) 0.99

(C) 1.5

(D) Choices (A) and (C)

(E) Choices (A), (B), and (C)

806. If you compute the correlation between the heights and weights of a group of students and then compute the correlation between their weights and heights, how will the two correlations compare?

Digging Deeper into How Correlations Can Change

807–810 You conduct a study to see whether changing the font size of the text displayed on a computer screen influences reading comprehension.

807. Which variable corresponds to "reading comprehension" in this study?

(A) the X variable

(B) the Y variable

(C) the response variable

(D) Choices (A) and (C)

(E) Choices (B) and (C)

808. Which variable corresponds to "font size" in this study?

(A) the X variable

(B) the Y variable

(C) the response variable

(D) Choices (A) and (C)

(E) Choices (B) and (C)

809. How will the correlation change if you switch the designation of the two variables — that is, if you make the X variable the Y variable and make the Y variable the X variable?

810. How does the correlation change if text font size is measured in centimeters rather than inches?

Making Conclusions about Correlations

811–813 Draw conclusions about the correlations in the following problems.

811. If a correlation between two quantitative variables is calculated to be 1.2, what do you conclude?

812. If you compute a correlation of –0.86 between two quantitative variables, what do you conclude?

813. If you compute a correlation of 0.27 between two quantitative variables, what do you conclude?

Getting More Practice with Scatter Plots and Correlation Changes

814–817 You conduct a study to see whether the amount of time spent studying per week is related to GPA for a group of college computer science majors.

814. How do you designate the "time spent studying" variable on a scatter plot of your data?

815. How do you designate the variable "GPA" on a scatter plot of your data?

(A)   the X variable

(B)   the Y variable

(C)   the response variable

(D)   Choices (A) and (C)

(E)   Choices (B) and (C)

816. How does the correlation change if you switch the measurement of study time from minutes to hours?

817. How does the correlation change if you switch the designation of the two variables — that is, make the X variable the Y variable and make the Y variable the X variable?

Making More Conclusions about Correlations

818–820 Draw conclusions about the correlations in the following problems.

818. If you compute a correlation of –0.23 between two quantitative variables, what do you conclude?

819. If you compute the correlation between two quantitative variables to be 1.05, what do you conclude?

820. If you compute a correlation of –0.87 between two quantitative variables, what do you conclude?

Chapter 16

Simple Linear Regression

• •

With simple linear regression, you look for a certain type of relationship between two quantitative (numerical) variables (like high-school GPA and college GPA.) This special relationship is a *linear relationship* — one whose pairs of data resemble a straight line. After you find that right relationship, you fit a line to it and use the line to make predictions for future values. Sounds romantic, doesn't it?

The Problems You'll Work On

Your job in this chapter is to find and interpret the results of a regression line and its elements and to carefully check exactly how well your line fits. *Note:* Regression assumes you've found that a strong relationship exists (see Chapter 15 for the details of correlation and scatter plots).

✔ Find the best fitting (regression) line to describe a linear relationship.

✔ Understand and/or interpret the slope and *y*-intercept of the regression line.

✔ Use the regression line to make predictions for one variable given another, where appropriate.

✔ Assess line fit and look for anomalies, such as offbeat patterns or points that stand out.

What to Watch Out For

It's easy to get caught up in all the calculations of regression. Always remember that understanding and interpreting your results is just as important as calculating them!

✔ Slope measures change and are used all over the place in regression — be sure to know them cold.

✔ Equations aren't smart — you have to know when they can be used and applied and when they can't.

✔ Many techniques exist for determining how well a line fits and for pointing out problems. Know all the tools available and the specifics of what their results tell you.

Introducing the Regression Line

821 Solve the following problem about regression line basics.

821. What conditions must be met before it's appropriate to find the least-squares regression line between two quantitative variables?

    (A) Both variables are numeric.

    (B) The scatter plot indicates a linear relationship.

    (C) The correlation is at least moderate.

    (D) Choices (A) and (C)

    (E) Choices (A), (B), and (C)

Knowing the Conditions for Regression

822–823 In this scatter plot, the two variables plotted are quantitative (numerical). The correlation is $r = 0.75$.

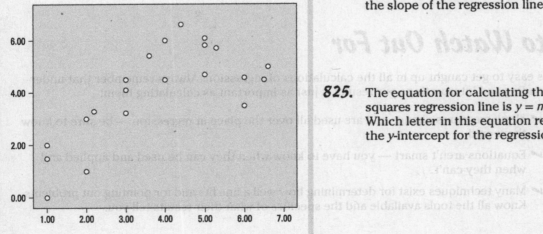

Illustration by Ryan Sneed

822. In looking at this scatter plot, which of the following violations of a necessary condition for fitting a regression line is observed?

    (A) The variables aren't numeric.

    (B) Their correlation isn't strong enough.

    (C) Their relationship isn't linear.

    (D) Choices (B) and (C)

    (E) None of the above.

823. The equation for calculating the least-squares regression line is $y = mx + b$. If two variables have a negative relationship, which letter will be preceded by the negative sign?

Examining the Equation for Calculating the Least-Squares Regression Line

824–825 Solve the following problems about the equation for the least-squares regression line.

824. The equation for calculating the least-squares regression line is $y = mx + b$. Which letter in this equation represents the slope of the regression line?

825. The equation for calculating the least-squares regression line is $y = mx + b$. Which letter in this equation represents the y-intercept for the regression line?

Finding the Slope and y-Intercept of a Regression Line

826–829 The linear relationship between two variables is described by the regression line $y = 3x + 1$. (Assume that the correlation is strong and that the scatter plot shows a strong linear relationship.)

826. What is the slope of the regression line?

827. What is the y-intercept of the regression line?

828. If $x = 3.5$, what is the expected value of y?

829. If $x = 0.4$, what is the expected value of y?

Seeing How Variables Can Change in a Regression Line

830–832 Suppose that the linear relationship between two quantitative variables is described by the regression line $y = -1.2x + 0.74$. (Assume that the correlation is strong and that the scatter plot shows a strong linear relationship.)

830. If x increases by 1.5, how does the value of y change?

831. If x decreases by 2.3, how does the value of y change?

832. Where does this regression line intersect the y-axis?

Finding a Regression Line

833–842 This scatter plot shows the relationship between the GRA verbal (GRA_V) and GRA math (GRA_M) scores of a group of high-school seniors.

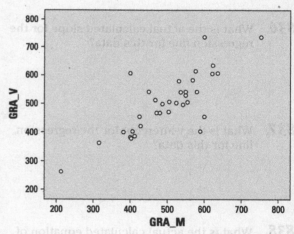

Illustration by Ryan Sneed

For the purpose of this analysis, consider GRA_M to be the X variable and GRA_V to be the Y variable. These variables and their relationship are characterized by the following statistics:

$$\bar{x} = 502.9; \bar{y} = 506.1$$

$$s_x = 103.2; s_y = 103.2$$

$$r = 0.792$$

833. How would you characterize the linear relationship between X and Y in this case?

834. What is the approximate range of values for each of these variables?

835. Without doing any calculations, what is the best estimate for the slope of the regression line for these variables?

836. What is the actual calculated slope for the regression line for this data?

837. What is the y-intercept for the regression line for this data?

838. What is the actual calculated equation of the regression line for this data?

839. The linear relationship between two quantitative variables is described by the equation $y = 0.792x + 107.8$. Using this equation, if $x = 230$, what is the expected value of y?

840. Student A has a math score 210 points higher than Student B. How much higher do you expect Student A's verbal score to be, compared to Student B?

841. Student C has a math score 50 points lower than Student D. Using the regression equation, how do you expect their verbal scores to compare?

842. There is a strong, positive correlation between GRA_M and GRA_V in this data. Why can't you assume that this is a cause-and-effect relationship?

Digging Deeper into Finding a Regression Line

843–852 The scatter plot represents data on home size (in square feet) and selling price (in thousands of dollars) for 35 recent sales in an American community. Home size is the X variable, and selling price is the Y variable.

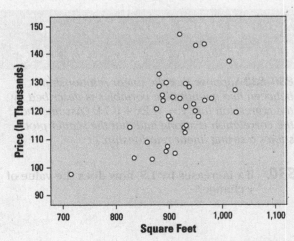

Illustration by Ryan Sneed

The following statistics describe these two variables and their relationship in the data set:

$\bar{x} = 915.1; \bar{y} = 121.1$

$s_x = 58.5; s_y = 11.8$

$r = 0.527$

843. How do you describe the linear relationship between these two variables?

844. Given the following home sizes in square feet, which size home do you expect to sell for the highest price?

    (A)   800 square feet

    (B)   850 square feet

    (C)   870 square feet

    (D)   890 square feet

    (E)   910 square feet

845. What is the slope of the regression line for this data?

846. What is the *y*-intercept of the regression line for this data?

847. What is the equation of the regression line for this data?

848. What do you expect a house of 1,000 square feet to sell for (in dollars)?

849. What do you expect a house of 1,500 square feet to sell for (in dollars)?

850. What do you expect a house of 890 square feet to sell for (in dollars)?

851. House A is 90 square feet larger than House B. What do you expect their price difference to be (in dollars)?

852. House C is 54 square feet smaller than House D. What do you expect their price difference to be (in dollars)?

Connecting to Correlation and Linear Relationships

853–856 Use the following scatter plot to answer the following problems.

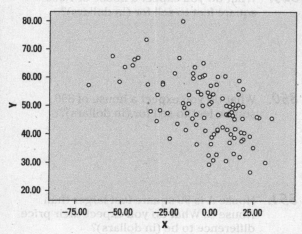

Illustration by Ryan Sneed

853. In words, how would you describe the general linear relationship between the X and Y variables in this scatter plot?

854. In terms of numbers, what is the most plausible value for the correlation between X and Y?

855. If the variables X and Y were switched in this scatter plot, how would the correlation be affected?

856. Does the scatter plot suggest that X and Y are good candidates for a linear regression analysis?

Determining Whether Variables Are Candidates for a Linear Regression Analysis

857–859 Figure out whether you can use the variables in the following problems in a linear regression analysis.

857. Suppose that you're considering running a linear regression with two variables, X and Y. As a preliminary check, you compute the correlation and find that it's 0.54. At this point, can you say that these variables are good candidates for a linear regression analysis? Why or why not?

858. Suppose that you're considering running a linear regression with two variables, X and Y. As a preliminary check, you compute the correlation and find that it's 0.05. Are these variables good candidates for a linear regression analysis? Why or why not?

859. Which of the following correlations indicates the strongest linear relationship between two variables? (Assume that the scatter plots match the correlations, respectively.)

(A) –0.9

(B) –0.5

(C) 0.0

(D) 0.9

(E) Choices (A) and (D)

Digging Deeper into Correlations and Linear Relationships

860–865 As part of a health behaviors study, you collect data on heights and weights for a group of adults. The scatter plot of the data indicates a possible linear relationship.

860. If the correlation between height and weight is 0.65, what is the correlation between weight and height?

861. Originally, you collected height in inches and weight in pounds. You decide to convert your measurements to centimeters and kilograms, respectively, and recalculate the correlation. How will it change?

862. For females in this data set, the correlation coefficient between weight and height is 0.50. For males, it's 0.80. What do these two correlations tell you about the linear relationship between weight and height for females and males in this data set? (Assume that the scatter plots agree with the correlations in both cases.)

863. Which of the following doesn't represent a possible value for a correlation in general?

(A) −1.5

(B) −0.6

(C) 0.0

(D) 0.2

(E) 0.8

864. Do a strong correlation and a strong linear pattern on a scatter plot allow you to conclude a cause-and-effect relationship between two variables *X* and *Y*? Why or why not?

865. If you want to predict weight using height, which variable is designated as the *X* variable: height or weight?

Describing Linear Relationships

866–872 As part of a study of academic performance, you collect data about GPA, major, minutes spent studying per week, and minutes spent watching TV per week from a random sample of college juniors. A scatter plot of minutes studying and GPA suggests a linear relationship.

866. For the entire sample, the correlation between minutes studying and GPA is 0.74. How would you describe the linear relationship between these two variables?

867. Suppose that the responses for minutes spent studying per week range from 0 to 480 minutes. Further, the responses average 250 minutes with a standard deviation of 60 minutes. A scatter plot between minutes studying and GPA demonstrates a linear trend, and the correlation coefficient of the two variables is 0.84. Among the five choices given, how many minutes spent studying is the most likely prediction for students with the highest GPAs, on average? Choose the best answer.

(A)   20 minutes

(B)   60 minutes

(C)   250 minutes

(D)   260 minutes

(E)   450 minutes

868. For the entire sample, the correlation between minutes watching TV and GPA is –0.38. How would you describe the linear relationship between these two variables?

869. Given a correlation of –0.68 between minutes watching TV and GPA for five students, which of the following minutes would you expect the student with the highest GPA to have spent watching TV, on average? (Assume that the scatter plot suggests a linear relationship.)

(A)   30

(B)   80

(C)   120

(D)   250

(E)   500

870. For the purpose of a regression analysis, which is the more logical choice for the Y variable: time spent studying or GPA?

871. Among English majors, the correlation between minutes spent studying and GPA is 0.48. Among engineering majors, it's 0.78. Given this information, which of the following statements about the linear relationship between time spent studying and GPA is correct for this data set? Assume that scatter plots suggest some type of linear relationship between the variables in each case.

(A)   The linear relationship is stronger for English majors.

(B)   The linear relationship is stronger for engineering majors.

(C)   The linear relationships are equal between English and engineering majors.

(D)   Correlation can't be used to compare two groups.

(E)   Not enough information to tell.

872. You calculate a correlation of –2.56 between minutes spent studying per week and minutes spent watching TV per week. What do you conclude?

Getting More Practice with Finding a Regression Line

873–877 You conduct a survey including variables for current income (measured in thousands of dollars) and life satisfaction (measured on a scale of 0 to 100, with 100 being most satisfied and 0 being least satisfied). A scatter plot of the data suggests a linear relationship.

873. If you want to predict life satisfaction using income, which variable would you designate as X?

874. For this data, the mean of satisfaction is 60.4, and the standard deviation of satisfaction is 12.5. The mean of income is 80.5, and the standard deviation of income is 16.7. The correlation coefficient between satisfaction and income is 0.77. What is the slope for the regression line predicting satisfaction from income?

875. For this data, the mean of satisfaction is 60.4, and the standard deviation of satisfaction is 12.5. The mean of income is 80.5, and the standard deviation of income is 16.7. The correlation coefficient between satisfaction and income is 0.77. What is the y-intercept for the regression line when predicting satisfaction from income?

876. The slope and y-intercept for the linear relationship between income (X) and life satisfaction (Y) are 0.58 and 13.7, respectively. What is the equation describing the linear relationship between income and life satisfaction based on this data?

877. In this data set, incomes (x values) range from $50,000 to $150,000. Which of the following would qualify as extrapolation when attempting to predict life satisfaction (Y), using income (X)?

(A) predicting satisfaction for someone with an income of $75,000

(B) predicting satisfaction for someone with an income of $45,000

(C) predicting satisfaction for someone with an income of $200,000

(D) Choices (A) and (B)

(E) Choices (B) and (C)

Making Predictions

878–884 A building contractor examines the cost of having carpentry work done in some of his buildings in the current year. He finds that the cost for a given job can be predicted by this equation:

$$y = 50x + 65$$

Here, y is the cost of a job (in dollars), and x is the number of hours a job takes to complete. So the cost of a given job can be predicted by a base fee of $65 per job plus a cost of $50 per hour. Assume that the scatter plot and correlation both indicate strong linear relationships.

878. What is the predicted cost of a job that takes 2.5 hours to complete?

879. What is the predicted cost for a job that takes 4.75 hours to complete?

880. How much more money do you predict a job taking 3.75 hours to complete will cost, as compared to a job taking 3.5 hours to complete?

881. Suppose that in a different city, a similar equation predicts carpentry costs, but the intercept is $75 (the slope remains the same). What is the predicted cost for a job taking 2 hours in this city?

882. Suppose you're comparing the predicted cost of a job lasting 3.5 hours in the two cities (the first with an intercept of $65 and the second with an intercept of $75). How much more do you predict a job lasting 3.5 hours will cost in the first city versus the second city?

883. In a third city, carpentry costs are predicted by an equation with a slope of $60 and an intercept of $65. How much do you predict a job lasting 2.3 hours will cost?

884. Looking at data from an earlier year, the equation predicting costs in the first city has an intercept of $65 and a slope of $60. Compared to the current year, how much more did a job lasting 3.6 hours cost in the previous year?

Figuring Out Expected Values and Differences

885–893 You conduct a survey of job satisfaction rating (measured on a scale of 0 to 100 points, with 100 being the most satisfied) and years of experience among employees in a large company. The linear relationship between these two variables is described by the regression line $y = 1.4x + 62$. Here, y is job satisfaction rating, and x is years of employment. Assume that the scatter plot and correlation project a linear relationship.

885. What are the y-intercept and slope for this equation?

886. What does the slope mean in a simple regression equation?

887. What does the y-intercept mean in a simple regression equation?

888. For this data, what is the expected job satisfaction rating for someone with 20 years of experience, on average?

889. For this data, what is the expected job satisfaction rating for someone with two years of experience, on average?

890. What is the expected difference in job satisfaction ratings when comparing someone with 15 years of experience to someone with 8 years of experience?

891. What is the expected difference in job satisfaction ratings, on average, comparing someone with 15 years of experience to a new hire (with 0 years of experience)?

892. What is the expected difference, on average, in job satisfaction ratings, comparing someone with 11.5 years of experience to a new hire (0 years of experience)?

893. For a second company, the equation predicting job satisfaction from years of employment, on average, is $y = 1.4x + 67$. Comparing employees with 10 years of experience, how much does the predicted job satisfaction rating for an employee from the second company differ from that of the first, on average?

Digging Deeper into Expected Values and Differences

894–902 You collect data on square footage and market value for homes in two communities, and you want to use square footage to predict market value. The variables x_1 and x_2 represent the square footage of homes in Community 1 and Community 2, respectively, and y_1 and y_2 represent the market value of homes in Community 1 and Community 2, respectively. The scatter plots and correlations both indicate strong linear relationships for each community. The regression equations describing these linear relationships are

$$y_1 = 77x_1 - 15,400$$
$$y_2 = 74x_2 - 11,300$$

894. What is the expected market value for a home of 1,500 square feet in Community 1, on average?

895. What is the expected market value for a home of 1,840 square feet in Community 1, on average?

896. What is the expected market value for a home of 1,500 square feet in Community 2, on average?

897. What is the expected market value for a home of 980 square feet in Community 2, on average?

898. Comparing two homes with 1,000 square feet each, one in Community 1 and one in Community 2, how do their expected values differ, on average?

899. Comparing two homes with 1,620 square feet each, one in Community 1 and one in Community 2, how do their expected values differ, on average?

900. Comparing two homes with 1,930 square feet each, one in Community 1 and one in Community 2, how do their expected values differ, on average?

901. Which has a larger expected market value: a home in Community 1 with 1,100 square feet or a home in Community 2 with 1,200 square feet, on average?

902. Suppose that the home sizes for Community 1 in your sample range from 800 square feet to 2,780 square feet. Which of the following would constitute extrapolation?

    (A)   estimating the market value of a home in Community 1 with 2,700 square feet

    (B)   estimating the market value of a home in Community 1 with 2,900 square feet

    (C)   estimating the market value of a home in Community 1 with 750 square feet

    (D)   Choices (A) and (B)

    (E)   Choices (B) and (C)

Digging Deeper into Predictions

903–905 For a group of 100 high-school students, the following equation relates their SAT math score (ranging from 200 to 800) with minutes per week watching TV (denoted by x, ranging from 0 to 720):
$$SAT = 725 - 0.5x$$

903. What is the predicted SAT math score for a student who watches 360 minutes of TV per week?

904. What is the predicted SAT math score for a student who watches 600 minutes of TV per week?

905. Student A watches 60 minutes of TV per week, Student B watches 800 minutes of TV per week, and Student C watches 220 minutes of TV per week. Rank the students from high to low in terms of their expected SAT math scores.

Getting More Practice with Expected Values and Differences

906–915 You collect data on income (in thousands of dollars) and years of experience of part-time employees working for two companies, and you calculate separate regression equations to explore these linear relationships. The regression equations describing these linear relationships are

$$y_1 = 6.7x_1 + 2.5$$
$$y_2 = 7.2x_2 + 1.2$$

Here, y_1 and x_1 are the expected salary and years of experience for an employee in Company 1, and y_2 and x_2 are the expected salary and years of experience for an employee in Company 2.

906. What is the expected salary for a part-time employee in Company 1 with 6 years of experience?

907. What is the expected salary for a part-time employee in Company 1 with 17 years of experience?

908. What is the expected salary for a part-time employee in Company 2 with 2.5 years of experience?

909. How much more do you expect a part-time employee in Company 2 with 13 years of experience to make when compared to a part-time employee in Company 2 with 7 years of experience?

910. How much more do you expect a part-time employee in Company 2 with 6.5 years of experience to make when compared to a part-time employee in Company 2 with 1.2 years of experience?

911. Which company has the higher expected starting salary for its part-time employees?

912. Which company has the higher rate of expected increase of salary for its part-time employees?

913. Compare expected salaries between an employee at Company 1 with 3 years of experience and an employee at Company 2 with 5.5 years of experience. Which employee has the higher expected salary?

914. Compare expected salaries between an employee at Company 1 with 3.8 years of experience and an employee at Company 2 with 4.3 years of experience. Which employee has the higher expected salary?

915. Under which circumstances could you declare that a strong relationship between x and y leads you to conclude that a change in x *causes* a change in y?

(A) replication of this study at other companies

(B) a longitudinal study tracking growth in individual employee salaries as years of employment increase

(C) adding additional variables to the model to control for other influences on salary

(D) Choices (A) and (B)

(E) Choices (A), (B), and (C)

Chapter 17

Two-Way Tables and Independence

. . . .

Many applications of statistics involve categorical variables, such as gender (male/female), opinion (yes/no/undecided), home ownership (yes/no), or blood type. One common statistical application is to look for relationships between two categorical variables. In this chapter, you focus on pairs of categorical variables: how to organize and interpret their data (the two-way tables part), and how to describe findings and look for relationships (the checking for independence part).

The Problems You'll Work On

In this chapter, you work on all the ins and outs of two-way tables and how to interpret and use them, including

- ✔ Being able to read and interpret all parts of a two-way table by using either counts or percentages
- ✔ Finding marginal, joint, and conditional probabilities from a two-way table or related graph
- ✔ Using probabilities from a two-way table to look for and describe relationships between two categorical variables

What to Watch Out For

Numbers sitting in a little table seem easy enough, but you'd be surprised at all the information you can get out of a table, and how many equations, formulas, and notations that you can squeeze out of them. Be ready and keep the following in mind:

- ✔ The numbers within a two-way table represent intersections of two characteristics.
- ✔ What goes into the denominator is the key for all two-way table probabilities. Pay attention to what group you're looking at; the total number in that group becomes your denominator.
- ✔ The wording of the problems for two-way tables can be extremely tricky; one small change in wording can lead to a totally different answer. Practice as many problems as you can.

Introducing Variables and Two-Way Tables

916–920 Solve the following problems about variables in two-way tables.

916. Which of the following variables are categorical (that is, their possible values fall into nonnumerical categories)?

(A) blood type

(B) country of origin

(C) annual income

(D) Choices (A) and (B)

(E) Choices (A), (B), and (C)

917. Which of the following variables is categorical?

(A) gender

(B) hair color

(C) zip code

(D) Choices (A) and (B)

(E) Choices (A), (B), and (C)

918. Which of the following variables would be suitable for a two-way table?

(A) years of education

(B) height in centimeters

(C) homeownership (yes/no)

(D) gender

(E) Choices (C) and (D)

919. For a two-way table including education as one of the variables, which of the following would be an appropriate way to categorize the data?

(A) whether someone is a high-school graduate

(B) whether someone is a college graduate

(C) highest level of school completed

(D) Choices (A) and (B)

(E) Choices (A), (B), and (C)

920. How many cells does a 2-x-2 table contain? (A cell is any of the possible combinations of the two variables being studied.)

Reading a Two-Way Table

921–933 This 2-x-2 table displays results from a poll of randomly selected male and female college students at a certain college, asking whether they were in favor of increasing student fees to expand the college's athletics program. The results of their opinions are broken down by gender in the following table.

	Favor Fee Increase	Do Not Favor Fee Increase
Male	72	108
Female	48	132

Illustration by Ryan Sneed

921. What does the value 72 represent in this table?

922. What does the value 132 represent in this table?

923. How many of the students are female and favor the fee increase?

924. How many male students were included in this poll?

925. How many female students were included in this poll?

926. How many students favor the fee increase?

927. How many students do not favor the fee increase?

928. What is the total number of students who took part in the poll?

929. What proportion of male students favors the fee increase?

930. What proportion of female students does not favor the fee increase?

931. What proportion of all students does not favor the fee increase?

932. The following pie chart was calculated based only on the 180 females who were polled. In the poll, the females were asked for their opinions on whether they favored a fee increase. What does the 27% represent?

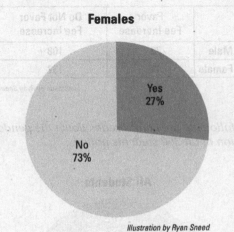

Illustration by Ryan Sneed

933. In this poll, 360 students were asked for their opinions on whether they favor a fee increase. A breakdown of the results is shown in the following pie chart. What does the 33% represent?

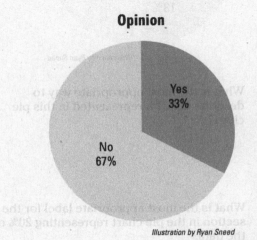

Illustration by Ryan Sneed

Interpreting a Two-Way Table by Using Percentages

934–937 This 2-x-2 table displays results from a poll of 360 randomly selected male and female college students at a certain college, asking whether they were in favor of increasing student fees to expand the college's athletics program. The results of their opinions are broken down by gender in the following table.

	Favor Fee Increase	Do Not Favor Fee Increase
Male	72	108
Female	48	132

Illustration by Ryan Sneed

The following pie chart breaks down the gender and opinion of all 360 students polled.

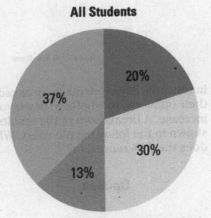

All Students

Illustration by Ryan Sneed

934. What is the most appropriate way to describe the 37% represented in this pie chart?

935. What is the most appropriate label for the section in the pie chart representing 20% of the data?

936. What is the most appropriate label for the section in the pie chart representing 30% of the data?

937. What is the most appropriate label for the section in the pie chart representing 13% of the data?

Interpreting a Two-Way Table by Using Counts

938–950 The following table displays information about cigarette smoking and diagnosis with hypertension for a group of patients at a medical clinic.

	Hypertension Diagnosis	No Hypertension Diagnosis
Smoker	48	24
Nonsmoker	26	50

Illustration by Ryan Sneed

938. How many of the patients are smokers?

939. How many of the patients have a hypertension diagnosis?

940. How many of the patients are both nonsmokers and have a hypertension diagnosis?

941. How many of the patients are smokers and have a hypertension diagnosis?

942. What is the total number of patients in this study?

943. How many of the patients do not have a hypertension diagnosis and are smokers?

944. How many of the patients do not have a hypertension diagnosis and are nonsmokers?

945. What proportion of patients with a hypertension diagnosis are smokers?

946. What proportion of patients with a hypertension diagnosis are nonsmokers?

947. What proportion of nonsmokers has a hypertension diagnosis?

948. What proportion of nonsmokers does not have a hypertension diagnosis?

949. What proportion of all the patients in this study are smokers and have no hypertension diagnosis?

950. What proportion of all the patients in this study are nonsmokers and have no hypertension diagnosis?

Connecting Conditional Probabilities to Two-Way Tables

951–955 The following table displays information about cigarette smoking and diagnosis with hypertension for a group of patients at a medical clinic.

	Hypertension Diagnosis	No Hypertension Diagnosis
Smoker	48	24
Nonsmoker	26	50

Illustration by Ryan Sneed

The following stacked bar graph displays the smoking and hypertension data for the group of patients at the medical clinic.

Note: *This graph shows two bars, one for the group with a hypertension diagnosis, and one for the group with no hypertension diagnosis. The total percentages within each bar sum to 100%. Any percentage within a group represents a conditional probability for that group — that is, the percentage within that group with a certain characteristic. Use conditional probability terms and notation to answer these problems.*

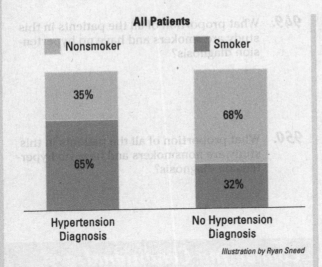

All Patients

Nonsmoker   Smoker

35%

68%

65%

32%

Hypertension
Diagnosis

No Hypertension
Diagnosis

Illustration by Ryan Sneed

951.   What does the area labeled 35% represent?

952.   What does the area labeled 65% represent?

953.   What does the area labeled 68% represent?

954.   What does the area labeled 32% represent?

955.   Based on this data, and understanding that you are working with only a single sample of data, which of the following statements appears to be true?

(A)   Patients with a hypertension diagnosis are more likely to be smokers than nonsmokers.

(B)   Patients with a hypertension diagnosis are less likely to be smokers than nonsmokers.

(C)   Patients without a hypertension diagnosis are more likely to be smokers than nonsmokers.

(D)   Patients without a hypertension diagnosis are more likely to be nonsmokers than smokers.

(E)   Choices (A) and (D)

Investigating Independent Variables

956–960 Solve the following problems about independent variables.

956.   If variables A and B are independent, which of the following must be true?

(A) $P(A) = P(B)$

(B) $P(A) \neq P(B)$

(C) $P(A)$ does not depend on whether or not B occurs.

(D) $P(A)$ depends on $P(B)$.

(E)   Choices (A) and (C)

957. You collect data on the choice of major among a random sample of students from a large university. You find that among engineering majors, 70% are male, and among English majors, 80% are female. Which of the following statements is true?

(A) Gender and choice of major are independent.

(B) Gender and choice of major are not independent.

(C) There are more males than females enrolled in the university.

(D) There are fewer males than females enrolled in the university.

(E) Impossible to say anything without further information

958. Suppose that in a population of high-school seniors, the choice to enroll in higher education after graduation is independent of gender. Which of the following statements would be true?

(A) The same number of males and females choose to enroll in higher education.

(B) The same proportion of males and females choose to enroll in higher education.

(C) More males enlist in the military, and more females go directly to full-time work.

(D) Choices (B) and (C)

(E) None of the above.

959. A small town has 300 male registered voters and 350 female registered voters. Overall, 60% of voters voted for a bond initiative. If voting is independent of gender in this sample, how many women voted for the bond initiative?

960. A primary school has 200 male students and 190 female students. Overall, 40% of students participate in after-school activities. If participation is independent of gender in this sample, how many boys did not participate in after-school activities?

Calculating Marginal Probability and More

961–970 The following table represents data from a survey on the type of diet and cholesterol level among a group of adults ages 50 to 75 years.

Note: *A vegetarian diet excludes meat, and a vegan diet excludes any animal-derived products in addition to meat. Regular diet in this table refers to those who are neither vegan nor vegetarian.*

	High Cholesterol	No High Cholesterol	
Vegetarian			100
Vegan			100
Regular Dieter			100
	100	200	300

Illustration by Ryan Sneed

961. What is the marginal probability of being a vegetarian?

962. What is the marginal probability of not being a vegan?

963. What is the marginal probability of having high cholesterol?

964. What is the marginal probability of not having high cholesterol?

965. If diet and cholesterol level are independent, which of the following would you expect to be true in this sample?

(A) The same percentage of vegetarians, vegans, and regular dieters will have high cholesterol.

(B) The percentage of vegans, vegetarians, and regular dieters with high cholesterol will differ.

(C) Among those with high cholesterol, equal numbers will be vegetarians, vegans, and regular dieters.

(D) Different numbers of people with high cholesterol will be vegetarians, vegans, and regular dieters.

(E) Choices (A) and (C)

966. If diet and cholesterol level are independent, how many adults in this data set would you expect to be vegetarians without high cholesterol?

967. If being a vegetarian and having high cholesterol are independent, how many adults in this data set would you expect to be vegetarians with high cholesterol?

968. Suppose that in this data, 10 vegetarians and 20 vegans have high cholesterol. How many regular dieters have high cholesterol?

969. Suppose that in this data, ten vegetarians have high cholesterol. How many vegetarians do not have high cholesterol?

970. Suppose that in this data, 35 regular dieters do not have high cholesterol. How many regular dieters do have high cholesterol?

Adding Joint Probability to the Mix

971–978 This table contains data from a survey that asked a sample of adults what type of phone they used most commonly. The adults were classified into three age categories: 18 to 40 years old, 41 to 65 years old, and 66 years old or older.

	Smartphone	Other Mobile Phone	Landline
Age 18–40	60	30	10
Age 41–65	40	40	20
Age 66 or older	20	30	50

Illustration by Ryan Sneed

971. How many people took part in this survey?

972. What is the marginal probability that an individual in the sample is 41 to 65 years old?

973. What is the marginal probability of a respondent's most commonly used type of phone not being a landline?

974. What is the joint probability of being between the ages of 18 and 40 and most commonly using a smartphone?

975. What is the joint probability of being age 66 or older and most commonly using a land-line phone?

976. Assume that the marginal frequencies are correct but that the cell entries of the table are unknown. If age and phone preference were independent, how many people ages 18 to 40 would prefer a smartphone?

977. Assume that the marginal frequencies are correct but that the cell entries of the table are unknown. If age and phone preference were independent, how many people age 66 or older would prefer a mobile phone?

978. Comparing the observed values in this table and the expected values under independence, what can you conclude about age and phone preference?

    (A) Age and phone preference are independent.

    (B) People ages 18 to 40 are less likely to prefer smartphones than would be expected if age and phone preference were independent.

    (C) People age 66 or older are less likely to prefer smartphones than would be expected if age and phone preference were independent.

    (D) People age 66 or older are more likely to prefer smartphones than would be expected if age and phone preference were independent.

    (E) Choices (B) and (D)

Digging Deeper into Conditional and Marginal Probabilities

979 Answer the following problem.

979. Suppose that you have a 2-x-2 table displaying values on gender (male or female) and laptop computer ownership (yes or no) for 100 male and 100 female college students. Overall, 75% of the students own laptops; 85% of the male students own laptops, as do 65% of the female students. Are gender and laptop ownership independent in this data set?

    (A) Yes, because the overall ownership rate is 75%.

    (B) Yes, because the conditional and marginal probabilities are equal.

    (C) No, because the ownership rates differ by gender.

    (D) No, because the marginal ownership rate differs from the conditional ownership rates.

    (E) Choices (C) and (D)

Figuring Out the Number of Cells in a Two-Way Table

980 Determine how many cells a two-way table will have in the following problem.

980. You are constructing a two-way table showing the responses to two questions in a survey: type of residence (private home, rented apartment, or condominium) and annual income category ($20,000 or less; $21,000 to $45,000; $46,000 to $75,000; and $76,000 or more). How many cells will this table have?

Including Conditional Probability

981–992 This bar chart displays frequency results from a survey conducted with a random sample of adults, asking their gender and whether they own a car. All four combinations are shown in the graph, similar to a two-way table.

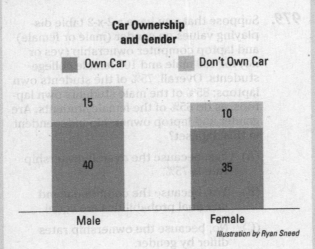

Car Ownership and Gender

Own Car   Don't Own Car

Male: 15, 40
Female: 10, 35

Illustration by Ryan Sneed

981. What was the total sample size for this survey?

982. If a person is male, what is the probability of him owning a car?

983. Given a person is female, what is the probability of her owning a car?

984. What is the marginal probability of car ownership?

985. What is the conditional probability of being male, given car ownership?

986. What is the conditional probability of being female, given car ownership?

987. Assuming that the marginal probability of car ownership among all adults also applies to the group of males, how many of the males in this sample would be car owners?

988. Assuming that the probability of car ownership among all adults also applies to the group of females, how many of the females in this sample would be car owners?

989. What is the marginal probability of being male?

990. What is the marginal probability of being female?

991. Which of the following statements is true when you compare the prevalence of car ownership between females and males?

(A) In this sample, more males than females own cars.

(B) In this sample, more females than males own cars.

(C) In this sample, the conditional probability of car ownership is higher for males.

(D) In this sample, the conditional probability of car ownership is higher for females.

(E) Choices (A) and (D)

992. Which of the following would bolster your ability to draw conclusions about the relationship between car ownership and gender?

(A) replication of the survey in other locations

(B) replication of the survey with a larger sample

(C) replicating the survey with a nationally representative sample

(D) Choices (A) and (B)

(E) Choices (A), (B), and (C)

Digging into Research Designs

993 Solve the following problem about research designs.

993. Which of the following research designs offers the best justification for making cause-and-effect statements from statistical results?

(A) a survey

(B) a randomized clinical trial

(C) an observational study

(D) Choices (A) and (B)

(E) Choices (A) and (C)

Digging Deeper into Two-Way Tables

994–1,001 This table displays results from a study evaluating the effectiveness of a new screening instrument for depression. All participants in the study were given the screening test and were also clinically evaluated for depression.

	Evaluated Positive	Evaluated Negative
Screened Positive	25	20
Screened Negative	10	60

Illustration by Ryan Sneed

994. Suppose that a third category, labeled "Inconclusive," was added to each of the two variables. How many cells would the table then have?

995. What is the marginal probability of screening positive for depression?

996. What is the marginal probability of evaluating positive for depression?

997. What is the conditional probability of evaluating negative for depression, given a positive screening result?

998. What is the conditional probability of evaluating positive for depression, given a negative screening result?

999. Of the four possible outcomes (screened positive and evaluated positive; screened positive and evaluated negative; screened negative and evaluated positive; screened negative and evaluated negative), which has the highest joint probability?

	Evaluated Positive	Evaluated Negative
Screened Positive	25	20
Screened Negative	10	90

1,000. Do the results from this sample suggest that screening positive for depression and being diagnosed for depression are independent? Why or why not?

1,001. Which of the following could strengthen your confidence in drawing causal conclusions from this study?

(A) The study sample was randomly selected from the population.

(B) The people doing the evaluation had no knowledge of the screening results.

(C) This study replicated an earlier study that produced similar results.

(D) Choices (A) and (B)

(E) Choices (A), (B), and (C)

Part II
The Answers

In this part . . .

This part provides fully worked-out solutions and discussions for each of the problems in Part I. Be sure to work through a problem as best as you can before peeking at the solutions — that way you'll be able to gauge which ideas you've already got under your belt and which ones you need to focus more time on.

As you're working through the practice problems and reading the answer explanations, you may decide that you could use a little brushing up in certain areas or a little more supporting material to give you an edge. No worries! The following books are easily available courtesy of your *For Dummies* friends (written by Deborah J. Rumsey, PhD, and published by Wiley):

- ✔ *Statistics For Dummies,* 2nd Edition
- ✔ *Statistics Workbook For Dummies*
- ✔ *Statistics Essentials For Dummies*

Visit www.dummies.com for more information.

Chapter 18

Answers

- -

1. all households in the city

A *population* is the entire group you're interested in studying. The goal here is to esti-mate what percent of *all households in a large city* have a single woman as the head of the household. The population is all households, and the variable is whether a single woman runs the household.

2. the 200 households selected

The *sample* is a subset drawn from the entire population you're interested in studying. So in this example, the subset is the 200 households selected out of all the households in the city.

3. the percent of households headed by single women in the city

A *parameter* is some characteristic of the population. Because studying a population directly isn't usually possible, parameters are usually estimated by using statistics (numbers calculated from sample data).

4. the percent of households headed by single women among the 200 selected households

The *statistic* is a number describing some characteristic that you calculate from your sample data; the statistic is used to estimate the parameter (the same characteristic in the population).

5. D.   a person's height, recorded in inches

Quantitative variables are measured and expressed numerically, have numeric mean-ing, and can be used in calculations. (That's why another name for them is numerical variables.) Although zip codes are written in numbers, the numbers are simply conve-nient labels and don't have numeric meaning (for example, you wouldn't add together two zip codes).

6.   E.   Choices (B) and (C) (college major; high-school graduate or not)

A categorical variable doesn't have numerical or quantitative meaning but simply describes a quality or characteristic of something. College major (such as English or mathematics) and high-school graduate (yes or no) both describe non-numerical qualities.

The numbers used in categorical or qualitative data designate a quality rather than a measurement or quantity. For example, you can assign the number 1 to a person who's married and the number 2 to a person who isn't married. The numbers themselves don't have meaning — that is, you wouldn't add the numbers together.

7.   E.   All of these choices are true.

Bias is systematic favoritism in the data. You want to get data that represents all customers at the store, no matter what day or what time they shop, whether they shop in couples or alone, and so on. You can't assume that the people who shopped during those three hours on that Saturday morning are representative of the store's total clientele. This sample wasn't drawn randomly — everyone who walked in was counted.

8.   E.   Choices (C) and (D) (the gender of each shopper who comes in during the time period; the number of men entering the store during the time period)

A variable is a characteristic or measurement on which data is collected and whose result can change from one individual to the next. That means gender is a variable, and the number of men entering the store is also a variable. The day you collect data and the store you observe are just part of the design of your study and were determined beforehand.

9.   gender; number of shoppers

Gender is a categorical variable (the categories are male or female), and *number of shoppers* is a quantitative variable (because it represents a count). The day you collect data and the store you observe are just part of the design of your study and were determined beforehand.

10.   D.   Choices (A) and (C) (a bar graph; a pie chart)

Gender is a categorical variable, so both bar graphs and pie charts are appropriate to display the proportion of males versus females among the shoppers. You could use a time plot only if you knew how many males and how many females were in the store at each individual time period.

11.   Add together the total shoppers from each observer and divide this total by 3.

The mean number of shoppers per hour is calculated by dividing the total number of shoppers (found by adding together the total from each observer) and dividing by the number of hours (3).

12.   E.   Choices (A) and (C) (3, 3, 3, 3, 3; 1, 2, 3, 4, 5)

To find the median, put the data in order from lowest to highest, and find the value in the middle. It doesn't matter how many times a number is repeated. In this case, the data sets 3, 3, 3, 3, 3 and 1, 2, 3, 4, 5 each have a median of 3.

13.   It means that 90% of students who took the exam had scores less than or equal to Susan's.

A *percentile* shows the relative standing of a score in a population by identifying the percent of values below that score. Susan scored in the 90th percentile, so 90% of the students' scores are less than or equal to Susan's.

14.   E.   all of the above

All of the choices are correct. A distribution is basically a list of all possible values of the variable and how often they occur. In a sentence, you can say, "60% of the 100 people surveyed said they like chocolate, and 40% said they don't" — this sentence gives the distribution. You can also make a table with rows labeled "Like chocolate" and "Don't like chocolate" and show the percents, or you can use a pie chart or a bar graph to visually describe the distribution.

15.   Your score is 0.70 standard deviations above the mean.

A *z*-score tells you how many standard deviations a data value is below or above the mean. If your *z*-score is 0.70, your exam score is 0.70 standard deviations above the mean. It doesn't tell you your actual score or how many students scored above or below you, but it does tell you where a data value stands, compared to the *average* exam score.

16.   It means that it's likely that between 59% and 71% of all Americans approve of the president.

The *margin of error* tells you how much your sample results are likely to change from sample to sample. It's measured as "plus or minus a certain amount." In this case, the margin of error of 6% tells you that the result from this sample (65% approving of the president) could change by as much as 6% on either side. Therefore, in using the sample results to draw conclusions about the whole population, the best you can say is, "Based on the data, the percentage of all Americans who approve of the president is likely between 59% (65% − 6%) and 71% (65% + 6%)."

17.   a confidence interval

You use a confidence interval when you want to estimate a population parameter (a number describing the population) when you have no prior information about it. In a confidence interval, you take a sample, calculate a statistic, and add/subtract a margin of error to come up with your estimate.

18. a hypothesis test

You use a hypothesis test when someone reports or claims that a population parameter (such as the population mean) is equal to a certain value and you want to challenge that claim. Here, the claim is that the percentage of all high-school graduates who participated in sports is equal to 60%. You think it's higher than that, so you're challenging that claim.

19. E. $p = 0.001$

A p-value measures how strong your evidence is against the other person's reported value. A small p-value means your evidence is strong against them; a large p-value says your evidence is weak against them. In this case, the smallest p-value is 0.001, which in any statistician's book is deemed highly significant, meaning your data and test results show strong evidence against the report.

20. C. 1, 1, 4, 4

The standard deviation measures how much variability (diversity) is in the data set, compared to the mean. If all the data values are the same, the standard deviation is 0. To increase the standard deviation, move the values farther and farther away from the mean. The choice that moves them the farthest from the mean here is 1, 1, 4, 4.

21. 25.7

Use the formula for calculating the mean

$$\bar{x} = \frac{\sum x}{n}$$

where $\bar{x}$ is the mean, $\sum$ represents the sum of the data values, and n is the number of values in the data set.

In this case, $x = 14 + 14 + 15 + 16 + 28 + 28 + 32 + 35 + 37 + 38 = 257$, and $n = 10$. So the mean is

$$\frac{257}{10} = 25.7$$

22. 52.4

Use the formula for calculating the mean

$$\bar{x} = \frac{\sum x}{n}$$

where $\bar{x}$ is the mean, $\sum$ represents the sum of the data values, and n is the number of values in the data set.

In this case, $x = 15 + 25 + 35 + 45 + 50 + 60 + 70 + 72 + 100 = 472$, and $n = 9$. So the mean is

$$\frac{472}{9} = 52.4444$$

The question asks for the nearest tenth, so you round to 52.4.

23.   7.5

Use the formula for calculating the mean

$$\bar{x} = \frac{\sum x}{n}$$

where $\bar{x}$ is the mean, $\sum$ represents the sum of the data values, and n is the number of values in the data set.

In this case, $x = 0.8 + 1.8 + 2.3 + 4.5 + 4.8 + 16.1 + 22.3 = 52.6$, and $n = 7$. So the mean is

$$\frac{52.6}{7} = 7.5143$$

The question asks for the nearest tenth, so you round to 7.5.

24.   4.525

Use the formula for calculating the mean

$$\bar{x} = \frac{\sum x}{n}$$

where $\bar{x}$ is the mean, $\sum$ represents the sum of the data values, and n is the number of values in the data set.

In this case, $x = 0.003 + 0.045 + 0.58 + 0.687 + 1.25 + 10.38 + 11.252 + 12.001 = 36.198$, and $n = 8$. So the mean is

$$\frac{36.198}{8} = 4.52475$$

The questions asks for the nearest thousandth, so you round to 4.525.

25.   15.0

To find the median, put the numbers in order from smallest to largest:

4, 5, 6, 12, 15, 16, 18, 20, 22

Because this data set has an odd number of values (nine), the median is simply the middle number in the data set: 15.

26.   17.0

To find the median, put the numbers in order from smallest to largest:

10, 12, 15.5, 16, 17, 17, 17, 18, 18, 21, 21

Because this data set has an odd number of values (11), the median is simply the middle number in the data set: 17.

27. 9.0

To find the median, put the numbers in order from smallest to largest:

1, 2, 6, 7, 8, 10, 14, 15, 21, 30

Because this data set has an even number of values (ten), the median is the average of the two middle numbers:

$$\frac{8+10}{2} = 9.0$$

28. 4.04

To find the median, put the numbers in order from smallest to largest.

0.001, 0.1, 0.25, 1.22, 6.85, 8.2, 13.2, 25.2

Because this data set has an even number of values (eight), the median is the average of the two middle numbers:

$$\frac{1.22+6.85}{2} = 4.035$$

The question asks for the nearest hundredth, so round to 4.04.

29. The mean will have a higher value than the median.

A data set distribution that is skewed right is asymmetrical and has a large number of values at the lower end and few numbers at the high end. In this case, the median, which is the middle number when you sort the data from smallest to largest, lies in the lower range of values (where most of the numbers are). However, because the mean finds the average of all the values, both high and low, the few outlying data points on the high end cause the mean to increase, making it higher than the median.

30. The mean will have a lower value than the median.

A data set distribution that is skewed left is asymmetrical and has a large number of values at the high end and few numbers at the low end. In this case, the median, which is the middle number when you sort the data from smallest to largest, lies in the upper range of values (where most of the numbers are). However, because the mean finds the average of all the values, both high and low, the few outlying data points on the low end cause the mean to decrease, making it lower than the median.

31. The mean and median will be fairly close together.

When a data set has a symmetrical distribution, the mean and the median are close together because the middle value in the data set, when ordered smallest to largest, resembles the balancing point in the data, which occurs at the average.

32. median

The median is the middle value of the data points when ordered from smallest to largest. When the data is ordered, it no longer takes into account the values of any of the other data points. This makes it resistant to being influenced by outliers. (In other words, outliers don't really affect the median.) In contrast, the mean takes every specific data value into account. If the data points contain some outliers that are extreme values to one side, the mean will be pulled toward those outliers.

33. how concentrated the data is around the mean

A standard deviation measures the amount of variability among the numbers in a data set. It calculates the typical distance of a data point from the mean of the data. If the standard deviation is relatively large, it means the data is quite spread out away from the mean. If the standard deviation is relatively small, it means the data is concentrated near the mean.

34. approximately 68%

According to the empirical rule, the bell-shaped curve of a normal distribution will have 68% of the data points within one standard deviation of the mean.

35. E. Choice (A) or (C) (standard deviation or variance)

The standard deviation is a way of measuring the typical distance that data is from the mean and is in the same units as the original data. The variance is a way of measuring the typical *squared* distance from the mean and isn't in the same units as the original data. Both the standard deviation and variance measure variation in the data, but the standard deviation is easier to interpret.

36. E. Choices (A) and (C) (margin of error; standard deviation)

The standard deviation measures the typical distance from the data to the mean (using all the data to calculate). Outliers are far from the mean, so the more outliers there are, the higher the standard deviation will be. You calculate the margin of error by using the sample standard deviation so it's also sensitive to outliers. The interquartile range is the range of the middle 50% of the data, so outliers won't be included, making it less sensitive to outliers than the standard deviation or margin of error.

37. 5 years

The formula for the sample standard deviation of a data set is

$$s = \sqrt{\frac{\sum (x - \bar{x})^2}{n - 1}}$$

where x is a single value, $\bar{x}$ is the mean of all the values, $\sum (x - \bar{x})^2$ represents the sum of the squared differences from the mean, and n is the sample size.

First, find the mean of the data set by adding together the data points and then dividing by the sample size (in this case, $n = 10$):

$$\bar{x} = \frac{0+1+2+4+8+3+10+17+2+7}{10}$$

$$= \frac{54}{10} = 5.4$$

Then, subtract the mean from each number in the data set and square the differences, $(x - \bar{x})^2$:

$$(0 - 5.4)^2 = (-5.4)^2 = 29.16$$

$$(1 - 5.4)^2 = (-4.4)^2 = 19.36$$

$$(2 - 5.4)^2 = (-3.4)^2 = 11.56$$

$$(4 - 5.4)^2 = (-1.4)^2 = 1.96$$

$$(8 - 5.4)^2 = (2.6)^2 = 6.76$$

$$(3 - 5.4)^2 = (-2.4)^2 = 5.76$$

$$(10 - 5.4)^2 = (4.6)^2 = 21.16$$

$$(17 - 5.4)^2 = (11.6)^2 = 134.56$$

$$(2 - 5.4)^2 = (-3.4)^2 = 11.56$$

$$(7 - 5.4)^2 = (1.6)^2 = 2.56$$

Next, add up the results from the squared differences:

$$29.16 + 19.36 + 11.56 + 1.96 + 6.76 + 5.76 + 21.16 + 134.56 + 11.56 + 2.56 = 244.4$$

Finally, plug the numbers into the formula for the sample standard deviation:

$$s = \sqrt{\frac{\sum (x - \bar{x})^2}{n - 1}}$$

$$= \sqrt{\frac{244.4}{10 - 1}}$$

$$= \sqrt{27.156}$$

$$= 5.21$$

The question asks for the nearest year, so round to 5 years.

38. 8 years

The formula for the sample standard deviation of a data set is

$$s = \sqrt{\frac{\sum (x - \bar{x})^2}{n - 1}}$$

where x is a single value, $\bar{x}$ is the mean of all the values, $\sum (x - \bar{x})^2$ represents the sum of the squared differences from the mean, and n is the sample size.

First, find the mean of the data set by adding together the data points and then dividing by the sample size (in this case, $n = 12$):

$$\bar{x} = \frac{12+10+16+22+24+18+30+32+19+20+35+26}{12}$$

$$= \frac{264}{12} = 22$$

Then, subtract the mean from each number in the data set and square the differences, $(x - \bar{x})^2$:

$$(12 - 22)^2 = (-10)^2 = 100$$
$$(10 - 22)^2 = (-12)^2 = 144$$
$$(16 - 22)^2 = (-6)^2 = 36$$
$$(22 - 22)^2 = (0)^2 = 0$$
$$(24 - 22)^2 = (2)^2 = 4$$
$$(18 - 22)^2 = (4)^2 = 16$$
$$(30 - 22)^2 = (8)^2 = 64$$
$$(32 - 22)^2 = (10)^2 = 100$$
$$(19 - 22)^2 = (-3)^2 = 9$$
$$(20 - 22)^2 = (-2)^2 = 4$$
$$(35 - 22)^2 = (13)^2 = 169$$
$$(26 - 22)^2 = (4)^2 = 16$$

Next, add up the results from the squared differences:

$$100 + 144 + 36 + 0 + 4 + 16 + 64 + 100 + 9 + 4 + 169 + 16 = 662$$

Finally, plug the numbers into the formula for the sample standard deviation:

$$s = \sqrt{\frac{\sum (x - \bar{x})^2}{n-1}}$$

$$= \sqrt{\frac{662}{12-1}}$$

$$= \sqrt{60.1818}$$

$$= 7.76$$

The question asks for the nearest year, so round to 8 years.

39.   8.7 points

The formula for the sample standard deviation of a data set is

$$s = \sqrt{\frac{\sum (x - \bar{x})^2}{n-1}}$$

where x is a single value, $\bar{x}$ is the mean of all the values, $\sum (x - \bar{x})^2$ represents the sum of the squared differences from the mean, and n is the sample size.

First, find the mean of the data set. Although you don't have a list of all the individual values, you do know the test score for each student in the sample. For example, you know that three students scored 92 points, so if you listed every student's score individually, you'd see 92 three times, or (92)(3). To find the mean this way, multiply each exam score by the number of students who received that score, add the products together, and then divide by the number of students in the sample ($n = 20$):

$$(98)(2) = 196$$
$$(95)(1) = 95$$
$$(92)(3) = 276$$
$$(88)(4) = 352$$
$$(87)(2) = 174$$
$$(85)(2) = 170$$
$$(81)(1) = 81$$
$$(78)(2) = 156$$
$$(73)(1) = 73$$
$$(72)(1) = 72$$
$$(65)(1) = 65$$

$$\bar{x} = \frac{196 + 95 + 276 + 352 + 174 + 170 + 81 + 156 + 73 + 72 + 65}{20}$$

$$= \frac{1,710}{20} = 85.5$$

Next, subtract the mean from each different exam score in the data set and square the differences, $(x - \bar{x})^2$. **Note:** There are 11 different exam scores here — 98, 95, 92, 88, 87, 85, 81, 78, 73, 72, and 65 — but 20 students. First, work with the 11 exam scores.

$$(98 - 85.5)^2 = (12.5)^2 = 156.25$$
$$(95 - 85.5)^2 = (9.5)^2 = 90.25$$
$$(92 - 85.5)^2 = (6.5)^2 = 42.25$$
$$(88 - 85.5)^2 = (2.5)^2 = 6.25$$
$$(87 - 85.5)^2 = (1.5)^2 = 2.25$$
$$(85 - 85.5)^2 = (-0.5)^2 = 0.25$$
$$(81 - 85.5)^2 = (-4.5)^2 = 20.25$$
$$(78 - 85.5)^2 = (-7.5)^2 = 56.25$$
$$(73 - 85.5)^2 = (-12.5)^2 = 156.25$$
$$(72 - 85.5)^2 = (-13.5)^2 = 182.25$$
$$(65 - 85.5)^2 = (-20.5)^2 = 420.25$$

Now, multiply each value by the number of students who got that score:

$$(156.25)(2) = 312.5$$
$$(90.25)(1) = 90.25$$
$$(42.25)(3) = 126.75$$

$$(6.25)(4) = 25$$

$$(2.25)(2) = 4.5$$

$$(0.25)(2) = 0.5$$

$$(20.25)(1) = 20.25$$

$$(56.25)(2) = 112.5$$

$$(156.25)(1) = 156.25$$

$$(182.25)(1) = 182.25$$

$$(420.25)(1) = 420.25$$

Then, add up those results:

$$312.5 + 90.25 + 126.75 + 25 + 4.5 + 0.5 + 20.25 + 112.5 + 156.25 + 182.25 + 420.25 = 1,451$$

Finally, plug the numbers into the formula for the sample standard deviation:

$$s = \sqrt{\frac{\sum (x - \bar{x})^2}{n-1}}$$

$$= \sqrt{\frac{1,451}{20-1}}$$

$$= \sqrt{76.37}$$

$$= 8.74$$

The question asks for the nearest tenth of a point, so round to 8.7.

---

40.   0.0036 cm

The formula for the sample standard deviation of a data set is

$$s = \sqrt{\frac{\sum (x - \bar{x})^2}{n-1}}$$

where x is a single value, $\bar{x}$ is the mean of all the values, $\sum (x - \bar{x})^2$ represents the sum of the squared differences from the mean, and n is the sample size.

First, find the mean of the data set by adding together the data points and then dividing by the sample size (in this case, $n = 10$):

$$\bar{x} = \frac{5.001 + 5.002 + 5.005 + 5.010 + 5.009 + 5.003 + 5.002 + 5.001 + 5.000}{10}$$

$$= \frac{50.033}{10} = 5.0033$$

Then, subtract the mean from each number in the data set and square the differences, $(x - \bar{x})^2$:

$$(5.001 - 5.0033)^2 = (-0.0023)^2 = 0.00000529$$

$$(5.002 - 5.0033)^2 = (-0.0013)^2 = 0.00000169$$

$$(5.005 - 5.0033)^2 = (0.0017)^2 = 0.00000289$$

$$(5.000 - 5.0033)^2 = (-0.0033)^2 = 0.00001089$$

$$(5.010 - 5.0033)^2 = (0.0067)^2 = 0.00004489$$

$$(5.009 - 5.0033)^2 = (0.0057)^2 = 0.00003249$$

$$(5.003 - 5.0033)^2 = (-0.0003)^2 = 0.00000009$$

$$(5.002 - 5.0033)^2 = (-0.0013)^2 = 0.00000169$$

$$(5.001 - 5.0033)^2 = (-0.0023)^2 = 0.00000529$$

$$(5.000 - 5.0033)^2 = (-0.0033)^2 = 0.00001089$$

Next, add up the results from the squared differences:

$$0.00000529 + 0.00000169 + 0.00000289 + 0.00001089 + 0.00004489 + 0.00003249 + 0.00000009 + 0.00000169 + 0.00000529 + 0.00001089 = 0.0001161$$

Finally, plug the numbers into the formula for the sample standard deviation:

$$s = \sqrt{\frac{\sum (x - \bar{x})^2}{n - 1}}$$

$$= \sqrt{\frac{0.0001161}{10 - 1}}$$

$$= \sqrt{0.0000129}$$

$$= 0.0036$$

The sample standard deviation for the jet engine turbine part is 0.0036 centimeters.

41. There is more variation in salaries in Magna Company than in Ace Corp.

The larger standard deviation in Magna Company shows a greater variation of salaries in both directions from the mean than Ace Corp. The standard deviation measures on average how spread out the data is (for example, the high and low salaries at each company).

42. B. measuring the variation in circuitry components when manufacturing computer chips

The quality of the vast majority of manufacturing processes depends on reducing variation to as little as possible. If a manufacturing process has a large standard deviation, it indicates a lack of predictability in the quality and usefulness of the end product.

43. Lake Town has a lower average temperature and less variability in temperatures than Sunshine City.

Lake Town has a much smaller standard deviation than Sunshine City, so its temperatures change (or vary) less. You don't know the actual range of temperatures for either city.

44. There will be no change in the standard deviation.

All the data points will shift up $2,000, and as a result, the mean will also increase by $2,000. But each individual salary's distance (or deviation) from the mean will be the same, so the standard deviation will stay the same.

45.

The sample variance is 2.3 ounces². The standard deviation is 1.5 ounces.

You find the sample variance with the following formula:

$$s^2 = \frac{\sum (x - \bar{x})^2}{n-1}$$

where x is a single value, $\bar{x}$ is the mean of all the values, $\sum (x - \bar{x})^2$ represents the sum of the squared difference scores, and n is the sample size.

First, find the mean by adding together the data points and dividing by the sample size (in this case, $n = 5$):

$$\bar{x} = \frac{7+6+5+6+9}{5}$$

$$= \frac{33}{5} = 6.6$$

Then, subtract the mean from each data point and square the differences, $(x - \bar{x})^2$:

$(7 - 6.6)^2 = (0.4)^2 = 0.16$

$(6 - 6.6)^2 = (-0.6)^2 = 0.36$

$(5 - 6.6)^2 = (-1.6)^2 = 2.56$

$(6 - 6.6)^2 = (-0.6)^2 = 0.36$

$(9 - 6.6)^2 = (2.4)^2 = 5.76$

Next, plug the numbers into the formula for the sample variance:

$$s^2 = \frac{\sum (x - \bar{x})^2}{n-1}$$

$$= \frac{0.16 + 0.36 + 2.56 + 0.36 + 5.76}{5-1}$$

$$= \frac{9.2}{4} = 2.3$$

The sample variance is 2.3 ounces². But these units don't make sense because there's no such thing as "square ounces." However, the standard deviation is the square root of the variance, so it can then be expressed in the original units: $s = 1.5$ ounces (rounded). For this reason, standard deviation is preferred over the variance when it comes to measuring and interpreting variability in a data set.

46.

The sample variance is 15 minutes². The standard deviation is 4 minutes.

You find the sample variance with the following formula:

$$s^2 = \frac{\sum (x - \bar{x})^2}{n-1}$$

where x is a single value, $\bar{x}$ is the mean of all the values, $\sum (x - \bar{x})^2$ represents the sum of the squared difference scores, and n is the sample size.

First, find the mean by adding together the data points and dividing by the sample size (in this case, $n = 5$):

$$\bar{x} = \frac{15+16+18+10+9}{5}$$

$$= \frac{68}{5} = 13.6$$

Then, subtract the mean from each data point and square the differences, $(x-\bar{x})^2$:

$$(15-13.6)^2 = (1.4)^2 = 1.96$$

$$(16-13.6)^2 = (2.4)^2 = 5.76$$

$$(18-13.6)^2 = (4.4)^2 = 19.36$$

$$(10-13.6)^2 = (-3.6)^2 = 12.96$$

$$(9-13.6)^2 = (-4.6)^2 = 21.16$$

Next, plug the numbers into the formula for the sample variance:

$$s^2 = \frac{\sum (x-\bar{x})^2}{n-1}$$

$$= \frac{1.96+5.76+19.36+12.96+21.16}{5-1}$$

$$= \frac{61.2}{4} = 15.3$$

The sample variance is 15.3 minutes². But these units don't make sense because there's no such thing as "square minutes." However, the standard deviation is the square root of the variance, so it can then be expressed in the original units: $s = 3.91$ minutes (rounded up to 4). For this reason, standard deviation is preferred over the variance when it comes to measuring and interpreting variability in a data set.

47. The standard deviation is 13 kilometers/hour.

You find the standard deviation with the following formula:

$$s = \sqrt{\frac{\sum (x-\bar{x})^2}{n-1}}$$

where x is a single value, $\bar{x}$ is the mean of all the values, $\sum (x-\bar{x})^2$ represents the sum of the squared difference scores, and n is the sample size.

First, find the mean by adding together the data points and dividing by the sample size (in this case, $n = 5$):

$$\bar{x} = \frac{10+15+35+40+30}{5}$$

$$= \frac{130}{5} = 26$$

Then, subtract the mean from each data point and square the differences, $(x-\bar{x})^2$:

$$(10-26)^2 = (-16)^2 = 256$$

$$(15-26)^2 = (-11)^2 = 121$$

$$(35-26)^2 = (9)^2 = 81$$

$$(40-26) = (14)^2 = 196$$

$$(30-26) = (4)^2 = 16$$

Next, plug the numbers into the formula for the standard deviation:

$$s = \sqrt{\frac{\sum (x - \bar{x})^2}{n-1}}$$

$$= \sqrt{\frac{256 + 121 + 81 + 196 + 16}{5 - 1}}$$

$$= \sqrt{\frac{670}{4}} = 12.942$$

Rounded to a whole number, the standard deviation is 13 kilometers/hour.

---

48.   D.   Choices (A) and (B) (Data Set 1; Data Set 2)

The original data set contains the numbers 1, 2, 3, 4, 5. Data Set 1 just shifts those numbers up by five units to get 6, 7, 8, 9, 10. Standard deviation represents typical (or average) distance from the mean, and although the mean in Data Set 1 changes from 3 to 8, the distances from each point to that new mean stay the same as they were for the original data set, so the average distance from the mean is the same.

Data Set 2 contains the numbers –2, –1, 0, 1, 2. These numbers shift the original data set's values down by three units. For example, 1 – 3 = –2, 2 – 3 = –1, and so forth. Therefore, the standard deviation doesn't change from the original data set.

Data Set 3 divides all the numbers in the original data set by 10, making them closer to the mean, on average, than the original data set. Therefore, the standard deviation is smaller.

---

49.   approximately 68%

The empirical rule states that in a normal (bell-shaped) distribution, approximately 68% of values are within one standard deviation of the mean.

---

50.   about 95%

The empirical rule states that in a normal (bell-shaped) distribution, approximately 95% of values are within two standard deviations of the mean.

---

51.   about 57 to 71 years

The empirical rule states that in a normal distribution, 95% of values are within two standard deviations of the mean. "Within two standard deviations" means two standard deviations below the mean *and* two standard deviations above the mean.

In this case, the mean is 64 years, and the standard deviation is 3.5 years. So two standard deviations is (3.5)(2) = 7 years.

To find the lower end of the range, subtract two standard deviations from the mean: 64 – 7 years = 57 years. And then to find the upper end of the range, add two standard deviations to the mean: 64 + 7 years = 71 years.

So about 95% of people who retire do so between the ages of about 57 to 71 years.

52. about 16%

The empirical rule states that approximately 68% of values are within one standard deviation of the mean. "Within one standard deviation" means one standard deviation below the mean *and* one standard deviation above the mean.

In this case, the mean is $48,000, so about 50% of the engineers made less than $48,000. In a normal distribution, half of the values are above the mean and half are below the mean. The standard deviation is $7,000.

To find the lower end of the range within one standard deviation of the mean, subtract the standard deviation from the mean: $48,000 – $7,000 = $41,000; to find the upper end of the range, add the standard deviation to the mean: $48,000 + $7,000 = $55,000.

Because the normal distribution is symmetrical, 34% of the values will be between $41,000 and $48,000, and 34% will be between $48,000 and $55,000.

Therefore, 50% + 34% = 84% of the data is $55,000 and below, which leaves 16% of the data above $55,000.

53. if a population has a normal distribution

You can use the empirical rule only if the distribution of the population is normal. Note that the rule says that *if* the distribution is normal, *then* approximately 68% of the values lie within one standard deviation of the mean, not the other way around. Many distributions have 68% of the values within one standard deviation of the mean that don't look like a normal distribution.

54. the mean and the standard deviation of the population

The empirical rule describes the distribution of the data in a population in terms of the mean and the standard deviation. For example, the first part of the empirical rule says that about 68% of the values lie within one standard deviation of the mean, so all you need to know is the mean and the standard deviation to use the rule.

55. 30

If you assume that the 600 lenses tested come from a population with a normal distribution (which they do), you can apply the empirical rule (also known as the 68-95-99.7 rule).

Using the empirical rule, approximately 95% of the data lies within two standard deviations of the mean, and 5% of the data lies outside this range. Because the lenses that are more than two standard deviations from the mean are rejected, you can expect about 5% of the 600 lenses, or (0.05)(600) = 30 lenses to be rejected.

56. cannot be determined with the information given

You could use the empirical rule (also known as the 68-95-99.7 rule) if the shape of the distribution of fish lengths was normal; however, this distribution is said to be "very much skewed left," so you can't use this rule. With the information given, you can't answer the question.

57. percentile

A percentile splits the data into two parts, the percentage below the value and the percentage above the value. In other words, a percentile measures where an individual data value stands compared to the rest of the data values. For example, the 90th percentile is the value where 90% of the values lie below it and 10% of the values lie above it.

58. median

The 50th percentile is the value where 50% of the data fall below it and 50% fall above it. This is the same as the definition of the median.

59. It means that 15% of the scores were better than your score.

Percentile is the relative standing in a set of data from the lowest values to highest values. If your score is in the 85th percentile, it means that 85% of the scores are below your score and 15% are above your score.

60. Bob's score is above 70.

A score in the 90th percentile means that 90% of the scores were lower. With 60 scores and an approximately normal distribution, the 90th percentile will certainly be above the mean, here given as 70.

61. It means that 30% of the students scored above you and that you correctly answered 82% of the test questions.

The 70th percentile means that 70% of the scores were below your score, and 30% were above your score. Your actual score was 82%, which means that you answered 82% of the test questions correctly.

62. 65%

The 50th percentile doesn't mean a score of 50%; it's the median (or middle number) of the data set. The middle number is 65%, so that is the 50th percentile.

63. 95

A percentage score is different than a percentile. In this case, a percentage score is the percent of questions answered correctly; a percentile expresses the relative standing of a score in terms of the other scores.

Use the procedure for calculating a percentile. To calculate the kth percentile (where k is any number between 0 and 100), follow these steps:

1. Put all the numbers in the data set in order from smallest to largest:

   80, 80, 82, 84, 85, 86, 88, 90, 91, 92, 92, 94, 96, 98, 100

2. Multiply k percent by the total number of values, n:

   $(0.80)(15) = 12$

3. Because your result in Step 2 is a whole number, count the numbers in the data set from left to right until you reach the number you found in Step 2 (in this case, the 12th number). The kth percentile is the average of that corresponding value in the data set and the value that directly follows it.

Find the average of the 12th and 13th numbers in the data set:

$$\frac{94+96}{2} = \frac{190}{2} = 95$$

The 80th percentile is 95.

64. Bill, Mary, Jose, Lisa, then Paul

If your score is at the kth percentile, that means k percent of the students scored less than you did, and the rest scored better than you did. For example, someone scoring at the 95th percentile knows that 95% of the other students scored lower than he did, and 5% scored higher.

When talking about exam scores, the person that scores at the highest percentile scored better than everyone else on the list. So in ranking from highest to lowest in terms of where their grades stand compared to each other, you have Bill first, followed by Mary, then Jose, then Lisa, and then Paul.

65. B. the median

The median of a data set is the middle value after you've put the data in order from smallest to largest (or the average of the two middle values if your data set contains an even number of values). Because the median concerns only the very middle of the data set, adding an outlier won't affect its value much (if any). It adds only one more value to one end or the other of the sorted data set.

The mean is based on the sum of all the data values, which includes the outlier, so the mean will be affected by adding an outlier. The standard deviation involves the mean in its calculation; hence it's also affected by outliers. The range is perhaps the most affected by an outlier, because it's the distance between the minimum and maximum values, so adding an outlier makes either the minimum value smaller or the maximum value larger. Either way, the distance between the minimum and maximum increases.

66. E. None of the above.

It's strange but true that all the scenarios are possible. You can use one data set as an example where all four scenarios occur at the same time: 5, 5, 5, 5, 5, 5, 5. In this case, the minimum and maximum are both 5, and the median (middle value) is 5. The median cuts the data set in half, creating an upper half and a lower half of the data set. To find the 1st quartile, take the median of the lower half of the data set, which gives you 5 in this case; to find the 3rd quartile, take the median of the upper half of the data set (also 5). The range is the distance from the minimum to the maximum, which is $5 - 5 = 0$. The IQR is the distance from the 1st to the 3rd quartile, which is $5 - 5 = 0$. Hence, the range and IQR are the same.

67.

1st quartile = 79, median = 90, 3rd quartile = 96

The 1st quartile is the 25th percentile, the median is the 50th percentile, and the 3rd quartile is the 75th percentile.

To find the values for these numbers, use the procedure for calculating a percentile. To calculate the kth percentile (where k is any number between 1 and 100), follow these steps:

1. Put all the numbers in the data set in order from smallest to largest:

   72, 74, 75, 77, 79, 82, 83, 87, 88, 90, 91, 91, 91, 92, 96, 97, 97, 98, 100

2. Multiply k percent times the total number of values, n.

   For the 1st quartile (or 25%): (0.25)(19) = 4.75.

   For the median (or 50%): (0.50)(19) = 9.5.

   For the 3rd quartile (or 75%): (0.75)(19) = 14.25.

3. Because the results in Step 2 aren't whole numbers, round up to the nearest whole number, and then count the numbers in the data set from left to right (from the smallest to the largest number) until you reach the value of the rounded number. The corresponding value in the data set is the kth percentile.

   For the 1st quartile, round 4.75 up to 5 and then find the fifth number in the data set: 79.

   For the median, round 9.5 up to 10 and then find the tenth number in the data set: 90.

   For the 3rd quartile, round 14.25 up to 15 and then find the 15th number in the data set: 96.

68.

The five-number summary can't be found.

The five-number summary of a data set includes the minimum value, the 1st quartile, the median, the 3rd quartile, and the maximum value. You're not given the minimum value or the maximum value here, so you can't fill out the five-number summary.

Note that even though you're given the range, which is the distance between the maximum and minimum values, you can't determine the actual values of the minimum and maximum.

69.

The median can't be greater in value than the 3rd quartile.

The five numbers in the five-number summary are the minimum (smallest) number in the data set, the 25th percentile (also known as the 1st quartile, or Q_1), the median (50th percentile), the 75th percentile (also known as the 3rd quartile, or Q_3), and the maximum (largest) number in the data set.

Because the median is at the 50th percentile, its value must be between the value of the 1st and 3rd quartiles, inclusive. In this case, the 1st quartile is 50, and the 3rd quartile is 80, so a median of 85 isn't possible.

70.

B. 15, 15, 15

Many data sets containing three numbers can have a mean of 15. However, if you force the standard deviation to be 0, you have only one choice: 15, 15, 15. A standard deviation of 0 means the average distance from the data values to the mean is 0. In other words, the data values don't deviate from the mean at all, and hence they have to be the same value.

71.

The median can't be less than Q_1.

Q_1 represents the 25th percentile, indicating that 25% of the salaries lie below that value. The median represents the 50th percentile, indicating that 50% of the salaries lie below that value. That means that the median has to be at least as large as Q_1.

Regarding the variance, the fact that it's large isn't alarming; variance represents the approximate average squared distance from the data values to the mean and isn't related to the other statistics directly.

The median doesn't have to be halfway between Q_1 and Q_3; it does have to lie somewhere between them (inclusive), but it can lie anywhere in that range.

The mean and median don't have to be equal, and the other descriptive statistics shown don't require a certain relationship between them.

72.

E. All of the above.

A data set is divided into four parts, each containing 25% of the data: (1) the minimum value to the 1st quartile, (2) the 1st quartile to the median, (3) the median to the 3rd quartile, and (4) the 3rd quartile to the maximum value. Each statement represents a distance that covers two adjacent parts out of the four, which gives a total percentage of 25%(2) = 50% in every case.

73.

E. None of the above.

The mean and median can have any type of relationship depending on the data set. The range, being the distance between the minimum and maximum values in the data set, must be greater than or equal to the IQR, which is the distance between the 1st and 3rd quartiles — that is, the range of the middle 50% of the data. The variance is the square of the standard deviation and is larger than the standard deviation only if the standard deviation is greater than 1. (If you pick a standard deviation less than 1, say 0.50, then the variance equals 0.50 squared, which is 0.25, a smaller value.)

74.

the variance of the weights

You calculate the variance by taking the distances from the mean and squaring each of them and then taking (an approximate) average. During the squaring process, the units become squared also. In this case, the units for the variance are square pounds, which doesn't make sense, so you typically don't report units for variance.

The standard deviation is the square root of the variance, which takes you from square units (square pounds) back to the original units (pounds).

The other statistics, such as the mean, median, and range, all stay in the original units because nothing is being done during their calculations to affect the units of the data.

75. none of the above

Measuring spread (variability) in a data set involves measuring distances of various types. The range is the distance from the minimum value to the maximum value, the IQR is the distance from the 1st quartile to the 3rd quartile, the standard deviation is the (approximate) average distance from the data points to the mean, and the variance is the (approximate) average squared distance from the data points to the mean. Different measures of spread are more appropriate than others in different situations, but they all involve something to do with distances.

76. C. 2, 2, 3, 4, 4

To begin, the standard deviation must be 1, so the (approximate) average from the data values to the mean must be 1. The data set 2, 2, 3, 4, 4 meets both criteria — it has a mean of 3, and its standard deviation (average distance from the data values to the mean) equals 1. (You find the standard deviation by subtracting the mean from each value in the set, squaring the results, adding them together, dividing by $n - 1$, where n is the number of values, and then taking the square root.)

77. IQR and median

The distribution is skewed right, meaning that there are a large number of low values concentrated over a relatively small range, and a long tail to the right, indicating a large number of high values spread over a wide range. When a distribution isn't approximately normal, the spread and center are best measured with the IQR and median, respectively. (The mean and standard deviation are sensitive to the outliers in the lower end of the test scores.)

78. standard deviation and mean

The distribution is approximately normal because the data is approximately symmetrical. The best measures of spread and center are the standard deviation and mean, respectively.

79. IQR and median

At least one outlier exists at the upper end of the salary range — $2 million — and the standard deviation is large. The mean is pulled up to $78,000 by the high outlier(s), whereas the median shows that half the company salaries are below $45,000, and half are above.

All these factors show that the distribution isn't symmetric and that the IQR and median are the best measures of spread and center, respectively.

80. IQR and median

The mean is much lower than the median, which tells you that the data is skewed (either the data is shifted off-center or at least one outlier exists). Because the median is highest, the data is skewed to the left and isn't symmetric.

The best measures for center and spread for skewed data are the median and the interquartile range (IQR) because they're less affected by the skewness than the mean and standard deviation are, and they represent the "typical" center and spread for most of the data.

81. Business

The Business College has 25% of the student enrollment, which is the largest proportion of any college.

82. D. Choices (B) and (C) (a separate pie chart for each college showing what percentage are enrolled and what percentage aren't; a bar graph where each bar represents a college and the height shows what percentage of students are enrolled)

To get the information across correctly, you want to show the results from each college separately by using separate pie charts or one bar graph (note that the heights of all the bars in a bar graph don't need to sum to 100%).

83. 39%

To find the percentage of students enrolled in these two colleges, simply add the percentages together: 23% + 16% = 39%.

84. 86%

To find the percentage of students *not* enrolled in the College of Engineering, subtract the percentage of students in the College of Engineering from the whole: 100% − 14% = 86%.

85. impossible to tell without further information

This pie chart contains information only about percentages, not the total number of students nor the number in each category.

86. 5,500

To find how many students are enrolled in the College of Arts & Sciences, you multiply the total number of students by the percentage enrolled in that college, which is 22%: (25,000)(0.22) = 5,500.

87. They distort the proportions of the slices.

Three-dimensional pie charts distort the proportions of the slices by making those in the front look larger than they should.

88. Attend a university.

University is the tallest bar in the graph and therefore represents the greatest number of students of any category.

89. Go into the military.

Military is the shortest bar in the graph and therefore represents the fewest number of students.

90. 140

Judging by the height of the bars, about 120 students plan to attend a university, and about 20 plan to take a gap year. So 120 + 20 = 140 students.

91. 322

You can find the number of students in each category from the height of the bars and then add them together: 120 + 82 + 18 + 82 + 20 = 322.

92. 25%

You can find the number of students in each category from the height of the bars. The total number of students is 322 (120 + 82 + 18 + 82 + 20 = 322). The number going to a community college is 82.

To find the percentage of students going to a community college, divide the number of students in that category by the total number of students: 82/322 = 0.2546, or about 25%.

93. 63%

Judging from the height of the bars, there are 322 students total (120 + 82 + 18 + 82 + 20 = 322), and 120 plan to attend a university. To find the percentage of students not planning to attend a university, subtract the number that do plan to attend from the total number of students (322 – 120 = 202), and then divide by the total: 202/322 = 0.627, or about 63%.

94. The *y*-axis has been stretched far beyond the range of the data.

The largest value is 120 but the *y*-axis runs to 300, making it difficult to compare the height of the bars.

95. histogram

Because the data is continuous with many possible values, a histogram would be the best choice for displaying this type of data.

96. Yes, each bar corresponds to a specific interval of values on the real number line.

For example, a histogram bar covering the range 5 to 10 on the *x*-axis would represent all values in the data set in the range of 5 through 10.

97. because the bars represent the distribution of continuous data

The bars in a histogram represent cutoff values of continuous data that takes on real numbers. Bars for categorical data stand for groups like male and female, which can be shown in either order.

98. the BMI scores

The x-axis represents the values of the variable: BMI scores, a continuous variable that, in this case, runs from 18 to 31.

99. the range of BMI scores for each group in the bar

The BMI scores were divided into groups on the number line. The width of a bar represents the range of the BMI scores in that group.

100. the number of adults with a BMI score in a particular range

The y-axis of a histogram represents how many individuals are in each group, either as a count (frequency) or as a percentage (relative frequency). In this case, the y-axis represents the number of adults (frequency) with a BMI score in the given range.

101. approximately normal

The shape of this distribution is approximately normal because it has bell-shaped characteristics.

102. 18 to 31

You can see from the x-axis that the lowest bar has a lower bound of 18 and the highest bar has an upper bound of 31, so no data is outside that range.

103. the seventh bar

The data is approximately symmetric (similar looking on each side of the mean) and is likely centered in the range of 24 to 25. This corresponds to the seventh bar on the histogram.

104. 21

You can determine the number of people in each BMI range by the height of the bar. In this case, 8 are in the 22 to 23 range, and 13 are in the 23 to 24 range. Add these numbers to get your answer: 8 + 13 = 21.

105. 15%

You can judge the number of people in each BMI range by the height of the bar. In this case, 9 are in the 28 to 29 range, 4 are in the 29 to 30 range, and 2 are in the 30 to 31 range. Add these numbers to get your answer: $9 + 4 + 2 = 15$. Find the proportion by dividing by the total sample size (101) to get 0.15, then multiply by 100 to get the percentage (15%).

106. right-skewed

The data is right-skewed because it has a large number of values in the lower income range and a long tail extending to the right, representing a few cases with high incomes.

107. median

The median is the best measure of the center when the data is highly skewed, and it's less affected by extreme values than the mean. Because the mean uses the value of every number in its calculation, the few high values pull the mean upward, while the median stays in the middle and is a better representative of the "center" of the data.

108. The mean will be higher.

In a right-skewed distribution, the mean will be greater than the median because it has a few high values compared to the rest, pulling the average up. The median is just the middle number when the data is ordered, so the few high values can't pull up the median.

109. 0

The lowest possible value is 0 because the values in the first bar (looking at the *x*-axis) go from 0 to 5,000. *Note:* You can't tell whether 0 is actually in the data set, so you know only that the lowest *possible* value is 0.

110. 85,000

Each bar represents a range of 5,000 (looking at the *x*-axis). The lower bound of the highest bar is 80,000, so its upper bound is 85,000.

111. 35

The height of a bar represents the number of people in that range. Each bar covers a range of $5,000 in income, so the first two bars represent incomes less than $10,000. The first bar has a height of about 11 and the second has a height of about 24. Together, that's 35 ($11 + 24 = 35$).

112. $10,000 to $15,000

The sample includes 110 adults, and if you sort their incomes from lowest to highest, the median is the number in the middle (between the 55th and 56th numbers). To find the bar that contains the median, count the heights of the bars until you reach 55 and 56. The third bar contains the median and ranges from $10,000 to $15,000.

113. Section 1 is approximately normal; Section 2 is approximately uniform.

Section 1 is clearly close to normal because it has an approximate bell shape. Section 2 is close to uniform because the heights of the bars are roughly equal all the way across.

114. They are the same.

The range of values lets you know where the highest and lowest values are. The grades are shown on the x-axis of each graph. Section 1's grades go from 70 to 90, and Section 2's grades go from 70 to 90, so they are the same.

115. They will be similar.

In both cases, the data appear to be fairly symmetric, which means that if you draw a line right down the middle of each graph, the shape of the data looks about the same on each side. For symmetric data, no skewness or outliers exist, so the average and the middle value (median) are similar.

116. They will be similar.

In both cases, the data appear to be fairly symmetric, which means that if you draw a line right down the middle of each graph, the shape of the data looks about the same on each side. For symmetric data, no skewness or outliers exist, so the average and the middle value (median) are similar.

117. 78.75 to 80

Because the sample size is 100, the median will be between the 50th and 51st data value when the data is sorted from lowest to highest. To find the bar that contains the median, count the heights of the bars until you reach 50 and 51. The bar containing the median has the range 78.75 to 80.

118. 77.5 to 82.5

Because the sample size is 100, the median will be between the 50th and 51st data value when the data is sorted from lowest to highest. To find the bar that contains the median, count the heights of the bars until you reach or pass 50 and 51. The bar containing the 50th data value has the range 77.5 to 80. The bar containing the 51st data value has the range 80 to 82.5. Thus the median is approximately 80 (the value that borders both intervals).

119.

Section 2, because a flat histogram has more variability than a bell-shaped histogram of a similar range.

Standard deviation is the average distance the data is from the mean. When the data is flat, it has a large average distance from the mean, overall, but if the data has a bell shape (normal), much more data is close to the mean, and the standard deviation is lower.

120.

the median

NBA salaries are highly skewed to the right — a few players make very high salaries, while most earn considerably less. The median, which is the salary in the middle when the salaries are arranged in order, is less affected by extreme values than the mean.

121.

2.5

The range of data is from 1.5 to 4.0, which is $4.0 - 1.5 = 2.5$.

122.

3.0

The thick line within the box indicates the median (or middle number) for the data.

123.

1.125

The interquartile range (IQR) is the distance between the 1st and 3rd quartiles (Q_1 and Q_3). In this case, IQR $= 3.5 - 2.375 = 1.125$.

124.

the GPA values

The numerical axis is a scale showing the GPAs of individual students ranging from 1.5 to 4.0.

125.

cannot tell

A box plot includes five values: the minimum value, the 25th percentile (Q_1), the median, the 75th percentile (Q_3), and the maximum value. The value of the mean isn't included on a box plot.

126.

skewed left

You can't tell the exact distribution of data from a box plot. But because the median is located above the center of the box and the lower tail is longer than the upper tail, this data is skewed left.

127.

50%

The actual box part of a box plot includes the middle 50% of the data, so the remaining 50% of the total must be outside the box.

128. 50%

The definition of a median is that half the data in a distribution is below it and half is above it. In a box plot, the median is indicated by the location of the line inside the box part of the box plot.

129. E. Choices (A), (B), and (C) (the total sample size; the number of students in each college; the mean of each data set)

The sample size isn't accessible from a box plot. You know that 25% of the data lies within each section, but you don't know the total sample size. You also don't know the mean; you see the median (the line inside the box), but the mean isn't included on a box plot.

130. College 1

The median is indicated by the line within the actual box part of the box plot. Comparing the medians, you can see College 1's median has a greater value than College 2's.

131. College 2

The interquartile range (IQR) is the distance between the 3rd and 1st quartiles and represents the length of the box. If you compare the IQR of the two box plots, the IQR for College 2 is larger than the IQR for College 1.

132. Impossible to tell without further information.

Just because one box plot has a longer box than another one doesn't mean it has more data in it. It just means that the data inside the box (the middle 50% of the data) is more spread out for that group. Each section marked off on a box plot represents 25% of the data; but you don't know how many values are in each section without knowing the total sample size.

133. The two data sets have the same percentage of GPAs above their medians.

The median is the place in the data set that divides the data in half: 50% above and 50% below. So both data sets have 50% of their GPAs above their median.

134. City 2, City 1, City 3

The bar in the center of the box represents the median of each distribution; City 2 is the highest, followed by City 1 and City 3.

135. City 2

To find the number of homes that sold for more than $72,000, look at the numerical axis, where you can see that City 2 has three-quarters of the data lying past 72 (because Q_1 for City 2 is greater than 72). The other cities don't.

136. City 2

The lower edge of the box, which represents Q_1, is above 72 for City 2, whereas the medians for City 1 and City 3 are below 72. Therefore, if all three cities had the same number of home sales in 2012, City 2 must have had the most above $72,000.

137. City 1

The range is the maximum value minus the minimum value of a data set (shown in the very top line and the very bottom line of the box plot).

City 1 range: 80 − 62 = 18

City 2 range: 125 − 43 = 82

City 3 range: 80 − 38 = 42

City 1 has the smallest range, 18.

138. D. Choices (A) and (B) (More than half of the homes in City 1 sold for more than $50,000; more than half of the homes in City 2 sold for more than $75,000.)

As indicated by the median lines within the boxes for each city, more than half of the homes in City 1 sold for more than $50,000, and more than half of the homes in City 2 sold for more than $75,000.

139. E. Choices (A) and (C) (About 25% of homes in City 1 sold for $75,000 or more; about 25% of homes in City 2 sold for $98,000 or more.)

The cutoff for the upper 25% of the values in a box plot is indicated by the largest number in the box. For City 1, about 25% of homes sold for $75,000 or more, and about 25% of homes in City 2 sold for $100,000 or more.

140. decreasing

Although not every year shows a decrease from the previous year, the clear overall pattern of the dropout rate is one of a steady decrease.

141. 3.8%

Find where the value for 2005 on the *x*-axis intersects with the value for the dropout rate on the *y*-axis; it's at 3.8%.

142. drop of 2%

The dropout rate in 2001 was about 5%, and in 2011, it was about 3%. So the dropout rate from 2001 to 2011 is 3% − 5% = −2%.

A change of −2% is the same as a drop of 2 percentage points.

143. rise of 0.2%

The dropout rate in 2003 was about 4.0%, and in 2004, it was 4.2%. So the dropout rate from 2003 to 2004 changed by 4.2% – 4.0% = 0.2%. This is the same as an increase (rise) of 0.2 percentage points.

144. The y-axis scale has been stretched (the range of y-values goes well outside the data).

The change in dropout rates is made to look less significant in this chart because the y-axis scale runs from 0 to 14, while the data runs from 3 to 5.

145. because you need to take into account the number of students enrolled each year, not just the number who dropped out

Suppose that one year, out of 1,000 students in the system, 10 of them dropped out, so 10/1,000 = 0.01, or 1% of the students dropped out. However, if the number of students in the system was only 500, but the number of dropouts was the same, then 10/500 = 0.02, or 2% of the students dropped out. Dividing by the total number of students calculates a dropout rate, which allows for a fair comparison.

146. Some points have been omitted from the scale of the x-axis.

Data for the years 2002 and 2009 have been omitted, but there's no indication of this omission on the chart (that is, no broken line calls attention to the missing years).

147. normal, uniform, bimodal

Income 1 has the bell shape of a normal distribution. Income 2 has a roughly flat shape. Income 3 has two peaks in it and is bimodal.

148. Income 3, Income 2, Income 1

Because you measure variability in terms of average distance from the mean, graphs with values more concentrated around the mean (such as the bell shape) have less variability than graphs with values not concentrated around the mean (like the uniform). So of the three graphs, Income 1 has the lowest variability, and Income 3 has the most. (Income 3 is relatively flat but has peaks at the extremes; these peaks increase the variability.)

149. D. the number of books purchased by a student in a year

A discrete random variable is one that can take on only values that are integers (positive and negative whole numbers and 0). The values of a discrete random variable can have a finite stopping point, like –1, 0, and 1, or they can go to infinity (for example, 1, 2, 3, 4, . . .).

A continuous random variable takes on all possible values within an interval on the real number line (such as all real numbers between –2 and 2, written as [–2, 2]).

A number of books takes on only positive integer values, such as 0, 1, or 2, and thus is a discrete random variable.

150.   B.   the annual rainfall in a city

A discrete random variable is one that can take on only values that are integers (positive and negative whole numbers and 0). The values of a discrete random variable can have a finite stopping point, like –1, 0, and 1, or they can go to infinity (for example, 1, 2, 3, 4, . . .).

A continuous random variable takes on all possible values within an interval on the real number line (such as all real numbers between –2 and 2, written as [–2, 2]).

The amount of rain that falls on a city in a year can take on any non-negative value on the real number line, such as 11.45 inches or 37.9 inches, and therefore is continuous rather than discrete.

151.   C.   the number of cars registered in a state

A discrete random variable is one that can take on only values that are integers (positive and negative whole numbers and 0). The values of a discrete random variable can have a finite stopping point, like –1, 0, and 1, or they can go to infinity (for example, 1, 2, 3, 4, . . .).

A continuous random variable takes on all possible values within an interval on the real number line (such as all real numbers between –2 and 2, written as [–2, 2]).

The number of cars registered in a state must be a non-negative integer, such as 0, 1, or 2, and thus is a discrete random variable.

152.   C.   the proportion of the American population that believes in ghosts

A discrete random variable is one that can take on only values that are integers (positive and negative whole numbers and 0). The values of a discrete random variable can have a finite stopping point, like –1, 0, and 1, or they can go to infinity (for example, 1, 2, 3, 4, . . .).

A continuous random variable takes on all possible values within an interval on the real number line (such as all real numbers between –2 and 2, written as [–2, 2]).

A proportion can take on any real number between 0 and 1 on the real number line and is therefore continuous.

153.   E.   the amount of gasoline used in the United States in 2012

A discrete random variable is one that can take on only values that are integers (positive and negative whole numbers and 0). The values of a discrete random variable can have a finite stopping point, like –1, 0, and 1, or they can go to infinity (for example, 1, 2, 3, 4, . . .).

A continuous random variable takes on all possible values within an interval on the real number line (such as all real numbers between –2 and 2, written as [–2, 2]).

The amount of gasoline can take on any value on the real number line that's greater than or equal to 0 and is therefore continuous.

154.   D.   the number of bird species observed in an area

A discrete random variable is one that can take on only values that are integers (positive and negative whole numbers and 0). The values of a discrete random variable can have a finite stopping point, like –1, 0, and 1, or they can go to infinity (for example, 1, 2, 3, 4, . . .).

A continuous random variable takes on all possible values within an interval on the real number line (such as all real numbers between –2 and 2, written as [–2, 2]).

The number of bird species takes on non-negative integer values, such as 0, 1, or 2, and thus is a discrete random variable.

Note: The amount of money spent by a household for food during a year is considered continuous because of all the possible values it can take on (even though the amounts would be rounded to dollars and cents).

155.   0.10

From the table, you see that 0.10 or 10% of the adults in the city are employed part-time. Using notation, this means that $P(part\text{-}time) = 0.10$.

156.   0.82

Because total probability is always equal to 1, the probability that someone isn't retired is 1 minus the probability that the person is retired (which, according to the table, is 0.18 in this case). So the probability that the adult isn't retired is $1 - 0.18 = 0.82$, or 82%. Using notation, this means that $P(not\ retired) = 0.82$.

157.   0.75

Because the categories don't overlap, the probability that someone is working either part-time or full-time is the sum of their individual probabilities. You can see from the table that the probability for part-time employment is 0.10 and for full-time employment, 0.65. Add these two probabilities to get your answer: $0.10 + 0.65 = 0.75$, or 75%. Using notation, this means that $P(part\text{-}time\ or\ full\text{-}time) = 0.75$.

158.   3.4

In this case, X represents the number of classes. The possible values of X are 4 and 3, denoted x_1 and x_2, respectively; their proportions (probabilities) are equal to 0.40 and 0.60 (denoted p_1 and p_2, respectively).

To find the average number of classes, or the mean of X (denoted by μ_x), multiply each value, x_i, by its probability, p_i, and then add the products:

$$\mu_x = \sum x_i p_i$$
$$= (4)(0.40) + (3)(0.60)$$
$$= 1.6 + 1.8$$
$$= 3.4$$

159.   18.75

In this case, X represents the age of a student. The possible values of X are 18, 19, and 20, denoted x_1, x_2, and x_3, respectively; their proportions (probabilities) are equal to 0.50, 0.25, and 0.25 (denoted p_1, p_2, and p_3, respectively).

To find the mean of X, or the average age of the students in the class (denoted by μ_x), multiply each value, x, by its probability, p, and then add the products:

$$\mu_x = \sum x_i p_i$$
$$= (18)(0.50) + (19)(0.25) + (20)(0.25)$$
$$= 9 + 4.75 + 5$$
$$= 18.75$$

160.   0.10

The total probability must equal 1, so you can subtract the sum of the known values in the table from 1 to find the missing value: $1 - (0.25 + 0.60 + 0.05) = 1 - 0.90 = 0.10$.

161.   0.95

In this case, X represents the number of automobiles. The possible values of X are 0, 1, 2, and 3, denoted x_1, x_2, x_3, and x_4, respectively; their proportions (probabilities) are equal to 0.25, 0.60, 0.10, and 0.05 (denoted p_1, p_2, p_3, and p_4, respectively).

To find the mean of X (denoted by μ_x), multiply each value, x, by its probability, p, and then add the products:

$$\mu_x = \sum x_i p_i$$
$$= (0)(0.25) + (1)(0.60) + (2)(0.10) + (3)(0.05)$$
$$= 0 + 0.60 + 0.20 + 0.15$$
$$= 0.95$$

162.   1.20

In this case, X represents the number of automobiles owned. You want the mean of X, designated as μ_x.

If every family not currently owning a car bought a car, then the total proportion of families owning one car would increase from 0.25 to $0.25 + 0.60 = 0.85$, and the proportion of families with no car would now be 0. You still have 0.10, or 10%, of the families owning 2 cars and 0.05, or 5%, of the families owning 3 cars. So the values for X are 0, 1, 2, and 3, (denoted x_1, x_2, x_3, and x_4, respectively) and their proportions (probabilities) are 0, 0.85, 0.10, and 0.05 (denoted p_1, p_2, p_3, and p_4, respectively).

To find the mean of X, multiply each value, x, by its probability, p, and then add the products:

$$\mu_x = \sum x_i p_i$$
$$= (0)(0) + (1)(0.85) + (2)(0.10) + (3)(0.05)$$
$$= 0 + 0.85 + 0.20 + 0.15$$
$$= 1.20$$

163. 1.00

In this case, X represents the number of automobiles owned. You want the mean of X, designated as μ_x.

If every family currently owning three cars bought a fourth car, the values of X would be 0, 1, 2, 3, and 4. The proportion of families owning three cars would be 0, and the total proportion of families owning four cars would be 0.05. You still have 0.25 of the families owning 0 cars, 0.60 of the families owning 1 car, and 0.10 of the families owning 2 cars.

So the values of X are 0, 1, 2, 3, and 4 (denoted $x_1, x_2, x_3, x_4,$ and x_5, respectively), and their proportions (probabilities) are 0.25, 0.60, 0.10, 0, and 0.05 (denoted $p_1, p_2, p_3, p_4,$ and p_5, respectively).

To find the mean of X, multiply each value, x_i, by its probability, p_i, and then add the products:

$$\mu_x = \sum x_i p_i$$
$$= (0)(0.25) + (1)(0.60) + (2)(0.10) + (3)(0) + (4)(0.05)$$
$$= 0 + 0.60 + 0.20 + 0 + 0.20$$
$$= 1.00$$

164. 1.73

The standard deviation is the square root of the variance, so if the variance of X is 3, the standard deviation of X is $\sqrt{3} = 1.73$ (rounded to two decimal places).

165. 0.42

The variance is the square of the standard deviation, so if the standard deviation of X is 0.65, the variance of X is $(0.65)^2 = 0.4225$. Rounded to two decimal places, the answer is 0.42.

166. 0.80

In this case, X represents the number of siblings a student has. The question asks for the mean of X, designated as μ_x.

The possible values of X are 0, 1, and 2, denoted $x_1, x_2,$ and x_3, respectively; their proportions (probabilities) are equal to 0.34, 0.52, and 0.14 (denoted $p_1, p_2,$ and p_3, respectively).

To find the mean of X, multiply each value, x_i, by its probability, p_i, and then add the products:

$$\mu_x = \sum x_i p_i$$
$$= (0)(0.34) + (1)(0.52) + (2)(0.14)$$
$$= 0 + 0.52 + 0.28$$
$$= 0.80$$

167. 0.44

To find the variance of X, denoted by σ_x^2, you take the first value of X, call it x_1, subtract the mean of X (denoted by μ_x), square that result, and then multiply it by the probability for x_1 (denoted p_1). Do the same thing for every other possible value of X, and then sum up all the results.

In this case, X represents the number of siblings. The values of X are 0, 1, and 2, denoted x_1, x_2, and x_3, respectively. Their probabilities are 0.34, 0.52, and 0.14, respectively.

You need to first find the mean of X because it's part of the formula to calculate the variance. Multiply each value, x_i by its probability, p_i and then add the products:

$$\mu_x = \sum x_i p_i$$
$$= (0)(0.34) + (1)(0.52) + (2)(0.14)$$
$$= 0 + 0.52 + 0.28$$
$$= 0.80$$

Now, plug this value into the formula for finding the variance:

$$\sigma_x^2 = \sum (x_i - \mu_x)^2 p_i$$
$$= (0 - 0.8)^2 (0.34) + (1 - 0.8)^2 (0.52) + (2 - 0.8)^2 (0.14)$$
$$= 0.2176 + 0.0208 + 0.2016$$
$$= 0.4400$$

168. 0.66

To find the standard deviation of X, (denoted by σ_x), you first find the variance of X (denoted by σ_x^2), and then take the square root of that result.

To find the variance of X, you take the first value of X, call it x_1, subtract the mean of X (denoted by μ_x), and square the result. Then, you multiply that result by the probability for x_1, denoted p_1. Do this for each possible value of X, and then add up all the results.

In this case, X represents the number of siblings. The values of X are 0, 1, and 2, denoted x_1, x_2, and x_3, respectively. Their probabilities are 0.34, 0.52, and 0.14, respectively.

You need to first find the mean of X because it's part of the formula to calculate the variance. Multiply each value, x_i by its probability, p_i and then add the products:

$$\mu_x = \sum x_i p_i$$
$$= (0)(0.34) + (1)(0.52) + (2)(0.14)$$
$$= 0 + 0.52 + 0.28$$
$$= 0.80$$

Now, plug this value into the formula to calculate the variance of X:

$$\sigma_x^2 = \sum (x_i - \mu_x)^2 p_i$$
$$= (0 - 0.8)^2 (0.34) + (1 - 0.8)^2 (0.52) + (2 - 0.8)^2 (0.14)$$
$$= 0.2176 + 0.0208 + 0.2016$$
$$= 0.4400$$

Finally, find the standard deviation of X. The standard deviation is the square root of the variance, or $\sqrt{0.44} = 0.66$.

169. The variance would be four times as large, and the standard deviation would be twice as large.

If you double all the values of X, their average distance from the mean (and hence the standard deviation) doubles as well. And because the variance of X is the square of the standard deviation, the variance of X becomes larger by a factor of $2^2 = 4$.

170. 1.45

In this case, X represents the number of books required. The question asks for the mean of X, designated as μ_x.

From the table, the possible values of X are 0, 1, 2, 3, and 4, denoted x_1, x_2, x_3, x_4, and x_5, respectively; their proportions (probabilities) are 0.30, 0.25, 0.25, 0.10, and 0.10, respectively.

To find the mean of X, multiply each value of X by its probability, and then add the products:

$$\mu_x = \sum x_i p_i$$
$$= (0)(0.30) + (1)(0.25) + (2)(0.25) + (3)(0.10) + (4)(0.10)$$
$$= 0 + 0.25 + 0.50 + 0.30 + 0.40$$
$$= 1.45$$

171. 1.65

To find the variance of X, you take the first value of X, call it x_1, subtract the mean of X (denoted by μ_x), and square the result. Then, you multiply that result by the probability for x_1, denoted p_1. Do this for each possible value of X, and then add up all the results.

In this case, X represents the number of books required. The values of X are 0, 1, 2, 3, and 4, denoted x_1, x_2, x_3, x_4, and x_5, respectively; their probabilities are 0.30, 0.25, 0.25, 0.10, and 0.10, respectively.

You need to first find the mean of X because it's part of the formula to calculate the variance. Multiply each value of X by its probability, and then add up the results:

$$\mu_x = \sum x_i p_i$$
$$= (0)(0.30) + (1)(0.25) + (2)(0.25) + (3)(0.10) + (4)(0.10)$$
$$= 0 + 0.25 + 0.50 + 0.30 + 0.40$$
$$= 1.45$$

Now, plug this value into the formula to calculate the variance of X:

$$\sigma_x^2 = \sum (x_i - \mu_x)^2 p_i$$
$$= (0-1.45)^2(0.30) + (1-1.45)^2(0.25) + (2-1.45)^2(0.25)$$
$$+ (3-1.45)^2(0.10) + (4-1.45)^2(0.10)$$
$$= 0.63075 + 0.050625 + 0.075625 + 0.24025 + 0.65025$$
$$= 1.6475$$

Rounded to two decimal places, the answer is 1.65.

172.   1.28

To find the standard deviation of X, you first find the variance, and take the square root of the result.

To find the variance of X (denoted σ_x^2), you take the first value of X, call it x_1, subtract the mean of X (denoted by μ_x), and square the result. Then, you multiply that result by the probability for x_1, denoted p_1. Do this for each possible value of X, and then sum up all the results.

In this case, X represents the number of books required. The values of X are 0, 1, 2, 3, and 4, denoted x_1, x_2, x_3, x_4, and x_5, respectively; their probabilities are 0.30, 0.25, 0.25, 0.10, and 0.10, respectively.

You need to first find the mean of X because it's part of the formula to calculate the variance. To find the mean of X, multiply each value of X by its probability, and then add up the results.

$$\mu_x = \sum x_i p_i$$
$$= (0)(0.30) + (1)(0.25) + (2)(0.25) + (3)(0.10) + (4)(0.10)$$
$$= 0 + 0.25 + 0.50 + 0.30 + 0.40$$
$$= 1.45$$

Now, plug this value into the formula to find the variance of X:

$$\sigma_x^2 = \sum (x_i - \mu_x)^2 p_i$$
$$= (0-1.45)^2(0.30) + (1-1.45)^2(0.25) + (2-1.45)^2(0.25)$$
$$+ (3-1.45)^2(0.10) + (4-1.45)^2(0.10)$$
$$= 0.63075 + 0.050625 + 0.075625 + 0.24025 + 0.65025$$
$$= 1.6475$$

The standard deviation, σ, is the square root of the variance:

$$\sigma = \sqrt{1.6475} = 1.28.$$

173.   Both would increase.

With the changes described, fewer students have the number of books near the mean (middle), and some students are getting a higher number of books. The mean number of books overall would increase a bit, but the number of books students have will be more spread out (compared to the mean) than they were before. That means that the variance increases and so does the standard deviation.

174.

E. all of the above

For a random variable to be binomial, it must have a fixed number of trials, with exactly two outcomes possible on each trial, a constant probability of success on all trials, and each trial must be independent.

175.

the total number of heads

For a random variable to be binomial, it must have a fixed number of trials (n), with exactly two outcomes possible on each trial, a constant probability of success on all trials, and each trial must be independent. Then you define X as the number of "successes" (what you're interested in counting).

In this case, you have 25 trials (coin flips), with exactly two possible outcomes on each flip: heads or tails. According to this problem, a success is a heads, the flips are independent, and the probability of a heads on each trial is the same for each flip (0.5), so the random variable X is binomial. Here, X represents the total number of heads.

176.

because each trial has more than two possible outcomes

In this case, you have six possible outcomes on each trial, but a binomial trial may have only two possible outcomes: success or failure. Here, X represents the outcome of one roll of the die (1, 2, 3, 4, 5, or 6), not the total number of die with a certain outcome (such as the total number of 6's rolled).

177.

because the number of trials isn't fixed

For a binomial experiment, the number of trials must be specified in advance. In this case, although you know in the end that you need 30 employees who say they graduated from high school, you don't know how many employees you'll have to ask before finding 30 who graduated from high school. Because the total number of trials, n, is unknown, X isn't binomial.

178.

because the trials aren't independent

If one sibling has the mutation, a higher chance exists that the other one will, too, so the results for each person aren't independent. Instead of recruiting 30 pairs of siblings for this test, you should recruit 60 people at random.

179.

7.2

The mean of a binomial random variable X is represented by the symbol μ. A binomial distribution has a special formula for the mean, which is $\mu = np$. Here, $n = 18$ and $p = 0.4$, so $\mu = (18)(0.4) = 7.2$.

180.

8.75

The mean of a binomial random variable X is represented by the symbol μ. A binomial distribution has a special formula for the mean, which is $\mu = np$. Here, $n = 25$ and $p = 0.35$, so $\mu = (25)(0.35) = 8.75$.

181. 2.08

The standard deviation of X is represented by σ and represents the square root of the variance. If X has a binomial distribution, the formula for the standard deviation is $\sigma = \sqrt{np(1-p)}$, where n is the number of trials and p is the probability of success on each trial. For this situation, $n = 18$ and $p = 0.4$, so

$$\sigma = \sqrt{18(0.4)(1-0.4)}$$
$$= \sqrt{4.32} = 2.08$$

182. 5.69

The variance is represented by σ^2 and represents the typical squared distance from the mean for all values of X.

For a binomial distribution, the variance has its own formula: $\sigma^2 = np(1-p)$. In this case, $n = 25$ and $p = 0.35$, so

$$\sigma^2 = 25(0.35)(1-0.35)$$
$$= 25(0.35)(0.65)$$
$$= 5.6875$$

Rounded to two decimal places, the answer is 5.69.

183. 130

The mean of a random variable X is denoted μ. For a binomial distribution, the mean has a special formula: $\mu = np$. In this case, $p = 0.14$ and μ is 18.2, so you need to find n. Plug the known values into the formula for the mean, so $18.2 = n(0.14)$, and then divide both sides by 0.14 to get $n = 18.2/0.14 = 130$.

184. 8

The value of $\binom{n}{x}$, called "n choose x" tells you the number of ways you can get X successes in n trials. The more ways to get those X successes, the higher the probability becomes. The formula for calculating "n choose x" is

$$\binom{n}{x} = \frac{n!}{x!(n-x)!}$$

The $n!$ stands for "n factorial." To calculate $n!$, you do a string of multiplications, starting from n and going down to 1. For example $5! = (5)(4)(3)(2)(1) = 120$; $2! = (2)(1) = 2$; $1! = 1$; and by convention, $0! = 1$.

To find $\begin{pmatrix} n \\ x \end{pmatrix}$ in this problem, where $n = 8$ and $x = 1$, plug the numbers into the formula:

$$\begin{pmatrix} 8 \\ 1 \end{pmatrix} = \frac{8!}{1!(8-1)!}$$

$$= \frac{(8)(7)(6)(5)(4)(3)(2)(1)}{(1)[(7)(6)(5)(4)(3)(2)(1)]}$$

$$= \frac{40{,}320}{5{,}040} = 8$$

---

185.   0.0164

The formula for calculating a probability for a binomial distribution is

$$P(X = x) = \begin{pmatrix} n \\ x \end{pmatrix} p^x (1-p)^{n-x}$$

Here, $\begin{pmatrix} n \\ x \end{pmatrix} = \dfrac{n!}{x!(n-x)!}$ and $n!$ means $n(n-1)(n-2)\ldots(3)(2)(1)$. For example $5! = (5)(4)$
$(3)(2)(1) = 120$; $2! = (2)(1) = 2$; $1! = 1$; and by convention, $0! = 1$.

To find the probability of exactly one success in eight trials, you need $P(X = 1)$, where $n = 8$ (remember that $p = 0.55$ here):

$$P(X = 1) = \begin{pmatrix} 8 \\ 1 \end{pmatrix} (0.55)^1 (1 - 0.55)^{8-1}$$

$$= \frac{8!}{1!(8-1)!} (0.55)^1 (1 - 0.55)^{8-1}$$

$$= \frac{(8)(7)(6)(5)(4)(3)(2)(1)}{(1)[(7)(6)(5)(4)(3)(2)(1)]} (0.55)(0.45)^7$$

$$= (8)(0.55)(0.00373669453125)$$

$$= 0.0164414559375$$

Rounded to four decimal places, the answer is 0.0164.

---

186.   0.0703

The formula for calculating a probability for a binomial distribution is

$$P(X = x) = \begin{pmatrix} n \\ x \end{pmatrix} p^x (1-p)^{n-x}$$

Here, $\begin{pmatrix} n \\ x \end{pmatrix} = \dfrac{n!}{x!(n-x)!}$ and $n!$ means $n(n-1)(n-2)\ldots(3)(2)(1)$. For example $5! = (5)(4)$
$(3)(2)(1) = 120$; $2! = (2)(1) = 2$; $1! = 1$; and by convention, $0! = 1$.

To find the probability of exactly two successes in eight trials, you want $P(X = 2)$, where $n = 8$ (remember that $p = 0.55$ here):

$$P(X=2) = \binom{8}{2}(0.55)^2(1-0.55)^{8-2}$$

$$= \frac{8!}{2!(8-2)!}(0.55)^2(1-0.55)^{8-2}$$

$$= \frac{(8)(7)(6)(5)(4)(3)(2)(1)}{[(2)(1)][(6)(5)(4)(3)(2)(1)]}(0.55)^2(0.45)^6$$

$$= (28)(0.3025)(0.008303765625)$$

$$= 0.07033289484375$$

Rounded to four decimal places, the answer is 0.0703.

---

187.   1

The formula for calculating "n choose x" is

$$\binom{n}{x} = \frac{n!}{x!(n-x)!}$$

The $n!$ stands for "n factorial." To calculate $n!$, you do a string of multiplications, starting from n and going down to 1. For example $5! = (5)(4)(3)(2)(1) = 120$; $2! = (2)(1) = 2$; $1! = 1$; and by convention, $0! = 1$.

So find $\binom{n}{x}$ in this problem, plugging in the numbers for $n = 8$ and $x = 0$:

$$\binom{8}{0} = \frac{8!}{0!(8-0)!}$$

$$= \frac{(8)(7)(6)(5)(4)(3)(2)(1)}{(1)[(8)(7)(6)(5)(4)(3)(2)(1)]}$$

$$= 1$$

Incidentally, when $x = 0$, $\binom{n}{x}$ is always 1.

---

188.   0.9983

The formula for calculating a probability for a binomial distribution is

$$P(X=x) = \binom{n}{x}p^x(1-p)^{n-x}$$

Here, $\binom{n}{x} = \frac{n!}{x!(n-x)!}$ and $n!$ means $n(n-1)(n-2)\dots(3)(2)(1)$. For example $5! = (5)(4)$ $(3)(2)(1) = 120$; $2! = (2)(1) = 2$; $1! = 1$; and by convention, $0! = 1$.

In this case, X is the number of successes in n trials. You want $P(X \geq 1)$ because "at least one" means the same as "one or more." The easiest way to answer this question is to take 1 minus $P(X = 0)$, because that's the opposite of $P(X \geq 1)$ and easier to find.

$$P(X=0) = \binom{8}{0}(0.55)^0(1-0.55)^{8-0}$$

$$= \frac{8!}{0!(8-0)!}(0.55)^0(0.45)^8$$

$$= \frac{(8)(7)(6)(5)(4)(3)(2)(1)}{(1)[(8)(7)(6)(5)(4)(3)(2)(1)]}(0.55)^0(0.45)^8$$

$$= (1)(1)(0.0016815125390625)$$

$$= 0.0016815125390625$$

Rounded to four decimal places, this answer is 0.017. Now, plug the value of $P(X = 0)$ in the formula to find $P(X > 0)$:

$$P(X>0) = 1 - P(X=0)$$

$$= 1 - 0.0017$$

$$= 0.9983$$

189.   0.0439

The formula for calculating a probability for a binomial distribution is

$$P(X=x) = \binom{n}{x}p^x(1-p)^{n-x}$$

Here, $\binom{n}{x} = \dfrac{n!}{x!(n-x)!}$ and $n!$ means $n(n-1)(n-2) \ldots (3)(2)(1)$. For example $5! = (5)(4)$ $(3)(2)(1) = 120$; $2! = (2)(1) = 2$; $1! = 1$; and by convention, $0! = 1$.

To find the probability of exactly eight successes in ten trials, you want $P(X = 8)$, where $n = 10$ (remember that $p = 0.50$ here):

$$P(X=8) = \binom{10}{8}(0.50)^8(1-0.50)^{10-8}$$

$$= \frac{10!}{8!(10-8)!}(0.50)^8(1-0.50)^{10-8}$$

$$= \frac{(10)(9)(8)(7)(6)(5)(4)(3)(2)(1)}{[(8)(7)(6)(5)(4)(3)(2)(1)][(2)(1)]}(0.50)^8(0.50)^2$$

$$= (45)(0.00390625)(0.25)$$

$$= 0.0439453125$$

Rounded to four decimal places, the answer is 0.0439.

190. 0.3125

The formula for calculating a probability for a binomial distribution is

$$P(X=x) = \binom{n}{x} p^x (1-p)^{n-x}$$

Here, $\binom{n}{x} = \dfrac{n!}{x!(n-x)!}$ and $n!$ means $n(n-1)(n-2) \ldots (3)(2)(1)$. For example $5! = (5)(4)$ $(3)(2)(1) = 120$; $2! = (2)(1) = 2$; $1! = 1$; and by convention, $0! = 1$.

In this case, $n = 5$ trials, $x = 3$ successes, and $p = 0.5$, the probability of success on each trial. You want $P(X = 3)$:

$$P(X=3) = \binom{5}{3} (0.50)^3 (1-0.50)^{5-3}$$

$$= \frac{5!}{3!(5-3)!} (0.50)^3 (1-0.50)^{5-3}$$

$$= \frac{(5)(4)(3)(2)(1)}{[(3)(2)(1)][(2)(1)]} (0.50)^3 (0.50)^2$$

$$= (10)(0.125)(0.25)$$

$$= 0.3125$$

191. 0.1563

The formula for calculating a probability for a binomial distribution is

$$P(X=x) = \binom{n}{x} p^x (1-p)^{n-x}$$

Here, $\binom{n}{x} = \dfrac{n!}{x!(n-x)!}$ and $n!$ means $n(n-1)(n-2) \ldots (3)(2)(1)$. For example $5! = (5)(4)$ $(3)(2)(1) = 120$; $2! = (2)(1) = 2$; $1! = 1$; and by convention, $0! = 1$.

In this case, $n = 5$ trials, $x = 4$ successes, and $p = 0.50$. You want $P(X = 4)$:

$$P(X=4) = \binom{5}{4} (0.50)^4 (1-0.50)^{5-4}$$

$$= \frac{5!}{4!(5-4)!} (0.50)^4 (1-0.50)^{5-4}$$

$$= \frac{(5)(4)(3)(2)(1)}{[(4)(3)(2)(1)](1)} (0.50)^4 (0.50)^1$$

$$= (5)(0.0625)(0.50)$$

$$= 0.15625$$

Rounded to four decimal places, the answer is 0.1563.

192. 1

The formula for calculating "n choose x" is

$$\binom{n}{x} = \frac{n!}{x!(n-x)!}$$

The $n!$ stands for "n factorial." To calculate $n!$, you do a string of multiplications, starting from n and going down to 1. For example $5! = (5)(4)(3)(2)(1) = 120$; $2! = (2)(1) = 2$; $1! = 1$; and by convention, $0! = 1$.

So find $\binom{n}{x}$ in this problem, plugging in the numbers for $n = 5$ and $x = 5$:

$$\binom{5}{5} = \frac{5!}{5!(5-5)!}$$

$$= \frac{(5)(4)(3)(2)(1)}{[(5)(4)(3)(2)(1)](1)}$$

$$= 1$$

Incidentally, $\binom{n}{x}$ is always 1 when n and x are the same number.

193. 0.4688

The formula for calculating a probability for a binomial distribution is

$$P(X = x) = \binom{n}{x} p^x (1-p)^{n-x}$$

Here, $\binom{n}{x} = \frac{n!}{x!(n-x)!}$ and $n!$ means $n(n-1)(n-2) \ldots (3)(2)(1)$. For example $5! = (5)(4)(3)(2)(1) = 120$; $2! = (2)(1) = 2$; $1! = 1$; and by convention, $0! = 1$.

In this problem, $n = 5$ trials, $x = 3$ or 4 successes, and $p = 0.50$, the probability of success on each trial. You want $P(X = 3 \text{ or } 4) = P(X = 3) + P(X = 4)$. First, find the probability of each outcome separately:

$$P(X = 3) = \binom{5}{3}(0.50)^3(1-0.50)^{5-3}$$

$$= \frac{5!}{3!(5-3)!}(0.50)^3(1-0.50)^{5-3}$$

$$= \frac{(5)(4)(3)(2)(1)}{[(3)(2)(1)][(2)(1)]}(0.50)^3(0.50)^2$$

$$= (10)(0.125)(0.25)$$

$$= 0.3125$$

$$P(X=4) = \binom{5}{4}(0.50)^4(1-0.50)^{5-4}$$

$$= \frac{5!}{4!(5-4)!}(0.50)^4(1-0.50)^{5-4}$$

$$= \frac{(5)(4)(3)(2)(1)}{[(4)(3)(2)(1)](1)}(0.50)^4(0.50)^1$$

$$= (5)(0.0625)(0.50)$$

$$= 0.15625$$

Then, add the results together:

$$P(X=3 \text{ or } 4) = P(X=3) + P(X=4)$$

$$= 0.3125 + 0.15625$$

$$= 0.46875$$

Rounded to four decimal places, the answer is 0.4688.

194.   0.5001

The formula for calculating a probability for a binomial distribution is

$$P(X=x) = \binom{n}{x}p^x(1-p)^{n-x}$$

Here, $\binom{n}{x} = \frac{n!}{x!(n-x)!}$ and $n!$ means $n(n-1)(n-2)\ldots(3)(2)(1)$. For example 5! = (5)(4)(3)(2)(1) = 120; 2! = (2)(1) = 2; 1! = 1; and by convention, 0! = 1.

In this case, $n = 5$ trials, $x = 3$, 4, or 5 successes, and $p = 0.50$, the probability of success on each trial. *Note:* The probability of *at least three* successes means the probability of three, four, or five successes. In other words, you need to find $P(X = 3 \text{ or } 4 \text{ or } 5) = P(X = 3) + P(X = 4) + P(X = 5)$. First, find the probability of each outcome separately:

$$P(X=3) = \binom{5}{3}(0.50)^3(1-0.50)^{5-3}$$

$$= \frac{5!}{3!(5-3)!}(0.50)^3(1-0.50)^{5-3}$$

$$= \frac{(5)(4)(3)(2)(1)}{[(3)(2)(1)][(2)(1)]}(0.50)^3(0.50)^2$$

$$= (10)(0.125)(0.25)$$

$$= 0.3125$$

$$P(X=4) = \binom{5}{4}(0.50)^4(1-0.50)^{5-4}$$

$$= \frac{5!}{4!(5-4)!}(0.50)^4(1-0.50)^{5-4}$$

$$= \frac{(5)(4)(3)(2)(1)}{[(4)(3)(2)(1)](1)}(0.50)^4(0.50)^1$$

$$= (5)(0.0625)(0.50)$$

$$= 0.15625$$

$$P(X=5) = \binom{5}{5}(0.50)^5(1-0.50)^{5-5}$$

$$= \frac{5!}{5!(5-0)!}(0.50)^5(1-0.50)^{5-5}$$

$$= \frac{(5)(4)(3)(2)(1)}{[(5)(4)(3)(2)(1)][(5)(4)(3)(2)(1)]}(0.50)^5(0.50)^0$$

$$= 0.0313$$

Then add the results together:

$$P(X=3 \text{ or } 4 \text{ or } 5) = P(X=3) + P(X=4) + P(X=5)$$

$$= 0.3125 + 0.15625 + 0.0313$$

$$= 0.50005$$

Rounded to four decimal places, the answer is 0.5001.

195. 0.4999

The formula for calculating a probability for a binomial distribution is

$$P(X=x) = \binom{n}{x}p^x(1-p)^{n-x}$$

Here, $\binom{n}{x} = \frac{n!}{x!(n-x)!}$ and $n!$ means $n(n-1)(n-2)\ldots(3)(2)(1)$. For example $5! = (5)(4)$

$(3)(2)(1) = 120$; $2! = (2)(1) = 2$; $1! = 1$; and by convention, $0! = 1$.

In this case, $n = 5$ trials, $x = $ no more than 2 successes, and $p = 0.50$, the probability of success on each trial. To find the probability of no more than two successes, you can either find $P(X \le 2)$ or find the probability of at least three successes, $P(X \ge 3)$ and subtract the sum of these probabilities from 1. For example, you find $P(X = 3)$, $P(X = 4)$, and $P(X = 5)$:

$$P(X=3) = \binom{5}{3}(0.50)^3(1-0.50)^{5-3}$$

$$= \frac{5!}{3!(5-3)!}(0.50)^3(1-0.50)^{5-3}$$

$$= \frac{(5)(4)(3)(2)(1)}{[(3)(2)(1)][(2)(1)]}(0.50)^3(0.50)^2$$

$$= (10)(0.125)(0.25)$$

$$= 0.3125$$

$$P(X=4) = \binom{5}{4}(0.50)^4(1-0.50)^{5-4}$$

$$= \frac{5!}{4!(5-4)!}(0.50)^4(1-0.50)^{5-4}$$

$$= \frac{(5)(4)(3)(2)(1)}{[(4)(3)(2)(1)](1)}(0.50)^4(0.50)^1$$

$$= (5)(0.0625)(0.50)$$

$$= 0.15625$$

$$P(X=5) = \binom{5}{5}(0.50)^5(1-0.50)^{5-5}$$

$$= \frac{5!}{5!(5-0)!}(0.50)^5(1-0.50)^{5-5}$$

$$= \frac{(5)(4)(3)(2)(1)}{[(5)(4)(3)(2)(1)][(5)(4)(3)(2)(1)]}(0.50)^5(0.50)^0$$

$$= 0.0313$$

And then you add the results:

$$P(X \geq 3) = P(X=3) + P(X=4) + P(X=5)$$

$$= 0.3125 + 0.15625 + 0.0313$$

$$= 0.50005$$

You can round this answer to four decimal places: 0.5001. Finally, you subtract the probability of $P(X \geq 3)$ from 1: $1 - 0.5001 = 0.4999$.

196. 0.012

The binomial table (Table A-3 in the appendix) has a series of mini-tables inside of it, one for each selected value of n. To find $P(X = 6)$, where $n = 15$ and $p = 0.7$, locate the mini-table for $n = 15$, find the row for $x = 6$, and follow across to where it intersects with the column for $p = 0.7$. This value is 0.012.

197. 0.219

The binomial table (Table A-3 in the appendix) has a series of mini-tables inside of it, one for each selected value of n. To find $P(X = 11)$, where $n = 15$ and $p = 0.7$, locate the mini-table for $n = 15$, find the row for $x = 11$, and follow across to where it intersects with the column for $p = 0.7$. This value is 0.219.

198. 0.995

In this case, 15 is the greatest possible value of X (because there are only 15 total trials), so to find $P(X < 15)$, you can first find $P(X = 15)$ and subtract this result from 1 to get what you need. (This makes the calculations much easier.)

The binomial table (Table A-3 in the appendix) has a series of mini-tables inside of it, one for each selected value of n. To find $P(X = 15)$, where $n = 15$ and $p = 0.7$, locate the mini-table for $n = 15$, find the row for $x = 15$, and follow across to where it intersects with the column for $p = 0.7$. This value is 0.005.

Now, subtract that from 1 so you have $P(X \geq 15) = 1 - 0.005 = 0.995$.

199. 0.015

You want the probability between 4 and 7, but you don't want to include 4 and 7. So you want only the probabilities for $X = 5$ and $X = 6$. You know that $n = 15$ and $p = 0.7$, which is the probability of success on each trial.

To find each of these probabilities, use the binomial table (Table A-3 in the appendix), which has a series of mini-tables inside of it, one for each selected value of n. To find $P(X = 5)$, where $n = 15$ and $p = 0.7$, locate the mini-table for $n = 15$, find the row for $x = 5$, and follow across to where it intersects with the column for $p = 0.7$. This value is 0.003. Now do the same for $P(X = 6)$ to get 0.012. Then add these probabilities together:

$$P(4 < X < 7) = P(X = 5) + P(X = 6)$$
$$= 0.003 + 0.012$$
$$= 0.015$$

200. 0.051

Here, you want to find the probability equal to 4 and 7 and everything in between. In other words, you want the probabilities for $X = 4$, $X = 5$, $X = 6$, and $X = 7$. You know that $n = 15$ and $p = 0.7$, which is the probability of success on each trial.

To find each of these probabilities, use the binomial table (Table A-3 in the appendix), which has a series of mini-tables inside of it, one for each selected value of n. To find $P(X = 4)$, where $n = 15$ and $p = 0.7$, locate the mini-table for $n = 15$, find the row for $x = 4$, and follow across to where it intersects with the column for $p = 0.7$. This value is 0.001. Now do the same for the other probabilities: $P(X = 5) = 0.003$; $P(X = 6) = 0.012$; and $P(X = 7) = 0.035$. Finally, add these probabilities together:

$$P(4 \leq X \leq 7) = P(X = 4) + P(X = 5) + P(X = 6) + P(X = 7)$$
$$= 0.001 + 0.003 + 0.012 + 0.035$$
$$= 0.051$$

201.   0.221

The binomial table (Table A-3 in the appendix) has a series of mini-tables inside of it, one for each selected value of n. To find $P(X = 5)$, where $n = 11$ and $p = 0.4$, locate the mini-table for $n = 11$, find the row for $x = 5$, and follow across to where it intersects with the column for $p = 0.4$. This value is 0.221.

202.   0.996

To find the probability that X is greater than 0, find the probability that X is equal to 0, and then subtract that probability from 1. This makes the calculations much easier.

The binomial table (Table A-3 in the appendix) has a series of mini-tables inside of it, one for each selected value of n. To find $P(X = 0)$, where $n = 11$ and $p = 0.4$, locate the mini-table for $n = 11$, find the row for $x = 0$, and follow across to where it intersects with the column for $p = 0.4$. This value is 0.004. Now subtract that from 1:

$$P(X > 0) = 1 - P(X = 0)$$
$$= 1 - 0.004$$
$$= 0.996$$

203.   0.120

To find the probability that X is less than or equal to 2, you first need to find the probability of each possible value of X less than 2. In other words, you find the values for $P(X = 0)$, $P(X = 1)$, and $P(X = 2)$.

To find each of these probabilities, use the binomial table (Table A-3 in the appendix), which has a series of mini-tables inside of it, one for each selected value of n. To find $P(X = 0)$, where $n = 11$ and $p = 0.4$, locate the mini-table for $n = 11$, find the row for $x = 0$, and follow across to where it intersects with the column for $p = 0.4$. This value is 0.004. Now do the same for the other probabilities: $P(X = 1) = 0.027$ and $P(X = 2) = 0.089$. Finally, add these probabilities together:

$$P(X \le 2) = P(X = 0) + P(X = 1) + P(X = 2)$$
$$= 0.004 + 0.027 + 0.089$$
$$= 0.120$$

204.   0.001

To find the probability that X is greater than 9, first find the probability that X is equal to 10 or 11 (in this case, 11 is the greatest possible value of x because there are only 11 total trials).

To find each of these probabilities, use the binomial table (Table A-3 in the appendix), which has a series of mini-tables inside of it, one for each selected value of n. To find $P(X = 10)$, where $n = 11$ and $p = 0.4$, locate the mini-table for $n = 11$, find the row for $x = 10$, and follow across to where it intersects with the column for $p = 0.4$. This value is 0.001. Now do the same for $P(X = 11)$, which gives you 0.000. (*Note:* $P(X = 11)$ isn't exactly

0.000 here; it's just a smaller probability than can be expressed in the four decimal places used in this table.) Finally, add the two probabilities together:

$$P(X > 9) = P(X = 10) + P(X = 11)$$
$$= 0.001 + 0.000$$
$$= 0.001$$

---

205. 0.634

Here, you want to find the probability equal to 3 and 5 and everything in between. In other words, you want the probabilities for $X = 3$, $X = 4$, and $X = 5$. You know that $n = 11$ and $p = 0.4$, which is the probability of success on each trial.

To find each of these probabilities, use the binomial table (Table A-3 in the appendix), which has a series of mini-tables inside of it, one for each selected value of n. To find $P(X = 3)$, where $n = 11$ and $p = 0.4$, locate the mini-table for $n = 11$, find the row for $x = 3$, and follow across to where it intersects with the column for $p = 0.4$. This value is 0.177. Now do the same for the other probabilities: $P(X = 4) = 0.236$ and $P(X = 5) = 0.221$. Finally, add these probabilities together:

$$P(3 \leq X \leq 5) = P(X = 3) + P(X = 4) + P(X = 5)$$
$$= 0.177 + 0.236 + 0.221$$
$$= 0.634$$

---

206. $n = 30$, $p = 0.4$

Two conditions must be met to use the normal approximation to the binomial: Both np and $n(1 - p)$ must be at least 10. Using the choices you're given, the only one that works is $n = 30$ and $p = 0.4$: $np = 30(0.4) = 12$, and $n(1 - p) = 30(1 - 0.4) = 30(0.6) = 18$.

---

207. 20

Two conditions must be met to use the normal approximation to the binomial: Both np and $n(1 - p)$ must be at least 10. So you need the smallest value of n that meets both of these conditions, knowing that $p = 0.5$ here.

First, take $np \geq 10$, or $n(0.5) \geq 10$. To get n by itself, divide both sides by 0.5, which gives you $n \geq 20$. Then, take $n(1 - p) \geq 10$, or $n(1 - 0.5) \geq 10$. Again, you get $n \geq 20$. The minimum sample size (n) that meets both of these requirements is 20.

Note: Sometimes p is very small or very large, which changes the values of np and $n(1 - p)$, so you must always check and meet both conditions every time.

---

208. 34

To use the normal approximation to the binomial, both np and $n(1 - p)$ must be at least 10. Here the value of p is 0.30 ("success" = green marble).

First, take $np \geq 10$, or $n(0.3) \geq 10$. To get n by itself, divide both sides by 0.3, which gives you $n \geq 33.33$ (round up to 34 to make sure the condition is met). Then, take $n(1 - p) \geq 10$, or $n(1 - 0.3) \geq 10$. That gives you $n \geq 14.29$ (round up to 15 to make sure the condition is met).

For the first condition, you need $n \geq 34$; for the second condition, you need $n \geq 15$. To meet both conditions, you need the larger n, which is 34.

209.

$P(X \geq 50)$

You can rephrase the probability of getting at least 50 heads out of 80 flips (that is, X is at least 50) as the probability that X is greater than or equal to 50 (because 50 is the lower limit of possible values): $P(X \geq 50)$.

210.

$P(X \leq 30)$

You can rephrase the probability of getting no more than 30 heads out of 80 flips (that is, X is no more than 30) as the probability that X is less than or equal to 30 (because 30 is the upper limit of possible values): $P(X \leq 30)$.

211.

40

For a binomial distribution, the mean, μ, is equal to np. In this case, $n = 80$ trials (flips of the coin) and $p = 0.50$ (chance of a heads/success on each flip). Therefore, $\mu = (80)(0.50) = 40$.

212.

4.47

For a binomial distribution, the standard deviation, σ, is equal to $\sqrt{np(1-p)}$. In this case, you have $n = 80$ trials (flips of the coin) and $p = 0.5$ (chance of a heads/success on each flip). Therefore,

$$\sigma = \sqrt{80(0.50)(1-0.50)}$$
$$= \sqrt{20} = 4.47$$

213.

1.12

To find the z-value for an x-value, subtract the population mean from x, and divide by the population standard deviation:

$$z = \frac{x - \mu}{\sigma}$$

The question is what do you use for μ and σ? Because X has a binomial distribution, you use the mean and standard deviation of the binomial distribution. The mean of a binomial distribution is $\mu = np$. In this case, $n = 80$ and $p = 0.50$, so $\mu = (80)(0.50) = 40$. And the formula for the standard deviation of a binomial distribution is $\sigma = \sqrt{np(1-p)}$. So you get

$$\sigma = \sqrt{80(0.50)(1-0.50)}$$
$$= \sqrt{20} = 4.47$$

Now, plug these numbers into the formula for the z-value, where $x = 45$:

$$z = \frac{x - \mu}{\sigma}$$
$$= \frac{(45-40)}{4.47} = 1.12$$

214.   0.1314

Here, you want the probability that X is at least 45, or $P(X \geq 45)$. In this case, $n = 80$ trials (flips of the coin) and $p = 0.50$ (chance of a heads/success on each flip).

Because n is so large, you may want to use the normal approximation to the binomial to solve this problem. But first you need to determine whether the two conditions are met — that is, both np and $n(1 - p)$ must be at least 10. So plug in the numbers: $np = (80)(0.50) = 40$, and $n(1 - p) = 80(1 - 0.50) = (80)(0.50) = 40$. Both conditions are at least 10, so you can move forward.

The first step in doing the normal approximation to find a binomial probability is to find the z-value. To find the z-value for an x-value, subtract the population mean from x, and divide by the population standard deviation:

$$z = \frac{x - \mu}{\sigma}$$

Because X has a binomial distribution, you use the mean and standard deviation of the binomial distribution. The mean of a binomial distribution is $\mu = np$. In this case, $n = 80$ and $p = 0.50$, so $\mu = (80)(0.50) = 40$. The formula for the standard deviation of a binomial distribution is $\sigma = \sqrt{np(1-p)}$. So you get

$$\sigma = \sqrt{80(0.50)(1 - 0.50)}$$
$$= \sqrt{20} = 4.47$$

Now, plug these numbers into the formula for the z-value, where $x = 45$:

$$z = \frac{x - \mu}{\sigma}$$
$$= \frac{(45 - 40)}{4.47} = 1.12$$

Use a Z-table, such as Table A-1 in the appendix, to find $P(Z \leq 1.12)$, and then subtract that from 1 to find $P(Z \geq 1.12)$. Locate the row for $z = 1.1$ and follow it across to where it intersects with the column for 0.02, which gives you 0.8686. This also corresponds to $P(Z \leq 1.12) = P(X \leq 45)$. Then subtract from 1 to get $P(X \geq 45)$: $1 - 0.8686 = 0.1314$.

215. $\mu = 45, \sigma = 4.97$

For a binomial distribution, the mean is $\mu = np$, and the standard deviation is $\sigma = \sqrt{np(1-p)}$. In this case, $n = 100$ trials and $p = 0.45$, the probability of success on each trial. Therefore, the mean of X is $\mu = (100)(0.45) = 45$, and the standard deviation is

$$\sigma = \sqrt{np(1-p)}$$
$$= \sqrt{100(0.45)(1 - 0.45)}$$
$$= \sqrt{24.75} = 4.97$$

216.

−1.00

To find the z-value for an x-value, subtract the population mean from x, and divide by the population standard deviation:

$$z = \frac{x - \mu}{\sigma}$$

For a binomial distribution, the mean is $\mu = np$, and the standard deviation is $\sigma = \sqrt{np(1-p)}$. In this case, $n = 100$ and $p = 0.45$, so $\mu = (100)(0.45) = 45$, and the standard deviation is

$$\sigma = \sqrt{np(1-p)}$$
$$= \sqrt{100(0.45)(1-0.45)}$$
$$= \sqrt{24.75} = 4.97$$

Now, plug these numbers into the formula for the z-value, where $x = 40$:

$$z = \frac{x - \mu}{\sigma}$$
$$= \frac{(40-45)}{4.97} = -1.00$$

217.

0.1587

Because n is so large, you may want to use the normal approximation to the binomial to solve this problem. But first you need to determine whether the two conditions are met — that is, both np and $n(1-p)$ must be at least 10. In this case, $np = 100(0.45) = 45$, and $n(1-p) = 100(1-0.45) = (100)(0.55) = 55$. Both conditions are at least 10, so you can move forward.

To find a "less than" probability for an x-value from a normal distribution, you convert the x-value to a z-value and then find the corresponding probability for that z-value by using a Z-table, such as Table A-1 in the appendix.

To find the z-value for an x-value, subtract the population mean from x, and divide by the population standard deviation:

$$z = \frac{x - \mu}{\sigma}$$

For a binomial distribution, the mean is $\mu = np$, and the standard deviation is $\sigma = \sqrt{np(1-p)}$. In this case, $n = 100$ and $p = 0.45$, so $\mu = (100)(0.45) = 45$, and the standard deviation is

$$\sigma = \sqrt{np(1-p)}$$
$$= \sqrt{100(0.45)(1-0.45)}$$
$$= \sqrt{24.75} = 4.97$$

Now, plug these numbers into the formula for the z-value, where $x = 40$:

$$z = \frac{x - \mu}{\sigma}$$

$$= \frac{(40 - 45)}{4.97} = -1.00$$

Then find $P(Z \le -1.00)$, using Table A-1. Look in the row for $z = -1.0$ and follow across to where it intersects with the column for 0.00, which gives you 0.1587. Remember this is only an approximation because you started with a binomial and were able to use the normal approximation to find the probability.

---

218.   0.8413

Because n is so large, you may want to use the normal approximation to the binomial to solve this problem. But first you need to determine whether the two conditions are met — that is, both np and $n(1 - p)$ must be at least 10. In this case, $np = 100(0.45) = 45$, and $n(1 - p) = 100(1 - 0.45) = (100)(0.55) = 55$. Both conditions are at least 10, so you can move forward.

To find a "greater than" probability for an x-value from a normal distribution, you convert the x-value to a z-value and find the corresponding probability for that z-value by using a Z-table, such as Table A-1 in the appendix; then, you subtract that result from 1 (because Table A-1 gives you "less than" probabilities only).

To find the z-value for an x-value, subtract the population mean from x, and divide by the population standard deviation:

$$z = \frac{x - \mu}{\sigma}$$

For a binomial distribution, the mean is $\mu = np$, and the standard deviation is $\sigma = \sqrt{np(1-p)}$. In this case, $n = 100$ and $p = 0.45$, so $\mu = (100)(0.45) = 45$, and the standard deviation is

$$\sigma = \sqrt{np(1-p)}$$

$$= \sqrt{100(0.45)(1-0.45)}$$

$$= \sqrt{24.75} = 4.97$$

Now, plug these numbers into the formula for the z-value, where $x = 40$:

$$z = \frac{x - \mu}{\sigma}$$

$$= \frac{(40 - 45)}{4.97} = -1.00$$

Then, find $P(Z \le -1.00)$, using Table A-1. Look in the row for $z = -1.0$ and follow across to where it intersects with the column for 0.00, which gives you 0.1587. To get $P(Z \ge -1.00)$, subtract that value from 1: $1 - 0.1587 = 0.8413$. Remember this is only an approximation because you started with a binomial random variable and were able to use the normal approximation to find the probability.

---

219. 0

Even though X is binomial and discrete, the Z-distribution is continuous, so the probability at a single value is 0. Probability for a continuous random variable is represented by the area under a curve. There's no area under the curve at a single point. You would have to use continuity correction to solve $P(X = 40)$ using a normal approximation.

---

220. 2.21

To find the z-value for an x-value, subtract the population mean from x, and divide by the population standard deviation:

$$z = \frac{x - \mu}{\sigma}$$

For a binomial distribution, the mean is $\mu = np$, and the standard deviation is $\sigma = \sqrt{np(1-p)}$. In this case, $n = 100$ and $p = 0.45$, so $\mu = (100)(0.45) = 45$, and the standard deviation is

$$\sigma = \sqrt{np(1-p)}$$
$$= \sqrt{100(0.45)(1-0.45)}$$
$$= \sqrt{24.75} = 4.97$$

Now, plug these numbers into the formula for the z-value, where $x = 56$:

$$z = \frac{x - \mu}{\sigma}$$
$$= \frac{(56 - 45)}{4.97} = 2.21$$

---

221. 0.0136

Because n is so large, you may want to use the normal approximation to the binomial to solve this problem. But first you need to determine whether the two conditions are met — that is, both np and $n(1 - p)$ must be at least 10. In this case, $np = 100(0.45) = 45$, and $n(1 - p) = 100(1 - 0.45) = (100)(0.55) = 55$. Both conditions are at least 10, so you can move forward.

To find a "greater than" probability for an x-value from a normal distribution, you convert the x-value to a z-value and then find the corresponding probability for that z-value by using a Z-table, such as Table A-1 in the appendix. You then subtract that result from 1 (because Table A-1 gives you "less than" probabilities only).

To find the z-value for an x-value, subtract the population mean from x, and divide by the population standard deviation:

$$z = \frac{x - \mu}{\sigma}$$

For a binomial distribution, the mean is $\mu = np$ and the standard deviation is $\sigma = \sqrt{np(1-p)}$. In this case, $n = 100$ and $p = 0.45$, so $\mu = (100)(0.45) = 45$, and the standard deviation is

$$\sigma = \sqrt{np(1-p)}$$
$$= \sqrt{100(0.45)(1-0.45)}$$
$$= \sqrt{24.75} = 4.97$$

Now, plug these numbers into the formula for the z-value, where $x = 56$:

$$z = \frac{x - \mu}{\sigma}$$
$$= \frac{(56-45)}{4.97} = 2.21$$

Then, find $P(Z \leq 2.21)$, using Table A-1. Look in the row for $z = 2.2$ and follow across to where it intersects with the column for 0.01, which gives you 0.9864. To get $P(Z \geq 2.21)$, subtract that value from 1: $1 - 0.9864 = 0.0136$. This corresponds to $P(X \geq 56)$.

Remember this is only an approximation because you started with a binomial random variable and were able to use the normal approximation to find the probability.

222.   0.9864

Because n is so large, you may want to use the normal approximation to the binomial to solve this problem. But first you need to determine whether the two conditions are met — that is, both np and $n(1 - p)$ must be at least 10. In this case, $np = 100(0.45) = 45$, and $n(1 - p) = 100(1 - 0.45) = (100)(0.55) = 55$. Both conditions are at least 10, so you can move forward.

To find a "less than" probability for an x-value from a normal distribution, you convert the x-value to a z-value and then find the corresponding probability for that z-value by using a Z-table, such as Table A-1 in the appendix.

To find the z-value for an x-value, subtract the population mean from x, and divide by the population standard deviation:

$$z = \frac{x - \mu}{\sigma}$$

For a binomial distribution, the mean is $\mu = np$, and the standard deviation is $\sigma = \sqrt{np(1-p)}$. In this case, $n = 100$ and $p = 0.45$, so $\mu = (100)(0.45) = 45$, and the standard deviation is

$$\sigma = \sqrt{np(1-p)}$$
$$= \sqrt{100(0.45)(1-0.45)}$$
$$= \sqrt{24.75} = 4.97$$

Now, plug these numbers into the formula for the z-value, where $x = 56$:

$$z = \frac{x - \mu}{\sigma}$$
$$= \frac{(56-45)}{4.97} = 2.21$$

Then find $P(Z \leq 2.21)$, using Table A-1. Look in the row for $z = 2.2$ and follow across to where it intersects with the column for 0.01, which gives you 0.9864. Remember this is only an approximation because you started with a binomial and were able to use the normal approximation to find the probability.

223.   0.0023

Because n is so large, you may want to use the normal approximation to the binomial to solve this problem. But first you need to determine whether the two conditions are met — that is, both np and $n(1-p)$ must be at least 10. In this case, $np = 100(0.45) = 45$, and $n(1-p) = 100(1-0.45) = (100)(0.55) = 55$. Both conditions are at least 10, so you can move forward.

To find the probability of "being between" two x-values from a normal distribution, convert each x-value to a z-value, and find the corresponding probability for each z-value by using a Z-table, such as Table A-1 in the appendix. Then subtract the lowest probability from the highest probability.

To find the z-value for an x-value, subtract the population mean from x, and divide by the population standard deviation:

$$z = \frac{x - \mu}{\sigma}$$

For a binomial distribution, the mean is $\mu = np$, and the standard deviation is $\sigma = \sqrt{np(1-p)}$. In this case, $n = 100$ and $p = 0.45$, so $\mu = (100)(0.45) = 45$, and the standard deviation is

$$\sigma = \sqrt{np(1-p)}$$
$$= \sqrt{100(0.45)(1-0.45)}$$
$$= \sqrt{24.75} = 4.97$$

Now, plug these numbers into the formula for the z-value, where $x = 56$:

$$z = \frac{x - \mu}{\sigma}$$
$$= \frac{(56-45)}{4.97} = 2.21$$

Then find $P(Z \le 2.21)$, using Table A-1. Look in the row for $z = 2.2$ and follow across to where it intersects with the column for 0.01, which gives you 0.9864.

Follow these steps for $x = 60$:

$$z = \frac{x - \mu}{\sigma}$$
$$= \frac{(60-45)}{4.97} = 3.02$$

Find $P(Z \le 3.02)$ in Table A-1, which is 0.9987.

Finally, to find the probability of being between the two z-values, subtract the lowest probability from the highest probability to get $0.9987 - 0.9864 = 0.0023$. Remember this is only an approximation because you started with a binomial and were able to use the normal approximation to find the probability because the necessary conditions were met.

224.   All of the above.

The properties of the normal distribution are that it's symmetrical, mean and median are the same, the most common values are near the mean and less common values are farther from it, and the standard deviation marks the distance from the mean to the inflection point.

225. 68%

The empirical rule (also known as the 68-95-99.7 rule) says that about 68% of the values in a normal distribution are within one standard deviation of the mean.

226. 95%

The empirical rule (also known as the 68-95-99.7 rule) says that about 95% of the values in a normal distribution are within two standard deviations of the mean.

227. 99.7%

The empirical rule (also known as the 68-95-99.7 rule) says that about 99.7% of the values in a normal distribution are within three standard deviations of the mean.

228. the mean and the standard deviation

You can re-create any normal distribution if you know two parameters: the mean and the standard deviation. The mean is the center of the bell-shaped picture, and the standard deviation is the distance from the mean to the inflection point (the place where the concavity of the curve changes on the graph).

229. C. $\mu = 5, \sigma = 1.75$

The larger the standard deviation (σ), the greater the spread for a normal distribution. The value of the mean (μ) doesn't affect the spread of the normal distribution; it just shows you where the center is.

230. 6.2

You want to find a value of X where 34% of the values lie between the mean (5) and x (and x is in the right side of the mean). First, note that the normal distribution has a total probability of 100%, and each half takes up 50%. You'll use the 50% idea to do this problem.

Because this is a normal distribution, according to the empirical rule, about 68% of values are within one standard deviation from the mean on either side. That means that about 34% are within one standard deviation above the mean.

If x is one standard deviation above its mean, x equals the mean (5) plus 1 times the standard deviation (1.2): $x = 5 + 1(1.2) = 6.2$.

231. 7.4

You want to find a value of X where 2.5% of the values are greater than x. First, note that the normal distribution has a total probability of 100%, and each half takes up 50%. You'll use this 50% idea to do this problem.

Because this is a normal distribution, according to the empirical rule, about 95% of values are within two standard deviations from the mean on either side. That means that about 47.5% are within two standard deviations above the mean, and beyond that point, you have about (50 – 47.5) = 2.5% of the values. (Remember that the total percentage above the mean equals 50%.) The value of X where this occurs is the one that's two standard deviations above its mean.

If x is two standard deviations above its mean, x equals the mean (5) plus 2 times the standard deviation (1.2): x = 5 + 2(1.2) = 7.4.

232. 3.8

You want to find a value of X where 16% of the values are less than x. First, note that the normal distribution has a total probability of 100%, and each half takes up 50%. You'll use this 50% idea to do this problem.

Because this is a normal distribution, according to the empirical rule (or the 68-95-99.7 rule), about 68% of values are within one standard deviation from the mean on either side. That means that about 34% are within one standard deviation below the mean, and 16% of the values lie below that value. (Remember the total percentage below the mean equals 50%, so you have 50 – 34 = 16%.) The value of X where all this occurs is one standard deviation below its mean.

If x is one standard deviation below its mean, x equals the mean (5) minus 1 times the standard deviation (1.2): x = 5 – 1(1.2) = 3.8.

233. 1.5

The empirical rule (or the 68-95-99.7 rule) says that in a normal distribution, about 99.7% of the values lie within three standard deviations above and below the mean (8), which includes the numbers between $8 - 3\sigma$ and $8 + 3\sigma$.

If the upper and lower values of this range are supposed to be 3.5 and 12.5, you know that $3.5 = 8 - 3\sigma$ and $12.5 = 8 + 3\sigma$. Solving for σ in the first equation, you get $3\sigma = 4.5$, so

$$\sigma = \frac{4.5}{3} = 1.5$$

The same answer works in the second equation.

234. $\mu = 0, \sigma = 1$

The Z-distribution, also called the standard normal distribution, has a mean (μ) of 0 and a standard deviation (σ) of 1.

235. 1.2

To calculate the z-score for a value of X, subtract the population mean from x and then divide by the standard deviation:

$$z = \frac{x - \mu}{\sigma}$$

$$= \frac{21.2 - 17}{3.5}$$

$$= 1.2$$

236.   −1.0

To calculate the z-score for a value of X, subtract the population mean from x and then divide by the standard deviation:

$$z = \frac{x - \mu}{\sigma}$$

$$= \frac{13.5 - 17}{3.5}$$

$$= -1.0$$

237.   2.5

To calculate the z-score for a value of X, subtract the population mean from x and then divide by the standard deviation:

$$z = \frac{x - \mu}{\sigma}$$

$$= \frac{25.75 - 17}{3.5}$$

$$= 2.5$$

238.   15.6

The question gives you a z-score and asks for its corresponding x-value. The z-formula contains both x and z, so as long as you know one of them you can always find the other:

$$z = \frac{x - \mu}{\sigma}$$

You know that $z = -0.4$, $\mu = 17$, and $\sigma = 3.5$, so you just plug these numbers into the z-formula and then solve for x:

$$-0.4 = \frac{x - 17}{3.5}$$

$$x = 17 - 0.4(3.5)$$

$$= 15.6$$

239.   24.7

The question gives you a z-score and asks for its corresponding x-value. The z-formula contains both x and z, so as long as you know one of them you can always find the other:

$$z = \frac{x - \mu}{\sigma}$$

You know that $z = 2.2$, $\mu = 17$, and $\sigma = 3.5$, so you just plug these numbers into the z-formula and then solve for x:

$$2.2 = \frac{x - 17}{3.5}$$

$$x = 17 + 2.2(3.5)$$

$$= 24.7$$

240.   Form A = 85, Form B = 86

The exam scores from the two forms have different means and different standard deviations, so you can convert both of them to Z-distributions so they're on the same scale and then do the work from there.

The formula that changes an x-value to a z-value is

$$z = \frac{x - \mu}{\sigma}$$

For Form A, you want the x-value corresponding to a z-score of 1.5. Form A has a mean of 70 and standard deviation of 10, so the x-value is

$$1.5 = \frac{x - 70}{10}$$
$$x = 70 + 1.5(10)$$
$$= 85$$

Similarly for Form B, with a mean of 74 and standard deviation of 8, you have

$$1.5 = \frac{x - 74}{8}$$
$$x = 74 + 1.5(8)$$
$$= 86$$

241.   Form A = 50, Form B = 58

The exam scores from the two forms have different means and different standard deviations, so you can convert both of them to Z-distributions so they're on the same scale and then do the work from there.

The formula that changes an x-value to a z-value is

$$z = \frac{x - \mu}{\sigma}$$

For Form A, you want the x-value corresponding to a z-score of –2.0. Form A has a mean of 70 and standard deviation of 10, so the x-value is

$$-2.0 = \frac{x - 70}{10}$$
$$x = 70 - 2.0(10)$$
$$= 50$$

Similarly for Form B, with a mean of 74 and standard deviation of 8, you have

$$-2.0 = \frac{x - 74}{8}$$
$$x = 74 - 2.0(8)$$
$$= 58$$

242.   82

To find and/or use x-values on two different normal distributions, you use the z-formula to get everything on the same scale and then work from there. For this question, first find the z-score that goes with a score of 80 on Form A, and then find the score on Form B that corresponds to that same z-score.

To find the z-score for Form A, use the z-formula with an x-score of 80, a mean of 70, and a standard deviation of 10:

$$z = \frac{x - \mu}{\sigma}$$

$$= \frac{80 - 70}{10}$$

$$= 1$$

So a score of 80 on Form A has a z-score of 1, meaning its score is one standard deviation above its mean. To find the corresponding score on Form B, add 1 standard deviation (8) to its mean (74): $74 + (1)(8) = 82$.

---

243.   86

To find and/or use x-values on two different normal distributions, you use the z-formula to get everything on the same scale and then work from there. For this question, first find the z-score that goes with a score of 85 on Form A, and then find the score on Form B that corresponds to that same z-score.

To find the z-score for Form A, use the z-formula with an x-score of 85, a mean of 70, and a standard deviation of 10:

$$z = \frac{x - \mu}{\sigma}$$

$$= \frac{85 - 70}{10}$$

$$= 1.5$$

So a score of 85 on Form A has a z-score of 1.5, meaning its score is one and a half standard deviations above its mean. To find the corresponding score on Form B, add 1.5 standard deviations (8) to its mean (74): $74 + (1.5)(8) = 86$.

---

244.   75

To find and/or use x-values on two different normal distributions, you use the z-formula to get everything on the same scale and then work from there. For this question, first find the z-score that goes with a score of 78 on Form B, and then find the score on Form A that corresponds to that same z-score.

To find the z-score for Form B, use the z-formula with an x-score of 78, a mean of 74, and a standard deviation of 8:

$$z = \frac{x - \mu}{\sigma}$$

$$= \frac{78 - 74}{8}$$

$$= 0.5$$

So a score of 78 on Form B has a z-score of 0.5, meaning its score is half a standard deviation above its mean. To find the corresponding score on Form A, add 0.5 standard deviations (10) to its mean (70): $70 + (0.5)(10) = 75$.

245. 62.5

To find and/or use x-values on two different normal distributions, you use the z-formula to get everything on the same scale and then work from there. For this question, first find the z-score that goes with a score of 68 on Form B, and then find the score on Form A that corresponds to that same z-score.

To find the z-score for Form B, use the z-formula with an x-score of 68, a mean of 74, and a standard deviation of 8:

$$z = \frac{x - \mu}{\sigma}$$
$$= \frac{68 - 74}{8}$$
$$= -0.75$$

So a score of 68 on Form B is 0.75 standard deviations below its mean. To find the corresponding score on Form A, add −0.75 of its standard deviations (10) to its mean (70): $70 + (-0.75)(10) = 62.5$.

246. $P(Z \le 2)$

Looking at the graph, you see that the shaded area represents the probability of all z-values of 2 or less. The probability notation for this is $P(Z \le 2)$.

247. $P(0 \le Z \le 2)$

Looking at the graph, you see that the shaded area represents the probability that Z is between 0 and 2, expressed as $P(0 \le Z \le 2)$.

248. $P(Z \ge -2)$

Looking at the graph, you see that the shaded area represents the probability of a z-value of −2 or higher, expressed as $P(Z \ge -2)$.

249. 0.9332

To find $P(Z \le 1.5)$, using the Z-table (Table A-1 in the appendix), find where the row for 1.5 intersects with the column for 0.00; this value is 0.9332. The Z-table shows only "less than" probabilities so it gives you exactly what you need for this question. *Note:* No probability is exactly at one single point, so $P(Z \le 1.5) = P(Z < 1.5)$.

250. 0.0668

You want $P(Z \ge 1.5)$, so use the Z-table (Table A-1 in the appendix) to find where the row for 1.5 intersects with the column for 0.00, which is 0.9332. Because the Z-table gives you only "less than" probabilities, subtract $P(Z < 1.5)$ from 1 (remember that the total probability for the normal distribution is 1.00, or 100%):

$$P(Z \ge 1.5) = 1 - P(Z < 1.5)$$
$$= 1 - 0.9332 = 0.0668$$

251. 0.7734

You want $P(Z \geq -0.75)$, so use the Z-table (Table A-1 in the appendix) to find where the row for –0.7 intersects with the column for 0.05, which is 0.2266. Because the Z-table gives you only "less than" probabilities, subtract $P(Z < -0.75)$ from 1 (remember that the total probability for the normal distribution is 1.00, or 100%):

$$P(Z \geq -0.75) = 1 - P(Z < -0.75)$$
$$= 1 - 0.2266 = 0.7734$$

252. 0.5328

To find the probability that Z is between two values, use the Z-table (Table A-1 in the appendix) to find the probabilities corresponding to each z-value, and then find the difference between the probabilities.

Here, you want the probability that Z is between –0.5 and 1.0. First, use the Z-table to find the value where the row for –0.5 intersects with the column for 0.00, which is 0.3085. Then, find the value where the row for 1.0 intersects with the column for 0.00, which is 0.8413. Because the Z-table gives you only "less than" probabilities, find the difference between the probability less than 1.0 (written as $P[Z \leq 1.0]$) and the probability less than –0.5 (written as $P[Z \leq -0.5]$):

$$P(-0.5 \leq Z \leq 1.0) = P(Z \leq 1.0) - P(Z \leq -0.50)$$
$$= 0.8413 - 0.3085 = 0.5328$$

253. 0.6826

To find the probability that Z is between two values, use the Z-table (Table A-1 in the appendix) to find the probabilities corresponding to each z-value, and then find the difference between the probabilities.

Here, you want the probability that Z is between –1.0 and 1.0. First, use the Z-table to find the value where the row for –1.0 intersects with 0.00, which is 0.1587. Then, find the value where the row for 1.0 intersects with the column for 0.00, which is 0.8413. Because the Z-table gives you only "less than" probabilities, find the difference between probability less than 1.0 (written as $P[Z \leq 1.0]$) and the probability less than –1.0 (written as $P[Z \leq -1.0]$):

$$P(-1.0 \leq Z \leq 1.0) = P(Z \leq 1.0) - P(Z \leq -1.0)$$
$$= 0.8413 - 0.1587 = 0.6826$$

254. 1.5

To find a z-score for a particular value of X, subtract the population mean from x, and then divide by the population standard deviation:

$$z = \frac{x - \mu}{\sigma}$$
$$= \frac{13 - 10}{2}$$
$$= 1.5$$

255. 0.0668

To find a "greater than" probability for an x-value, you first convert the x-value to a z-score and then find the corresponding probability for that z-score by using a z-table, such as Table A-1 in the appendix. Then, you subtract that result from 1 (because Table A-1 gives you "less than" probabilities only).

To find the z-score for an x-value, subtract the population mean from x, and then divide by the population standard deviation:

$$z = \frac{x - \mu}{\sigma}$$

Here, $x = 13$ centimeters in diameter, and you want $P(X \geq 13)$, the mean, μ, is 10, and the standard deviation, σ, is 2. Plug these numbers into the z-formula to convert to a z-score:

$$z = \frac{13 - 10}{2} = 1.5$$

Using Table A-1, find where the row for 1.5 and the column for 0.00 intersect; you get $P(Z < 1.5) = 0.9332$. Now, subtract this value from 1 to get $P(Z > 1.5) = 1 - 0.9332 = 0.0668$.

256. 0.9332

To find the probability of a value "no greater than 13" means that the value must be "less than or equal to 13." So, first, convert 13 into a z-score, and then use the Z-table (Table A-1 in the appendix) to find the probability (because Table A-1 provides "less than" probabilities only).

To find the z-score for an x-value, subtract the population mean from x, and then divide by the population standard deviation:

$$z = \frac{x - \mu}{\sigma}$$

Here, $x = 13$ centimeters in diameter, and you want $P(X \leq 13)$, the mean, μ, is 10, and the standard deviation, σ, is 2. Plug these numbers into the z-formula to convert to a z-score:

$$z = \frac{13 - 10}{2} = 1.5$$

Using Table A-1, find where the row for 1.5 and the column for 0.00 intersect; you get $P(Z \leq 1.5) = 0.9332$.

257. 0.4332

To find the probability that X is between two values, change both values to z-scores and then use the Z-table (Table A-1 in the appendix) to find the probabilities corresponding to each z-value; finally, find the difference between the probabilities.

Here, you want the probability that X is between 10 and 13. To find the z-score for an x-value, subtract the population mean from x, and then divide by the population standard deviation:

$$z = \frac{x - \mu}{\sigma}$$

Change $x = 10$ to a z-score with a mean of 10 and a standard deviation of 2:

$$z = \frac{10 - 10}{2} = 0$$

Then, do the same for $x = 13$:

$$z = \frac{13 - 10}{2} = 1.5$$

Now, find the probabilities that Z is between 0 and 1.5. First, use the Z-table to find the value where the row for 0.0 intersects with the column for 0.00, which is 0.5000. Then, find the value where the row for 1.5 intersects with the column for 0.00, which is 0.9332. Because the Z-table gives you only "less than" probabilities, find the difference between the probability less than 1.5 (written as $P[Z \leq 1.5]$) and the probability less than 0 (written as $P[Z \leq 0]$):

$$P(Z \leq 1.5) - P(Z \leq 0) = 0.9332 - 0.5000 = 0.4332$$

---

258. 0.4322

To find the probability that X is between two values, change both values to z-scores and then use the Z-table (Table A-1 in the appendix) to find the probabilities corresponding to each z-value; finally, find the difference between the probabilities.

Here, you want the probability that X is between 7 and 10. To find the z-score for an x-value, subtract the population mean from x, and then divide by the population standard deviation:

$$z = \frac{x - \mu}{\sigma}$$

Change $x = 7$ to a z-score with a mean of 10 and a standard deviation of 2:

$$z = \frac{7 - 10}{2} = -1.5$$

Then, do the same for $x = 10$:

$$z = \frac{10 - 10}{2} = 0$$

Now, find the probabilities that Z is between -1.5 and 0. First, use the Z-table to find the value where the row for -1.5 intersects with the column for 0.00, which is 0.0668. Then, find the value where the row for 0 intersects with the column for 0.00, which is 0.5000. Because the Z-table gives you only "less than" probabilities, find the difference between the probability less than 0 (written as $P[Z \leq 0]$) and the probability less than -1.5 (written as $P[Z \leq -1.5]$):

$$P(Z \leq 0) - P(Z \leq -1.5) = 0.5000 - 0.0668 = 0.4332$$

---

259. -1.25

To find a z-score for a value of X, subtract the population mean (μ) from x, and then divide by the population standard deviation (σ):

$$z = \frac{x - \mu}{\sigma}$$

$$= \frac{135 - 160}{20} = -1.25$$

260.   0.5

To find a z-score for a value of X, subtract the population mean (μ) from x, and then divide by the population standard deviation (σ):

$$z = \frac{x - \mu}{\sigma}$$

$$= \frac{170 - 160}{20} = 0.5$$

261.   −2.25

To find a z-score for a value of X, subtract the population mean (μ) from x, and then divide by the population standard deviation (σ):

$$z = \frac{x - \mu}{\sigma}$$

$$= \frac{115 - 160}{20} = -2.25$$

262.   3

To find a z-score for a value of X, subtract the population mean (μ) from x, and then divide by the population standard deviation (σ):

$$z = \frac{x - \mu}{\sigma}$$

$$= \frac{220 - 160}{20} = 3$$

263.   2.25

To find a z-score for a value of X, subtract the population mean (μ) from x, and then divide by the population standard deviation (σ):

$$z = \frac{x - \mu}{\sigma}$$

$$= \frac{205 - 160}{20} = 2.25$$

264.   0.0013

To find a "greater than" probability for an x-value, first convert the x-value to a z-score and then find the corresponding probability for that z-score by using a Z-table, such as Table A-1 in the appendix; finally, you subtract that result from 1 (because Table A-1 gives you "less than" probabilities only).

To find the z-score for an x-value, subtract the population mean from x, and then divide by the population standard deviation:

$$z = \frac{x - \mu}{\sigma}$$

Here, x is 220 pounds, and you want $P(X > 220)$; the mean, μ, is 160, and the standard deviation, σ, is 20. Plug these numbers into the z-formula to convert to a z-score:

$$z = \frac{220 - 160}{20} = 3$$

Using Table A-1, find where the row for 3.0 and the column for 0.00 intersect; you get $P(Z \le 3.0) = 0.9987$. Now, subtract this value from 1 to get $P(Z > 3) = 1 - 0.9987 = 0.0013$.

---

265.   0.9987

To find a "less than" probability for a value of x from a normal distribution, first convert the value into a z-score, and then use the Z-table (Table A-1 in the appendix) to find the probability.

To find the z-score for an x-value, subtract the population mean from x, and then divide by the population standard deviation:

$$z = \frac{x - \mu}{\sigma}$$

Here, x is 220 pounds, the mean, μ, is 160, and the standard deviation, σ, is 20. Plug these numbers into the z-formula to convert to a z-score:

$$z = \frac{220 - 160}{20} = 3$$

Using Table A-1, find where the row for 3.0 and the column for 0.00 intersect; you get $P(Z < 3.0) = 0.9987$.

---

266.   0.3944

To find the probability that X is between two values, change both values to z-scores and then use the Z-table (Table A-1 in the appendix) to find the probabilities corresponding to each z-value; finally, find the difference between the probabilities.

Here, you want the probability that X is between 135 and 160. To find the z-score for an x-value, subtract the population mean from x, and then divide by the population standard deviation:

$$z = \frac{x - \mu}{\sigma}$$

Change $x = 135$ to a z-score with a mean of 160 and a standard deviation of 20:

$$z = \frac{135 - 160}{20} = -1.25$$

Then, do the same for $x = 160$:

$$z = \frac{160 - 160}{20} = 0$$

Now, find the probabilities that Z is between -1.25 and 0. First, use the Z-table to find the value where the row for -1.2 intersects with the column for 0.05, which is 0.1056. Then, find the value where the row for 0.0 intersects with the column for 0.00, which is 0.5000. Because the Z-table gives you only "less than" probabilities, find the difference between the probability less than 0 (written as $P[Z \le 0]$) and the probability less than –1.25 (written as $P[Z \le -1.25]$):

$$P(135 \le X \le 160) = P(-1.25 \le Z \le 0)$$
$$= P(Z \le 0) - P(Z \le -1.25)$$
$$= 0.5000 - 0.1056 = 0.3944$$

---

267.   0.0109

To find the probability that X is between two values, change both values to z-scores and then use the Z-table (Table A-1 in the appendix) to find the probabilities corresponding to each z-value; finally, find the difference between the probabilities.

Here, you want the probability that X is between 205 and 220, written as $P(205 \le X \le 220)$. To find the z-score for an x-value, subtract the population mean from x, and then divide by the population standard deviation:

$$z = \frac{x - \mu}{\sigma}$$

Change $x = 205$ to a z-score with a mean of 160 and a standard deviation of 20:

$$z = \frac{205 - 160}{20} = 2.25$$

Then, do the same for $x = 220$:

$$z = \frac{220 - 160}{20} = 3$$

Now, find the probabilities that Z is between 2.25 and 3. First, use the Z-table to find the value where the row for 2.2 intersects with the column for 0.05, which is 0.9878. Then, find the value where the row for 3.0 intersects with the column for 0.00, which is 0.9987. Because the Z-table gives you only "less than" probabilities, find the difference between the probability less than 3.0 (written as $P[Z \le 3.0]$) and the probability less than 2.25 (written as $P[Z \le 2.25]$). In essence, you're starting with everything below 3.0 and taking off what you don't want, which is everything below 2.25:

$$P(205 \le X \le 220) = P(2.25 \le Z \le 3.0)$$
$$= P(Z \le 3.0) - P(Z \le 2.25)$$
$$= 0.9987 - 0.9878 = 0.0109$$

---

268.   0.0934

To find the probability that X is between two values, change both values to z-scores and then use the Z-table (Table A-1 in the appendix) to find the probabilities corresponding to each z-value; finally, find the difference between the probabilities.

Here, you want the probability that X is between 115 and 135, written as $P(115 \le X \le 135)$. To find the z-score for an x-value, subtract the population mean from x, and then divide by the population standard deviation:

$$z = \frac{x - \mu}{\sigma}$$

Change $x = 115$ to a z-score with a mean of 160 and a standard deviation of 20:

$$z = \frac{115 - 160}{20} = -2.25$$

Then, do the same for $x = 135$:

$$z = \frac{135 - 160}{20} = -1.25$$

Now, find the probabilities that Z is between -2.25 and -1.25. First, use the Z-table to find the value where the row for -2.2 intersects with the column for 0.05, which is 0.0122. Then, find the value where the row for -1.2 intersects with the column for 0.05, which is 0.1056. Because the Z-table gives you only "less than" probabilities, find the difference between the probability less than -1.25 (written as $P[Z < -1.25]$) and the probability less than -2.25 (written as $P[Z < -2.25]$). In essence, you're starting with everything below -1.25 and taking off what you don't want, which is everything below -2.25:

$$P(115 \leq X \leq 135) = P(-2.25 \leq Z \leq -1.25)$$
$$= P(Z \leq -1.25) - P(Z \leq -2.25)$$
$$= 0.1056 - 0.0122 = 0.0934$$

---

269.   0.5859

To find the probability that X is between two values, change both values to z-scores and then use the Z-table (Table A-1 in the appendix) to find the probabilities corresponding to each z-value; finally, find the difference between the probabilities.

Here, you want the probability that X is between 135 and 170, written as $P(135 \leq X \leq 170)$. To find the z-score for an x-value, subtract the population mean from x, and then divide by the population standard deviation:

$$z = \frac{x - \mu}{\sigma}$$

Change $x = 135$ to a z-score with a mean of 160 and a standard deviation of 20:

$$z = \frac{135 - 160}{20} = -1.25$$

Then, do the same for $x = 170$:

$$z = \frac{170 - 160}{20} = 0.50$$

Now, find the probabilities that Z is between -1.25 and -0.50. First, use the Z-table to find the value where the row for -1.2 intersects with the column for 0.05, which is 0.1056. Then, find the value where the row for 0.5 intersects with the column for 0.00, which is 0.6915. Because the Z-table gives you only "less than" probabilities, find the difference between the probability less than 0.50 (written as $P[Z < 0.50]$) and the probability less than -1.25 (written as $P[Z < -1.25]$). In essence, you're starting with everything below 0.50 and taking off what you don't want, which is everything below -1.25:

$$P(135 \leq X \leq 170) = P(-1.25 \leq Z \leq 0.50)$$
$$= P(Z \leq 0.50) - P(Z \leq -1.25)$$
$$= 0.6915 - 0.1056 = 0.5859$$

---

270.   0.3072

To find the probability that X is between two values, change both values to z-scores and then use the Z-table (Table A-1 in the appendix) to find the probabilities corresponding to each z-value; finally, find the difference between the probabilities.

Here, you want the probability that X is between 170 and 220, written as $P(170 \leq X \leq 220)$. To find the z-score for an x-value, subtract the population mean from x, and then divide by the population standard deviation:

$$z = \frac{x - \mu}{\sigma}$$

Change $x = 170$ to a z-score with a mean of 160 and a standard deviation of 20:

$$z = \frac{170 - 160}{20} = 0.5$$

Then, do the same for $x = 220$:

$$z = \frac{220 - 160}{20} = 3$$

Now, find the probabilities that Z is between 0.5 and 3.0. First, use the Z-table to find the value where the row for 0.5 intersects with the column for 0.00, which is 0.6915. Then, find the value where the row for 3.0 intersects with the column for 0.00, which is 0.9987. Because the Z-table gives you only "less than" probabilities, find the difference between the probability less than 3.0 (written as $P[Z \leq 3.0]$) and the probability less than 0.50 (written as $P[Z \leq 0.50]$). In essence, you're starting with everything below 3.0 and taking off what you don't want, which is everything below 0.50:

$$P(170 \leq X \leq 220) = P(0.50 \leq Z \leq 3.0)$$
$$= P(Z \leq 3.0) - P(Z \leq 0.50)$$
$$= 0.9987 - 0.6915 = 0.3072$$

---

271.   27

In this case, using intuition is very helpful. If you have a normal distribution for the population, then half of the values lie below the mean (because it's symmetrical and the total percentage is 100%). Here, the mean is 27, so 50%, or half, of the population of adults has a BMI lower than 27.

---

272.   23.65

You want to find the value of X (BMI) where 25% of the population lies below it. In other words, you want to find the 25th percentile of X. First, you need to find the 25th percentile for Z (using the Z-table, or Table A-1 in the appendix) and then change the z-value to an x-value by using the z-formula:

$$z = \frac{x - \mu}{\sigma}$$

To find the 25th percentile for Z (or the cutoff point where 25% of the population lies below it), look at the Z-table and find the probability that's closest to 0.25. (**Remember:** The probabilities for the Z-table are the values *inside* the table. The numbers on the outsides that tell which row/column you're in are actual z-values, not probabilities.) Searching Table A-1, you see that the closest probability to 0.25 is 0.2514.

Next, find what z-score this probability corresponds to. After you've located 0.2514 inside the table, find its corresponding row (–0.6) and column (0.07). Put these numbers together and you get the z-score of –0.67. This is the 25th percentile for Z. In other words, 25% of the z-values lie below –0.67.

To find the corresponding BMI that marks the 25th percentile, use the z-formula and solve for x. You know that $z = -0.67$, $\mu = 27$, and $\sigma = 5$:

$$z = \frac{x - \mu}{\sigma}$$

$$-0.67 = \frac{x - 27}{5}$$

$$x = 27 - 0.67(5)$$

$$= 23.65$$

So 25% of the population has a BMI lower than 23.65.

---

273. 18.80

You want to find the value of X (BMI) where 5% of the population lies below it. In other words, you want to find the 5th percentile of X. First, you need to find the 5th percentile for Z (using the Z-table, or Table A-1 in the appendix) and then change the z-value to an x-value by using the z-formula:

$$z = \frac{x - \mu}{\sigma}$$

To find the 5th percentile for Z (or the cutoff point where 5% of the population lies below it), look at the Z-table and find the probability that's closest to 0.05. (**Remember:** The probabilities for the Z-table are the values *inside* the table. The numbers on the outsides that tell which row/column you're in are actual z-values, not probabilities.) Searching Table A-1, you see that the closest probability to 0.05 is either 0.0495 or 0.0505 (use 0.0505 in this case).

Next, find what z-score this probability corresponds to. After you've located 0.0505 inside the table, find its corresponding row (–1.6) and column (0.04). Put these numbers together and you get the z-score of –1.64. This is the 5th percentile for Z. In other words, 5% of the z-values lie below –1.64.

To find the corresponding BMI that marks the 5th percentile, use the z-formula and solve for x. You know that $z = -1.64$, $\mu = 27$, and $\sigma = 5$:

$$z = \frac{x - \mu}{\sigma}$$

$$-1.64 = \frac{x - 27}{5}$$

$$x = 27 - 1.64(5)$$

$$= 18.80$$

So 5% of the population has a BMI lower than 18.80.

---

274. 20.60

You want to find the value of X (BMI) where 10% of the population lies below it. In other words, you want to find the 10th percentile of X. First, you need to find the 10th percentile for Z (using the Z-table, or Table A-1 in the appendix) and then change the z-value to an x-value by using the z-formula:

$$z = \frac{x - \mu}{\sigma}$$

To find the 10th percentile for *Z* (or the cutoff point where 10% of the population lies below it), look at the *Z*-table and find the probability that's closest to 0.10. (***Remember:*** The probabilities for the *Z*-table are the values *inside* the table. The numbers on the outsides that tell which row/column you're in are actual *z*-values, not probabilities.) Searching Table A-1, you see that the closest probability to 0.10 is 0.1003.

Next, find what *z*-score this probability corresponds to. After you've located 0.1003 inside the table, find its corresponding row (–1.2) and column (0.08). Put these numbers together and you get the *z*-score of –1.28. This is the 10th percentile for *Z*. In other words, 10% of the *z*-values lie below –1.28.

To find the corresponding BMI that marks the 10th percentile, use the *z*-formula and solve for *x*. You know that $z = -1.28$, $\mu = 27$, and $\sigma = 5$:

$$z = \frac{x - \mu}{\sigma}$$

$$-1.28 = \frac{x - 27}{5}$$

$$x = 27 + (-1.28)(5)$$

$$= 20.60$$

So 10% of the population has a BMI lower than 20.60.

---

275.   33.40

You want to find the value of *X* (BMI) where 10% of the population lies above it. Because you need to use the *Z*-table to solve this problem and because the *Z*-table shows only "less than" probabilities, work this problem as if you wanted the cutoff for the lower 90%. In other words, you want to find the 90th percentile of *X* (don't worry; you'll get the same answer). First, you need to find the 90th percentile for *Z* (using the *Z*-table, or Table A-1 in the appendix) and then change the *z*-value to an *x*-value by using the *z*-formula:

$$z = \frac{x - \mu}{\sigma}$$

To find the 90th percentile for *Z*, look at the *Z*-table and find the probability that's closest to 0.90. (***Remember:*** The probabilities for the *Z*-table are the values *inside* the table. The numbers on the outsides that tell which row/column you're in are actual *z*-values, not probabilities.) Searching Table A-1, you see that the closest probability to 0.90 is 0.8997.

Next, find what *z*-score this probability corresponds to. After you've located 0.8997 inside the table, find its corresponding row (1.2) and column (0.08). Put these numbers together and you get the *z*-score of 1.28. This is the 90th percentile for *Z*. In other words, 90% of the *z*-values lie below 1.28 (and 10% are above it).

To find the corresponding BMI that marks the 90th percentile, use the *z*-formula and solve for *x*. You know that $z = 1.28$, $\mu = 27$, and $\sigma = 5$:

$$z = \frac{x - \mu}{\sigma}$$

$$1.28 = \frac{x - 27}{5}$$

$$x = 27 + (1.28)(5)$$

$$= 33.40$$

So the BMI marking the upper 10% for this population is 33.40.

276. 35.25

You want to find the value of X (BMI) where 5% of the population lies above it. Because you need to use the Z-table to solve this problem and because the Z-table shows only "less than" probabilities, work this problem as if you wanted the cutoff for the lower 95%. In other words, you want to find the 95th percentile of X (don't worry; you'll get the same answer). First, you need to find the 95th percentile for Z (using the Z-table, or Table A-1 in the appendix) and then change the z-value to an x-value by using the z-formula:

$$z = \frac{x - \mu}{\sigma}$$

To find the 95th percentile for Z, look at the Z-table and find the probability that's closest to 0.95. (**Remember:** The probabilities for the Z-table are the values *inside* the table. The numbers on the outsides that tell which row/column you're in are actual z-values, not probabilities.) In Table A-1, use the probability 0.9505.

Next, find what z-score this probability corresponds to. After you've located 0.9505 inside the table, find its corresponding row (1.6) and column (0.05). Put these numbers together and you get the z-score of 1.65. This is the 95th percentile for Z. In other words, 95% of the z-values lie below 1.65 (and 5% are above it).

To find the corresponding BMI that marks the 95th percentile, use the z-formula and solve for x. You know that $z = 1.65$, $\mu = 27$, and $\sigma = 5$:

$$z = \frac{x - \mu}{\sigma}$$

$$1.65 = \frac{x - 27}{5}$$

$$x = 27 + 1.65(5)$$

$$= 35.25$$

So the BMI marking the upper 5% for this population is 35.25.

277. 29.60

You want to find the value of X (BMI) where 30% of the population lies above it. Because you need to use the Z-table to solve this problem and because the Z-table shows only "less than" probabilities, work this problem as if you wanted the cutoff for the lower 70%. In other words, you want to find the 70th percentile of X (don't worry; you'll get the same answer). First, find the 70th percentile for Z (using the Z-table, or Table A-1 in the appendix) and then change the z-value to an x-value by using the z-formula:

$$z = \frac{x - \mu}{\sigma}$$

To find the 70th percentile for Z, look at the Z-table and find the probability that's closest to 0.70. (**Remember:** The probabilities for the Z-table are the values *inside* the table. The numbers on the outsides that tell which row/column you're in are actual z-values, not probabilities.) Searching Table A-1, you see that the closest probability to 0.70 is 0.6985.

Next, find what z-score this probability corresponds to. After you've located 0.6985 inside the table, find its corresponding row (0.5) and column (0.02). Put these numbers together and you get the z-score of 0.52. This is the 70th percentile for Z. In other words, 70% of the z-values lie below 0.52 (and 30% are above it).

To find the corresponding BMI that marks the 70th percentile, use the z-formula and solve for x. You know that $z = 0.52$, $\mu = 27$, and $\sigma = 5$:

$$z = \frac{x - \mu}{\sigma}$$

$$0.52 = \frac{x - 27}{5}$$

$$x = 27 + (0.52)(5)$$

$$= 29.60$$

So the BMI marking the upper 30% for this population is 29.60.

---

278. 23.65, 30.35

The 1st quartile (Q_1) is the value with 25% of the distribution below it, and the 3rd quartile (Q_3) is the value with 75% of the values below it. Using the Z-table (Table A-1 in the appendix), you can find that the value closest to 0.25 is 0.2514, corresponding to a z-score of –0.67 (the value where the row for –0.6 and the column for 0.07 intersect).

To find the 1st quartile (Q_1) of X (a BMI score) corresponding to $z = -0.67$, use the z-formula and solve for x:

$$z = \frac{x - \mu}{\sigma}$$

$$-0.67 = \frac{x - 27}{5}$$

$$x = 27 + (-0.67)(5)$$

$$= 23.65$$

To find the 3rd quartile (Q_3) for X, follow the same procedure: First, find the value in the Z-table that's closest to 0.75, which is 0.67 (note the symmetry in the values). Then, use the z-formula to solve for x:

$$z = \frac{x - \mu}{\sigma}$$

$$0.67 = \frac{x - 27}{5}$$

$$x = 27 + (0.67)(5)$$

$$= 30.35$$

---

279. 69.96

You want to find the value of X (exam score) where 20% of the population lies below it. In other words, you want to find the 20th percentile of X. First, find the 20th percentile for Z (using the Z-table, or Table A-1 in the appendix) and then change the z-value to an x-value by using the z-formula:

$$z = \frac{x - \mu}{\sigma}$$

To find the 20th percentile for Z (or the cutoff point where 20% of the population lies below it), look at the Z-table and find the probability that's closest to 0.20. (**Remember:** The probabilities for the Z-table are the values *inside* the table. The numbers on the outsides that tell which row/column you're in are actual z-values, not probabilities.) Searching Table A-1, you see that the closest probability to 0.20 is 0.2005.

Next, find what z-score this probability corresponds to. After you've located 0.2005 inside the table, find its corresponding row (–0.8) and column (0.04). Put these numbers together and you get the z-score of –0.84. This is the 20th percentile for Z. In other words, 20% of the z-values lie below –0.84.

To find the corresponding exam score that marks the 20th percentile, use the z-formula and solve for x. You know that $z = -0.84$, $\mu = 75$, and $\sigma = 6$:

$$z = \frac{x - \mu}{\sigma}$$

$$-0.84 = \frac{x - 75}{6}$$

$$x = 75 + (-0.84)(6)$$

$$= 69.96$$

So 20% of the students scored below 69.96.

280.   65.16

You want to find the value of X (exam score) where 5% of the population lies below it. In other words, you want to find the 5th percentile of X. First, you need to find the 5th percentile for Z (using the Z-table, or Table A-1 in the appendix) and then change the z-value to an x-value by using the z-formula:

$$z = \frac{x - \mu}{\sigma}$$

To find the 5th percentile for Z (or the cutoff point where 5% of the population lies below it), look at the Z-table and find the probability that's closest to 0.05. (*Remember:* The probabilities for the Z-table are the values *inside* the table. The numbers on the outsides that tell which row/column you're in are actual z-values, not probabilities.) Searching Table A-1, you see that the closest probability to 0.05 is either 0.0495 or 0.0505 (use 0.0505 in this case).

Next, find what z-score this probability corresponds to. After you've located 0.0505 inside the table, find its corresponding row (–1.6) and column (0.04). Put these numbers together and you get the z-score of –1.64. This is the 5th percentile for Z. In other words, 5% of the z-values lie below –1.64.

To find the corresponding exam score that marks the 5th percentile, use the z-formula and solve for x. You know that $z = -1.64$, $\mu = 75$, and $\sigma = 6$:

$$z = \frac{x - \mu}{\sigma}$$

$$-1.64 = \frac{x - 75}{6}$$

$$x = 75 - 1.64(6)$$

$$= 65.16$$

So 5% of the students scored below 65.16.

281.   82.68

You want to find the value of X (exam score) where 10% of the population lies above it. Because you need to use the Z-table to solve this problem and because the Z-table shows only "less than" probabilities, work this problem as if you wanted the cutoff for the lower 90%. In other words, you want to find the 90th percentile of X (don't worry; you'll get the same answer). First, you need to find the 90th percentile for Z (using the Z-table, or Table A-1 in the appendix) and then change the z-value to an x-value by using the z-formula:

$$z = \frac{x - \mu}{\sigma}$$

To find the 90th percentile for Z, look at the Z-table and find the probability that's closest to 0.90. (**Remember:** The probabilities for the Z-table are the values *inside* the table. The numbers on the outsides that tell which row/column you're in are actual z-values, not probabilities.) Searching Table A-1, you see that the closest probability to 0.90 is 0.8997.

Next, find what z-score this probability corresponds to. After you've located 0.8997 inside the table, find its corresponding row (1.2) and column (0.08). Put these numbers together and you get the z-score of 1.28. This is the 90th percentile for Z. In other words, 90% of the z-values lie below 1.28 (and 10% lie above it).

To find the corresponding exam score that marks the 90th percentile, use the z-formula and solve for x. You know that z = 1.28, $\mu = 75$, and $\sigma = 6$:

$$z = \frac{x - \mu}{\sigma}$$

$$1.28 = \frac{x - 75}{6}$$

$$x = 75 + 1.28(6)$$

$$= 82.68$$

So 10% of the students scored above 82.68.

282.   88.98

You want to find the value of X (exam score) where 1% of the population lies above it. Because you need to use the Z-table to solve this problem and because the Z-table shows only "less than" probabilities, work this problem as if you wanted the cutoff for the lower 99%. In other words, you want to find the 99th percentile of X (don't worry; you'll get the same answer). First, you need to find the 99th percentile for Z (using the Z-table, or Table A-1 in the appendix) and then change the z-value to an x-value by using the z-formula:

$$z = \frac{x - \mu}{\sigma}$$

To find the 99th percentile for Z, look at the Z-table and find the probability that's closest to 0.99. (**Remember:** The probabilities for the Z-table are the values *inside* the table. The numbers on the outsides that tell which row/column you're in are actual z-values, not probabilities.) Searching Table A-1, you see that the closest probability to 0.99 is 0.9901.

Next, find what z-score this probability corresponds to. After you've located 0.9901 inside the table, find its corresponding row (2.3) and column (0.03). Put these numbers together and you get the z-score of 2.33. This is the 99th percentile for Z. In other words, 99% of the z-values lie below 2.33 and 1% lie above it.

To find the corresponding exam score that marks the 99th percentile, use the z-formula and solve for x. You know that z = 2.33, $\mu = 75$, and $\sigma = 6$:

$$z = \frac{x - \mu}{\sigma}$$

$$2.33 = \frac{x - 75}{6}$$

$$x = 75 + 2.33(6)$$

$$= 88.98$$

So 1% of the students scored above 88.98.

283.   86.76

You want to find the value of X (exam score) where 2.5% of the population lies above it. Because you need to use the Z-table to solve this problem and because the Z-table shows only "less than" probabilities, work this problem as if you wanted the cutoff for the lower 97.5%. In other words, you want to find the 97.5th percentile of X (don't worry; you'll get the same answer). First, you need to find the 97.5th percentile for Z (using the Z-table, or Table A-1 in the appendix) and then change the z-value to an x-value by using the z-formula:

$$z = \frac{x - \mu}{\sigma}$$

To find the 97.5th percentile for Z, look at the Z-table and find the probability that's closest to 0.975. (*Remember:* The probabilities for the Z-table are the values *inside* the table. The numbers on the outsides that tell which row/column you're in are actual z-values, not probabilities.) Searching Table A-1, you see that the closest probability to 0.975 is exactly 0.9750.

Next, find what z-score this probability corresponds to. After you've located 0.9750 inside the table, find its corresponding row (1.9) and column (0.06). Put these numbers together and you get the z-score of 1.96. This is the 97.5th percentile for Z. In other words, 97.5% of the z-values lie below 1.96 (and 2.5% lie above it).

To find the corresponding exam score that marks the 97.5th percentile, use the z-formula and solve for x. You know that $z = 1.96$, $\mu = 75$, and $\sigma = 6$:

$$z = \frac{x - \mu}{\sigma}$$

$$1.96 = \frac{x - 75}{6}$$

$$x = 75 + (1.96)(6)$$

$$= 86.76$$

So 2.5% of the students scored above 86.76.

284.   84.84

You want to find the value of X (exam score) where 5% of the population lies above it. Because you need to use the Z-table to solve this problem and because the Z-table shows only "less than" probabilities, work this problem as if you wanted the cutoff for the lower 95%. In other words, you want to find the 95th percentile of X (don't worry; you'll get the same answer). First, you need to find the 95th percentile for Z (using the Z-table, or Table A-1 in the appendix) and then change the z-value to an x-value by using the z-formula:

$$z = \frac{x - \mu}{\sigma}$$

To find the 95th percentile for Z, look at the Z-table and find the probability that's closest to 0.95. (*Remember:* The probabilities for the Z-table are the values *inside* the table. The numbers on the outsides that tell which row/column you're in are actual z-values, not probabilities.) In Table A-1, use the probability 0.9495.

Next, find what z-score this probability corresponds to. After you've located 0.9495 inside the table, find its corresponding row (1.6) and column (0.04). Put these numbers together and you get the z-score of 1.64. This is the 95th percentile for Z. In other words, 95% of the z-values lie below 1.64 (and 5% are above it).

To find the corresponding exam score that marks the 95th percentile, use the z-formula and solve for x. You know that $z = 1.64$, $\mu = 75$, and $\sigma = 6$:

$$z = \frac{x - \mu}{\sigma}$$

$$1.64 = \frac{x - 75}{6}$$

$$x = 75 + (1.64)(6)$$

$$= 84.84$$

So 5% of the students scored above 84.84.

---

285.   310.8

The fastest (and best) times are at the lower end of the distribution. Using the Z-table (Table A-1 in the appendix), find the value where only 5% of the times are below it. The closest table value to 0.05 is 0.0505, which corresponds to a z-value of -1.64.

To find the time corresponding to a particular z-score, use the z-formula to solve for x:

$$z = \frac{x - \mu}{\sigma}$$

$$-1.64 = \frac{x - 360}{30}$$

$$x = 360 + (-1.64)(30)$$

$$= 310.8$$

This means a time of 310.8 seconds is the cutoff for the fastest 5% of the times.

---

286.   360

In this case, using intuition is very helpful. If you have a normal distribution for the population, then half of the values lie below the mean (because it's symmetrical and the total percentage is 100%). Here, the mean is 360, so 50%, or half, of the military recruits have a time of 360 seconds.

---

287.   398.4

The slowest (and worst) times are at the upper end of the distribution. Using the Z-table (Table A-1 in the appendix), find the value where 90% of the times are below it. The closest table value to 0.90 is 0.8997, corresponding to a z-value of 1.28.

To find the time corresponding to a particular z-score, use the z-formula to solve for x:

$$z = \frac{x - \mu}{\sigma}$$

$$1.28 = \frac{x - 360}{30}$$

$$x = 360 + (1.28)(30)$$

$$= 398.4$$

So the slowest 10% of the recruits had a time of 398.4 seconds or more.

288. 321.6

The fastest (and best) times are at the lower end of the distribution. Using the Z-table (Table A-1 in the appendix), find the value where only 10% of the times are below it. The closest table value to 0.10 is 0.1003, which corresponds to a z-value of –1.28.

To find the time corresponding to a particular z-score, use the z-formula to solve for x:

$$z = \frac{x - \mu}{\sigma}$$

$$-1.28 = \frac{x - 360}{30}$$

$$x = 360 + (-1.28)(30)$$

$$= 321.6$$

So the fastest 10% of the recruits had a time of 321.6 seconds or less.

289. 339.9

The fastest (and best) times are at the lower end of the distribution. Using the Z-table (Table A-1 in the appendix), find the value where only 25% of the times are below it. The closest table value to 0.25 is 0.2514, which corresponds to a z-value of –0.67.

To find the time corresponding to a particular z-score, use the z-formula to solve for x:

$$z = \frac{x - \mu}{\sigma}$$

$$-0.67 = \frac{x - 360}{30}$$

$$x = 360 + (-0.67)(30)$$

$$= 339.9$$

So the fastest 25% of the recruits had times of 339.9 seconds or less.

290. 380.1

The slowest (and worst) times are at the upper end of the distribution. Using the Z-table (Table A-1 in the appendix), you first have to rewrite what you're looking for in terms of a "less than" probability; so you find the value where 75% of the times are below it. The closest table value to 0.75 is 0.7486, corresponding to a z-value of 0.67.

To find the time corresponding to a particular z-score, use the z-formula to solve for x:

$$z = \frac{x - \mu}{\sigma}$$

$$0.67 = \frac{x - 360}{30}$$

$$x = 360 + (0.67)(30)$$

$$= 380.1$$

So the slowest 25% of the recruits had times of 380.1 seconds or more.

291.　E.　Choices (A), (B), and (C) (The t-distribution has thicker tails than the Z-distribution; the t-distribution has a proportionately larger standard deviation than the Z-distribution; the t-distribution is bell-shaped but has a lower peak than the Z-distribution.)

Compared to the Z-distribution, the t-distribution has thicker tails and a proportionately larger standard deviation. It's still bell-shaped, but it has a lower peak than the Z-distribution.

292.　t_{29}

A t-distribution for a study with one population with a sample size of 30 has $n - 1 = 30 - 1 = 29$ degrees of freedom, so the correct distribution is t_{29}.

293.　25

A t_{24} distribution has $n - 1 = 24$ degrees of freedom, so the sample size, n, is 25.

294.　The peak of the Z-distribution would be higher.

In general, the t-distribution is bell-shaped but is flatter and has a lower peak than the standard normal (Z-) distribution, particularly with smaller degrees of freedom for the t-distribution.

295.　The t-distribution would have thicker tails.

The t-distribution is flatter, has a lower peak, and has thicker tails compared to the standard normal (Z-) distribution, particularly with smaller degrees of freedom for the t-distribution.

296.　100

As the degrees of freedom increase, the t-distribution tends to look more like the Z-distribution. So the t-distribution with the highest degrees of freedom most resembles the Z-distribution.

297.　the degrees of freedom

A t-distribution is defined by its degrees of freedom, unlike the normal distribution, which is defined by its mean and standard deviation. The t-distribution always has a mean of 0 (like the Z-distribution), and the more degrees of freedom that a t-distribution has, the smaller its standard deviation gets (because the tails aren't as thin).

298.　the one-sample t-test

You're testing the mean of one population, so the answer has to be a one-sample test. You can't use the one-sample Z-test without knowing the population standard deviation, so in this instance, you'd use the one-sample t-test.

299. the paired t-test

You use the paired t-test to study mean differences among paired subjects according to some variable — for example, the mean difference in weight before and after a weight loss program or the mean difference in weight loss in study participants who are matched according to similar characteristics.

300. t_{24}

This type of study is called a matched-pairs design. A matched-pairs design with 50 observations from the two samples combined has 25 pairs of data, so $n = 25$, and the degrees of freedom is $n - 1 = 25 - 1 = 24$. So the t-distribution corresponding to this scenario is t_{24}.

301. $df = 17$

The study involving one population and a sample size of 18 has $n - 1 = 18 - 1 = 17$ degrees of freedom.

302. $df = 21$

A matched-pairs design with 44 total observations has 22 pairs. The degrees of freedom is one less than the number of pairs: $n - 1 = 22 - 1 = 21$.

303. $p = 0.025$

The column headings of Table A-2 display upper-tail ("greater than") probabilities for specified t-values, so you can read the value for an upper-tail probability of 0.025 directly from the column heading for 0.025.

304. $p = 0.005$

For a hypothesis test, α is the level of significance; if the p-value for the test is less than α, then H_0 (the null hypothesis) is rejected. (The p-value is the probability of being beyond your test statistic.)

The column headings of Table A-2 display upper-tail ("greater than") probabilities for specified t-values. For a two-tailed test (where H_a, or the alternative hypothesis, is "not equal to") with significance level α, you select the column for $\alpha/2$, which gives you the probability for each tail. So in this case, you need the column for $\alpha/2 = 0.01/2 = 0.005$.

305. $p = 0.025$

For a hypothesis test, α is the level of significance; if the p-value for the test is less than α, then H_0 (the null hypothesis) is rejected. (The p-value is the probability of being beyond your test statistic.)

The column headings of Table A-2 display upper-tail ("greater than") probabilities for specified t-values. For a two-tailed test with significance level α, you select the column for $\alpha/2$, which gives you the probability for each tail. So in this case, you need the column for $\alpha/2 = 0.05/2 = 0.025$.

306.   0.05

The column headings of Table A-2 display upper-tail ("greater than") probabilities for specified t-values. To find the upper-tail probability for $t_{10} \geq 1.81$, locate the row for $df = 10$ and follow it across until you find the t-value 1.81. The column heading ("greater than" probability) for this value is 0.05.

307.   0.01

The column headings in Table A-2 display upper-tail ("greater than") probabilities for specified t-values. To find the upper-tail probability for $t_{25} \geq 2.49$, locate the row for $df = 25$ and follow it across until you find the t-value 2.49. The column heading ("greater than" probability) for this value is 0.01.

308.   0.10

The column headings in Table A-2 display upper-tail ("greater than") probabilities for specified t-values. To find the upper-tail probability for $t_{15} \geq 1.34$, locate the row for $df = 15$ and follow it across until you find the t-value 1.34. The column heading ("greater than" probability) for this value is 0.10.

309.   0.05 and 0.025

Using Table A-2, locate the row with 22 degrees of freedom and look for 1.80. However, this exact value doesn't lie in this row, so look for the values on either side of it: 1.717144 and 2.07387. The upper-tail probabilities in Table A-2 appear in the column headings; the column heading for 1.717144 is 0.05, and the column heading for 2.07387 is 0.025. Hence, the upper-tail probability for a t-value of 1.80 must lie between 0.05 and 0.025.

310.   0.025 and 0.01

Using Table A-2, locate the row with 14 degrees of freedom and look for 2.35. However, this exact value doesn't lie in this row, so look for the values on either side of it: 2.14479 and 2.62449. The upper-tail probabilities in Table A-2 appear in the column headings; the column heading for 2.14479 is 0.025, and the column heading for 2.62449 is 0.01. Hence, the upper-tail probability for a t-value of 2.35 must lie between 0.025 and 0.01.

311.   0.01

The t-distribution is symmetrical, so the probability of being in the upper tail ("greater than") with a positive value of t is the same as the probability of being in the lower tail (less than) with the corresponding negative value of t. Table A-2 gives you upper-tail probabilities for positive values of t, so locate the row for $df = 15$ and follow it across to the t-value of 2.60; you find that $P(t_{15} \geq 2.60)$ is 0.01 (the column heading). This means (by symmetry) that $P(t_{15} \leq -2.60)$ is also 0.01.

312. 0.025

The t-distribution is symmetrical, so the probability of being in the upper tail ("greater than") with a positive value of t is the same as the probability of being in the lower tail ("less than") with the corresponding negative value of t. Table A-2 gives you upper-tail probabilities for positive values of t, so locate the row for $df = 27$ and follow it across to the t-value of 2.05; you find that $P(t_{27} \geq 2.05)$ is 0.025 (the column heading). This means (by symmetry) that $P(t_{27} \leq -2.05)$ is also 0.025.

313. 0.05

The t-distribution is symmetrical, so the probability of being in the upper tail ("greater than") with a positive value of t is the same as the probability of being in the lower tail ("less than") with the corresponding negative value of t. Table A-2 gives you upper-tail probabilities for positive values of t, so locate the row for $df = 27$ and follow it across to the t-value of 2.05; you find that $P(t_{27} \geq 2.05)$ is 0.025 (the column heading). This means (by symmetry) that $P(t_{27} \leq -2.05)$ is also 0.025. To find $P(t_{27} \geq 2.05$ or $\leq -2.05)$, you sum the two individual probabilities: $0.025 + 0.025 = 0.05$.

314. 0.01

The t-distribution is symmetrical, so the probability of being in the upper tail ("greater than") with a positive value of t is the same as the probability of being in the lower tail ("less than") with the corresponding negative value of t. Table A-2 gives you upper-tail probabilities for positive values of t, so locate the row for $df = 9$, and follow it across to the t-value of 3.25; you find that $P(t_9 \geq 3.25)$ is 0.005 (the column heading). This means (by symmetry) that $P(t_9 \leq -3.25)$ is also 0.005. To find $P(t_9 \geq 3.25$ or $\leq -3.25)$, you sum the two individual probabilities: $0.005 + 0.005 = 0.01$.

315. 1.81

The 95th percentile of a distribution is the value that 95% of values are less than and 5% are greater than. The column headings in Table A-2 show "greater than" probabilities, so locate the row for $df = 10$, and follow it across to the t-value with a greater than probability of 0.05 (this is the value of t that intersects column 0.05 and row 10), which is 1.81. Because 5% of the values are greater than 1.81, you know that 95% are less than 1.81; so the 95th percentile is $t = 1.81$.

316. 0.26

The 60th percentile of a distribution is the value that 60% of values are less than and 40% are greater than. The column headings in Table A-2 show "greater than" probabilities, so locate the row for $df = 28$, and follow it across to the t-value with a greater than probability of 0.40 (this is the value of t that intersects column 0.40 and row 28), which is 0.26. Because 40% of the values are greater than 0.26, you know that 60% are less than 0.26; so the 60th percentile is $t = 0.26$.

317. −1.33

The 10th percentile of a distribution is the value that 10% of values are less than and 90% are greater than. The column headings in Table A-2 show "greater than" probabilities. Note that 0.90 isn't one of them, but 0.10 is there, and because the t-distribution is symmetrical, the t-value for the 10th percentile is the negative of the t-value for the 90th percentile.

So to find the 10th percentile value, locate the 90th percentile value in Table A-2 and then take its negative value. First, locate the row for $df = 20$, and follow it across to the value that intersects with the column *0.10,* which gives you 1.33 (the 90th percentile). So because the t-distribution is symmetrical, you know that 10% of values are less than −1.33. That means that the 10th percentile is $t = -1.33$.

318. −0.69

The 25th percentile of a distribution is the value that 25% of values are less than and 75% are greater than. The column headings in Table A-2 show "greater than" probabilities. Note that 0.75 isn't one of them, but 0.25 is there, and because the t-distribution is symmetrical, the t-value for 25th percentile is the negative of the t-value for the 75th percentile.

So to find the 25th percentile value, locate the 75th percentile in Table A-2 and then take its negative value. First, locate the row for $df = 20$, and follow it across to the value that intersects with the column *0.25,* which gives you 0.69 (the 75th percentile). So because the t-distribution is symmetrical, you know that 25% of values are less than −0.69. That means that the 25th percentile is $t = -0.69$.

319. −1.34

The 10th percentile of a distribution is the value that 10% of values are less than and 90% are greater than. The column headings in Table A-2 show "greater than" probabilities. Note that 0.90 isn't one of them, but 0.10 is there, and because the t-distribution is symmetrical, the t-value for the 10th percentile is the negative of the t-value for the 90th percentile.

So to find the 10th percentile value, locate the 90th percentile value in Table A-2 and then take its negative value. First, locate the row for $df = 16$, and follow it across to the value that intersects with the column *0.10,* which gives you 1.34 (the 90th percentile). So because the t-distribution is symmetrical, you know that 10% of values are less than −1.34. That means that the 10th percentile is $t = -1.34$.

320. −1.75

The 5th percentile of a distribution is the value that 5% of values are less than and 95% are greater than. The column headings in Table A-2 show "greater than" probabilities. Note that 0.95 isn't one of them, but 0.05 is there, and because the t-distribution is symmetrical, the t-value for the 5th percentile is the negative of the t-value for the 95th percentile.

So to find the 5th percentile value, locate the 95th percentile value in Table A-2 and then take its negative value. First, locate the row for $df = 16$, and follow it across to the value that intersects with the column 0.05, which gives you 1.75 (the 95th percentile). So because the t-distribution is symmetrical, you know that 5% of values are less than -1.75. That means that the 5th percentile is $t = -1.75$.

321. D. Choices (A) and (B) (You have a small sample size; you don't know the population standard deviation)

You use the t-distribution rather than the Z-distribution to calculate confidence intervals when you have a small sample size and/or when you don't know the population standard deviation (so you use the sample standard deviation instead). In both cases, you pay a penalty, hence the wider (flatter) t-distribution.

322. 99%

Here, you want the t-values for a 99% confidence interval, so look at the last row of Table A-2 (the row labeled CI), and find the value 99%. Using this column, you can find the t-values for a 99% confidence level by following it to where it intersects with the row for the degrees of freedom you want.

323. 0.005

A 99% confidence interval means that 99%, or 0.99, of all the values lie inside the confidence interval, and 10%, or 0.01, of all the values lie on the outside, with $0.01/2 = 0.005$ of the values above (greater than) the confidence interval and 0.005 of the values below (less than) the confidence interval.

The first row (column headings) of Table A-2 show upper-tail ("greater than") probabilities only. To find a t-value for a 99% confidence interval using the first row of Table A-2, look in the column for $0.01/2 = 0.005$. Then go to the row corresponding to the degrees of freedom to find the t-value you need.

324. 2.13

Here, you want a t-value for a 95% confidence interval, so look at the last row of Table A-2 (labeled CI), find the value 95%, and then intersect this column with the row for $df = 15$, which gives you 2.13.

325. 2.81

In Table A-2, find where the row for $df = 23$ and the column for 99% CI (in the last row of the table) intersect. The value at this intersection is 2.81.

326. 1.70

In Table A-2, find where the row for $df = 30$ and the column for 90% CI (in the last row of the table) intersect. The value at this intersection is 1.697261, which rounds to 1.70.

327. 95%

In Table A-2, you find the value 2.09 where the row for $df = 19$ and the column for 95% CI (in the last row of the table) intersect.

328. 99%

In Table A-2, you find the value 2.78 where the row for $df = 26$ and the column for 99% CI (in the last row of the table) intersect.

329. 80%

In Table A-2, you find the value 1.36 where the row for $df = 12$ and the column for 80% CI (in the last row of the table) intersect.

330. 50

As the degrees of freedom increase, the t-distribution resembles the Z-distribution more closely. The peak in the bell shape rises higher and higher, and the tails become thinner and thinner, until the two distributions are practically indistinguishable. So the t-distribution with the highest degrees of freedom most resembles the Z-distribution.

331. 10

The smaller the degrees of freedom, the less the t-distribution resembles the Z-distribution. The peak in the bell shape gets lower and lower, and the tails become thicker and thicker. So the t-distribution with the lowest degrees of freedom least resembles the Z-distribution.

332. 10

The column in Table A-2 with a heading of 0.05 shows all the different t-values with right-tail ("greater than") probabilities of 0.05 for various degrees of freedom. As the degrees of freedom decrease (moving from bottom to top in the column), the t-values increase because the tails in the t-distributions with fewer degrees of freedom are thicker, and you have to move out farther on the t-distribution to get to the 5% mark. Moving farther out requires a higher t-value; so the t-distribution with the lowest degrees of freedom is the one with the largest t-value.

333. 50

The column in Table A-2 with a heading of 0.10 shows all the different t-values with right-tail ("greater than") probabilities of 0.10 for various degrees of freedom. As the degrees of freedom increase (moving from top to bottom in the column), the t-values decrease because the tails in the t-distributions with higher degrees of freedom are thinner, and you don't have to move out as far on the t-distribution to get to the 10% mark. That means you'll have a smaller t-value; so the t-distribution with the highest degrees of freedom is the one with the smallest t-value.

334. 80%

For any distribution (including the t-distribution with 40 degrees of freedom), the greater the confidence level is for a confidence interval, the wider the confidence interval is; and the lower the confidence level is, the narrower the confidence interval is. Therefore, the confidence interval with the lowest confidence level (in this case, 80%) will be the narrowest.

335. 99%

For any distribution (including the t-distribution with 50 degrees of freedom), the greater the confidence level is for a confidence interval, the wider the confidence interval is; and the lower the confidence level is, the narrower the confidence interval is. Therefore, the confidence interval with the highest confidence level (in this case, 99%) will be the widest.

336. −2.74

Because you don't know the population standard deviation and are testing the mean of one population, you compute the one-sample t-test, using the following formula for the test statistic:

$$t = \frac{\bar{x} - \mu_0}{s/\sqrt{n}}$$

In this case, the sample mean, $\bar{x}$, is 4.8; the target population mean, μ_0, is 5 (this value goes in the null hypothesis H_0, hence the subscript 0 in both expressions); the sample standard deviation, s, is 0.4; the sample size, n, is 30; and the degrees of freedom, $n - 1$, is 29. Now, plug these numbers into the formula and solve:

$$t = \frac{4.8 - 5.0}{0.4/\sqrt{30}} = -2.74$$

337. between 0.01 and 0.005

In Table A-2, using the row for $df = 29$, the upper-tail ("greater than") probability for 2.46202 is 0.01 (found in the column heading), and the probability for 2.75639 is 0.005. Because the t-distribution is symmetrical, the lower-tail ("less than") probability for −2.46202 is also 0.01, and the probability for −2.75639 is also 0.005. The t-value of −2.74 lies between these two numbers, so the probability is between 0.01 and 0.005.

338. (4.65, 4.95)

The formula for the confidence interval for one population mean, using the t-distribution, is

$$\bar{x} \pm t_{n-1}\frac{s}{\sqrt{n}}$$

In this case, the sample mean, $\bar{x}$, is 4.8; the sample standard deviation, s, is 0.4; the sample size, n, is 30; and the degrees of freedom, $n-1$, is 29. That means $t_{n-1} = 2.05$ (from Table A-2).

Now, plug in the numbers:

$$\bar{x} \pm t_{n-1} \frac{s}{\sqrt{n}}$$

$$= 4.8 \pm 2.05 \frac{0.4}{\sqrt{30}}$$

$$= 4.8 \pm 0.1497$$

$$= 4.6503 \text{ to } 4.9497$$

Rounded to two decimal places, the answer is 4.65 to 4.95.

339. (4.68, 4.92)

The formula for the confidence interval for one population mean, using the t-distribution, is

$$\bar{x} \pm t_{n-1} \frac{s}{\sqrt{n}}$$

In this case, the sample mean, $\bar{x}$, is 4.8; the sample standard deviation, s, is 0.4; the sample size, n, is 30; and the degrees of freedom, $n-1$, is 29. That means that $t_{n-1} = 1.70$ (from Table A-2).

Now, plug in the numbers:

$$\bar{x} \pm t_{n-1} \frac{s}{\sqrt{n}}$$

$$= 4.8 \pm 1.70 \frac{0.4}{\sqrt{30}}$$

$$= 4.8 \pm 0.1242$$

$$= 4.6758 \text{ to } 4.9242$$

Rounded to two decimal places, the answer is 4.68 to 4.92.

340. (4.60, 5.00)

The formula for the confidence interval for one population mean, using the t-distribution, is

$$\bar{x} \pm t_{n-1} \frac{s}{\sqrt{n}}$$

In this case, the sample mean, $\bar{x}$, is 4.8; the sample standard deviation, s, is 0.4; the sample size, n, is 30; and the degrees of freedom, $n-1$, is 29. That means that $t_{n-1} = 2.76$ (from Table A-2).

Now, plug in the numbers:

$$\bar{x} \pm t_{n-1} \frac{s}{\sqrt{n}}$$

$$= 4.8 \pm 2.76 \frac{0.4}{\sqrt{30}}$$

$$= 4.8 \pm 0.2016$$

$$= 4.5984 \text{ to } 5.0016$$

Rounded to two decimal places, the answer is 4.60 to 5.00.

341. random variable

A *random variable* is an assignment of numbers to the outcome of some random (or partially random) event. Many things can be random variables. For example, on a coin toss, you can assign 1 to heads and 0 to tails, and the outcome of the coin toss would be a random variable. You can also toss a coin five times and count the number of times it comes up heads, and that number would be a random variable. If you rolled two dice and added the numbers that came up on both, the total of the roll would be a random variable.

342. a sampling distribution of the sample means

A *sampling distribution* is a collection of all the means from all possible samples of the same size taken from a population. In this case, the population is the 10,000 test scores, each sample is 100 test scores, and each sample mean is the average of the 100 test scores.

343. $\mu_x = 3.11$

Because you found the average GPA of every student in the university, you used a population value, which needs a Greek letter. μ_x refers to the mean of all individual values in the population.

344. $\mu_{\bar{x}} = 3.5$

Here, you take all possible samples (of the same size), find all their possible means, and treat those as a population. Then, you find the mean of that entire population of sample means. The notation for this is $\mu_{\bar{x}} = 3.5$.

345. $\bar{x} = 3.5$

Because the value is the result of only a sample of dice rools, and not the full population of all possible rolls, you must use the sample mean notation, $\bar{x} = 3.5$.

346. D. Each of the observations in the distribution must consist of a statistic that describes a collection of data points.

A sampling distribution is a set of all possible values in a population, except the values themselves represent statistics, like sample means or sample standard deviations.

The critical element in each case is that data points going into your distribution each represent a summary statistic for a sample.

347. E. a distribution showing the weight of each individual football fan entering a stadium on game day

A sampling distribution is a population of data points where each data point represents a summary statistic from one sample of individuals. A population distribution is a population of data points where each data point represents an individual.

348.
a random variable denoting the outcome from a single roll of the die

X is a random variable with possible values 1, 2, 3, 4, 5, and 6, denoting the outcome from a single roll of the die.

349.
a random variable denoting the average value when you roll the die n times (where n is some fixed number)

$\bar{X}$ is a random variable representing any calculated average from a certain number of rolls of the die. You just don't know what its value is yet because you haven't rolled the die yet.

350.
2.6

$\bar{x}$ represents the sample mean; you find it by adding the numbers and dividing by n (the sample size). Use the following formula:

$$\bar{x} = \frac{\sum\limits_{i=1}^{n} x_i}{n}$$

$$= \frac{3+4+2+3+1}{5}$$

$$= \frac{13}{5}$$

$$= 2.6$$

Here, each x_i represents a value in the data set — x_1 is the first number, x_2 is the second number, and so on, and then x_n is the nth, or last, number.

351.
4.2

$\bar{x}$ represents the sample mean; you find it by adding the numbers and dividing by n (the sample size). Use the following formula:

$$\bar{x} = \frac{\sum\limits_{i=1}^{n} x_i}{n}$$

$$= \frac{3+4+6+3+5}{5}$$

$$= \frac{21}{5}$$

$$= 4.2$$

Here, each x_i represents a value in the data set — x_1 is the first number, x_2 is the second number, and so on, and then x_n is the nth, or last, number.

352.
C. $\sigma_{\bar{X}} = \dfrac{\sigma_X}{\sqrt{n}}$

The formula for the standard error of a sample mean is

$$\sigma_{\bar{X}} = \frac{\sigma_X}{\sqrt{n}}$$

where σ_X is the population standard deviation and n is the sample size.

353.   standard deviation; standard error

The standard deviation represents the variability in the entire population, or the variability of X, while the standard error represents the variability of the sample means, or the variability of $\bar{X}$.

354.   B.   be approximately the same; be smaller

You don't expect the sample mean to change with the sample size. However, a larger sample size is expected to result in a smaller standard error because the formula for standard error includes dividing by the sample size:

$$\sigma_{\bar{x}} = \frac{\sigma_X}{\sqrt{n}}$$

where σ_X is the population standard deviation and n is the sample size.

Dividing the same population standard deviation by the square root of a larger n results in a smaller standard error. Larger samples have a smaller standard error because their mean changes less from sample to sample.

355.   3.68

To calculate the standard error, use the following formula:

$$\sigma_{\bar{x}} = \frac{\sigma_X}{\sqrt{n}}$$

where σ_X is the population standard deviation and n is the sample size.

Substitute the known values into the formula and solve:

$$\sigma_{\bar{x}} = \frac{26}{\sqrt{50}}$$
$$= \frac{26}{7.071}$$
$$= 3.677$$

This rounds to 3.68.

356.   3.36

To calculate the standard error, use the following formula:

$$\sigma_{\bar{x}} = \frac{\sigma_X}{\sqrt{n}}$$

where σ_X is the population standard deviation and n is the sample size.

Substitute the known values into the formula and solve:

$$\sigma_{\bar{x}} = \frac{26}{\sqrt{60}}$$
$$= \frac{26}{7.746}$$
$$= 3.356$$

This rounds to 3.36.

357. 4.75

To calculate the standard error, use the following formula:

$$\sigma_{\bar{x}} = \frac{\sigma_X}{\sqrt{n}}$$

where σ_X is the population standard deviation and n is the sample size.

Substitute the known values into the formula and solve:

$$\sigma_{\bar{x}} = \frac{26}{\sqrt{30}}$$
$$= \frac{26}{5.477}$$
$$= 4.747$$

This rounds to 4.75.

358. $\bar{x}$

A small x with a bar over it indicates the average of a set of individual scores.

359. μ_X

This notation represents the population parameter for the mean of the random variable X.

360. $\sigma_{\bar{x}}$

The standard error of the mean is also known as the standard deviation (σ) of the sampling distribution of the sample mean ($\bar{X}$), hence the subscript.

361. σ_X

Population parameters are represented by Greek letters — in this case, by the lower-case letter sigma (σ) with X as a subscript to indicate that it's the standard deviation of individual scores.

362. $\sigma_{\bar{x}}$

In this case, you're taking repeated samples of size $n = 5$, computing a sample average each time, and then computing the dispersion around the average of all sample averages. Such a procedure is conceptually a way of getting the standard error of the mean, which is denoted as $\sigma_{\bar{x}}$.

363. $\sigma_{\bar{x}}$

The fish boat owner is looking at the standard deviation of the average weights for his catches over time. He's basically looking at the standard deviation of the sample means (thinking of each catch as a sample). Statistically speaking, this term is the standard error of the sample mean and is denoted by $\sigma_{\bar{x}}$.

364. x_i

In this case, you're considering the sale price of an individual house, so you're dealing with an individual data point. The subscript indicates which data point (or home in this case) you're referring to in the data set.

365. D. a smaller sample size

The formula for calculating standard error is

$$\sigma_{\bar{x}} = \frac{\sigma_X}{\sqrt{n}}$$

where σ_X is the population standard deviation and n is the sample size.

As you can see, the population mean has no effect on the standard error. A smaller population standard deviation will produce a smaller standard error because the population standard deviation is the numerator of the standard error formula. Because sample size is the denominator of the formula, a smaller sample size will produce a larger standard error, while a larger sample size will produce a smaller standard error. Larger samples have a smaller standard error because their mean changes less from sample to sample.

366. It would lower the standard error of the sample mean.

Use the formula for calculating the standard error of the sample mean:

$$\sigma_{\bar{x}} = \frac{\sigma_X}{\sqrt{n}}$$

where σ_X is the population standard deviation and n is the sample size. Dividing the same population standard deviation by the square root of a larger n results in a smaller standard error. In other words, increasing the sample sizes reduces the amount of change (standard error) in the sample means.

367. Population B has a smaller standard error because of the smaller population standard deviation.

Use the formula for calculating the standard error of the mean:

$$\sigma_{\bar{x}} = \frac{\sigma_X}{\sqrt{n}}$$

where σ_X is the population standard deviation and n is the sample size. Dividing a smaller population standard deviation by the square root of the same n results in a smaller standard error. Samples drawn from a population that is less variable, as shown by a smaller standard deviation, are more likely to have means closer to the sample mean and hence a smaller standard error.

368. Quadruple the sample size.

Use the formula for calculating the standard error of the mean:

$$\sigma_{\bar{x}} = \frac{\sigma_X}{\sqrt{n}}$$

where σ_X is the population standard deviation and n is the sample size.

To double $\sigma_{\bar{x}}$, you have to divide $\sigma_{\bar{x}}$ by half as much. Because the divisor is the square root of n, you must quadruple the sample size to get a divisor twice as large.

369.   6.6667

Use the formula for calculating the standard error of the mean:

$$\sigma_{\bar{x}} = \frac{\sigma_X}{\sqrt{n}}$$

where σ_X is the population standard deviation and n is the sample size.

Substitute the known values into the formula and solve:

$$\sigma_{\bar{x}} = \frac{20}{\sqrt{9}}$$

$$= \frac{20}{3}$$

$$= 6.\bar{6}$$

This rounds to 6.6667

370.   5

Use the formula for calculating the standard error of the mean:

$$\sigma_{\bar{x}} = \frac{\sigma_X}{\sqrt{n}}$$

where σ_X is the population standard deviation and n is the sample size.

Substitute the known values into the formula and solve:

$$\sigma_{\bar{x}} = \frac{20}{\sqrt{16}}$$

$$= \frac{20}{4}$$

$$= 5$$

371.   E.   Choices (B) and (D) (a smaller sample size; a larger population standard deviation)

Given the formula for the standard error of the mean

$$\sigma_{\bar{x}} = \frac{\sigma_X}{\sqrt{n}}$$

where σ_X is the population standard deviation and n is the sample size, increasing the numerator or decreasing the denominator will both result in a larger standard error. A more variable population will result in more variable sample means, and a smaller sample size will also result in more variable sample means, in both cases resulting in a larger sample error.

372. 25

The formula for standard error can be rearranged to find the population standard deviation, given sample size and standard error. Multiply both sides by the square root of n, substitute the values given, and solve:

$$\sigma_{\bar{x}} = \frac{\sigma_X}{\sqrt{n}}$$

$$\sigma_{\bar{x}}(\sqrt{n}) = \sigma_X$$

$$5(\sqrt{25}) = \sigma_X$$

$$5(5) = 25$$

373. centimeter

Units for standard error are the same as for the original measurements.

374. A. 0.4856

A smaller standard error will give you a more precise estimate of the mean because the sample means will cluster more closely around the population mean.

375. No specific requirement for sample size is needed.

Because you know that the individual scores come from a normal distribution, the distribution of the sample means will also have a normal distribution, regardless of the sample size.

376. D. Individual scores x_i are normally distributed.

If the individual scores are normally distributed, then the sampling distribution of the sample means is normal. The magic of the central limit theorem is that as samples become sufficiently large (30 or more), the sampling distribution of the sample means becomes approximately normal.

377. It is exactly normal.

Because the individual data points are normally distributed, the sampling distribution of sample means is also normal, no matter what the size of each sample is. (You don't need the central limit theorem and the $n \geq 30$ requirement if you start with a normal distribution.)

378. C. right-skewed, 60

If the population's distribution is normal, the sampling distribution of the sample means is also normal, so the central limit theorem is required only for non-normal (that is, right-skewed) populations.

379. normally distributed

When data points are drawn from a normal population of data points, the sampling distribution of the sample mean is normal.

380. It is exactly normal for any sample size.

When a sample is drawn from a normal population, the sampling distribution of the sample means is normal.

381. the shape of a normal distribution

Repeatedly sampling from a population of scores and then forming a histogram of the means of the samples creates a sampling distribution. When individual scores are normally distributed, the sampling distribution of the sample means is also normal, which makes a bell shape.

382. It would be expected to be normally distributed.

Although the sample size is small (four), the distribution of sample means from a population with an underlying normal distribution is also expected to be normal.

383. It is exactly normal.

When all samples of a fixed size, even small ones, are drawn from a normally distributed population, the sampling distribution of sample means is normal.

384. a precise normal distribution

When samples are drawn from a normally distributed population, the sampling distribution of sample means is normal.

385. normal

The sampling distribution of the sample means is expected to be normal, although the underlying distribution isn't normal, because the sample size is sufficiently large (35) that the central limit theorem applies.

386. E. All of the above (Population A, Population B, Population C, Population D)

According to the central limit theorem, with reasonably large samples ($n \geq 30$), the sampling distribution of sample means is expected to be normally distributed, no matter what shape the underlying distribution of individual observations has (most situations work well if the sample size is at least 30).

387. Population D

A sample size of 20 is too small to expect that the sampling distribution of the sample means will be approximately normal, unless the population has a normal distribution.

388. normal for all four populations

Because the sample size is fairly large (40), the central limit theorem tells you that the sampling distribution of sample means is expected to be normal as well, no matter what kind of distribution the underlying individual observations have, as long as the sample size is at least 30.

389. not able to determine from the information given

Here, you're not told how large each sample is; you're told only how large each *population* is. The central limit theorem says that sampling distributions of sample means are normal if the sample sizes are sufficiently large ($n \geq 30$) or if the underlying distribution of observations is normal.

In this case, you may be talking about small samples (of less than 30 observations), and three of the four distributions involved are clearly not normal. Therefore, you can't make any general statement about the resulting shape of the sampling distributions of sample means.

390. $n \geq 30$

When data isn't drawn from a normal distribution, the sampling distribution of sample means becomes normal only when sufficiently large samples ($n \geq 30$) are used.

391. a precise normal distribution

In this case, you know that the observations themselves are normally distributed, so samples of any size will give rise to a normal sampling distribution of sample means (even samples of size 3).

392. an approximate normal distribution

Because the sample sizes are large ($n = 100$, which is much larger than the approximately 30 cases required by the central limit theorem), you know that the sampling distribution of sample means should be approximately normal.

393. No, because the sample sizes are too small to use the central limit theorem.

In this case, the original population distribution is unknown, so you can't assume that you have a normal distribution. The central limit theorem can't be invoked because the sample sizes are too small (less than 30).

394. $n = 30$

According to the central limit theorem, if you repeatedly take sufficiently large samples, the distribution of the means from those samples will be approximately normal. For most non-normal populations, you can choose sample sizes of at least 30 from the distribution, which usually leads to a normal sampling distribution of sample means no matter what the underlying shape of the distribution of scores is. In fact, if the underlying distribution of values approximates a normal distribution, it may be possible to achieve a normal sampling distribution of sample means with smaller samples.

For populations with several peaks, wild variation, and/or extreme outliers, you may need larger sample sizes.

395. The researcher has not violated the condition because the sampling distribution of sample means is approximately normal whenever the sample size is at least 30.

The central limit theorem says that the sampling distribution of sample means is approximately normal if the sample size is at least 30. The central limit theorem is used to ensure that studies meet the assumptions underlying other tests, which rely on a normal sampling distribution of sampling means. It isn't necessary to draw repeated samples from a distribution to invoke the central limit theorem. It's only necessary to have either an underlying normal distribution of observations or a sample size of $n \geq 30$ in most cases, for the condition to be met.

396. The sampling distribution of sample means is normal.

The sampling distribution of sample means is normal whenever the observations come from a normally distributed population of scores, which is true in this case.

397. C. The central limit theorem can be used because the sample size is large enough and the population distribution is unknown.

Because the distribution of the population isn't discussed, you can't assume that it's normal. You have to appeal to the central limit theorem. In this case, the condition of the central limit theorem is met: The sample size is well over 30 ($n = 150$), so you can use it to find probabilities about the sample mean.

398. No, the central limit theorem can't be used to infer a normal sampling distribution of sample means drawn from a non-normal population because there are too few observations in each sample.

To use the central limit theorem for samples drawn from a population that isn't normally distributed, the sample size must be relatively large ($n \geq 30$). Because $n = 10$ for each sample, the central limit theorem can't be used.

399. The central limit theorem can be used to infer a normal sampling distribution of sample means.

Although the observations have a skewed (and hence non-normal) distribution, the central limit theorem does allow you to conclude that the sampling distribution of sample means is normal, because the size of each sample is sufficiently large ($n = 150$ here, which is well more than the 30 or more suggested by the central limit theorem). Note that the number of samples taken isn't relevant here.

400. Yes, the central limit theorem can be used to infer a normal sampling distribution of sample means.

This case just barely qualifies to invoke the central limit theorem and infer that the sampling distribution of sample means is normal. The population of observations isn't normal, but the sample size is just large enough to use the central limit theorem. (The larger the sample size, the better the approximation in all situations.)

401.   9.6

Use the formula for finding the mean:

$$\bar{x} = \frac{\Sigma x_i}{n}$$

where Σx_i is the sum of the original observations, and n is the sample size. Substitute the known values into the formula and solve:

$$\bar{x} = \frac{7+6+7+14+9+9+11+11+11+11}{10}$$

$$= \frac{96}{10}$$

$$= 9.6$$

402.   D.   Larger samples tend to yield more precise estimates of the population mean.

The sample mean is the best unbiased estimate of the population mean, and a larger sample will generally produce a more precise estimate because larger samples change less from sample to sample.

403.   E.   Choices (B) and (C) (The sample mean is the best estimate of the population mean; larger samples yield more precise estimates of the population mean.)

By definition, the sample mean is the best estimator of the population mean. Larger samples yield more precise estimates of the sample mean because they vary less from one sample to another than do smaller samples.

404.   67.44%

Because the observations are drawn from a normally distributed population, the sample means are also normally distributed.

First, find the z-score equivalent of this sample mean by using the population mean and the standard error of the mean for samples of this size.

$$z = \frac{\bar{x} - \mu_X}{\sigma_X / \sqrt{n}}$$

Here, $\bar{x}$ is the sample mean, μ_X is the population mean, σ_X is the population standard deviation, and n is the sample size.

Now, substitute the known values into the formula and solve:

$$z = \frac{9.6 - 10}{3 / \sqrt{10}}$$

$$= \frac{-0.4}{3 / 3.1623}$$

$$= \frac{-0.4}{0.9487}$$

$$= -0.4216$$

Because the probabilities equate to tail areas in a normal distribution, you start by finding the area under the curve associated with that z-score by using a Z-table (Table A-1 in the appendix). So round the z-score to –0.42, and use Table A-1 to find the associated probability of 0.3372.

Now, figure out how much area is in the tails. In the Z-table, areas reflect the proportion of the normal curve in one tail. Because you want to find the probability of a score this far from the mean *in either direction,* you need to double the area you just found to account for the area in the other tail:

$$\text{Total tail area} = 2(\text{Single-tail area})$$
$$= 2(0.3372)$$
$$= 0.6744, \text{ or } 67.44\%$$

Finally, report that probability (the proportion of curve area contained by the two tails) as the probability of finding a mean score this extreme or more so.

---

405.   2.5%

Because the observations are drawn from a normally distributed population, the sample means are also normally distributed.

First, find the z-score equivalent of this sample mean by using the population mean and the standard error of the mean for samples of this size.

$$z = \frac{\bar{x} - \mu_X}{\sigma_X/\sqrt{n}}$$

Here, $\bar{x}$ is the sample mean, μ_X is the population mean, σ_X is the population standard deviation, and n is the sample size.

Now, substitute the known values into the formula and solve:

$$z = \frac{9.7 - 10}{3/\sqrt{500}}$$
$$= \frac{-0.3}{3/22.3607}$$
$$= \frac{-0.3}{0.1342}$$
$$= -2.2355$$

Because probabilities equate to tail areas in a normal distribution, you start by finding the area under the curve associated with that z-score by using a Z-table (Table A-1 in the appendix). So round the z-score to –2.24, and use Table A-1 to find the associated probability of 0.0125

Now, figure out how much area is in the tails. In the Z-table, areas reflect the proportion of the normal curve in one tail. Because you want to find the probability of a score this far from the mean *in either direction,* you need to double the area you just found to account for the area in the other tail:

$$\text{Total tail area} = 2(\text{Single-tail area})$$
$$= 2(0.0125)$$
$$= 0.025, \text{ or } 2.5\%$$

Finally, report that probability (the proportion of curve area contained by the two tails) as the probability of finding a mean score this extreme or more so.

406. 4.88%

Because the sample size is greater than 30, you can use the central limit theorem to solve this problem. Assume a score 10 units below the mean, to simplify calculations.

First, find the z-score equivalent of this sample mean by using the population mean and the standard error of the mean for samples of this size.

$$z = \frac{\bar{x} - \mu_X}{\sigma_X / \sqrt{n}}$$

Here, $\bar{x}$ is the sample mean, μ_X is the population mean, σ_X is the population standard deviation, and n is the sample size.

Now, substitute the known values into the formula and solve:

$$z = \frac{90 - 100}{30 / \sqrt{35}}$$

$$= \frac{-10}{30 / 5.916}$$

$$= \frac{-10}{5.071}$$

$$= -1.972$$

Because probabilities equate to tail areas in a normal distribution, you start by finding the area under the curve associated with that z-score by using a Z-table (Table A-1 in the appendix). So round the z-score to –1.97, and use Table A-1 to find the associated probability of 0.0244.

Now, figure out how much area is in the tails. In the Z-table, areas reflect the proportion of the normal curve in one tail. Because you want to find the probability of a score this far from the mean *in either direction,* you need to double the area you just found to account for the area in the other tail:

$$\text{Total tail area} = 2(\text{Single-tail area})$$

$$= 2(0.0244)$$

$$= 0.0488, \text{ or } 4.88\%$$

Finally, report that probability (the proportion of curve area contained by the two tails) as the probability of finding a mean score this extreme or more so.

407. 4.56%

Because the observations are drawn from a normally distributed population, the sample means are also normally distributed. Assume a score 10 units below the mean, to simplify calculations.

First, find the z-score equivalent of this sample mean by using the population mean and the standard error of the mean for samples of this size.

$$z = \frac{\bar{x} - \mu_X}{\sigma_X / \sqrt{n}}$$

Here, $\bar{x}$ is the sample mean, μ_X is the population mean, σ_X is the population standard deviation, and n is the sample size.

Now, substitute the known values into the formula and solve:

$$z = \frac{90 - 100}{40/\sqrt{64}}$$

$$= \frac{-10}{40/8}$$

$$= \frac{-10}{5}$$

$$= -2.0$$

Because probabilities equate to tail areas in a normal distribution, you start by finding the area under the curve associated with that z-score by using a Z-table (Table A-1 in the appendix). Using Table A-1, you find the associated probability of 0.0228

Now, figure out how much area is in the tails. In the Z-table, areas reflect the proportion of the normal curve in one tail. Because you want to find the probability of a score this far from the mean *in either direction,* you need to double the area you just found to account for the area in the other tail:

$$\text{Total tail area} = 2(\text{Single-tail area})$$

$$= 2(0.0228)$$

$$= 0.0456, \text{ or } 4.56\%$$

Finally, report that probability (the proportion of curve area contained by the two tails) as the probability of finding a mean score this extreme or more so.

---

408. less than 0.02%

Because the observations are drawn from a normally distributed population, the sample means are also normally distributed. Assume a score 10 units below the mean, to simplify calculations.

First, find the z-score equivalent of this sample mean by using the population mean and the standard error of the mean for samples of this size.

$$z = \frac{\bar{x} - \mu_X}{\sigma_X/\sqrt{n}}$$

Here, $\bar{x}$ is the sample mean, μ_X is the population mean, σ_X is the population standard deviation, and n is the sample size.

Now, substitute the known values into the formula and solve:

$$z = \frac{40 - 50}{16/\sqrt{64}}$$

$$= \frac{-10}{16/8}$$

$$= \frac{-10}{2}$$

$$= -5$$

Because probabilities equate to tail areas in a normal distribution, you start by finding the area under the curve associated with that z-score by using a Z-table (Table A-1 in the appendix).

Use Table A-1 to find the associated probability; because this value is more extreme than any value included in the table, the probability is less than the smallest table value of 0.0001.

Now, figure out how much area is in the tails. In the Z-table, areas reflect the proportion of the normal curve in one tail. Because you want to find the probability of a score this far from the mean *in either direction,* you need to double the area you just found to account for the area in the other tail:

$$\text{Total tail area} = 2(\text{Single-tail area})$$

$$= 2(0.0001)$$

$$= 0.0002, \text{ or } 0.02\%$$

Finally, report that probability (the proportion of curve area contained by the two tails) as the probability of finding a mean score this extreme or more so.

409.   1.24%

Because the observations are drawn from a normally distributed population, the sample means are also normally distributed. Assume a score 10 units below the mean, to simplify calculations.

First, find the z-score equivalent of this sample mean by using the population mean and the standard error of the mean for samples of this size.

$$z = \frac{\bar{x} - \mu_X}{\sigma_X / \sqrt{n}}$$

Here, $\bar{x}$ is the sample mean, μ_X is the population mean, σ_X is the population standard deviation, and n is the sample size.

Now, substitute the known values into the formula and solve:

$$z = \frac{40 - 50}{16 / \sqrt{16}}$$

$$= \frac{-10}{16/4}$$

$$= \frac{-10}{4}$$

$$= -2.5$$

Because probabilities equate to tail areas in a normal distribution, you start by finding the area under the curve associated with that z-score by using a Z-table (Table A-1 in the appendix). Using Table A-1, you find the associated probability of 0.0062

Now, figure out how much area is in the tails. In the Z-table, areas reflect the proportion of the normal curve in one tail. Because you want to find the probability of a score this far from the mean *in either direction,* you need to double the area you just found to account for the area in the other tail:

$$\text{Total tail area} = 2(\text{Single-tail area})$$

$$= 2(0.0062)$$

$$= 0.0124, \text{ or } 1.24\%$$

Finally, report that probability (the proportion of curve area contained by the two tails) as the probability of finding a mean score this extreme or more so.

410. 52.86%

Because the observations are drawn from a normally distributed population, the sample means are also normally distributed. Assume a score 10 units below the mean, to simplify calculations.

First, find the z-score equivalent of this sample mean by using the population mean and the standard error of the mean for samples of this size.

$$z = \frac{\bar{x} - \mu_X}{\sigma_X / \sqrt{n}}$$

Here, $\bar{x}$ is the sample mean, μ_X is the population mean, σ_X is the population standard deviation, and n is the sample size.

Now, substitute the known values into the formula and solve:

$$z = \frac{-10 - 0}{160 / \sqrt{100}}$$

$$= \frac{-10}{160 / 10}$$

$$= \frac{-10}{16}$$

$$= -0.625, \text{ rounded to } -0.63$$

Because probabilities equate to tail areas in a normal distribution, you start by finding the area under the curve associated with that z-score by using a Z-table (Table A-1 in the appendix). Using Table A-1, you find the associated probability of 0.2643.

Now, figure out how much area is in the tails. In the Z-table, areas reflect the proportion of the normal curve in one tail. Because you want to find the probability of a score this far from the mean *in either direction,* you need to double the area you just found to account for the area in the other tail:

$$\text{Total tail area} = 2(\text{Single-tail area})$$

$$= 2(0.2643)$$

$$= 0.5286, \text{ or } 52.86\%$$

Finally, report that probability (the proportion of curve area contained by the two tails) as the probability of finding a mean score this extreme or more so.

411. 4.65%

Before going through the steps for solving this problem, determine whether you can use the central limit theorem. Although the lab technician doesn't know whether the population of white blood cell counts is normally distributed, by taking a sample of 40 independent measurements, the technician sets up a case where the central limit theorem can be used.

First, find the z-score equivalent of this sample mean by using the population mean and the standard error of the mean for samples of this size.

$$z = \frac{\bar{x} - \mu_X}{\sigma_X / \sqrt{n}}$$

Here, $\bar{x}$ is the sample mean, μ_X is the population mean, σ_X is the population standard deviation, and n is the sample size.

Now, substitute the known values into the formula and solve:

$$z = \frac{7,616 - 7,250}{1,375/\sqrt{40}}$$

$$= \frac{366}{1,375/6.32456}$$

$$= \frac{366}{217.4064}$$

$$= 1.6835$$

Because probabilities equate to tail areas in a normal distribution, you start by finding the tail area associated with this z-score (rounding to 1.68), using a Z-table (Table A-1 in the appendix).

The table gives tail areas only for negative z-scores, but the normal distribution is symmetrical, so you can look up the left-tail area equivalent to $z = -1.68$, which turns out to be 0.0465.

Now, figure out the total tail area. The problem asks for the probability of getting a sample mean of 7,616 or larger. Because the problem specifically refers to only one tail of the distribution, you should *not* double the probability. There's only one tail specified by the question (scores of 7,616 *or larger*), so you can use 0.0465, or 4.65%, as your answer.

412. 49.08%

The distribution of the weights of the cookies is unknown, and you can't assume that it's normal. To answer this question, look to the central limit theorem. The central limit theorem can be applied to this problem because the sample size of 36 is large enough ($n \geq 30$).

First, find the z-score equivalent of this sample mean by using the population mean and the standard error of the mean for samples of this size.

$$z = \frac{\bar{x} - \mu_X}{\sigma_X/\sqrt{n}}$$

Here, $\bar{x}$ is the sample mean, μ_X is the population mean, σ_X is the population standard deviation, and n is the sample size.

Now, substitute the known values into the formula and solve:

$$z = \frac{12.011 - 12}{0.1/\sqrt{36}}$$

$$= \frac{0.011}{0.1/6}$$

$$= \frac{0.011}{0.01667}$$

$$= 0.6599$$

Because probabilities equate to tail areas in a normal distribution, you start by finding the tail area associated with this z-score (rounding to 0.66), using a Z-table (Table A-1 in the appendix).

The table gives tail areas only for negative z-scores, but the normal distribution is symmetrical, so you can look up the left-tail area equivalent to $z = -0.66$, which turns out to be 0.2546.

The question asks for the probability of a value at least this close to the mean, so you want the probability *excluding* the tail area. The total area under the curve is 1, so you double the tail area and subtract from 1:

$$1-(2)(0.2546)$$
$$=1-0.5092$$
$$=0.4908, \text{ or } 49.08\%$$

413.   22.06%

The distribution of the weights of the cookies is unknown, and you can't assume that it's normal. To answer this question, look to the central limit theorem. The central limit theorem can be applied to this problem because the sample size of 49 is large enough ($n \geq 30$).

First, find the z-score equivalent of this sample mean by using the population mean and the standard error of the mean for samples of this size.

$$z = \frac{\bar{x}-\mu_X}{\sigma_X/\sqrt{n}}$$

Here, $\bar{x}$ is the sample mean, μ_X is the population mean, σ_X is the population standard deviation, and n is the sample size.

Now, substitute the known values into the formula and solve:

$$z = \frac{12.004-12}{0.1/\sqrt{49}}$$
$$= \frac{0.004}{0.1/7}$$
$$= \frac{0.004}{0.01429}$$
$$= 0.2799$$

Because probabilities equate to tail areas in a normal distribution, you start by finding the tail area associated with this z-score (rounding to 0.28), using a Z-table (Table A-1 in the appendix).

The table gives tail areas only for negative z-scores, but the normal distribution is symmetrical, so you can look up the left-tail area equivalent to $z = -0.28$, which turns out to be 0.3897.

The question asks for the probability of a value at least this close to the mean, so you want the probability *excluding* the tail area.

The total area under the curve is 1, so double the tail area and subtract from 1:

$$1-(2)(0.3897)$$
$$=1-0.7794$$
$$=0.2206, \text{ or } 22.06\%$$

414.   76.98%

The distribution of the weights of the cookies is unknown, and you can't assume that it's normal. To answer this question, look to the central limit theorem. The central limit theorem can be applied to this problem because the sample size of 36 is large enough ($n \geq 30$).

First, find the z-score equivalent of this sample mean by using the population mean and the standard error of the mean for samples of this size.

$$z = \frac{\bar{x} - \mu_X}{\sigma_X / \sqrt{n}}$$

Here, $\bar{x}$ is the sample mean, μ_X is the population mean, σ_X is the population standard deviation, and n is the sample size.

Now, substitute the known values into the formula and solve:

$$z = \frac{12.02 - 12}{0.1 / \sqrt{36}}$$

$$= \frac{0.02}{0.1/6}$$

$$= \frac{0.02}{0.01667}$$

$$= 1.1998$$

Because probabilities equate to tail areas in a normal distribution, you start by finding the tail area associated with this z-score (rounding to 1.20), using a Z-table (Table A-1 in the appendix).

The table gives tail areas only for negative z-scores, but the normal distribution is symmetrical, so you can look up the left-tail area equivalent to $z = -1.20$, which turns out to be 0.1151.

The question asks for the probability of a value at least this close to the mean, so you want the probability *excluding* the tail area. The total area under the curve is 1, so double the tail area and subtract from 1:

$$1 - (2)(0.1151)$$

$$= 1 - 0.2302$$

$$= 0.7698, \text{ or } 76.98\%$$

415. 83.84%

The distribution of the weights of the cookies is unknown, and you can't assume that it's normal. To answer this question, look to the central limit theorem. The central limit theorem can be applied to this problem because the sample size of 49 is large enough ($n \geq 30$).

First, find the z-score equivalent of this sample mean by using the population mean and the standard error of the mean for samples of this size.

$$z = \frac{\bar{x} - \mu_X}{\sigma_X / \sqrt{n}}$$

Here, $\bar{x}$ is the sample mean, μ_X is the population mean, σ_X is the population standard deviation, and n is the sample size.

Now, substitute the known values into the formula and solve:

$$z = \frac{12.02 - 12}{0.1 / \sqrt{49}}$$

$$= \frac{0.02}{0.1/7}$$

$$= \frac{0.02}{0.014286}$$

$$= 1.39997$$

Because probability is associated with tail area when working with the normal distribution, you find the tail area associated with this z-score (rounding to 1.40), using a Z-table (Table A-1 in the appendix).

The table gives tail areas only for negative z-scores, but the normal distribution is symmetrical, so you can look up the left-tail area equivalent to $z = -1.40$, which turns out to be 0.0808.

The question asks for the probability of a value at least this close to the mean, so you want the probability *excluding* the tail area. The total area under the curve is 1, so double the tail area and subtract from 1:

$$1 - (2)(0.0808)$$
$$= 1 - 0.1616$$
$$= 0.8384, \text{ or } 83.84\%$$

---

416. 0.69

$\hat{p}$ (pronounced "p-hat") is the observed proportion of female zombies in the sample. Because 20 zombies were identified as female and 9 were identified as male, the proportion is

$$\hat{p} = \frac{\text{Number of female zombies in the sample}}{\text{Total number of zombies in the sample}}$$
$$= \frac{20}{20+9}$$
$$= \frac{20}{29}$$
$$= 0.689$$

---

417. 0.50

The population proportion, p, can be an actual proportion observed in a population or a theoretical proportion that should happen under some set of assumptions, such as the assumption that exactly half (or 0.50) of zombies would be female in the absence of bias.

---

418. 0.0928

$\sigma_{\hat{p}}$ stands for the standard error of a sample proportion. You calculate $\sigma_{\hat{p}}$ with the following formula:

$$\sigma_{\hat{p}} = \sqrt{\frac{p(1-p)}{n}}$$

where p is the population proportion and n is the sample size.

Now, substitute the known values into the formula and solve:

$$\sigma_{\hat{p}} = \sqrt{\frac{0.5(1-0.5)}{29}}$$
$$= \sqrt{\frac{0.25}{29}}$$
$$= 0.092848$$

This rounds to 0.0928.

419. Yes

The number of female zombies, X, has a binomial distribution with $p = 0.5$ and $n = 29$. You can apply the central limit theorem as long as both np and $n(1 - p)$ are greater than 10. (*Note:* You don't use the condition that n is at least 30 like you do for other central limit theorem problems. The binomial is very common and has its own special conditions to check.)

In this case, $np = (29)(0.5) = 14.5$ and $n(1 - p) = 29(1 - 0.5) = 14.5$, so the sample size is large enough to use the central limit theorem.

420. 0.03

Use the formula for calculating the standard error of a sample proportion

$$\sigma_{\hat{p}} = \sqrt{\frac{p\,(1-p)}{n}}$$

where p is the population proportion and n is the sample size.

In this case, $p = 0.9$ and $n = 100$, so you get

$$\sigma_{\hat{p}} = \sqrt{\frac{0.9(1-0.9)}{100}}$$

$$= \sqrt{\frac{0.09}{100}}$$

$$= 0.03$$

421. 0.03

$\sigma_{\hat{p}}$ is the standard error of a sample proportion, which you can find by using this formula:

$$\sigma_{\hat{p}} = \sqrt{\frac{p\,(1-p)}{n}}$$

where p is the population proportion and n is the sample size.
In this case, $p = 0.1$ and $n = 100$, so you get

$$\sigma_{\hat{p}} = \sqrt{\frac{0.1(1-0.1)}{100}}$$

$$= \sqrt{\frac{0.09}{100}}$$

$$= 0.03$$

422. 0.05

$\sigma_{\hat{p}}$ is the standard error of a sample proportion, which you can find by using this formula:

$$\sigma_{\hat{p}} = \sqrt{\frac{p\,(1-p)}{n}}$$

where p is the population proportion and n is the sample size.

In this case, $p = 0.5$ and $n = 100$, so you get

$$\sigma_p = \sqrt{\frac{0.5(1-0.5)}{100}}$$

$$= \sqrt{\frac{0.25}{100}}$$

$$= 0.05$$

423. 0.0607

σ_p is the standard error of the sample proportion, which you can find by using this formula:

$$\sigma_p = \sqrt{\frac{p(1-p)}{n}}$$

where p is the population proportion and n is the sample size.

In this case, $p = 0.67$ and $n = 60$, so you get

$$\sigma_p = \sqrt{\frac{0.67(1-0.67)}{n}}$$

$$= \sqrt{\frac{0.2211}{60}}$$

$$= 0.0607$$

424. 100

To use the central limit theorem for proportions, both np and $n(1-p)$ must be 10 or greater, where n is the sample size and p is the population proportion. (**Note:** You don't use the condition that n is at least 30 like you do for other central limit theorem problems. The binomial is very common and has its own special conditions to check.)

In this case, $p < (1-p)$, so you need to ensure that np is at least 10, which you can write as $np \geq 10$ and rearrange as

$$n \geq \frac{10}{p}$$

(Because $p > 0$, you don't need to reverse the inequality here.) Now substitute 0.1 for p to get

$$n \geq \frac{10}{0.1}$$

$$n \geq 100$$

The sample size must be at least 100.

425. 20

To use the central limit theorem for proportions, both np and $n(1-p)$ must be 10 or greater, where n is the sample size and p is the population proportion. (**Note:** You don't use the condition that n is at least 30 like you do for other central limit theorem problems. The binomial is very common and has its own special conditions to check.)

Because p and $(1 - p)$ are the same (0.5), solving for either np or $n(1 - p)$ will work.

So you can rearrange the statement $np \geq 10$ as

$$n \geq \frac{10}{p}$$

Now substitute 0.5 for p to get

$$n \geq \frac{10}{0.5}$$
$$n \geq 20$$

The sample size must be at least 20.

---

426.   −1.58

Use the formula for a z-score for proportions:

$$z = \frac{\hat{p} - p}{\sigma_p}$$

where $\hat{p}$ is the sample proportion, p is the population proportion, and σ_p is the standard error of the sample proportion.

Now, substitute the known values into the formula and solve:

$$z = \frac{0.25 - 0.5}{0.1581}$$
$$= \frac{-0.25}{0.1581}$$
$$= -1.5813$$

---

427.   −2.5

Use the formula for a z-score for proportions:

$$z = \frac{\hat{p} - p}{\sigma_p}$$

where $\hat{p}$ is the sample proportion, p is the population proportion, and σ_p is the standard error of the sample proportion.

Now, substitute the known values into the formula and solve:

$$z = \frac{0.25 - 0.5}{0.1}$$
$$= \frac{-0.25}{0.1}$$
$$= -2.5$$

---

428.   0

You can find the answer to this problem quickly, without doing the math. If you notice that the observed proportion (0.25) is the same as the population proportion (0.25), you know that the z-score is 0. You can see why by looking at the equation for a z-score:

$$z = \frac{\hat{p} - p}{\sigma_{\hat{p}}}$$

Here, $\hat{p}$ is the sample proportion, p is the population proportion, and $\sigma_{\hat{p}}$ is the standard error of the sample proportion. If the numerator is 0 ($0.25 - 0.25 = 0$), then z is also 0.

429. cannot use the central limit theorem here

For the central limit theorem to work here, you need to verify that np and $n(1-p)$ are both 10 or greater. In this case, $np = (10)(0.5) = 5$, so you can't use the central limit theorem.

Note: You can compute the exact probability by using the binomial probability formula or table.

430. 0.62%

For the central limit theorem to work here, you need to verify that np and $n(1-p)$ are both 10 or greater. In this case, $np = (25)(0.5) = 12.5$ and $n(1-p) = 25(1-0.5) = 12.5$, so you can use the central limit theorem.

First, use the formula for finding the standard error:

$$\sigma_{\hat{p}} = \sqrt{\frac{p(1-p)}{n}}$$

where p is the population proportion and n is the sample size.

$$\sigma_{\hat{p}} = \sqrt{\frac{0.5(1-0.5)}{25}}$$

$$= \sqrt{\frac{0.25}{25}}$$

$$= \sqrt{0.01}$$

$$= 0.1$$

Then convert the information you have to a z-score, using the following formula:

$$z = \frac{\hat{p} - p}{\sigma_{\hat{p}}}$$

where $\hat{p}$ is the sample proportion, p is the population proportion, and $\sigma_{\hat{p}}$ is the standard error of the sample proportion.

In this case, $\hat{p} = 0.25$, $p = 0.5$, and $\sigma_{\hat{p}} = 0.1$, so you get

$$z = \frac{0.25 - 0.5}{0.1}$$

$$= \frac{-0.25}{0.1}$$

$$= -2.5$$

Because probabilities equate to tail areas in a normal distribution, Table A-1 in the appendix shows the probability in the tail below any z-value you look up. From the table, you can see that the area under the curve below a z-score of –2.5 is 0.0062, or 0.62%, which is the approximate probability needed by the central limit theorem.

431. 50%

Before diving into the calculations with this problem, think it through a little. You can see that the sample probability and the population probability are both the same (0.25). So this question is really just asking you how likely it is that the sample probability is less than the population probability. Because the sample probability is an unbiased estimator of the population probability, it's equally likely to be above or below p, so the probability that it's below p is 50%. You should also confirm that the central limit theorem applies here by computing $np = (40)(0.25) = 10$ and $p(1 - p) = 40(1 - 0.25) = 30$. Both are at least 10, so the central limit theorem is applicable.

432. 15.87%

Because the coin is fair, the probability of heads on each toss is $p = 0.5$. The 36 tosses form a sample of $n = 36$.

Check to see whether you can use the normal approximation to the binomial by making sure that both np and $n(1 - p)$ equal at least 10. In this case, $n = 36$ and $p = 0.5$, so $np = (36)(0.5) = 18$ and $n(1 - p) = (36)(1 - 0.5) = (36)(0.5) = 18$.

Now convert 21 heads to proportions by dividing by the sample size:

$$\hat{p} = \frac{21}{36} = 0.5833$$

Next, find the standard error, using this formula:

$$\sigma_{\hat{p}} = \sqrt{\frac{p(1 - p)}{n}}$$

where p is the population proportion and n is the sample size.

Substitute the known values into the formula and solve:

$$\sigma_{\hat{p}} = \sqrt{\frac{0.5(1 - 0.5)}{36}}$$

$$= \sqrt{\frac{0.25}{36}}$$

$$= 0.0833$$

Then convert the sample proportion to a z-value with this formula:

$$z = \frac{\hat{p} - p}{\sigma_{\hat{p}}}$$

where $\hat{p}$ is the sample proportion, p is the population proportion, and $\sigma_{\hat{p}}$ is the standard deviation.

$$z = \frac{\hat{p} - p}{\sigma_{\hat{p}}}$$

$$= \frac{0.5833 - 0.5}{0.0833}$$

$$= \frac{0.0833}{0.0833}$$

$$= 1$$

Because probabilities equate to tail areas in a normal distribution, Table A-1 in the appendix shows the probability in the tail below the z-value you look up. From the table, you can see that the area under the curve with a z-score of -2.5 is 0.0092, or 0.92%, which is the approximate probability needed by the central limit theorem.

Because tail area is the same as probability when you're dealing with the normal distribution, use Table A-1 in the appendix to find the area to the left of $z = +1.0$ is 0.8413; to find the area to the right, subtract this from 1 (because total probability is always 1):

$$1 - 0.8413 = 0.1587, \text{ or } 15.87\%$$

433.   >99.99%

Because the coin is fair, the probability of heads on each toss is $p = 0.5$. The 50 tosses form a sample of $n = 50$.

Check to see whether you can use the normal approximation to the binomial by making sure that both np and $n(1-p)$ equal at least 10. In this case, $n = 50$ and $p = 0.5$, so $np = (50)(0.5) = 25$ and $n(1-p) = (50)(1-0.5) = (50)(0.5) = 25$.

Now convert ten heads to proportions by dividing by the sample size:

$$\hat{p} = \frac{10}{50} = 0.2$$

Next, find the standard error, using this formula:

$$\sigma_{\hat{p}} = \sqrt{\frac{p(1-p)}{n}}$$

where p is the population proportion and n is the sample size.

Substitute the known values into the formula and solve:

$$\sigma_{\hat{p}} = \sqrt{\frac{0.5(1-0.5)}{50}}$$

$$= \sqrt{\frac{0.25(0.25)}{50}}$$

$$= 0.07071$$

Then convert each sample proportion to a z-value with this formula:

$$z = \frac{\hat{p} - p}{\sigma_{\hat{p}}}$$

where $\hat{p}$ is the sample proportion, p is the population proportion, and $\sigma_{\hat{p}}$ is the standard deviation.

$$z = \frac{\hat{p} - p}{\sigma_{\hat{p}}}$$

$$= \frac{0.2 - 0.5}{0.07071}$$

$$= \frac{-0.3}{0.07071}$$

$$= -4.2427$$

To find the probability of these results, find the probability of the z-score using Table A-1. The smallest value for a z-score in Table A-1 is –3.69; the probability of a z-score lower than this value is 0.0001. The probability of a z-score of –4.2427 is less than this, because –4.2427 is farther from 0 than –3.69. So you can say that the probability of getting at least ten heads in 50 tosses of a fair coin is greater than the following:

$$1 - 0.0001 = 0.9999 \text{ or } 99.99\%$$

434.   2.28%

Start by finding the observed proportion of interest. The question asks about getting more than 60 heads, so you're interested in sample probabilities greater than $60/100 = 0.6$, or $\hat{p} = 0.6$.

Note that the value of the population proportion (p) used in the following formulas is 0.5, the chance of getting heads on one flip of a fair coin.

Next, find the standard error, using this formula:

$$\sigma_{\hat{p}} = \sqrt{\frac{p(1-p)}{n}}$$

where p is the population proportion and n is the sample size.

Substitute the known values into the formula and solve:

$$\sigma_{\hat{p}} = \sqrt{\frac{0.5(1-0.5)}{100}}$$

$$= \sqrt{\frac{0.25}{100}}$$

$$= 0.05$$

Then convert the sample proportion to a z-value with this formula:

$$z = \frac{\hat{p}-p}{\sigma_{\hat{p}}}$$

where $\hat{p}$ is the sample proportion, p is the population proportion, and $\sigma_{\hat{p}}$ is the standard deviation.

$$z = \frac{0.6-0.5}{0.05}$$

$$= 2.0$$

So you're interested in the probability of getting a z-score greater than 2.0. Table A-1 in the appendix gives the probability of getting a z-score less than 2.0 as 0.9722.

You can subtract this from 1 to get the probability of a score greater than 2.0:
$1 - 0.9722 = 0.0228$.

435.   5.59%

First, identify the information you have in symbolic notation, converting from percentages to proportions as necessary:

$p = 0.01$ (the population proportion)

$n = 1,000$ (the sample size)

$\hat{p} = 0.015$ (the sample proportion)

Find out whether you can use the central limit theorem by making sure that both np and $n(1-p)$ are at least 10. Because np is $(1,000)(0.01) = 10$ and $1,000(1-0.01) = 1,000(0.99) = 990$, you can proceed.

To get the appropriate probability, you need to find the standard error and then convert the sample proportion to a z-value. The formula for standard error is

$$\sigma_{\hat{p}} = \sqrt{\frac{p\,(1-p)}{n}}$$

where p is the population proportion and n is the sample size.

Substitute the known values in the formula and solve:

$$\sigma_{\hat{p}} = \sqrt{\frac{0.01(1-0.01)}{1,000}}$$

$$= \sqrt{\frac{0.0099}{1,000}}$$

$$= 0.003146$$

Then convert the sample proportion to a z-value, using this formula:

$$z = \frac{\hat{p}-p}{\sigma_{\hat{p}}}$$

where $\hat{p}$ is the sample proportion, p is the population proportion, and $\sigma_{\hat{p}}$ is the standard deviation.

$$z = \frac{0.015-0.01}{0.003146}$$

$$= \frac{0.005}{0.003146}$$

$$= 1.58931$$

You can find the probability of observing a z-value of 1.589 or less by using Table A-1 in the appendix (rounding z to 1.59): 0.9411.

To answer the question, you need the probability of observing a z-value this high or greater, which is found by subtracting from 1 (the total area under the curve):

$$1 - 0.9441 = 0.0559 = 5.59\%$$

436. 19.02%

First, find out whether you can use the normal approximation to the binomial by making sure that both np and $n(1-p)$ are at least 10. Here, $np = 36(0.3) = 10.8$, and $n(1-p) = 36(1-0.3) = 36(0.7) = 25.2$, so you can proceed.

To get the appropriate probability, you need to find the standard error, using this formula:

$$\sigma_{\hat{p}} = \sqrt{\frac{p\,(1-p)}{n}}$$

where p is the population proportion and n is the sample size.

Substitute the known values into the formula and solve:

$$\sigma_{\hat{p}} = \sqrt{\frac{0.3(1-0.3)}{36}}$$

$$= \sqrt{\frac{0.21}{36}}$$

$$= 0.07638$$

Then convert the observed proportions to z-scores with this formula:

$$z = \frac{\hat{p}-p}{\sigma_{\hat{p}}}$$

where $\hat{p}$ is the sample proportion, p is the population proportion, and $\sigma_{\hat{p}}$ is the standard deviation.

$$z_1 = \frac{\hat{p}_1 - p}{\sigma_{\hat{p}}}$$
$$= \frac{0.2 - 0.3}{0.07638}$$
$$= \frac{-0.1}{0.07638}$$
$$= -1.3092$$

$$z_2 = \frac{\hat{p}_2 - p}{\sigma_{\hat{p}}}$$
$$= \frac{0.4 - 0.3}{0.07638}$$
$$= \frac{0.1}{0.07638}$$
$$= 1.3092$$

Because probabilities equate to tail areas in a normal distribution, you can find the area in the left tail by using the Z-table (Table A-1 in the appendix) and looking up the area to the left of z = –1.3092, which is 0.0951.

Doubling that gives a total tail area of 0.1902, or 19.02%.

437. 100%

A proportion must be in the range of 0 to 1, so there's no probability of observing a proportion greater than or equal to 2. Therefore, the probability of an observed proportion of less than 2 is 100%.

438. 90.49%

First, determine whether you can use the central limit theorem by making sure that both np and $n(1-p)$ are at least 10. Because $np = 36(0.3) = 10.8$ and $n(1-p) = 36(1-0.3) = 36(0.7) = 25.2$, you can proceed.

Next, find the standard error, using this formula:

$$\sigma_{\hat{p}} = \sqrt{\frac{p(1-p)}{n}}$$

where p is the population proportion and n is the sample size.

Substitute the known values into the formula and solve:

$$\sigma_{\hat{p}} = \sqrt{\frac{0.3(1-0.3)}{36}}$$
$$= \sqrt{\frac{0.21}{36}}$$
$$= 0.07638$$

Then convert the observed proportions to z-scores with this formula:

$$z = \frac{\hat{p} - p}{\sigma_p}$$

where $\hat{p}$ is the sample proportion, p is the population proportion, and σ_p is the standard deviation.

$$z = \frac{\hat{p}_1 - p}{\sigma_p}$$

$$= \frac{0.4 - 0.3}{0.07638}$$

$$= \frac{0.1}{0.07638}$$

$$= 1.3092$$

Because probabilities equate to tail areas in a normal distribution, you can find the approximate area in the left tail by using the Z-table (Table A-1 in the appendix) and looking up the area to the left of $z = 1.3092$ (rounded to 1.31), which is 0.9049, or 90.49%.

439.   5%

First, determine whether you can use the central limit theorem by making sure that both np and $n(1-p)$ are at least 10. Because $np = 81(0.3) = 24.3$ and $n(1-p) = 81(1-0.3) = 81(0.7) = 56.7$, you can proceed.

Next, find the standard error, using this formula:

$$\sigma_p = \sqrt{\frac{p(1-p)}{n}}$$

where p is the population proportion and n is the sample size.

Substitute the known values into the formula and solve:

$$\sigma_p = \sqrt{\frac{0.3(1-0.3)}{81}}$$

$$= \sqrt{\frac{0.21}{81}}$$

$$= 0.05092$$

Then convert the observed proportions to z-scores with this formula:

$$z = \frac{\hat{p} - p}{\sigma_p}$$

where $\hat{p}$ is the sample proportion, p is the population proportion, and σ_p is the standard deviation.

$$z_1 = \frac{\hat{p}_1 - p}{\sigma_p}$$

$$= \frac{0.2 - 0.3}{0.05092}$$

$$= \frac{-0.1}{0.05092}$$

$$= -1.96386$$

$$z_2 = \frac{\hat{p}_2 - p}{\sigma_{\hat{p}}}$$

$$= \frac{0.4 - 0.3}{0.05092}$$

$$= \frac{0.1}{0.05092}$$

$$= 1.96386$$

Because probabilities equate to tail areas in a normal distribution, you can find the area in the left tail by rounding the smaller z-value to two decimal places and using the Z-table (Table A-1 in the appendix) and looking up the area to the left of z = –1.96, which is 0.025.

Doubling that gives you a total tail area of 0.05, or 5%.

440. 99.12%

First, determine whether you can use the normal approximation to the binomial by making sure that both np and $n(1-p)$ are at least 10. Because $np = 144(0.3) = 43.2$ and $n(1-p) = 144(1-0.3) = 144(0.7) = 100.8$, you can proceed.

Next, find the standard error, using this formula:

$$\sigma_{\hat{p}} = \sqrt{\frac{p(1-p)}{n}}$$

where p is the population proportion and n is the sample size.

Substitute the known values into the formula and solve:

$$\sigma_{\hat{p}} = \sqrt{\frac{0.3(1-0.3)}{144}}$$

$$= \sqrt{\frac{0.21}{144}}$$

$$= 0.03819$$

Then convert the observed proportions to z-scores with this formula:

$$z = \frac{\hat{p} - p}{\sigma_{\hat{p}}}$$

where $\hat{p}$ is the sample proportion, p is the population proportion, and $\sigma_{\hat{p}}$ is the standard deviation.

$$z_1 = \frac{\hat{p}_1 - p}{\sigma_{\hat{p}}}$$

$$= \frac{0.2 - 0.3}{0.03819}$$

$$= \frac{-0.1}{0.03819}$$

$$= -2.6185$$

$$z_2 = \frac{\hat{p}_2 - p}{\sigma_p}$$

$$= \frac{0.4 - 0.3}{0.03819}$$

$$= \frac{0.1}{0.03819}$$

$$= 2.6185$$

Because probabilities equate to tail areas in a normal distribution, you can find the area in the left tail by rounding the smaller z-value to two decimal places and using the Z-table (Table A-1 in the appendix) and looking up the area to the left of $z = -2.62$ which is 0.0044.

Doubling that gives a total tail area of 0.0088. This is the proportion of probability in the range. To find the proportion outside the range, subtract from 1:

$$1 - 0.0088 = 0.9912, \text{ or } 99.12\%.$$

---

441. It tells you how precise you can expect the results to be, across many random samples of the same size.

The basic idea in surveys is that sample results vary. *Margin of error* measures how much you expect your sample results could change if you took many different samples of the same size from this population. Here, the survey result of 60% is based on one sample, and the 4% margin of error means that, with a certain level of confidence, this value of 60% can change by as much as 4% on either side if different samples of the same size are taken.

Note: You assume that all samples were chosen at random in this case, or the margin of error means nothing. Margin of error assumes that the samples were randomly selected and only measures how much the results can change from sample to sample.

---

442. the confidence level

Any statistical result involving a margin of error is basically calculating a confidence interval, which is the sample statistic plus or minus the margin of error. The claim requires a sample result, the margin of error, and the level of confidence. Here, the sample is 1,000, the sample statistic is 0.93, but the confidence level is missing.

---

443. The election is too close to call.

You can use the poll to conclude that 54% of the voters in this sample would vote for Garcia, and when you project the results to the population, you add a margin of error of ± 5%. That means that the proportion voting for Garcia is estimated to be between 54% − 5% = 49% and 54% + 5% = 59% in the population with 95% confidence.

You can also use the poll to conclude that 46% of the voters in this sample would vote for Smith, and when you project the results to the population, you add a margin of error of ± 5%. That means that the proportion voting for Smith is estimated to be between 46% − 5% = 41% and 46% + 5% = 51% in the population with 95% confidence (over many samples).

Garcia's confidence interval is 49% to 59%, and Smith's confidence interval is 41% to 51%. Because the confidence intervals overlap, the election is too close to call.

444. ± 2.94

The formula to find the margin of error when estimating a population mean is

$$\text{MOE} = \pm z^* \left(\frac{\sigma}{\sqrt{n}} \right)$$

where z^* is the value from Table A-4 for a given confidence level (95% in this case, or 1.96), σ is the population standard deviation (15), and n is the sample size (100).

Now, substitute these values into the formula and solve:

$$\text{MOE} = \pm 1.96 \left(\frac{15}{\sqrt{100}} \right)$$

$$= \pm 1.96 (1.5)$$

$$= \pm 2.94$$

The margin of error for a 95% confidence interval for the mean is ± 2.94.

445. ± 0.438

The formula for margin of error when estimating a population mean is

$$\text{MOE} = \pm z^* \left(\frac{\sigma}{\sqrt{n}} \right)$$

where z^* is the value from Table A-4 for a given confidence level (95% in this case, or 1.96), σ is the population standard deviation (5), and n is the sample size (500).

Now, substitute the values into the formula and solve:

$$\text{MOE} = \pm 1.96 \left(\frac{5}{\sqrt{500}} \right)$$

$$= \pm 1.96 \, (0.2236)$$

$$= \pm 0.438$$

The margin of error for a 95% confidence interval for the population mean is ± 0.438.

446. ± $3,099.03

The formula for margin of error when estimating a population mean is

$$\text{MOE} = \pm z^* \left(\frac{\sigma}{\sqrt{n}} \right)$$

where z^* is the value from Table A-4 for a given confidence level (95% in this case, or 1.96), σ is the population standard deviation ($10,000), and n is the sample size (40).

Now, substitute the values into the formula and solve:

$$MOE = \pm 1.96 \left(\frac{\$10{,}000}{\sqrt{40}} \right)$$

$$= \pm 1.96(\$1{,}581.13883)$$

$$= \pm \$3{,}099.03$$

The margin of error for a 95% confidence interval for the population mean is ± $3,099.03.

---

447.   ± $3.72

The formula for margin of error when estimating a population mean is

$$MOE = \pm z^* \left(\frac{\sigma}{\sqrt{n}} \right)$$

where z^* is the value from Table A-4 for a given confidence level (99% in this case, or 2.58), σ is the standard deviation ($25), and n is the sample size (300).

Now, substitute the values into the formula and solve:

$$MOE = \pm 2.58 \left(\frac{\$25}{\sqrt{300}} \right)$$

$$= \pm 2.58(\$1.4434)$$

$$= \pm \$3.72$$

The margin of error for a 99% confidence interval for the population mean is ± $3.72.

---

448.   ± $2.83

The formula for margin of error when estimating a population mean is

$$MOE = \pm z^* \left(\frac{\sigma}{\sqrt{n}} \right)$$

where z^* is the value from Table A-4 for a given confidence level (95% in this case, or 1.96), σ is the standard deviation ($25), and n is the sample size (300).

Now, substitute the values into the formula and solve:

$$MOE = \pm 1.96 \left(\frac{\$25}{\sqrt{300}} \right)$$

$$= \pm 1.96(\$1.4434)$$

$$= \pm \$2.83$$

The margin of error for a 95% confidence interval for the population mean is ± $2.83.

449. $83.15

To find the lower limit for the 80% confidence interval, you first have to find the margin of error. The formula for the margin of error when estimating a population mean is

$$MOE = \pm z^* \left(\frac{\sigma}{\sqrt{n}} \right)$$

where z^* is the value from Table A-4 for a given confidence level (80% in this case, or 1.28), σ is the standard deviation ($25), and n is the sample size (300).

Now, substitute the values into the formula and solve:

$$MOE = \pm 1.28 \left(\frac{\$25}{\sqrt{300}} \right)$$

$$= \pm 1.28 (\$1.4434)$$

$$= \pm \$1.85$$

Next, subtract the MOE from the sample mean to find the lower limit: $85.00 – $1.85 = $83.15.

450. 1.96

First off, if you look at Table A-4, you see that the number you need for z^* for a 95% confidence interval is 1.96. However, when you look up 1.96 on Table A-1 in the appendix, you get a probability of 0.975. Why?

In a nutshell, Table A-1 shows only the probability below a certain z-value, and you want the probability between two z-values, $-z$ and z. If 95% of the values must lie between $-z$ and z, you expand this idea to notice that a combined 5% of the values lie above z and below $-z$. So 2.5% of the values lie above z, and 2.5% of the values lie below $-z$. To get the total area below this z-value, take the 95% between $-z$ and z plus the 2.5% below $-z$, and you get 97.5%. That's the z-value with 97.5% area below it. It's also the number with 95% lying between two z-values, $-z$ and z.

To avoid all these extra steps and headaches, Table A-4 has already done this conversion for you. So when you look up 1.96 on Table A-4, you automatically find 95% (not 97.5%).

451. 2.58

Table A-4 shows the answer: A 99% confidence level has a z^*-value of 2.58.

452. 1.28

Table A-4 shows the answer: An 80% confidence level has a z^*-value of 1.28.

453. ± 0.83 years

The formula for margin of error when estimating a population mean is

$$MOE = \pm z^* \left(\frac{\sigma}{\sqrt{n}} \right)$$

where z^* is the value from Table A-4 for a given confidence level (95% in this case, or 1.96), σ is the standard deviation (3 years), and n is the sample size (50).

Now, substitute the numbers into the formula and solve:

$$MOE = \pm 1.96 \left(\frac{3 \text{ years}}{\sqrt{50}} \right)$$

$$= \pm 1.96 \, (0.4243 \text{ years})$$

$$= \pm 0.83 \text{ years}$$

With 50 women in the sample, the sociologist has a margin of error of ± 0.83 years.

454. ± 0.59 years

The formula for margin of error when estimating a population mean is

$$MOE = \pm z^* \left(\frac{\sigma}{\sqrt{n}} \right)$$

where z^* is the value from Table A-4 for a given confidence level (95% in this case, or 1.96), σ is the standard deviation (3 years), and n is the sample size (100).

Now, substitute the numbers into the formula and solve:

$$MOE = \pm 1.96 \left(\frac{3 \text{ years}}{\sqrt{100}} \right)$$

$$\pm 1.96 (0.3 \text{ years})$$

$$= \pm 0.59 \text{ years}$$

With this sample size, the margin of error is ± 0.59 years.

455. 9

To solve this problem, you have to do a little algebra, using the general formula for margin of error when estimating a population mean:

$$MOE = \pm z^* \left(\frac{\sigma}{\sqrt{n}} \right)$$

Here, z^* is the value from Table A-4 for a given confidence level (95% in this case, or 1.96), σ is the population standard deviation (3 years), and n is the sample size.

To find the sample size, *n*, you have to rearrange the formula for margin of error: Multiply both sides by the square root of *n*, divide both sides by the margin of error, and square both sides (note that a sample size can only be positive). Here are the steps:

$$MOE = \pm z^* \left(\frac{\sigma}{\sqrt{n}} \right)$$

$$MOE\sqrt{n} = \pm z^*(\sigma)$$

$$(\sqrt{n}) = \pm \frac{z^*(\sigma)}{MOE}$$

$$n = \left(\frac{z^*(\sigma)}{MOE} \right)^2$$

Now to solve the problem at hand, substitute the values into the formula:

$$n = \left(\frac{(1.96)(3)}{2} \right)^2$$

$$= 8.643$$

You can't have a fraction of a participant and you need to make sure the margin of error is less than or equal to two years, so round up to 9 participants.

Note: Even if the result was a number that you'd normally round down (like 8.123), you still always round up to the next greatest integer when solving for sample size.

---

456. ± 0.00182 mm

To calculate the margin of error (MOE) for estimating a population mean, use this formula:

$$MOE = \pm z^* \left(\frac{\sigma}{\sqrt{n}} \right)$$

Here, z^* is the value from Table A-4 for a given confidence level (99% in this case, or 2.58), σ is the standard deviation (0.01 millimeters), and *n* is the sample size (200).

Now, substitute all the known values into the formula and solve:

$$MOE = \pm 2.58 \left(\frac{0.01 \, mm}{\sqrt{200}} \right)$$

$$= \pm 2.58(0.000707 \, mm)$$

$$= \pm 0.00182 \, mm$$

---

457. ± 0.00912 mm

To calculate the margin of error (MOE) for estimating a population mean, use this formula:

$$MOE = \pm z^* \left(\frac{\sigma}{\sqrt{n}} \right)$$

Here, z^* is the value from Table A-4 for a given confidence level (99% in this case, or 2.58), σ is the population standard deviation (0.05 millimeters), and n is the sample size (200).

Now, substitute the known values into the formula and solve:

$$MOE = \pm 2.58 \left(\frac{0.05 \text{mm}}{\sqrt{200}} \right)$$

$$= \pm 2.58 (0.003536 \text{ mm})$$

$$= \pm 0.00912 \text{ mm}$$

458. ± 0.0182 mm

To calculate the margin of error (MOE) for estimating a population mean, use this formula:

$$MOE = \pm z^* \left(\frac{\sigma}{\sqrt{n}} \right)$$

Here, z^* is the value from Table A-4 for a given confidence level (99% in this case, or 2.58), σ is the population standard deviation (0.10 millimeters), and n is the sample size (200).

Substitute the known values into the formula and solve:

$$MOE = \pm 2.58 \left(\frac{0.1 \text{ mm}}{\sqrt{200}} \right)$$

$$= \pm 2.58 (0.00707 \text{ mm})$$

$$= \pm 0.0182 \text{ mm}$$

459.   Quadruple the original sample size.

The basic formula for margin of error when estimating a population mean is

$$MOE = \pm z^* \left(\frac{\sigma}{\sqrt{n}} \right)$$

where z^* is the value from Table A-4 for a given confidence level (99% in this case, or 2.58), σ is the population standard deviation, and n is the sample size (200). You don't need to know the units of measurement or the value of the standard deviation to answer this question.

Note that the sample size, n, appears in the denominator here. If you want to cut the margin of error in half, the sample size will change. Halving the MOE is equivalent to dividing the MOE by 2; to maintain the integrity of the equation, you have to divide the other half of the equation by 2 also. Because the denominator of the right-hand side of the equation is the square root of n, dividing by the square root of 4 is equivalent to dividing by 2, because 2 is the square root of 4.

$$MOE = \pm z^* \left(\frac{\sigma}{\sqrt{n}} \right)$$

$$\frac{MOE}{2} = \pm z^* \left(\frac{\sigma}{2\sqrt{n}} \right)$$

$$\frac{MOE}{2} = \pm z^* \left(\frac{\sigma}{\sqrt{4n}} \right)$$

Therefore, with all else held constant, increasing n fourfold will cut the MOE in half.

Note that, because of the distributive law, the following is true:

$$\sqrt{4n} = (\sqrt{4})(\sqrt{n}) = 2\sqrt{n}$$

---

460. converted a margin of error at 99% confidence to a margin of error at 80% confidence

Note that you don't need to know what part or what dimension of the part is involved; you can solve this problem through your knowledge of the formula for margin of error. Switching from a 99% confidence level to an 80% confidence level will change the margin of error by a little more than half.

Consider the formula for calculating margin of error:

$$MOE = z^* \left(\frac{\sigma}{\sqrt{n}} \right)$$

where z^* is the value from Table A-1 corresponding to the margin of error, σ is the standard deviation, and n is the sample size. To preserve the equation but reduce the MOE by half, you must divide both sides by 2 (or multiply both by ½).

$$MOE = z^* \left(\frac{\sigma}{\sqrt{n}} \right)$$

$$\left(\frac{1}{2} \right) MOE = \left(\frac{1}{2} \right) z^* \left(\frac{\sigma}{\sqrt{n}} \right)$$

You can't change the sample size or standard deviation; the only thing you can change is z^*. You need two values, one of which is about twice the other. The value of z^* for a 99% confidence interval is 2.58, which is about twice that for an 80% confidence interval (1.28). The best conclusion is that the factory owner changed the width of the confidence interval.

---

461. 800

The sample size, n, appears in the denominator of the formula for margin of error.

$$MOE = \pm z^* \left(\frac{\sigma}{\sqrt{n}} \right)$$

The researcher wants to cut the MOE to ⅓ of its current value, which is equivalent to dividing by 3. To retain the integrity of the equation, you must divide both sides by 3. Note that the denominator of the right-hand side of the equation is the square root of n; dividing by the square root of 9 is equivalent to dividing by 3, because 3 is the square root of 9. So to divide the MOE by 3, holding everything else constant, you must increase the sample size to 9 times its current value.

$$MOE = \pm z^* \left(\frac{\sigma}{\sqrt{n}} \right)$$

$$\frac{MOE}{3} = \pm z^* \left(\frac{\sigma}{3\sqrt{n}} \right)$$

$$\frac{MOE}{3} = \pm z^* \left(\frac{\sigma}{\sqrt{9n}} \right)$$

Note that, because of the distributive law, the following is true:

$$\sqrt{9n} = (\sqrt{9})(\sqrt{n}) = 3\sqrt{n}$$

Because $n = 100$ and $9n = (9)(100) = 900$, the market researcher needs 800 more participants to get the desired margin of error (a total of 900 participants is required).

462. $\pm 0.978\%$

The formula to calculate the margin of error (MOE) for a population proportion is

$$MOE = \pm z^* \sqrt{\frac{\hat{p}(1-\hat{p})}{n}}$$

where z^* is the value from Table A-4 for a given confidence level (95% in this case, or 1.96), $\hat{p}$ is the sample proportion (0.53), and n is the sample size (10,000).

You convert 53% to the proportion 0.53 by dividing the percentage by 100: $53/100 = 0.53$.

Now, substitute the known values in the formula and solve:

$$MOE = \pm 1.96 \sqrt{\frac{0.53(1-0.53)}{10,000}}$$

$$= \pm 1.96 \sqrt{\frac{0.2491}{10,000}}$$

$$= \pm 1.96 \sqrt{0.00002491}$$

$$= \pm 1.96(0.00499)$$

$$= \pm 0.00978$$

Convert this proportion to a percentage by multiplying by 100%: $(0.00978)(100\%) = 0.978\%$

The margin of error is therefore $\pm 0.978\%$.

You estimate that 53% $\pm$ 0.978% of all Europeans are unhappy with the euro, based on these survey results with 95% confidence.

463. ± 7.69%

The formula for the margin of error (MOE) for a proportion is

$$MOE = \pm z^* \sqrt{\frac{\hat{p}(1-\hat{p})}{n}}$$

where z^* is the value from Table A-4 for a given confidence level (99% in this case, or 2.58), $\hat{p}$ is the sample proportion (0.48), and n is the sample size (281).

To compute the sample proportion, divide the number who responded that they intend to vote by the total number polled: 135/281 = 0.48.

Now, substitute the known values into the formula and solve:

$$MOE = \pm 2.58 \sqrt{\frac{0.48(1-0.48)}{281}}$$

$$= \pm 2.58 \sqrt{\frac{0.2496}{281}}$$

$$= \pm 2.58(0.0298)$$

$$= \pm 0.0769$$

Convert the proportion to a percentage by multiplying by 100%: (0.0769)(100%) = 7.69%.

Using a 99% confidence level, the county election office has achieved a margin of error of ± 7.69%.

464. 90% confidence

The confidence level tells you how many standard errors to add and subtract to get the margin of error you want, which is quantified by the z-value. To find the confidence level, you first need to solve for z by rearranging the margin of error formula for a population proportion:

$$MOE = \pm z^* \sqrt{\frac{\hat{p}(1-\hat{p})}{n}}$$

Here, z^* is the value from Table A-4 for a given confidence level, $\hat{p}$ is the sample proportion (0.46), and n is the sample size (922).

Rearrange this formula as follows:

$$MOE = \pm z^* \sqrt{\frac{\hat{p}(1-\hat{p})}{n}}$$

$$\frac{MOE}{\sqrt{\frac{\hat{p}(1-\hat{p})}{n}}} = \pm z^*$$

Note that you can drop the ± symbol because the standard normal distribution is symmetrical. Also, by convention, this formula is usually written with z^* on the left side (because the two sides are equivalent, it doesn't matter which way you write the equation):

$$z^* = \frac{\text{MOE}}{\sqrt{\dfrac{\hat{p}(1-\hat{p})}{n}}}$$

Then substitute the known values into the formula and solve. (First, convert the margin of error to a proportion by dividing by 100%: 2.7%/100% = 0.027.)

$$z^* = \frac{0.027}{\sqrt{\dfrac{0.46(1-0.46)}{922}}}$$

$$= \frac{0.027}{\sqrt{\dfrac{0.2484}{922}}}$$

$$= \frac{0.027}{0.01641}$$

$$= 1.64534$$

Rounding to three decimal places, you get 1.645. Now you can find the confidence level by looking at Table A-4 and seeing that 1.645 corresponds to the 90% confidence level z^*-value. So you can safely say that the confidence level is 90%.

465.   B.   If the same study were repeated many times, about 95% of the time, the confidence interval would contain the average money spent for all the customers.

The average money spent for all the customers is an unknown value, called a *population parameter*. The average money spent for the 100 customers in the sample is a known value, $45, which is called a *statistic*.

The store is using a sample statistic to estimate a population parameter. Because samples vary from sample to sample, they know the sample mean may not correspond exactly to the population mean, so they use confidence intervals to state a plausible range of values for the population mean. If the same experiment were repeated many times (drawing a sample of the same size from the same population and calculating the sample average), the population mean would be expected to be contained in 95% of the confidence intervals created.

466.   E.   Choices (A) and (C) (The store studied a sample of sales records rather than the entire population of sales records; because sample results vary, the sample mean is not expected to correspond exactly to the population mean, so a range of likely values is required.)

The store studied a sample of records to estimate a population parameter, and because sample results vary (called *sampling error*), the sample mean isn't expected to correspond exactly to the population mean. If another sample of the same size were drawn from the population, the sample mean would be expected to be somewhat different, so a range of likely values for the population mean (that is, a confidence interval) is required.

467.   D.   Choices (B) and (C) (The larger sample will produce a more precise estimate of the population mean; the 95% confidence interval calculated from the larger sample will be narrower.)

A larger sample drawn from the same population will tend to produce a narrower confidence interval and a more precise estimate of the population mean. The amount of bias isn't measured by the confidence interval.

468.

E. None of the above

The sample mean isn't expected to be exactly the same as the population mean nor is another sample of the same size drawn from the same population expected to have exactly the same mean. That number by itself isn't a "good" number to use to estimate the population mean; you need a margin of error to go with it. The sample mean changes if another sample were taken.

469.

Sample D

For samples of the same size, greater variability in the data will tend to produce a wider confidence interval. In this case, Sample A has no variability, Sample B has limited variability, Sample C has some variability, and Sample D has the greatest variability due to one extremely high value (20). You can conduct calculations on these data sets to confirm this notion.

470.

one with a confidence level of 99%

The confidence level is the amount of confidence you have that a confidence interval will contain the actual population parameter (in this case, the population mean) if you repeated the process over and over.

If the confidence level is higher, the interval needs to be wider to include a wider range of likely values for the population parameter. Therefore, the confidence interval with the highest confidence level will be the widest.

471.

The width of the confidence interval will increase.

Increasing the confidence level will increase the width of the confidence interval because to be more confident in your process of trying to estimate the mean of the population, you must include a wider range of possible values for it (assuming that all the other parts involved in the confidence interval stay the same).

472.

A higher confidence level means that the margin of error is increased, requiring a wider confidence interval.

With the factors remaining equal, increasing the confidence level means that you're offering a wider range of possible values for the population parameter. This increases the width of the confidence interval.

473.

± 5%

The margin of error is the number that is added or subtracted from the sample statistic to produce the confidence interval. In this case, the sample statistic is 65%. To start with 65% and end up with a confidence interval of 60% to 70%, you must add and subtract 5%:

$$65\% + 5\% = 70\%$$

$$65\% - 5\% = 60\%$$

So the margin of error is ± 5%.

474. A. $n = 100$

All else held constant, the smaller the sample size is, the larger the margin of error will be. You don't have as much precision when you have a smaller data set. Less precision means a larger margin of error, which means that the confidence interval will be wider. Sets of small samples have greater variability than sets of large samples.

475. E. $n = 5,000$

All else held constant, the larger the sample size is, the more precise your results and the smaller the margin of error will be. A smaller margin of error means that the confidence interval is narrower. Sets of large samples don't have as much variability as sets of small samples.

476. E. Choices (B) and (D) (The sample of 500 will have a narrower 95% confidence interval; the sample of 500 will produce a more precise estimate of the population mean.)

All else being equal, a larger sample will produce a narrower 95% confidence interval and a more precise estimate of the population mean because sets of large samples don't have as much variability as sets of small samples.

477. For a higher confidence level, the confidence interval will be wider.

When calculating confidence intervals for different confidence levels from the same sample, a higher confidence level will produce a wider confidence interval.

478. C. The confidence interval related to Population A is expected to be narrower.

All else being equal, a sample from a less-variable population can be expected to produce a narrower confidence interval. That's because the samples don't change as much when they come from a population whose values are more similar (less variable).

479. E. Choices (B) and (C) (A confidence level of 80% will produce a narrower confidence interval than a confidence level of 90%; a sample of 300 students will produce a narrower confidence interval than a sample of 150 students.)

A larger sample and a lower confidence level will both produce a narrower confidence interval when dealing with random samples drawn from the same population.

480. Sample B

For the same sample size and confidence level, greater variability in the sample will produce a wider confidence interval. Sample B has the greatest variability and will have the widest confidence interval for a given confidence level.

481.

D. all workers ages 22 to 30 who live in North America (Canada, the U.S., and Mexico), adjusted in U.S. dollars

The variability of the income of this population is the greatest because it includes workers from countries with substantially different mean incomes. This causes larger margin of error when calculating a confidence interval.

482.

E. adults ages 55 to 65

When sample size and confidence level are held constant, the sample from the least variable population will be expected to product the least variable sample and, hence, the smallest confidence interval. The population with the least variability is expected to be that of adults who have attained their full adult height (adults ages 55 to 65) but have not yet become subject to decreases in height due to osteoporosis.

483.

B. one with confidence level 95%, $n = 200$, and $\sigma = 12.5$

The margin of error increases as the confidence level and the standard deviation increase. The margin of error also increases with a smaller sample.

484.

D. Choices (A) and (B) (Increase the sample size from 200 to 1,000 subjects; decrease the confidence level from 95% to 90%.)

A larger sample size and a lower confidence level will both decrease the margin of error of a confidence interval.

485.

E. The 95% confidence interval for the mean height of all the boys is between 5 feet 5 inches and 6 feet 1 inch.

The confidence interval is constructed as the statistic (point estimate) plus or minus the margin of error. In this case, the 95% confidence interval is the point estimate of 5 feet 9 inches plus or minus the margin of error of 4 inches, which is 5 feet 5 inches to 6 feet 1 inch.

There can be a wide distribution of individual heights of the boys. A confidence interval and a confidence level don't say anything about the height of an individual boy or the total range of heights overall.

486.

The 95% confidence interval for the mean summer income of all college students is $4,100 to $4,900.

The confidence interval is constructed as the sample statistic (point estimate) plus or minus the margin of error. In this case, the 95% confidence interval is the point estimate of $4,500 plus or minus the margin of error of $400, giving a confidence interval of $4,100 to $4,900.

487.

C. The margin is used to calculate a range of likely values for a population parameter, based on a sample.

The margin of error isn't due to an error in the sample or survey. It factors in the fact that sample results vary from sample to sample and gives you a measure of how much you expect them to vary with a certain level of confidence.

488.   B.   The margin of error measures the amount by which your sample results could change, with 99% confidence.

The confidence interval is constructed as the point estimate plus or minus the margin of error. In this case, the 99% confidence interval is the point estimate of $450 plus or minus the margin of error of $50, giving a confidence interval of $400 to $500.

489.   The election is too close to predict. The actual support for either candidate could be above 50%.

The margin of error is used to construct the confidence interval, which is a range of likely values for the population parameter (here, the parameter is the percentage of all voters who would vote for a candidate). To calculate a confidence interval, you take the result from the sample and add and subtract the margin of error.

In this case, the 98% confidence interval for the proportion of all voters for Candidate Smith is 48% (from the sample) plus or minus 3% (the margin of error), which is a range of 45% to 51%. For Candidate Jones, the 98% confidence interval is 52% (from the sample) plus or minus 3% (the margin of error), which is a range of 49% to 55%. Both confidence intervals contain possible values above 50%, so either candidate could win; therefore, the results are too close to call.

490.   A.   The survey has a built-in bias.

This survey is biased because it wasn't conducted with a random sample of Americans but rather with a sample of fans who attended this particular football game. It's plausible that these fans don't represent the taste preferences of all Americans.

Taking a higher sample of fans won't make the survey less biased, and having a narrower confidence interval won't solve the problem, because bias isn't measured by the margin of error (it measures only how the results of a random sample would change from sample to sample).

491.   E.   Choices (B) and (D) (The survey is biased because it was based only on first-year employees, who may feel differently about their jobs than other employees. The sample size is only 30; the margin of error must be higher than 3% based on the size of the sample and the confidence level.)

First, the sample has bias because even though it was a random sample, it was based only on first-year employees, who may feel differently about their jobs than employees who have been on the job longer.

Second, with a sample of only 30 employees, the margin of error can't be as small as 3%. The formula for margin of error for a population proportion is

$$\text{MOE} = z^* \sqrt{\frac{\hat{p}(1-\hat{p})}{n}}$$

Where z^* is the value from Table A-1 for the selected confidence level, $\hat{p}$ is the sample proportion, and n is the sample size.

The sample size is 30, $\hat{p}$ is 0.8, and z^* is 1.96 for this problem.

$$MOE = 1.96\sqrt{\frac{0.8(1-0.8)}{30}}$$

$$= 1.96\sqrt{\frac{(0.8)(0.2)}{30}}$$

$$= 1.96(0.0730)$$

$$= 0.1431$$

Convert the proportion to a percent by multiplying by 100%:

$$0.1431(100\%) = 14.31\%$$

This value is much larger than the reported MOE of 3%.

492.   B.   The survey can be used only by the company as part of its analysis of the Internet spending habits of all visitors to its website during the last three months.

The random sample can be used only in analysis for that specific website for that specific time period. It can't be extended to a longer period of time or to a larger group of customers. But it can be used to draw conclusions to the visitors because that's what the random sample is based on. The statistics for the mean, margin of error, and confidence level are realistic with a sample size of 1,000.

493.   A.   The results are invalid because the survey was done at a movie theater.

The survey is asking only those teenagers who are already at a theater. This means that each response would be at least 1, but certainly it's possible that someone visited a theater 0 times in the past 12 months. A valid random sample would include all teenagers.

494.   E.   Choices (B), (C), and (D) (The sample isn't based on a representative sample of the magazine's readers. Because the magazine is based on Colorado, more readers are likely to buy the magazine who are from Colorado; therefore, they'd be more likely to vote it as the best place to live. The sample results are likely biased because the respondents had to make the effort to mail back the survey.)

The survey results are almost certain to be biased. A mail-in survey isn't a valid way to draw a random sample, because those who choose to take part in such a survey are probably not representative of the magazine's entire readership. It's also biased because the magazine is based on Colorado and, hence, is more likely to be purchased by people living in Colorado, who are more likely to choose Colorado as their favorite place to live.

A larger sized sample can't reduce bias that already exists.

495.   A.   With 95% confidence, the average points scored by all intramural basketball players is between 7.3 and 8.7 points.

Use the formula for finding the confidence interval for a population when the standard deviation is known:

$$\bar{x} \pm MOE = \bar{x} \pm z^* \left(\frac{\sigma}{\sqrt{n}} \right)$$

where $\bar{x}$ is the sample mean, σ is the population standard deviation, n is the sample size, and z^* represents the appropriate z^*-value from the standard normal distribution for your desired confidence level. The data has to come from a normal distribution, or n has to be large enough (a standard rule of thumb is at least 30 or so), for the central limit theorem to apply. You can find z^*-values from Table A-1 or Table A-4 in the appendix. The z^*-value is 1.96 for a two-tailed confidence interval with a confidence level of 95%.

Next, substitute the values into the formula:

$$MOE = 1.96\left(\frac{2.5}{\sqrt{50}}\right) \approx 0.693$$

The 95% confidence interval is 8 ± 0.7 (rounded to the nearest tenth), or 7.3 to 8.7 points scored.

---

496. The 99% confidence interval for the average SAT math score for all students at the high school is between 624.2 and 678.8.

Use the formula for finding the confidence interval for a population when the standard deviation is known:

$$\bar{x} \pm MOE = \bar{x} \pm z^*\left(\frac{\sigma}{\sqrt{n}}\right)$$

where $\bar{x}$ is the sample mean, σ is the population standard deviation, n is the sample size, and z^* represents the appropriate z^*-value from the standard normal distribution for your desired confidence level. The data has to come from a normal distribution, or n has to be large enough (a standard rule of thumb is at least 30 or so), for the central limit theorem to apply. You can find z^*-values from Table A-1 or Table A-4 in the appendix. The z^*-value for a two-tailed confidence interval with a confidence level of 99% is 2.58.

Next, substitute the values into the formula:

$$MOE = 2.58\left(\frac{100}{\sqrt{100}}\right) = 25.8$$

The confidence interval is 650 ± 25.8 (rounded to the nearest tenth), or 624.2 to 678.8.

---

497. The 99% confidence interval for the average weight of all apples from the ten trees is between 6.5 and 7.5 ounces.

Use the formula for finding the confidence interval for a population when the standard deviation is known:

$$\bar{x} \pm MOE = \bar{x} \pm z^*\left(\frac{\sigma}{\sqrt{n}}\right)$$

where $\bar{x}$ is the sample mean, σ is the population standard deviation, n is the sample size, and z^* represents the appropriate z^*-value from the standard normal distribution for your desired confidence level. The data has to come from a normal distribution, or n has to be large enough (a standard rule of thumb is at least 30 or so), for the central limit theorem to apply. You can find z^*-values from Table A-1 or Table A-4 in the appendix. The z^*-value for a two-tailed confidence interval with a confidence level of 99% is 2.58.

Next, substitute the values into the formula:

$$MOE = 2.58 \left(\frac{1.5}{\sqrt{50}} \right) \approx 0.5473$$

The confidence interval is 7 ± 0.5 (rounded to the nearest tenth), or 6.5 to 7.5 ounces.

---

498. The 90% confidence interval for the average time students at the university spend doing homework each day is between 2.88 and 3.12 hours.

Use the formula for finding the confidence interval for a population when the standard deviation is known:

$$\bar{x} \pm MOE = \bar{x} \pm z^* \left(\frac{\sigma}{\sqrt{n}} \right)$$

where $\bar{x}$ is the sample mean, σ is the population standard deviation, n is the sample size, and z^* represents the appropriate z^*-value from the standard normal distribution for your desired confidence level. The data has to come from a normal distribution, or n has to be large enough (a standard rule of thumb is at least 30 or so), for the central limit theorem to apply. You can find z^*-values from Table A-1 or Table A-4 in the appendix. The z^*-value for a two-tailed confidence interval with a confidence level of 90% is 1.645.

Next, substitute the values into the formula:

$$MOE = 1.645 \left(\frac{1}{\sqrt{200}} \right) \approx 0.1163$$

The confidence interval is 3 ± 0.12 (rounded to the nearest hundredth), or 2.88 to 3.12 hours of homework.

---

499. The 95% confidence interval for the average spending of people ages 18 to 22 on a typical outing with a friend is between $30.42 and $34.58.

Use the formula for finding the confidence interval for a population when the standard deviation is known:

$$\bar{x} \pm MOE = \bar{x} \pm z^* \left(\frac{\sigma}{\sqrt{n}} \right)$$

where $\bar{x}$ is the sample mean, σ is the population standard deviation, n is the sample size, and z^* represents the appropriate z^*-value from the standard normal distribution for your desired confidence level. The data has to come from a normal distribution, or n has to be large enough (a standard rule of thumb is at least 30 or so), for the central limit theorem to apply. You can find z^*-values from Table A-1 or Table A-4 in the appendix. The z^*-value for a two-tailed confidence interval with a confidence level of 95% is 1.96.

Next, substitute the values into the formula:

$$MOE = 1.96 \left(\frac{15.00}{\sqrt{200}} \right) \approx 2.0789$$

The confidence interval is 32.50 ± 2.08 (rounded to the nearest hundredth), or \$30.42 to \$34.58 spent.

---

500. The 90% confidence interval for the average time people above age 17 spend doing vigorous exercise is 28 to 32 minutes per day.

Use the formula for finding the confidence interval for a population when the standard deviation is known:

$$\bar{x} \pm MOE = \bar{x} \pm z^* \left(\frac{\sigma}{\sqrt{n}} \right)$$

where $\bar{x}$ is the sample mean, σ is the population standard deviation, n is the sample size, and z^* represents the appropriate z^*-value from the standard normal distribution for your desired confidence level. The data has to come from a normal distribution, or n has to be large enough (a standard rule of thumb is at least 30 or so), for the central limit theorem to apply. You can find z^*-values from Table A-1 or Table A-4 in the appendix. The z^*-value for a two-tailed confidence interval with a confidence level of 90% is 1.645.

Next, substitute the values into the formula:

$$MOE = 1.645 \left(\frac{15}{\sqrt{150}} \right) \approx 2.0147$$

The confidence interval is 30 ± 2.0 (rounded to the nearest tenth), or 28 to 32 minutes.

---

501. The 95% confidence interval for the average income of all first-year college graduates is between \$34,891 and \$37,109.

Use the formula for finding the confidence interval for a population when the standard deviation is known:

$$\bar{x} \pm MOE = \bar{x} \pm z^* \left(\frac{\sigma}{\sqrt{n}} \right)$$

where $\bar{x}$ is the sample mean, σ is the population standard deviation, n is the sample size, and z^* represents the appropriate z^*-value from the standard normal distribution for your desired confidence level. The data has to come from a normal distribution, or n has to be large enough (a standard rule of thumb is at least 30 or so), for the central limit theorem to apply. You can find z^*-values from Table A-1 or Table A-4 in the appendix. The z^*-value for a two-tailed confidence interval with a confidence level of 95% is 1.96.

Next, substitute the values into the formula:

$$MOE = 1.96 \left(\frac{8,000}{\sqrt{200}} \right) \approx 1,108.7434$$

The confidence interval is $36,000 \pm 1,109$, or \$34,891 to \$37,109.

502.

The 95% confidence interval for the average amount of time of all bus trips along this route is between 44:40 and 45:20 minutes.

Use the formula for finding the confidence interval for a population when the standard deviation is known:

$$\bar{x} \pm \text{MOE} = \bar{x} \pm z^* \left(\frac{\sigma}{\sqrt{n}} \right)$$

where $\bar{x}$ is the sample mean, σ is the population standard deviation, n is the sample size, and z^* represents the appropriate z^*-value from the standard normal distribution for your desired confidence level. The data has to come from a normal distribution, or n has to be large enough (a standard rule of thumb is at least 30 or so), for the central limit theorem to apply. You can find z^*-values from Table A-1 or Table A-4 in the appendix. The z^*-value for a two-tailed confidence interval with a confidence level of 95% is 1.96.

Next, substitute the values into the formula:

$$\text{MOE} = 1.96 \left(\frac{3}{\sqrt{300}} \right) \approx 0.3395$$

The confidence interval is 45 ± 0.34 minutes (rounded to the nearest hundredth), or 44.66 minutes to 45.34 minutes. Convert the fractional minutes to the easier-to-use unit of seconds: 44 minutes and 40 seconds to 45 minutes and 20 seconds, or 44:40 to 45:20.

503.

The 98% confidence interval for the average amount of time among all requests for an airline itinerary to be displayed online is between 4.36 and 4.64 seconds.

Use the formula for finding the confidence interval for a population when the standard deviation is known:

$$\bar{x} \pm \text{MOE} = \bar{x} \pm z^* \left(\frac{\sigma}{\sqrt{n}} \right)$$

where $\bar{x}$ is the sample mean, σ is the population standard deviation, n is the sample size, and z^* represents the appropriate z^*-value from the standard normal distribution for your desired confidence level. The data has to come from a normal distribution, or n has to be large enough (a standard rule of thumb is at least 30 or so), for the central limit theorem to apply. You can find z^*-values from Table A-1 or Table A-4 in the appendix. The z^*-value for a two-tailed confidence interval with a confidence level of 98% is 2.33.

Next, substitute the values into the formula:

$$\text{MOE} = 2.33 \left(\frac{2}{\sqrt{1,100}} \right) \approx 0.1405$$

The confidence interval is 4.5 ± 0.14 (rounded to the nearest hundredth), or 4.36 to 4.64 seconds.

504.

The 95% confidence interval for the average amount of time to assemble an MP3 player is between 11.95 and 12.55 minutes.

Use the formula for finding the confidence interval for a population when the standard deviation is known:

$$\bar{x} \pm \text{MOE} = \bar{x} \pm z^* \left(\frac{\sigma}{\sqrt{n}} \right)$$

where $\bar{x}$ is the sample mean, σ is the population standard deviation, n is the sample size, and z^* represents the appropriate z^*-value from the standard normal distribution for your desired confidence level. The data has to come from a normal distribution, or the n has to be large enough (a standard rule of thumb is at least 30 or so) for the central limit theorem to apply. You can find z^*-values from Table A-1 or Table A-4 in the appendix. The z^*-value for a two-tailed confidence interval with a confidence level of 95% is 1.96.

Next, substitute the values into the formula:

$$\text{MOE} = 1.96 \left(\frac{2.15}{\sqrt{200}} \right) \approx 0.2980$$

The confidence interval is 12.25 ± 0.30 (rounded to the nearest hundredth), or 11.95 to 12.55 minutes.

505.

The 90% confidence interval for the average of all such distances to the university from a student's hometown is between 121.2 and 128.8 miles.

Use the formula for finding the confidence interval for a population when the standard deviation is known:

$$\bar{x} \pm \text{MOE} = \bar{x} \pm z^* \left(\frac{\sigma}{\sqrt{n}} \right)$$

where $\bar{x}$ is the sample mean, σ is the population standard deviation, n is the sample size, and z^* represents the appropriate z^*-value from the standard normal distribution for your desired confidence level. The data has to come from a normal distribution, or n has to be large enough (a standard rule of thumb is at least 30 or so), for the central limit theorem to apply. You can find z^*-values from Table A-1 or Table A-4 in the appendix. The z^*-value for a two-tailed confidence interval with a confidence level of 90% is 1.645.

Next, substitute the values into the formula:

$$\text{MOE} = 1.645 \left(\frac{40}{\sqrt{300}} \right) \approx 3.7990$$

The confidence interval is 125 ± 3.8 (rounded to the nearest tenth), or 121.2 to 128.8 miles.

506. The 95% confidence interval for the average number of errors among data specialists is between 2.53 and 2.87 errors in 10,000 entries.

Use the formula for finding the confidence interval for a population when the standard deviation is known:

$$\bar{x} \pm \text{MOE} = \bar{x} \pm z^* \left(\frac{\sigma}{\sqrt{n}} \right)$$

where $\bar{x}$ is the sample mean, σ is the population standard deviation, n is the sample size, and z^* represents the appropriate z^*-value from the standard normal distribution for your desired confidence level. The data has to come from a normal distribution, or n has to be large enough (a standard rule of thumb is at least 30 or so), for the central limit theorem to apply. You can find z^*-values from Table A-1 or Table A-4 in the appendix. The z^*-value for a two-tailed confidence interval with a confidence level of 95% is 1.96.

Next, substitute the values into the formula:

$$\text{MOE} = 1.96 \left(\frac{0.75}{\sqrt{75}} \right) \approx 0.1697$$

The confidence interval is 2.7 ± 0.17 (rounded to the nearest hundredth), or 2.53 to 2.87 errors.

507. The 99% confidence level for the average length of all major league 38-inch bats is between 38.009 and 38.011 inches.

Use the formula for finding the confidence interval for a population when the standard deviation is known:

$$\bar{x} \pm \text{MOE} = \bar{x} \pm z^* \left(\frac{\sigma}{\sqrt{n}} \right)$$

where $\bar{x}$ is the sample mean, σ is the population standard deviation, n is the sample size, and z^* represents the appropriate z^*-value from the standard normal distribution for your desired confidence level. The data has to come from a normal distribution, or n has to be large enough (a standard rule of thumb is at least 30 or so), for the central limit theorem to apply. You can find z^*-values from Table A-1 or Table A-4 in the appendix. The z^*-value for a two-tailed confidence interval with a confidence level of 99% is 2.58.

Next, substitute the values into the formula:

$$\text{MOE} = 2.58 \left(\frac{0.01}{\sqrt{500}} \right) \approx 0.00115$$

The confidence interval is 38.01 ± 0.001 (rounded to the nearest thousandth), or 38.009 to 38.011 inches.

508. The 99% confidence interval for the average length of all special valve engine parts is between 3.2536 and 3.2564 centimeters.

Use the formula for finding the confidence interval for a population when the standard deviation is known:

$$\bar{x} \pm MOE = \bar{x} \pm z^* \left(\frac{\sigma}{\sqrt{n}} \right)$$

where $\bar{x}$ is the sample mean, σ is the population standard deviation, n is the sample size, and z^* represents the appropriate z^*-value from the standard normal distribution for your desired confidence level. The data has to come from a normal distribution, or n has to be large enough (a standard rule of thumb is at least 30 or so), for the central limit theorem to apply. You can find z^*-values from Table A-1 or Table A-4 in the appendix. The z^*-value for a two-tailed confidence interval with a confidence level of 99% is 2.58.

Next, substitute the values into the formula:

$$MOE = 2.58 \left(\frac{0.025}{\sqrt{2,000}} \right) \approx 0.001442$$

The confidence interval is 3.2550 ± 0.0014 (rounded to the nearest ten-thousandth), or 3.2536 to 3.2564 centimeters.

509. The 95% confidence interval for the average cost of medium-quality hardwood during the 12-month period had an average cost of between $0.743 and $0.817 per board foot.

Use the formula for finding the confidence interval for a population when the standard deviation is known:

$$\bar{x} \pm MOE = \bar{x} \pm z^* \left(\frac{\sigma}{\sqrt{n}} \right)$$

where $\bar{x}$ is the sample mean, σ is the population standard deviation, n is the sample size, and z^* represents the appropriate z^*-value from the standard normal distribution for your desired confidence level. The data has to come from a normal distribution, or n has to be large enough (a standard rule of thumb is at least 30 or so), for the central limit theorem to apply. You can find z^*-values from Table A-1 in the appendix or Table A-4. The z^*-value for a two-tailed confidence interval with a confidence level of 95% is 1.96.

Next, substitute the values into the formula:

$$MOE = 1.96 \left(\frac{0.12}{\sqrt{40}} \right) \approx 0.0371$$

The confidence interval is 0.78 ± 0.037 (rounded to the nearest thousandth), or $0.743 to $0.817.

510. The 95% confidence interval for the average first-year college student score for the math test is between 81.9% and 86.1%.

Use the formula for the confidence interval when the population's standard deviation is unknown and n is small (less than 30):

$$\bar{x} \pm \text{MOE} = \bar{x} \pm t^*_{n-1}\left(\frac{s}{\sqrt{n}}\right)$$

where $\bar{x}$ is the sample mean, t^*_{n-1} is the critical t^*-value from the t-distribution with $n-1$ degrees of freedom (where n is the sample size) and the confidence level desired, and s is the sample standard deviation (if the population's standard deviation is known, substitute it for s).

Use the t-table (Table A-2 in the appendix) to find the t^*-value for 95% with degrees of freedom for a sample size of 25 ($n-1 = 25 - 1 = 24$). Find 95% in the *CI* row at the bottom of the table, and move up the column to intersect with the *df/p* row labeled 24: $t^* = 2.06390$.

Next, substitute the values into the formula:

$$\text{MOE} = 2.06390\left(\frac{5}{\sqrt{25}}\right) = 2.0639$$

The confidence interval is 84 ± 2.1 (rounded to nearest tenth), or 81.9 and 86.1.

511. The 90% confidence interval for the average household size is between 3.13 and 3.67 people.

Use the formula for the confidence interval when the population's standard deviation is unknown and n is small (less than 30):

$$\bar{x} \pm \text{MOE} = \bar{x} \pm t^*_{n-1}\left(\frac{s}{\sqrt{n}}\right)$$

where $\bar{x}$ is the sample mean, t^*_{n-1} is the critical t^*-value from the t-distribution with $n-1$ degrees of freedom (where n is the sample size) and the confidence level desired, and s is the sample standard deviation (if the population's standard deviation is known, substitute it for s).

Use the t-table (Table A-2 in the appendix) to find the t^*-value for 90% with degrees of freedom for a sample size of 25 ($n-1 = 25 - 1 = 24$). Find 90% in the *CI* row at the bottom of the table, and move up the column to intersect with the *df/p* row labeled 24: $t^* = 1.710882$

Next, substitute the values into the formula:

$$\text{MOE} = 1.710882\left(\frac{0.8}{\sqrt{25}}\right) \approx 0.2737$$

The confidence interval is 3.4 ± 0.27 (rounded to the nearest hundredth), or 3.13 to 3.67 people.

512.

The 95% confidence interval for teenagers' average number of social network friends is 66 to 104.

Use the formula for the confidence interval when the population's standard deviation is unknown.

$$\bar{x} \pm \text{MOE} = \bar{x} \pm t^*_{n-1}\left(\frac{s}{\sqrt{n}}\right)$$

where $\bar{x}$ is the sample mean, t^*_{n-1} is the critical t^*-value from the t-distribution with $n-1$ degrees of freedom (where n is the sample size) and the confidence level desired, and s is the sample standard deviation (if the population's standard deviation is known, substitute it for s).

Use the t-table (Table A-2 in the appendix) to find the t^*-value for 95% with degrees of freedom for a sample size of 30 ($n-1 = 30 - 1 = 29$). Find 95% in the *CI* row at the bottom of the table, and move up the column to intersect with the *df/p* row labeled 29: $t^* = 2.04523$.

Next, substitute the values into the formula:

$$\text{MOE} = 2.04523\left(\frac{50}{\sqrt{30}}\right) \approx 18.6703$$

The confidence interval is 85 ± 19 (rounded to the nearest whole number), or 66 to 104 friends.

513.

The 95% confidence interval for the average length of the longest trip first-year college students took last year was between 273 and 527 miles.

Use the formula for the confidence interval when the population's standard deviation is unknown and n is small (less than 30):

$$\bar{x} \pm \text{MOE} = \bar{x} \pm t^*_{n-1}\left(\frac{s}{\sqrt{n}}\right)$$

where $\bar{x}$ is the sample mean, t^*_{n-1} is the critical t^*-value from the t-distribution with $n-1$ degrees of freedom (where n is the sample size) and the confidence level desired, and s is the sample standard deviation (if the population's standard deviation is known, substitute it for s).

Use the t-table (Table A-2 in the appendix) to find the t^*-value for 95% with a degree of freedom for a sample size of 24 ($n-1 = 24 - 1 = 23$). Find 95% in the *CI* row at the bottom of the table, and move up the column to intersect with the *df/p* row labeled 23: $t^* = 2.06866$.

Next, substitute the values into the formula:

$$\text{MOE} = 2.06866\left(\frac{300}{\sqrt{24}}\right) \approx 126.6790$$

The confidence interval is 400 ± 127 (rounded to the nearest whole number), or 273 to 527 miles.

514. The 95% confidence interval for the average amount of purchases by mall shoppers that day is between \$54.75 and \$102.25.

Use the formula for the confidence interval when the population's standard deviation is unknown and n is small (less than 30):

$$\bar{x} \pm \text{MOE} = \bar{x} \pm t^*_{n-1}\left(\frac{s}{\sqrt{n}}\right)$$

where $\bar{x}$ is the sample mean, t^*_{n-1} is the critical t^*-value from the t-distribution with $n-1$ degrees of freedom (where n is the sample size) and the confidence level desired, and s is the sample standard deviation (if the population's standard deviation is known, substitute it for s).

Use the t-table (Table A-2 in the appendix) to find the t^*-value for 95% with degrees of freedom for a sample size of 20 ($n-1 = 20-1 = 19$). Find 95% in the *CI* row at the bottom of the table, and move up the column to intersect with the *df/p* row labeled 19: $t^* = 2.09302$.

Next, substitute the values into the formula:

$$\text{MOE} = 2.09302\left(\frac{50.75}{\sqrt{20}}\right) = 23.7517$$

The confidence interval is 78.50 ± 23.75 (rounded to the nearest hundredth), or \$54.75 to \$102.25.

515. The 90% confidence interval for the average amount of time visitors spent in the museum that day is between 2.6 and 3.4 hours.

Use the formula for the confidence interval when the population's standard deviation is unknown and n is small (less than 30):

$$\bar{x} \pm \text{MOE} = \bar{x} \pm t^*_{n-1}\left(\frac{s}{\sqrt{n}}\right)$$

where $\bar{x}$ is the sample mean, t^*_{n-1} is the critical t^*-value from the t-distribution with $n-1$ degrees of freedom (where n is the sample size) and the confidence level desired, and s is the sample standard deviation (if the population's standard deviation is known, substitute it for s).

Use the t-table (Table A-2 in the appendix) to find the t^*-value for 90% with degrees of freedom for a sample size of 20 ($n-1 = 20-1 = 19$). Find 90% in the *CI* row at the bottom of the table, and move up the column to intersect with the *df/p* row labeled 19: $t^* = 1.729133$.

Next, substitute the values into the formula:

$$\text{MOE} = 1.729133\left(\frac{1}{\sqrt{20}}\right) \approx 0.3866$$

The confidence interval is 3 ± 0.4 (rounded to the nearest tenth), or 2.6 to 3.4 hours (or between 2 hours 36 minutes and 3 hours and 24 minutes).

516. The 90% confidence interval for the average weight of all 1-pound loaves of bread is between 17.65 and 18.35 ounces.

Use the formula for the confidence interval when the population's standard deviation is unknown and n is small (less than 30):

$$\bar{x} \pm \text{MOE} = \bar{x} \pm t^*_{n-1}\left(\frac{s}{\sqrt{n}}\right)$$

where $\bar{x}$ is the sample mean, t^*_{n-1} is the critical t^*-value from the t-distribution with $n-1$ degrees of freedom (where n is the sample size) and the confidence level desired, and s is the sample standard deviation (if the population's standard deviation is known, substitute it for s).

Use the t-table (Table A-2 in the appendix) to find the t^*-value for 90% with degrees of freedom for a sample size of 50 ($n-1 = 50-1 = 49$). Find 90% in the *CI* row at the bottom of the table, and move up the column to the z-value (because when $df > 30$, the t- and z-values are almost the same), or 1.644854. For ease of calculation, you can use 1.645 (which is also the t-table value rounded to three decimal places).

Next, substitute the values into the formula:

$$\text{MOE} = 1.644854\left(\frac{1.5}{\sqrt{50}}\right) \approx 0.3489$$

The confidence interval is 18 ± 0.35 (rounded to the nearest hundredth), or 17.65 to 18.35 ounces.

517. The 80% confidence limit for the average number of visits by a person to the store each month is between 1.93 and 3.67 visits.

Use the formula for the confidence interval when the population's standard deviation is unknown and n is small (less than 30):

$$\bar{x} \pm \text{MOE} = \bar{x} \pm t^*_{n-1}\left(\frac{s}{\sqrt{n}}\right)$$

where $\bar{x}$ is the sample mean, t^*_{n-1} is the critical t^*-value from the t-distribution with $n-1$ degrees of freedom (where n is the sample size) and the confidence level desired, and s is the sample standard deviation (if the population's standard deviation is known, substitute it for s).

Use the t-table (Table A-2 in the appendix) to find the t^*-value for 80% with degrees of freedom for a sample size of 10 ($n-1 = 10-1 = 9$). Find 80% in the *CI* row at the bottom of the table, and move up the column to intersect with the df/p row labeled 9: $t^* = 1.383029$.

Next, substitute the values into the formula:

$$\text{MOE} = 1.383029\left(\frac{2}{\sqrt{10}}\right) \approx 0.8747$$

The confidence interval is 2.8 ± 0.87 (rounded to the nearest hundredth), or 1.93 to 3.67 visits.

518.

The 90% confidence interval for the average number of movies per month watched by first-year university students is 3.8 to 6.2 movies.

Use the formula for the confidence interval when the population's standard deviation is unknown and n is small (less than 30):

$$\bar{x} \pm \text{MOE} = \bar{x} \pm t^*_{n-1}\left(\frac{\sigma}{\sqrt{n}}\right)$$

where $\bar{x}$ is the sample mean, t^*_{n-1} is the critical t^*-value from the t-distribution with $n-1$ degrees of freedom (where n is the sample size) and the confidence level desired, and s is the sample standard deviation (if the population's standard deviation is known, substitute it for s).

Use the t-table (Table A-2 in the appendix) to find the t^*-value for 90% with degrees of freedom for a sample size of 18 ($n-1 = 18-1 = 17$). Find 90% in the CI row at the bottom of the table, and move up the column to intersect with the df/p row labeled 17: $t^* = 1.739607$.

Next, substitute the values into the formula:

$$\text{MOE} = 1.739607\left(\frac{3}{\sqrt{18}}\right) \approx 1.2301$$

The confidence interval is 5 ± 1.2 (rounded to the nearest tenth), or 3.8 to 6.2 movies.

519.

The 99% confidence interval for the average spending by a park visitor that day is $28.64 to $35.36.

Use the formula for the confidence interval when the population's standard deviation is unknown and n is small (less than 30):

$$\bar{x} \pm \text{MOE} = \bar{x} \pm t^*_{n-1}\left(\frac{\sigma}{\sqrt{n}}\right)$$

where $\bar{x}$ is the sample mean, t^*_{n-1} is the critical t^*-value from the t-distribution with $n-1$ degrees of freedom (where n is the sample size) and the confidence level desired, and s is the sample standard deviation (if the population's standard deviation is known, substitute it for s).

Use the t-table (Table A-2 in the appendix) to find the t^*-value for 99% with degrees of freedom for a sample size of 25 ($n-1 = 25-1 = 24$). Find 99% in the CI row at the bottom of the table, and move up the column to intersect with the df/p row labeled 24: $t^* = 2.79694$.

Next, substitute the values into the formula:

$$\text{MOE} = 2.79694\left(\frac{6.00}{\sqrt{25}}\right) \approx 3.3563$$

The confidence interval is $32.00 \pm $3.36 (rounded to the nearest hundredth), or $28.64 to $35.36.

520.   D.   Choices (A) and (B) (117.2; 117.6)

Fractional sample sizes are always rounded up, even if the fractional is less than 0.5, so 117.2 and 117.6 are both rounded up to 118, while 118.1 is rounded up to 119. In other words, if you need a fraction of an individual to be included in the sample, no matter how small the fraction is, you need the whole individual to meet the conditions for the margin of error.

521.   122, 122, 132, 132

Fractional results in sample size calculations are always rounded up, even if the fractional part is less than 0.5. Therefore, 121.1 and 121.5 both round up to 122, and 131.2 and 131.6 both round up to 132. In other words, if you need a fraction of an individual to be included in the sample, no matter how small the fraction is, you need the whole individual to meet the conditions for the margin of error.

522.   E.   Choices (A), (B), and (C) (A larger sample often means greater costs; it may be difficult to recruit a larger sample [for example, if you're studying people with a rare disease]; at some point, increasing the sample size may not significantly improve precision [for instance, increasing the sample size from 3,000 to 3,500])

Although it's important to have an adequate sample size when conducting a study, costs and difficulties with recruiting subjects can both limit the size of samples a researcher can work with. In addition, increasing the size of a large sample produces proportionately less of an increase in precision than increasing the size of a small sample so that the relatively small gains in precision may not be worth the added effort and cost.

523.   35 records

The formula to calculate the sample size needed for a confidence interval for a sample mean is

$$n \geq \left(\frac{z^* \sigma}{\text{MOE}} \right)^2$$

where n is the sample size required, z^* is the value from Table A-1 for the chosen confidence level, σ is the population standard deviation, and MOE is the margin of error. For this example, z^* is 1.96, σ is 1.6, and the MOE is 1.5.

$$n \geq \left(\frac{z^* \sigma}{\text{MOE}} \right)^2$$

$$\geq \left(\frac{(1.96)(4.5)}{1.5} \right)^2$$

$$\geq 34.5744$$

Samples sizes are always rounded up to the nearest integer, so the sample size must be at least 35.

524. 50 records

The formula to calculate sample size to estimate a mean, when the standard deviation is known, is

$$n \geq \left(\frac{z^*\sigma}{MOE}\right)^2$$

where n is the sample size required, z^* is the value from Table A-1 for the chosen confidence level, σ is the population standard deviation, and MOE is the margin of error. For this example, z^* is 1.96, σ is 1.6, and the MOE is 0.5.

$$n \geq \left(\frac{z^*\sigma}{MOE}\right)^2$$

$$\geq \left(\frac{1.96(1.8)}{0.5}\right)^2$$

$$\geq 49.787136$$

She needs a sample of at least 50 to achieve the desired precision.

525. The sample size must be at least 35 to produce a margin of error of ± 1 inch.

The formula to find the required sample size based on a desired margin of error is

$$n \geq \left(\frac{z^*\sigma}{MOE}\right)^2$$

Here, MOE is the margin of error, z^* is the z^*-value corresponding to your desired confidence level, and σ is the population standard deviation. If σ is unknown, you can do a small pilot study to find the standard deviation of the sample (including making a conservative adjustment to the sample standard deviation to be safe).

Substitute the known values into the formula:

$$n \geq \left(\frac{1.96(3)}{1}\right)^2 \approx 34.57$$

Always round up the answer to the nearest whole number to be sure the sample size is large enough to give the margin of error needed. So $n \geq 35$. That means that you need at least 35 boys in your sample to get a margin of error of no more than 1 inch for average height.

Note that a sample size of 35 will give you the margin of error you want; a higher sample size will give you an even lower margin of error.

526. The sample size (number of students) must be at least 107 to produce a margin of error of ± $5.

The formula to find the required sample size based on a desired margin of error is

$$n \geq \left(\frac{z^* \sigma}{\text{MOE}} \right)^2$$

Here, MOE is the margin of error, z^* is the z^*-value corresponding to your desired confidence level, and σ is the population standard deviation. If σ is unknown, you can do a small pilot study to find the standard deviation of the sample (including making a conservative adjustment to the sample standard deviation to be safe).

Substitute the known values into the formula:

$$n \geq \left(\frac{2.58(20)}{5} \right)^2 = 106.5$$

Always round up the answer to the nearest whole number to be sure the sample size is large enough to give the margin of error needed. So $n \geq 107$. That means that you need at least 107 students in your sample to get a margin of error of ± $5 when finding the mean weekly earnings.

Note that a sample size of 107 will give you the margin of error you want; a higher sample size will give you an even lower margin of error.

527. The sample size (number of students) must be at least 82 to produce a margin of error no more than ± $10.

The formula to find the required sample size based on a desired margin of error is

$$n \geq \left(\frac{z^* \sigma}{\text{MOE}} \right)^2$$

Here, MOE is the margin of error, z^* is the z^*-value corresponding to your desired confidence level, and σ is the population standard deviation. If σ is unknown, you can do a small pilot study to find the standard deviation of the sample (including making a conservative adjustment to the sample standard deviation to be safe).

Substitute the known values into the formula:

$$n \geq \left(\frac{1.645(55)}{10} \right)^2 \approx 81.86$$

Always round up the answer to the nearest whole number to be sure the sample size is large enough to give the margin of error needed. So $n \geq 82$. That means that you need at least 82 students in your sample to have a margin of error of no more than ± $10 for the mean weekly earnings.

Note that a sample size of 82 will give you the margin of error you want; a higher sample size will give you an even lower margin of error.

528.

The sample size (number of students) must be at least 89 to produce a margin of error no more than ± 0.25 hours.

The formula to find the required sample size based on a desired margin of error is

$$n \geq \left(\frac{z^* \sigma}{\text{MOE}} \right)^2$$

Here, MOE is the margin of error, z^* is the z^*-value corresponding to your desired confidence level, and σ is the population standard deviation. If σ is unknown, you can do a small pilot study to find the standard deviation of the sample (including making a conservative adjustment to the sample standard deviation to be safe).

Substitute the known values into the formula:

$$n \geq \left(\frac{1.96(1.2)}{0.25} \right)^2 \approx 88.5105$$

Always round up the answer to the nearest whole number to be sure the sample size is large enough to give the margin of error needed. So $n \geq 89$. That means that you need at least 89 students in your sample to have a margin of error of ± 0.25 hours for the average amount of sleep per night.

Note that a sample size of 89 will give you the margin of error you want; a higher sample size will give you an even lower margin of error.

529.

The sample size (number of games) must be at least 32 to produce a margin of error of no more than ± 800 people.

The formula to find the required sample size based on a desired margin of error is

$$n \geq \left(\frac{z^* \sigma}{\text{MOE}} \right)^2$$

Here, MOE is the margin of error, z^* is the z^*-value corresponding to your desired confidence level, and σ is the population standard deviation. If σ is unknown, you can do a small pilot study to find the standard deviation of the sample (including making a conservative adjustment to the sample standard deviation to be safe).

Substitute the known values into the formula:

$$n \geq \left(\frac{1.96(2,300)}{800} \right)^2 \approx 31.7532$$

Always round up the answer to the nearest whole number to be sure the sample size is large enough to give the margin of error needed. So $n \geq 32$. That means that you need at least 32 games in your sample to have a margin of error of ± 800 for average game attendance.

Note that a sample size of 32 will give you the margin of error you want; a higher sample size will give you an even lower margin of error.

530.

D. Choices (B) and (C) (You are using sample data to estimate a parameter; if you drew a different sample of the same size, you would expect the results to be slightly different.)

Your goal is to estimate the proportion of all students in the university who are thinking of changing their major. However, you're working with a sample rather than the entire university population. Thus, you don't expect your parameter estimate to be exactly the same as the true parameter, and you realize that if you drew another sample of the same size, your parameter estimate would probably be slightly different. The confidence interval expresses this uncertainty.

531.

E. Choices (C) and (D) (an estimate of the proportion of all students at the university who are thinking of changing their major; the proportion of students in the sample of 100 who are thinking of changing their major)

The value 0.38 represents the proportion of students in the sample who are thinking of changing their major. In other words, in your sample of 100, 38 students indicated they were thinking about changing their major, so 38/100 = 0.38.

It is also an estimate of the proportion of all students in the university who are thinking of changing their major.

532.

Yes, because $n\hat{p}$ and $n(1-\hat{p})$ are both greater than 10.

To use the normal approximation to the binomial to calculate a confidence interval, both $n\hat{p}$ and $n(1-\hat{p})$ must be greater than 10, where n is the sample size and $\hat{p}$ is the sample proportion.

In this case, $n = 100$ and $\hat{p} = 0.38$. So plugging in the numbers, you get $n\hat{p} = (100)0.38 = 38$ and $n(1-\hat{p}) = (100)(1-0.38) = 62$.

Both 38 and 62 are greater than 10, so you can use the normal approximation.

533.

0.0485

The formula for the standard error of a proportion is

$$SE = \sqrt{\frac{\hat{p}(1-\hat{p})}{n}}$$

where n is the sample size and $\hat{p}$ is the sample proportion.

In this example, $n = 100$ and $\hat{p} = 0.38$. So simply plug in the numbers and solve for the standard error:

$$SE = \sqrt{\frac{0.38(1-0.38)}{100}}$$

$$= \sqrt{\frac{0.2356}{100}}$$

$$\approx 0.0485$$

534. E. 99%

Without doing any calculations, you know that the 99% confidence interval will be widest, because it includes the largest set of samples taken from the population. In other words, it includes the largest range of plausible guesses for the population parameter.

535. between 30% and 46%

To determine a confidence interval (CI) for one population proportion, use this formula:

$$CI = \hat{p} \pm z^* \sqrt{\frac{\hat{p}(1-\hat{p})}{n}}$$

Here, $\hat{p}$ is the sample proportion, n is the sample size, and z^* is the appropriate value from the standard normal distribution for your desired confidence level (see Table A-4 in the appendix for various confidence levels).

First, you have to find the sample proportion, $\hat{p}$, by dividing the number of "successes" (38 in this case) by the sample size (100):

$$\hat{p} = \frac{\text{Number of successes}}{n}$$

$$= \frac{38}{100} = 0.38$$

Then confirm whether you can use the normal approximation to the binomial. To use the normal approximation to the binomial to calculate a confidence interval, both $n\hat{p}$ and $n(1-\hat{p})$ must be greater than 10, where n is the sample size and $\hat{p}$ is the sample proportion.

In this case, $n = 100$ and $\hat{p} = 0.38$. So plugging in the numbers, you get $n\hat{p} = (100)0.38 = 38$ and $n(1-\hat{p}) = (100)(1-0.38) = 62$.

Both 38 and 62 are greater than 10, so you can use the normal approximation to the binomial.

Next, substitute the known values into the formula for the confidence interval and solve:

$$CI = 0.38 \pm 1.645 \sqrt{\frac{0.38(1-0.38)}{100}}$$

$$= 0.38 \pm 1.645 \sqrt{\frac{0.2356}{100}}$$

$$= 0.38 \pm 1.645 \sqrt{0.002356}$$

$$= 0.38 \pm 1.645(0.04854)$$

$$= 0.38 \pm 0.0798$$

Add and subtract to/from the sample proportion to find the range:

$$0.38 - 0.0798 = 0.3002$$

$$0.38 + 0.0798 = 0.4598$$

Then convert to percentages by multiplying by 100%:

$$0.3040(100\%) = 30.02\%$$

$$0.4598(100\%) = 45.98\%$$

Round to the nearest whole percentage point, so the 90% confidence interval for the proportion of all students thinking of changing their major is 30% to 46%.

536. between 28% and 48%

To determine a confidence interval (CI) for one population proportion, use this formula:

$$CI = \hat{p} \pm z^* \sqrt{\frac{\hat{p}(1-\hat{p})}{n}}$$

Here, $\hat{p}$ is the sample proportion, n is the sample size, and z^* is the appropriate value from the standard normal distribution for your desired confidence level (see Table A-4 in the appendix for various confidence levels).

First, you have to find the sample proportion, $\hat{p}$, by dividing the number of "successes" (38 in this case) by the sample size (100):

$$\hat{p} = \frac{\text{Number of successes}}{n}$$

$$= \frac{38}{100} = 0.38$$

Then confirm whether you can use the normal approximation to the binomial. To use the normal approximation to the binomial to calculate a confidence interval, both $n\hat{p}$ and $n(1-\hat{p})$ must be greater than 10, where n is the sample size and $\hat{p}$ is the sample proportion.

In this case, $n = 100$ and $\hat{p} = 0.38$. So plugging in the numbers, you get $n\hat{p} = (100)0.38 = 38$ and $n(1-\hat{p}) = (100)(1-0.38) = 62$.

Both 38 and 62 are greater than 10, so you can use the normal approximation to the binomial.

Next, substitute the known values into the formula for the confidence interval and solve:

$$CI = 0.38 \pm 1.96 \sqrt{\frac{0.38(1-0.38)}{100}}$$

$$= 0.38 \pm 1.96 \sqrt{\frac{0.38(0.62)}{100}}$$

$$= 0.38 \pm 1.96 \sqrt{\frac{0.2356}{100}}$$

$$= 0.38 \pm 1.96 \sqrt{0.002356}$$

$$= 0.38 \pm 1.96(0.04854)$$

$$= 0.38 \pm 0.0951$$

Add and subtract to/from the sample proportion to find the range:

$0.38 + 0.0951 = 0.4751$

$0.38 - 0.0951 = 0.2849$

Then convert to percentages by multiplying by 100%:

$0.4751(100\%) = 47.51\%$

$0.2849(100\%) = 28.49\%$

Round to the nearest percentage point, so the 95% confidence interval for the proportion of all students thinking of changing their major is 28% to 48%.

537. 0.75, 0.03

Find the sample proportion, $\hat{p}$, by dividing the number of "successes" (75 in this case) by the sample size (200):

$$\hat{p} = \frac{\text{Number of successes}}{n}$$

$$= \frac{150}{200} = 0.75$$

The sample proportion represents the proportion of customers in the sample who are satisfied with their online purchases.

Then use the following formula to find the standard error (SE):

$$SE = \sqrt{\frac{\hat{p}(1-\hat{p})}{n}}$$

where $\hat{p}$ is the sample proportion and n is the sample size:

$$SE = \sqrt{\frac{0.75(1-0.75)}{200}}$$

$$= \sqrt{\frac{0.75(0.25)}{200}}$$

$$= \sqrt{\frac{0.1875}{200}}$$

$$\approx 0.03$$

So the standard error for the sample proportion in this example is 0.03.

538. 0.06

Use the formula for finding the margin of error (MOE):

$$MOE = z^* \sqrt{\frac{\hat{p}(1-\hat{p})}{n}}$$

Here, $\hat{p}$ is the sample proportion, n is the sample size, and z^* is the appropriate value from the standard normal distribution for your desired confidence level (see Table A-4 in the appendix for various confidence levels). For a 95% confidence level, the z^*-value is 1.96. Note that the margin of error is the z^*-value times the standard error.

Now, plug in the known values and solve:

$$MOE = 1.96 \sqrt{\frac{0.75(0.25)}{200}}$$

$$= 1.96 \sqrt{\frac{0.1875}{200}}$$

$$= 1.96(0.0306)$$

$$\approx 0.06$$

With 95% confidence, the margin of error is ± 0.06 for estimating the proportion of all customers who purchased products online in the past 12 months.

539. 0.67 to 0.83

To determine a confidence interval (CI) for one population proportion, use this formula:

$$CI = \hat{p} \pm z^* \sqrt{\frac{\hat{p}(1-\hat{p})}{n}}$$

Here, $\hat{p}$ is the sample proportion, n is the sample size, and z^* is the appropriate value from the standard normal distribution for your desired confidence level (see Table A-4 for various confidence levels). For a confidence level of 99%, $z^* = 2.58$.

First, you have to find the sample proportion, $\hat{p}$, by dividing the number of "successes" (150 in this case) by the sample size (200):

$$\hat{p} = \frac{\text{Number having the characteristic}}{n}$$

$$= \frac{150}{200} = 0.75$$

Then confirm whether you can use the normal approximation to the binomial. To use the normal approximation to the binomial to calculate a confidence interval, both $n\hat{p}$ and $n(1-\hat{p})$ must be greater than 10, where n is the sample size and $\hat{p}$ is the sample proportion.

In this case, $n = 200$ and $\hat{p} = 0.75$. So plugging in the numbers, you get $n\hat{p} = (200)(0.75) = 150$ and $n(1-\hat{p}) = (200)(1-0.75) = 50$.

Both 150 and 50 are greater than 10, so you can use the normal approximation to the binomial.

Next, substitute the known values into the formula for the confidence interval and solve:

$$CI = 0.75 \pm 2.58 \sqrt{\frac{0.75(1-0.75)}{200}}$$

$$= 0.75 \pm 2.58 \sqrt{\frac{0.75(0.25)}{200}}$$

$$= 0.75 \pm 2.58 \sqrt{\frac{0.1875}{200}}$$

$$= 0.75 \pm 2.58 \sqrt{0.0009375}$$

$$= 0.75 \pm 2.58(0.0306)$$

$$\approx 0.75 \pm 0.0789$$

Add and subtract to/from the sample proportion to find the range:

$$0.75 - 0.0789 = 0.6711$$

$$0.75 + 0.0789 = 0.8289$$

Round to two decimal places, so the 99% confidence interval for the proportion of all customers who purchased products online in the past 12 months is 0.67 to 0.83.

540. Yes, because $n\hat{p}$ and $n(1-\hat{p})$ are both greater than 10.

To use the normal approximation to the binomial, both $n\hat{p}$ and $n(1-\hat{p})$ must be greater than 10.

For this example, $n = 80$ and $\hat{p} = 0.15$. So, plugging in the numbers, you find that $n\hat{p} = 80(0.15) = 12$ and $n(1-\hat{p}) = 80(0.85) = 68$.

Both 12 and 68 are greater than 10, so you can use the normal approximation for this data.

541. 0.0989 to 0.2011

To use the normal approximation to the binomial, both $n\hat{p}$ and $n(1-\hat{p})$ must be greater than 10.

For this example, $n = 80$ and $\hat{p} = 0.15$. So, plugging in the numbers, you find that $n\hat{p} = 80(0.15) = 12$ and $n(1-\hat{p}) = 80(0.85) = 68$.

Both 12 and 68 are greater than 10, so you can use the normal approximation.

To calculate a confidence interval (CI) for a proportion, use this formula:

$$CI = \hat{p} \pm z^* \sqrt{\frac{\hat{p}(1-\hat{p})}{n}}$$

where $\sqrt{\frac{\hat{p}(1-\hat{p})}{n}}$ is the standard error, and z^* is the appropriate value from the standard normal distribution for your desired confidence level (see Table A-4 in the appendix for various confidence levels). For a confidence level of 80%, $z^* = 1.28$.

Calculate the standard error:

$$SE = \sqrt{\frac{0.15(1-0.15)}{80}}$$

$$= \sqrt{\frac{0.1275}{80}}$$

$$= 0.03992$$

Plug in the known values into the formula for the confidence interval and solve:

$$CI = 0.15 \pm 1.28(0.03992)$$

$$= 0.15 \pm 0.0510976$$

$$= 0.15 \pm 0.0511$$

To find the limits of the confidence interval, add and subtract 0.0511 from the sample proportion:

$$0.15 - 0.0511 = 0.0989$$

$$0.15 + 0.0511 = 0.2011$$

The 80% confidence interval for the population proportion is 0.0989 to 0.2011.

542. 0.0843 to 0.2175

To use the normal approximation to the binomial, both $n\hat{p}$ and $n(1-\hat{p})$ must be greater than 10.

For this example, $n = 80$ and $\hat{p} = 0.15$. So, plugging in the numbers, you find that $n\hat{p} = 80(0.15) = 12$ and $n(1-\hat{p}) = 80(0.85) = 68$.

Both 12 and 68 are greater than 10, so you can use the normal approximation.

To calculate a confidence interval (CI) for a proportion, use this formula:

$$CI = \hat{p} \pm z^* \sqrt{\frac{\hat{p}(1-\hat{p})}{n}}$$

where $\sqrt{\frac{\hat{p}(1-\hat{p})}{n}}$ is the standard error, and z^* is the appropriate value from the standard normal distribution for your desired confidence level (see Table A-4 in the appendix for various confidence levels). For a confidence level of 90%, $z^* = 1.645$.

Calculate the standard error:

$$SE = \sqrt{\frac{0.15(1-0.15)}{80}}$$

$$= \sqrt{\frac{0.1275}{80}}$$

$$= 0.03992$$

Plug in the known values into the formula for the confidence interval and solve:

$$CI = 0.15 \pm 1.645(0.03992)$$

$$\approx 0.15 \pm 0.0657$$

To find the limits of the confidence interval, add and subtract 0.0657 from the sample proportion:

$$0.15 - 0.0657 = 0.0843$$

$$0.15 + 0.0657 = 0.2175$$

The 90% confidence interval for the population proportion is 0.0843 to 0.2175.

543. 0.0718 to 0.2282

To use the normal approximation to the binomial, both $n\hat{p}$ and $n(1-\hat{p})$ must be greater than 10.

For this example, $n = 80$ and $\hat{p} = 0.15$. So, plugging in the numbers, you find that $n\hat{p} = 80(0.15) = 12$ and $n(1-\hat{p}) = 80(0.85) = 68$.

Both 12 and 68 are greater than 10, so you can use the normal approximation.

To calculate a confidence interval (CI) for a proportion, use this formula:

$$CI = \hat{p} \pm z^* \sqrt{\frac{\hat{p}(1-\hat{p})}{n}}$$

where $\sqrt{\dfrac{\hat{p}(1-\hat{p})}{n}}$ is the standard error, and z^* is the appropriate value from the standard normal distribution for your desired confidence level (see Table A-4 in the appendix for various confidence levels). For a confidence level of 95%, $z^* = 1.96$.

Calculate the standard error:

$$SE = \sqrt{\frac{0.15(1-0.15)}{80}}$$

$$= \sqrt{\frac{0.1275}{80}}$$

$$= 0.03992$$

Plug in the known values into the formula for the confidence interval and solve:

$$CI = 0.15 \pm 1.96(0.03992)$$

$$\approx 0.15 \pm 0.0782$$

To find the limits of the confidence interval, add and subtract 0.0782 from the sample proportion:

$$0.15 - 0.0782 = 0.0718$$

$$0.15 + 0.0782 = 0.2282$$

The 95% confidence interval for the population proportion is 0.0718 to 0.2282.

---

544.   0.0570 to 0.2430

To use the normal approximation to the binomial, both $n\hat{p}$ and $n(1-\hat{p})$ must be greater than 10.

For this example, $n = 80$ and $\hat{p} = 0.15$. So, plugging in the numbers, you find that $n\hat{p} = 80(0.15) = 12$ and $n(1-\hat{p}) = 80(0.85) = 68$.

Both 12 and 68 are greater than 10, so you can use the normal approximation.

To calculate a confidence interval (CI) for a proportion, use this formula:

$$CI = \hat{p} \pm z^* \sqrt{\frac{\hat{p}(1-\hat{p})}{n}}$$

where $\sqrt{\dfrac{\hat{p}(1-\hat{p})}{n}}$ is the standard error, and z^* is the appropriate value from the standard normal distribution for your desired confidence level (see Table A-4 in the appendix for various confidence levels). For a confidence level of 98%, $z^* = 2.33$.

Calculate the standard error:

$$SE = \sqrt{\frac{0.15(1-0.15)}{80}}$$

$$= \sqrt{\frac{0.1275}{80}}$$

$$= 0.03992$$

Plug in the known values into the formula for the confidence interval and solve:

$$CI = 0.15 \pm 2.33(0.03992)$$
$$\approx 0.15 \pm 0.0930$$

To find the limits of the confidence interval, add and subtract 0.0930 from the sample proportion:

$$0.15 - 0.0930 = 0.0570$$
$$0.15 + 0.0930 = 0.2430$$

The 98% confidence interval for the population proportion is 0.0570 to 0.2430.

---

545.   0.0470 to 0.2530

To use the normal approximation to the binomial, both $n\hat{p}$ and $n(1-\hat{p})$ must be greater than 10.

For this example, $n = 80$ and $\hat{p} = 0.15$. So, plugging in the numbers, you find that $n\hat{p} = 80(0.15) = 12$ and $n(1-\hat{p}) = 80(0.85) = 68$.

Both 12 and 68 are greater than 10, so you can use the normal approximation.

To calculate a confidence interval (CI) for a proportion, use this formula:

$$CI = \hat{p} \pm z^* \sqrt{\frac{\hat{p}(1-\hat{p})}{n}}$$

where $\sqrt{\dfrac{\hat{p}(1-\hat{p})}{n}}$ is the standard error, and z^* is the appropriate value from the standard normal distribution for your desired confidence level (see Table A-4 in the appendix for various confidence levels). For a confidence level of 99%, $z^* = 2.58$.

Calculate the standard error:

$$SE = \sqrt{\frac{0.15(1-0.15)}{80}}$$
$$= \sqrt{\frac{0.1275}{80}}$$
$$= 0.03992$$

Plug in the known values into the formula for the confidence interval and solve:

$$CI = 0.15 \pm 2.58(0.03992)$$
$$\approx 0.15 \pm 0.1030$$

To find the limits of the confidence interval, add and subtract 0.1030 from the sample proportion:

$$0.15 - 0.1030 = 0.0470$$
$$0.15 + 0.1030 = 0.2530$$

The 99% confidence interval for the population proportion is 0.0470 to 0.2530.

546. A. 0.15 to 0.35

For the same data, a 98% confidence interval will be wider than a 95% confidence interval.

547. D. Choice (A) or (B) (80% or 90%)

For the same data, a 99% confidence interval will be wider than a 95% confidence interval, while an 80% and 90% confidence interval will be narrower. The confidence interval 0.22 to 0.28 is narrower than the 95% confidence interval of 0.20 to 0.30, so it must represent a lower level of confidence.

548. 0.55

If you had only one number to use to estimate the population proportion, you'd use the sample proportion. To find the sample proportion, $\hat{p}$, divide the number of "successes" (88 in this case) by the sample size (160):

$$\hat{p} = \frac{\text{Number of successes}}{n}$$

$$= \frac{88}{160} = 0.55$$

However, note that a confidence interval is actually the best estimate of a population proportion because you know that the sample proportion changes with each new sample. So using the sample proportion of 0.55, plus or minus a margin of error, gives you the best possible estimate.

549. 0.0393

The formula for the standard error of a sample proportion is

$$SE = \sqrt{\frac{\hat{p}(1-\hat{p})}{n}}$$

where n is the sample size, and $\hat{p}$ is the sample proportion. In this example, $n = 160$ and $\hat{p} = 88/160$ (number of success divided by sample size) = 0.55.

Now, plug in the known values and solve for the standard error:

$$SE = \sqrt{\frac{0.55(1-0.55)}{160}}$$

$$= \sqrt{\frac{0.2475}{160}}$$

$$\approx 0.0393$$

550. 0.0770

To find the margin of error (MOE), use the formula

$$MOE = z^* \sqrt{\frac{\hat{p}(1-\hat{p})}{n}}$$

where $\hat{p}$ is the sample proportion, n is the sample size, and z^* is the appropriate value from the standard normal distribution for your desired confidence level (see Table A-4 in the appendix for various confidence levels). For a 95% confidence level, the z^*-value is 1.96. Note that the margin of error is the z^*-value times the standard error.

First, you have to find the sample proportion, $\hat{p}$, by dividing the number of "successes" (in this case, 88) by the sample size (160):

$$\hat{p} = \frac{\text{Number of successes}}{n}$$

$$= \frac{88}{160} = 0.55$$

Then, plug in the known values to the formula for the margin of error:

$$\text{MOE} = 1.96\sqrt{\frac{0.55(1-0.55)}{160}}$$

$$= 1.96\sqrt{\frac{0.55(0.45)}{160}}$$

$$= 1.96\sqrt{\frac{0.2475}{160}}$$

$$= 1.96(0.0393)$$

$$\approx 0.0770$$

With a 95% confidence level, 0.0770 is the margin of error for estimating the proportion of all adults in the city who favor the new tax.

---

551.   0.1014

To find the margin of error (MOE), use the formula

$$\text{MOE} = z^*\sqrt{\frac{\hat{p}(1-\hat{p})}{n}}$$

where $\hat{p}$ is the sample proportion, n is the sample size, and z^* is the appropriate value from the standard normal distribution for your desired confidence level (see Table A-4 in the appendix for various confidence levels). For a 99% confidence level, the z^*-value is 2.58. Note that the margin of error is the z^*-value times the standard error.

First, you have to find the sample proportion, $\hat{p}$, by dividing the number of "successes" (in this case, 88) by the sample size (160):

$$\hat{p} = \frac{\text{Number of successes}}{n}$$

$$= \frac{88}{160} = 0.55$$

Then, plug in the known values to the formula for the margin of error:

$$MOE = 2.58\sqrt{\frac{0.55(1-0.55)}{160}}$$

$$= 2.58\sqrt{\frac{0.55(0.45)}{160}}$$

$$= 2.58\sqrt{\frac{0.2475}{160}}$$

$$= 2.58(0.0393)$$

$$\approx 0.1014$$

With a 99% confidence level, 0.1014 is the margin of error for estimating the proportion of all adults in the city who favor the new tax.

---

552.   0.0503

To find the margin of error (MOE), use the formula

$$MOE = z^*\sqrt{\frac{\hat{p}(1-\hat{p})}{n}}$$

where $\hat{p}$ is the sample proportion, n is the sample size, and z^* is the appropriate value from the standard normal distribution for your desired confidence level (see Table A-4 for various confidence levels). For an 80% confidence level, the z^*-value is 1.28. Note that the margin of error is the z^*-value times the standard error.

First, you have to find the sample proportion, $\hat{p}$, by dividing the number of "successes" (in this case, 88) by the sample size (160):

$$\hat{p} = \frac{\text{Number of successes}}{n}$$

$$= \frac{88}{160} = 0.55$$

Then, plug in the known values to the formula for the margin of error:

$$MOE = 1.28\sqrt{\frac{0.55(1-0.55)}{160}}$$

$$= 1.28\sqrt{\frac{0.55(0.45)}{160}}$$

$$= 1.28\sqrt{\frac{0.2475}{160}}$$

$$= 1.28(0.0393)$$

$$\approx 0.0503$$

With an 80% confidence level, the margin of error for estimating the proportion of all adults in the city who favor the new tax is 0.0503.

553. 0.30

If you could use only one number to estimate the difference between two population proportions, you'd use the difference in the two sample proportions. So calling the population of males Population 1 and the population of females Population 2, for this specific data set from these samples taken from the populations, you get

$$\hat{p}_1 - \hat{p}_2 = 0.55 - 0.25$$
$$= 0.30$$

Note, however, that a confidence interval is the best estimate for the difference in two population proportions because you know the sample proportions change as soon as the sample changes, and a confidence interval provides a range of likely values rather than just one number for the population parameter. So using 0.30 plus or minus a margin of error gives you the best possible estimate.

554. 0.0660

To calculate the standard error for the estimated difference in two population proportions, use the formula

$$SE = \sqrt{\frac{\hat{p}_1(1-\hat{p}_1)}{n_1} + \frac{\hat{p}_2(1-\hat{p}_2)}{n_2}}$$

where $\hat{p}_1$ and n_1 are the sample proportion and sample size of the sample from Population 1, and $\hat{p}_2$ and n_2 are the sample proportion and sample size of the sample from Population 2.

Treating the sample of males from Population 1 as Sample 1 and the sample of females from Population 2 as Sample 2, plug in the numbers and solve:

$$SE = \sqrt{\frac{0.55(1-0.55)}{100} + \frac{0.25(1-0.25)}{100}}$$
$$= \sqrt{\frac{0.55(0.45)}{100} + \frac{0.25(0.75)}{100}}$$
$$= \sqrt{0.002475 + 0.001875}$$
$$= \sqrt{0.00435}$$
$$\approx 0.0660$$

So the standard error for the estimate of the difference in proportions in the male and female populations is 0.0660.

555. between 17% and 43%

To find a confidence interval when estimating the difference of two population proportions, use the formula

$$CI = (\hat{p}_1 - \hat{p}_2) \pm z^* \sqrt{\frac{\hat{p}_1(1-\hat{p}_1)}{n_1} + \frac{\hat{p}_2(1-\hat{p}_2)}{n_2}}$$

where $\hat{p}_1$ and n_1 are the sample proportion and sample size of the sample taken from Population 1, $\hat{p}_2$ and n_2 are the sample proportion and sample size of the sample taken from Population 2, and z^* is the appropriate value from the standard normal distribution for your desired confidence level (see Table A-4 in the appendix for various confidence levels).

To solve, follow these steps:

1. Use the confidence level to find the appropriate z^*-value by referring to Table A-4. The z^*-value for a confidence level of 95% is 1.96.

2. To make the calculations somewhat easier, label the group of males as "Sample 1" and the group of females as "Sample 2."

3. For each proportion, divide the number having the attribute by the sample size:

$$\hat{p} = \frac{\text{Number having the characteristic}}{n}$$

Sample 1: $\frac{55}{100} = 0.55$

Sample 2: $\frac{25}{100} = 0.25$

4. Substitute the values into the formula for the confidence interval and solve:

$$CI = (0.55 - 0.25) \pm 1.96 \sqrt{\frac{0.55(1 - 0.55)}{100} + \frac{0.25(1 - 0.25)}{100}}$$

$$= 0.30 \pm 1.96 \sqrt{\frac{0.55(0.45)}{100} + \frac{0.25(0.75)}{100}}$$

$$= 0.30 \pm 1.96 \sqrt{0.002475 + 0.001875}$$

$$= 0.30 \pm 1.96 \sqrt{0.00435}$$

$$= 0.30 \pm 1.96(0.0660), \text{rounded}$$

$$\approx 0.30 \pm 0.12936$$

5. Subtract and add the margin of error:

$$0.30 - 0.12936 = 0.17064$$

$$0.30 + 0.12936 = 0.42936$$

6. Convert to percentages by multiplying by 100%:

$$0.17064(100\%) = 17.064\%$$

$$0.42936(100\%) = 42.936\%$$

Round to the nearest whole percentage point so the 95% confidence interval is 17% to 43%.

This is a 95% confidence interval for the difference in the percentage of all males and females favoring Johnson among all likely voters. Because you subtracted the sample proportion of females from the sample proportion of males to get these results, you can conclude that the males are the ones with a higher likelihood to vote for candidate Johnson.

556. between 19% and 41%

To find a confidence interval when estimating the difference of two population proportions, use the formula

$$CI = (\hat{p}_1 - \hat{p}_2) \pm z^* \sqrt{\frac{\hat{p}_1(1 - \hat{p}_1)}{n_1} + \frac{\hat{p}_2(1 - \hat{p}_2)}{n_2}}$$

where $\hat{p}_1$ and n_1 are the sample proportion and sample size of the sample taken from Population 1, $\hat{p}_2$ and n_2 are the sample proportion and sample size of the sample taken from Population 2, and z^* is the appropriate value from the standard normal distribution for your desired confidence level (see Table A-4 in the appendix for various confidence levels).

To solve, follow these steps:

1. Use the confidence level to find the appropriate z^*-value by referring to Table A-4. The z^*-value for a confidence level of 90% is 1.645.

2. To make the calculations somewhat easier, label the sample of males taken from Population 1 as "Sample 1" and the sample of females taken from Population 2 as "Sample 2."

3. For each proportion, divide the number having the attribute by the sample size:

$$\hat{p} = \frac{\text{Number having the characteristic}}{n}$$

Sample 1: $\dfrac{55}{100} = 0.55$

Sample 2: $\dfrac{25}{100} = 0.25$

4. Substitute the values into the formula for the confidence interval and solve:

$$CI = (0.55 - 0.25) \pm 1.645 \sqrt{\frac{0.55(1-0.55)}{100} + \frac{0.25(1-0.25)}{100}}$$

$$= 0.30 \pm 1.645 \sqrt{\frac{0.55(0.45)}{100} + \frac{0.25(0.75)}{100}}$$

$$= 0.30 \pm 1.645 \sqrt{0.002475 + 0.001875}$$

$$= 0.30 \pm 1.645 \sqrt{0.00435}$$

$$= 0.30 \pm 1.645(0.0660), \text{ rounded}$$

$$\approx 0.30 \pm 0.10857$$

5. Subtract and add the margin of error:

$0.30 - 0.10857 = 0.19143$

$0.30 + 0.10857 = 0.40857$

6. Convert to percentages by multiplying by 100%:

$0.19143(100\%) = 19.143\%$

$0.40857(100\%) = 40.857\%$

Round to the nearest whole percentage point so the 90% confidence interval is 19% to 41%.

This is a 90% confidence interval for the difference in the percentage of all males and females favoring Johnson among all likely voters. Because you subtracted the sample proportion of females from the sample proportion of males to get these results, you can conclude that the males are the ones with a higher likelihood to vote for candidate Johnson.

557. between 13% and 47%

To find a confidence interval when estimating the difference of two population proportions, use the formula

$$CI = (\hat{p}_1 - \hat{p}_2) \pm z^* \sqrt{\frac{\hat{p}_1(1-\hat{p}_1)}{n_1} + \frac{\hat{p}_2(1-\hat{p}_2)}{n_2}}$$

where $\hat{p}_1$ and n_1 are the sample proportion and sample size of the sample taken from Population 1, $\hat{p}_2$ and n_2 are the sample proportion and sample size of the sample taken from Population 2, and z^* is the appropriate value from the standard normal distribution for your desired confidence level (see Table A-4 in the appendix for various confidence levels).

To solve, follow these steps:

1. Use the confidence level to find the appropriate z^*-value by referring to Table A-4. The z^*-value for a confidence level of 99% is 2.58.

2. To make the calculations somewhat easier, label the sample of males taken from Population 1 as "Sample 1" and the sample of females taken from Population 2 as "Sample 2."

3. For each proportion, divide the number having the attribute by the sample size:

$$\hat{p} = \frac{\text{Number having the characteristic}}{n}$$

Sample 1: $\frac{55}{100} = 0.55$

Sample 2: $\frac{25}{100} = 0.25$

4. Substitute the values into the formula and solve:

$$CI = (0.55 - 0.25) \pm 2.58 \sqrt{\frac{0.55(1-0.55)}{100} + \frac{0.25(1-0.25)}{100}}$$

$$= 0.30 \pm 2.58 \sqrt{\frac{0.55(0.45)}{100} + \frac{0.25(0.75)}{100}}$$

$$= 0.30 \pm 2.58 \sqrt{0.002475 + 0.001875}$$

$$= 0.30 \pm 2.58 \sqrt{0.00435}$$

$$= 0.30 \pm 2.58(0.0660), \text{ rounded}$$

$$\approx 0.30 \pm 0.17028$$

5. Subtract and add the margin of error:

$$0.30 - 0.17028 = 0.12972$$

$$0.30 + 0.17028 = 0.47028$$

6. Convert to percentages by multiplying by 100%:

$$0.12972(100\%) = 12.972\%$$

$$0.47028(100\%) = 47.028\%$$

Round to the nearest whole percentage point so the 99% confidence interval is 13% to 47%.

This is a 99% confidence interval for the difference in the percentage of all males and females favoring Johnson among all likely voters. Because you subtracted the sample proportion of females from the sample proportion of males to get these results, you can conclude that the males are the ones with a higher likelihood to vote for candidate Johnson.

---

558. 0.3333

If you could choose only one number to estimate the difference in two population proportions, you'd use the difference between two sample proportions (one from the large cites and one from the small cities).

For cities with more than 1 million in population, 220 out of 300 adults wanted increased funding, so the proportion wanting increased funding is $220/300 \approx 0.7333$.

For cities with fewer than 100,000 in population, 120 of 300 adults wanted increased funding, so the proportion wanting increased funding is $120/300 = 0.4$.

The difference in the sample proportions is $0.7333 - 0.4 = 0.3333$ for this sample of data. Because of the order of subtraction (large cities minus small cities), the value of 0.3333 means the proportion in favor from large cities is larger than the proportion in favor for small cities for these samples.

Note, however, that a confidence interval is the best estimate for the difference in population proportions because you know that the sample proportions change as soon as the sample changes, and a confidence interval provides a range of likely values rather than just one number for the population parameter. So adding and subtracting a margin of error to the value of 0.3333 gives you the best possible estimate.

---

559. 0.2706 to 0.3960

To find a confidence interval for the difference of two population proportions, use the formula

$$CI = (\hat{p}_1 - \hat{p}_2) \pm z^* \sqrt{\frac{\hat{p}_1(1-\hat{p}_1)}{n_1} + \frac{\hat{p}_2(1-\hat{p}_2)}{n_2}}$$

where $\hat{p}_1$ and n_1 are the sample proportion and sample size of the sample taken from Population 1, $\hat{p}_2$ and n_2 are the sample proportion and sample size of the sample taken from Population 2, and z^* is the appropriate value from the standard normal distribution for your desired confidence level (see Table A-4 in the appendix for various confidence levels).

To solve, follow these steps:

1. Use the confidence level to find the appropriate z^*-value by referring to Table A-4. The z^*-value for a confidence level of 90% is 1.645.

2. To make the calculations somewhat easier, label the sample of adults taken from large cities as "Sample 1" and the sample of adults taken from small cities as "Sample 2."

3. For each proportion, divide the number having the attribute by the sample size:

$$\hat{p} = \frac{\text{Number having the characteristic}}{n}$$

Sample 1: $\frac{220}{300} \approx 0.7333$

Sample 2: $\frac{120}{300} = 0.4$

4. Substitute the values into the formula and solve:

$$CI = (\hat{p}_1 - \hat{p}_2) \pm z^* \sqrt{\frac{\hat{p}_1(1-\hat{p}_1)}{n_1} + \frac{\hat{p}_2(1-\hat{p}_2)}{n_2}}$$

$$= (0.7333 - 0.4) \pm 1.645 \sqrt{\frac{0.7333(1-0.7333)}{300} + \frac{0.4(1-0.4)}{300}}$$

$$= 0.3333 \pm 1.645 \sqrt{\frac{0.19557}{300} + \frac{0.24}{300}}$$

$$= 0.3333 \pm 1.645 \sqrt{0.0006519 + 0.0008}$$

$$= 0.3333 \pm 1.645(0.038104)$$

$$= 0.3333 \pm 0.06268$$

This rounds to 0.0627.

5. Subtract and add the margin of error:

$$0.3333 - 0.0627 = 0.2706$$

$$0.3333 + 0.0627 = 0.3960$$

This is a 90% confidence interval for the difference in the proportion of all adults who are in favor of public transportation, comparing large cities and small cities.

Note that because you took the sample proportion from large cities (as Population 1) minus the sample proportion from small cities (as Population 2) and your confidence interval contains all positive values, you can conclude that the large cities are the ones more in favor of public transportation.

560. 0.2485 to 0.3821

To find a confidence interval for the difference of two population proportions, use the formula

$$CI = (\hat{p}_1 - \hat{p}_2) \pm z^* \sqrt{\frac{\hat{p}_1(1-\hat{p}_1)}{n_1} + \frac{\hat{p}_2(1-\hat{p}_2)}{n_2}}$$

where $\hat{p}_1$ and n_1 are the sample proportion and sample size of the sample taken from Population 1, $\hat{p}_2$ and n_2 are the sample proportion and sample size of the sample taken from Population 2, and z^* is the appropriate value from the standard normal distribution for your desired confidence level (see Table A-4 in the appendix for various confidence levels).

To solve, follow these steps:

1. Use the confidence level to find the appropriate z^*-value by referring to Table A-4. The z^*-value for a confidence level of 80% is 1.28.

2. To make the calculations somewhat easier, label the sample of adults taken from large cities as "Sample 1" and the sample of adults taken from small cities as "Sample 2."

3. For each proportion, divide the number having the attribute by the sample size:

$$\hat{p} = \frac{\text{Number having the characteristic}}{n}$$

Sample 1: $\frac{220}{300} \approx 0.7333$

Sample 2: $\frac{120}{300} = 0.4$

4. Substitute the known values into the formula for the confidence interval and solve.

    (**Remember:** The standard error, $\sqrt{\dfrac{\hat{p}_1(1-\hat{p}_1)}{n_1}+\dfrac{\hat{p}_2(1-\hat{p}_2)}{n_2}}$, is 0.0381.)

    $$CI = (0.7333 - 0.4) \pm 1.28(0.0381)$$
    $$\approx 0.3333 \pm 0.0488$$

5. Subtract and add the margin of error:

    $$0.3333 - 0.0488 = 0.2845$$
    $$0.3333 + 0.0488 = 0.3821$$

    This is an 80% confidence interval for the difference in the proportion of all adults who are in favor of public transportation, comparing large cities and small cities.

    Note that because you took the sample proportion from large cities (as Population 1) minus the sample proportion from small cities (as Population 2) and your confidence interval contains all positive values, you can conclude that the large cities are the ones more in favor of public transportation.

---

561.   3.35 to 4.65 inches

To find the confidence interval for the difference of two population means, where the population standard deviations are known, use the following formula:

$$CI = (\bar{x}_1 - \bar{x}_2) \pm z^* \sqrt{\dfrac{\sigma_1^2}{n_1} + \dfrac{\sigma_2^2}{n_2}}$$

Here, $\bar{x}_1$ and n_1 are the mean and the size of the sample taken from Population 1, whose population standard deviation, σ_1, is given (known); $\bar{x}_2$ and n_2 are the mean and the size of the sample taken from Population 2, whose population standard deviation, σ_2, is given (known).

Follow these steps to solve:

1. Use the confidence level to find the appropriate z^*-value by referring to Table A-4 in the appendix for various confidence levels. The z^*-value for a 95% confidence level is 1.96.

2. Substitute the known values into the equation and solve, making sure to follow the order of operations:

    $$CI = (71 - 67) \pm 1.96 \sqrt{\dfrac{2^2}{70} + \dfrac{1.8^2}{60}}$$
    $$= 4 \pm 1.96 \sqrt{\dfrac{4}{70} + \dfrac{3.24}{60}}$$
    $$= 4 \pm 1.96 \sqrt{0.05714 + 0.054}$$
    $$= 4 \pm 1.96 \sqrt{0.11114}$$
    $$= 4 \pm 1.96(0.3334)$$
    $$= 4 \pm 0.653464$$

3. Find the _lower end_ of the confidence interval by subtracting the margin of error from the difference of the means:

    $$4 - 0.653464 = 3.346536$$

4. Find the *upper end* of the confidence interval by adding the margin of error to the difference of the means:

    $$4 + 0.653464 = 4.653464$$

5. Round to the nearest hundredth, so the 95% confidence interval is 3.35 to 4.65 inches.

This is a 95% confidence interval for the difference in heights of all boys and girls in these populations.

Note that because you took the sample mean of the sample of boys taken from Population 1 minus the sample mean of the sample of girls taken from Population 2, and the confidence interval contains all positive values, you can conclude that the boys are the ones with the higher average height.

562.   3.57 to 4.43 inches

To find the confidence interval for the difference of two population means, where the population standard deviations are known, use the following formula:

$$\text{CI} = (\bar{x}_1 - \bar{x}_2) \pm z^* \sqrt{\frac{\sigma_1^2}{n_1} + \frac{\sigma_2^2}{n_2}}$$

Here, $\bar{x}_1$ and n_1 are the mean and the size of the sample taken from Population 1, whose populations standard deviation, σ_1, is given (known); $\bar{x}_2$ and n_2 are the mean and the size of the sample taken from Population 2, whose population standard deviation, σ_2, is given (known).

Follow these steps to solve:

1. Use the confidence level to find the appropriate z^*-value by referring to Table A-4 in the appendix for various confidence levels. The z^*-value for an 80% confidence level is 1.28.

2. Substitute the known values into the equation and solve, making sure to follow the order of operations:

    $$\text{CI} = (71 - 67) \pm 1.28\sqrt{\frac{2^2}{70} + \frac{1.8^2}{60}}$$

    $$= 4 \pm 1.28\sqrt{\frac{4}{70} + \frac{3.24}{60}}$$

    $$= 4 \pm 1.28\sqrt{0.05714 + 0.054}$$

    $$= 4 \pm 1.28\sqrt{0.11114}$$

    $$= 4 \pm 1.28(0.3334)$$

    $$= 4 \pm 0.426752$$

3. Find the *lower end* of the confidence interval by subtracting the margin of error from the difference of the means:

    $$4 - 0.426752 = 3.573248$$

4. Find the *upper end* of the confidence interval by adding the margin of error to the difference of the means:

    $$4 + 0.426752 = 4.426752$$

5. Round to the nearest hundredth, so the 80% confidence interval is 3.57 to 4.43 inches.

This is a 80% confidence interval for the difference in heights of all boys and girls in these populations.

Note that because you took the sample mean of the sample of boys taken from Population 1 minus the sample mean of the sample of girls taken from Population 2, and the confidence interval contains all positive values, you can conclude that the boys are the ones with the higher average height.

---

563. $\qquad$ 3.14 to 4.86 inches

To find the confidence interval for the difference of two population means, where the population standard deviations are known, use the following formula:

$$CI = (\bar{x}_1 - \bar{x}_2) \pm z^* \sqrt{\frac{\sigma_1^2}{n_1} + \frac{\sigma_2^2}{n_2}}$$

Here, $\bar{x}_1$ and n_1 are the mean and the size of the sample taken from Population 1, whose population standard deviation, σ_1, is given (known); $\bar{x}_2$ and n_2 are the mean and the size of the sample taken from Population 2, whose population standard deviation, σ_2, is given (known).

Follow these steps to solve:

1. Use the confidence level to find the appropriate z^*-value by referring to Table A-4 in the appendix for various confidence levels. The z^*-value for a 99% confidence level is 2.58.

2. Substitute the known values into the equation and solve, making sure to follow the order of operations:

$$CI = (71 - 67) \pm 2.58 \sqrt{\frac{2^2}{70} + \frac{1.8^2}{60}}$$

$$= 4 \pm 2.58 \sqrt{\frac{4}{70} + \frac{3.24}{60}}$$

$$= 4 \pm 2.58 \sqrt{0.05714 + 0.054}$$

$$= 4 \pm 2.58 \sqrt{0.11114}$$

$$= 4 \pm 2.58 (0.3334)$$

$$= 4 \pm 0.860172$$

3. Find the *lower end* of the confidence interval by subtracting the margin of error from the difference of the means:

$$4 - 0.860172 = 3.139828$$

4. Find the *upper end* of the confidence interval by adding the margin of error to the difference of the means:

$$4 + 0.860172 = 4.860172$$

5. Round to the nearest hundredth, so the 99% confidence interval is 3.14 to 4.86 inches.

This is a 99% confidence interval for the difference in heights of all boys and girls in these populations.

Note that because you took the sample mean of the sample of boys taken from Population 1 minus the sample mean of the sample of girls taken from Population 2, and the confidence interval contains all positive values, you can conclude that the boys are the ones with the higher average height.

564. 3.22 to 4.78 inches

To find the confidence interval for the difference of two population means, where the population standard deviations are known, use the following formula:

$$CI = (\bar{x}_1 - \bar{x}_2) \pm z^* \sqrt{\frac{\sigma_1^2}{n_1} + \frac{\sigma_2^2}{n_2}}$$

Here, $\bar{x}_1$ and n_1 are the mean and the size of the sample taken from Population 1, whose population standard deviation, σ_1, is given (known); $\bar{x}_2$ and n_2 are the mean and the size of the sample taken from Population 2, whose population standard deviation, σ_2, is given (known).

Follow these steps to solve:

1. Use the confidence level to find the appropriate z^*-value by referring to Table A-4 in the appendix for various confidence levels. The z^*-value for a 98% confidence level is 2.33.

2. Substitute the known values into the equation and solve, making sure to follow the order of operations:

$$CI = (71 - 67) \pm 2.33 \sqrt{\frac{2^2}{70} + \frac{1.8^2}{60}}$$

$$= 4 \pm 2.33 \sqrt{\frac{4}{70} + \frac{3.24}{60}}$$

$$= 4 \pm 2.33 \sqrt{0.05714 + 0.054}$$

$$= 4 \pm 2.33 \sqrt{0.11114}$$

$$= 4 \pm 2.33 (0.3334)$$

$$= 4 \pm 0.776822$$

3. Find the *lower end* of the confidence interval by subtracting the margin of error from the difference of the means:

$$4 - 0.776822 = 3.223178$$

4. Find the *upper end* of the confidence interval by adding the margin of error to the difference of the means:

$$4 + 0.776822 = 4.776822$$

5. Round to the nearest hundredth, so the 98% confidence interval is 3.22 to 4.78 inches.

This is a 98% confidence interval for the difference in heights of all boys and girls in these populations.

Note that because you took the sample mean of the sample of boys taken from Population 1 minus the sample mean of the sample of girls taken from Population 2, and the confidence interval contains all positive values, you can conclude that the boys are the ones with the higher average height.

565. If you switch the order of the populations (treating the population of girls as Population 1 and population of boys as Population 2), the mean difference would be negative, but the margin of error would be the same.

This result is clear from the formula for a confidence interval for the difference in two means:

$$CI = (\bar{x}_1 - \bar{x}_2) \pm z^* \sqrt{\frac{\sigma_1^2}{n_1} + \frac{\sigma_2^2}{n_2}}$$

You use the means only to calculate the estimated difference in means, not the margin of error. So switching girls and boys switches the order in which the means are subtracted, changing the difference from positive (boys – girls) to negative (girls – boys). A negative difference means the results from Group 1 are smaller than the results from Group 2. (For example, if you take 67 – 71 you get a negative number, meaning 2 is smaller than 4.)

However, switching Populations 1 and 2 doesn't change the margin of error, mainly because the values are squared and summed in this formula rather than subtracted. So in the end, when you switch the population names, the difference in means changes sign, but the margin of error stays the same. The confidence intervals don't change in their width, but their possible values have different signs. Bottom line: It's important to know which population is designated Population 1 and which population is designated Population 2.

566. 1.2

The margin of error (MOE) is the quantity added or subtracted from the sample mean when calculating a confidence interval. For a confidence interval of the difference in two population means, when the population standard deviations are known, the formula for the MOE is

$$MOE = z^* \sqrt{\frac{\sigma_1^2}{n_1} + \frac{\sigma_2^2}{n_2}}$$

where n_1 is the sample size of the sample taken from Population 1, whose population standard deviation is σ_1, and n_2 is the sample size of the sample taken from Population 2, whose population standard deviation is σ_2.

To solve, follow these steps:

1. Use the confidence level to find the appropriate z^*-value by referring to Table A-4 in the appendix for various confidence levels. The z^*-value for a 90% confidence level is 1.645.

2. Substitute the known values into the equation and solve, making sure to follow the order of operations:

$$MOE = 1.645 \sqrt{\frac{7^2}{120} + \frac{4^2}{130}}$$

$$= 1.645 \sqrt{\frac{49}{120} + \frac{16}{130}}$$

$$= 1.645 \sqrt{0.4083 + 0.1231}$$

$$= 1.645 \sqrt{0.5314}$$

$$= 1.645(0.7290)$$

$$= 1.199205$$

Rounded to one decimal place, the margin of error is 1.2 hours.

This is the margin of error for the estimated difference in average time spent on homework for college physics majors versus college English majors for a 90% confidence level.

567.   0.9

The margin of error (MOE) is the quantity added or subtracted from the sample mean when calculating a confidence interval. For a confidence interval of the difference in two population means, when the population standard deviations are known, the formula for the MOE is

$$MOE = z^* \sqrt{\frac{\sigma_1^2}{n_1} + \frac{\sigma_2^2}{n_2}}$$

where n_1 is the sample size of the sample taken from Population 1, whose population standard deviation is σ_1, and n_2 is the sample size of the sample taken from Population 2, whose population standard deviation is σ_2.

To solve, follow these steps:

1. Use the confidence level to find the appropriate z^*-value by referring to Table A-4 in the appendix for various confidence levels. The z^*-value for an 80% confidence level is 1.28.

2. Substitute the known values into the equation and solve, making sure to follow the order of operations:

$$MOE = 1.28 \sqrt{\frac{7^2}{120} + \frac{4^2}{130}}$$

$$= 1.28 \sqrt{\frac{49}{120} + \frac{16}{130}}$$

$$= 1.28 \sqrt{0.4083 + 0.1231}$$

$$= 1.28 \sqrt{0.5314}$$

$$= 1.28(0.7290)$$

$$= 0.93312$$

Rounded to one decimal place, the margin of error is 0.9 hours.

This is the margin of error for the estimated difference in average time spent on homework for college physics majors versus college English majors with an 80% confidence level.

568.   5.6 to 8.4 hours

To find the confidence interval for the difference of two population means, where the population standard deviations are known, use the following formula:

$$CI = (\bar{x}_1 - \bar{x}_2) \pm z^* \sqrt{\frac{\sigma_1^2}{n_1} + \frac{\sigma_2^2}{n_2}}$$

Here, $\bar{x}_1$ and n_1 are the mean and the size of the sample taken from Population 1, whose population standard deviation, σ_1, is given (known); $\bar{x}_2$ and n_2 are the mean and size of the sample taken from Population 2, whose population standard deviation, σ_2, is given (known).

To solve, follow these steps:

1. Use the confidence level to find the appropriate z^*-value by referring to Table A-4 in the appendix for various confidence levels. The z^*-value for a 95% confidence level is 1.96.

2. Substitute the known values into the equation and solve, making sure to follow the order of operations:

$$CI = (25 - 18) \pm 1.96 \sqrt{\frac{7^2}{120} + \frac{4^2}{130}}$$

$$= 7 \pm 1.96 \sqrt{\frac{49}{120} + \frac{16}{130}}$$

$$= 7 \pm 1.96 \sqrt{0.4083 + 0.1231}$$

$$= 7 \pm 1.96 \sqrt{0.5314}$$

$$= 7 \pm 1.96 (0.7290)$$

$$= 7 \pm 1.42884$$

3. Find the *lower end* of the confidence interval by subtracting the margin of error from the difference of the two sample means:

$$7 - 1.42884 = 5.57116$$

4. Find the *upper end* of the confidence interval by adding the margin of error to the difference of the two means:

$$7 + 1.42884 = 8.42884$$

5. Round to the nearest tenth to get 5.6 to 8.4 hours.

So a 95% confidence interval for the difference in average study time is 5.6 to 8.4 hours. Because you treated the population of physics majors as Population 1, and all the values in the confidence interval are positive, you can conclude that the physics majors are the ones with the higher average homework time.

569.   5.1 to 8.9 hours

To find the confidence interval for the difference of two population means, where the population standard deviations are known, use the following formula:

$$CI = (\bar{x}_1 - \bar{x}_2) \pm z^* \sqrt{\frac{\sigma_1^2}{n_1} + \frac{\sigma_2^2}{n_2}}$$

Here, $\bar{x}_1$ and n_1 are the mean and the size of the sample taken from Population 1, whose population standard deviation, σ_1, is given (known); $\bar{x}_2$ and n_2 are the mean and size of the sample taken from Population 2, whose population standard deviation, σ_2, is given (known).

To solve, follow these steps:

1. Use the confidence level to find the appropriate z^*-value by referring to Table A-4 for various confidence levels. The z^*-value for a 99% confidence level is 2.58.

2. Substitute the known values into the equation and solve, making sure to follow the order of operations:

$$CI = (25 - 18) \pm 2.58 \sqrt{\frac{7^2}{120} + \frac{4^2}{130}}$$

$$= 7 \pm 2.58 \sqrt{\frac{49}{120} + \frac{16}{130}}$$

$$= 7 \pm 2.58 \sqrt{0.4083 + 0.1231}$$

$$= 7 \pm 2.58 \sqrt{0.5314}$$

$$= 7 \pm 2.58(0.7290)$$

$$= 7 \pm 1.88082$$

3. Find the *lower end* of the confidence interval by subtracting the MOE from the difference in sample means:

$$7 - 1.88082 = 5.11918$$

4. Find the *upper end* of the confidence interval by adding the margin of error to the difference in sample means:

$$7 + 1.88082 = 8.88082$$

5. Round to the nearest tenth to get 5.1 to 8.9 hours.

So a 99% confidence interval for the difference in average homework time is 5.1 to 8.9 hours. Because you treated the population of physics majors as Population 1, and all the values in the confidence interval are positive, you can conclude that the physics majors are the ones with the higher average homework time.

---

570. D. Choices (A) and (B) (You would use t^* from a t-distribution rather than z^* from the standard normal distribution; you would use the sample standard deviations rather than the population standard deviations.)

If you're estimating the difference in two population means and don't know the population standard deviations, you use t^* rather than z^* and the sample standard deviations when calculating a confidence interval.

---

571. You would use a t^*-value rather than a z^*-value.

If you're estimating the difference in two population means and one or both of your sample sizes are less than 30, you use a t^*-value from a t-distribution rather than a z^*-value from a standard normal distribution when calculating a confidence interval.

572. 0.6

The margin of error (MOE) is the quantity added or subtracted from the sample mean when calculating a confidence interval. For a confidence interval of the difference in two population means, when the population standard deviations are known, the formula for the MOE is

$$MOE = z^* \sqrt{\frac{\sigma_1^2}{n_1} + \frac{\sigma_2^2}{n_2}}$$

where n_1 is the size of the sample taken from Population 1, whose population standard deviation is σ_1, and n_2 is the size of the sample taken from Population 2, whose population standard deviation is σ_2.

To solve, follow these steps:

1. Use the confidence level to find the appropriate z^*-value by referring to Table A-4 in the appendix for various confidence levels. The z^*-value for an 80% confidence level is 1.28.

2. Substitute the known values into the equation and solve, making sure to follow the order of operations:

$$MOE = 1.28 \sqrt{\frac{6^2}{200} + \frac{4^2}{220}}$$

$$= 1.28 \sqrt{\frac{36}{200} + \frac{16}{220}}$$

$$= 1.28 \sqrt{0.18 + 0.0727}$$

$$= 1.28 \sqrt{0.2527}$$

$$= 1.28 \, (0.5027)$$

$$= 0.643456$$

Rounded to one decimal place, the margin of error is 0.6 years.

For an 80% confidence level, the margin of error for the estimate of the difference in average age at first marriage for men and women is ± 0.6 years.

573. 0.8

The margin of error (MOE) is the quantity added or subtracted from the sample mean when calculating a confidence interval. For a confidence interval of the difference in two population means, when the population standard deviations are known, the formula for the MOE is

$$MOE = z^* \sqrt{\frac{\sigma_1^2}{n_1} + \frac{\sigma_2^2}{n_2}}$$

where n_1 is the size of the sample taken from Population 1, whose population standard deviation is σ_1, n_2 is the size of the sample taken from Population 2, whose population standard deviation is σ_2, and z^* is found by using Table A-4.

To solve, follow these steps:

1. Use the confidence level to find the appropriate z^*-value by referring to Table A-4 for various confidence levels. The z^*-value for a 90% confidence level is 1.645.

2. Substitute the known values into the equation and solve, making sure to follow the order of operations:

$$MOE = 1.645\sqrt{\frac{6^2}{200} + \frac{4^2}{220}}$$

$$= 1.645\sqrt{\frac{36}{200} + \frac{16}{220}}$$

$$= 1.645\sqrt{0.18 + 0.0727}$$

$$= 1.645\sqrt{0.2527}$$

$$= 1.645\,(0.5027)$$

$$= 0.8269415$$

Round to one decimal place to get 0.8 years.

So the margin of error (MOE) for the difference in average ages between men and women at the time of their first marriage at a confidence level of 90% is ± 0.8 years.

574.   2.0 to 4.0 years

To find the confidence interval for the difference of two population means, where the population standard deviations are known, use the following formula:

$$CI = (\bar{x}_1 - \bar{x}_2) \pm z^* \sqrt{\frac{\sigma_1^2}{n_1} + \frac{\sigma_2^2}{n_2}}$$

Here, $\bar{x}_1$ and n_1 are the mean and the size of the sample taken from Population 1, whose population standard deviation, σ_1, is given (known); $\bar{x}_2$ and n_2 are the mean and size of the sample taken from Population 2, whose population standard deviation, σ_2, is given (known).

To solve, follow these steps:

1. Use the confidence level to find the appropriate z^*-value by referring to Table A-4 in the appendix for various confidence levels. The z^*-value for a 95% confidence level is 1.96.

2. Substitute the known values into the equation and solve, making sure to follow the order of operations:

$$CI = (29 - 26) \pm 1.96\sqrt{\frac{6^2}{200} + \frac{4^2}{220}}$$

$$= 3 \pm 1.96\sqrt{\frac{36}{200} + \frac{16}{220}}$$

$$= 3 \pm 1.96\sqrt{0.18 + 0.0727}$$

$$= 3 \pm 1.96\sqrt{0.2527}$$

$$= 3 \pm 1.96\,(0.5027)$$

$$= 3 \pm 0.9853$$

3. Find the *lower end* of the confidence interval by subtracting the margin of error from the difference of the two means:

$$3 - 0.9853 = 2.0147$$

4. Find the *upper end* of the confidence interval by adding the margin of error to the difference of the two means:

$$3 + 0.9853 = 3.9853$$

5. Round to the nearest tenth to get 2.0 to 4.0 years.

For a 95% confidence level, the confidence interval for the difference in average age at first marriage between men and women is 2.0 to 4.0 years. Because you treated men as Population 1, and all the values in the confidence interval are positive, you can conclude that the males are the ones with the higher average age at first marriage.

---

575.   38

Because both sample sizes are less than 30, you'll use a t^*-value rather than a z^*-value to calculate the confidence interval. The formula to calculate the degrees of freedom (*df*) for a difference in means is

$$df = n_1 + n_2 - 2$$

where n_1 is the first sample size and n_2 is the second sample size. In this example, n_1 (12th graders) is 20 and n_2 is also 20, so you get $20 + 20 - 2 = 38$.

---

576.   12

Because both sample sizes are less than 30, you'll use a t^*-value to calculate the margin of error. The formula for the margin of error (MOE) is

$$MOE = t *_{n_1+n_2-2} \left(\sqrt{\frac{(n_1-1)\,s_1^2 + (n_2-1)s_2^2}{n_1+n_2-2}} \right) \left(\sqrt{\frac{1}{n_1} + \frac{1}{n_2}} \right)$$

Here, t^* is the critical value from the t-table (Table A-2 in the appendix) with $n_1 + n_2 - 2$ degrees of freedom, n_1 and n_2 are the two sample sizes respectively, and s_1 and s_2 are the two sample standard deviations.

First, calculate the degrees of freedom, using the formula $df = n_1 + n_2 - 2 = 20 + 20 - 2 = 38$.

This value is larger than any *df* value in Table A-2, so use the *z* row. For confidence intervals, use the *CI* row at the bottom of the table. So for a 99% level of confidence, the t^*-value (as estimated by a *z*-value) is 2.57583. Because you're rounding to whole numbers, two decimal places are sufficient for your calculations, so this rounds to 2.58.

Now, plug the values into the formula and solve:

$$\text{MOE} = 2.58 \left(\sqrt{\frac{(20-1)18^2 + (20-1)12^2}{20+20-2}} \right) \left(\sqrt{\frac{1}{20} + \frac{1}{20}} \right)$$

$$= 2.58 \left(\sqrt{\frac{(19)(324) + (19)(144)}{38}} \right) \sqrt{0.05 + 0.05}$$

$$= 2.58 \left(\sqrt{\frac{6,156 + 2,736}{38}} \right) (0.3162)$$

$$= 2.58 \left(\sqrt{\frac{8,892}{38}} \right) (0.3162)$$

$$= 2.58 \left(\sqrt{234} \right) (0.3162)$$

$$= 2.58 \, (15.297) \, (0.3162)$$

$$= 12.4792$$

Rounded to the nearest whole number of pounds, the margin of error is 12 for a 99% confidence level.

---

577.   6

Because both sample sizes are less than 30, you'll use a t^*-value to calculate the margin of error. The formula for the margin of error (MOE) is

$$\text{MOE} = t^*_{n_1+n_2-2} \left(\sqrt{\frac{(n_1-1)\,s_1^2 + (n_2-1)\,s_2^2}{n_1+n_2-2}} \right) \left(\sqrt{\frac{1}{n_1} + \frac{1}{n_2}} \right)$$

Here, t^* is the critical value from the t-table (Table A-2 in the appendix) with $n_1 + n_2 - 2$ degrees of freedom, n_1 and n_2 are the two sample sizes respectively, and s_1 and s_2 are the two sample standard deviations.

First, calculate the degrees of freedom, using the formula $df = n_1 + n_2 - 2 = 20 + 20 - 2 = 38$.

This value is larger than any df value in Table A-2, so use the z row. For confidence intervals, use the CI row at the bottom of the table. For an 80% level of confidence, the t^*-value (as estimated by a z-value) is 1.281552. Because you're rounding to whole numbers, two decimal places are sufficient for your calculations, so this rounds to 1.28.

Now, plug the values into the formula and solve:

$$MOE = 1.28\left(\sqrt{\frac{(20-1)18^2+(20-1)12^2}{20+20-2}}\right)\left(\sqrt{\frac{1}{20}+\frac{1}{20}}\right)$$

$$=1.28\left(\sqrt{\frac{(19)(324)+(19)(144)}{38}}\right)\left(\sqrt{0.05+0.05}\right)$$

$$=1.28\left(\sqrt{\frac{6,156+2,736}{38}}\right)(0.3162)$$

$$=1.28\left(\sqrt{\frac{8,892}{38}}\right)(0.3162)$$

$$=1.28\left(\sqrt{234}\right)(0.3162)$$

$$=1.28(15.297)(0.3162)$$

$$=6.1912$$

Rounded to the nearest whole number of pounds, the margin of error is 6 for an 80% confidence level.

578. 8

Because both sample sizes are less than 30, you'll use a t^*-value to calculate the margin of error. The formula for the margin of error (MOE) is

$$MOE = t^*_{n_1+n_2-2}\left(\sqrt{\frac{(n_1-1)s_1^2+(n_2-1)s_2^2}{n_1+n_2-2}}\right)\left(\sqrt{\frac{1}{n_1}+\frac{1}{n_2}}\right)$$

Here, t^* is the critical value from the t-table (Table A-2 in the appendix) with n_1+n_2-2 degrees of freedom, n_1 and n_2 are the two sample sizes respectively, and s_1 and s_2 are the two sample standard deviations.

First, calculate the degrees of freedom, using the formula $df = n_1 + n_2 - 2 = 20 + 20 - 2 = 38$.

This value is larger than any df value in Table A-2, so use the z row. For confidence intervals, use the CI row at the bottom of the table. For a 90% level of confidence, the t^*-value (as estimated by a z-value) is 1.644854. This rounds to 1.645.

Now, plug the values into the formula and solve:

$$MOE = 1.645 \left(\sqrt{\frac{(20-1)18^2 + (20-1)12^2}{20+20-2}} \right) \left(\sqrt{\frac{1}{20} + \frac{1}{20}} \right)$$

$$= 1.645 \left(\sqrt{\frac{(19)(324) + (19)(144)}{38}} \right) \left(\sqrt{0.05 + 0.05} \right)$$

$$= 1.645 \left(\sqrt{\frac{6,156 + 2,736}{38}} \right) (0.3162)$$

$$= 1.645 \left(\sqrt{\frac{8,892}{38}} \right) (0.3162)$$

$$= 1.645 \left(\sqrt{234} \right) (0.3162)$$

$$= 1.645 (15.297)(0.3162)$$

$$= 7.9567$$

Rounded to the nearest whole number of pounds, the margin of error is 8 for a 90% confidence level.

---

579. 21 to 39 pounds

Use the formula for creating a confidence interval for the difference of two population means when the population standard deviation isn't known and/or the sample sizes are small (less than 30) and you can't be sure whether your data came from a normal distribution.

$$CI = (\bar{x}_1 - \bar{x}_2) \pm t^*_{n_1+n_2-2} \left(\sqrt{\frac{(n_1-1)s_1^2 + (n_2-1)s_2^2}{n_1+n_2-2}} \right) \left(\sqrt{\frac{1}{n_1} + \frac{1}{n_2}} \right)$$

Here, t^* is the critical value from the t-table (Table A-2 in the appendix) with $n_1 + n_2 - 2$ degrees of freedom, n_1 and n_2 are the two sample sizes respectively, $\bar{x}_1$ and $\bar{x}_2$ are the two sample means, and s_1 and s_2 are the two sample standard deviations.

Follow these steps to solve:

1. Determine the t^*-value in the t-table by finding the number in the df row that intersects with the given confidence level (or CI).

In this question, you have a 95% confidence level and $df = n_1 + n_2 - 2 = 20 + 20 - 2 = 38$. Because the degrees of freedom is more than 30, use the number in row z, so $t^* \approx 1.95996$. Two decimal places are sufficient because you're rounding to whole numbers, so use 1.96.

2. Substitute all the values into the formula and solve:

$$CI = (170 - 140) \pm 1.96 \left(\sqrt{\frac{(20-1)18^2 + (20-1)12^2}{20+20-2}} \right) \left(\sqrt{\frac{1}{20} + \frac{1}{20}} \right)$$

$$= 30 \pm 1.96 \left(\sqrt{\frac{(19)(324) + (19)(144)}{38}} \right) \left(\sqrt{0.05 + 0.05} \right)$$

$$= 30 \pm 1.96 \left(\sqrt{\frac{6{,}156 + 2{,}736}{38}} \right) (0.3162)$$

$$= 30 \pm 1.96 \left(\sqrt{\frac{8{,}892}{38}} \right) (0.3162)$$

$$= 30 \pm 1.96 \left(\sqrt{234} \right) (0.3162)$$

$$= 30 \pm 1.96 \, (15.297) \, (0.3162)$$

$$= 30 \pm 9.4803$$

3. Subtract and add the margin of error:

   $30 - 9.4803 = 20.5197$

   $30 + 9.4803 = 39.4803$

4. Round to the nearest whole number to get 21 to 39 pounds.

So a 95% confidence interval for the true difference in the mean weights of all the 12th-grade and 9th-grade boys at this school is 21 to 39 pounds. Because you treated the population of 12th graders as Population 1 and all the values in the confidence interval are positive, you can conclude that the 12th graders are the ones with the higher average weight.

580.   19 to 41 pounds

Use the formula for creating a confidence interval for the difference of two population means when the population standard deviation isn't known and/or the sample sizes are small (less than 30) and you can't be sure whether your data came from a normal distribution.

$$CI = (\bar{x}_1 - \bar{x}_2) \pm t^*_{n_1 + n_2 - 2} \left(\sqrt{\frac{(n_1 - 1)s_1^2 + (n_2 - 1)s_2^2}{n_1 + n_2 - 2}} \right) \left(\sqrt{\frac{1}{n_1} + \frac{1}{n_2}} \right)$$

Here, t^* is the critical value from the t-table (Table A-2 in the appendix) with $n_1 + n_2 - 2$ degrees of freedom, n_1 and n_2 are the two sample sizes respectively, $\bar{x}_1$ and $\bar{x}_2$ are the two sample means, and s_1 and s_2 are the two sample standard deviations.

Follow these steps to solve:

1. Determine the t^*-value in the t-table by finding the number in the df row that intersects with the given confidence level (or CI).

   In this question, you have a 98% confidence level and $df = n_1 + n_2 - 2 = 20 + 20 - 2 = 38$. Because the degrees of freedom is more than 30, use the number in row z, so $t^* \approx 2.32635$. Two decimal places are sufficient because you're rounding to whole numbers, so use 2.33.

2. Substitute all the values into the formula and solve:

$$CI = (170 - 140) \pm 2.33 \left(\sqrt{\frac{(20-1)18^2 + (20-1)12^2}{20+20-2}} \right) \left(\sqrt{\frac{1}{20} + \frac{1}{20}} \right)$$

$$= 30 \pm 2.33 \left(\sqrt{\frac{(19)(324) + (19)(144)}{38}} \right) \left(\sqrt{0.05 + 0.05} \right)$$

$$= 30 \pm 2.32635 \left(\sqrt{\frac{6,156 + 2,736}{38}} \right) (0.3162)$$

$$= 30 \pm 2.33 \left(\sqrt{\frac{8,892}{38}} \right) (0.3162)$$

$$= 30 \pm 2.33 \left(\sqrt{234} \right) (0.3162)$$

$$= 30 \pm 2.33 \, (15.297) \, (0.3162)$$

$$= 30 \pm 11.2700$$

3. Subtract and add the margin of error:

$$30 - 11.27 = 18.73$$

$$30 + 11.27 = 41.27$$

4. Round to the nearest whole number to get a confidence interval of 19 to 41 pounds.

---

581. 43

Because both sample sizes are less than 30, you'll use a t^*-value rather than a z^*-value to calculate the confidence interval. The formula to calculate the degrees of freedom (*df*) for a difference in means is

$$df = n_1 + n_2 - 2$$

where n_1 is the size of the sample taken from Population 1, and n_2 is the size of the sample taken from Population 2. In this example, n_1 (men) is 20 and n_2 (women) is 25, so you get $20 + 25 - 2 = 43$.

---

582. 1.6

Because both sample sizes are less than 30, you'll use a t^*-value to calculate the margin of error. The formula for the margin of error (MOE) is

$$MOE = t^*_{n_1 + n_2 - 2} \left(\sqrt{\frac{(n_1 - 1)s_1^2 + (n_2 - 1)s_2^2}{n_1 + n_2 - 2}} \right) \left(\sqrt{\frac{1}{n_1} + \frac{1}{n_2}} \right)$$

Here, t^* is the critical value from the *t*-table (Table A-2 in the appendix) with $n_1 + n_2 - 2$ degrees of freedom, n_1 and n_2 are the two sample sizes respectively, and s_1 and s_2 are the two sample standard deviations.

First, calculate the degrees of freedom using the formula $df = n_1 + n_2 - 2 = 20 + 25 - 2 = 43$.

This value is larger than any *df* value in Table A-2, so use the z row. For confidence intervals, use the *CI* row at the bottom of the table. For a 90% level of confidence, the t^*-value (as estimated by a z-value) is 1.644854. This rounds to 1.645.

Now, plug the values into the formula and solve:

$$\text{MOE} = 1.645 \left(\sqrt{\frac{(20-1)3.5^2 + (25-1)3^2}{20+25-2}} \right) \left(\sqrt{\frac{1}{20} + \frac{1}{25}} \right)$$

$$= 1.645 \left(\sqrt{\frac{(19)(12.25) + (24)(9)}{43}} \right) \left(\sqrt{0.09} \right)$$

$$= 1.645 \left(\sqrt{\frac{232.75 + 216}{43}} \right) (0.3)$$

$$= 1.645 \left(\sqrt{\frac{448.75}{43}} \right) (0.3)$$

$$= 1.645 \left(\sqrt{10.436} \right) (0.3)$$

$$= 1.645 (3.230)(0.3)$$

$$= 1.594$$

Rounded to one decimal place, the margin of error is 1.6 thousand dollars, which is the difference in average salaries between men and women for a 90% confidence level.

---

583. 1.9

Because both sample sizes are less than 30, you'll use a t^*-value to calculate the margin of error. The formula for the margin of error (MOE) is

$$\text{MOE} = t^*_{n_1+n_2-2} \left(\sqrt{\frac{(n_1-1)s_1^2 + (n_2-1)s_2^2}{n_1+n_2-2}} \right) \left(\sqrt{\frac{1}{n_1} + \frac{1}{n_2}} \right)$$

Here, t^* is the critical value from the *t*-table (Table A-2 in the appendix) with $n_1 + n_2 - 2$ degrees of freedom, n_1 and n_2 are the two sample sizes respectively, and s_1 and s_2 are the two sample standard deviations.

First, calculate the degrees of freedom using the formula $df = n_1 + n_2 - 2 = 20 + 25 - 2 = 43$.

This value is larger than any *df* value in Table A-2, so use the z row. For confidence intervals, use the *CI* row at the bottom of the table. For a 95% level of confidence, the t^*-value (as estimated by a z-value) is 1.95996. This rounds to 1.96.

Now, plug the values into the formula and solve:

$$MOE = 1.96 \left(\sqrt{\frac{(20-1)3.5^2 + (25-1)\,3^2}{20+25-2}} \right) \left(\sqrt{\frac{1}{20} + \frac{1}{25}} \right)$$

$$= 1.96 \left(\sqrt{\frac{(19)(12.25) + (24)(9)}{43}} \right) \left(\sqrt{0.09} \right)$$

$$= 1.96 \left(\sqrt{\frac{232.75 + 216}{43}} \right) (0.3)$$

$$= 1.96 \left(\sqrt{\frac{448.75}{43}} \right) (0.3)$$

$$= 1.96 \left(\sqrt{10.436} \right) (0.3)$$

$$= 1.96(3.230)(0.3)$$

$$= 1.899$$

Rounded to one decimal place, the margin of error is 1.9 thousand dollars, which is the difference in average salaries between men and women for a 95% confidence level.

584. 4.5 to 9.5

Use the formula for creating a confidence interval for the difference of two population means when the population standard deviation isn't known and/or the sample sizes are small (less than 30) and you can't be sure whether your data came from a normal distribution.

$$CI = (\bar{x}_1 - \bar{x}_2) \pm t*_{n_1+n_2-2} \left(\sqrt{\frac{(n_1-1)\,s_1^2 + (n_2-1)\,s_2^2}{n_1+n_2-2}} \right) \left(\sqrt{\frac{1}{n_1} + \frac{1}{n_2}} \right)$$

Here, $t*$ is the critical value from the t-table (Table A-2 in the appendix) with $n_1 + n_2 - 2$ degrees of freedom, n_1 and n_2 are the two sample sizes respectively, $\bar{x}_1$ and $\bar{x}_2$ are the two sample means, and s_1 and s_2 are the two sample standard deviations.

Follow these steps to solve:

1. Determine the $t*$-value in the t-table by finding the number in the df row that intersects with the given confidence level (or CI).

   In this question, you have a 99% confidence level and $df = n_1 + n_2 - 2 = 20 + 25 - 2 = 43$. Because the degrees of freedom is more than 30, use the number in row z, so $t* = 2.57583$. This rounds to 2.58.

2. Substitute all the values into the formula and solve:

$$CI = (37-30) \pm 2.58 \left(\sqrt{\frac{(20-1)3.5^2 + (25-1)\,3^2}{20+25-2}} \right) \left(\sqrt{\frac{1}{20} + \frac{1}{25}} \right)$$

$$= 7 \pm 2.58 \left(\sqrt{\frac{(19)\,(12.25) + (24)\,(9)}{43}} \right) \left(\sqrt{0.09} \right)$$

$$= 7 \pm 2.58 \left(\sqrt{\frac{232.75 + 216}{43}} \right) (0.3)$$

$$= 7 \pm 2.58 \,(3.230)\,(0.3)$$

$$= 7 \pm 2.500$$

3. Subtract and add the margin of error:

$$7 - 2.500 = 4.500$$

$$7 + 2.500 = 9.500$$

4. Round to one decimal place to get 4.5 to 9.5 (thousands of dollars).

So a 99% confidence interval for the true difference in average income between all North American men and women after five years of employment is 4.5 and 9.5, in thousands of dollars.

---

585.   **E.**   Choices (A) and (C) (An automobile factory claims 99% of its parts meet stated specifications; an automobile factory claims that it can assemble 500 automobiles an hour when the assembly line is fully staffed.)

A hypothesis test is a statistical procedure undertaken to test a quantifiable claim. "The best-quality cars" isn't quantifiable on its own and can't be tested in this manner, while the other two claims can.

If the "best quality" is defined as having "an average of 30 amenities per vehicle," then a hypothesis test could be devised.

---

586.   **C.**   A school gives its students standardized tests to measure levels of achievement compared to prior years.

A hypothesis test is a statistical procedure undertaken to test a quantifiable claim. No quantifiable claim is clearly defined in the preceding statement, and therefore no hypothesis test is possible without further clarification.

If you're told that last year's standardized test yielded an average math subscore of 80, you could test whether this year's students perform significantly better. But as currently stated, you're not given a clear definition of the population value (parameter) of interest.

---

587. $H_0\colon p = 0.75$

The null hypothesis is the prior claim that you want to test — in this case, that "75% of voters conclude the bond issue." The null and alternative hypotheses are always stated in terms of a population parameter (p in this case).

588.

$H_a: p \neq 0.75$

The alternative hypothesis is the statement about the world that you will conclude if you have statistical evidence to reject the null hypothesis, based on the data. The null and alternative hypotheses are always stated in terms of a population parameter (in this case p).

589.

impossible to tell without further information

The null and alternative hypotheses are defined by a real-world situation. Suppose that 132 is the intended mean amount of M&M's in a bag of set weight. A consumer advocacy group may want to solely test whether $\mu < 132$ to ensure that customers aren't being cheated. A quality control manager may want to test that $\mu \neq 132$ so that bags aren't being underfilled or overfilled. You can't know the alternative hypothesis without knowing the context of the situation.

590.

$H_0: \mu = 10.50$

The null hypothesis is the original claim or current "best guess" at the value of interest. The null hypothesis is always written in terms of a population parameter (in this case, μ) being equivalent to a specific value.

591.

$H_0: \mu = 52$

The null hypothesis states a current claim about the condition of the world. The null hypothesis is always written in terms of a population parameter (in this case, μ) being equivalent to a specific value.

592.

The average number of songs on an MP3 player owned by a college student is 228.

A null hypothesis states the specific value of a population parameter, using an equal sign. In this case, you claim that the population mean (μ) equals 228.

593.

$H_0: p = 0.78$

A null hypothesis states the specific value of a population parameter, using an equal sign. In this case, you define that the parameter is the proportion (p) of the entire teenage population who owns cellphones.

594.

$H_0: p_1 - p_2 = 0$

Both proportions are assumed to be the same, so their difference is assumed to be zero. Another way to represent this statement is $H_0: p_1 = p_2$.

Null hypotheses are always written using population parameters, in this case p_1 and p_2.

595. $H_a: p < 0.70$

The alternative hypothesis is the hypothesis that you conclude if you have sufficient evidence to reject the null hypothesis. In this case, you hope to reject the null hypothesis that 70% of Americans think Congress is doing a good job ($H_0: p = 0.70$) and thus conclude the alternative hypothesis that the true percentage is lower, based on the data. The hypothesis test will help you decide.

596. $H_a: \mu > 2.5$

The alternative hypothesis is the hypothesis that you will conclude if you have sufficient evidence to reject the null hypothesis. In this case, you're hoping to reject the null hypothesis that the train trip takes an average of 2.5 hours ($H_0: \mu = 2.5$) and conclude the alternative that it takes longer than 2.5 hours, based on the data.

597. $H_a: p < 0.92$

The alternative hypothesis is the hypothesis that you will conclude if you have sufficient evidence to reject the null hypothesis. In this case, you're hoping to reject the null hypothesis that the airline's planes arrive early 92% of the time ($H_0: p = 0.92$) and conclude the alternative that the true proportion is lower, based on the data.

598. $H_a: \mu < 39$

The alternative hypothesis is the hypothesis that you will conclude if you have sufficient evidence to reject the null hypothesis. In this case, you're hoping to reject the null hypothesis that the car averages 39 miles per gallon ($H_0: \mu = 39$) and conclude the alternative that the true average is lower, based on the data.

599. $H_a: p > 0.005$

The alternative hypothesis is the hypothesis that you will conclude if you have sufficient evidence to reject the null hypothesis. In this case, you're hoping to reject the null hypothesis that only 1 in 200 computers has a mechanical malfunction ($H_0: p = 0.005$) and conclude the alternative hypothesis that the true proportion is higher, based on the data.

600. $H_a: p > 0.05$

The alternative hypothesis is the hypothesis that you will conclude if you have sufficient evidence to reject the null hypothesis. In this case, you're hoping to reject the null hypothesis that only 5% of patients are dissatisfied with their care ($H_0: p = 0.05$) and conclude the alternative hypothesis that the true proportion is higher, based on the data.

601. $H_a: \mu \neq 3{,}300$

The alternative hypothesis is the hypothesis that you will conclude if you have sufficient evidence to reject the null hypothesis, based on the data. In this case, you're hoping to reject the null hypothesis that American adults consume an average of 3,300 calories per day ($H_0: \mu = 3{,}300$) and conclude the alternative hypothesis that the true average is different, based on the data.

You're not given specific direction as to whether you believe the statistic underestimates or overestimates the truth, so you need the two-sided alternative.

602. $H_a: \mu > 1.8$

The alternative hypothesis is the hypothesis that you will conclude if you have sufficient evidence to reject the null hypothesis. In this case, you're hoping to reject the null hypothesis that adults watch an average of 1.8 hours of television per day ($H_0: \mu = 1.8$) and conclude the alternative hypothesis that the true average is higher, based on the data.

603. $H_a: \mu < 0.08$

The alternative hypothesis is the hypothesis that you will conclude if you have sufficient evidence to reject the null hypothesis. In this case, you're hoping to reject the null hypothesis that the investment company's clients make an average 8% return each year ($H_0: \mu = 0.08$) and conclude the alternative hypothesis that the true average return is lower, based on the data.

Note that interest rates are continuous measures. They're not binomial proportions, as if to say "8% of the time you get a return." As such, the 8% return is an average, not a proportion.

604. $H_a: p_1 - p_2 \neq 0.25$

The alternative hypothesis is the hypothesis that you will conclude if you have sufficient evidence to reject the null hypothesis. In this case, you're hoping to reject the null hypothesis that the difference in college attendance between the two cities is 25% ($H_0: p_1 - p_2 = 0.25$) and conclude the alternative that the true difference in proportions is some other value, based on the data.

You're not given specific direction as to whether you believe the statistic underestimates or overestimates the truth. The possibility exists that the true difference is greater than 25%, or perhaps that it is significantly less than that. So you need the two-sided alternative.

605. -2

You calculate the test statistic by subtracting the claimed value (from the null hypothesis) from the sample statistic and dividing by the standard error. In this example, the claimed value is 4, the sample statistic is 3, and the standard error is 0.5, so the test statistic is

$$\frac{3-4}{0.5} = -2$$

606. 1

You calculate the test statistic by subtracting the claimed value (from the null hypothesis) from the sample statistic and dividing by the standard error. In this example, the claimed value is 4, the sample statistic is 4.5, and the standard error is 0.5, so the test statistic is

$$\frac{4.5 - 4}{0.5} = 1$$

607. 2.4

You calculate the test statistic by subtracting the claimed value (from the null hypothesis) from the sample statistic and dividing by the standard error. In this example, the claimed value is 4, the sample statistic is 5.2, and the standard error is 0.5, so the test statistic is

$$\frac{5.2 - 4}{0.5} = 2.4$$

608. −0.8

You calculate the test statistic by subtracting the claimed value (from the null hypothesis) from the sample statistic and dividing by the standard error. In this example, the claimed value is 4, the sample statistic is 3.6, and the standard error is 0.5, so the test statistic is

$$\frac{3.6 - 4}{0.5} = -0.8$$

609. 0.1556

The *p*-value tells you the probability of a result being at or beyond your test statistic, if the null hypothesis is true. Because you have a two-tailed test, you will double the table probability to account for test results both above and below the claimed value. Note that the *p*-value is a probability and can never be negative.

In Table A-1 in the appendix, find 1.4 in column *z*. Then read across the *1.4* row to find the column labeled *0.02*. The number where row *1.4* intersects with column *0.02* is 0.9222.

Table A-1 shows the probability of a value below a given *z*-score. You want the probability of a value above 1.42, so subtract the table value from 1 (because total probability always equals 1): 1 − 0.9222 = 0.0778.

To get the *p*-value in this case, double this number because H$_a$ is "not equal to" and both the upper and the lower ends of the distribution must be included: 2(0.0778) = 0.1556.

610. 0.1188

The *p*-value tells you the probability of a result being at or beyond your test statistic, if the null hypothesis is true. Because you have a two-tailed test, you will double the table probability to account for test results both above and below the claimed value.

In Table A-1 in the appendix, find –1.5 in column z. Then read across the –1.5 row to find the column labeled 0.06. The number where row –1.5 intersects with column 0.06 is 0.0594.

To get the p-value in this case, double this number because H_a is "not equal to" and both the upper and the lower ends of the distribution must be included: 2(0.0594) = 0.1188.

611.   0.4532

The p-value tells you the probability of a result being at or beyond your test statistic, if the null hypothesis is true. Because you have a two-tailed test, you will double the table probability to account for test results in the extremes both above and below the assumed mean. Note that the p-value is a probability and can never be negative.

In Table A-1 in the appendix, find 0.7 in column z. Then read across the 0.7 row to find the column labeled 0.05. The number where row 0.7 intersects with column 0.05 is 0.7734.

Table A-1 shows the probability of a value below a given z-score. You want the probability of a value above 0.75, so subtract the table value from 1 (because total probability always equals 1): 1 – 0.7734 = 0.2266.

To get the p-value in this case, double this number because H_a is "not equal to" (two-sided) and both the upper and the lower ends of the distribution must be included: 2(0.2266) = 0.4532.

612.   0.4180

The p-value tells you the probability of a result being at or beyond your test statistic, if the null hypothesis is true. Because you have a two-tailed test, you will double the table probability to account for test results in the extremes both above and below the assumed mean. Note that the p-value is a probability and can never be negative.

In Table A-1 in the appendix, find –0.8 in column z. Then read across the –0.8 row to find the column labeled 0.01. The number where row –0.8 intersects with column 0.01 is 0.2090.

To get the p-value in this case, double this number because H_a is "not equal to" (two-sided) and both the upper and the lower ends of the distribution must be included: 2(0.2090) = 0.4180.

613.   Fail to reject H_0 because the p-value of your result is greater than alpha.

To reject the null hypothesis at the alpha = 0.05 level, you need a test statistic with a p-value of less than 0.05.

Your alternative hypothesis is directional (>), so you'll reject the null hypothesis only if your test statistic is positive and has a p-value of less than the alpha level of 0.05.

Using Table A-1 in the appendix, find the value where the 1.5 row intersects the 0.01 column. This value of 0.9345 is the probability of a value being less than 1.51. To find the probability of a value being greater than 1.51, subtract the table value from 1 (because total probability is always 1):

$$1 - 0.9345 = 0.0655$$

This value is the p-value of your test statistic of 1.51, if the null hypothesis is true and you use a single-tailed test (your alternative hypothesis is that the population proportion is greater than 0.45).

This p-value is greater than your alpha level of 0.05, so you will fail to reject the null hypothesis when the alpha level is 0.05. You data didn't provide enough evidence to reject the null hypothesis.

---

614. Reject H_0 because the p-value of your result is less than alpha.

To reject the null hypothesis at the alpha = 0.10 level, you need a test statistic with a p-value of less than 0.10.

Your alternative hypothesis is directional (>), so you'll reject the null hypothesis only if your test statistic is positive and has a p-value of less than the alpha level of 0.10.

Using Table A-1 in the appendix, find the value where the *1.5* row intersects the *0.01* column. This value of 0.9345 is the probability of a value being less than 1.51. To find the probability of a value being greater than 1.51, subtract the table value from 1 (because total probability is always 1):

$$1 - 0.9345 = 0.0655$$

This value is the p-value of your test statistic of 1.51, if the null hypothesis is true and you use a single-tailed test (your alternative hypothesis is that the population proportion is greater than 0.45).

This p-value is less than your alpha level of 0.10, which means that your observed test statistic was unlikely, assuming that the original claim was true. So you will reject the null hypothesis at the 0.10 alpha level.

---

615. Fail to reject H_0 because the p-value of your result is greater than alpha.

To reject the null hypothesis at the alpha = 0.01 level, you need a test statistic with a p-value of less than 0.01.

Your alternative hypothesis is directional (>), so you'll reject the null hypothesis only if your test statistic is positive and has a p-value of less than the alpha level of 0.01.

Using Table A-1 in the appendix, find the value where the *1.9* row intersects the *0.08* column. This value of 0.9761 is the probability of a value being less than 1.98. To find the probability of a value being greater than 1.98, subtract the table value from 1 (because total probability is always 1):

$$1 - 0.9761 = 0.0239$$

This value is the p-value of your test statistic of 1.98, if the null hypothesis is true and you use a single-tailed test (your alternative hypothesis is that the population proportion is greater than 0.45).

This p-value is greater than your alpha level of 0.01, so you will fail to reject the null hypothesis at the alpha level of 0.01. Your data didn't provide enough evidence to reject the null hypothesis.

616.

Reject H_0 because the p-value of your result is less than alpha.

To reject the null hypothesis at the alpha = 0.05 level, you need a test statistic with a p-value of less than 0.05.

Your alternative hypothesis is directional (>), so you'll reject the null hypothesis only if your test statistic is positive and has a p-value of less than the alpha level of 0.05.

Using Table A-1 in the appendix, find the value where the *1.9* row intersects the *0.08* column. This value of 0.9761 is the probability of a value being less than 1.98. To find the probability of a value being greater than 1.98, subtract the table value from 1 (because total probability is always 1):

$$1 - 0.9761 = 0.0239$$

This value is the p-value of your test statistic of 1.98, if the null hypothesis is true and you use a single-tailed test (your alternative hypothesis is that the population proportion is greater than 0.45).

This p-value is less than your alpha level of 0.05, which means that your observed test statistic was unlikely, assuming the original claim was true. So you will reject the null hypothesis at the alpha level of 0.05.

617.

Fail to reject H_0 because your test statistic is negative, while the alternative hypothesis is that the population probability is greater than 0.45.

You don't have to do any calculations for this problem. Your alternative hypothesis ($p > 0.45$) indicates that your test statistic must also be positive to reject the null hypothesis. That is, you can reject the claim only if your observed value is significantly greater than 0.45. However, in this example, the test statistic is negative, so you won't reject the null hypothesis.

You could still calculate the p-value to verify. To reject the null hypothesis at the alpha = 0.05 level, you need a test statistic with a p-value of less than 0.05. Using Table A-1 in the appendix, find the value where the -1.9 row intersects the *0.08* column. This value of 0.0239 is the probability of a value being less than −1.98. To find the probability of a value being greater than −1.98, subtract the table value from 1 (because total probability is always 1):

$$1 - 0.0239 = 0.9761$$

This value is the p-value of your test statistic of −1.98, if the null hypothesis is true and you use a single-tailed test (your alternative hypothesis is that the population proportion is greater than 0.45).

This p-value is much greater than your alpha level of 0.05, so you will fail to reject the null hypothesis at the 0.05 alpha level. Your data didn't provide enough evidence to reject the null hypothesis.

618.

Fail to reject H_0 because your test statistic is negative, while the alternative hypothesis is that the population probability is greater than 0.45.

You don't have to do any calculations for this problem. Your alternative hypothesis ($p > 0.45$) indicates that your test statistic must also be positive to reject the null hypothesis. That is, you can reject the claim only if your observed value is significantly greater than 0.45. However, in this example, the test statistic is negative, so you won't reject the null hypothesis.

You could still calculate the p-value to verify. To reject the null hypothesis at the alpha = 0.01 level, you need a test statistic with a p-value of less than 0.01. Using Table A-1 in the appendix, find the value where the –3.0 row intersects the 0.00 column. This value of 0.0013 is the probability of a value being less than –3.0. To find the probability of a value being greater than –3.0, subtract the table value from 1 (because total probability is always 1):

$$1 - 0.0013 = 0.9987$$

This value is the p-value of your test statistic of –3.0, if the null hypothesis is true and you use a single-tailed test (your alternative hypothesis is that the population proportion is greater than 0.45).

This p-value is much greater than your alpha level of 0.01, so you will fail to reject the null hypothesis at the 0.01 alpha level. Your data didn't provide enough evidence to reject the null hypothesis.

619. There is a 1% chance of getting a value at least that extreme, if the null hypothesis is true.

The p-value of a test statistic tells you the probability of getting a specific, observed value or a value more extreme (farther from the claimed value), assuming the null hypothesis is true.

620. Choices (A) and (C) (There is a 10% chance that you will reject the null hypothesis when it is true; you should reject the null hypothesis if your test statistic has a p-value of 0.10 or less.)

An alpha level of 0.10 means that you have a 10% chance that you'll randomly reject the null hypothesis when it's actually true and also that you'll reject the null hypothesis if your test statistic has a p-value of 0.10 or less.

621. Fail to reject the null hypothesis.

You reject the null hypothesis if the p-value for your test statistic is less than the alpha level. It's never correct to say that you accept the null hypothesis (it implies that you know the null hypothesis is factually true) or that you reject the alternative hypothesis (because it's the null, not the alternative hypothesis, that you're putting to the test).

622. highly statistically significant

A test result with a p-value of 0.001, when the alpha level is 0.05, is typically stated as being "highly statistically significant," because it's so much smaller than the required value of 0.05.

623. Reject H_0 because the results are statistically significant.

The p-value is well below the significance level $\alpha = 0.05$. The results support rejecting H_0. Alpha values and hypotheses should always be set before results are calculated. It's incorrect to adjust either after the data has been gathered, in the hope of getting a significant result.

624. B. 0.01

H_0 is rejected if the p-value is less than the significance level ($\alpha = 0.02$).

625. E. 0.04

H_0 is rejected if the p-value is less than the significance level ($\alpha = 0.05$).

626. Reject H_0.

You reject H_0 because the p-value is less than the significance level (α level) of 0.01.

627. Reject H_0.

You reject H_0 because the p-value is less than the significance level (α level) of 0.03.

628. Fail to reject H_0.

The p-value of 0.06 is greater than the significance level of $\alpha = 0.05$. You must fail to reject H_0.

629. Fail to reject H_0.

The p-value of 0.42 is much greater than the significance level of $\alpha = 0.05$, so you fail to reject H_0.

630. Fail to reject H_0.

The p-value of 0.2 is much greater than the significance level of $\alpha = 0.02$, so you fail to reject H_0.

631. Fail to reject H_0.

The p-value of 0.018 is greater than the significance level of $\alpha = 0.01$, so you fail to reject H_0.

632. impossible to determine from the given information

Even though the p-value is less than the significance level of $\alpha = 0.05$, notice that the test statistic is negative. This means that the sample mean is *less* than the claimed population mean of 9.65. However, the alternative hypothesis states that the sample mean is expected to be *greater* than the population mean. This is a contradiction that tells you either the test statistic or the p-value was calculated incorrectly. You should return to the original data and double-check the calculations.

633. You should fail to reject H_0.

If the alpha level is 0.05 and the p-value of the test statistic is 0.07, you should fail to reject H_0. ***Note:*** It's never correct to say that you *accept* H_0.

634.   C. rejecting the null hypothesis when it is true

You make a Type I error when the null hypothesis is true but you reject it. This error is just by random chance, because if you knew for a fact that the null was true, you certainly wouldn't reject it. But there's a slim chance (alpha level) that it could happen.

A Type I error is sometimes referred to as a "false alarm," because rejecting the null hypothesis is like sounding an alarm to change an established value. If the null is true, then there's no need for such a change.

635.   D. failing to reject the null hypothesis when it is false

You make a Type II error when the null hypothesis is false but you fail to reject it because your data couldn't detect it, just by chance.

This error is sometimes referred to as "missing out on a detection." The claim really was wrong, but you didn't get a random sample that would provide enough evidence (small enough p-value) to reject it.

636.   0.01

The alpha level (or significance level) of 0.01 indicates the probability of a Type I error — that is, the error of rejecting the null hypothesis when it's actually true.

637.   impossible to tell without further information

The probability of a Type II error isn't directly related to the alpha level in terms of its formula or calculations. Instead, it's determined by a combination of several factors, including sample size. However, the behavior of alpha is related to the behavior of beta (probability of Type II error). In general, they have an indirect relationship: The larger alpha is, the more likely you are to reject (regardless of the truth) and thus the less likely you are to make a Type II error.

638.   D. the probability of rejecting the null hypothesis when it is false

The power of the test describes the probability of rejecting the null hypothesis when it's false — that is, of making the correct decision when the null hypothesis is false. In that way, it's the opposite (or complement) of making a Type II error.

639.   E. Choices (B) and (C) (Having a random sample of data; having a large sample size.)

A random sample is necessary to try to get the most representative and unbiased data possible. Otherwise, a sample could be tampered with to have only data values that support the null hypothesis, forcing you to fail to reject H_0 no matter what.

A large sample size increases the power of the test and makes it more likely that you'll be able to correctly detect when H_0 is false.

640.

E. Choices (A), (B), and (C) (having a low significance level; having a random sample of data; having a large sample size)

A low significance level, such as $\alpha = 0.01$ or $\alpha = 0.05$, means that you'll reject only if the observed data provides a result that you'd consider very unlikely under the condition of the null hypothesis.

A random sample is necessary to try to get the most representative and unbiased data possible. Otherwise, a sample could be tampered with to have only extreme data values far from the claimed value in the null hypothesis, forcing you to reject H_0 regardless of the truth.

A large sample will reduce the variation by giving you a more accurate estimate of the truth than a smaller sample would.

The value of the p-value being low or high is out of your hands and up to the data. But a large, random sample will help ensure a more reliable p-value.

641.

the population standard deviation and a statement that soda consumption is normally distributed among U.S. teens

To run a z-test to see whether a sample mean differs from a population mean, you need the population standard deviation; you also need to know either that the characteristic of interest is normally distributed in the population or that the sample size is at least $n = 30$. Because the sample is small ($n = 15$), you need to know that soda consumption is normally distributed among U.S. teens and the population standard deviation of soda consumption.

642.

E. Choices (A) and (C) (whether the characteristic of interest is normally distributed in the population; the population mean and standard deviation)

In addition to needing the sample mean and sample size, you can run a z-test if you have information about the population of interest. You need the population mean and standard deviation and some knowledge of the behavior of the population. In this scenario, because the sample size is small ($n = 20$), you need to know that the characteristic of interest is normally distributed in the population.

643.

$H_0: \mu = 25$

The null hypothesis is always that the population parameter is *equal to* some specific value.

For a test of one population mean, the null hypothesis is that the population mean of interest is equal to a certain claimed value, which is 25 in this case.

644.

$H_a: \mu \neq 10$

The null hypothesis is always that the population mean is *equal to* the stated value. In this case, the researcher believes that the null hypothesis is wrong, but the researcher doesn't have any theory about whether the true population mean is higher or lower than that, so the alternative hypothesis is a *not equal to* hypothesis.

645.

$H_a: \mu > 5$

The computer store owner believes that her customers buy more than five flash drives per year on average. In other words, she believes that her customers' population average, μ, is greater than the claimed value of five. Therefore, the alternative hypothesis is a *greater than* hypothesis.

646.

$H_a: \mu < 3$

The man believes that the cost of dry-cleaning a shirt in his town is lower than the average amount of 3 dollars. Therefore, the alternative hypothesis is a *less than* hypothesis.

647.

0.0668

Using Table A-1, find −1.5 in the left-hand column, and then go across the row to the column for 0.00, where the value is 0.0668. This is the proportion of the curve area that's to the left of (less than) the test statistic value of z that you're looking up. In this case, the alternative hypothesis is a *less than* hypothesis, so you can read the p-value from the table without doing further calculations.

648.

0.1336

Using Table A-1, find −1.5 in the left-hand column, and then go across the row to the column for 0.00, where the value is 0.0668. This is the proportion of the curve area that's to the left of (less than) the value of z you're looking up. In this case, the alternative hypothesis is a *not equal to* hypothesis, so you double the outlying tail quantity (area below the z-value of −1.5) to get the p-value.

649.

0.0456

Using Table A-1, find −2.0 in the left-hand column, and then go across the row to the column for 0.0, where the value is 0.0228. This is the proportion of the curve area that's to the left of (less than) the value of z you're looking up. In this case, the alternative hypothesis is a *not equal to* hypothesis, so you double the outlying tail quantity (area below the z-value of −2.0) to get the p-value.

650.

0.2714

Using Table A-1, find 1.1 in the left-hand column, then go across the row to the column for 0.0, where the value is 0.8643. This is the area under the curve to the left of the z value of 1.1. Because total area under the curve equals 1, the area above z in this case is $1 - 0.8643 = 0.1357$.

For a *not equal to* alternative hypothesis, you double the value of the outlying tail area: $p = 2(0.1357) = 0.2714$.

651.

3.54

Start by identifying the sample mean $\bar{x}$. In this case, you're told that the same mean is 115 degrees.

Next, calculate the standard error. For a one-sample z-test, the standard error is the population standard deviation, σ_X, divided by the square root of the sample size, n:

$$\sigma_{\bar{x}} = \frac{\sigma_X}{\sqrt{n}}$$

$$= \frac{10}{\sqrt{50}}$$

$$= 1.4142$$

To get the z-test statistic, find the difference between the sample mean, $\bar{x}$, and the claimed population mean, μ_0, and divide that by the standard error, $\sigma_{\bar{x}}$:

$$z = \frac{\bar{x} - \mu_0}{\sigma_{\bar{x}}}$$

$$= \frac{115 - 110}{1.4142}$$

$$= 3.5355678, \text{ or } 3.54 \text{ (rounded)}$$

652.

−3.4031

First, find the standard error by dividing the population standard deviation, σ_X, by the square root of the sample size, n:

$$\sigma_{\bar{x}} = \frac{\sigma_X}{\sqrt{n}}$$

$$= \frac{26.52}{\sqrt{40}}$$

$$= 4.19318$$

Then, calculate the z-statistic by subtracting the claimed population mean, μ_0, from the sample mean, $\bar{x}$, and dividing the result by the standard error, $\sigma_{\bar{x}}$:

$$z = \frac{\bar{x} - \mu_0}{\sigma_{\bar{x}}}$$

$$= \frac{172.12 - 186.39}{4.19318}$$

$$= \frac{-14.27}{4.19318}$$

$$= -3.403145, \text{ or } 3.4031 \text{ (rounded)}$$

653.

0.0132

First, set up the null and alternative hypotheses. The null hypothesis is always an *equal to* hypothesis:

$$H_0: \mu = 40$$

Because the research director believes that customers hope to use a pen for fewer than 40 days, you use a *less than* alternative hypothesis:

$$H_a: \mu < 40$$

Next, identify the sample mean, which is 36 in this case.

Then, compute the standard error by dividing the population standard deviation, σ_X, by the square root of the sample size, n:

$$\sigma_{\bar{x}} = \frac{\sigma_X}{\sqrt{n}}$$

$$= \frac{9}{\sqrt{25}}$$

$$= 1.8$$

Now, find the z-statistic by subtracting the claimed population mean, μ_0, from the sample mean, $\bar{x}$, and dividing the result by the standard error, $\sigma_{\bar{x}}$:

$$z = \frac{\bar{x} - \mu_0}{\sigma_{\bar{x}}}$$

$$= \frac{36 - 40}{1.8}$$

$$= -2.222\bar{2}$$

Rounded to two decimal places (the degree of precision in Table A-1 in the appendix), you get –2.22. Using Table A-1, find –2.2 in the left-hand column, and then go across the row to the column for 0.02, where the value is 0.0132. This is the area to the left of –2.22. Because the alternative hypothesis is a *less than* hypothesis, the p-value is the same as the value you find in the table: 0.0132.

---

654.   0.1587

First, set up the null hypothesis and alternative hypotheses. The null hypothesis is that the mean of the farmer's population of bush yields will be the same as the claimed mean:

$$H_0: \mu = 3$$

Because the farmer expects to get less fruit than she read about, you use a *less than* alternative hypothesis:

$$H_a: \mu < 3$$

Next, identify the sample mean, which is 2.9 in this case.

Then, compute the standard error by dividing the population standard deviation, σ_X, by the square root of the sample size, n:

$$\sigma_{\bar{x}} = \frac{\sigma_X}{\sqrt{n}}$$

$$= \frac{1}{\sqrt{100}}$$

$$= 0.1$$

Now, find the z-statistic by subtracting the claimed population mean, μ_0, from the sample mean, $\bar{x}$, and dividing the result by the standard error, $\sigma_{\bar{x}}$:

$$z = \frac{\bar{x} - \mu_0}{\sigma_{\bar{x}}}$$

$$= \frac{2.9 - 3}{0.1}$$

$$= -1$$

Use Table A-1 in the appendix, find the value −1.0 in the left-hand column, and then go across the row to the column for 0.0. This value is the area under the curve to the left (less than) of this value of z, 0.1587. Because you have a *less than* alternative hypothesis, this value is also the p-value.

655.

Reject the null hypothesis.

Any time the p-value is less than the alpha level (α, also known as the significance level), you reject the null hypothesis. In this case, you're given a p-value of 0.02 and a significance level of α = 0.05, which is enough information to reject the null hypothesis for this study.

656.

The null hypothesis can't be rejected.

First, set up the null and alternative hypotheses. The null hypothesis is that business travelers will have the same population mean as the average airplane passenger:

$$H_0: \mu = 45$$

The alternative hypothesis is that business travelers carry less that the claimed value of luggage, so you use a *less than* hypothesis:

$$H_a: \mu < 45$$

Next, identify the sample mean, which is 44.5 in this case.

Then, compute the standard error by dividing the population standard deviation, σ_X, by the square root of the sample size, n:

$$\sigma_{\bar{x}} = \frac{\sigma_X}{\sqrt{n}}$$

$$= \frac{10}{\sqrt{250}}$$

$$= 0.6325$$

Now, find the z-statistic by subtracting the claimed population mean, μ_0, from the sample mean, $\bar{x}$, and dividing the result by the standard error, $\sigma_{\bar{x}}$:

$$z = \frac{\bar{x} - \mu_0}{\sigma_{\bar{x}}}$$

$$= \frac{44.5 - 45}{0.6325}$$

$$= -0.790513834$$

Use a Z-table, such as Table A-1 in the appendix, to find the area under the normal curve to the left of the computed test statistic value. Round the test statistic value to two decimal places (to −0.79). Find the value of −0.7 in the left-hand column, and then go across the row to the column for 0.09. Table A-1 specifies that an area of 0.2148 lies to the left of this value. For a *less than* alternate hypothesis, the p-value is the same as the table value: 0.2148.

Finally, compare the p-value (0.2148) to the stated significance level (0.05). In this case, the p-value is bigger than the significance level, so you *can't reject* the null hypothesis on the basis of this data.

657.

The shopkeeper doesn't have enough evidence to conclude that his rent is cheaper on average.

The shopkeeper can't conclude that the rent on average is lower than $2.00 per month because the mean in his sample ($3.00) is higher than that amount.

658.

0.0082

First, set up the null and alternative hypotheses. The null hypothesis is that the doctor's patients have the same average temperature as humans in general:

$H_0: \mu = 98.6$

Because the doctor believes that her patients have a higher temperature than humans on average, you use a *greater than* alternative hypothesis:

$H_a: \mu > 98.6$

Next, identify the sample mean, which is 98.8 in this case.

Then, compute the standard error by dividing the population standard deviation, σ_X, by the square root of the sample size, n:

$$\sigma_{\bar{x}} = \frac{\sigma_X}{\sqrt{n}}$$

$$= \frac{0.5}{\sqrt{36}}$$

$$= \frac{0.5}{6}$$

$$= 0.0833$$

Now, find the z-statistic by subtracting the claimed population mean, μ_0, from the sample mean, $\bar{x}$, and dividing the result by the standard error, $\sigma_{\bar{x}}$:

$$z = \frac{\bar{x} - \mu_0}{\sigma_{\bar{x}}}$$

$$= \frac{98.8 - 98.6}{0.0833}$$

$$= 2.40096$$

Use a Z-table, such as Table A-1 in the appendix, find the value of 2.4 in the left-hand column, and then go across to the column for 0.00. The value of 0.9918 is the area to the left of the z-value of 2.40. Because you have a *greater than* alternative hypothesis, you need to subtract the table value from 1 (the total area under the curve) to get the p-value: $1 - 0.9918 = 0.0082$.

659.

0.0606

First, set up the null and alternative hypotheses. The null hypothesis is that the mean of the population of interest (the Northeastern division customers) is the same as the claimed mean:

$H_0: \mu = 5$

In this case, the researcher suspects that the population mean for the Northeastern division is higher than 5, so you use a *greater than* alternative hypothesis:

$H_a: \mu > 5$

Next, identify the sample mean, which is 5.1 in this case.

Then, compute the standard error by dividing the population standard deviation, σ_X, by the square root of the sample size, n:

$$\sigma_{\bar{x}} = \frac{\sigma_X}{\sqrt{n}}$$

$$= \frac{0.5}{\sqrt{60}}$$

$$= \frac{0.5}{7.746}$$

$$= 0.0645$$

Now, find the z-statistic by subtracting the claimed population mean, μ_0, from the sample mean, $\bar{x}$, and dividing the result by the standard error, $\sigma_{\bar{x}}$:

$$z = \frac{\bar{x} - \mu_0}{\sigma_{\bar{x}}}$$

$$= \frac{5.1 - 5.0}{0.0645}$$

$$= \frac{0.1}{0.0645}$$

$$= 1.5504$$

Use a Z-table, such Table A-1 in the appendix, find the value of 1.5 in the left-hand column, and then go across to the column for 0.05. The value of 0.9394 is the area to the left of the z-value of 1.55. Because you have a *greater than* alternative hypothesis, you need to subtract the curve area from 1 to get the p-value: $1 - 0.9394 = 0.0606$.

---

660. Fail to reject the null hypothesis that the average words per minute is equal to 20.

First, set up the null and alternative hypotheses. The null hypothesis is that the mean of the population of interest (the employees of the manager's branch) is equal to the claimed mean:

$$H_0: \mu = 20$$

In this case, you use a *greater than* alternative hypothesis because the manager believes his employees have a speed greater than the claimed average:

$$H_a: \mu > 20$$

Next, identify the sample mean, which is 20.5 words per minute.

Then, compute the standard error by dividing the population standard deviation, σ_X, by the square root of the sample size, n:

$$\sigma_{\bar{x}} = \frac{\sigma_X}{\sqrt{n}}$$

$$= \frac{3}{\sqrt{30}}$$

$$= 0.5477$$

Now, find the z-statistic by subtracting the claimed population mean, μ_0, from the sample mean, $\bar{x}$, and dividing the result by the standard error, $\sigma_{\bar{x}}$:

$$z = \frac{\bar{x} - \mu_0}{\sigma_{\bar{x}}}$$

$$= \frac{20.5 - 20}{0.5477}$$

$$= 0.9129$$

Use a Z-table, such Table A-1 in the appendix, find the value of 0.9 in the left-hand column, and then go across to the column for 0.01. The value of 0.8186 is the area to the left of the z-value of 0.91. Because you have a *greater than* alternative hypothesis, the p-value is 1 minus the table value: $1 - 0.8186 = 0.1814$.

Compare the p-value to the significance level and reject the null hypothesis only if the p-value is less than the significance level. Here, p-value $= 0.1814$, which is greater than the significance level of 0.05. This means you *fail to reject the null hypothesis*.

661.

Fail to reject H_0.

First, set up the null and alternative hypotheses. In this case, μ_0 represents the current average amount of glaze put on a single vase by workers in the workshop. The target value of 2 ounces plays the same role as a claimed value might. The potter wants to know whether her current value is significantly above 2 ounces.

$$H_0: \mu = 2$$

Because the potter believes her workshop uses more than 2 ounces, the alternative hypothesis is a *greater than* hypothesis:

$$H_a: \mu > 2$$

Next, identify the sample mean, which is 2.3 ounces in this case.

Then, compute the standard error by dividing the population standard deviation, σ_X, by the square root of the sample size, n:

$$\sigma_{\bar{x}} = \frac{\sigma_X}{\sqrt{n}}$$

$$= \frac{0.8}{\sqrt{30}}$$

$$= 0.14606$$

Now, find the z-statistic by subtracting the claimed population mean, μ_0, from the sample mean, $\bar{x}$, and dividing the result by the standard error, $\sigma_{\bar{x}}$:

$$z = \frac{\bar{x} - \mu_0}{\sigma_{\bar{x}}}$$

$$= \frac{2.3 - 2}{0.14606}$$

$$= 2.054$$

Use a Z-table, such as Table A-1 in the appendix, find the value of 2.0 in the left-hand column, and then go across to the column for 0.05. The value of 0.9798 is the area to the left of the z-value of 2.05.

Now, find the p-value by subtracting the curve area from 1, because you have a *greater than* alternative hypothesis: $1 - 0.9798 = 0.0202$.

Finally, compare this value to your significance level. The p-value (0.0202) is greater than the significance level (0.01), so you *fail to reject the null hypothesis*.

662. Reject H_0.

First, set up the null and alternative hypotheses. The null hypothesis is that the mean of the population of interest (the density of the tissue cultures) is equal to the claimed mean:

$$H_0: \mu = 0.0047$$

The researcher suspects that her samples are heavier than the norm, so you have a *greater than* alternative hypothesis:

$$H_a: \mu > 0.0047$$

Next, identify the sample mean, 0.005 in this case, which is greater than the claimed value, so proceed with testing.

Then, compute the standard error by dividing the population standard deviation, σ_X, by the square root of the sample size, n:

$$\sigma_{\bar{X}} = \frac{\sigma_X}{\sqrt{n}}$$
$$= \frac{0.00047}{\sqrt{40}}$$
$$= 0.00007431$$

Now, find the z-statistic by subtracting the claimed population mean, μ_0, from the sample mean, $\bar{x}$, and dividing the result by the standard error, $\sigma_{\bar{X}}$:

$$z = \frac{\bar{x} - \mu_0}{\sigma_{\bar{X}}}$$
$$= \frac{0.005 - 0.0047}{0.00007431}$$
$$= 4.037$$

This rounds to 4.04, which is above the highest value in Table A-1 in the appendix, so the probability of a value at least this extreme is less than 0.0001 (the area under the curve above the highest table value of 3.69). A p-value of 0.0001 is less than the significance level of 0.001, so you should reject the null hypothesis.

The researcher has enough evidence to conclude that the tissue cultures in her lab are denser than average. And because the p-value is so small (0.0001), she can say these results are highly significant.

663. The farmer can't reject the null hypothesis at a significance level of 0.05.

First, set up the null and alternative hypotheses. The null hypothesis is that the farmer's hens lay an average of 15 eggs per month:

$$H_0: \mu = 15$$

The alternative hypothesis is that the average farmer's hens is a different value (a *not equal to* hypothesis):

$$H_a: \mu \neq 15$$

Then, compute the standard error by dividing the population standard deviation, σ_X, by the square root of the sample size, n:

$$\sigma_{\bar{x}} = \frac{\sigma_X}{\sqrt{n}}$$

$$= \frac{5}{\sqrt{30}}$$

$$= 0.9129$$

Now, find the z-statistic by subtracting the claimed population mean, μ_0, from the sample mean, $\bar{x}$, and dividing the result by the standard error, $\sigma_{\bar{x}}$:

$$z = \frac{\bar{x} - \mu_0}{\sigma_{\bar{x}}}$$

$$= \frac{16.5 - 15}{0.9129}$$

$$= 1.6431$$

Now, find the associated area under the normal curve to the left of the value you got for z, using a Z-table, such as Table A-1 in the appendix. Find the value of 1.6 in the left-hand column, and go across the row to the column for 0.04. The value is 0.9495.

Because the alternative hypothesis is a *not equal to* hypothesis and the test statistic value is positive, you need to subtract the value you found in the table from 1 $(1 - 0.9495 = 0.0505)$ and then double that result to get the p-value: $2(0.0505) = 0.101$.

Because the p-value is larger than the significance level, the farmer can't reject the null hypothesis. In other words, he can't say that his hens lay eggs any differently than the norm.

664. Reject the null hypothesis.

First, set up the null and alternative hypotheses. The null hypothesis is that the average number of boxes used by Chicago families is equal to the national average:

$$H_0: \mu = 110$$

Because the Chicago company just wants to see how it compares to the national average in terms of boxes used when moving, it's interested in being either higher or lower on average. So you use a *not equal to* alternative hypothesis:

$$H_a: \mu \neq 110$$

Then, compute the standard error by dividing the population standard deviation, σ_X, by the square root of the sample size, n:

$$\sigma_{\bar{x}} = \frac{\sigma_X}{\sqrt{n}}$$

$$= \frac{30}{\sqrt{80}}$$

$$= 3.3541$$

Now, find the z-statistic by subtracting the claimed population mean, μ_0, from the sample mean, $\bar{x}$, and dividing the result by the standard error, $\sigma_{\bar{x}}$:

$$z = \frac{\bar{x} - \mu_0}{\sigma_{\bar{x}}}$$

$$= \frac{103 - 110}{3.3541}$$

$$= -2.087$$

Now find the associated area under the normal curve to the left of the value you got for z, using a Z-table, such as Table A-1 in the appendix. Find the value of –2.0 in the left-hand column, and go across the row to the column for 0.09 (rounding to two decimals). The value is 0.0183, which is the area under the curve to the left of this z-value.

Because you have a *not equal to* alternative hypothesis and the test statistic value is negative, you need to double the curve area to get the p-value: 2(0.0183) = 0.0366.

This p-value is lower than the significance level of 0.05, so you *reject the null hypothesis*. In other words, you reject the claim that the average number of boxes used by Chicago families is equal to the national average. (Because the z-value is negative, the average box count is probably less than the national average.)

---

665. Fail to reject the null hypothesis.

First, set up the null and alternative hypotheses. The null hypothesis is that the average American family who uses a car to go on vacation travels an average of 382 miles from home:

$$H_0: \mu = 382$$

Because the researcher is interested in seeing whether a difference occurs in average travel distance, the alternative hypothesis is a *not equal to* hypothesis:

$$H_a: \mu \neq 382$$

Then, compute the standard error by dividing the population standard deviation, σ_X, by the square root of the sample size, n:

$$\sigma_{\bar{x}} = \frac{\sigma_X}{\sqrt{n}}$$

$$= \frac{150}{\sqrt{30}}$$

$$= 27.3861$$

Now, find the z-statistic by subtracting the claimed population mean, μ_0, from the sample mean, $\bar{x}$, and dividing the result by the standard error, $\sigma_{\bar{x}}$:

$$z = \frac{\bar{x} - \mu_0}{\sigma_{\bar{x}}}$$

$$= \frac{398 - 382}{27.3861}$$

$$= 0.5842$$

Now find the associated area under the normal curve to the left of the value you got for z, using a Z-table, such as Table A-1 in the appendix. Find the value of 0.5 in the left-hand column, and go across the row to the column for 0.08. The value is 0.7190, which is the area to the left of this value. Because you have a *not equal to* alternative hypothesis, you find the p-value by subtracting the area under the curve from 1 and doubling the result: p-value = 2(1 − 0.7190) = 0.562.

This value is greater than the significance level of 0.05, so you fail to reject the null hypothesis. You can't say families who vacation by car with dogs travel a different distance on average than other families, based on this data.

666. $H_0: \mu = 120; H_a: \mu < 120$

The question is about the average number of minutes U.S. teenagers spend texting, which represents a population. So the symbol for the population mean, μ, is in the null and alternative hypotheses (and not the symbol for the sample mean, $\bar{x}$). The claim is that the mean equals 120, so this is the value in the null hypothesis. You believe it's less than that, so your alternative hypothesis is the *less than* alternative.

667. $H_0: \mu = 250; H_a: \mu > 250$

The null and alternative hypotheses are always about a population value (in this case, the population mean), not about a sample value. So the symbols in the null and alternative hypothesis should be μ (population mean), not $\bar{x}$ (sample mean). In addition, the actual numbers in the null and alternative hypotheses should reference the original claim regarding the population mean (250 calories). Finally, because the researcher believes that the pizza on the college campus where he works has more calories, the alternative hypothesis is a *greater than* hypothesis.

668.   A.   The population standard deviation is unknown.

If the population has a normal distribution but the population standard deviation is unknown, you use a *t*-test (otherwise, use a *z*-test). The sample standard deviation comes from your data, so you'll always know its value.

669.   A.   The population standard deviation isn't known.

When the population standard deviation isn't known but can be estimated from the sample standard deviation, you should use a *t*-test instead of using the *Z*-distribution to test a hypothesis about a single population mean.

670.   The student should do a single population *t*-test.

Whenever the population standard deviation isn't known but the sample standard deviation allows an estimate of it, you should do a *t*-test for a single population mean rather than use the *Z*-distribution.

671. $H_0: \mu_1 = \mu_0$

The null hypothesis is that the mean of the population from which the sample was drawn is equal to the claimed mean.

672. $H_a: \mu_1 < \mu_0$

Because the student believes that his friends spend less time than is claimed, the alternative hypothesis is a *less than* hypothesis.

673. She could make the conclusion under other conditions, but the sample from her store doesn't represent all stores in this retail chain.

An important condition of any hypothesis test is that the data going into the calculations is reliable and valid. In this case, because the question resolves around the average price of a certain hair product from a national chain, whose stores are throughout the United States, the data collected must also represent this population. In this case, the data represent 30 bottles of this product taken from her store only. So any conclusions she makes from a hypothesis test should be rendered invalid.

674. $H_0: \mu = 3.5; H_a: \mu < 3.5$

The null hypothesis is that the population mean from which the sample was drawn is equal to the claimed population mean. The alternative hypothesis in this case is that the dentist's patients experience less pain than average, so the alternative is a *less than* hypothesis.

675. −2

To find the value for the test statistic, t, use the t-test formula:

$$t = \frac{\bar{x} - \mu_0}{s/\sqrt{n}}$$

where $\bar{x}$ is the sample mean, μ_0 is the claimed population mean, s is the sample standard deviation, and n is the sample size. Substitute these values from the question into the formula and solve:

$$t = \frac{30 - 35}{10/\sqrt{16}}$$

$$= \frac{-5}{2.5}$$

$$= -2$$

676. −2.733

To find the value for the test statistic, t, use the t-test formula:

$$t = \frac{\bar{x} - \mu_0}{s/\sqrt{n}}$$

where $\bar{x}$ is the sample mean, μ_0 is the claimed population mean, s is the sample standard deviation, and n is the sample size. Substitute these values from the question into the formula and solve:

$$t = \frac{5.2 - 6.3}{1.8/\sqrt{20}}$$

$$= \frac{-1.1}{0.4025}$$

$$= -2.733$$

677. 1.761310

First, figure out the degrees of freedom, which is one less than the sample size n: $df = 15 - 1 = 14$.

Because the researcher believes that population of interest has a mean greater than 6.1, you use a *greater than* alternative hypothesis, so 0.05 of the area under the curve should be in the upper tail. Using a *t*-table, such as Table A-2 in the appendix, find the row for 14 degrees of freedom and the column for 0.05. The critical value is 1.761310.

678. 2.26216

First, figure out the degrees of freedom, which is one less than the sample size n: $df = 10 - 1 = 9$.

Because the research believes that the sampled population of interest has a mean that differs from the claimed value, you use a *not equal to* alternative hypothesis, so half of the confidence level will be in each tail: $0.05/2 = 0.025$.

Using a *t*-table, such as Table A-2 in the appendix, find the row for 9 degrees of freedom and the column for 0.025. The critical values are 2.26216 and –2.26216; you reject the null hypothesis if your test statistic is outside the range of –2.26216 to 2.26216.

679. She would reject H_0.

The test statistic is –2.733, which is in the correct direction for a *less than* alternative hypothesis. Compare it to the critical value for a *less than* alternative hypothesis with a confidence level of 0.05, with $n - 1 = 20 - 1 = 19$ degrees of freedom. Using a *t*-table, such as Table A-2 in the appendix, find the row for 19 degrees of freedom and the column for 0.05. The value you find is 1.729133. However, because this is a left-tailed test, the critical value is –1.729133 in this case.

Because Table A-2 contains only positive values, compare the absolute value of the test statistic, 2.733, with the critical value. The absolute value of the test statistic is greater, so reject the null hypothesis. The researcher has significant evidence to suggest that mothers get less sleep than the average person.

680. The test statistic t is larger than the positive critical value.

First, figure out the degrees of freedom, which is one less than the sample size, n: $df = 17 - 1 = 16$.

Because this is a *not equal to* alternative hypothesis (the question specified that the researcher is interested in a difference "in either direction,") half the confidence level will be in the upper tail and half in the lower tail (0.05 in each).

Then, find the critical value from a *t*-table, such as Table A-2 in the appendix. Find the row for 16 degrees of freedom and the column for 0.05. The critical values are –1.745884 and 1.745884.

Next, find the test statistic t, using the basic formula for t:

$$t = \frac{\bar{x} - \mu_0}{s/\sqrt{n}}$$

where $\bar{x}$ is the sample mean, μ_0 is the population mean, s is the sample standard deviation, and n is the sample size.

$$t = \frac{\bar{x} - \mu_0}{s/\sqrt{n}}$$

$$= \frac{0.0123 - 0.0112}{0.0019/\sqrt{17}}$$

$$= 2.387061$$

This is larger than the positive critical value of t.

---

681. 0.20 < p-value < 0.50

First, find the degrees of freedom, which is one less than the sample size, n: $df =$ 11 − 1 = 10.

Then, using Table A-2 in the appendix, find the row for 10 degrees of freedom. Read from the left to find the last value that's *smaller than* your test statistic t. In this case, that's 0.699812, which corresponds to a single-tail area of 0.25. Because the alternative is two-sided, you have to double that to get 0.50 as one bound of the p-value.

On the same row, read from the left to find the first value that's *greater than* the test statistic t, which is 1.372184. Now, look at the column head to find the single-tail area, 0.10. Double that to get the other bound of the p-value, 0.20.

You can say that the p-value is between 0.20 and 0.50, or that 0.20 < p-value < 0.50.

---

682. 0.10 < p-value < 0.25

First, find the degrees of freedom, which is one less than the sample size, n: $df =$ 29 − 1 = 28.

Then, find the test statistic t, using the basic formula for t:

$$t = \frac{\bar{x} - \mu_0}{s/\sqrt{n}}$$

where $\bar{x}$ is the sample mean, μ_0 is the population mean, s is the sample standard deviation, and n is the sample size.

$$t = \frac{\bar{x} - \mu_0}{s/\sqrt{n}}$$

$$= \frac{89.8 - 90}{1/\sqrt{29}}$$

$$= \frac{-0.2}{0.1857}$$

$$= -1.077$$

You have a *less than* alternative hypothesis because the researcher "believes that her sample mean is smaller than 90." Because of this, and because the sample mean is below the hypothesized population mean, you must translate the t-statistic into an absolute value of 1.077. Using Table A-2 in the appendix, go to the row for 28 degrees of freedom, and find the closest values to your test statistic. This value falls between the 0.25 and 0.10 columns, so the p-value is between 0.10 and 0.25.

683.

0.01 < p-value < 0.02

First, find the degrees of freedom, which is one less than the sample size, n:
$df = 12 - 1 = 11$.

Because the warden has no hypothesis about the direction her inmate costs may differ, the alternative hypothesis is a *not equal to* hypothesis.

Calculate the test statistic t, using the formula for a one-sample t-test,

$$t = \frac{\bar{x} - \mu_0}{s/\sqrt{n}}$$

where $\bar{x}$ is the sample mean, μ_0 is the population mean, s is the sample standard deviation, and n is the sample size.

$$t = \frac{\bar{x} - \mu_0}{s/\sqrt{n}}$$

$$= \frac{58,660 - 50,000}{10,000/\sqrt{12}}$$

$$= \frac{8,660}{2,886.75}$$

$$= 2.9999$$

Because you have a positive value of t, you can use the t-table without worrying about the sign on t. Using Table A-2 in the appendix, read across the row for 11 degrees of freedom. The nearest numbers in the table to the test statistic are 2.71808, which has a right-tail area of 0.01 (from the column heading), and 3.10581, which has a right-tail area of 0.005.

You have a *not equal to* hypothesis because no direction is stated in the problem. Double the right-tail areas (because there's an equal area in the left tail), so the p-value is between 2(0.005) and 2(0.01), or between 0.01 and 0.02.

684.

The p-value is between 0.01 and 0.025.

Using Table A-2 in the appendix, find the row for 14 degrees of freedom. The test statistic, 2.5, falls between 2.14479 (corresponding to a right-tail area of 0.025 as given in the column heading) and 2.62449 (with a right-tail area of 0.01).

Because you have a *greater than* alternative hypothesis, the p-values are the same as right-tail areas. So the p-value is between 0.01 and 0.025.

685.

0.02 < p-value < 0.05

Because the test statistic is negative but the t-table deals with positive values, use the absolute value of the test statistic, 2.5.

On Table A-2 in the appendix, read across the row for 20 degrees of freedom. The nearest numbers in the table to the test statistic are 2.08596, which has a right-tail area of 0.025, and 2.52798, which has a right-tail area of 0.01. Because you have a *not equal to* alternative hypothesis, double both values, giving you a range of 0.02 to 0.05.

686. There is enough evidence to conclude that the average exam time is more than 45 minutes.

The claim is that the average test-taking time is 45 minutes (the null hypothesis). You believe that the average test-taking time is more than 45 minutes (the alternative hypothesis). Because the p-value from your sample is smaller than your significance level, you reject the null hypothesis, but you can go further and say that you can conclude that the average time is more than 45 minutes, based on your data. *Note:* You could be wrong, so your results don't "prove" anything; however, they do provide strong evidence against the claim.

687. Fail to reject H_0.

Because the sample size is small and you're testing the population mean with unknown population standard deviation (only the sample standard deviation is given), a t-test is in order.

The null hypothesis is that the music students have average verbal ability. The scientist believes that it's lower, so the alternative hypothesis is a *less than* alternative.

First, note the degrees of freedom, which is one less than the sample size n: $df = 8 - 1 = 7$.

Then, to find the critical value, use Table A-2 in the appendix to get the value of 1.894579. However, because this is a left-tailed test, the critical value is –1.894579.

Next, calculate the test statistic t, using this formula:

$$t = \frac{\bar{x} - \mu_0}{s/\sqrt{n}}$$

where $\bar{x}$ is the sample mean, μ_0 is the population mean, s is the sample standard deviation, and n is the sample size.

$$t = \frac{\bar{x} - \mu_0}{s/\sqrt{n}}$$

$$= \frac{97.5 - 100}{5/\sqrt{8}}$$

$$= \frac{-2.5}{1.7678}$$

$$= -1.4142$$

The value of this test statistic (–1.4142) is closer to zero than the critical value (–1.894579), which was found earlier. Therefore, you *can't reject* the null hypothesis. There isn't enough evidence to say that people who play musical instruments have below average verbal ability.

688. Reject H_0.

Because the sample size is small and you're testing the population mean with unknown population standard deviation (only the sample standard deviation is given), a t-test is in order.

The null hypothesis is that the employees donate the target amount of $50 per year on average. The president believes it's lower than that, so the alternative hypothesis is a less than alternative.

First, note the degrees of freedom, which is one less than the sample size *n*: *df* = 10 – 1 = 9.

Then, find the critical value in Table A-2 in the appendix, which is –1.833113 (because this is a left-tailed test).

Next, calculate the test statistic *t*, using this formula:

$$t = \frac{\bar{x} - \mu_0}{s/\sqrt{n}}$$

where $\bar{x}$ is the sample mean, μ_0 is the population mean, *s* is the sample standard deviation, and *n* is the sample size.

$$t = \frac{\bar{x} - \mu_0}{s/\sqrt{n}}$$

$$= \frac{43.40 - 50}{5.2/\sqrt{10}}$$

$$= \frac{-6.6}{1.6444}$$

$$= -4.0136$$

Because you have a negative *t*-test statistic and a *less than* alternative hypothesis, you reject the null hypothesis because the value of the *t*-test statistic is less than the critical value (–4.0136 < –1.833113).

So the president has a point and can conclude that the donations of her employees to charity are lower than the targeted value on average, based on her data.

689. Reject H$_0$.

Because the sample size is small and you're testing the population mean with unknown population standard deviation (only the sample standard deviation is given), a *t*-test is in order.

The null hypothesis is that the coats are protected to an average of –5 degrees Centigrade, on average. But the coat maker believes that the average temperature is lower than that; so the alternative hypothesis is a *less than* alternative.

First, note the degrees of freedom, which is one less than the sample size *n*: *df* = 15 – 1 = 14.

Then, to find the critical value, use Table A-2 in the appendix, going to the row for 14 degrees of freedom and the column for 0.10; the value is 1.345030. Because this is a left-tailed test, the critical value is –1.345030.

Next, calculate the test statistic *t*, using this formula:

$$t = \frac{\bar{x} - \mu_0}{s/\sqrt{n}}$$

where $\bar{x}$ is the sample mean, μ_0 is the population mean, s is the sample standard deviation, and n is the sample size.

$$t = \frac{\bar{x} - \mu_0}{s/\sqrt{n}}$$

$$= \frac{-6.5 - (-5)}{1/\sqrt{15}}$$

$$= \frac{-1.5}{0.2582}$$

$$= -5.8095$$

Because you have a *less than* alternative hypothesis and the test result is negative, you reject the null hypothesis because the test statistic is less than the critical value ($-5.8095 < -1.345030$).

Based on his data, the coat maker has enough evidence to reject the advertisement's claim and conclude that the average protection temperature is lower than that.

---

690. Fail to reject H_0.

Because the sample size is small and you're testing the population mean with unknown population standard deviation (only the sample standard deviation is given), a *t*-test is in order.

The null hypothesis is that the average teacher evaluations at the teacher's school are the same as those in the district. But the teacher believes that the average is lower than that; so the alternative hypothesis is a *less than* alternative.

First, note the degrees of freedom, which is one less than the sample size n: $df = 6 - 1 = 5$.

The significance level is $\alpha = 0.05$. Because you have a *less than* alternative hypothesis and the sample mean is less than the hypothesized value, you can just use the column heading of the *t*-table (Table A-2 in the appendix) to find the significance level (0.05) and read down to find the row corresponding to the degrees of freedom (5). You get a value of 2.015048. Because this is a left-tailed test, the critical value is –2.015048.

Now, calculate the test statistic t, using this formula:

$$t = \frac{\bar{x} - \mu_0}{s/\sqrt{n}}$$

where $\bar{x}$ is the sample mean, μ_0 is the population mean, s is the sample standard deviation, and n is the sample size.

$$t = \frac{\bar{x} - \mu_0}{s/\sqrt{n}}$$

$$= \frac{6.667 - 7.2}{2/\sqrt{6}}$$

$$= \frac{-0.533}{0.8165}$$

$$= -0.6528$$

The fact that the alternative hypothesis is *less than* (because the teacher "believes that other teachers in her school get lower evaluations compared to teachers in other schools in the district") and the test statistic is not less than (is closer to zero than) the critical value (−0.6528 > −2.015048), you *fail to reject* the null hypothesis. There isn't enough evidence to conclude that the average of the teacher evaluations in the teacher's school is lower than the district.

691. Fail to reject H_0.

Because the sample size is small and you're testing the population mean with unknown population standard deviation (only the sample standard deviation is provided), a *t*-test is in order.

The null hypothesis is that the average popsicle temperature is 1.92 degrees Centigrade. The president believes that the average is lower than that; so the alternative hypothesis is a *less than* alternative.

First, note the degrees of freedom, which is one less than the sample size n:
$df = 5 − 1 = 4$.

Then, to find the critical value, use Table A-2 in the appendix. You get a value of 3.74695 by going to the row for 4 degrees of freedom and the column for 0.01. Because this is a left-tailed test, the value is −3.74695.

Next, calculate the test statistic t, using this formula:

$$t = \frac{\bar{x} - \mu_0}{s/\sqrt{n}}$$

where $\bar{x}$ is the sample mean, μ_0 is the population mean, s is the sample standard deviation, and n is the sample size.

$$t = \frac{\bar{x} - \mu_0}{s/\sqrt{n}}$$
$$= \frac{-2.25 - (-1.92)}{1.62/\sqrt{5}}$$
$$= \frac{-0.33}{0.7245}$$
$$\doteq -0.4555$$

Because you have a *less than* alternative hypothesis, you fail to reject the null hypothesis because the test statistic is not less than (is closer to zero than) the critical value (−0.4555 > −3.74695). The president doesn't have enough evidence to conclude that the temperature for the popsicles is set too low on average, based on this data.

692. Reject H_0.

Because the sample size is small and you're testing the population mean with unknown population standard deviation, a *t*-test is in order.

The null hypothesis is that the average weight is 50 grams per object. The alternative hypothesis is that the average weight is more than 50 grams, so here you have a *greater than* alternative hypothesis.

First, note the degrees of freedom, which is one less than the sample size n:
$df = 16 − 1 = 15$.

Then, find the critical value in Table A-2 in the appendix, which is 1.753050, by going to the row for 15 degrees of freedom and the column for 0.05.

Next, calculate the test statistic t, using this formula:

$$t = \frac{\bar{x} - \mu_0}{s/\sqrt{n}}$$

where $\bar{x}$ is the sample mean, μ_0 is the population mean, s is the sample standard deviation, and n is the sample size.

$$t = \frac{\bar{x} - \mu_0}{s/\sqrt{n}}$$

$$= \frac{54 - 50}{8/\sqrt{16}}$$

$$= \frac{4}{2} = 2$$

Because the test statistic (2) is larger than the critical value (1.753050), you reject the null hypothesis. In other words, you reject the null hypothesis that the average weight is 50 grams per object in favor of the alternative hypothesis that the average weight is more than 50 grams.

---

693. Fail to reject H_0.

Because you're testing the population mean with unknown population standard deviation, a t-test is in order.

The null hypothesis is that the average number of beads per 1-pound bag is 1,200. The alternative hypothesis is that the average weight is more than that, so you have a *greater than* alternative hypothesis.

First, note the degrees of freedom, which is one less than the sample size n: $df = 30 - 1 = 29$.

Then, find the critical value in Table A-2 in the appendix by going to the row for 29 degrees of freedom and the column for 0.01; the critical value is 2.46202.

Next, calculate the test statistic t, using this formula:

$$t = \frac{\bar{x} - \mu_0}{s/\sqrt{n}}$$

where $\bar{x}$ is the sample mean, μ_0 is the population mean, s is the sample standard deviation, and n is the sample size.

$$t = \frac{\bar{x} - \mu_0}{s/\sqrt{n}}$$

$$= \frac{1,350 - 1,200}{500/\sqrt{30}}$$

$$= \frac{150}{91.2871}$$

$$= 1.6432$$

The critical value of 2.46202 is larger than the test statistic value of 1.6432, so the conclusion is that you fail to reject the null hypothesis. There isn't enough evidence for the retailer to say that the average number of beads in a 1-pound bag is more than 1,200.

694.

Fail to reject H_0.

Because you're testing the population mean with unknown population standard deviation, a *t*-test is in order.

The null hypothesis is that the average number of beads per 1-pound bag is 1,200. The alternative hypothesis is that the average weight is more than that, so you have a *greater than* alternative.

First, note the degrees of freedom, which is one less than the sample size *n*: $df = 30 - 1 = 29$.

Then, find the critical value in Table A-2 in the appendix by going to the row for 29 degrees of freedom and the column for 0.05; the critical value is 1.699127.

Next, calculate the test statistic *t*, using this formula:

$$t = \frac{\bar{x} - \mu_0}{s/\sqrt{n}}$$

where $\bar{x}$ is the sample mean, μ_0 is the population mean, s is the sample standard deviation, and n is the sample size.

$$t = \frac{\bar{x} - \mu_0}{s/\sqrt{n}}$$

$$= \frac{1,350 - 1,200}{500/\sqrt{30}}$$

$$= \frac{150}{91.2871}$$

$$= 1.6432$$

The test statistic of 1.6432 is less than the critical value of 1.699127 so you fail to reject the null hypothesis. There isn't enough evidence to say that the average number of beads in a 1-pound bag is more than 1,200.

695.

Reject H_0.

Because you're testing the population mean with unknown population standard deviation (only the sample standard deviation is given), a *t*-test is in order.

The null hypothesis is that the average amount of money spent on entertainment is $100 (maximum). The alternative hypothesis is that the average money spent is more than that, so you have a *greater than* alternative hypothesis.

First, note the degrees of freedom, which is one less than the sample size *n*: $df = 25 - 1 = 24$.

Then, find the critical value in Table A-2 in the appendix by going to the row for 24 degrees of freedom and the column for 0.01; the critical value is 2.49216.

Next, calculate the test statistic *t*, using this formula:

$$t = \frac{\bar{x} - \mu_0}{s/\sqrt{n}}$$

where $\bar{x}$ is the sample mean, μ_0 is the population mean, s is the sample standard deviation, and n is the sample size.

$$t = \frac{\bar{x} - \mu_0}{s/\sqrt{n}}$$

$$= \frac{118.44 - 100}{35/\sqrt{25}}$$

$$= \frac{18.44}{7}$$

$$= 2.6343$$

The test statistic is greater than the critical value, so you reject H_0. There's enough evidence based on this data that the average money spent on entertainment is more than $100 per month.

---

696. Reject H_0.

Because you're testing the population mean with unknown population standard deviation (only the sample standard deviation is given), a t-test is in order.

The null hypothesis is that the average amount of cheese in the United States matches that of Europe, which is 25.83 kilograms per person per year. The alternative hypothesis is that the average amount of cheese eaten in the United States is more than that, so you have a greater than alternative hypothesis.

First, note the degrees of freedom, which is one less than the sample size n: $df = 30 - 1 = 29$.

Then, find the critical value in Table A-2 in the appendix by going to the row for 29 degrees of freedom and the column for 0.05; the critical value is 1.699127.

Next, calculate the test statistic t, using this formula:

$$t = \frac{\bar{x} - \mu_0}{s/\sqrt{n}}$$

where $\bar{x}$ is the sample mean, μ_0 is the population mean, s is the sample standard deviation, and n is the sample size.

$$t = \frac{\bar{x} - \mu_0}{s/\sqrt{n}}$$

$$= \frac{27.86 - 25.83}{6.46/\sqrt{30}}$$

$$= \frac{2.03}{1.1794}$$

$$= 1.7212$$

The test statistic is larger than the critical value, so you reject the null hypothesis. There's enough evidence based on this data to reject the claim that the average cheese consumption in the United States is 25.83 kilograms per person and conclude that it's actually higher than that.

697. Reject H_0.

Because you're testing the population mean with unknown population standard deviation (only the sample standard deviation is given), a *t*-test is in order.

The null hypothesis is that his students match the ideal meditation time of 20 minutes. The alternative hypothesis is that the average amount differs from that, so you have a *not equal to* alternative hypothesis.

First, note the degrees of freedom, which is one less than the sample size *n*: $df = 9 - 1 = 8$.

Using Table A-2 in the appendix, find the critical values, with 8 degrees of freedom and a right-tail area of 0.05 (half of 0.10 because the alternative is two-sided); the value is 1.859548. Because this is a two-tailed test, the critical values are 1.859548 and −1.859548.

Next, calculate the test statistic *t*, using this formula:

$$t = \frac{\bar{x} - \mu_0}{s/\sqrt{n}}$$

where $\bar{x}$ is the sample mean, μ_0 is the population mean, *s* is the sample standard deviation, and *n* is the sample size.

$$t = \frac{\bar{x} - \mu_0}{s/\sqrt{n}}$$

$$= \frac{24 - 20}{5/\sqrt{9}}$$

$$= \frac{4}{1.6667}$$

$$= 2.39995$$

The test statistic of 2.4 (rounded up) is greater than the positive critical value of 1.859548, so you reject the null hypothesis. There's sufficient evidence to conclude that this instructor's students don't meditate the ideal amount on average.

698. Reject H_0.

Because you're testing the population mean with unknown population standard deviation (only the sample standard deviation is given), a *t*-test is in order.

The null hypothesis is that the average output of this type of laser printer before servicing is 20,000 pages. The alternative hypothesis is that the average output differs from that, so you have a *not equal to* alternative hypothesis.

First, note the degrees of freedom, which is one less than the sample size *n*: $df = 16 - 1 = 15$.

Next, find the critical values. Table A-2 in the appendix gives right-tail probabilities in the column headings. The right-tail probability is half of the significance level when the alternative hypothesis is *not equal to*. This works out to half of $\alpha = 0.05$, that is, 0.025;

find the value by going to the row for 15 degrees of freedom and the column for 0.025. The value is 2.13145. Because this is a two-tailed test, the critical values are 2.13145 and −2.13145.

You'll reject the null hypothesis if the test statistic value of t falls outside of the range of −2.13145 to 2.13145.

Next, calculate the test statistic t, using this formula:

$$t = \frac{\bar{x} - \mu_0}{s/\sqrt{n}}$$

where $\bar{x}$ is the sample mean, μ_0 is the population mean, s is the sample standard deviation, and n is the sample size.

$$t = \frac{\bar{x} - \mu_0}{s/\sqrt{n}}$$

$$= \frac{18,356 - 20,000}{2,741/\sqrt{16}}$$

$$= \frac{-1,644}{685.25}$$

$$= -2.39912$$

The test statistic value of −2.39912 is outside the range defined by the critical value of t (−2.13145 to +2.13145). This means that you reject the null hypothesis. There's significant evidence that this type of laser printer doesn't require servicing at 20,000 pages on average. It appears (because the test statistic is negative) that it's likely to be less than that.

---

699. Fail to reject H_0.

Because you're testing the population mean with unknown population standard deviation (only the sample standard deviation is given), a t-test is in order.

The null hypothesis is that the students' dissertations in the doctoral program match the norm of 90 pages. The alternative hypothesis is that the average amount differs from that, so you have a *not equal to* alternative hypothesis.

First, note the degrees of freedom, which is one less than the sample size n: $df = 10 - 1 = 9$.

Then, find the critical value. Because this is a *not equal to* test, the 0.05 significance level means that 5% of curve area is split between two tails, leaving 2.5% (or 0.025) in each tail. Use Table A-2 in the appendix to find the critical value of t by going to the row for 9 degrees of freedom and the column for 0.025; the value is 2.26216. Because this is a *not equal to* test, you'll fail to reject the null hypothesis if the test statistic is in the range of −2.26216 to +2.26216.

Next, calculate the test statistic t, using this formula:

$$t = \frac{\bar{x} - \mu_0}{s/\sqrt{n}}$$

where $\bar{x}$ is the sample mean, μ_0 is the population mean, s is the sample standard deviation, and n is the sample size.

$$t = \frac{\bar{x} - \mu_0}{s/\sqrt{n}}$$

$$= \frac{85.2 - 90}{7.59/\sqrt{10}}$$

$$= \frac{-4.8}{2.40}$$

$$= -2.0$$

The test value is within the range defined by the critical values –2.26216 and +2.26216, so you fail to reject the null hypothesis. The student doesn't have enough support for the idea that students in her doctoral program write dissertations that differ in length from the supposed mean of 90 pages.

---

700. Reject H_0.

Because you're testing the population mean with unknown population standard deviation (only the sample standard deviation is given), a *t*-test is in order.

The null hypothesis is that the average age of the trees in the forest is 30 years. The alternative hypothesis is that the average age differs from that, so here you have a *not equal to* alternative hypothesis.

First, note the degrees of freedom, which is one less than the sample size n: $df = 5 - 1 = 4$.

Then, find the critical value. In Table A-2 in the appendix, find the critical value for 4 degrees of freedom that will leave 0.25 area in each tail (because this is a *not equal to* alternative hypothesis) by going to the row for 4 degrees of freedom and the column for 0.25; the value is 0.740697. You'll reject the null hypothesis if the test statistic is outside the range of –0.740697 to +0.740697.

Next, calculate the test statistic t, using this formula:

$$t = \frac{\bar{x} - \mu_0}{s/\sqrt{n}}$$

where $\bar{x}$ is the sample mean, μ_0 is the population mean, s is the sample standard deviation, and n is the sample size.

$$t = \frac{\bar{x} - \mu_0}{s/\sqrt{n}}$$

$$= \frac{33 - 30}{5.6/\sqrt{5}}$$

$$= \frac{3}{2.5044}$$

$$= 1.19789$$

The computed test statistic *t*-value of 1.19789 is clearly outside the range of –0.740697 to +0.740697, so you reject the null hypothesis. The logger has evidence to conclude that the average age of the trees in this forest isn't equal to 30 years. (And because the test statistic is positive, it is likely that the average age is higher than that.)

701. The null hypothesis should be rejected; the bank should open the new branch.

First, set up the null and alternative hypotheses:

$H_0: p_0 = 0.10$

$H_a: p_0 > 0.10$

Note: Because the bank wants to ensure that *at least* 10% of the residents will use the new branch, it's implicitly interested in a *greater than* alternative hypothesis.

Next, determine whether the sample is large enough to run a z-test by checking that both np_0 and $n(1 - p_0)$ equal at least 10. In this case, $n = 100$ and $p_0 = 0.10$, so $np_0 = (100)(0.10) = 10$, and $n(1 - np_0) = 100(1 - 0.10) = 100(0.90) = 90$.

Then, compute the standard error with this formula:

$$SE = \sqrt{\frac{p_0 (1-p_0)}{n}}$$

where p_0 is the population proportion and n is the sample size. Substitute the known values into the formula to get

$$SE = \sqrt{\frac{0.10(1-0.10)}{100}}$$

$$= \sqrt{\frac{0.10(0.90)}{100}}$$

$$= \sqrt{\frac{0.09}{100}}$$

$$= \sqrt{0.0009}$$

$$= 0.03$$

Next, find the observed proportion by dividing the number who said they would consider banking at the new branch by the sample size of 100: $19/100 = 0.19$.

Then, calculate the z-test statistic, using this formula:

$$z = \frac{\hat{p} - p_0}{SE}$$

where $\hat{p}$ is the observed proportion, p_0 is the hypothesized proportion, and SE is the standard error.

$$z = \frac{0.19 - 0.1}{0.03}$$

$$= 3$$

Now, use a Z-table, such as Table A-1 in the appendix, to determine the probability of observing a z-score this high or higher. Unfortunately, the table shows the probability of observing a score of $z = 3.0$ or lower, so you have to subtract the table probability from 1 to get the probability: $1 - 0.9987 = 0.0013$.

Finally, compare the probability (that is, the p-value) with the alpha (α) level. The bank wanted to use a significance level of 0.05, so $\alpha = 0.05$, and the p-value of 0.0013 is much lower than that. So you reject the null hypothesis.

702. The null hypothesis should be rejected; the standard has been met and exceeded.

First, set up the null and alternative hypotheses:

$$H_0: p_0 = 0.75$$

$$H_a: p_0 > 0.75$$

Note: The alternative is a *greater than* hypothesis because the call center sets a minimum threshold for performance and is asking only whether the system meets or exceeds the threshold with a high degree of confidence.

Then, determine whether the sample is large enough to run a z-test by checking that both np_0 and $n(1 - p_0)$ equal at least 10. In this case, $n = 50$ and $p_0 = 0.75$, so $np_0 = 50(0.75) = 37.5$, and $n(1 - p_0) = 50(1 - 0.75) = 50(0.25) = 12.5$.

Compute the standard error with this formula:

$$SE = \sqrt{\frac{p_0(1 - p_0)}{n}}$$

where p_0 is the population proportion and n is the sample size. Substitute the known values into the formula to get

$$SE = \sqrt{\frac{0.75(1 - 0.75)}{50}}$$

$$= \sqrt{\frac{0.75(0.25)}{50}}$$

$$= \sqrt{\frac{0.1875}{50}}$$

$$= \sqrt{0.00375}$$

$$= 0.061237$$

Next, find the observed proportion by dividing the number of "successes" by the sample size: $45/50 = 0.90$.

Then, calculate the z-test statistic, using this formula:

$$z = \frac{\hat{p} - p_0}{SE}$$

where $\hat{p}$ is the observed proportion, p_0 is the hypothesized proportion, and SE is the standard error.

$$z = \frac{0.90 - 0.75}{0.061237}$$

$$= 2.4495, \text{ or } 2.45 \text{ (rounded)}$$

Now, use a Z-table, such as Table A-1 in the appendix, to figure out the probability of observing a value this high or higher (because the alternative is a *greater than* hypothesis). The table gives you the proportion of the area under the curve that's less than a given value of z, which is 0.9929. To get the desired area (proportion above this z-value), you have to subtract that number from 1: $1 - 0.9929 = 0.0071$.

Finally, identify the desired alpha (α) level (0.05) and compare the probability you found from the Z-table with that. Because 0.0071 is less than 0.05, you reject the null hypothesis.

703. The null hypothesis can't be rejected; the clerk shouldn't buy the books.

First, set up the null and alternative hypotheses:

$H_0: p_0 = 0.50$

$H_a: p_0 > 0.50$

Note: The store sets a minimum threshold (50%) and is interested in finding whether a collection of books is at least that marketable or more, so the alternative is a *greater than* hypothesis.

Next, determine whether the sample is large enough to run a z-test by checking that both np_0 and $n(1 - p_0)$ equal at least 10. In this case, $n = 30$ and $p_0 = 0.5$, so $np_0 = 30(0.5) = 15$, and $n(1 - p_0) = 30(1 - 0.5) = 15$.

Calculate the observed proportion by dividing the number of books likely to sell by the number of books offered: $17/30 = 0.5667$.

Then, compute the standard error with this formula:

$$SE = \sqrt{\frac{p_0 (1-p_0)}{n}}$$

where p_0 is the population proportion and n is the sample size. Substitute the known values into the formula to get

$$SE = \sqrt{\frac{0.5 (1-0.5)}{30}}$$

$$= \sqrt{\frac{0.50 (0.50)}{30}}$$

$$= \sqrt{\frac{0.25}{30}}$$

$$= \sqrt{0.008\overline{3}}$$

$$= 0.091287$$

Next, calculate the z-test statistic, using this formula:

$$z = \frac{\hat{p} - p_0}{SE}$$

where $\hat{p}$ is the observed proportion, p_0 is the hypothesized proportion, and SE is the standard error.

$$z = \frac{0.5667 - 0.5}{0.091287}$$

$$= 0.7307, \text{ or } 0.73 \text{ (rounded)}$$

Now, find this z-score in a Z-table, such as Table A-1 in the appendix. The table value is 0.7673, which is the area under the curve to the left, the z-test statistic you observed in the sample. However, you want to know the probability of getting a z-score this high or higher (because the alternative is a *greater than* hypothesis), so you have to subtract the table value from 1 to get the p-value for the z-statistic: $1 - 0.7673 = 0.2327$.

Finally, compare this value to the α level (0.05), and you find that the p-value is greater than α. Thus, your conclusion is that you can't reject the null hypothesis, and the clerk shouldn't offer to buy the collection of books.

704. The null hypothesis shouldn't be rejected; the owner can't conclude that the defect rate is less than 1%.

First, set up the null and alternative hypotheses:

$$H_0: p_0 = 0.01$$
$$H_a: p_0 < 0.01$$

Note: The alternative hypothesis is a *less than* hypothesis because the factory owner has a maximum acceptable error rate, and she wants to ensure that the process makes an error 1% or less of the time.

Then, determine whether the sample is large enough to run a z-test by checking that both np_0 and $n(1-p_0)$ equal at least 10. In this case, $n = 1,000$ and $p_0 = 0.01$, so $np_0 = 1,000(0.01) = 10$, and $n(1-p_0) = 1,000(1-0.01) = 1,000(0.99) = 999$.

Calculate the observed proportion by dividing the number of defective ball bearing by the sample size. If 6 out of 1,000 ball bearings are found to be defective, the proportion is $6/1,000 = 0.006$.

Next, compute the standard error with this formula:

$$SE = \sqrt{\frac{p_0(1-p_0)}{n}}$$

where p_0 is the population proportion and n is the sample size. Substitute the known values into the formula to get

$$SE = \sqrt{\frac{0.01(0.99)}{1,000}}$$
$$= \sqrt{0.0000099}$$
$$= 0.0031464$$

Then, calculate the z-test statistic, using this formula:

$$z = \frac{\hat{p} - p_0}{SE}$$

where $\hat{p}$ is the observed proportion, p_0 is the hypothesized proportion, and SE is the standard error.

$$z = \frac{0.006 - 0.01}{0.0031464}$$
$$= -1.2712, \text{ or} -1.27 \text{ (rounded)}$$

Now, find the probability of getting a test statistic value of z at least this far from the claimed proportion under the null hypothesis. In this case, the Z-table (Table A-1 in the appendix) gives you the proportion under the curve to the left of the z-value you look up, and the alternative hypothesis is *less than,* so you get the probability you need directly from the table. That probability is 0.1020 and represents the p-value.

Compare the p-value with the desired level of $\alpha = 0.05$. Because the p-value is greater than α, you fail to reject the null hypothesis.

In practical terms, the factory owner doesn't have significant evidence that the process is working correctly and is keeping the error rate to 1% or less.

705. $\hat{p}$ is the sample proportion, and p_0 is the claimed value for the population proportion.

When you do a hypothesis test for a proportion, the null hypothesis (H_0) makes a claim about what the population proportion is; this claimed value for the population proportion is denoted by p_0. For example, a newspaper article may say that 30% (or 0.30) of Americans wear glasses; so $p_0 = 0.30$ is a claimed value for the proportion of all Americans in the population who wear glasses. You test the claim by taking a sample of Americans and finding the proportion of people in the sample who wear glasses. This result is called the sample proportion and is denoted $\hat{p}$. So $\hat{p}$ is a value that comes from your sample (the sample proportion), while p_0 is a value that someone claims to be the population proportion.

706. The null hypothesis can't be rejected; the manufacturer shouldn't accept the shipment.

First, set up the null and alternative hypotheses:

$$H_0: p_0 = 0.01$$

$$H_a: p_0 < 0.01$$

Note: The alternative hypothesis is *less than* because the company wants to work only with suppliers whose parts are less than 1% defective.

Then, determine whether the sample is large enough to run a z-test by checking that both np_0 and $n(1 - p_0)$ equal at least 10. In this case, $n = 10,000$ and $p_0 = 0.01$, so $np_0 = 10,000(0.01) = 100$, and $n(1 - p_0) = 10,000(1 - 0.01) = 9,900$.

Calculate the observed proportion by dividing the number of defective items by the sample size: $90/10,000 = 0.009$.

Next, compute the standard error with this formula:

$$SE = \sqrt{\frac{p_0(1-p_0)}{n}}$$

where p_0 is the population proportion and n is the sample size. Substitute the known values into the formula to get

$$SE = \sqrt{\frac{0.01(1-0.01)}{10,000}}$$

$$= \sqrt{\frac{0.01(0.99)}{10,000}}$$

$$= \sqrt{\frac{0.0099}{10,000}}$$

$$= 0.000994987$$

Then, calculate the z-test statistic, using this formula:

$$z = \frac{\hat{p} - p_0}{SE}$$

where $\hat{p}$ is the observed proportion, p_0 is the hypothesized proportion, and SE is the standard error.

$$z = \frac{0.009 - 0.01}{0.000994987}$$

$$= -1.005038, \text{ or } -1.01 \text{ (rounded)}$$

Now, find the area under the curve to the left of the test statistic value of z and convert this to a p-value. According to Table A-1, the nearest value for $z = -1.01$, which gives an area of 0.1562. Because you had a *less than* alternative hypothesis and the value of z is negative, the table value is the same as the p-value.

Finally, compare the p-value with the $\alpha = 0.01$ level of significance. In this case, the p-value is greater than α, so you can't reject the null hypothesis. The company can't be reasonably sure that the shipment keeps the defect rate low enough.

---

707.   E.   None of the above.

The true value for the population proportion is denoted by p. Because p is typically unknown, it's often given a claimed value (denoted by p_0). The claimed value for p is challenged by comparing it to the results of a sample, using a hypothesis test. In the end, a probability is reported, which is called the p-value. The p-value is the probability that the results in the sample happened by random chance, while the claimed value for p is true. A small p-value indicates the sample results were unlikely to be due to chance. It gives evidence against the claimed value of p, resulting in possibly rejecting H_0, depending on how small it is.

---

708.   There is insufficient evidence to draw a conclusion.

In this case, the dealer didn't draw a sufficient sample size. To determine whether the sample is large enough to run a z-test, you check that both np_0 and $n(1 - p_0)$ equal at least 10. In this case, $n = 10$ and $p_0 = 0.05$, so $np_0 = 10(0.05) = 0.5$. This is less than 10, so you can't use the normal approximation to the binomial.

Ultimately, there's insufficient evidence to draw a conclusion in this case when sampling only ten items.

---

709.   100% confidence

It's important to remember basic concepts before starting to run statistical tests. A sample is a part of a population. If impurities exist in a sample of blood specimens, then impurities certainly exist in the population of specimens from which it's drawn. The blood bank can say with 100% confidence that the blood donation population of specimens has disease in it. The z-test is irrelevant here.

---

710.   The null hypothesis can't be rejected; the city council should work to control the pigeon population.

In this case, a little thought will save a lot of statistical testing. If the city council wants the infection rate to be 3% or less and it observes an infection rate of 3% in a sample (6/200 is 3%), then the null hypothesis can't be rejected, regardless of the significance level. Or simply consider that the numerator of the test statistic z would be 0 because the sample proportion and assumed population proportion are the same. The p-value would be 0.50, far from significant. You fail to reject the null hypothesis, and the pigeons need to be controlled better, given the policy.

711. The null hypothesis can be rejected; the designer shouldn't accept the shipment.

First, set up the null and alternative hypotheses:

$$H_0: p_0 = 0.25$$

$$H_a: p_0 \neq 0.25$$

Note: The alternative hypothesis is *not equal to* because the designer aims for a defect rate of 0.25 and neither significantly more or less.

Next, determine whether the sample is large enough to run a z-test by checking that both np_0 and $n(1 - p_0)$ equal at least 10. In this case, $n = 50$ and $p_0 = 0.25$, so $np_0 = 50(0.25) = 12.5$ and $n(1 - p_0) = 50(1 - 0.25) = 50(0.75) = 37.5$.

The sample proportion is 0.12, as stated in the problem.

Then, compute the standard error with this formula:

$$SE = \sqrt{\frac{p_0(1 - p_0)}{n}}$$

where p_0 is the population proportion and n is the sample size. Substitute the known values into the formula to get

$$SE = \sqrt{\frac{0.25(1 - 0.25)}{50}}$$

$$= \sqrt{\frac{0.25(0.75)}{50}}$$

$$= \sqrt{\frac{0.1875}{50}}$$

$$= \sqrt{0.00375}$$

$$= 0.061237$$

Next, calculate the z-test statistic, using this formula:

$$z = \frac{\hat{p} - p_0}{SE}$$

where $\hat{p}$ is the observed proportion, p_0 is the hypothesized proportion, and SE is the standard error.

$$z = \frac{0.12 - 0.25}{0.061237}$$

$$= -2.1229, \text{ or } -2.12 \text{ (rounded)}$$

Find the area under the curve to the left of the test statistic value of z, and convert this to a p-value. Using Table A-1 in the appendix (or another Z-table), you find that 0.0170 of the curve lies to the left of $z = -2.12$. The alternative hypothesis in this case is *not equal to,* so that curve area must be doubled: $2(0.0170) = 0.0340$.

Now, compare the p-value with the desired α level of 0.05. Because the p-value is less than α, the null hypothesis can be rejected. In this case, the designer should reject the shipment because it has too few defects for his taste.

---

712. $H_0: p = 0.50$; $H_a: p \neq 0.50$

You're testing to see whether a coin is fair. If the coin is fair, the probability of getting a heads (or the proportion of heads in an infinite number of coin flips) is $p = 0.50$. If the coin isn't fair, there could be either a significantly smaller proportion of heads than 0.50 or a significantly larger proportion of heads than 0.50. In other words, if the coin isn't fair, $p \neq 0.50$. In a hypothesis test, you start by claiming that the coin is fair unless evidence shows otherwise. That means that H_0 is $p = 0.50$ and H_a is $p \neq 0.50$.

---

713. $H_0: p = 0.45$; $H_a: p > 0.45$

In this case, p is the proportion of correct answers Joe would get if you played this game an infinite number of times. If Joe were just guessing at the suit of each card, he'd be expected to guess 25% of the card suits correctly (because there are four possible suits, and each suit occurs equally often among each deck). In other words, if Joe is guessing, p would equal 0.25 in the long run. You decided that he has to be at least 20 percentage points above the accuracy you would expect by chance: $0.25 + 0.20 = 0.45$. So the null and alternative hypotheses in this case are $H_0: p = 0.45$ and $H_a: p > 0.45$.

---

714. The null hypothesis can't be rejected, and the sample shouldn't be rejected either.

First, set up the null and alternative hypotheses:

$$H_0: p_0 = 0.25$$
$$H_a: p_0 \neq 0.25$$

Next, determine whether the sample is large enough to run a z-test by checking that both np_0 and $n(1 - p_0)$ equal at least 10. In this case, $n = 1{,}000{,}000$ and $p_0 = 0.25$, so $np_0 = 1{,}000{,}000(0.25) = 250{,}000$, and $n(1 - p_0) = 1{,}000{,}000(1 - 0.25) = 1{,}000{,}000(0.75) = 750{,}000$.

Calculate the observed proportion by dividing the cells with the phenotype by sample size: $250{,}060/1{,}000{,}000 = 0.25006$.

Then, compute the standard error with this formula:

$$SE = \sqrt{\frac{p_0(1-p_0)}{n}}$$

where p_0 is the population proportion and n is the sample size. Substitute the known values into the formula to get

$$SE = \sqrt{\frac{p_0(1-p_0)}{n}}$$

$$= \sqrt{\frac{0.25(1-0.25)}{1{,}000{,}000}}$$

$$= \sqrt{\frac{0.25(0.75)}{1{,}000{,}000}}$$

$$= \sqrt{\frac{0.1875}{1{,}000{,}000}}$$

$$= \sqrt{0.0000001875}$$

$$= 0.000433$$

Next, calculate the z-test statistic, using this formula:

$$z = \frac{\hat{p} - p_0}{SE}$$

where $\hat{p}$ is the observed proportion, p_0 is the hypothesized proportion, and SE is the standard error.

$$z = \frac{0.25006 - 0.25}{0.000433}$$

$$= 0.1386, \text{ or } 0.14 \text{ (rounded)}$$

Find the area under the curve to the right of the test statistic value of z (that is, away from the claimed proportion), and convert this to a p-value. The closest z-value is 0.14, which corresponds to an area of 0.5557. This, however, is the area to the left of z = 0.14, so you find 1 – 0.5557 = 0.4443 to be the area to the right of z. The *not equal to* alternative hypothesis means that you need the total area in two tails. To get that, multiply the result by 2:

p-value = 2(0.4443) = 0.8886

Now, compare this with the α level. The p-value says that there's an 88.86% probability of seeing a sample as or more extreme that what the biologist observed. This is evidence that a reasonably common proportion was seen. This far exceeds the threshold of α = 0.10. Because the p-value is greater than α, the biologist should fail to reject the null hypothesis and accept this sample for further study.

715. You know you can't reject H_0 because 0.20 isn't one of the values in H_a.

Bob's theory is that 30% of the customers in the checkout line buy something (so H_0 is p = 0.30). You believe it could even be higher than that (so H_a is p > 0.30); in this case, any values less than 0.30 don't matter. Given this situation, the only hope you have of rejecting H_0 is if your sample results are higher than 30%; then it would be a matter of calculating out the hypothesis test to determine whether your sample results are high enough above 30% to reject Bob's claim. However, because your sample results (20%) are lower than 30% right out of the box, you don't have to go any further; you know you won't be able to reject H_0.

716. $H_0: \mu_1 - \mu_2 = 0; H_a: \mu_1 - \mu_2 > 0$

The null hypothesis is always a statement of equality. The alternative in this case is that the population mean of Group 1 will be larger than that of Group 2.

717. 1.645

First, determine whether a z-test for independent groups is appropriate. Because the population of scores is normally distributed and you have population standard deviation information for both groups, you can proceed to do the test.

This is a *greater than* alternative hypothesis with an alpha level of 0.05. Using Table A-1 in the appendix, find the critical value of z such that 0.05 of the probability lies above it. This value falls between 1.64 and 1.65 and just happens to round to 1.645.

718. 4.472

To calculate the standard error, use this formula:

$$SE = \sqrt{\frac{\sigma_1^2}{n_1} + \frac{\sigma_2^2}{n_2}}$$

where σ_1^2 and σ_2^2 are the variances of the two populations, and n_1 and n_2 are the two sample sizes. So for this test, the standard error is

$$SE = \sqrt{\frac{17.32^2}{30} + \frac{17.32^2}{30}}$$

$$= \sqrt{\frac{299.98}{30} + \frac{299.98}{30}}$$

$$= 4.472$$

719. 4.2263

First, find the standard error, using this formula:

$$SE = \sqrt{\frac{\sigma_1^2}{n_1} + \frac{\sigma_2^2}{n_2}}$$

where σ_1^2 and σ_2^2 are the variances of the two populations, and n_1 and n_2 are the two sample sizes. So for this test, the standard error is

$$SE = \sqrt{\frac{17.32^2}{30} + \frac{17.32^2}{30}}$$

$$= \sqrt{\frac{299.98}{30} + \frac{299.98}{30}}$$

$$= 4.472$$

Then, use the formula to find the test statistic:

$$z = \frac{(\bar{x}_1 - \bar{x}_2) - (\mu_1 - \mu_2)}{SE}$$

where $\bar{x}_1$ and $\bar{x}_2$ are the sample means, and μ_1 and μ_2 are the population means. The null hypothesis is that the two populations have the same mean (in other words, that the difference in population means is 0).

$$z = \frac{(33.3 - 14.4) - (0)}{4.472}$$

$$= 4.2263$$

720. The null hypothesis can be rejected; it appears that workers who smile more are more productive.

Set up the null and alternative hypotheses. Because the manager believes that happy workers are more productive, it makes sense to set up a *greater than* alternative hypothesis.

$$H_0: \mu_1 - \mu_2 = 0$$
$$H_a: \mu_1 - \mu_2 > 0$$

Next, identify the alpha level. The question says to use $\alpha = 0.05$.

Use a Z-table, such as Table A-1 in the appendix, to find a critical value for the z-test. Because the alternative hypothesis is *greater than,* the critical value for z will be positive and occur at the point where 0.05 of the curve area lies to the right of the z-score, putting $1 - 0.05 = 0.95$ of the area to the left of that z-score. As a result, you look for a table value of 0.95 and identify the z-value corresponding to that. It turns out to be between 1.64 and 1.65, so you can call it 1.645.

Now, calculate the standard error, using this formula:

$$SE = \sqrt{\frac{\sigma_1^2}{n_1} + \frac{\sigma_2^2}{n_2}}$$

where σ_1^2 and σ_2^2 are the variances of the two populations, and n_1 and n_2 are the two sample sizes. So for this test, the standard error is

$$SE = \sqrt{\frac{17.32^2}{30} + \frac{17.32^2}{30}}$$

$$= \sqrt{\frac{299.98}{30} + \frac{299.98}{30}}$$

$$= 4.472$$

Then use the formula to find the test statistic:

$$z = \frac{(\bar{x}_1 - \bar{x}_2) - (\mu_1 - \mu_2)}{SE}$$

where $\bar{x}_1$ and $\bar{x}_2$ are the sample means, and μ_1 and μ_2 are the population means. The null hypothesis is that the two populations have the same mean (in other words, that the difference in population means is 0).

$$z = \frac{(33.3 - 14.4) - (0)}{4.472}$$

$$= 4.2263$$

Finally, determine whether the test statistic value for z is greater than the critical value for z, and reject the null hypothesis if so. In this case, the test statistic value (4.2263) is greater than the critical value (1.645), so you reject the null hypothesis. It appears that checkout clerks who smile more are more productive.

721. $H_0: \mu_1 = \mu_2; H_a: \mu_1 \neq \mu_2$

The null hypothesis is that the two population means don't differ, while the alternative hypothesis is that they do differ (a *not equal to* alternative hypothesis).

722. 1.96

Because this is a *not equal to* alternative hypothesis, half the alpha value of 0.05 will be in the upper tail and 0.05 in the lower tail. Using Table A-1, you can see that 1.96 is the z-value that has 0.025 of the total area above it. Doubling this probability (to include the lower tail) gives a significance level of 0.05.

723. −1.2123

To find the test statistic, use this formula:

$$z = \frac{\bar{x}_1 - \bar{x}_2}{\sqrt{(\sigma_1^2/n_1) + (\sigma_2^2/n_2)}}$$

where $\bar{x}_1$ and $\bar{x}_2$ are the sample means, σ_1^2 and σ_2^2 are the population variances, and n_1 and n_2 are the sample sizes.

Then, substitute the known values into the formula and solve:

$$z = \frac{51.9 - 52.6}{\sqrt{(5/30) + (5/30)}}$$
$$= \frac{-0.7}{\sqrt{0.1667 + 0.1667}}$$
$$= \frac{-0.7}{\sqrt{0.3334}}$$
$$= -1.2123$$

724. The null hypothesis can't be rejected. This study doesn't support the idea that smokers and nonsmokers differ in IQ.

First, find the test statistic, using this formula:

$$z = \frac{\bar{x}_1 - \bar{x}_2}{\sqrt{(\sigma_1^2/n_1) + (\sigma_2^2/n_2)}}$$

where $\bar{x}_1$ and $\bar{x}_2$ are the sample means, σ_1^2 and σ_2^2 are the population variances, and n_1 and n_2 are the sample sizes.

Then, substitute the known values into the formula and solve:

$$z = \frac{51.9 - 52.6}{\sqrt{(5/30) + (5/30)}}$$
$$= \frac{-0.7}{\sqrt{0.1667 + 0.1667}}$$
$$= \frac{-0.7}{\sqrt{0.3334}}$$
$$= -1.2123$$

As you can see in Table A-1 in the appendix, the test statistic for a *not equal to* hypothesis with an alpha level of 0.10 is 1.645. The z-statistic is only 1.2123 standard deviations away from the claimed mean of 0. In other words, because the absolute value of z is less than the critical value (1.2123 < 1.645), you don't have sufficient evidence to reject the null hypothesis.

725.
The null hypothesis can't be rejected. This study doesn't support the idea that smokers and nonsmokers differ in IQ.

First, find the test statistic, using this formula:

$$z = \frac{\bar{x}_1 - \bar{x}_2}{\sqrt{(\sigma_1^2/n_1) + (\sigma_2^2/n_2)}}$$

where $\bar{x}_1$ and $\bar{x}_2$ are the sample means, σ_1^2 and σ_2^2 are the population variances, and n_1 and n_2 are the sample sizes.

Then, substitute the known values into the formula and solve:

$$z = \frac{51.9 - 52.6}{\sqrt{(5/30) + (5/30)}}$$

$$= \frac{-0.7}{\sqrt{0.1667 + 0.1667}}$$

$$= \frac{-0.7}{\sqrt{0.3334}}$$

$$= -1.2123$$

As you can see from Table A-1 in the appendix, the test statistic for a *not equal to* hypothesis with an alpha level of 0.5 is 1.96. The z-statistic is only 1.2123 standard deviations away from the claimed mean of 0. In other words, because the absolute value of z is less than the critical value (1.2123 < 1.96), you don't have sufficient evidence to reject the null hypothesis.

726.
$H_0: \mu_1 = \mu_2; H_a: \mu_1 \neq \mu_2$

You may think that more sleep would lead to a better score on the memory test. But because this isn't explicitly stated, you must assume that the researcher is interested in whether the two groups differ at all. Thus, the alternative hypothesis must be *not equal to*, while the null hypothesis is that the population mean scores are equal.

727.
$H_0: \mu_1 = \mu_2; H_a: \mu_1 > \mu_2$

If the researcher is interested only in whether Group 1 (the group allowed to sleep five hours) performs better than Group 2 (the group allowed to sleep only three hours), this is a *less than* alternative hypothesis. The null hypothesis is always a statement of equality.

728.
2.8803

To calculate the test statistic, use this formula:

$$z = \frac{\bar{x}_1 - \bar{x}_2}{\sqrt{(\sigma_1^2/n_1) + (\sigma_2^2/n_2)}}$$

where $\bar{x}_1$ and $\bar{x}_2$ are the sample means, σ_1^2 and σ_2^2 are the population variances, and n_1 and n_2 are the sample sizes.

Then, substitute the known values into the formula and solve:

$$z = \frac{62-58}{\sqrt{(6^2/40)+(6^2/35)}}$$

$$= \frac{4}{\sqrt{(36/40)+(36/35)}}$$

$$= \frac{4}{\sqrt{0.9+1.0286}}$$

$$= \frac{4}{\sqrt{1.3887}}$$

$$= 2.8803$$

729. Reject the null hypothesis and conclude that a difference in sleep is associated with a difference in performance.

First, find the test statistic, using this formula:

$$z = \frac{\bar{x}_1 - \bar{x}_2}{\sqrt{(\sigma_1^2/n_1)+(\sigma_2^2/n_2)}}$$

where $\bar{x}_1$ and $\bar{x}_2$ are the sample means, σ_1^2 and σ_2^2 are the population variances, and n_1 and n_2 are the sample sizes.

Then, substitute the known values into the formula and solve:

$$z = \frac{62-58}{\sqrt{(6^2/40)+(6^2/35)}}$$

$$= \frac{4}{\sqrt{(36/40)+(36/35)}}$$

$$= \frac{4}{\sqrt{0.9+1.0286}}$$

$$= \frac{4}{\sqrt{1.3887}}$$

$$= 2.8803$$

For a two-tailed z-test with a significance level of $\alpha = 0.01$, alpha must be split between the two tails. Using Table A-1 and looking for a probability of $0.01/2 = 0.005$, you find the critical value to be roughly 2.58. The test statistic is greater than this so the researcher will reject the null hypothesis. Based on the two-sided alternative hypothesis, you can conclude only that a difference in sleep is associated with a difference in performance.

730. Reject the null hypothesis and conclude that more sleep is associated with better perfor

First, find the test statistic, using this formula:

$$z = \frac{\bar{x}_1 - \bar{x}_2}{\sqrt{(\sigma_1^2/n_1)+(\sigma_2^2/n_2)}}$$

where $\bar{x}_1$ and $\bar{x}_2$ are the sample means, σ_1^2 and σ_2^2 are the population variances n_1 and n_2 are the sample sizes.

Then, substitute the known values into the formula and solve:

$$z = \frac{62 - 58}{\sqrt{(6^2/40) + (6^2/35)}}$$

$$= \frac{4}{\sqrt{(36/40) + (36/35)}}$$

$$= \frac{4}{\sqrt{0.9 + 1.0286}}$$

$$= \frac{4}{\sqrt{1.3887}}$$

$$= 2.8803$$

Using Table A-1 in the appendix, find the critical value for a one-tailed z-test at $\alpha = 0.05$, which is roughly 1.64. The test statistic is greater than this so the researcher will reject the null hypothesis and conclude that more sleep is associated with better performance.

---

731. a t-test of paired population means

The households are observed twice, once for each appeal, so the two measurements on each household are paired rather than independent. The sample size of ten is too small for a z-test, so the t-test must be used.

---

732. A. The wallet appeal was more successful.

For the paired t-test, you calculate differences by subtracting the second group from the first. Because the mean difference score, $\bar{d}$, is negative, on average, energy consumption was lower in the wallet condition as compared to the moral condition. However, to test whether the two conditions differed significantly, you need to conduct a paired t-test.

9

The appropriate test is the paired t-test. The degrees of freedom is one less than the number of pairs: $n_{\text{pairs}} - 1 = 10 - 1 = 9$.

14.67

To calculate the standard error for a paired t-test, use this formula:

$$SE = \frac{s_{\bar{d}}}{\sqrt{n_{\text{pairs}}}}$$

where $s_{\bar{d}}$ is the standard deviation of the differences from the sample, and n_{pairs} is the number of pairs.

The standard error is

$$SE = \frac{46.39}{\sqrt{10}}$$

$$= 14.67$$

735. −5.69

First, find the standard error with this formula:

$$SE = \frac{s_{\bar{d}}}{\sqrt{n_{pairs}}}$$

where $s_{\bar{d}}$ is the standard deviation of the differences from the sample, and n_{pairs} is the number of pairs.

So the standard error is

$$SE = \frac{46.39}{\sqrt{10}}$$

$$= 14.67$$

Then, find the test statistic, using this formula:

$$t = \frac{\bar{d}}{SE}$$

where $\bar{d}$ is the average of the paired difference from the sample. So, substituting the numbers, that makes the test statistic

$$t = \frac{-83.5}{14.67}$$

$$= -5.69189$$

736. Reject the null hypothesis and conclude that moral appeal is less successful than the wallet appeal.

You will conduct a paired t-test. This is a *not equal to* alternative hypothesis, so the null and alternative hypotheses are

$$H_0: \mu_1 - \mu_2 = 0$$

$$H_a: \mu_1 - \mu_2 \neq 0$$

To calculate the standard error for a paired t-test, use this formula:

$$SE = \frac{s_{\bar{d}}}{\sqrt{n_{pairs}}}$$

where $s_{\bar{d}}$ is the standard deviation of the differences from the sample, and n_{pairs} is the number of pairs.

$$SE = \frac{46.39}{\sqrt{10}}$$

$$= 14.67$$

Then, find the test statistic:

$$t = \frac{\bar{d}}{SE}$$

$$= \frac{-83.5}{14.67}$$

$$= -5.69189$$

where $\bar{d}$ is the average of the paired differences from the sample. This result rounds to −5.69.

The degrees of freedom are $n - 1 = 10 - 1 = 9$. You can find the critical value for an alpha level of 0.10 by first splitting the level of significance in half for each of the two tails, because this test has a two-sided alternative. Examining the column for $0.10/2 = 0.05$ and the row for 9 df in Table A-2 in the appendix, you find the critical value is 1.833113.

The test statistic is greater than the critical value, so you reject the null hypothesis. The direction of the difference in means indicates that less power was used following the wallet appeal, making the moral appeal less successful.

737. a t-test of paired population means

The data is paired because both brands are tested on each toy, and the lengths of time each brand of battery operated a given toy are compared.

738. $H_0: \mu_d = 0; H_a: \mu_d \neq 0$

In a paired t-test, the null hypothesis is that the mean of the difference scores is 0. As a *not equal to* hypothesis, the alternative hypothesis is that the mean of the difference scores isn't 0.

739. -0.9286

You calculate the mean of the difference scores, $\bar{d}$, by averaging the individual difference scores, with this formula:

$$\bar{d} = \frac{\sum_{i=1}^{n} d_i}{n}$$

where d_i represents the individual difference scores, and n is the number of pairs. So, substituting the numbers into the formula, you get

$$\bar{d} = \frac{-1.7 + (-2.2) + (-0.5) + (-0.8) + (-1.1) + 0.7 + (-0.9)}{7}$$

$$= \frac{-6.5}{7}$$

$$= -0.9286$$

740. 0.3483

To find the standard error, use this formula:

$$SE = \frac{s_d}{\sqrt{n}}$$

where s_d is the standard deviation of the difference scores in the sample, and n is the sample size.

Substitute the known values into the formula to get

$$SE = \frac{0.9214}{\sqrt{7}}$$

$$= 0.3483$$

---

741. 2.44691

To find the critical value, you need the degrees of freedom and the alpha level. The degrees of freedom are one less than the number of pairs: $7 - 1 = 6$. The alpha level is given.

For a *not equal to* alternative hypothesis, you want half the alpha value of 0.05 in the lower tail and half in the upper tail. Looking in the column for 0.025 and the row for 6 *df* in the *t*-table (Table A-2 in the appendix), you find the critical value 2.44691.

---

742. 3.70743

To find the critical value, you need the degrees of freedom and the alpha level. The degrees of freedom are one less than the number of pairs: $7 - 1 = 6$. The alpha level is given.

For a *not equal to* alternative hypothesis, you want half the alpha value of 0.01 in the lower tail and half in the upper tail. Looking in the column for 0.005 and the row for 6 *df* in the *t*-table (Table A-2 in the appendix), you find the critical value 3.70743.

---

743. −2.6661

Calculate the test statistic for a paired *t*-test using this formula:

$$t = \frac{\bar{d}}{SE}$$

You calculate the mean of the difference scores, $\bar{d}$, by averaging the individual difference scores with this formula.

$$\bar{d} = \frac{\sum\limits_{i=1}^{n} d_i}{n}$$

where the d_i are the individual difference scores, and n is the number of pairs. So, substituting the numbers into the formula, you get

$$\bar{d} = \frac{-1.7 + (-2.2) + (-0.5) + (-0.8) + (-1.1) + 0.7 + (-0.9)}{7}$$

$$= \frac{-6.5}{7}$$

$$= -0.9286$$

To find the standard error, use this formula:

$$SE = \frac{s_d}{\sqrt{n}}$$

where s_d is the standard deviation of the difference scores in the sample, and n is the sample size.

Substitute the known values into the formula to get

$$SE = \frac{0.9214}{\sqrt{7}}$$

$$= 0.3483$$

Plugging in the values, the test statistic is therefore

$$t = \frac{\bar{d}}{SE}$$

$$= \frac{-0.9286}{0.3483}$$

$$= -2.6661$$

744. Fail to reject the null hypothesis.

You calculate the mean of the difference scores, $\bar{d}$, by averaging the individual difference scores, with this formula:

$$\bar{d} = \frac{\sum\limits_{i=1}^{n} d_i}{n}$$

where d_i represents the individual difference scores, and n is the number of pairs. So, substituting the numbers into the formula, you get

$$\bar{d} = \frac{-1.7 + (-2.2) + (-0.5) + (-0.8) + (-1.1) + 0.7 + (-0.9)}{7}$$

$$= \frac{-6.5}{7}$$

$$= -0.9286$$

To find the standard error, use this formula:

$$SE = \frac{s_d}{\sqrt{n}}$$

where s_d is the standard deviation of the difference scores in the sample, and n is the sample size.

Substitute the known values into the formula to get

$$SE = \frac{0.9214}{\sqrt{7}}$$

$$= 0.3483$$

Then calculate the test statistic:

$$t = \frac{\bar{d}}{SE}$$

$$= \frac{-0.9286}{0.3483}$$

$$= -2.6661$$

To find the critical value, you need the degrees of freedom and the alpha level. The degrees of freedom are one less than the number of pairs: $7 - 1 = 6$, and the alpha level is given as 0.01.

For a *not equal to* alternative hypothesis, you want half the alpha value of 0.01 in the lower tail and half in the upper tail. Looking in the column for 0.005 and the row for 6 *df* in the *t*-table (Table A-2 in the appendix), you find the critical value 3.70743.

Because you have a *not equal to* alternative hypothesis, you'll reject the null hypothesis if the absolute value of your test statistic exceeds the critical value. However, the absolute value of your test statistic, 2.6661, is less than the critical value, so you don't reject the null hypothesis.

Your test statistic is closer to the mean than the critical value is, so you fail to reject the null hypothesis.

745. Reject the null hypothesis and conclude that there's a significant difference in battery life between the two brands.

You need to calculate a *t*-statistic and compare it to the critical value to decide whether to accept or reject the null hypothesis. The degrees of freedom for the test statistic are one less than the number of pairs: $7 - 1 = 6$.

For a *not equal to* alternative hypothesis, you want half the alpha value of 0.01 in the lower tail and half in the upper tail. Looking in the column for 0.005 and the row for 6 *df* in the *t*-table (Table A-2 in the appendix), you find the critical value 2.44691.

You calculate the mean of the difference scores, $\bar{d}$, by averaging the individual difference scores with this formula:

$$\bar{d} = \frac{\sum_{i=1}^{n} d_i}{n}$$

where the d_i are the individual difference scores, and n is the number of pairs. So, substituting the numbers into the formula, you get

$$\bar{d} = \frac{-1.7 + (-2.2) + (-0.5) + (-0.8) + (-1.1) + 0.7 + (-0.9)}{7}$$

$$= \frac{-6.5}{7}$$

$$= -0.9286$$

To find the standard error, use this formula:

$$SE = \frac{s_d}{\sqrt{n}}$$

where s_d is the standard deviation of the difference scores in the sample, and n is the sample size.

Substitute the known values into the formula to get

$$SE = \frac{0.9214}{\sqrt{7}}$$

$$= 0.3483$$

Then calculate the test statistic:

$$t = \frac{\bar{d}}{SE}$$

$$= \frac{-0.9286}{0.3483}$$

$$= -2.6661$$

Your test statistic is farther from the mean than the critical value is, so you will reject the null hypothesis. Your sample data provides you with sufficient evidence to reject the null hypothesis and conclude that there is a difference between the two brands.

746. a z-test of two population proportions

The research question is whether the proportion of security breaches differs between two independent populations, and the sample size is large enough to support a z-test.

747. $H_0: p_1 = p_2$; $H_a: p_1 \neq p_2$

The null and alternative hypotheses are always stated in terms of population parameters — in this case, population proportions p_1 and p_2. The null hypothesis is always a statement of equality; when the researcher has no initial hunch about the direction of population differences, the alternative hypothesis is written as *not equal to*, using the $\neq$ symbol.

748. 0.0211 and 0.05144

You calculate the sample proportions, $\hat{p}_1$ and $\hat{p}_2$, for each group by dividing the number of security breaches by the number of accounts observed.

Group 1 had 1,055 breaches in 50,000 cases, so the observed proportion is

$$\hat{p}_1 = \frac{1,055}{50,000} = 0.02110$$

Group 2 had 2,572 breaches in 50,000 cases, so the observed proportion is

$$\hat{p}_2 = \frac{2,572}{50,000} = 0.05144$$

749. 0.03627

You calculate the overall sample proportion, $\hat{p}$, by dividing the total number of security breaches by the total number of accounts observed. In this example, Group 1 had 1,055 breaches, and Group 2 had 2,572 breaches. Each group had 50,000 accounts.

$$\hat{p} = \frac{1,055 + 2,572}{50,000 + 50,000} = 0.03627$$

750.

0.0012

Calculate the standard error using the following formula, where $\hat{p}$ is the population proportion and n_1 and n_2 are the sample sizes:

$$SE = \sqrt{\hat{p}\,(1-\hat{p})\left(\frac{1}{n_1}+\frac{1}{n_2}\right)}$$

$$= \sqrt{0.03627(1-0.03627)\left(\frac{1}{50{,}000}+\frac{1}{50{,}000}\right)}$$

$$= \sqrt{0.034954(0.00004)}$$

$$= 0.0011824$$

751.

−25.66

Calculate the z-statistic using the following formula, where $\hat{p}$ is the population proportion and n_1 and n_2 are the sample sizes:

$$z = \frac{\hat{p}_1 - \hat{p}_2}{\sqrt{\hat{p}\,(1-\hat{p})\left(\dfrac{1}{n_1}+\dfrac{1}{n_2}\right)}}$$

$$= \frac{0.02110 - 0.05144}{\sqrt{0.03627(1-0.03627)\left(\dfrac{1}{50{,}000}+\dfrac{1}{50{,}000}\right)}}$$

$$= \frac{-0.03034}{\sqrt{0.034954(0.00004)}}$$

$$= \frac{-0.03034}{0.0011824}$$

$$= -25.6597, \text{ or } -25.66 \text{ (rounded)}$$

752.

Reject the null hypothesis and conclude that there's a significant difference in the security of the two types of passwords.

You will conduct a z-test for two population proportions, for a _not equal to_ question, using the null and alternative hypotheses:

$$H_0: p_1 = p_2$$

$$H_a: p_1 \neq p_2$$

To calculate the test statistic, you use this formula:

$$z = \frac{\hat{p}_1 - \hat{p}_2}{\sqrt{\hat{p}\,(1-\hat{p})\left(\dfrac{1}{n_1}+\dfrac{1}{n_2}\right)}}$$

where $\hat{p}_1$ and $\hat{p}_2$ are the two sample proportions, and n_1 and n_2 are the sample sizes.

You calculate the overall sample proportion, $\hat{p}$, by dividing the total number of security breaches by the total number of accounts observed. In this example, Group 1 had 1,055 breaches, and Group 2 had 2,572 breaches. Each group had 50,000 accounts.

$$\hat{p} = \frac{1,055 + 2,572}{50,000 + 50,000}$$
$$= \frac{3,627}{100,000}$$
$$= 0.03627$$

Then you find the sample proportions, $\hat{p}_1$ and $\hat{p}_2$, for each group by dividing the number of security breaches by the number of accounts observed.

Group 1 had 1,055 breaches in 50,000 cases, so the observed proportion is

$$\hat{p}_1 = \frac{1,055}{50,000}$$
$$= 0.02110$$

Group 2 had 2,572 breaches in 50,000 cases, so the observed proportion is

$$\hat{p}_2 = \frac{2,572}{50,000}$$
$$= 0.05144$$

Substitute the values in the formula and solve:

$$z = \frac{\hat{p}_1 - \hat{p}_2}{\sqrt{\hat{p}(1-\hat{p})\left(\frac{1}{n_1} + \frac{1}{n_2}\right)}}$$

$$= \frac{0.02110 - 0.05144}{\sqrt{0.03627(1 - 0.03627)\left(\frac{1}{50,000} + \frac{1}{50,000}\right)}}$$

$$= \frac{-0.03034}{\sqrt{0.034954(0.00004)}}$$

$$= \frac{-0.03034}{0.0011824}$$

$$= -25.6597, \text{ or } -25.66 \text{ (rounded)}$$

The critical value, using Table A-1, a significance level of 0.05, and a *not equal to* alternative hypothesis, means that the critical value for z is ± 1.96 (the value that leaves 2.5%, or 0.025, of the curve area in each tail). You'll reject the null hypothesis if the test statistic value of z is outside the range of -1.96 to $+1.96$.

Your test statistic of -25.66 is outside the range of -1.96 to 1.96, so you reject the null hypothesis. Thus, the more plausible explanation is the two-sided alternative hypothesis that there's a difference between the two sets of security rules. The alternative hypothesis didn't favor one set of security rules over the other, but because the test statistic is negative, you know that the first group had fewer security breaches than the second, so the extra rule about passwords seems to increase security.

753. 0.5 and 0.7

$\hat{p}_1$ and $\hat{p}_2$ are the sample proportions for Group 1 and Group 2. You calculate them by dividing the number of cells with cases of interest (behavioral disturbance) in each group by the sample size for each group.

$$\hat{p}_1 = \frac{50}{100} = 0.5$$

$$\hat{p}_2 = \frac{70}{100} = 0.7$$

754. 0.6

You find the overall sample proportion, $\hat{p}$, by dividing the total number of cells with cases of interest (those with a behavioral disturbance) by the total number of cases in the study.

$$\hat{p} = \frac{50+70}{100+100}$$

$$= \frac{120}{200}$$

$$= 0.6$$

755. 0.0693

Calculate the standard error by using this formula:

$$SE = \sqrt{\hat{p}\,(1-\hat{p})\left(\frac{1}{n_1} + \frac{1}{n_2}\right)}$$

where n_1 and n_2 are the two sample sizes, and the overall sample proportion, $\hat{p}$, is calculated by dividing the total number of cases of interest (those with a behavioral disturbance) by the total number of cases in the study.

$$\hat{p} = \frac{50+70}{100+100}$$

$$= \frac{120}{200}$$

$$= 0.6$$

Then, plug in the numbers for the standard error formula:

$$SE = \sqrt{\hat{p}\,(1-\hat{p})\left(\frac{1}{n_1} + \frac{1}{n_2}\right)}$$

$$= \sqrt{0.6(1-0.6)\left(\frac{1}{100} + \frac{1}{100}\right)}$$

$$= \sqrt{0.24(0.02)}$$

$$= 0.06928, \text{ or } 0.0693 \text{ (rounded)}$$

756. 2.58

This is a *not equal to* alternative hypothesis, so you'll split the probability of 0.01 between the upper and lower tails of the distribution, resulting in a value of 0.01/2 = 0.005 in each tail. Using Table A-1, you find that the critical value is roughly 2.58.

757. 2.33

This is a *greater than* alternative hypothesis, so you want the entire probability of 0.01 in the upper tail of the distribution. Using Table A-1, you find that the critical value is about 2.33.

758. −2.8868

To find the test statistic, use this formula:

$$z = \frac{\hat{p}_1 - \hat{p}_2}{\sqrt{\hat{p}(1-\hat{p})\left(\frac{1}{n_1} + \frac{1}{n_2}\right)}}$$

where n_1 and n_2 are the samples sizes for the two groups and $\hat{p}$ is the sample proportion. You find the value of $\hat{p}$ by dividing the total number of cells with a behavioral disturbance by the total number of cells in the study.

$$\hat{p} = \frac{50+70}{100+100}$$
$$= \frac{120}{200}$$
$$= 0.6$$

Now, plug in the numbers and solve:

$$z = \frac{\hat{p}_1 - \hat{p}_2}{\sqrt{\hat{p}(1-\hat{p})\left(\frac{1}{n_1} + \frac{1}{n_2}\right)}}$$

$$= \frac{0.5-0.7}{\sqrt{0.6(1-0.6)\left(\frac{1}{100} + \frac{1}{100}\right)}}$$

$$= \frac{-0.2}{\sqrt{(0.24)(0.02)}}$$

$$= \frac{-0.2}{\sqrt{0.0048}}$$

$$= -2.8868$$

759.

Reject the null hypothesis and conclude that there's a significant difference in behavioral disturbances between the two groups.

Compute the test statistic, using this formula:

$$z = \frac{\hat{p}_1 - \hat{p}_2}{\sqrt{\hat{p}(1-\hat{p})\left(\frac{1}{n_1} + \frac{1}{n_2}\right)}}$$

where n_1 and n_2 are the samples sizes for the two groups. You calculate the overall sample proportion, $\hat{p}$, by dividing the total number of cells with a behavioral disturbance by the total number of cells in the study.

$$\hat{p} = \frac{50+70}{100+100}$$
$$= \frac{120}{200}$$
$$= 0.6$$

Now, plug the numbers into the formula and solve:

$$z = \frac{\hat{p}_1 - \hat{p}_2}{\sqrt{\hat{p}(1-\hat{p})\left(\frac{1}{n_1} + \frac{1}{n_2}\right)}}$$

$$= \frac{0.5 - 0.7}{\sqrt{0.6(1-0.6)\left(\frac{1}{100} + \frac{1}{100}\right)}}$$

$$= \frac{-0.2}{\sqrt{(0.24)(0.02)}}$$

$$= \frac{-0.2}{\sqrt{0.0048}}$$

$$= -2.8868$$

The test statistic of –2.8868 is outside the range of –1.96 to 1.96, which are the critical values (from Table A-1) for a z-test for a *not equal to* alternative hypothesis with a significance level of 0.05. You therefore reject the null hypothesis. Thus, the more plausible explanation is the two-sided alternative hypothesis that there's a difference in behavior between allowing Internet use and not allowing it. As stated, the alternative hypothesis doesn't favor a particular course of action, but because the test statistic is negative, this shows that the second group (those without Internet access) had more behavioral disturbances than the first.

760. Reject the null hypothesis and conclude that there's a significant difference in behavioral disturbances between the two groups.

You calculate the overall sample proportion, $\hat{p}$, by dividing the total number of cells with a behavioral disturbance by the total number of cells in the study.

$$\hat{p} = \frac{50+70}{100+100}$$

$$= \frac{120}{200}$$

$$= 0.6$$

To find the test statistic, use this formula:

$$z = \frac{\hat{p}_1 - \hat{p}_2}{\sqrt{\hat{p}(1-\hat{p})\left(\frac{1}{n_1} + \frac{1}{n_2}\right)}}$$

where n_1 and n_2 are the samples sizes for the two groups.

Now, plug in the numbers and solve:

$$z = \frac{\hat{p}_1 - \hat{p}_2}{\sqrt{\hat{p}(1-\hat{p})\left(\frac{1}{n_1} + \frac{1}{n_2}\right)}}$$

$$= \frac{0.5-0.7}{\sqrt{0.6(1-0.6)\left(\frac{1}{100} + \frac{1}{100}\right)}}$$

$$= \frac{-0.2}{\sqrt{(0.24)(0.02)}}$$

$$= \frac{-0.2}{\sqrt{0.0048}}$$

$$= -2.8868$$

The test statistic of –2.8868 is outside the range of –2.58 to 2.58, which are the critical values (from Table A-1) for a z-test with a *not equal to* alternative hypothesis. You therefore reject the null hypothesis. Thus, the more plausible explanation is the two-sided alternative hypothesis that there's a difference in behavior between allowing Internet use and not allowing it. As stated, the alternative hypothesis doesn't favor a particular course of action, but because the test statistic is negative, this shows that the second group (those without Internet access) had more behavioral disturbances than the first.

761.   all adult drivers in the metro area

The target population is the people you want your results to apply to, which, in this example, is all adult drivers in the metro area.

762.   bias

Bias means that the results from your sample are unlikely to be true for the population as a whole in some systematic way.

763.   Choices (A), (B), and (C) (Not everyone has a phone or a listed phone number; not everyone is at home during the day on weekdays; not everyone is willing to participate in telephone polls.)

Using published phone directories as a sampling frame and scheduling calls during only one part of the day can introduce several types of bias to a study. First, people without a phone or a published phone number have no possibility of being selected for the sample. Second, people who aren't at home during the scheduled time can't supply data. And third, phone surveys in general are subject to non-response bias because a high proportion of people contacted may refuse to participate in the survey. All these factors may bias the sample and the data, so your results don't represent the target population.

764.   It indicates that one answer is preferred and may introduce bias.

In responsible polling, questions should be stated in a neutral manner.

765.   non-response bias

Non-response bias occurs when those among the sample who defer to take part in a study differ in some significant way from those who do take part.

766.   response bias

In this case, it's likely that some of the respondents weren't being truthful. This is when response bias occurs. For example, most people believe that voting in elections is a positive characteristic, so they're more likely to report having voted, even if they didn't.

767.   because it may be impossible to detect and, thus, impossible to correct

Many potential causes of bias in survey research exist, and they're particularly problematic because the survey results alone may not give any indication of what kinds of bias, if any, affected the results. Thus, it may be impossible to compensate or correct for any biases present in the data.

768. A. calculating the mean age of all the students by using their official records

You can reasonably assume that all students in the school have official records, and the records include age. This method is the only one that guarantees gathering of information from the entire target population, which is the definition of a census.

769. E. numbering the students by using the school's official roster and selecting the sample by using a random number generator

This method is the only one described that will result in a simple random sample. The other methods suggested are a stratified sample (classifying the students as male or female and drawing a random sample from each), a systematic sample (using an alphabetized student roster and selecting every 15th name, starting with the first one), a cluster sample (selecting three tables at random from the cafeteria during lunch hour and asking the students at those tables for their age), and a snowball sample (selecting one student at random, asking him or her to suggest three friends to participate and continuing in this fashion until you have your sample size).

770. undercoverage bias

Undercoverage bias results when part of the target population is excluded from the possibility of being selected for the sample. In this case, the exclusion is due to the sampling frame not including all the firm's current employees.

771. volunteer sample bias

You'll receive responses only from people who happen to be watching the program and then volunteer to participate in the survey. Because you didn't select them yourself beforehand, they don't make up a statistical sample, and they probably don't represent any real population of interest.

772. convenience sample bias

You're sampling and interviewing a sample of people that's the most convenient for you, and there's no way to know what population they represent (if any).

773. to reduce bias caused by always using positive or negative statements

Some people may have a tendency to agree to positive statements rather than the equivalent question written as a negative statement, so by including both types of statements, this bias can be reduced.

774. It asks two questions at once.

Because this survey item asks two questions in one (should everyone go to college, and should everyone seek gainful employment), some respondents may be confused about how to answer, especially if they agree with only one part of the question. For example, they may think that everyone should seek gainful employment but not necessarily go to college.

775. Choices (A) and (B) (because the respondent may be uninformed about the topic; because the respondent may not remember enough information to answer the question)

Including "don't know" as a category is needed for people who don't know the answer to a question, whether because it requires knowledge that they don't possess (about a world political situation, for example) or because they don't remember the information needed to answer the question (for example, what their typical diet was 30 years ago). If you're concerned about respondents finding a question offensive, include a separate category, such as "choose not to answer."

776. E. Choices (A) and (C) (strong; positive)

This scatter plot displays a strong, positive linear relationship ($r = 0.77$) between high-school and college GPA.

777. E. 4.0

The strong, positive linear relationship between high-school and college GPA suggests that, without having any further information, the student with the highest high-school GPA will also have the highest college GPA.

778. The points slope upward from left to right.

That the points slope upward from left to right means that lower values on the X variable tend to have lower values on the Y variable as well, and higher values on the X variable tend to have higher values on the Y variable.

779. The points cluster closely around a straight line.

This relationship isn't perfect (if it was, all the points would lie on a perfectly straight line), but the points do cluster closely around a line, so the relationship is fairly strong.

780. The points would all lie on a perfectly straight line sloping upward.

With a correlation of 1.0, all the points would lie perfectly on a straight line instead of just clustering around it. The line would slope upward from left to right.

781. D. Choices (A) and (C) (The points would all lie on a straight line; all the points would slope downward from left to right.)

The scatter plot of two variables with a correlation of –1.0 would have all the points lying perfectly on a line sloping downward from left to right.

782. C. weight

Scatter plots display the relationship between two quantitative variables. Of these choices, only weight is quantitative; the others are categorical. Zip codes can't be treated as a quantitative variable; for example, you can't find the average zip code for the United States.

783. E. Choices (A) and (D) (height and age; height and weight)

Scatter plots display the relationship between two quantitative variables. Of these variable pairs, only height and age and height and weight are quantitative.

784. B. −0.8

Among the choices, only −0.8 indicates a relationship that's both negative and strong.

785. C. 0.2

Among the choices, only 0.2 indicates a relationship that's both positive and weak.

786. E. 0.9

Among the choices, only 0.9 indicates a relationship that's both positive and very strong.

787. A. −0.2

Among the choices, only −0.2 indicates a relationship that's both negative and weak.

788. nonlinear

Linear relationships resemble a straight line. This relationship resembles a curve (in fact, it's a quadratic relationship) and is therefore nonlinear.

789. 0

Although these variables are strongly related, the relationship isn't linear. Correlation measures only linear relationships. In this case, there's no linear relationship.

790. Their relationship is nonlinear.

Correlation expresses the degree of linear relationship between two variables and isn't appropriate for nonlinear relationships.

791. A. –0.85

The strongest linear relationship is indicated by the largest absolute value of the correlation; a correlation of –0.85 and 0.85 represents equally strong relationships between the variables. The sign just indicates whether the relationship is uphill or downhill.

792. C. 0.1

The weakest relationship is indicated by the smallest absolute value of the correlation; a correlation of –0.1 and 0.1 represents equally weak relationships between the variables. The sign just indicates whether the relationship is uphill or downhill.

793. D. Choices (A) and (C) (–1; 1)

Both 1 and –1 represent perfect correlations (one uphill and one downhill).

794. 2.5

To calculate the mean, add together all the values and then divide by the total number of values:

$$\bar{x} = \frac{1+2+3+4}{4}$$
$$= \frac{10}{4} = 2.5$$

795. 2.75

To calculate the mean, add together all the values and then divide by the total number of values:

$$\bar{y} = \frac{2+2+4+3}{4}$$
$$= \frac{11}{4} = 2.75$$

796. 3

n is the number of cases (or pairs of data in this case). In this example, $n = 4$, so $n - 1 = 4 - 1 = 3$.

797. 1.29

To calculate the standard deviation of the X values, use this formula:

$$s_x = \sqrt{\frac{\sum (x-\bar{x})^2}{n-1}}$$

where x is a single value, $\bar{x}$ is the mean of all the values, Σ represents the sum of the squared differences from the mean, and n is the sample size.

$$s_x = \sqrt{\frac{(1-2.5)^2+(2-2.5)^2+(3-2.5)^2+(4-2.5)^2}{4-1}}$$

$$= \sqrt{\frac{2.25+0.25+0.25+2.25}{3}}$$

$$= \sqrt{\frac{5}{3}}$$

$$= 1.29099$$

798. 0.96

To calculate the standard deviation of the Y values, use this formula:

$$s_y = \sqrt{\frac{\Sigma(y-\bar{y})^2}{n-1}}$$

where y is a single value, $\bar{y}$ is the mean of all the values, Σ represents the sum of the squared differences from the mean, and n is the sample size.

$$s_y = \sqrt{\frac{(2-2.75)^2+(2-2.75)^2+(4-2.75)^2+(3-2.75)^2}{4-1}}$$

$$= \sqrt{\frac{0.5625+0.5625+1.5625+0.0625}{3}}$$

$$= \sqrt{\frac{2.75}{3}}$$

$$= 0.95743$$

799. 0.67

To calculate the correlation between X and Y, divide the sum of cross products by the standard deviations of x and y, and then divide the result by $n-1$.

For this example, the sum of cross products is 2.5, n is 4, the standard deviation of X is 1.29, and the standard deviation of Y is 0.96.

$$r = \frac{1}{n-1}\left(\frac{\sum_x\sum_y(x-\bar{x})(y-\bar{y})}{s_x s_y}\right)$$

$$= \frac{1}{3}\left[\frac{2.5}{(1.29)(0.96)}\right]$$

$$= 0.6729$$

800. 0.82

To calculate the correlation between X and Y, divide the sum of cross products by the standard deviations of X and Y, and then divide the result by $n-1$. In this case, the sum of cross products is 274, the standard deviation of X is 4.47, the standard deviation of Y is 5.36, and n is 15:

$$r = \frac{1}{n-1}\left(\frac{\sum_x\sum_y(x-\bar{x})(y-\bar{y})}{s_x s_y}\right)$$

$$= \frac{1}{14}\left[\frac{274}{(4.47)(5.36)}\right]$$

$$= 0.81686$$

801. It will decrease.

The correlation will decrease because you'd divide by a larger number ($n-1=19$) rather than $n-1=14$.

802. It will increase.

The correlation will increase because you'd divide by a smaller number (4.82 rather than 5.36).

803. It will increase.

The correlation will increase because the numerator would be larger (349 rather than 274).

804. It will stay the same.

The correlation is a unitless measure, so change in the units in which variables are measured won't change their correlation.

805. D.   Choices (A) and (C) (−2.64; 1.5)

Correlations are always between −1 and 1.

806. The two correlations will be the same.

In a correlation, it doesn't matter which variable is designated as X and which as Y; the correlation will be the same either way.

807. E. Choices (B) and (C) (the Y variable; the response variable)

This study examines whether text font size may influence reading comprehension, so it's logical to designate reading comprehension as the Y, or response, variable.

808. A. the X variable

This study examines whether text font size may influence reading comprehension, so it's logical to designate font size as the X, or independent, variable.

809. It won't change.

In a correlation, it doesn't matter which variable is designated as X and which as Y; the correlation will be the same either way.

810. It stays the same.

The correlation is a unitless measure, so a change in the units won't change the correlation.

811. You made a mistake in your calculations.

Correlations are always between −1 and 1, so if you get a value outside this range, you made a mistake in your calculations.

812. They have a strong, negative linear relationship.

A correlation of −0.86 indicates a strong, negative linear relationship between two variables.

813. They have a weak, positive linear relationship.

A correlation of 0.27 indicates a weak, positive linear relationship between two variables.

814. the X variable

The logical assumption in this study is that time spent studying influences grades (GPA), so it makes sense to designate "time spent studying" as the X, or explanatory, variable.

815. E. Choices (B) and (C) (the Y variable; the response variable)

The logical assumption in this study is that time spent studying influences GPA, so it's logical to designate "GPA" as the Y, or response, variable.

816. It doesn't change.

The correlation is a unitless measure, so changing the units in which variables are measured won't change their correlation.

817. It doesn't change.

In a correlation, it doesn't matter which variable is designated as X and which as Y; the correlation will be the same either way.

818. They have a weak, negative linear relationship.

A correlation of –0.23 indicates a weak, negative linear relationship between two variables.

819. You made a mistake in your calculations.

A correlation is always between –1 and 1, so a value outside that range indicates an error in calculation.

820. They have a strong, negative linear relationship.

A correlation of –0.87 indicates a strong, negative relationship.

821. E. Choices (A), (B), and (C) (Both variables are numeric; the scatter plot indicates a linear relationship; the correlation is at least moderate.)

Before computing a regression between two variables, you should determine that both variables are quantitative (numeric), that they're at least moderately correlated, and that the scatter plot indicates a linear relationship.

822. C. Their relationship isn't linear.

The variables are numeric, and a correlation of 0.75 is sufficient to perform linear regression. However, the relationship of the points in the scatter plot isn't linear. The points initially have a positive relationship but then curve downward into a negative relationship.

823. m

The slope of the equation is m. If two variables have a negative relationship, they will have a negative slope.

824. m

In the equation for the least-squares regression line, m designates the slope of the regression line.

825. *b*

In the equation for the least-squares regression line, *b* designates the *y*-intercept for the regression line.

826. 3

The equation for calculating the regression line is $y = mx + b$, and *m* represents the slope. So in the regression line $y = 3x + 1$, the slope is 3.

827. 1

The equation for calculating the regression line is $y = mx + b$, and *b* represents the *y*-intercept. So in the regression line $y = 3x + 1$, the *y*-intercept is 1.

828. 11.5

The equation for the regression line is $y = 3x + 1$. To find the value for *y* when $x = 3.5$, substitute 3.5 for *x* in the equation:

$$y = 3x + 1 = (3)(3.5) + 1 = 11.5$$

829. 2.2

The equation for the regression line is $y = 3x + 1$. To find the value for *y* when $x = 0.4$, substitute 0.4 for *x* in the equation:

$$y = 3x + 1 = (3)(0.4) + 1 = 2.2$$

830. It decreases by 1.8.

The equation for calculating the regression line is $y = mx + b$, and *m* represents the slope. In this case, the slope is –1.2. If *x* increases by 1.5, *y* changes by $(-1.2)(1.5) = -1.8$. This is a decrease of 1.8.

831. It increases by 2.76.

The equation for calculating the regression line is $y = mx + b$, and *m* represents the slope. In this case, the slope is –1.2. If *x* decreases by 2.3, *y* changes by $(-1.2)(-2.3) = 2.76$. This is an increase of 2.76.

832. (0, 0.74)

To find the *y*-intercept, or the point where the line intersects the *y*-axis, find the value of *y* when $x = 0$. To do this, substitute 0 for *x* in the equation:

$$y = -1.2x + 0.74 = (-1.2)(0) + 0.74 = 0.74$$

833. strong and positive

The correlation of 0.792 and the scatter plot showing the points clustering fairly closely around a line running upward from left to right indicate a strong, positive linear relationship.

834. 200 to 800 points

Looking at the scatter plot, you see that no values are outside the range of 200 to 800 for either variable.

835. 0.79

Because s_x and s_y are the same, the slope will be equal to the correlation (a rare occurrence.).

836. 0.792

The equation to calculate the slope is

$$m = r\left(\frac{s_y}{s_x}\right)$$

In this case, the correlation is 0.792, the standard deviation of y is 103.2, and the standard deviation of x is 103.2. Plug these numbers into the formula and solve:

$$m = r\left(\frac{s_y}{s_x}\right)$$

$$= 0.792\left(\frac{103.2}{103.2}\right) = 0.792$$

Note: Because s_x and s_y are the same, the slope will be equal to the correlation (a rare occurrence).

837. 107.8 points

The equation to calculate the y-intercept is $b = \bar{y} - m\bar{x}$.

In this case, you know that the mean of y is 506.1 and the mean of x is 502.9. To find the slope, divide the standard deviation of y by the standard deviation of x and then multiply by the correlation. In this case, the correlation is 0.792, the standard deviation of y is 103.2, and the standard deviation of x is 103.2.

$$m = r\left(\frac{s_y}{s_x}\right)$$

$$= 0.792\left(\frac{103.2}{103.2}\right) = 0.792$$

Now, plug the values into the formula for the y-intercept:

$$b = \bar{y} - m\bar{x}$$
$$= 506.1 - (0.792)(502.9)$$
$$= 506.1 - 398.2968$$
$$= 107.8032$$

838. $y = 0.792x + 107.8$

The equation for a regression line is $y = mx + b$. To find the calculated equation of this regression line, you first need to find the slope and y-intercept.

The equation to calculate the slope is

$$m = r\left(\frac{s_y}{s_x}\right)$$

In this case, the correlation (r) is 0.792, the standard deviation of y is 103.2, and the standard deviation of x is 103.2.

$$m = r\left(\frac{s_y}{s_x}\right)$$

$$= 0.792\left(\frac{103.2}{103.2}\right) = 0.792$$

The equation to calculate the y-intercept is $b = \bar{y} - m\bar{x}$. In this case, the mean of y is 506.1, the mean of x is 502.9, and the slope is 0.792:

$$b = \bar{y} - m\bar{x}$$
$$= 506.1 - (0.792)(502.9)$$
$$= 506.1 - 398.2968$$
$$= 107.8032$$

Now, having the values of m and b, you simply plug them into the equation of the regression line to get $y = 0.792x + 107.8$.

839. 289.96 points

To find the expected value of y (verbal score) when x (math score) is 230 points, substitute 230 for x in the equation and solve for y:

$$y = 0.792(230) + 107.8 = 289.96 \text{ points}$$

840.   166.32 points

First, you need to find the slope of this equation by dividing the standard deviation of y by the standard deviation of x and then multiplying by the correlation. In this case, the correlation is 0.792, the standard deviation of y is 103.2, and the standard deviation of x is 103.2.

$$m = r\left(\frac{s_y}{s_x}\right)$$

$$= 0.792\left(\frac{103.2}{103.2}\right) = 0.792$$

So for every one unit increase in x, you expect to see a 0.792 unit increase in y. In other words, if x (math score) is higher by 1 point, then y (verbal score) is expected to be 0.792 points higher. Here, Student A's math score (x) is 210 points higher, so you expect Student A's verbal score (y) to be $(0.792)(210) = 166.32$ points higher (on average).

841.   Student C's verbal score will be 39.6 points lower than Student D's verbal score.

First, you need to find the slope of this equation by dividing the standard deviation of y by the standard deviation of x and then multiplying by the correlation. In this case, the correlation is 0.792, the standard deviation of y is 103.2, and the standard deviation of x is 103.2.

$$m = r\left(\frac{s_y}{s_x}\right)$$

$$= 0.792\left(\frac{103.2}{103.2}\right) = 0.792$$

So for every one point increase in x (math score), you expect to see a 0.792 point increase in y (verbal score). The opposite also applies: A one unit decrease in x results in a 0.792 decrease in y. Here, Student C's math score (x) decreases by 50 points compared to Student D, so you expect Student C's verbal score (y) to decrease by $(0.792)(50) = 39.6$ points compared to Student D. (*Note:* This value isn't the actual verbal score; it's the amount of drop in Student C's verbal score.)

842.   A third variable could be causing the observed relationship between GRA_V and GRA_M.

Finding a correlation between two variables doesn't automatically establish a causal relationship between them. For example, an increase in drug dosage may cause a change in blood pressure, but an increase in shoe size doesn't cause an increase in height. A third variable could be related to the relationship. For example, some research shows that students who are good in music are more likely to be good in both math and in verbal abilities.

843. moderate and positive

The scatter plot and the computed correlation of 0.527 both indicate a moderate, positive linear relationship between home size in square feet and selling price.

844. E. 910 square feet

Although the linear relationship between home size and selling price is moderate ($r = 0.527$), it's positive. Therefore, unless you have further information to contradict the pattern, you can expect that larger houses will generally sell for higher prices.

845. 0.106

To calculate the slope of a regression line for x and y, divide the standard deviation of y by the standard deviation of x, and then multiply by the correlation. In this case, the standard deviation of y is 11.8, the standard deviation of x is 58.5, and the correlation is 0.527.

$$m = r\left(\frac{s_y}{s_x}\right)$$

$$= 0.527\left(\frac{11.8}{58.5}\right)$$

$$= 0.10630$$

846. 24.1

To calculate the intercept of the regression line for x and y, multiply the mean of x by the slope, and then subtract that product from the mean of y. In this case, you know that the mean of x is 915.1 and the mean of y is 121.1. But you need to find the slope. To calculate the slope of a regression line for x and y, divide the standard deviation of y by the standard deviation of x, and then multiply by the correlation. In this case, the standard deviation of y is 11.8, the standard deviation of x is 58.5, and the correlation is 0.527.

$$m = r\left(\frac{s_y}{s_x}\right)$$

$$= 0.527\left(\frac{11.8}{58.5}\right)$$

$$= 0.10630$$

Now, plug the values into the equation for the intercept:

$$b = \bar{y} - m\bar{x}$$

$$= 121.1 - (0.106)(915.1)$$

$$= 24.0994$$

847. $y = 0.106x + 24.1$

The equation for a regression line is $y = mx + b$. You can find the slope by dividing the standard deviation of y (11.8) by the standard deviation of x (58.5) and then multiplying by the correlation (0.527).

$$m = r\left(\frac{s_y}{s_x}\right)$$

$$= 0.527\left(\frac{11.8}{58.5}\right)$$

$$= 0.10630$$

To find the intercept, you multiply the mean of x (915.1) by the slope (0.106), and then subtract that product from the mean of y (121.1).

$$b = \bar{y} - m\bar{x}$$
$$= 121.1 - (0.106)(915.1)$$
$$= 24.0994$$

Now, plug these numbers into the equation for the regression line: $y = 0.106x + 24.1$.

848. $130,100

First, you have to find the regression equation for relating square feet (x) to selling price (y). You do that by calculating the slope and the intercept, using the information provided:

$$m = r\left(\frac{s_y}{s_x}\right)$$

$$= 0.527\left(\frac{11.8}{58.5}\right)$$

$$= 0.10630$$
$$b = \bar{y} - m\bar{x}$$
$$= 121.1 - (0.106)(915.1)$$
$$= 24.0994$$

Then you plug these numbers into the regression line equation: $y = 0.106x + 24.1$.

To find the expected selling price measured in thousands of dollars for a house of 1,000 square feet, substitute 1,000 for x in the equation:

$$y = 0.106(1,000) + 24.1 = 130.1$$

Finally, you convert that to whole dollars by multiplying by $1,000: 130.1($1,000) = $130,100.

849. Cannot make an appropriate price prediction for a house of 1,500 square feet.

The sizes of the houses in this data set range from 700 to 1,000 square feet on the scatter plot. Therefore, making predictions for price is only appropriate for houses whose square footage lies within this range or is close to it. (This house has 1,500 square feet, so you can't make an appropriate price prediction.) If you did make a prediction for a house outside of the range of the data, you'd be committing an error called extrapolation.

850. $118,440

First, you have to find the regression equation for relating square feet (x) to selling price (y). You do that by calculating the slope and the intercept, using the information provided:

$$m = r\left(\frac{s_y}{s_x}\right)$$

$$= 0.527\left(\frac{11.8}{58.5}\right)$$

$$= 0.10630$$

$$b = \bar{y} - m\bar{x}$$

$$= 121.1 - (0.106)(915.1)$$

$$= 24.0994$$

Then, you plug these numbers into the regression line equation: $y = 0.106x + 24.1$.

To find the expected selling price measured in thousands of dollars for a house of 890 square feet, substitute 890 for x in the equation:

$$y = 0.106(890) + 24.1 = 118.44$$

Finally, convert to whole dollars by multiplying by $1,000: 118.44($1,000) = $118,440.

851. $9,540 more

First, you have to find the slope by dividing the standard deviation of y by the standard deviation of x and then multiplying by the correlation. In this case, the standard deviation of y is 11.8, the standard deviation of x is 58.5, and the correlation is 0.527:

$$m = r\left(\frac{s_y}{s_x}\right)$$

$$= 0.527\left(\frac{11.8}{58.5}\right)$$

$$= 0.10630$$

Due to the slope, for every unit change in square footage, price increases by 0.106 thousand dollars: $90(0.106) = 9.54$. So a house that's 90 square feet larger is expected to cost $9,540 more.

852. House C should cost about $5,720 less than House D.

First, you have to find the slope by dividing the standard deviation of y by the standard deviation of x and then multiplying by the correlation. In this case, the standard deviation of y is 11.8, the standard deviation of x is 58.5, and the correlation is 0.527:

$$m = r \left(\frac{s_y}{s_x} \right)$$

$$= 0.527 \left(\frac{11.8}{58.5} \right)$$

$$= 0.10630$$

Due to the slope, for every unit change in square footage, price increases by 0.106 thousand dollars, and for every unit decrease in square footage, price decreases by 0.106 thousand dollars. In this case, the difference in size is a decrease in 54 square feet; the difference in price is a decrease of $(54)(0.106) = 5.724$. Convert to whole dollars by multiplying by $1,000: $5.724(\$1,000) = \$5,724$, or about $5,720.

853. moderate negative

These points are mostly clustered around a line running from the upper left to lower right, indicating a moderate negative linear relationship. They're not tightly packed enough around a line to consider the linear relationship to be "strong."

854. −0.5

These variables have a moderate negative correlation, as evidenced by their loose clustering around a line running from the upper left to the lower right. In fact, their correlation is −0.54.

855. It wouldn't change.

Correlation measures the strength of the pattern around a line as well as the direction of the line (uphill or downhill). When you switch X and Y, you don't change the strength of their relationship or the direction of the relationship. For example, if the correlation between height and weight is −0.5, the correlation between weight and height is still −0.5.

856. Yes, because they're moderately correlated and the points suggest a linear trend.

The scatter plot indicates a possible linear relationship between the variables, and the correlation coefficient of −0.5 typically has an absolute value high enough to justify beginning a linear regression analysis.

857. impossible to tell without looking at the scatter plot

A moderate correlation alone isn't sufficient to indicate that two variables are good candidates for linear regression. You also need to make a scatter plot to confirm whether their relationship is at least approximately linear. In some cases, the scatter plot shows a bit of a curve, but the correlation is still moderately high; or the correlation is weak, but the scatter plot is made to look like a strong relationship exists.

858. no, because the correlation isn't high enough to justify a linear regression analysis

A correlation of 0.05 is barely better than chance and, by itself, doesn't indicate that two variables are good candidates for linear regression. You don't even have to look at the scatter plot to know that this isn't a good linear relationship.

859. E. Choices (A) and (D) (–0.9; 0.9)

The numerical part (absolute value) of the correlation is the part that measures the strength of the relationship; the sign on the correlation determines only the direction (uphill or downhill). The correlations 0.9 and –0.9 have the same strength and are the strongest of the choices on the list (because their absolute values are closest to 1).

860. 0.65

Correlation measures the strength and direction of the linear relationship between two variables and nothing more. Switching the X and Y variables has no influence on the value of a correlation, so the correlation of weight and height is the same as the correlation of height and weight.

861. It won't change.

Correlation has a universal interpretation — in other words, it doesn't depend on the units of the variables. By design, changing units has no effect on the size or direction of a correlation. Correlation is a "unit-free" statistic.

862. The linear relationship is stronger for males than for females.

The linear relationship between height and weight is stronger for males, as indicated by the higher correlation, and is weaker for females, as indicated by the lower correlation. (Notice that you can't add correlations together for two groups; the correlation must be between –1 and 1.)

863. A. –1.5

Correlations are always between –1 and +1, and –1.5 is outside of this range.

864. no, because correlation alone doesn't establish a causal linear relationship between two variables

An observed correlation, however strong, isn't sufficient to establish a causal linear relationship. The observed correlation could be due to numerous reasons other than a causal linear relationship, including confounding variables (variables not included in the study that could affect the results), such as age or gender. For example, if you want to lower your weight, you can't lower your height.

865. height

Height is the X variable because you're using height to predict weight. Height is the independent (explanatory) variable in the equation, and weight is the dependent (response) variable.

866. strong positive linear

Because a linear trend exists in the scatter plot of two variables, a correlation of 0.74 indicates a strong positive linear relationship between the two variables. In this case, larger numbers of minutes studying are associated with higher GPAs, and smaller numbers of minutes studying correspond with lower GPAs.

867. E. 450 minutes

Given a strong positive correlation between minutes studying and GPA, you can predict that students who spend the most minutes studying would have the highest GPAs on average. This isn't necessarily true for each student; it's a prediction based on the information given.

868. weak negative

A correlation of –0.38 indicates a weak negative linear relationship between two variables. In this case, high numbers of minutes watching TV are generally associated with low GPA, and low numbers of minutes watching TV generally correspond with high GPA, but the linear relationship is fairly weak.

869. A. 30

The linear relationship between minutes watching TV and GPA is moderately strong and negative; because the scatter plot looks good and the correlation is moderately strong, you'd predict that the student who spends the least amount of time watching TV would have the highest GPA.

870. GPA

Although this study can't establish a causal linear relationship merely from the linear relationship between these two variables, it's interesting to see whether time spent studying signifies a higher GPA instead of the other way around.

871. B. . The linear relationship is stronger for engineering majors.

The closer the absolute value of the correlation is to 1, the stronger the linear relationship becomes. The correlation of 0.48 for English majors is lower than that of 0.78 for engineering majors, indicating a stronger linear relationship for engineering majors and a weaker linear relationship for English majors.

872. You made a mistake in your calculations.

The value of a correlation must be between –1 and +1. The value of –2.56 is outside this range and thus indicates that you made a mistake in your calculations.

873. income

Income is the X variable because you're using income to predict life satisfaction. Income is the independent (explanatory) variable in the equation, and life satisfaction is the dependent (response) variable.

874. 0.58

To find the slope of a regression line, use this formula:

$$m = \left(r \frac{s_y}{s_x} \right)$$

where s_y is the standard deviation of y, s_x is the standard deviation of x, and r is the correlation. In this case, the X variable is income because it's being used to predict life satisfaction (Y). Substitute the values into the formula to solve:

$$m = \left(r \frac{s_y}{s_x} \right)$$

$$= \left[0.77 \left(\frac{12.5}{16.7} \right) \right]$$

$$\approx 0.58$$

875. 13.7

To find the y-intercept of a regression line, use this formula:

$$b = \bar{y} - m\bar{x}$$

where $\bar{x}$ is the mean of x, m is the slope, and $\bar{y}$ is the mean of y. In this case, the X variable is income because it's being used to predict life satisfaction (Y). Substitute the values into the formula to solve:

$$b = \bar{y} - m\bar{x}$$

$$= 60.4 - 0.58 (80.5)$$

$$= 60.4 - 46.69$$

$$= 13.7$$

876.

$y = 0.58x + 13.7$

This equation predicts life satisfaction (Y) from income (X), with a slope of 0.58 and a y-intercept of 13.7.

877.

E. Choices (B) and (C) (predicting satisfaction for someone with an income of $45,000; predicting satisfaction for someone with an income of $200,000)

In regression, extrapolation means using an equation to make predictions outside of the range of data used to make the equation. In this case, incomes of $45,000 and $200,000 fall outside the range of data used to find the regression equation. Because you don't have data that extends that far, you can't be sure that the same linear trend exists for those values.

878.

$190.00

To figure out the predicted cost, use the equation $y = 50x + 65$, replacing x with the given number of hours to complete the job. In this case, $x = 2.5$, so $y = 50(2.5) + 65 = 190.00$.

879.

$302.50

To figure out the predicted cost of a job, use the equation $y = 50x + 65$, replacing x with the given number of hours to complete the job. In this case, $x = 4.75$, so $y = 50(4.75) + 65 = 302.50$.

880.

$12.50

You can solve this problem in two ways.

First, the slope measures the change in cost (Y) for a given change in the number of hours (X). So you can simply calculate the change in hours (3.75 − 3.50 = 0.25), and then multiply by slope (50) to get the difference in cost, $(0.25)(50) = \$12.50$.

Second, you can calculate the costs based on both number of hours, and then take the difference. So substitute $x = 3.75$ (hours) into the equation, and substitute $x = 3.50$ (hours) into the equation, calculate their y values (costs), and subtract. So you have

$y = 50(3.75) + 65 = 252.50$

$y = 50(3.50) + 65 = 240.00$

Subtract these two values to get $252.50 − \$240.00 = \$12.50$.

This means the job is predicted to cost $12.50 more if the hours increase from 3.50 to 3.75.

881.

$175

If the intercept is $75 while the slope remains the same, the new equation for predicting costs will be $y = 50x + 75$.

In this case, $x = 2$, so $y = 50(2) + 75 = 175$.

882. $10

Because the slope is the same in the two equations, you need to consider only the difference in intercept: $75 – $65 = $10.

883. $203

If the slope is $60 and the intercept is $65, the equation to predict costs is $y = 60x + 65$. In this case, $x = 2.3$, so $y = 60(2.3) + 65 = 203$.

884. $36

You can solve this problem in two ways.

The equation of the regression line predicting costs in the current year is $y = 50x + 65$. The equation for the earlier year is $y = 60x + 65$.

First, the slope (50) measures the change in cost (y) for a given change in the number of hours (x). So you can simply calculate the change in slope ($60 – 50 = 10$), then multiply by hour (3.6) to get the difference in cost, $(10)(3.6) = 36.00$.

Second, you can calculate the costs based on both number of hours, then take the difference. So substitute $x = 3.6$ (hours) into both equations, calculate the cost in each year, and subtract.

$$y = 50(3.6) + 65 = 245$$

$$y = 60(3.6) + 65 = 281$$

Subtract these two values to get $281 – 245 = 36.

Costs in the previous year would have been $36 higher than in the current year for a job requiring 3.6 hours to complete.

885. y-intercept = 62; slope = 1.4

This equation is written in the form $y = mx + b$, where m is the slope and b is the y-intercept. Therefore, in the regression equation $y = 1.4x + 62$, 62 is the y-intercept and 1.4 is the slope.

886. the amount of change expected in y for a one-unit change in x

The slope is the amount of change you can expect in the value of the y variable when the x variable changes by one unit.

887. the point where the regression line crosses the y-axis

The y-intercept is the value of y when $x = 0$. This is the same as the point where the regression line crosses the y-axis. If the value of $x = 0$ isn't within the range of observed values of x, then you can't interpret the y value at that point. However, you still use the y-intercept as part of the equation of the regression line.

888. 90 points

To find the expected job satisfaction rating, use the regression equation for predicting job satisfaction (y), substitute the value for x (years of experience), and calculate y:

$$y = 1.4x + 62$$

In this case, $x = 20$, so $y = 1.4(20) + 62 = 90$.

So the expected job satisfaction rating for an employee with 20 years of experience, on a scale of 0 to 100, is 90 points.

889. 64.8 points

To find the expected job satisfaction rating, use the regression equation for predicting job satisfaction (y), substitute the value for x (years of experience), and calculate y:

$$y = 1.4x + 62$$

In this case, $x = 2$, so $y = 1.4(2) + 62 = 64.8$.

So the expected job satisfaction rating for an employee with 2 years of experience, on a scale of 0 to 100, is 64.8 points.

890. 9.8 points

To find the expected difference in job satisfaction ratings, use the regression equation for predicting job satisfaction (y), substitute the value for x (years of experience), and calculate y:

$$y = 1.4x + 62$$

In this case, you substitute both years of experience ($x = 15$ years and $x = 8$ years) into the equation, calculate their y values (satisfaction), and then find the difference, like so:

$$y = 1.4(15) + 62 = 83$$

$$y = 1.4(8) + 62 = 73.2$$

$$83 - 73.2 = 9.8$$

Or, because the slope (1.4) measures the change in satisfaction for a given change in years of experience (x), you can simply calculate the change in job satisfaction rating ($15 - 8 = 7$) and then multiply by slope (1.4) to get the difference: $1.4(7) = 9.8$.

Whatever method you choose, the answers are the same: The person with 15 years of experience is expected to rate 9.8 points higher on job satisfaction than someone with 8 years of experience.

891. 21.0 points

To find the expected difference in job satisfaction ratings, use the regression equation for predicting job satisfaction (y), substitute the value for x (years of experience), and calculate y:

$$y = 1.4x + 62$$

In this case, you substitute both years of experience ($x = 15$ years and $x = 0$ years) into the equation, calculate their y values (satisfaction) and then find the difference, like so:

$$y = 1.4(15) + 62 = 83$$

$$y = 1.4(0) + 62 = 62$$

$$83 - 62 = 21$$

Or, because the slope (1.4) measures the change in satisfaction for a given change in years of experience (x), you can simply calculate the change in job satisfaction rating ($15 - 0 = 15$) and then multiply by slope (1.4) to get the difference: $1.4(15) = 21$.

Whatever method you choose, the answers are the same: The person with 15 years of experience is expected to rate 21 points higher on job satisfaction than someone with 0 years of experience.

892. 16.1 points

To find the expected difference in job satisfaction ratings, use the regression equation for predicting job satisfaction (y), substitute the value for x (years of experience), and calculate y:

$$y = 1.4x + 62$$

In this case, you substitute both years of experience ($x = 11.5$ years and $x = 0$ years) into the equation, calculate their y values (satisfaction), and then find the difference, like so:

$$y = 1.4(11.5) + 62 = 78.1$$

$$y = 1.4(0) + 62 = 62$$

$$78.1 - 62 = 16.1$$

Or, because the slope (1.4) measures the change in satisfaction for a given change in years of experience (x), you can simply calculate the change in job satisfaction rating ($11.5 - 0 = 11.5$) and then multiply by slope (1.4) to get the difference: $1.4(11.5) = 16.1$.

Whatever method you choose, the answers are the same: The person with 15 years of experience is expected to rate 16.1 points higher on job satisfaction than someone with 0 years of experience.

893. 5

The regression lines have the same slope but different y-intercepts. Therefore, for employees from the two companies with the same years of experience, the only difference in their predicted job satisfaction ratings is the difference in y-intercepts: $67 - 62 = 5$.

894. $100,100

To find the expected market value, use the regression equation for Community 1 and substitute the value for x (square footage):

$$y_1 = 77x_1 - 15,400$$
$$= 77(1,500) - 15,400$$
$$= 115,500 - 15,400$$
$$= 100,100$$

So the expected market value for a home of 1,500 square feet in Community 1 is $100,100.

895. $126,280

To find the expected market value, use the regression equation for Community 1 and substitute the value for x (square footage):

$$y_1 = 77x_1 - 15,400$$
$$= 77(1,840) - 15,400$$
$$= 141,680 - 15,400$$
$$= 126,280$$

So the expected market value for a home of 1,840 square feet in Community 1 is $126,280.

896. $99,700

To find the expected market value, use the regression equation for Community 2 and substitute the value for x (square footage):

$$y_2 = 74x_2 - 11,300$$
$$= 74(1,500) - 11,300$$
$$= 111,000 - 11,300$$
$$= 99,700$$

So the expected market value for a home of 1,500 square feet in Community 2 is $99,700.

897. $61,220

To find the expected market value, use the regression equation for Community 2 and substitute the value for x (square footage):

$$y_2 = 74x_2 - 11,300$$
$$= 74(980) - 11,300$$
$$= 72,520 - 11,300$$
$$= 61,220$$

So the expected market value for a home of 980 square feet in Community 2 is $61,220.

898. The home in Community 2 has an expected market value of $1,100 greater than the home in Community 1.

Because the slope and y-intercept are both different for the two equations, you need to calculate the expected value for each home and then find their difference. In both cases, $x = 1,000$.

For the home in Community 1, the expected market value is

$$y_1 = 77x - 15,400$$
$$= 77(1,000) - 15,400$$
$$= 77,000 - 15,400$$
$$= 61,600$$

For the home in Community 2, the expected market value is

$$y_2 = 74x - 11,300$$
$$= 74(1,000) - 11,300$$
$$= 74,000 - 11,300$$
$$= 62,700$$

The difference in these two values is $62,700 - 61,600 = 1,100$, with the home in Community 2 having the greater value.

899. The home in Community 2 has an expected market value of $760 less than the home in Community 1.

Because the slope and y-intercept are both different for the two equations, you need to calculate the expected value for each home and then find their difference. In both cases, $x = 1,620$.

For the home in Community 1, the expected market value is

$$y_1 = 77x - 15,400$$
$$= 77(1,620) - 15,400$$
$$= 124,740 - 15,400$$
$$= 109,340$$

For the home in Community 2, the expected market value is

$$y_2 = 74x - 11,300$$
$$= 74(1,620) - 11,300$$
$$= 119,880 - 11,300$$
$$= 108,580$$

The difference in these two values is $108,580 - 109,340 = -760$, with the home in Community 2 having the lesser value.

900. The home in Community 1 has an expected market value of $1,690 more than the home in Community 2.

Because the slope and y-intercept are both different for the two equations, you need to calculate the expected value for each home and then find their difference. In both cases, $x = 1,930$.

For the home in Community 1, the expected market value is

$y_1 = 77x - 15,400$

$= 77(1,930) - 15,400$

$= 148,610 - 15,400$

$= 133,210$

For the home in Community 2, the expected market value is

$y_2 = 74x - 11,300$

$= 74(1,930) - 11,300$

$= 142,820 - 11,300$

$= 131,520$

The difference in these two values is $131,520 - 133,210 = -1,690$, with the home in Community 1 having the greater value.

901. The home in Community 1 has an expected market value of $8,200 less than the home in Community 2.

Because the slope and y-intercept are both different for the two equations, you need to calculate the expected value for each home and then calculate their difference. In this case, $x_1 = 1,100$ and $x_2 = 1,200$.

For the home in Community 1, the expected market value is

$y_1 = 77x - 15,400$

$= 77(1,100) - 15,400$

$= 84,700 - 15,400$

$= 69,300$

For the home in Community 2, the expected market value is

$y_2 = 74x - 11,300$

$= 74(1,200) - 11,300$

$= 88,800 - 11,300$

$= 77,500$

The difference in these two values is $77,500 - 69,300 = 8,200$, with the home in Community 1 having the lesser value.

902. E. Choices (B) and (C) (estimating the market value of a home in Community 1 with 2,900 square feet; estimating the market value of a home in Community 1 with 750 square feet)

Extrapolation means applying a regression equation for values beyond the range included in the data set used to create the regression equation. In this case, both 750 square feet and 2,900 square feet are outside the range of the data set used to create the equation. Because you don't have data that extends that far, you can't be sure that the same linear trend exists for those values.

903. 545

To find the answer, substitute the given x-value of 360 into the equation and solve:

$SAT = 725 - 0.5(360) = 545$

904. 425

To find the answer, substitute the given x-value of 600 into the equation and solve:

$$SAT = 725 - 0.5(600) = 425$$

905. Student A, Student C, Student B

Because the coefficient for minutes of TV watching is negative, students who watch the least amount of TV will have the highest predicted SAT scores. In this example, Student A watches the least amount of TV, followed by Student C and Student B.

906. $42,700

Use the regression equation for Company 1 and substitute the x_1 value of 6:

$$y_1 = 6.7(6) + 2.5$$
$$= 40.2 + 2.5$$
$$= 42.7$$

This answer is in thousands of dollars, so multiply by $1,000: 42.7($1,000) = $42,700.

907. $116,400

Use the regression equation for Company 1 and substitute the x_1 value of 17:

$$y_1 = 6.7(17) + 2.5$$
$$= 113.9 + 2.5$$
$$= 116.4$$

This answer is in thousands of dollars, so multiply by $1,000: 116.4($1,000) = $116,400.

908. $19,200

Use the regression equation for Company 2 and substitute the x_2 value of 2.5:

$$y_2 = 7.2(2.5) + 1.2$$
$$= 18 + 1.2$$
$$= 19.2$$

This answer is in thousands of dollars, so multiply by $1,000: 19.2($1,000) = $19,200.

909. $43,200

You can find the exact difference in two ways.

First, you can calculate the salary of each employee based on his/her respective years of experience and then find the difference, like so:

$$7.2(13) + 1.2 = 94.8$$
$$7.2(7) + 1.2 = 51.6$$
$$94.8 - 51.6 = 43.2$$

Convert this difference to thousands of dollars: 43.2($1,000) = $43,200.

Or, because the slope measures the change in salary (y) for a given change in the years of experience (x), you can simply calculate the difference in years (13 – 7 = 6) and then multiply by the slope to get the difference in salary:

$$y = (6)(7.2) = 43.2$$

Convert to whole dollars: 43.2($1,000) = $43,200.

Whatever method you choose, the answers are the same: You expect a part-time employee from Company 2 with 13 years of experience to make $43,200 more than one with 7 years of experience.

910. $38,160

You can find the exact difference in two ways.

First, you can calculate the salary of each employee based on his/her respective years of experience and then find the difference, like so:

$$7.2(6.5) + 1.2 = 48$$
$$7.2(1.2) + 1.2 = 9.84$$
$$48 - 9.84 = 38.16$$

Convert this difference to thousands of dollars: 38.16($1,000) = $38,160.

Or, because the slope measures the change in salary (y) for a given change in the years of experience (x), you can simply calculate the difference in years (6.5 – 1.2 = 5.3) and then multiply by the slope to get the difference in salary:

$$y = (5.3)(7.2) = 38.16$$

Convert to thousands of dollars: 38.16($1,000) = $38,160.

Whatever method you choose, the answers are the same: You expect a part-time employee from Company 2 with 6.5 years of experience to make $38,160 more than one with 1.2 years of experience.

911. Company 1

The starting salary is the salary associated with 0 years of experience. This means that the x terms fall out of the equations (being multiplied by 0, they'll both equal 0), and you find the answer by comparing the y-intercepts. A larger y-intercept means a larger starting salary.

For Company 1, the y-intercept is 2.5. For Company 2, the y-intercept is 1.2. Therefore, the starting salary is higher at Company 1.

912. Company 2

The rate of increase is the amount that salary is expected to increase with each additional year of employment. To find which is higher, compare the slopes; the company with the larger slope will have the higher rate of increase.

For Company 1, the slope is 6.7. For Company 2, the slope is 7.2. Therefore, the rate of increase is higher at Company 2.

913. The employee at Company 1 is expected to make $18,200 less than the other.

Because the slopes and y-intercepts both differ in these equations, you must calculate the expected salary for each case and then compare them. For this example, $x_1 = 3$ and $x_2 = 5.5$.

For the employee at Company 1, the expected salary (in thousands of dollars) is

$$y_1 = 6.7x + 2.5$$
$$= 6.7(3) + 2.5$$
$$= 20.1 + 2.5$$
$$= 22.6$$

For the employee at Company 2, the expected salary (in thousands of dollars) is

$$y_2 = 7.2x + 1.2$$
$$= 7.2(5.5) + 1.2$$
$$= 39.6 + 1.2$$
$$= 40.8$$

The difference between the two expected salaries is $40.8 - 22.6 = 18.2$.

This answer is in thousands of dollars, so multiply by $1,000: 18.2($1,000) = $18,200. The employee at Company 2 has the higher expected salary.

914. The employee at Company 1 is expected to make $4,200 less than the other.

Because the slopes and y-intercepts both differ in these equations, you must calculate the expected salary for each case and then compare them. For this example, $x_1 = 3.8$ and $x_2 = 4.3$.

For the employee at Company 1, the expected salary (in thousands of dollars) is

$$y_1 = 6.7x + 2.5$$
$$= 6.7(3.8) + 2.5$$
$$= 25.46 + 2.5$$
$$= 27.96$$

For the employee at Company 2, the expected salary (in thousands of dollars) is

$$y_2 = 7.2x + 1.2$$
$$= 7.2(4.3) + 1.2$$
$$= 30.96 + 1.2$$
$$= 32.16$$

The difference between the two expected salaries is $32.16 - 27.96 = 4.2$.

This answer is in thousands of dollars, so multiply by $1,000: 4.2($1,000) = $4,200. The employee at Company 2 has the higher expected salary.

915. E. Choices (A), (B), and (C) (replication of this study at other companies; a longitudinal study tracking growth in individual employee salaries as years of employment increase; adding additional variables to the model to control for other influences on salary)

Although observing a strong linear relationship between salary and years of experience isn't sufficient to draw causal inferences, additional information can strengthen your ability to draw such inferences. Examples of studies that can strengthen causal inference include replication studies, longitudinal studies, and studies including additional variables to control for other influences on the outcome variable.

916. D. Choices (A) and (B) (blood type; country of origin)

Blood type (A, AB, B, or O) and country of origin are both categorical because the data can take on a limited set of values, and the values have no numeric meaning. Annual income is continuous rather than categorical because the data can take on an infinite number of values, and the values have numeric meaning.

917. E. Choices (A), (B), and (C) (gender; hair color; zip code)

Gender, hair color, and zip code are all categorical because the data can take on a limited set of values, and the values have no numeric meaning. Although zip codes are written with numeric symbols (for example, 10024), the digits are symbols rather than numbers that can be added, subtracted, and so forth.

918. E. Choices (C) and (D) (homeownership [yes/no]; gender)

For a two-way table, the data must be categorical. Years of education and height are numerical and can take on an undetermined number of possible values and can even be considered continuous, not categorical. Gender and homeownership are categorical and can be used in a two-way table.

919. E. Choices (A), (B), and (C) (whether someone is a high-school graduate; whether someone is a college graduate; highest level of school completed)

Whether someone is a high-school graduate and *whether someone is a college graduate* both clearly have only two possible values — yes or no — and can be considered appropriate for a two-way table. *Highest level of school completed* is also a categorical variable whose possible values can be listed (for example, "less than high school," "high-school graduate," "some college," "college graduate," and "graduate school"). Therefore, it's also appropriate for a two-way table.

920. 4

A 2-x-2 table is a term that means a two-way table with exactly two rows and two columns. The two rows represent the two possible categories for one of the variables, and the two columns represent the two possible categories for the other variable. To find the number of cells in a table, multiply the number of rows by the number of columns. A 2-x-2 table has two rows and two columns, so it has four cells: (2)(2) = 4. These represent the four possible combinations of the values of the two variables.

921. the number of all students polled who are both male and favor an increase

In a cross-tabulation table, the number in an individual cell represents the cases that have the characteristics described by the row and column intersecting at that cell. In this example, 72 is in the cell at the intersection of the row for *Male* and the column for *Favor Fee Increase* so 72 represents the number of students who are both male and favor a fee increase.

922. the number of all students polled who are female and not in favor of the increase

In a cross-tabulation table, the number in an individual cell represents the cases that have the characteristics described by the row and column intersecting at that cell. In this example, 132 is in the cell at the intersection of the row for *Female* and the column for *Do Not Favor Fee Increase* so 132 represents the number of students who are both female and do not favor a fee increase.

923. 48

To find the number of students polled who are both female and favor the fee increase, find the cell where the row for *Female* and the column for *Favor Fee Increase* intersect. The value in this cell is 48. The keyword in this question is *and*, which means intersection.

924. 180

To find the total number of male students included in this poll (out of all 360 students polled), add the values in the row labeled *Male:* 72 + 108 = 180.

This sum represents all the students who are male and in favor of the increase plus all the students who are male and not in favor of the increase.

925. 180

To find the total number of female students included in this poll (out of all 360 students polled), add the values in the row labeled *Female:* 48 + 132 = 180.

This sum represents all the students who are female and in favor of the increase plus all the students who are female and not in favor of the increase.

926. 120

To find the total number of students who favor the fee increase (out of all 360 students polled), add the values in the column labeled *Favor Fee Increase:* 72 + 48 = 120.

This sum represents all the students who are in favor of the increase and male plus all the students who are in favor of the increase and female.

927. 240

To find the total number of students who don't favor the fee increase, add the values in the column labeled *Do Not Favor Fee Increase:* 108 + 132 = 240.

This sum represents all the students who are not in favor of the increase and male plus all the students who are not in favor of the increase and female.

928. 360

To find the total number of students who took part in the poll, add the values of all four cells in the 2-x-2 table: $72 + 108 + 48 + 132 = 360$.

929. 0.40

A proportion involves a fraction, including a numerator and denominator. The denominator is the key because it represents the total number of individuals in the group you're looking at, and you may be looking at different groups from problem to problem.

To find the proportion of male students who favor the fee increase, you divide the number of students who are male and in favor of the fee increase (72) by the total number of male students in the poll ($72 + 108 = 180$):

$$\frac{72}{180} = 0.4$$

Note that you don't divide by 360 (all students polled) because the question asks you to find the *proportion of male students,* not both male and female students, who favor the increase. So you divide by the total number of male students polled (180).

930. 0.73

A proportion involves a fraction, including a numerator and denominator. The denominator is the key because it represents the total number of individuals in the group you're looking at, and you may be looking at different groups from problem to problem.

To find the proportion of female students who don't favor the fee increase, you divide the number of students who are female and don't favor the fee increase (132) by the total number of female students in the poll ($132 + 48 = 180$):

$$\frac{132}{180} \approx 0.73$$

Note that you don't divide by 360 (all students polled) because the question asks you to find the *proportion of female students,* not both female and male students, who don't favor the increase. So you divide by the total number of female students polled (180).

931. 0.67

A proportion involves a fraction, including a numerator and denominator. The denominator is the key because it represents the total number of individuals in the group you're looking at, and you may be looking at different groups from problem to problem.

To find the proportion of all students who don't favor the fee increase, divide the total number of students who don't favor the fee increase ($108 + 132 = 240$) by the total number of students in the poll ($108 + 132 + 72 + 48 = 360$):

$$\frac{240}{360} \approx 0.67$$

Note that you divide by 360 here because you want the proportion of *all* students.

932. the percentage of females polled who favor the fee increase

This pie chart represents the opinions of only the 180 females who were polled, not all the students who were polled, so you'll use 180 as the denominator to calculate the percentage of females polled. According to the table, of the 180 females polled, 48 of them favor the increase, so the proportion of females who favor the fee increase is

$$\frac{48}{180} \approx 0.27, \text{ or } 27\%$$

933. the percentage of all students who favor the fee increase

This pie chart represents the opinions of all 360 students polled, so you divide by 360 to do your calculations.

Of the 360 students, 120 of them favor the increase, so the proportion of all students who favor the fee increase is

$$\frac{120}{360} \approx 0.33, \text{ or } 33\%$$

934. the percentage of all students who are female and not in favor of the increase

Looking at the survey data from the 2-x-2 table, you see that there are 360 total students. To find the number corresponding to 37% of these students, convert 37% to a proportion by dividing by 100, and then multiply that result by 360:

$$\frac{37}{100} = 0.37$$

$$(0.37)(360) = 133.2$$

The difference between 133.2 and 132 is due to rounding:

$$\frac{132}{360} = 0.3\bar{6}$$

which rounds up to 0.37. The closest value in the table is 132, so 37% represents the percentage of all students who are female and not in favor of the increase.

Note: If the results were based only on the females in the poll, you'd divide by 180, the total number of females. If the results were based only on those students not in favor of the fee increase, you'd divide by 240, the total number of students polled not in favor of the fee increase.

You can check this by finding what percentage of students in each other cell in the table represents:

$$\frac{48}{360} = 0.1\bar{3}, \text{ or } 13\%$$

$$\frac{72}{360} = 0.2, \text{ or } 20\%$$

$$\frac{108}{360} = 0.3, \text{ or } 30\%$$

935. the percentage of all students who are male and in favor of the fee increase

Looking at the survey data from the 2-x-2 table, you see that there are 360 total students. To find the number corresponding to 20% of these students, convert 20% to a proportion by dividing by 100, and then multiply that result by 360:

$$\frac{20}{100} = 0.2$$

$$(0.2)(360) = 72$$

There are 72 students who are male and favor the fee increase, so that's who the 20% represents.

936. the percentage of all students who are male and not in favor of the fee increase

Looking at the survey data from the 2-x-2 table, you see that there are 360 total students. To find the number corresponding to 30% of these students, convert 30% to a proportion by dividing by 100, and then multiply that result by 360:

$$\frac{30}{100} = 0.3$$

$$(0.3)(360) = 108$$

There are 108 students who are male and do favor the fee increase, so that's who the 30% represents.

937. percentage of all students who are female and in favor of the fee increase

Looking at the survey data from the 2-x-2 table, you see that there are 360 total students. To find the number corresponding to 13% of these students, convert 13% to a proportion by dividing by 100, and then multiply that result by 360:

$$\frac{13}{100} = 0.13$$

$$(0.13)(360) = 46.8$$

The closest value in the table is 48, so 13% represents the percentage of students who are female and in favor of the fee increase.

The difference between 46.8 and 48 is due to rounding:

$$\frac{48}{360} = 0.1\overline{3}$$

This rounds down to 0.13.

938. 72

To find the total number of smokers, add the cells in the row labeled *Smoker*:
$48 + 24 = 72$.

939. 74

To find the total number of patients with a hypertension diagnosis, add the cells in the column labeled *Hypertension Diagnosis*: $48 + 26 = 74$.

940. 26

To find the number of patients who are nonsmokers and have a hypertension diagnosis, find the cell at the intersection of the row labeled *Nonsmoker* and the column labeled *Hypertension Diagnosis:* 26.

941. 48

To find the number of patients who are smokers and have a hypertension diagnosis, find the cell at the intersection of the row labeled *Smoker* and the column labeled *Hypertension Diagnosis:* 48.

942. 148

To find the total number of patients in the study, add the numbers in each cell of the table: 48 + 24 + 26 + 50 = 148.

943. 24

To find the number of patients who don't have a hypertension diagnosis and are smokers, find the cell at the intersection of the column labeled *Hypertension Diagnosis* and the row labeled *Smoker:* 24.

944. 50

To find the number of patients who don't have a hypertension diagnosis and are nonsmokers, find the cell at the intersection of the column labeled *No Hypertension Diagnosis* and the row labeled *Nonsmoker:* 50.

945. 0.65

To find the proportion of patients with a hypertension diagnosis who are smokers, you focus on the first column of the table. The total number of patients with a hypertension diagnosis is 74: 48 of them are smokers, and 26 are nonsmokers. So the proportion of the hypertension patients who are smokers is

$$\frac{48}{74} \approx 0.65$$

946. 0.35

To find the proportion of patients with a hypertension diagnosis who are nonsmokers, you focus on the first column of the table. The total number of patients with a hypertension diagnosis is 74: 48 of them are smokers, and 26 are nonsmokers. The proportion of the hypertension patients who are nonsmokers is

$$\frac{26}{74} = 0.35$$

947. 0.34

To find the proportion of nonsmokers with a hypertension diagnosis, you focus on the second row of the table, because you're limited to the 76 nonsmokers. Of that group, 26 of them have a hypertension diagnosis. So you divide the number of nonsmokers with a hypertension diagnosis by the total number of nonsmokers:

$$\frac{26}{76} \approx 0.34$$

948. 0.66

To find the proportion of nonsmokers with no hypertension diagnosis, you focus on the second row of the table, because you're limited to the 76 nonsmokers. Of that group, 50 of them don't have a hypertension diagnosis. So you divide the number of nonsmokers with no hypertension diagnosis by the total number of nonsmokers:

$$\frac{50}{76} = 0.66$$

949. 0.16

To find the proportion of all patients who are smokers with no hypertension diagnosis, divide the number of patients who are smokers and have no hypertension diagnosis (24) by the total number of patients (148):

$$\frac{24}{148} \approx 0.16$$

950. 0.34

To find the proportion of all patients who are nonsmokers with no hypertension diagnosis, divide the number of patients who are nonsmokers and have no hypertension diagnosis (50) by the total number of patients (148):

$$\frac{50}{148} \approx 0.34$$

951. the conditional probability of not smoking, given a hypertension diagnosis

The 35% portion is in the bar labeled *Hypertension Diagnosis*, which sums to 100%, so this area is a conditional probability for those who have a hypertension diagnosis. You know that in the data from the table, among people with hypertension, smoking is more common than not smoking. So the smaller area of this bar must be the conditional probability of not smoking, given a hypertension diagnosis.

You can also calculate the conditional probability of not smoking, given a hypertension diagnosis, by dividing the number of respondents who don't smoke and have a hypertension diagnosis (26) by the total number with a hypertension diagnosis (74):

$$P(\text{nonsmoker} \mid \text{hypertension}) = \frac{26}{74} \approx 0.35$$

This probability notation has *hypertension* in the back part of the parentheses because that's the subgroup you're looking at (and why you divide by 74). The *nonsmoker* goes in the front part of the parentheses because you want to know what proportion of that subgroup are nonsmokers.

952. the conditional probability of smoking, given a hypertension diagnosis

The 65% portion is in the bar labeled *Hypertension Diagnosis*, which sums to 100%, so this area is a conditional probability for those who have a hypertension diagnosis. You know that in the data from the table, among people with hypertension, smoking is more common than not smoking. So the larger area of this bar must be the conditional probability of smoking, given a hypertension diagnosis.

You can also calculate the conditional probability of smoking, given a hypertension diagnosis, by dividing the number of respondents who do smoke and have a hypertension diagnosis (48) by the total number with a hypertension diagnosis (74):

$$P(\text{smoker} \mid \text{hypertension}) = \frac{48}{74} \approx 0.65$$

This probability notation has *hypertension* in the back part of the parentheses because that's the subgroup you're looking at (and why you divide by 74). The *smoker* goes in the front part of the parentheses because you want to know what proportion of that subgroup are smokers.

953. the conditional probability of not smoking, given no hypertension diagnosis

The 68% portion is in the bar labeled *No Hypertension Diagnosis*, which sums to 100%, so this area is a conditional probability for those who don't have a hypertension diagnosis. You know that in the data from the table, among people with no hypertension diagnosis, not smoking is more common than smoking. So the larger area of this bar must be the conditional probability of not smoking, given no hypertension diagnosis.

You can also calculate the conditional probability of not smoking, given no hypertension diagnosis, by dividing the number of respondents who don't smoke and don't have a hypertension diagnosis (50) by the total number who don't have a hypertension diagnosis (74):

$$P(\text{nonsmoker} \mid \text{no hypertension}) = \frac{50}{74} \approx 0.68$$

This probability notation has *no hypertension* in the back part of the parentheses because that's the subgroup you're looking at (and why you divide by 74). The *nonsmoker* goes in the front part of the parentheses because you want to know what proportion of that subgroup are nonsmokers.

954. the conditional probability of smoking, given no hypertension diagnosis

The 32% portion is in the bar labeled *No Hypertension Diagnosis*, which sums to 100%, so this area is a conditional probability for those who don't have a hypertension diagnosis. You know that in the data from the table, among people with no hypertension diagnosis, smoking is less common than not smoking. So the smaller area of this bar must be the conditional probability of smoking, given no hypertension diagnosis.

You can also calculate the conditional probability of smoking, given no hypertension diagnosis, by dividing the number of respondents who smoke and don't have a hypertension diagnosis (24) by the total number of patients who don't have a hypertension diagnosis (74):

$$P(\text{smoker} \mid \text{no hypertension}) = \frac{24}{74} \approx 0.32$$

This probability notation has *no hypertension* in the back part of the parentheses because that's the subgroup you're looking at (and why you divide by 74). The *smoker* goes in the front part of the parentheses because you want to know what proportion of that subgroup are smokers.

955. E. Choices (A) and (D) (Patients with a hypertension diagnosis are more likely to be smokers than nonsmokers; patients without a hypertension diagnosis are more likely to be non-smokers than smokers.)

Although you don't want to generalize too broadly from a single sample of data, the patterns found in this data set indicate that patients with a hypertension diagnosis are more likely to be smokers (65%) rather than nonsmokers (35%), and patients without a hypertension diagnosis are more likely to be nonsmokers (68%) than smokers (32%).

You can find these percentages in the two bar graphs.

956. C. $P(A)$ does not depend on whether or not B occurs.

The question states that the variables A and B are independent. Two variables are independent if the probability of one event occurring doesn't depend on whether the other event occurs; therefore, their probabilities aren't affected by the occurrence of the other event.

957. B. Gender and choice of major are not independent.

You don't know anything about the *number* of students in either group; you're given only percentages. If gender and choice of major were independent, you'd expect to see the same proportions of men and women enrolled in each major. In engineering, 70% of students are male, but in English 20% are male. And in engineering, 30% of students are female, while in English 80% are female.

958. B. The same proportion of males and females choose to enroll in higher education.

Note that although the same proportion of males and females will choose to enroll, it may not be true that the same number of males and females choose to enroll, because the senior class may not have the same number of males and females.

959. 210

Given this data, if 60% of voters voted for the bond initiative and voting was independent of gender, you'd also expect 60% of female voters to vote for the bond initiative. To find the expected number of women who voted for the bond initiative, you multiply the total number of female registered voters by 60%: 350(0.6) = 210.

960. 120

Given this data, if 40% of students participate in after-school activities, then 60% don't participate, because $1.0 - 0.4 = 0.6$, or 60%.

If 60% of students don't participate in after-school activities and participation is independent of gender, you'd expect that 60% of boys wouldn't participate in after-school activities. To find out how many boys 60% is, you multiply the total number of male students by 60%: $200(0.6) = 120$.

961. 0.33

Marginal probability is the probability of having a certain characteristic of one variable, without regard to the other variable(s).

In this case, the marginal probability of a person being a vegetarian is the probability of being a vegetarian, without regard to whether that person has high cholesterol. In this data, 100 adults are vegetarian out of a total of 300 adults. To find the marginal probability, divide the number of vegetarians by the total number of adults:

$$\frac{100}{300} \approx 0.33$$

962. 0.67

Marginal probability is the probability of having a certain characteristic of one variable, without regard to the other variable(s).

In this table, there are three dietary categories: vegetarian, vegan, and regular dieter. The marginal probability of a person not being a vegan is the probability of being either vegetarian or a regular dieter, without regard to whether that person has high cholesterol. In this data, 200 adults aren't vegans, out of a total of 300 adults. To find the marginal probability, divide the number of non-vegans by the total number of adults:

$$\frac{200}{300} = 0.67$$

963. 0.33

Marginal probability is the probability of having a certain characteristic of one variable, without regard to the other variable(s).

In this example, the marginal probability of a person having high cholesterol is the probability of having high cholesterol, without regard to that person's dietary habits. In this data, 100 adults have high cholesterol out of a total of 300 adults. To find the marginal probability, divide the number of adults with high cholesterol by the total number of adults:

$$\frac{100}{300} \approx 0.33$$

964. 0.67

Marginal probability is the probability of having a certain characteristic of one variable, without regard to the other variable(s).

In this example, the marginal probability of a person not having high cholesterol is the probability of not having high cholesterol, without regard to that person's dietary habits. In this data, 200 adults don't have high cholesterol out of a total of 300 adults. To find the marginal probability, divide the number of adults without high cholesterol by the total number of adults:

$$\frac{200}{300} \approx 0.67$$

965. E. Choices (A) and (C) (The same percentage of vegetarians, vegans, and regular dieters will have high cholesterol; among those with high cholesterol, equal numbers will be vegetarians, vegans, and regular dieters.)

This data set includes equal percentages of vegans, vegetarians, and regular dieters (100/300 = 33.3% each). If diet and cholesterol level are unrelated, then the probability of one variable occurring doesn't affect the probability of the other variable occurring. Therefore, the same percentage of vegans, vegetarians, and regular dieters should have high cholesterol, and vegans, vegetarians, and regular dieters should be equally represented among those with high cholesterol.

966. 67

If diet and cholesterol level are independent, you can find the joint probability by multiplying the marginal frequencies and dividing by the sample size.

In this data, there are 100 vegetarians and 200 adults without high cholesterol, out of 300 total adults:

$$\frac{100(200)}{300} \approx 67$$

967. 33

If being vegetarian and having high cholesterol are independent, you can find the joint probability by multiplying the marginal frequencies and dividing by the sample size.

In this data, there are 100 vegetarians and 100 adults with high cholesterol, out of 300 total adults:

$$\frac{100(100)}{300} \approx 33$$

968. 70

The marginal total of adults with high cholesterol is 100. If 10 of these are vegetarians and 20 are vegans, the remainder must be regular dieters: 100 − (10 + 20) = 70.

969. 90

The marginal total of vegetarians is 100. If ten of these have high cholesterol, the remainder must not have high cholesterol: $100 - 10 = 90$.

970. 65

The marginal total of regular dieters is 100. If 35 of these don't have high cholesterol, the remainder must have high cholesterol: $100 - 35 = 65$.

971. 300

To find the total number of participants, add the numbers from each cell in the table: $60 + 30 + 10 + 40 + 40 + 20 + 20 + 30 + 50 = 300$.

972. 0.33

Marginal probability is the probability of having a certain characteristic of one variable, without regard to the other variable(s).

To find the marginal probability of an individual being 41 to 65 years old, divide the number of participants ages 41 to 65 ($40 + 40 + 20 = 100$) by the total number of participants ($60 + 30 + 10 + 40 + 40 + 20 + 20 + 30 + 50 = 300$):

$$\frac{100}{300} \approx 0.33$$

973. 0.73

Marginal probability is the probability of having a certain characteristic of one variable, without regard to the other variable(s).

To find the marginal probability of a respondent's most commonly used type of phone not being a landline, divide the number of participants who said they most commonly used a smartphone or other mobile phone ($60 + 40 + 20 + 30 + 40 + 30 = 220$) by the total number of participants ($60 + 30 + 10 + 40 + 40 + 20 + 20 + 30 + 50 = 300$):

$$\frac{220}{300} \approx 0.73$$

974. 0.20

The joint probability refers to the probability of having two or more characteristics — in this case, being in a particular age group and using a particular type of phone.

To find the joint probability of a respondent being between 18 and 40 years old and most commonly using a smartphone, divide the number of participants ages 18 to 40 who most commonly use a smartphone (60) by the total number of participants ($60 + 30 + 10 + 40 + 40 + 20 + 20 + 30 + 50 = 300$):

$$\frac{60}{300} = 0.20$$

975.

0.17

The joint probability refers to the probability of having two or more characteristics — in this case, being in a particular age group and using a particular type of phone.

To find the joint probability of a respondent being age 66 or older and most commonly using a landline phone, divide the number of participants age 66 and older who most commonly use a landline (50) by the total number of participants (60 + 30 + 10 + 40 + 40 + 20 + 20 + 30 + 50 = 300):

$$\frac{50}{300} \approx 0.17$$

976.

40

To find the expected number of people ages 18 to 40 who prefer a smartphone, if age and phone preference are independent, multiply the marginal frequencies for a respondent being between 18 and 40 years old (60 + 30 + 10 = 100) and for preferring a smartphone (60 + 40 + 20 = 120), and divide by the total number of participants (60 + 30 + 10 + 40 + 40 + 20 + 20 + 30 + 50 = 300):

$$\frac{100(120)}{300} = \frac{12,000}{300} = 40$$

977.

33

To find the expected number of people age 66 or older who prefer a mobile phone, if age and phone preference are independent, multiply the marginal frequencies for a respondent being age 66 or older (20 + 30 + 50 = 100) and for preferring a mobile phone (30 + 40 + 30 = 100), and divide by the total number of participants (60 + 30 + 10 + 40 + 40 + 20 + 20 + 30 + 50 = 300):

$$\frac{100(100)}{300} = \frac{10,000}{300} \approx 33$$

978.

C. People age 66 or older are less likely to prefer smartphones than would be expected if age and phone preference were independent.

You can calculate the expected number of people in each joint category (age group and phone preference) by multiplying the marginal frequencies and dividing by the total number of participants. If age and phone preference are independent, these expected values will be the same as the observed number in each category.

In this example, 20 people age 66 or older said they most commonly used a smartphone (the observed value). You can find the expected value by multiplying the number of people age 66 or older (20 + 30 + 50 = 100) by the number of people who prefer a smartphone (60 + 40 + 20 = 120) and then dividing by the total number of participants (60 + 30 + 10 + 40 + 40 + 20 + 20 + 30 + 50 = 300):

$$\frac{(100)(120)}{300} = \frac{12,000}{300} = 40$$

Because the observed number is lower than the expected number, it's correct to say that people age 66 or older are less likely to prefer smartphones than would be expected if age and phone preference were independent.

979. E. Choices (C) and (D) (No, because the ownership rates differ by gender; no, because the marginal ownership rate differs from the conditional ownership rates.)

If two variables were independent, all three percentages given would have to be equal. The conditional probability of ownership (0.75) would be the same as the marginal probabilities of ownership (0.85 for males; 0.65 for females), and the marginal probabilities of ownership would be the same for both genders.

980. 12

To find the number of cells in a two-way table, multiply the number of categories for each variable. In this case, type of residence has three categories, and annual income has four categories: $(3)(4) = 12$.

981. 100

To find the total sample size, add together the frequencies for each category: $15 + 40 + 10 + 35 = 100$.

982. 0.73

A *conditional probability* represents the percentage of individuals within a given group who have a certain characteristic. The number of individuals in the given group always goes in the denominator.

To find the conditional probability of owning a car, given that the person is male, divide the number of male car owners (40) by the total number of males ($40 + 15 = 55$):

$$\frac{40}{55} \approx 0.73$$

983. 0.78

A *conditional probability* represents the percentage of individuals within a given group who have a certain characteristic. The number of individuals in the given group always goes in the denominator.

To find the conditional probability of owning a car, given that the person is female, divide the number of female car owners (35) by the total number of females ($35 + 10 = 45$):

$$\frac{35}{45} \approx 0.78$$

984. 0.75

Marginal probability is the probability of having a certain characteristic of one variable, without regard to the other variable(s).

To find the marginal probability of owning a car, divide the number of car owners ($40 + 35 = 75$) by the total number of survey participants ($40 + 15 + 35 + 10 = 100$):

$$\frac{75}{100} \approx 0.75$$

985. 0.53

A *conditional probability* represents the percentage of individuals within a given group who have a certain characteristic. The number of individuals in the given group always goes in the denominator.

To find the conditional probability of being male, given car ownership, divide the number of male car owners (40) by the total number of car owners (40 + 35 = 75):

$$\frac{40}{75} \approx 0.53$$

986. 0.47

A *conditional probability* represents the percentage of individuals within a given group who have a certain characteristic. The number of individuals in the given group always goes in the denominator.

To find the conditional probability of being female, given car ownership, divide the number of female car owners (35) by the total number of car owners (40 +35 = 75):

$$\frac{35}{75} \approx 0.47$$

987. 41.25

Marginal probability is the probability of having a certain characteristic of one variable, without regard to the other variable(s).

To find the marginal probability of car ownership, divide the number of car owners (40 + 35) by the sample size (40 + 15 +35 + 10 = 100):

$$\frac{75}{100} \approx 0.75$$

To apply this probability to males, multiply the number of males (40 + 15 = 55) by the marginal probability of car ownership:

$$55(0.75) = 41.25$$

988. 33.75

Marginal probability is the probability of having a certain characteristic of one variable, without regard to the other variable(s).

To find the marginal probability of car ownership, divide the number of car owners (40 + 35) by the sample size (40 + 15 +35 + 10 = 100):

$$\frac{75}{100} \approx 0.75$$

To apply this probability to females, multiply the number of females (35 + 10 = 45) by the marginal probability of car ownership:

$$45(0.75) = 33.75$$

989.

0.55

Marginal probability is the probability of having a certain characteristic of one variable, without regard to the other variable(s).

The marginal probability of being male is the probability that someone chosen at random from the sample is a male. It also represents the total percentage of females in the group.

To find the marginal probability of being male, divide the number of males in the sample (40 + 15 = 55) by the total sample size (40 + 15 + 35 + 10 = 100):

$$\frac{55}{100} = 0.55$$

990.

0.45

Marginal probability is the probability of having a certain characteristic of one variable, without regard to the other variable(s).

The marginal probability of being female is the probability that someone chosen at random from the sample is a female. It also represents the total percentage of females in the group.

To find the marginal probability of being female, divide the number of females in the sample (35 + 10 = 45) by the total sample size (40 + 15 + 35 + 10 = 100):

$$\frac{45}{100} = 0.45$$

991.

E.   Choices (A) and (D) (In this sample, more males than females own cars; in this sample, the conditional probability of car ownership is higher for females.)

In this data, more males (40) than females (35) own cars, but the conditional probability of car ownership is higher for females.

To find the conditional probability of owning a car, given male gender, divide the number of male car owners (40) by the total number of males (15 + 40 = 55):

$$P(\text{car} \mid \text{male}) = \frac{40}{55} \approx 0.73$$

To find the conditional probability of owning a car, given female gender, divide the number of female car owners (35) by the total number of females (35 + 10 = 45):

$$P(\text{car} \mid \text{female}) = \frac{35}{45} \approx 0.78$$

992.

E.   Choices (A), (B), and (C) (replication of the survey in other locations; replication of the survey with a larger sample; replicating the survey with a nationally representative sample)

Drawing cause-and-effect conclusions from surveys is difficult, but replicating the same survey with different samples, including larger samples, samples from different locations, and with a nationally representative sample, would all help to strengthen the argument that a relationship exists between car ownership and gender.

993. B. a randomized clinical trial

Although statistical evidence of a relationship between two variables isn't enough to establish causality, the randomized clinical trial design can strengthen your ability to draw cause-and-effect conclusions by minimizing bias, using sufficient sample sizes, and controlling for other variables that may affect the outcome.

994. 9

To find the number of cells in a table, multiply the number of rows by the number of columns. With the addition of the "Inconclusive" category to each of the two variables, this table would have three rows and three columns: $(3)(3) = 9$.

995. 0.39

Marginal probability is the probability of having a certain characteristic of one variable, without regard to the other variable(s).

To calculate the marginal probability for screening positive for depression, divide the total number of participants who screened positive for depression $(25 + 20 = 45)$ by the total sample size $(25 + 20 + 10 + 60 = 115)$:

$$\frac{45}{115} \approx 0.39$$

996. 0.30

Marginal probability is the probability of having a certain characteristic of one variable, without regard to the other variable(s).

To calculate the marginal probability for evaluating positive for depression, divide the total number of participants who evaluated positive for depression $(25 + 10 = 35)$ by the total sample size $(25 + 20 + 10 + 60 = 115)$:

$$\frac{35}{115} \approx 0.30$$

997. 0.44

A *conditional probability* represents the percentage of individuals within a given group who have a certain characteristic. The number of individuals in the given group always goes in the denominator.

To calculate the conditional probability of evaluating negative for depression, given a positive screening result, divide the number of participants with a positive screening result and negative evaluation (20) by the total number of participants with a positive screening result $(25 + 20 = 45)$:

$$\frac{20}{45} \approx 0.44$$

998. 0.14

A *conditional probability* represents the percentage of individuals within a given group who have a certain characteristic. The number of individuals in the given group always goes in the denominator.

To calculate the conditional probability of evaluating positive for depression, given a negative screening result, divide the number of participants with a negative screening result and positive evaluation (10) by the total number of participants with a negative screening result (10 + 60 = 70):

$$\frac{10}{70} \approx 0.14$$

999. screened negative and evaluated negative

The joint probability refers to the probability of having two or more characteristics — in this case, having a particular screening result and having a particular evaluation result.

To find the joint probabilities of these four possible outcomes, divide the number of participants with each particular combination (screening outcome and evaluation outcome) by the total sample size:

$$P(\text{screened positive and evaluated positive}) = \frac{25}{115} \approx 0.22$$

$$P(\text{screened positive and evaluated negative}) = \frac{20}{115} \approx 0.17$$

$$P(\text{screened negative and evaluated positive}) = \frac{10}{115} \approx 0.09$$

$$P(\text{screened negative and evaluated negative}) = \frac{60}{115} \approx 0.52$$

Screened negative and evaluated negative has the highest joint probability at 0.52.

Note: Because all these joint probability calculations share the total sample size as the denominator, the outcome with the highest frequency will also have the highest probability.

1,000. No, because the conditional probabilities for evaluating positive are different depending on the screening results.

If screening results and positive evaluation for depression were independent, you'd expect the conditional probability for evaluating positive to be the same whether a person screened positive or negative. This is not the case. As you find out when you calculate the conditional probabilities, they're quite different.

To calculate the conditional probability of evaluating positive for depression, given a positive screening result, divide the number of participants with a positive screening result and positive evaluation (25) by the total number of participants with a positive screening result (25 + 20 = 45):

$$P(\text{evaluated positive} \mid \text{screened positive}) = \frac{25}{45} \approx 0.56$$

To calculate the conditional probability of evaluating positive for depression, given a negative screening result, divide the number of participants with a negative screening result and positive evaluation (10) by the total number of participants with a negative screening result (10 + 60 = 70):

$$P(\text{evaluated positive} \mid \text{screened negative}) = \frac{10}{70} \approx 0.14$$

These two conditional probabilities aren't equal, which means that the screening process has an effect on whether the person is diagnosed as being depressed. Here, you see that if someone is screened positive, he has a higher chance of being diagnosed (0.56) than if he's not screened (0.14). Because screening has an effect on the outcome of the diagnosis, screening and diagnosis aren't independent. Their outcomes are related.

1,001. E. Choices (A), (B), and (C) (The study sample was randomly selected from the population; the people doing the evaluation had no knowledge of the screening results; this study replicated an earlier study that produced similar results.)

Although you must observe caution when drawing causal conclusions from statistical results, several factors could increase your confidence in doing so, including working with a study sample randomly selected from the population, blinding those doing the evaluation from the screening results, and replicating results of a previous study with similar results.

Appendix

Tables for Reference

*E*xcerpted from Statistics For Dummies, *2nd Edition, by Deborah J. Rumsey, PhD (2011, Wiley). This material is reproduced with permission of John Wiley & Sons, Inc.*

This appendix includes tables for finding probabilities for three distributions used in this book: the *Z*-distribution (standard normal), the *t*-distribution, and the binomial distribution. It also includes a table listing *z**-values for selected (percentage) confidence levels.

The Z-Table

Table A-1 shows less-than-or-equal-to probabilities for the *Z*-distribution; that is, $p(Z \leq z)$ for a given *z*-value. To use Table A-1, do the following:

1. **Determine the *z*-value for your particular problem.**

   The *z*-value should have one leading digit before the decimal point (positive, negative, or zero) and two digits after the decimal point — for example, $z = 1.28$, -2.69, or 0.13.

2. **Find the row of the table corresponding to the leading digit and first digit after the decimal point.**

   For example, if your *z*-value is 1.28, look in the *1.2* row; if $z = -1.28$, look in the *–1.2* row.

3. **Find the column corresponding to the second digit after the decimal point.**

   For example, if your *z*-value is 1.28 or –1.28, look in the *0.08* column.

4. **Intersect the row and column from Steps 2 and 3.**

   This is the probability that *Z* is less than or equal to your *z*-value. In other words, you've found $p(Z \leq z)$. For example, if $z = 1.28$, you see $p(Z \leq 1.28) = 0.8997$. For $z = -1.28$, you see $p(Z \leq -1.28) = 0.1003$.

Table A-1   The Z-Table

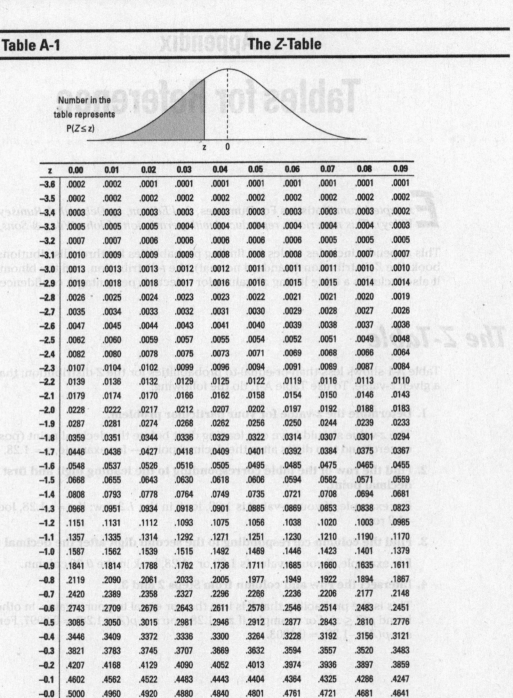

Number in the table represents $P(Z \le z)$

z	0.00	0.01	0.02	0.03	0.04	0.05	0.06	0.07	0.08	0.09
−3.6	.0002	.0002	.0001	.0001	.0001	.0001	.0001	.0001	.0001	.0001
−3.5	.0002	.0002	.0002	.0002	.0002	.0002	.0002	.0002	.0002	.0002
−3.4	.0003	.0003	.0003	.0003	.0003	.0003	.0003	.0003	.0003	.0002
−3.3	.0005	.0005	.0005	.0004	.0004	.0004	.0004	.0004	.0004	.0003
−3.2	.0007	.0007	.0006	.0006	.0006	.0006	.0006	.0005	.0005	.0005
−3.1	.0010	.0009	.0009	.0009	.0008	.0008	.0008	.0008	.0007	.0007
−3.0	.0013	.0013	.0013	.0012	.0012	.0011	.0011	.0011	.0010	.0010
−2.9	.0019	.0018	.0018	.0017	.0016	.0016	.0015	.0015	.0014	.0014
−2.8	.0026	.0025	.0024	.0023	.0023	.0022	.0021	.0021	.0020	.0019
−2.7	.0035	.0034	.0033	.0032	.0031	.0030	.0029	.0028	.0027	.0026
−2.6	.0047	.0045	.0044	.0043	.0041	.0040	.0039	.0038	.0037	.0036
−2.5	.0062	.0060	.0059	.0057	.0055	.0054	.0052	.0051	.0049	.0048
−2.4	.0082	.0080	.0078	.0075	.0073	.0071	.0069	.0068	.0066	.0064
−2.3	.0107	.0104	.0102	.0099	.0096	.0094	.0091	.0089	.0087	.0084
−2.2	.0139	.0136	.0132	.0129	.0125	.0122	.0119	.0116	.0113	.0110
−2.1	.0179	.0174	.0170	.0166	.0162	.0158	.0154	.0150	.0146	.0143
−2.0	.0228	.0222	.0217	.0212	.0207	.0202	.0197	.0192	.0188	.0183
−1.9	.0287	.0281	.0274	.0268	.0262	.0256	.0250	.0244	.0239	.0233
−1.8	.0359	.0351	.0344	.0336	.0329	.0322	.0314	.0307	.0301	.0294
−1.7	.0446	.0436	.0427	.0418	.0409	.0401	.0392	.0384	.0375	.0367
−1.6	.0548	.0537	.0526	.0516	.0505	.0495	.0485	.0475	.0465	.0455
−1.5	.0668	.0655	.0643	.0630	.0618	.0606	.0594	.0582	.0571	.0559
−1.4	.0808	.0793	.0778	.0764	.0749	.0735	.0721	.0708	.0694	.0681
−1.3	.0968	.0951	.0934	.0918	.0901	.0885	.0869	.0853	.0838	.0823
−1.2	.1151	.1131	.1112	.1093	.1075	.1056	.1038	.1020	.1003	.0985
−1.1	.1357	.1335	.1314	.1292	.1271	.1251	.1230	.1210	.1190	.1170
−1.0	.1587	.1562	.1539	.1515	.1492	.1469	.1446	.1423	.1401	.1379
−0.9	.1841	.1814	.1788	.1762	.1736	.1711	.1685	.1660	.1635	.1611
−0.8	.2119	.2090	.2061	.2033	.2005	.1977	.1949	.1922	.1894	.1867
−0.7	.2420	.2389	.2358	.2327	.2296	.2266	.2236	.2206	.2177	.2148
−0.6	.2743	.2709	.2676	.2643	.2611	.2578	.2546	.2514	.2483	.2451
−0.5	.3085	.3050	.3015	.2981	.2946	.2912	.2877	.2843	.2810	.2776
−0.4	.3446	.3409	.3372	.3336	.3300	.3264	.3228	.3192	.3156	.3121
−0.3	.3821	.3783	.3745	.3707	.3669	.3632	.3594	.3557	.3520	.3483
−0.2	.4207	.4168	.4129	.4090	.4052	.4013	.3974	.3936	.3897	.3859
−0.1	.4602	.4562	.4522	.4483	.4443	.4404	.4364	.4325	.4286	.4247
−0.0	.5000	.4960	.4920	.4880	.4840	.4801	.4761	.4721	.4681	.4641

Number in the
table represents
$P(Z \leq z)$

z	0.00	0.01	0.02	0.03	0.04	0.05	0.06	0.07	0.08	0.09
0.0	.5000	.5040	.5080	.5120	.5160	.5199	.5239	.5279	.5319	.5359
0.1	.5398	.5438	.5478	.5517	.5557	.5596	.5636	.5675	.5714	.5753
0.2	.5793	.5832	.5871	.5910	.5948	.5987	.6026	.6064	.6103	.6141
0.3	.6179	.6217	.6255	.6293	.6331	.6368	.6406	.6443	.6480	.6517
0.4	.6554	.6591	.6628	.6664	.6700	.6736	.6772	.6808	.6844	.6879
0.5	.6915	.6950	.6985	.7019	.7054	.7088	.7123	.7157	.7190	.7224
0.6	.7257	.7291	.7324	.7357	.7389	.7422	.7454	.7486	.7517	.7549
0.7	.7580	.7611	.7642	.7673	.7704	.7734	.7764	.7794	.7823	.7852
0.8	.7881	.7910	.7939	.7967	.7995	.8023	.8051	.8078	.8106	.8133
0.9	.8159	.8186	.8212	.8238	.8264	.8289	.8315	.8340	.8365	.8389
1.0	.8413	.8438	.8461	.8485	.8508	.8531	.8554	.8577	.8599	.8621
1.1	.8643	.8665	.8686	.8708	.8729	.8749	.8770	.8790	.8810	.8830
1.2	.8849	.8869	.8888	.8907	.8925	.8944	.8962	.8980	.8997	.9015
1.3	.9032	.9049	.9066	.9082	.9099	.9115	.9131	.9147	.9162	.9177
1.4	.9192	.9207	.9222	.9236	.9251	.9265	.9279	.9292	.9306	.9319
1.5	.9332	.9345	.9357	.9370	.9382	.9394	.9406	.9418	.9429	.9441
1.6	.9452	.9463	.9474	.9484	.9495	.9505	.9515	.9525	.9535	.9545
1.7	.9554	.9564	.9573	.9582	.9591	.9599	.9608	.9616	.9625	.9633
1.8	.9641	.9649	.9656	.9664	.9671	.9678	.9686	.9693	.9699	.9706
1.9	.9713	.9719	.9726	.9732	.9738	.9744	.9750	.9756	.9761	.9767
2.0	.9772	.9778	.9783	.9788	.9793	.9798	.9803	.9808	.9812	.9817
2.1	.9821	.9826	.9830	.9834	.9838	.9842	.9846	.9850	.9854	.9857
2.2	.9861	.9864	.9868	.9871	.9875	.9878	.9881	.9884	.9887	.9890
2.3	.9893	.9896	.9898	.9901	.9904	.9906	.9909	.9911	.9913	.9916
2.4	.9918	.9920	.9922	.9925	.9927	.9929	.9931	.9932	.9934	.9936
2.5	.9938	.9940	.9941	.9943	.9945	.9946	.9948	.9949	.9951	.9952
2.6	.9953	.9955	.9956	.9957	.9959	.9960	.9961	.9962	.9963	.9964
2.7	.9965	.9966	.9967	.9968	.9969	.9970	.9971	.9972	.9973	.9974
2.8	.9974	.9975	.9976	.9977	.9977	.9978	.9979	.9979	.9980	.9981
2.9	.9981	.9982	.9982	.9983	.9984	.9984	.9985	.9985	.9986	.9986
3.0	.9987	.9987	.9987	.9988	.9988	.9989	.9989	.9989	.9990	.9990
3.1	.9990	.9991	.9991	.9991	.9992	.9992	.9992	.9992	.9993	.9993
3.2	.9993	.9993	.9994	.9994	.9994	.9994	.9994	.9995	.9995	.9995
3.3	.9995	.9995	.9995	.9996	.9996	.9996	.9996	.9996	.9996	.9997
3.4	.9997	.9997	.9997	.9997	.9997	.9997	.9997	.9997	.9997	.9998
3.5	.9998	.9998	.9998	.9998	.9998	.9998	.9998	.9998	.9998	.9998
3.6	.9998	.9998	.9999	.9999	.9999	.9999	.9999	.9999	.9999	.9999

© John Wiley & Sons, Inc.

The t-Table

Table A-2 shows right-tail probabilities for selected *t*-distributions. Follow these steps to use Table A-2 to find right-tail probabilities and *p*-values for hypothesis tests involving *t*:

1. **Find the *t*-value for which you want the right-tail probability (call it *t*), and find the sample size (for example, *n*).**

2. **Find the row corresponding to the degrees of freedom (*df*) for your problem (for example, *n* – 1). Follow the row across to find the two *t*-values between which your *t* falls.**

   For example, if your *t* is 1.60, and your *n* is 7, you look in the row for *df* = 7 – 1 = 6. Across that row, you find your *t* lies between *t*-values 1.44 and 1.94.

3. **Look at the top of the columns containing the two *t*-values from Step 2.**

   The right-tail (greater-than) probability for your *t*-value is somewhere between the two values at the top of these columns. For example, your *t* = 1.60 is between *t*-values 1.44 and 1.94 (*df* = 6), so the right-tail probability for your *t* is between 0.10 (column heading for *t* = 1.44) and 0.05 (column heading for *t* = 1.94).

The row near the bottom with *z* in the *df* column gives right-tail (greater-than) probabilities from the *Z*-distribution.

To use Table A-2 to find *t**-values (critical values) for a confidence interval involving *t*, do the following

1. **Determine the confidence level you need (as a percentage).**

2. **Determine the sample size (for example, *n*).**

3. **Look at the bottom row of the table where the percentages are shown. Find your % confidence level there.**

4. **Intersect this column with the row representing *n* – 1 degrees of freedom (*df*).**

   This is the *t*-value you need for your confidence interval. For example, a 95% confidence interval with *df* = 6 has *t** = 2.45. (Find 95% on the last line and follow it up to row *6*.)

Table A-2   The *t*-Table

Numbers in each row of the table are values on a *t*-distribution with
(*df*) degrees of freedom for selected right-tail (greater-than) probabilities (*p*).

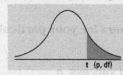

t (p, df)

df/p	0.40	0.25	0.10	0.05	0.025	0.01	0.005	0.0005
1	0.324920	1.000000	3.077684	6.313752	12.70620	31.82052	63.65674	636.6192
2	0.288675	0.816497	1.885618	2.919986	4.30265	6.96456	9.92484	31.5991
3	0.276671	0.764892	1.637744	2.353363	3.18245	4.54070	5.84091	12.9240
4	0.270722	0.740697	1.533206	2.131847	2.77645	3.74695	4.60409	8.6103
5	0.267181	0.726687	1.475884	2.015048	2.57058	3.36493	4.03214	6.8688
6	0.264835	0.717558	1.439756	1.943180	2.44691	3.14267	3.70743	5.9588
7	0.263167	0.711142	1.414924	1.894579	2.36462	2.99795	3.49948	5.4079
8	0.261921	0.706387	1.396815	1.859548	2.30600	2.89646	3.35539	5.0413
9	0.260955	0.702722	1.383029	1.833113	2.26216	2.82144	3.24984	4.7809
10	0.260185	0.699812	1.372184	1.812461	2.22814	2.76377	3.16927	4.5869
11	0.259556	0.697445	1.363430	1.795885	2.20099	2.71808	3.10581	4.4370
12	0.259033	0.695483	1.356217	1.782288	2.17881	2.68100	3.05454	43178
13	0.258591	0.693829	1.350171	1.770933	2.16037	2.65031	3.01228	4.2208
14	0.258213	0.692417	1.345030	1.761310	2.14479	2.62449	2.97684	4.1405
15	0.257885	0.691197	1.340606	1.753050	2.13145	2.60248	2.94671	4.0728
16	0.257599	0.690132	1.336757	1.745884	2.11991	2.58349	2.92078	4.0150
17	0.257347	0.689195	1.333379	1.739607	2.10982	2.56693	2.89823	3.9651
18	0.257123	0.688364	1.330391	1.734064	2.10092	2.55238	2.87844	3.9216
19	0.256923	0.687621	1.327728	1.729133	2.09302	2.53948	2.86093	3.8834
20	0.256743	0.686954	1.325341	1.724718	2.08596	2.52798	2.84534	3.8495
21	0.256580	0.686352	1.323188	1.720743	2.07961	2.51765	2.83136	3.8193
22	0.256432	0.685805	1.321237	1.717144	2.07387	2.50832	2.81876	3.7921
23	0.256297	0.685306	1.319460	1.713872	2.06866	2.49987	2.80734	3.7676
24	0.256173	0.684850	1.317836	1.710882	2.06390	2.49216	2.79694	3.7454
25	0.256060	0.684430	1.316345	1.708141	2.05954	2.48511	2.78744	3.7251
26	0.255955	0.684043	1.314972	1.705618	2.05553	2.47863	2.77871	3.7066
27	0.255858	0.683685	1.313703	1.703288	2.05183	2.47266	2.77068	3.6896
28	0.255768	0.683353	1.312527	1.701131	2.04841	2.46714	2.76326	3.6739
29	0.255684	0.683044	1.311434	1.699127	2.04523	2.46202	2.75639	3.6594
30	0.255605	0.682756	1.310415	1.697261	2.04227	2.45726	2.75000	3.6460
z	0.253347	0.674490	1.281552	1.644854	1.95996	2.32635	2.57583	3.2905
CI	———	———	80%	90%	95%	98%	99%	99.9%

© John Wiley & Sons, Inc.

The Binomial Table

Table A-3 shows probabilities for the binomial distribution. To use Table A-3, do the following:

1. **Find these three numbers for your particular problem:**
   - The sample size, n
   - The probability of success, p
   - The x-value for which you want $p(X = x)$

2. **Find the section of Table A-3 that's devoted to your n.**

3. **Look at the row for your x-value and the column for your p.**

4. **Intersect that row and column.** You have found $p(X = x)$.

5. **To get the probability of being less than, greater than, greater than or equal to, less than or equal to, or between two values of X, you add the appropriate values of Table A-3.**

   For example, if $n = 10$, $p = 0.6$, and you want $p(X = 9)$, go to the $n = 10$ section, the $x = 9$ row, and the $p = 0.6$ column to find 0.04.

Table A-3   The Binomial Table

Numbers in the table represent $p(X=x)$ for a binomial distribution with n trials and probability of success p.

Binomial probabilities:

$$\binom{n}{x} p^x (1-p)^{n-x}$$

							p					
n	x	0.1	0.2	0.25	0.3	0.4	0.5	0.6	0.7	0.75	0.8	0.9
1	0	0.900	0.800	0.750	0.700	0.600	0.500	0.400	0.300	0.250	0.200	0.100
	1	0.100	0.200	0.250	0.300	0.400	0.500	0.600	0.700	0.750	0.800	0.900
2	0	0.810	0.640	0.563	0.490	0.360	0.250	0.160	0.090	0.063	0.040	0.010
	1	0.180	0.320	0.375	0.420	0.480	0.500	0.480	0.420	0.375	0.320	0.180
	2	0.010	0.040	0.063	0.090	0.160	0.250	0.360	0.490	0.563	0.640	0.810
3	0	0.729	0.512	0.422	0.343	0.216	0.125	0.064	0.027	0.016	0.008	0.001
	1	0.243	0.384	0.422	0.441	0.432	0.375	0.288	0.189	0.141	0.096	0.027
	2	0.027	0.096	0.141	0.189	0.288	0.375	0.432	0.441	0.422	0.384	0.243
	3	0.001	0.008	0.016	0.027	0.064	0.125	0.216	0.343	0.422	0.512	0.729
4	0	0.656	0.410	0.316	0.240	0.130	0.063	0.026	0.008	0.004	0.002	0.000
	1	0.292	0.410	0.422	0.412	0.346	0.250	0.154	0.076	0.047	0.026	0.004
	2	0.049	0.154	0.211	0.265	0.346	0.375	0.346	0.265	0.211	0.154	0.049
	3	0.004	0.026	0.047	0.076	0.154	0.250	0.346	0.412	0.422	0.410	0.292
	4	0.000	0.002	0.004	0.008	0.026	0.063	0.130	0.240	0.316	0.410	0.656
5	0	0.590	0.328	0.237	0.168	0.078	0.031	0.010	0.002	0.001	0.000	0.000
	1	0.328	0.410	0.396	0.360	0.259	0.156	0.077	0.028	0.015	0.006	0.000
	2	0.073	0.205	0.264	0.309	0.346	0.313	0.230	0.132	0.088	0.051	0.008
	3	0.008	0.051	0.088	0.132	0.230	0.313	0.346	0.309	0.264	0.205	0.073
	4	0.000	0.006	0.015	0.028	0.077	0.156	0.259	0.360	0.396	0.410	0.328
	5	0.000	0.000	0.001	0.002	0.010	0.031	0.078	0.168	0.237	0.328	0.590
6	0	0.531	0.262	0.178	0.118	0.047	0.016	0.004	0.001	0.000	0.000	0.000
	1	0.354	0.393	0.356	0.303	0.187	0.094	0.037	0.010	0.004	0.002	0.000
	2	0.098	0.246	0.297	0.324	0.311	0.234	0.138	0.060	0.033	0.015	0.001
	3	0.015	0.082	0.132	0.185	0.276	0.313	0.276	0.185	0.132	0.082	0.015
	4	0.001	0.015	0.033	0.060	0.138	0.234	0.311	0.324	0.297	0.246	0.098
	5	0.000	0.002	0.004	0.010	0.037	0.094	0.187	0.303	0.356	0.393	0.354
	6	0.000	0.000	0.000	0.001	0.004	0.016	0.047	0.118	0.178	0.262	0.531
7	0	0.478	0.210	0.133	0.082	0.028	0.008	0.002	0.000	0.000	0.000	0.000
	1	0.372	0.367	0.311	0.247	0.131	0.055	0.017	0.004	0.001	0.000	0.000
	2	0.124	0.275	0.311	0.318	0.261	0.164	0.077	0.025	0.012	0.004	0.000
	3	0.023	0.115	0.173	0.227	0.290	0.273	0.194	0.097	0.058	0.029	0.003
	4	0.003	0.029	0.058	0.097	0.194	0.273	0.290	0.227	0.173	0.115	0.023
	5	0.000	0.004	0.012	0.025	0.077	0.164	0.261	0.318	0.311	0.275	0.124
	6	0.000	0.000	0.001	0.004	0.017	0.055	0.131	0.247	0.311	0.367	0.372
	7	0.000	0.000	0.000	0.000	0.002	0.008	0.028	0.082	0.133	0.210	0.478

(continued)

Table A-3 *(continued)*

Numbers in the table represent $p(X=x)$ for a binomial distribution with n trials and probability of success p.

Binomial probabilities:

$$\binom{n}{x} p^x (1-p)^{n-x}$$

n	x	0.1	0.2	0.25	0.3	0.4	0.5	0.6	0.7	0.75	0.8	0.9
8	0	0.430	0.168	0.100	0.058	0.017	0.004	0.001	0.000	0.000	0.000	0.000
	1	0.383	0.336	0.267	0.198	0.090	0.031	0.008	0.001	0.000	0.000	0.000
	2	0.149	0.294	0.311	0.296	0.209	0.109	0.041	0.010	0.004	0.001	0.000
	3	0.033	0.147	0.208	0.254	0.279	0.219	0.124	0.047	0.023	0.009	0.000
	4	0.005	0.046	0.087	0.136	0.232	0.273	0.232	0.136	0.087	0.046	0.005
	5	0.000	0.009	0.023	0.047	0.124	0.219	0.279	0.254	0.208	0.147	0.033
	6	0.000	0.001	0.004	0.010	0.041	0.109	0.209	0.296	0.311	0.294	0.149
	7	0.000	0.000	0.000	0.001	0.008	0.031	0.090	0.198	0.267	0.336	0.383
	8	0.000	0.000	0.000	0.000	0.001	0.004	0.017	0.058	0.100	0.168	0.430
9	0	0.387	0.134	0.075	0.040	0.010	0.002	0.000	0.000	0.000	0.000	0.000
	1	0.387	0.302	0.225	0.156	0.060	0.018	0.004	0.000	0.000	0.000	0.000
	2	0.172	0.302	0.300	0.267	0.161	0.070	0.021	0.004	0.001	0.000	0.000
	3	0.045	0.176	0.234	0.267	0.251	0.164	0.074	0.021	0.009	0.003	0.000
	4	0.007	0.066	0.117	0.172	0.251	0.246	0.167	0.074	0.039	0.017	0.001
	5	0.001	0.017	0.039	0.074	0.167	0.246	0.251	0.172	0.117	0.066	0.007
	6	0.000	0.003	0.009	0.021	0.074	0.164	0.251	0.267	0.234	0.176	0.045
	7	0.000	0.000	0.001	0.004	0.021	0.070	0.161	0.267	0.300	0.302	0.172
	8	0.000	0.000	0.000	0.000	0.004	0.018	0.060	0.156	0.225	0.302	0.387
	9	0.000	0.000	0.000	0.000	0.000	0.002	0.010	0.040	0.075	0.134	0.387
10	0	0.349	0.107	0.056	0.028	0.006	0.001	0.000	0.000	0.000	0.000	0.000
	1	0.387	0.268	0.188	0.121	0.040	0.010	0.002	0.000	0.000	0.000	0.000
	2	0.194	0.302	0.282	0.233	0.121	0.044	0.011	0.001	0.000	0.000	0.000
	3	0.057	0.201	0.250	0.267	0.215	0.117	0.042	0.009	0.003	0.001	0.000
	4	0.011	0.088	0.146	0.200	0.251	0.205	0.111	0.037	0.016	0.006	0.000
	5	0.001	0.026	0.058	0.103	0.201	0.246	0.201	0.103	0.058	0.026	0.001
	6	0.000	0.006	0.016	0.037	0.111	0.205	0.251	0.200	0.146	0.088	0.011
	7	0.000	0.001	0.003	0.009	0.042	0.117	0.215	0.267	0.250	0.201	0.057
	8	0.000	0.000	0.000	0.001	0.011	0.044	0.121	0.233	0.282	0.302	0.194
	9	0.000	0.000	0.000	0.000	0.002	0.010	0.040	0.121	0.188	0.268	0.387
	10	0.000	0.000	0.000	0.000	0.000	0.001	0.006	0.028	0.056	0.107	0.349
11	0	0.314	0.086	0.042	0.020	0.004	0.000	0.000	0.000	0.000	0.000	0.000
	1	0.384	0.236	0.155	0.093	0.027	0.005	0.001	0.000	0.000	0.000	0.000
	2	0.213	0.295	0.258	0.200	0.089	0.027	0.005	0.001	0.000	0.000	0.000
	3	0.071	0.221	0.258	0.257	0.177	0.081	0.023	0.004	0.001	0.000	0.000
	4	0.016	0.111	0.172	0.220	0.236	0.161	0.070	0.017	0.006	0.002	0.000
	5	0.002	0.039	0.080	0.132	0.221	0.226	0.147	0.057	0.027	0.010	0.000
	6	0.000	0.010	0.027	0.057	0.147	0.226	0.221	0.132	0.080	0.039	0.002
	7	0.000	0.002	0.006	0.017	0.070	0.161	0.236	0.220	0.172	0.111	0.016
	8	0.000	0.000	0.001	0.004	0.023	0.081	0.177	0.257	0.258	0.221	0.071
	9	0.000	0.000	0.000	0.001	0.005	0.027	0.089	0.200	0.258	0.295	0.213
	10	0.000	0.000	0.000	0.000	0.001	0.005	0.027	0.093	0.155	0.236	0.384
	11	0.000	0.000	0.000	0.000	0.000	0.000	0.004	0.020	0.042	0.086	0.314

Numbers in the table represent $p(X=x)$ for a binomial distribution with n trials and probability of success p.

Binomial probabilities:

$$\binom{n}{x} p^x(1-p)^{n-x}$$

n	x	0.1	0.2	0.25	0.3	0.4	0.5	0.6	0.7	0.75	0.8	0.9
12	0	0.282	0.069	0.032	0.014	0.002	0.000	0.000	0.000	0.000	0.000	0.000
	1	0.377	0.206	0.127	0.071	0.017	0.003	0.000	0.000	0.000	0.000	0.000
	2	0.230	0.283	0.232	0.168	0.064	0.016	0.002	0.000	0.000	0.000	0.000
	3	0.085	0.236	0.258	0.240	0.142	0.054	0.012	0.001	0.000	0.000	0.000
	4	0.021	0.133	0.194	0.231	0.213	0.121	0.042	0.008	0.002	0.001	0.000
	5	0.004	0.053	0.103	0.158	0.227	0.193	0.101	0.029	0.011	0.003	0.000
	6	0.000	0.016	0.040	0.079	0.177	0.226	0.177	0.079	0.040	0.016	0.000
	7	0.000	0.003	0.011	0.029	0.101	0.193	0.227	0.158	0.103	0.053	0.004
	8	0.000	0.001	0.002	0.008	0.042	0.121	0.213	0.231	0.194	0.133	0.021
	9	0.000	0.000	0.000	0.001	0.012	0.054	0.142	0.240	0.258	0.236	0.085
	10	0.000	0.000	0.000	0.000	0.002	0.016	0.064	0.168	0.232	0.283	0.230
	11	0.000	0.000	0.000	0.000	0.000	0.003	0.017	0.071	0.127	0.206	0.377
	12	0.000	0.000	0.000	0.000	0.000	0.000	0.002	0.014	0.032	0.069	0.282
13	0	0.254	0.055	0.024	0.010	0.001	0.000	0.000	0.000	0.000	0.000	0.000
	1	0.367	0.179	0.103	0.054	0.011	0.002	0.000	0.000	0.000	0.000	0.000
	2	0.245	0.268	0.206	0.139	0.045	0.010	0.001	0.000	0.000	0.000	0.000
	3	0.100	0.246	0.252	0.218	0.111	0.035	0.006	0.001	0.000	0.000	0.000
	4	0.028	0.154	0.210	0.234	0.184	0.087	0.024	0.003	0.001	0.000	0.000
	5	0.006	0.069	0.126	0.180	0.221	0.157	0.066	0.014	0.005	0.001	0.000
	6	0.001	0.023	0.056	0.103	0.197	0.209	0.131	0.044	0.019	0.006	0.000
	7	0.000	0.006	0.019	0.044	0.131	0.209	0.197	0.103	0.056	0.023	0.001
	8	0.000	0.001	0.005	0.014	0.066	0.157	0.221	0.180	0.126	0.069	0.006
	9	0.000	0.000	0.001	0.003	0.024	0.087	0.184	0.234	0.210	0.154	0.028
	10	0.000	0.000	0.000	0.001	0.006	0.035	0.111	0.218	0.252	0.246	0.100
	11	0.000	0.000	0.000	0.000	0.001	0.010	0.045	0.139	0.206	0.268	0.245
	12	0.000	0.000	0.000	0.000	0.000	0.002	0.011	0.054	0.103	0.179	0.367
	13	0.000	0.000	0.000	0.000	0.000	0.000	0.001	0.010	0.024	0.055	0.254
14	0	0.229	0.044	0.018	0.007	0.001	0.000	0.000	0.000	0.000	0.000	0.000
	1	0.356	0.154	0.083	0.041	0.007	0.001	0.000	0.000	0.000	0.000	0.000
	2	0.257	0.250	0.180	0.113	0.032	0.006	0.001	0.000	0.000	0.000	0.000
	3	0.114	0.250	0.240	0.194	0.085	0.022	0.003	0.000	0.000	0.000	0.000
	4	0.035	0.172	0.220	0.229	0.155	0.061	0.014	0.001	0.000	0.000	0.000
	5	0.008	0.086	0.147	0.196	0.207	0.122	0.041	0.007	0.002	0.000	0.000
	6	0.001	0.032	0.073	0.126	0.207	0.183	0.092	0.023	0.008	0.002	0.000
	7	0.000	0.009	0.028	0.062	0.157	0.209	0.157	0.062	0.028	0.009	0.000
	8	0.000	0.002	0.008	0.023	0.092	0.183	0.207	0.126	0.073	0.032	0.001
	9	0.000	0.000	0.002	0.007	0.041	0.122	0.207	0.196	0.147	0.086	0.008
	10	0.000	0.000	0.000	0.001	0.014	0.061	0.155	0.229	0.220	0.172	0.035
	11	0.000	0.000	0.000	0.000	0.003	0.022	0.085	0.194	0.240	0.250	0.114
	12	0.000	0.000	0.000	0.000	0.001	0.006	0.032	0.113	0.180	0.250	0.257
	13	0.000	0.000	0.000	0.000	0.000	0.001	0.007	0.041	0.083	0.154	0.356
	14	0.000	0.000	0.000	0.000	0.000	0.000	0.001	0.007	0.018	0.044	0.229

(continued)

© John Wiley & Sons, Inc.

Table A-3 (continued)

Numbers in the table represent $p(X=x)$ for a binomial distribution with n trials and probability of success p.

Binomial probabilities:

$$\binom{n}{x} p^x (1-p)^{n-x}$$

n	x	0.1	0.2	0.25	0.3	0.4	0.5	0.6	0.7	0.75	0.8	0.9
15	0	0.206	0.035	0.013	0.005	0.000	0.000	0.000	0.000	0.000	0.000	0.000
	1	0.343	0.132	0.067	0.031	0.005	0.000	0.000	0.000	0.000	0.000	0.000
	2	0.267	0.231	0.156	0.092	0.022	0.003	0.000	0.000	0.000	0.000	0.000
	3	0.129	0.250	0.225	0.170	0.063	0.014	0.002	0.000	0.000	0.000	0.000
	4	0.043	0.188	0.225	0.219	0.127	0.042	0.007	0.001	0.000	0.000	0.000
	5	0.010	0.103	0.165	0.206	0.186	0.092	0.024	0.003	0.001	0.000	0.000
	6	0.002	0.043	0.092	0.147	0.207	0.153	0.061	0.012	0.003	0.001	0.000
	7	0.000	0.014	0.039	0.081	0.177	0.196	0.118	0.035	0.013	0.003	0.000
	8	0.000	0.003	0.013	0.035	0.118	0.196	0.177	0.081	0.039	0.014	0.000
	9	0.000	0.001	0.003	0.012	0.061	0.153	0.207	0.147	0.092	0.043	0.002
	10	0.000	0.000	0.001	0.003	0.024	0.092	0.186	0.206	0.165	0.103	0.010
	11	0.000	0.000	0.000	0.001	0.007	0.042	0.127	0.219	0.225	0.188	0.043
	12	0.000	0.000	0.000	0.000	0.002	0.014	0.063	0.170	0.225	0.250	0.129
	13	0.000	0.000	0.000	0.000	0.000	0.003	0.022	0.092	0.156	0.231	0.267
	14	0.000	0.000	0.000	0.000	0.000	0.000	0.005	0.031	0.067	0.132	0.343
	15	0.000	0.000	0.000	0.000	0.000	0.000	0.000	0.005	0.013	0.035	0.206
20	0	0.122	0.012	0.003	0.001	0.000	0.000	0.000	0.000	0.000	0.000	0.000
	1	0.270	0.058	0.021	0.007	0.000	0.000	0.000	0.000	0.000	0.000	0.000
	2	0.285	0.137	0.067	0.028	0.003	0.000	0.000	0.000	0.000	0.000	0.000
	3	0.190	0.205	0.134	0.072	0.012	0.001	0.000	0.000	0.000	0.000	0.000
	4	0.090	0.218	0.190	0.130	0.035	0.005	0.000	0.000	0.000	0.000	0.000
	5	0.032	0.175	0.202	0.179	0.075	0.015	0.001	0.000	0.000	0.000	0.000
	6	0.009	0.109	0.169	0.192	0.124	0.037	0.005	0.000	0.000	0.000	0.000
	7	0.002	0.055	0.112	0.164	0.166	0.074	0.015	0.001	0.000	0.000	0.000
	8	0.000	0.022	0.061	0.114	0.180	0.120	0.035	0.004	0.001	0.000	0.000
	9	0.000	0.007	0.027	0.065	0.160	0.160	0.071	0.012	0.003	0.000	0.000
	10	0.000	0.002	0.010	0.031	0.117	0.176	0.117	0.031	0.010	0.002	0.000
	11	0.000	0.000	0.003	0.012	0.071	0.160	0.160	0.065	0.027	0.007	0.000
	12	0.000	0.000	0.001	0.004	0.035	0.120	0.180	0.114	0.061	0.022	0.000
	13	0.000	0.000	0.000	0.001	0.015	0.074	0.166	0.164	0.112	0.055	0.002
	14	0.000	0.000	0.000	0.000	0.005	0.037	0.124	0.192	0.169	0.109	0.009
	15	0.000	0.000	0.000	0.000	0.001	0.015	0.075	0.179	0.202	0.175	0.032
	16	0.000	0.000	0.000	0.000	0.000	0.005	0.035	0.130	0.190	0.218	0.090
	17	0.000	0.000	0.000	0.000	0.000	0.001	0.012	0.072	0.134	0.205	0.190
	18	0.000	0.000	0.000	0.000	0.000	0.000	0.003	0.028	0.067	0.137	0.285
	19	0.000	0.000	0.000	0.000	0.000	0.000	0.000	0.007	0.021	0.058	0.270
	20	0.000	0.000	0.000	0.000	0.000	0.000	0.000	0.001	0.003	0.012	0.122

z*-Values for Selected Confidence Levels

Table A-4 gives you the particular z^*-value that you need to get the confidence level (also known as percentage confidence) you want when you're calculating two types of confidence intervals in this book:

- Confidence intervals for a population mean where the population standard deviation σ is known

- Confidence intervals for a population proportion where the two conditions are met to use the normal approximation
    - $n\hat{p} \geq 10$
    - $n(1 - \hat{p}) \geq 10$

Note: You don't use Table A-4 if you're calculating confidence intervals for a population mean when the population standard deviation, σ, is unknown. For this type of confidence interval, you use the *t*-table (Table A-2).

For the two appropriate scenarios in the preceding list, some of the more commonly used confidence levels, along with their corresponding z^*-values, are in Table A-4. Here is how you find the z^*-value you need:

1. **Determine the confidence level needed for the confidence interval you're doing (this is typically given in the problem).**

    Find the row pertaining to this confidence level. For example, you may be asked to find a 95% confidence interval for the mean. In that case, the confidence level is 95%, so look in that row.

2. **Find the corresponding z^*-value in the second column of that same row in the table.**

    For example, for a 95% confidence level, the z^*-value is 1.96.

3. **Take the z^*-value from the table and plug it into the appropriate confidence interval formula you need.**

Note: To find a z^*-value for a confidence level that isn't included in Table A-4, you use the Z-table (Table A-1) with a modification. The Z-table shows the z-value corresponding to the percentage *below* a number. For a confidence interval, you want a z^*-value corresponding to the percentage *between* two numbers. To modify the Z-table to find what you need, take your original between percentage (confidence level) and convert it to a *less-than* percentage. Do this by taking your original percentage (confidence level) and adding half of what remains when you subtract it from 1. Look up this new percentage in the body of the table, and see what z-value it belongs to in the matching row/column of the table. That's the z^*-value you use in your appropriate confidence interval formula.

For example, a 95% confidence level means the between probability is 95%, so the less-than probability is 95% plus 2.5% (half of what's left), or 97.5%. Look up 0.975 in the body of the Z-table and find $z^* = 1.96$ for a 95% confidence level.

Table A-4	z*-Values for Selected (Percentage) Confidence Levels
Percentage Confidence	**z*-Value**
80	1.28
90	1.645
95	1.96
98	2.33
99	2.58

Illustration by Ryan Sneed

Index

Publisher's Acknowledgments

Executive Editor: Lindsay Sandman Lefevere
Senior Project Editor: Georgette Beatty
Copy Editor: Jennette ElNaggar
Contributor: LearningMate Solutions
Contributing Editor: Deborah J. Rumsey, PhD
Technical Editors: Jason Molitierno, David Unger
Art Coordinator: Alicia B. South

Project Coordinator: Erin Zeltner
Illustrator: Ryan Sneed
Cover Image: ©iStockphoto.com/Natalia Silych

Math & Science

Algebra I For Dummies,
2nd Edition
978-0-470-55964-2

Anatomy and Physiology
For Dummies, 2nd Edition
978-0-470-92326-9

Astronomy For Dummies,
3rd Edition
978-1-118-37697-3

Biology For Dummies,
2nd Edition
978-0-470-59875-7

Chemistry For Dummies,
2nd Edition
978-1-118-00730-3

1001 Algebra II Practice
Problems For Dummies
978-1-118-44662-1

Microsoft Office

Excel 2013 For Dummies
978-1-118-51012-4

Office 2013 All-in-One
For Dummies
978-1-118-51636-2

PowerPoint 2013
For Dummies
978-1-118-50253-2

Word 2013 For Dummies
978-1-118-49123-2

Music

Blues Harmonica
For Dummies
978-1-118-25269-7

Guitar For Dummies,
3rd Edition
978-1-118-11554-1

iPod & iTunes For Dummies,
10th Edition
978-1-118-50864-0

Programming

Beginning Programming
with C For Dummies
978-1-118-73763-7

Excel VBA Programming
For Dummies, 3rd Edition
978-1-118-49037-2

Java For Dummies,
6th Edition
978-1-118-40780-6

Religion & Inspiration

The Bible For Dummies
978-0-7645-5296-0

Buddhism For Dummies,
2nd Edition
978-1-118-02379-2

Catholicism For Dummies,
2nd Edition
978-1-118-07778-8

Self-Help & Relationships

Beating Sugar Addiction
For Dummies
978-1-118-54645-1

Meditation For Dummies,
3rd Edition
978-1-118-29144-3

Seniors

Laptops For Seniors
For Dummies, 3rd Edition
978-1-118-71105-7

Computers For Seniors
For Dummies, 3rd Edition
978-1-118-11553-4

iPad For Seniors
For Dummies, 6th Edition
978-1-118-72826-0

Social Security For Dummies
978-1-118-20573-0

Smartphones & Tablets

Android Phones
For Dummies, 2nd Edition
978-1-118-72030-1

Nexus Tablets For Dummies
978-1-118-77243-0

Samsung Galaxy S 4
For Dummies
978-1-118-64222-1

Samsung Galaxy Tabs
For Dummies
978-1-118-77294-2

Test Prep

ACT For Dummies,
5th Edition
978-1-118-01259-8

ASVAB For Dummies,
3rd Edition
978-0-470-63760-9

GRE For Dummies,
7th Edition
978-0-470-88921-3

Officer Candidate Tests
For Dummies
978-0-470-59876-4

Physician's Assistant Exam
For Dummies
978-1-118-11556-5

Series 7 Exam For Dummies
978-0-470-09932-2

Windows 8

Windows 8.1 All-in-One
For Dummies
978-1-118-82087-2

Windows 8.1 For Dummies
978-1-118-82121-3

Windows 8.1 For Dummies
Book + DVD Bundle
978-1-118-82107-7

Available in print and e-book formats.

Take Dummies with you everywhere you go!

Whether you are excited about e-books, want more from the web, must have your mobile apps, or are swept up in social media, Dummies makes everything easier.

Leverage the Power

For Dummies is the global leader in the reference category and one of the most trusted and highly regarded brands in the world. No longer just focused on books, customers now have access to the For Dummies content they need in the format they want. Let us help you develop a solution that will fit your brand and help you connect with your customers.

Advertising & Sponsorships

Connect with an engaged audience on a powerful multimedia site, and position your message alongside expert how-to content.

Targeted ads • Video • Email marketing • Microsites • Sweepstakes sponsorship

21 Million Monthly Page Views & 13 Million Unique Visitors